Details and Materials for Resilient Sites

Providing comprehensive information for closing material loops and reducing carbon impacts of site construction, this ground-breaking book is an essential resource for landscape architects and engineers to meet the environmental challenges of the 21st century.

It provides strategies, tools, and detailed information for the practice of climate positive design. Reduced carbon, resource efficient materials, and material mixes are addressed for both standard and alternative materials including stone, concrete, asphalt pavement, brick, earth-based materials, wood, biobased materials, metals, and plastics. Construction considerations are discussed and illustrated, and strategies to reduce environmental impacts are offered. Emphasis is placed on durable, resource efficient structures, and reduced embodied carbon approaches for pavements, porous pavements, green roofs, bioengineered embankments, retaining walls freestanding walls, decks and boardwalks, and rails, fences, and screens. Exemplary case studies from around the world are highlighted, and key resources such as assessment tools, information transparency tools, and benchmarking measures are provided.

Responding to the need for landscape architecture to account for climate change, resource consumption, and habitat destruction, this important book is key reading for both professionals and students alike.

Meg Calkins, FCELA, FASLA, is Professor of Landscape Architecture and Environmental Planning at North Carolina State University. She holds an MLA and MArch from the University of California at Berkeley and has taught and written about resilient site design and sustainable construction for over 25 years.

Meg is also the author of *Materials for Sustainable Sites* (2008) and editor of the *Sustainable Sites Handbook* (2012). She has been a frequent contributing editor to *Landscape Architecture* magazine, writing several articles on site construction materials and exemplary designed works, and she won the Bradford Williams Medal for Journalistic Excellence in 2003.

Meg has taken an active leadership role in development and implementation of the Sustainable Sites Initiative (SITES) since 2003. In 2023–24, Meg received a Landscape Architecture Foundation Fellowship for Leadership and Innovation where she developed key themes for this book. She currently serves on the LAF Board of Directors.

"Meg Calkins has provided cutting-edge material research, specification best practices, and outlined detailing and construction strategies that will accelerate the shift to climate positive practices and sites. This book challenges readers to think beyond short-term gains within project boundaries and instead focus on long-term sustainable practices and decisions at the macro level."

—**Pamela Conrad**, *Founder and Executive Director, Climate Positive Design*

"The science on the effects of carbon is clear. Calkins has given landscape architects, architects, engineers, developers, and government officials tangible tools for taking action. Now get to work and use them – there are no more excuses."

—**Lucinda Sanders**, *CEO and Partner, OLIN*

"Meg Calkins continues the revolution she started with *The Sustainable Sites Handbook*. Here, backed by over a decade of fact-finding and scrutiny, she asks designers and engineers to be leaders in reducing the globe's embodied carbon footprint, then provides the thinking and steps to get there. She is not asking for sacrifice in quality or aesthetics, but for persistence in expanding our environmental fluency and reach. Let's answer the call."

—**Laura Solano**, *Partner, Michael Van Valkenburgh Associates, Landscape Architects*

"Professor Calkins' new book prepares practitioners to deliver climate positive projects through their design and material choices and leads with the necessary sufficiency-first mindset. This is a must-have comprehensive resource for practitioners to leverage their material palettes for climate positive outcomes."

—**Vincent Martinez**, *President & COO, Architecture 2030*

Details and Materials for Resilient Sites

A Climate Positive Approach

Meg Calkins

Routledge
Taylor & Francis Group
NEW YORK AND LONDON

Designed cover image: Michal Szlaga

First published 2025
by Routledge
605 Third Avenue, New York, NY 10158

and by Routledge
4 Park Square, Milton Park, Abingdon, Oxon, OX14 4RN

Routledge is an imprint of the Taylor & Francis Group, an informa business

ISBN: 978-1-032-53856-3 (hbk)
ISBN: 978-1-032-53857-0 (pbk)
ISBN: 978-1-003-41397-4 (ebk)

DOI: 10.4324/9781003413974

Typeset in OfficinaSans
by Apex CoVantage, LLC

To Jack and Annie
For your future

Contents

Foreword

Today our world finds itself at a pivotal moment with design and construction central to the challenge and the solutions. Approximately 75% of global greenhouse gas emissions come from the urban built environment, including buildings, transportation, infrastructure, and sites. As nations around the world commit to global decarbonization goals, the architecture and engineering professions and the construction industry face enormous pressure. While emissions from building design and construction have been understood for some time, it wasn't until recently that we discovered that close to 80% of the emissions in the exterior built environment come from materials – the very building blocks of our cities and communities.

These emissions, often referred to as "embodied carbon", result from the extraction, transportation, manufacturing, and on-site construction of materials and built elements. A shift is necessary, not only for what we specify, but fundamentally in how we design, produce, and use these construction materials. The current global context sets the backdrop for the critical importance and timely manner of this book. While municipalities, governments, corporations, and private landowners set targets for meeting aggressive carbon reduction goals, exterior built environment professionals such as landscape architects, engineers, and planners can play a major role in this transformation. Taking steps to reduce these construction material emissions is not only an opportunity, but also a responsibility for a more sustainable future.

Though we are currently on a path to decarbonizing the built environment – we must ask ourselves – is it too late? In response to this time-sensitive situation, efforts to accelerate a low-carbon transition are underway. Since the launch of the Climate Positive Design Challenge and Pathfinder application in 2019, the landscape architecture profession has joined forces with allied professionals from around the world who are passionate about taking climate action. Landscape architects committed change for the first time at the United Nations Framework Convention on Climate Change (UNFCCC) COP 26 in 2021. The launch of the International Federation of Landscape Architects (IFLA) Climate Action Commitment saw over 75,000 professionals from 77 nations pledge to meet the 1.5°C Communique led by Architecture 2030, which seeks to cut emissions in half by 2030 and reach zero emissions by 2040. The American Society of Landscape Architects (ASLA) launched its supporting Climate Action Plan in 2022 alongside several other professional organizations.

With this guidance in place, various built environment professions – including architecture, landscape architecture, planning, and engineering – are now working together to align reporting standards through the Embodied Carbon Harmonization and Optimization (ECHO) project. Collaboration across industries and disciplines is critical to driving meaningful and lasting change. We must continue to strive for a shared vision of a sustainable future that unites designers, policymakers, manufacturers, and the public.

However, goals and targets are meaningless if they are not implemented. This is a multi-dimensional challenge that requires rethinking how designers, manufacturers, and contractors interact. Increasing dialogue is leading to new low-carbon material choices becoming more readily available, deeper awareness of energy intensity and ability to measure, and heightened ability and intentionality for material reuse. Without a fundamental understanding and rigor behind how materials are produced and constructed, unfortunately these efforts may be futile. The deep dive of this book will ground designers, from technical experts to students, on the basic knowledge needed to shift business-as-usual design to a climate positive future.

Meg Calkins has created much more than just a standard "how-to" guide. This book challenges readers to think beyond short-term gains within project boundaries and instead focus on long-term sustainable practices and decisions at the macro level. She has provided cutting-edge material research and specification best practices and outlined detailing and construction strategies that will accelerate the shift to climate positive practices and sites. Decisions made today will have long-term either positive or negative impacts on our planet.

No single industry or entity can solve the climate crisis, but landscape architects and the construction industry are uniquely positioned to make a significant impact. This book offers practical tools and insights to help make that happen and respond to the call for action. The nudge toward a shift in thinking is needed to change the way professionals and industries operate and has the potential for supporting a lasting, lifelong regenerative future for generations to come. And the time to act is now.

By Pamela Conrad, Founder and
Executive Director of Climate Positive Design

Preface

Sixteen years ago, I wrote my first book about the impacts of production and disposal of site construction materials on ecosystems around the world. I offered solutions to reduce those impacts, and while our practices have started to shift, exponential growth in the built environment has resulted in even greater use of material resources and release of greenhouse gas emissions. The challenge that confronts us is daunting yet achievable: Less than 16 years from now, we must meet our 2040 target for climate positive sites, infrastructure, and buildings.

At the time of my first book, embodied carbon was a consideration, but climate change was not the urgent issue that it is today. In fact, if we could return to the 20% lower 2008 greenhouse gas emissions (GHG) levels, we would be in a better place. Since that time our planet has warmed to over 1.4°C, and six of the nine planetary boundaries to support life on earth have been crossed.

Unprecedented global growth has consumed more and more resources, generating massive amounts of waste, degrading ecosystems, and releasing pollutants. This growth is projected to accelerate over the next 25 years, particularly in less economically advanced countries.

The production of construction materials has grown into a global industry where raw materials can be extracted in Brazil, fabricated in China, then used in a project in Amsterdam. Transport of materials is inexpensive and labor in some countries even more so. And this global supply chain has a dark side that is increasingly exposed. In some countries, forced and child labor abuses occur in material production industries, further subsidizing the low-cost of imported materials.

Given that 75% of global GHG emissions stem from the construction, operations, and transportation in the built environment, we have both a serious responsibility and a great opportunity to slow climate change. And since an estimated 70% of global infrastructure that will be in place in 2050 has not even been built yet, we have a chance to get it right this time. **We must substantially shift our practices to build carbon positive sites and infrastructure, close materials loops with a circular economy, and source equitably produced, healthy construction materials.**

Unlike 16 years ago, professionals, product manufacturers, material industries, and legislators are increasingly committed to addressing climate change. Concrete, steel, aluminum, and other industries have developed detailed plans for carbon neutrality by 2050. Professional organizations for landscape architects, architects, and engineers have adopted climate action plans; and perhaps most importantly, national governments have signed the Paris Climate Agreement and are providing financial incentives and legislative support for abating greenhouse gas emissions.

But these commitments to reduce GHG emissions are not enough. They must be followed by action on every construction project and with every decision about the materials and structures to build it. First, it may be baby steps, but quickly it must be substantive change. Construction industries must shift substantially toward **climate positive design** within a few years. This must be prioritized as highly as cost and aesthetics in our decision-making.

Substantive, even radical change is not easy. Designers and engineers may wish to change their specs and details, but standards, clients, material industries, and policies may get in the way. Therefore, this change requires collaboration among a wide array of stakeholders from designers and engineers to policymakers and regulators. Everyone must work together. And with the climate crisis at the forefront, we are at a critical juncture. We can respond accordingly, or we can continue with business as usual. **Now is the time we must work together for substantive change.**

What is climate positive design for site and infrastructure projects? Climate positive design employs carbon sequestration and carbon reduction strategies to ultimately sequester more carbon than is released for the materials, construction, and operation of a site. From a construction materials standpoint, that means we need to use low-carbon materials and material mixes and, where possible, natural, renewable, and even living materials.

But it's not just about materials, it is how we assemble them. Our site structures must be durable, flexible, repairable, and living structures. Where possible, we should employ bioengineered embankments instead of retaining walls, and stone dust pavements instead of concrete. This approach may result in a different appearance than we currently expect from certain types of landscapes, but we must adjust our aesthetic expectations for climate positive design.

Climate positive design will be simplest to achieve when structures on a site are reused, rigid walls and pavement are minimized, and renewable materials and living structures employed. But the programmatic requirements of many new site and infrastructure projects require rigid pavements, retaining walls, guard rails, and other carbon-intensive structures.

In some projects, these structures can be achieved using biobased or earth-based materials and structures; but in most instances, we will still need to use standard construction materials such as concrete, brick, steel, plastics, and asphalt pavement. Reducing the embodied carbon and environmental impacts of these standard construction materials is possible, sometimes substantially, with modifications to mixes, incorporation of high percentages of recycled content, and optimized structures.

This book provides strategies, tools, and detailed information for the next practices of climate positive design. Reduced carbon and resource efficient materials and material mixes, both standard and alternative, are addressed along with a variety of approaches to optimize the design of site structures.

We have less than 16 years to transform our practices to meet 2040 targets for climate positive design. **We can get there if the next practices provided in this book are employed to affect substantive change in the way we specify construction materials and design structures for the built environment.** We must embrace this great opportunity and important responsibility to slow climate change.

Acknowledgements

This book took shape with the advice, inspiration, and skills of many people. I am grateful for discussions with colleagues on the ASLA Biodiversity and Climate Action Committee, the Sustainable Sites Initiative (SITES), the *Landscape Architecture* magazine Editorial Advisory Board, and the Landscape Architecture Foundation (LAF). In particular, I appreciate the conversations and collaborations I've had over the past two years with Chris Hardy about embodied carbon calculations for site construction materials and the inspiration I've drawn from conversations with Pamela Conrad of Climate Positive Design.

Funds from the LAF Fellowship for Innovation and Leadership provided a research assistant to help with book graphics and a reduced teaching load. More importantly, fellowship advisors Laura Solano and Cindy Sanders facilitated an atmosphere for growth and development of many of the book's themes. Cohort Seven members and guest reviewers also provided invaluable feedback.

I'd like to sincerely thank chapter reviewers who generously gave of their time and expertise to ensure the accuracy of the book's content. They are Laura Solano, Chris Hardy, Stefan Faber, Stanton Jones, Sarah Gregg, Vincent Martinez, Jennifer Mitchell, Satyam Maharaj, Howard Marks, Richard Willis, and Zac Tolbert. A very special thank you to Laura Solano who thoughtfully reviewed and provided detailed feedback on three dense chapters.

Kathleen Elliott produced all of the details and other graphics and also provided research support for this project. Her high-quality work and positive attitude, despite changes and changes back, made this process so much smoother. Other graphic assistance was provided by Drake Bruner and Lizandro Marcial-Armas. All of their graphic talents are very much appreciated.

While they may not have contributed directly to this book, I would like to thank my mentors from across the decades. Linda Jewell inspired my passion for materials and compelling construction details, and she has supported me in positive ways at every stage of my career. Bruce Ferguson and Linda inspired my approach to disseminating information to our profession to move it toward change.

The hundreds of students who have taken my landscape construction classes over the years have pushed my work forward by asking tough questions and embracing many of the ideas. It has not been easy for them to go out into the profession and affect change, but I thank them for their efforts and their progress. Colleagues in LAEP at North Carolina State University have fostered a collegial and rigorous work environment that I enjoy being a part of. Thank you.

Thank you to my children, Jack and Annie, for their constant humorous antics, love, and support. I appreciate their readiness to slam on the brakes on a winding country road in the Cotswolds so I could take pictures of a beautiful willow fence, venture on a 75 kilometer detour to see reclaimed building materials from ancient Greece in a Middle Ages church in Croatia, and get their hands dirty modeling soil testing for book photos. They have always supported and engaged in my work, and as young adults they are following their own paths working on environmental and social issues.

Last but not least, thank you to my husband Steve Badger, who tolerated weekend after weekend of my work without complaint. I am so grateful for his love, emotional support, and supreme talents in the kitchen. My weekends are free now. Let's go explore this beautiful world.

CHAPTER 1

Introduction

Details and Materials for Resilient Sites

Our current approach to the design and construction of sites and infrastructure contributes to climate change, non-renewable resource use, and waste. The materials we use to build sites and infrastructure contribute about 7.2% of annual greenhouse gas emissions and use billions of metric tons of material resources each year (Thacker et al. 2021). With the UN Environmental Program Sustainable Development Goals and significant population growth in Africa and Asia, material use is expected to more than double by 2060 with a third of this being used for construction (UNEP 2022).

We are at critical juncture in the design and construction of our built environment. We can choose to proceed with business as usual by designing resource- and carbon-intensive built landscapes that we call sustainable because they provide ecosystem services, or we can take a hard look at the impacts of our standard construction materials and our standard construction details to build differently. **Now is the time to design and build climate positive sites.**

On a typical project, hundreds of choices are made about materials, details, structures, and the site design. A variety of criteria factor into those decisions, including cost, aesthetics, performance, and sometimes carbon footprint. But this optional, fringe approach of reducing the embodied carbon of our sites is no longer enough. Embodied carbon reductions must be at the top of our list to avoid the worst impacts of climate change, habitat loss, and resource consumption.

Our current linear systems of extraction, production, use, and waste to produce construction materials are causing immeasurable damage to the people and the planet. These practices have led to the *overshoot* of five of the nine planetary boundaries within which humanity can safely operate: climate change, biosphere integrity (biodiversity loss and extinction), land system change, freshwater use, and biogeochemical flows (phosphorous and nitrogen cycles) (Figures 1.1 and 1.2).

Scientists at the Stockholm Resilience Center emphasize:

> *Crossing boundaries increases the risk of generating large-scale abrupt or irreversible environmental changes. Drastic changes will not necessarily happen overnight, but together the boundaries mark a critical threshold for increasing risks to people and the ecosystems we are part of.*
>
> (Richardson et al. 2023)

And these impacts are not equitable. Populations in developing countries bear the brunt of the climate and biodiversity impacts, yet their consumption has not caused it.

Impacts From Site Construction Materials and Opportunities to Reduce Them

Our current development practices for sites, infrastructures, buildings, and communities are not safe for people or the planet. Greenhouse gas emissions, depletion of finite resources and production of waste, health impacts of material production and disposal, and inequities and labor abuses in the material supply chain make our current development unsustainable (Figure 1.3).

Greenhouse Gas Emissions Across the Life Cycle of Materials and Products

Current GHG emissions from raw material extraction, manufacture, and disposal of construction materials and products are excessive with billions of metric tons of planet warming CO_2e released each year for their production. Emissions from cement and concrete, steel and aluminum, and plastics are the main construction materials to decarbonize as their production accounts for nearly 20% of annual greenhouse gas emissions (McGhee and Olano 2023).

We can work to reduce use of these standard materials, but they will continue to be important for structures and infrastructure. Therefore, these industries must be

DOI: 10.4324/9781003413974-1

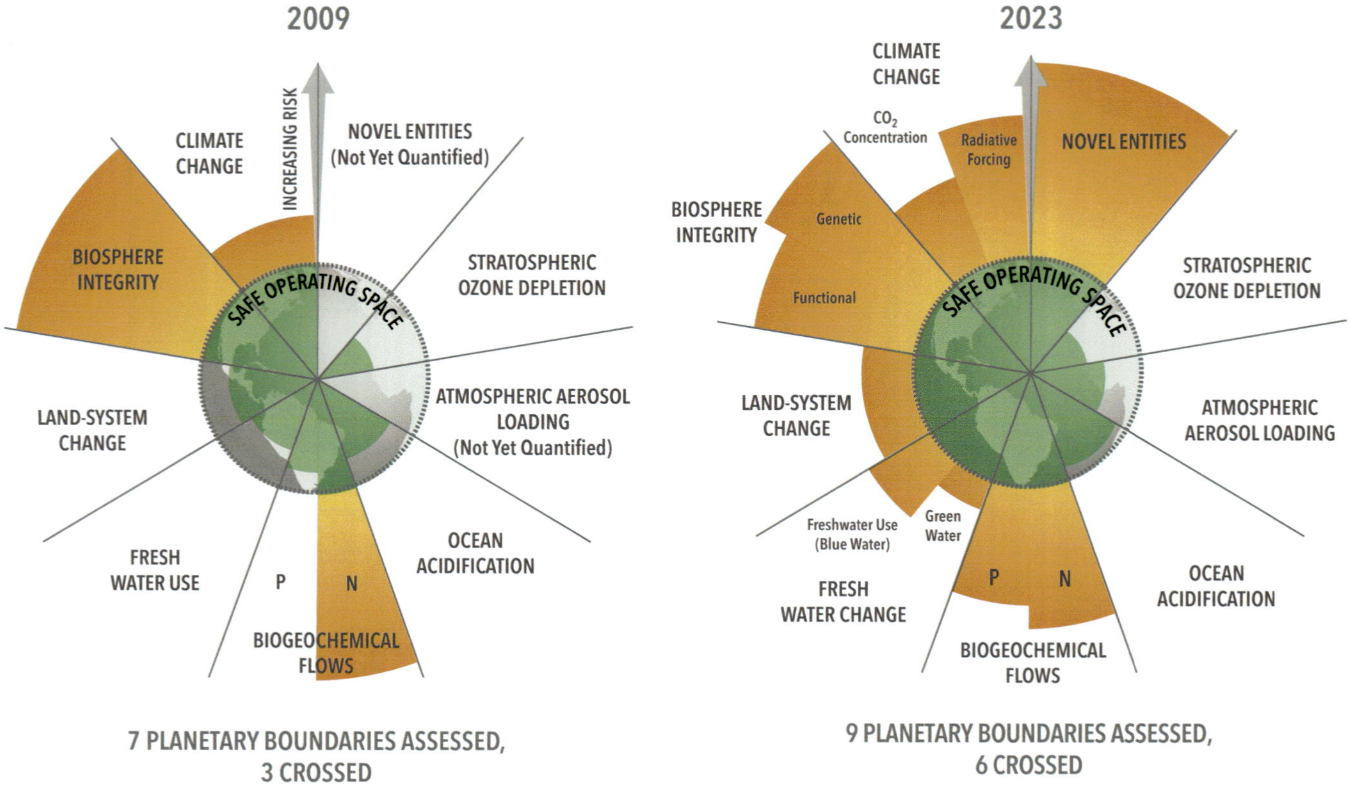

Figure 1.1 and Figure 1.2 Planetary boundaries are nine critical processes that together sustain a stable and resilient earth. In 2009, we had already crossed three of the nine boundaries; and in 2023, we had crossed a total of six. Based on Richardson et al. 2023; Steffen et al. 2015; Rockström et al. 2009. Image credit: Azote for Stockholm Resilience Centre, Stockholm University, redrawn by Meg Calkins

decarbonized, and low-carbon mixes of concrete and asphalt can be specified. Safer plastics and metals with high recycled content should be used.

Building codes and standards can be adapted to facilitate use of lower carbon mixes and materials, and engineers and designers can innovate with how these materials are used and assembled. At the same time, we can scale up use of biobased and alternative materials and structures. We can employ living retention and fencing structures wherever possible – even in urban landscapes.

Current global materials markets result in substantive emissions released from the transportation of products moved hundreds and even thousands of miles along the supply chain to reach consumers. Given the magnitude of GHG impacts from constructing sites and infrastructure, reducing the embodied carbon of construction materials and assemblies is key to achieving climate positive development by 2040.

We can move away from the global materials economy to source hyper-local materials including "mining" project sites and surrounding built environments; and we can design sites with vernacular structures and materials. We can specify alternative structures made with earth-based masonry products or bamboo; and we can use the rapidly growing body of materials produced with the help of biological organisms instead of heat (Hio 2022; Alemu et al. 2022).

Chapter 2 "Decarbonization of Materials and Structures" addresses decarbonization of construction materials and site structures. Chapter 5 "Assessing the Impacts of Materials, Products, and Structures" provides information about tools, techniques, and resources for identifying the embodied carbon of materials and material assemblies. Decarbonization approaches are covered in every chapter of this book as they relate to the chapter's topic.

Depletion of Finite Resources and Production of Waste for Materials and Products

Supplies of the earth's minerals, fossil fuels, metals, soil, and water resources are limited. Between 2015 and 2021, the global economy consumed 70% more resources than the earth could replenish (Circle Economy 2023). In fact, there is now more human produced mass than all biomass on earth, and waste is clogging our oceans and accumulating unchecked on land.

Non-renewable, energy-intensive, mineral-based materials currently make up most of the resources used to build our environment. Three to twenty metric tons of raw material can be consumed to produce one metric ton of metal or

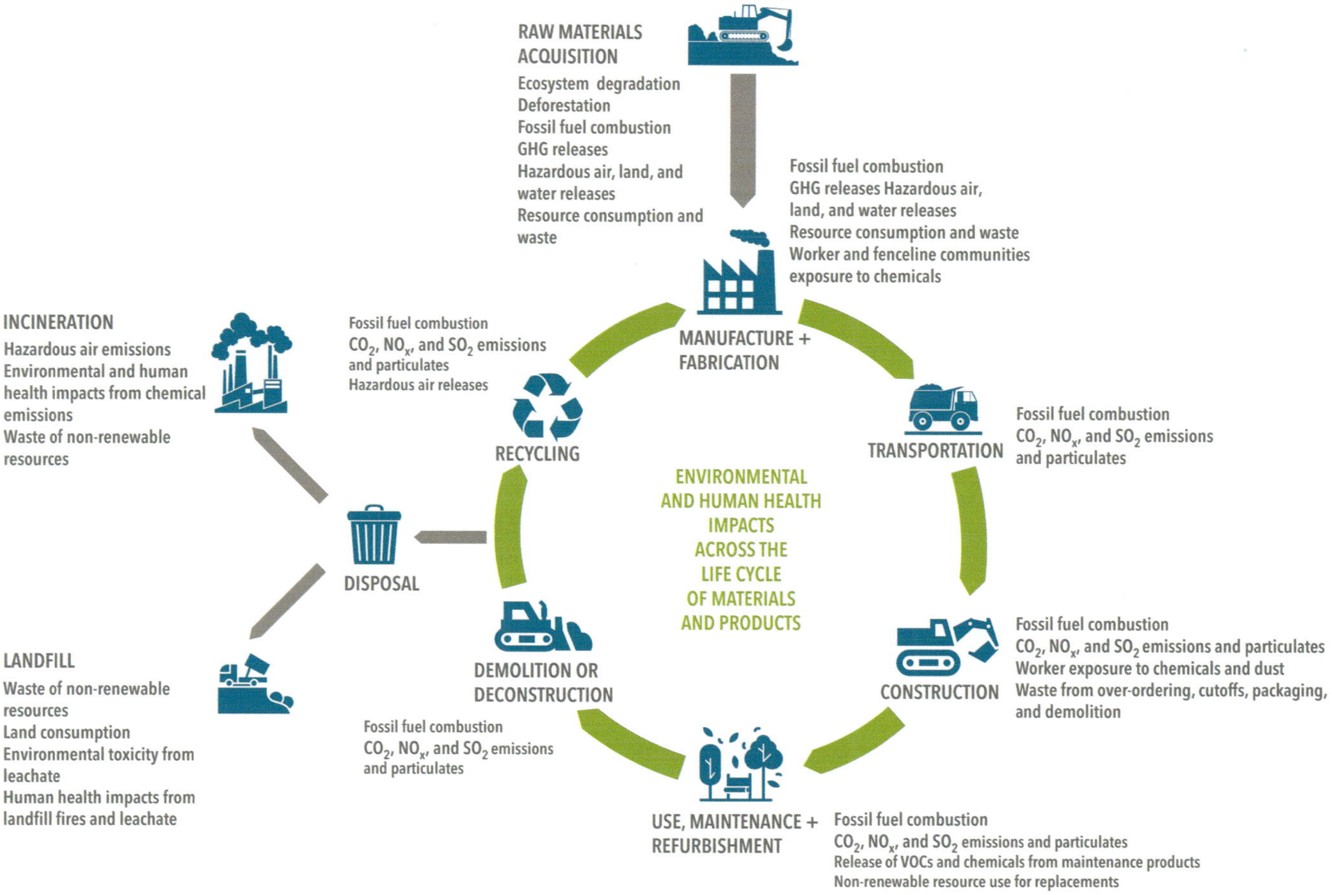

Figure 1.3 Environmental and human health impacts throughout material and product life cycles. Circular life cycles where waste from one project, product, or process is food for another can reduce these impacts. Image credit: Meg Calkins

dimension stone. And up to 30% of materials brought to a construction site can become waste before the construction is complete (Osmani 2011).

Chapter 3 "Resource Efficiency and the Circular Economy" defines resource consumption and waste throughout the material life cycle. It offers detailed strategies for reducing resource use and waste and shifting to material circularity to support circular economies. Resource efficiency approaches are covered in every chapter of this book as they relate to the chapter's topic.

Human and Environmental Health Impacts of Materials and Products

The production, use, and disposal of site construction materials poses health impacts to humans and the environment. Extraction of raw materials impacts ecosystems far away from a project site through removal of forests, landcover, soils, and overburden; sedimentation and pollution released to receiving waters; and air pollution from toxic chemicals, heavy metals, and particulate matter that are carried on air currents to nearby communities and receiving waters. Material manufacturing and fabrication facilities release solid, liquid, and airborne wastes to the environment, and disposal of materials and products releases additional toxins and waste to air, land, and water.

Understanding these impacts can be very challenging given the complex life cycles of materials and products. Chapter 4 "Healthy and Ethical Materials and Products" provides detailed information about human health impacts of materials and products across the life cycle, and tools and resources for identifying health risks are also explained. In Chapter 5, Table 5.2 defines typical environmental impact indicators.

Inequities and Human Rights Challenges During the Life Cycle of Materials and Products

An estimated 7% of the world's population is employed in the production of construction materials or in the construction sector. Alarming numbers of these workers are in inequitable employment situations where they are underpaid, work long hours, or are in unsafe conditions. Some of these laborers are children or adults trapped in forced labor situations.

The International Labor Organization estimates that 28 million people around the world are in forced labor situations with several times this number working as illegal laborers (ILO 2024). Nearly 160 million children ages 5–17 work in situations that are harmful to their physical and mental development (Prince et al. 2020).

Chapter 4 "Healthy and Ethical Materials and Products" explains these troubling issues of our construction material production and provides tools and resources to source equitably produced materials. Ethical sourcing is addressed in greater detail in the materials chapters of this book.

While designers, engineers, and specifiers play an important role in affecting this change and reducing environmental and human health impacts, several other stakeholders also have responsibilities. The material and product manufacturing sectors must change their resource and carbon-intensive production processes, source local materials, participate in the circular economy by reducing and reusing waste, and shift to renewable energy and heat sources. The recycling industries must catch up to meet demand for recycled content.

Policymakers and regulators must set carbon goals and write them into policies, regulations, and building codes. Codes and design standards must be modified to support resource efficient and low-carbon materials, materials mixes, and structures. Standards and industry organizations must develop and expand standards for cement substitutes and cement alternatives, alternative materials and structures, and practices that support circularity.

In addition to the extreme carbon impacts of the global materials economy, the supply chains of some building materials are at high risk for unethical and unsafe working conditions. Bricks, steel, rubber, plastics, timber, stone, copper, iron, and minerals are extracted and processed in some countries using millions of forced, illegal, and child laborers (Prince et al. 2020). While this activity is increasingly well documented, it is out of view, and the lure of inexpensive materials and labor from overseas is difficult to resist. Fair labor conditions must become an important criterion in our decision-making. Becoming aware of supply chains and specifying regionally and locally produced materials and products can help address these issues.

Materials and Details for Resilient Sites Defined

Sites and infrastructure are uniquely positioned to offer solutions to these environmental problems rather than contributing to them. Landscapes can be designed to provide multiple ecosystem services including carbon sequestration, habitat health, flood control, water purification, clean air, pollination, and human health and well-being. Site structures can receive reclaimed, reprocessed, and recycled materials from deconstruction of the built environment for use in concrete, asphalt, and aggregate applications.

With 75% of the infrastructure required by 2050 still unbuilt, there is a tremendous opportunity to change course and focus on building low, carbon resource efficient sites (UN Climate Change 2021). And with the focus on designing sites to provide ecosystem services, it is critically important to ensure that the materials and structures used to build these sites are not destroying ecosystems elsewhere.

A carbon positive approach to site materials and structures means building less and building with less, using circular material strategies, accelerating the use of lower carbon construction materials and material mixes, and designing and maintaining sites for reduced operational emissions.

There is no single strategy to achieve climate positive sites. Instead, the design and construction of sites can employ the multiple strategies discussed next. Some are low hanging fruit while others will take time and experience to achieve. Now is the time to implement a few, then share lessons learned. On the next project implement more, and then even more (Figure 1.4; Table 1.1).

Lower Embodied Carbon Materials, Products, and Structures

Reducing embodied carbon of materials, products, structures, and whole sites, along with maximizing use of carbon sequestering materials and vegetation, will facilitate the achievement of the ASLA goal of zero carbon from built sites by 2040. The following strategies to reduce embodied carbon of materials and structures are detailed in Chapter 2 "Decarbonization of Materials and Structures" and in all chapters of the book.

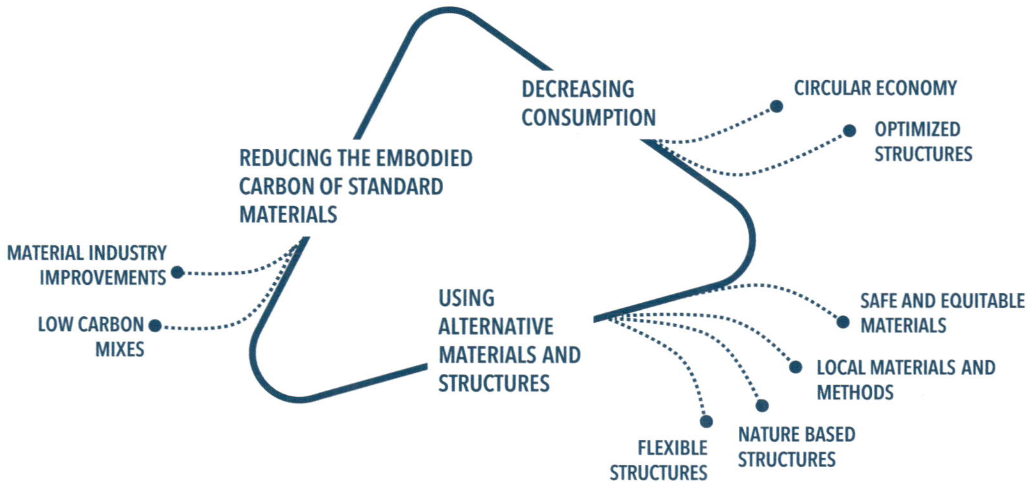

Figure 1.4 The solutions to reduce embodied carbon, conserve resources, and source safe and equitable materials and products for site construction. Image credit: Meg Calkins

Table 1.1 General Resources for Reducing the Environmental and Human Health Impacts of Site Materials, Products, and Structures

Information Resources	
SITES	Sustainable Sites Initiative by GBCI
International Living Future Institute	Living Building Challenge Living Building Challenge Red List Living Product Challenge Living Community Challenge Zero Carbon Certification
US Green Building Council (USGBC)	LEED BD+C: New Construction LEED Cities and Communities
Building Research Establishment (BRE) BREEAM	BREEAM Infrastructure BREEAM New Construction BREEAM Communities
Green Building Council of Australia	Green Star Buildings Green Start Communities
INVEST	Infrastructure Voluntary Evaluation Sustainability Tool
Sustainable Transportation Council	Greenroads Rating System
Institute for Sustainable Infrastructure	Envision
UNEP resources	UNEP Sustainable Development Goals UNEP Sustainable Public Procurement Implementation Guidelines UNEP Sustainable Buildings and Construction UNEP Circular Economy Roadmap
One Planet Network	Sustainable Consumption and Production. Very extensive Database of policies and standards.
Global Reporting Initiative (GRI)	Global Reporting Initiative
European Union Buying Green Handbook	EU Buying Green Handbook EU Guidance for Bio-Based Products in Construction
Individual Product Standards	Forest Stewardship Council Responsible Mining Initiative Sustainable Stone Standard Concrete Sustainability Council Sustainable Steel Principles Responsible Steel Aluminum Stewardship Initiative

Note: Refer to Tables 2.9, 3.5, and 4.6 for additional resources, standards, and certifications.

It is important to note that most resource efficiency strategies such as optimized structures, reused materials, and use of high recycled content in materials and products will reduce embodied carbon as well. These strategies are discussed in the section entitled "Resource Efficiency for the Circular Economy".

Low-Carbon Materials and Products

Materials that are less processed and avoid extensive periods of heating during manufacture are lower in embodied carbon. Examples of low-carbon materials include air-dried brick, minimally processed stone, lumber, compressed earth blocks with low cement content, and reclaimed or high recycled content materials.

The production processes of some materials such as cement, steel, and plastics give off process emissions from CO_2 released during the chemical conversion of the materials. These process emissions are more challenging to reduce than emissions from fossil fuel combustion to power manufacturing equipment. Fossil fuel sources of energy can be replaced by renewable energy sources. Some process emissions can be captured through carbon capture, utilization, and storage (CCUS), although this is a new and relatively expensive technology.

Higher embodied carbon materials, such as fired brick, can be considered low carbon if they have a very long service life. Bricks can be used over and over in multiple structures if mortar is avoided or it is removed for reuse.

Embodied carbon of materials and products is calculated in life cycle assessments (LCAs) and communicated in environmental product declarations (EPDs). Most commonly, embodied carbon is only quantified for the product stage (phases A1–A3) of a materials life cycle, but GHG emissions also occur during use and disposal phases as well.

Material sectors such as steel, aluminum, and cement and concrete have published plans to transition their industries to carbon neutrality by 2050. These plans propose production efficiencies and even new production processes, as well as a transition to renewable energy sources.

The embodied carbon of materials and products can be substantially reduced by altering some material mixes as well. For instance, the embodied carbon of concrete can be reduced by 30–45% or more through use of cement substitutes, alternative cements, use of fiber reinforcing in slabs, and incorporation of recycled aggregates. Smaller, thinner walls and slabs will reduce it further.

Low-Carbon Structures and Assemblies

Comparing the embodied carbon of materials for a given site structure does not produce equivalent results. Just identifying the embodied carbon of concrete and steel does not account for how these two materials would be used very differently in a structure, therefore calculating the embodied carbon of structures and assemblies is a more accurate approach to understand tradeoffs and work toward climate positive sites (Chapter 5).

When calculating the embodied carbon of site structures, it is very important to factor in the expected service life of a material or product. If it will not last the expected life of the structure, a replacement factor must be employed. For instance, ACQ pressure-treated pine deck boards will last for 10–15 years on average. If the deck will be in use for 30 years, a replacement factor of 2.5 must be used in the calculations.

Local Materials and Products

Use of local materials and products will reduce carbon emissions from transportation of heavy and bulky site construction materials. This can also boost local economies and reference the vernacular materials of a place or a natural landscape.

The carbon impacts of freight trucking can be substantial at around seven times the GHG emissions of train transportation. For example, trucking base aggregates 15 kilometers will release more greenhouse gas emissions than their production.

While shipping construction materials and products have a lower carbon footprint than trucking, shipping materials and products long distances across oceans adds up. In today's global materials economy, it is not unusual for dimension stone to be quarried by in India, worked in China, and installed in Boston; or for tropical hardwood to be sourced in Southeast Asia, then used in a project in Chicago.

Even greater impacts stem from transporting large amounts of material that will become waste prior to the product reaching consumers. The bauxite for aluminum is likely mined in the tropics of Southeast Asia, then sent unprocessed to Europe for refining, leaving 4 metric tons of waste in Europe for every metric ton of aluminum ore obtained.

It is important to note that some materials might appear to be local given a local distributor or company, but questions about origin of materials and point(s) of manufacturing must be asked. It is not unusual for some materials, such as dimension stone, to be quarried locally then sent overseas for fabrication due to lower labor costs. Heavy and bulky materials should be sourced within 80 kilometers of the project site; and lighter and/or more specialized products should be obtained within 200 kilometers of the site.

Materials and Products That Store and Sequester Carbon

Biobased materials such as wood lumber and engineered wood products store carbon that was in cells of the tree or plant when it was living. The product will store some of this carbon until it is burned or decays. If the forest or field is replanted, the carbon storage in the material and the carbon sequestration in the new trees and plants can result in a net increase in carbon sequestration (Kriegh et al. 2021).

Concrete and lime absorb small amounts of carbon over several years in a process called natural carbonation. A conservative estimate is that concrete will absorb 20% of the carbon released during manufacture of its cement over its service life (The Concrete Centre 2016).

Carbon from process emissions can be captured and sequestered in concrete at concrete plants or in cement. Carbon sequestration in concrete is a new technology, yet it is available in concrete plants around the United States and Europe. CCUS in cement does not require major modifications to most kilns and is already being implemented in some plants; however it is a relatively new technology and only stores a fraction of the process emissions from producing cement.

Products and Structures That Reduce Operating Carbon

While only estimated to be 10 to 20% of the total carbon footprint of a built site, there are products that can help reduce operational carbon during the use phase of the site. Lighting with built-in solar panels, high efficiency lights, energy star pumps, and irrigation controllers can reduce operational carbon impacts. Maintenance equipment such as power washers, mowers, and blowers that are electric hold potential to reduce carbon by using renewable energy sources for their electricity.

Resource Efficiency for the Circular Economy

Resource efficiency measures to support a circular economy are key considerations in material use and the design and detailing of site structures. Resource efficiency measures will also decarbonize site construction.

Building Nothing New

Far and away the most resource efficient strategy is to build nothing new. Avoiding new construction will extend the service life of a previously developed site and maximize the life cycle of the existing materials and structures. Designing sites for adaptability with multi-use spaces and open, flexible plans can maximize the service life of a site and its structures.

Reuse of Structures in Place

Retrofitting and adaptively reusing existing structures on a site can extend their service life with minimal use of new resources. This approach also reduces embodied carbon and the environmental impacts associated with material extraction. Reuse of existing structures can reference and celebrate the former identity of the site. Site and building audits during early project stages can identify reusable structures.

Optimized Material Use for Structures and Assemblies

Optimizing the material use for new structures can reduce use of resources and the associated impacts of their extraction and manufacture. Thoughtful design can produce material efficiencies such as smaller, thinner structures and structural members. Walls, slabs, and structural members should not be sized larger than the recommendations by structural or geotechnical engineers or span tables.

Structures such as decks and rails should be designed to use standard lumber lengths with only minimal cut offs. Downsizing path, roadway, and parking drive aisle widths will reduce material use. Some examples of material optimization are as follows:

- Metal angles should be favored over square metal tubes as they will perform similarly structurally with just over half the material.
- Eighty-nine mm wood posts will use 60% less material than 140x140mm posts.
- Pin foundation systems for decks and fences can use less material and have lower embodied carbon than standard concrete footings.
- Flexible wall structures will not require extensive spread footings in areas with deeper frost lines (60cm or lower).
- Wall or pavement structures with cast stone, stone, or brick veneer are an inefficient use of material as they are not structural. The concrete or concrete masonry unit wall and pavement slabs are the structure, and the veneer is unnecessary. Brick, stone, or precast concrete walls should be the structure, not the veneer. Unit pavements should be set on well compacted, stable, and thicker aggregate bases.

Durable Materials

Use of durable assemblies, products, and materials will reduce resource use by prolonging the service life of a structure. Structures built with durable materials and products and removeable connections can be deconstructed, and the materials and products can be used in another structure.

Designing structures for easy repair will prolong the life of the structure. Flexible walls can be releveled whereas rigid walls with severe cracking must be removed. Flexible pavements and pavement bases such as asphalt pavement, aggregate, or stone dust pavement can be releveled and resurfaced without the removal of the entire structure. Unit materials such as brick, concrete pavers, or stone can be easily removed and repaired or reused in another landscape if mortar is not used.

It can be challenging to obtain information about the expected service life of materials and products. Many manufacturers will offer a "limited lifetime warranty" but not commit to the number of years their product is likely to be in use.

Reclaimed and Reused Materials

Reusing materials and products will reduce use of primary materials and their associated extraction and manufacturing impacts. Reuse of materials and products on a previously developed site will also save resource use and emissions from transportation. Performing site and building inventories prior to conceptual design can identify materials and products for reuse. This reduced waste can save landfill and procurement costs, which could partially or fully offset the additional cost of deconstruction (Figures 1.5–1.8).

Reclaimed materials should be procured or identified early in schematic design so they can inform design decisions. The cultural history of a place can be referenced with reclaimed materials and products. Storing the procured materials for several months can be challenging or expensive, but some materials will need protection from excessive moisture. If

Figure 1.6 At the old Texaco Depot, now Cosmic Saltillo in Austin, Texas, Ten Eyck Landscape Architects reused structures, concrete pavement slabs, and reclaimed steel members in an overhead structure. Image credit: Erika Rich

Figure 1.7 Ten Eyck Landscape Architects maintained existing bunker walls for outdoor dining spaces at Cosmic Saltillo, and a reclaimed sewer pipe is the water source for a nearby fountain. Image credit: Erika Rich

Figure 1.5 The mammoth steel structure of the last operating steel plant within Pittsburgh city limits was integrated into the Mill 19 landscape by Ten x Ten, revealing the site's rich history while providing new opportunities to gather, explore, and relax amid disturbance-adapted gardens. Image credit: Gaffer Photography

Figure 1.8 At Still Alive, a park in Torino County, Italy, Wagon Landscaping created The Rockery from on-site demolition waste to resemble ruins from romantic gardens. Image credit Yann Monel

materials are stored on-site, they should not be stockpiled in an area of healthy soils.

If the materials will be used in a structural capacity, they might need to be inspected and possibly regraded. Other materials may require additional labor such as nail pulling or mortar removal.

If not reused on-site, reclaimed materials should be obtained from local sources. Some online materials exchanges allow searching by state. Larger metropolitan areas have several outlets for reclaimed materials. Designers, contractors, or clients may be able to help locate materials from their other projects.

On-Site Reprocessing of Materials

Reprocessed materials are those that are broken, crushed, or size reduced from a structure or their original size. Reclaimed concrete is reprocessed as moving it will break it. It can be reprocessed into smaller slabs for use in pavements or walls. Or it can be crushed and graded for use in new concrete, asphalt, or base material. Asphalt pavement can also be partially or fully removed, reprocessed, and reused in place.

Reprocessing materials usually "downcycles" materials as they are not able to be used in their original form. Although downcycled, reprocessing materials uses less energy and produces fewer emissions than recycling for remanufacturing. On larger projects, crushing and grading equipment for concrete and asphalt can be brought to the site, saving hauling costs and impacts.

Designing for Deconstruction

Designing sites for deconstruction at the end of their service life will allow for reuse of materials and products in a new project. Structures should be designed to allow for access to removeable connections without breaking, cutting, or compromising structural members. Use of mortar and nailed and welded connections will limit reuse of materials. Removing mortar can be cost prohibitive, and welded or glued connections require cutting of the metal members. As-built drawings of a site can assist in deconstruction and reuse of materials and products.

Materials and Products with Recycled Content

Structures for sites and infrastructure are uniquely positioned to receive recycled materials from demolition, quarry waste, and manufacturing waste as they use large volumes of aggregates in concrete, asphalt pavement, and as base materials. Crushed concrete, recycled tires, glass, sludge, quarry overburden, metals manufacturing slags, and more can be used in site structures (Tables 12.5, 14.8 and 15.4).

Many site construction products incorporate "secondary materials". The US EPA's Comprehensive Procurement Guidelines set minimum recycled content percentages for common site construction products (Table 21.2). Post-consumer recycled content is preferable to pre-consumer content as the life cycle is longer and it is more likely to have been diverted from landfills. Pre-consumer waste should be minimized in manufacturing or reused in other processes.

Material recycling varies widely by region. In areas with waste reduction mandates and rigorous landfill policies, there will usually be robust recycling industries.

Use of secondary materials in some products will significantly reduce the embodied carbon of new materials. For instance, use of 100% recycled aluminum in the production of new aluminum uses just 6% of the embodied energy and produces just 7% of the carbon emissions.

But in some industries, such as metals, there is not enough secondary material to produce a 100% recycled content material. Despite high recycling rates, the demand for new metals is just too great.

Some materials are unsafe to recycle. PVC is a common fencing, piping, and decking material. While technically recyclable, it releases dioxins and sometimes phthalates when heated, therefore recyclers do not typically recycle it (Chapter 21 "Plastic and plastic products").

Mixed materials are not recyclable. Composite lumber is often 50% recycled plastic, and 50% agricultural waste is a high recycled content product, but after its service life, it can only be reused, not recycled, due to the mixed materials. Other product assemblies with multiple materials will pose the same challenges.

Renewable and Rapidly Renewable Materials

Renewable materials are biological materials such wood lumber that can be regrown within a generation or two. Rapidly renewable materials such as bamboo, jute, hemp, and coir can be regrown within five years. Renewable, biobased materials store carbon until they burn or decompose. Most renewable materials can be composted unless they are preserved, modified, or mixed with plastics.

There is a distinct shift in the building industry toward renewable, biobased materials due to their renewability, lower embodied carbon than many mineral-based materials, and carbon storage. This shift could also happen in site construction with two important considerations. First, biobased materials in exterior settings will decompose with exposure to moisture and decay organisms unless preserved, modified, or naturally decay resistant. In recent years, wood treatments such as thermal modification, acetylation, or furfurylation are used to produce durable lumber for exterior settings.

The second important consideration with use of biobased materials is their growth and harvest practices. Agricultural practices for growing plants for biobased materials and bioplastics can impact ecosystems around the fields and forests. Fertilizers, pesticides, and herbicides impact surrounding waterways, insect populations, and habitats. These inputs also add considerably to the embodied carbon of these materials as they are primarily fossil fuel based.

Ecosystems around tropical hardwoods can be severely impacted when foundation tree species are removed for lumber. Research shows decline of these ecosystems within a few years and conversion to pastureland (Bourgoin et al. 2024; Schulze et al. 2008).

Materials or Products From Manufacturers With Take Back Programs

Some manufacturers will offer take back programs for packaging or their products at the end of the service life. EU legislation has resulted in incentives and/or mandates for take back of products or packaging in some countries. Lighting products, pumps, and controllers are good site product candidates for these programs. Packaging takeback programs are more common and have resulted in more resource efficient packaging methods.

Materials and Products That Reduce Water Consumption

Water resources are consumed in raw material extraction, manufacturing, and use of material and products. Water use in quarrying and mining activities results in sludge and mud that is not easily reused because it may contain toxic metals and other constituents. For example, large quantities of red mud result from bauxite processing for aluminum, and there are few reuse possibilities for it (Chapter 20 "Metal and metal products"). Water use in manufacturing can be captured and reused, although some constituents in the wastewater will prohibit this practice.

Site products that use water efficiently, such as drip irrigation, irrigation sensors and timers, and stormwater collection cisterns can reduce water consumption during site operation. The US EPA's WaterSense program certifies water-efficient fixtures.

Responsibly Produced and Healthy Materials and Products

Responsible production of materials and products involves reduced GHG releases, resource use, and waste from extraction and manufacturing; non-toxic materials and material by-products; and safe, fair, and just working conditions at all phases of their life cycle.

Sustainably Extracted and Manufactured Materials and Products

Many material manufacturing sectors are taking steps to decarbonize their practices to reach net-zero carbon by 2050. They are also making resource and water efficiency improvements and reducing toxic constituents and by-products.

Some companies measure their impacts and publish their data in corporate sustainability reports and environmental product declarations. Others have developed remediation and waste management plans for their points of raw material extraction.

The following standards and certification programs encourage responsibly produced materials and products.

- The ***Natural Stone Sustainability Standard ANSI/NSC 373: Sustainable Production of Dimension Stone*** is a multi-attribute voluntary standard designed to encourage and recognize sustainable practices in the natural dimension stone industry.
- The ***Forest Stewardship Council (FSC)*** is a rigorous, voluntary forest certification used around the world. FSC requires sustainable forestry practices and a certified supply chain.
- The ***Initiative for Responsible Mining Assurance*** (IRMA) offers a third-party assessment of mine sites to improve their practices and protect the rights of workers.
- The ***Responsible Steel International Production Standard V2*** was developed to recognize, encourage, and reward steel sites operating in an environmentally and socially responsible manner.

Safe and Healthy Materials

Environmental and human health impacts from materials can occur at all phases of their life cycle. Workers can be impacted by toxic materials or hazardous air pollutants in raw material extraction, manufacturing, fabrication, and recycling phases (Chapter 4 "Healthy and Ethical Materials and Products").

Material sectors with environmental or human health concerns are metals mining, primary metal production, metal finishing, cement production, adhesives and finishes industries, and some plastics. Fence-line communities around manufacturing facilities and along transportation lines are at risk, particularly from chemical transport.

The US EPA maintains the Toxics Release Inventory (TRI) of manufacturer's self-reported data on their toxic releases by compound. Releases can be searched by sector, type of pollutants, and location.

Healthy materials are those that avoid chemicals on the Living Building Challenge Red List and other listings of carcinogens, persistent organic pollutants (POPs), mutagens, and reproductive toxins. Volatile organic compound (VOC) limits for architectural coatings are listed by the California Air Resources Board.

Materials and Products From Companies That Provide Transparent, Third-Party Verified Information

Transparent information about the environmental and human health impacts of materials and products can help designers and specifiers make informed decisions. EPDs communicate life cycle assessment data for environmental indicators such as global warming potential, resource use, ozone depletion potential, eutrophication potential, acidification potential, and photochemical ozone formation (Chapter 5 "Assessing the Impacts of Materials, Products, and Structures"). Other materials and products are certified by standards such as Cradle to Cradle or the Living Product Challenge.

Health product declarations (HPDs) disclose a product's materials and substances and any associated health hazards. They list substances by the Chemical Abstracts Service Registry Number (CASRN), the role the constituent plays in the product, the hazard data, and warnings. Substances included on lists such as the IARC carcinogen groups or the IFLI Red List are identified. The Declare label is another tool for companies to disclose the constituents of their products.

Materials and Products From Companies With Responsible Corporate Practices

Companies with responsible corporate practices protect the health, safety, and well-being of their employees; provide fair compensation and equal opportunity for all workers; protect consumer health and safety; and contribute positively to community health and well-being. These companies will avoid forced or child labor in their supply chain (Chapter 5 "Assessing the Impacts of Materials, Products, and Structures").

Companies with responsible corporate practices will document their goals and practices in corporate sustainability plans, ethics statements, and other publications. Specify materials and products from companies that share this information publicly.

Ethical and Equitable Material Supply Chains

The International Labor Organization's *Global Estimates of Modern Slavery: Forced Labor and Forced Marriage* (ILO et al. 2022) estimates that the manufacturing and construction sectors employ a combined 35% of all forced laborers. The site construction materials identified by the Design for Freedom toolkits as "at the highest risk of embedded slavery" are bricks, dimension stone, rubber, steel and iron, timber, and precursors (especially soda ash and limestone) (Table 4.5). If a material or product was manufactured in another country and is substantially less expensive than nationally produced materials, there is a good chance that it was made with forced, illegal, or even child labor.

A social product declaration (SPD) provides information on the social sustainability performance of a material product at all phases of the life cycle. Structured similarly to its much more common cousin, the EPD, SPDs disclose information about the social and health impacts of a material or product on workers who produce a product or material, communities along the supply and disposal chain, and users of the product or material. Social product declarations are in their infancy with standards developed by EPD International (EPD International 2024).

Materials, Products, and Structures That Provide Ecosystem Services

Ecosystem services are the benefits people obtain from ecosystems. While natural and built landscapes primarily provide ecosystem services, materials and structures can provide them as well leading to truly regenerative landscapes.

Living Structures

Living structures such as living fences and bioengineered embankments provide ecosystem services including carbon storage, photosynthesis, habitat provision, biodiversity, flood control, shade, air purification, and human health and well-being. They also reduce the use of non-renewable material resources and their associated impacts.

Bioengineered embankments, including live staking, live fascines, and brush layering, employing the roots of plants to retain slopes (Chapter 8 "Retaining Walls and Bioengineered Embankments"). Living fences are made from live plant material that is trained, pleached, or woven as it grows to form a natural barrier (Chapter 11 "Rails, Fences, and Screens").

Stormwater Structures

Stormwater structures such as green roofs, porous paving (Chapter 7 "Porous Pavements and Green Roofs"), bioretention basins, and bioswales provide ecosystem services of flood control and water purification along with other ecosystem services provided by plants.

Design of sites to maximize storm water infiltration and improve storm water quality can protect the hydrologic health of a site and a region. The materials from which stormwater structures are built such as geotextiles, drain mats, and paving grids can be considered materials for resilient sites because they facilitate the ecosystem services of these structures.

Erosion Control and Slope Stabilization Structures

Erosion control and slope stabilization products and structures will maintain topsoil and soil health promoting the ecosystem services offered by soil such as water purification, carbon storage, biodiversity, and fertility and food production. Bioengineered embankments stabilize slopes with the roots of plants that grow on the structures. Plants on these structures offer additional ecosystem services (Chapter 8 "Retaining Walls and Bioengineered Embankments").

Biophilic Materials and Structures

Biophilic materials and structures provide ecosystem service benefits of human health and well-being. A biophilic design approach using natural materials connects users of a site or building more closely to nature, leading to better health outcomes (Wilson 1986). Biophilic materials include natural stone, earthen materials, wood, and other biobased materials. Biophilic structures include stone walls, earthen walls, and stone dust and flagstone pavements.

Materials and Products That Reduce the Urban Heat Island Effect

Pavement materials can either contribute to or reduce the urban heat island effect (Chapter 6 "Pavements"). The urban heat island is caused by the retention of solar energy on dark colored surfaces in urban and suburban areas. The heat radiates out of the surfaces elevating temperatures as much as 3.9°C above surrounding areas (EPA 2024). Air quality is negatively impacted by these elevated temperatures.

Use of lighter colored, reflective paving materials or open grid pavement structures with vegetation in the cells can help to reduce the heat island effect. Pervious pavements will cool pavement by allowing air and water to circulate through them (Chapter 7 "Porous Pavements and Green Roofs").

Shifts in Site Structures and Materials

A climate positive, resilient approach to the materials and structures of built landscapes will necessitate a shift in priorities, performance, and expectations for sites and infrastructure. Climate positive sites must be prioritized to address the climate crisis; and sites that are considered regenerative should be constructed from renewable materials and living structures. This shift will lead to a new aesthetic that prioritizes biological materials over technical materials and local, biophilic materials leading to regionally distinctive sites (Figures 1.9 and 1.10).

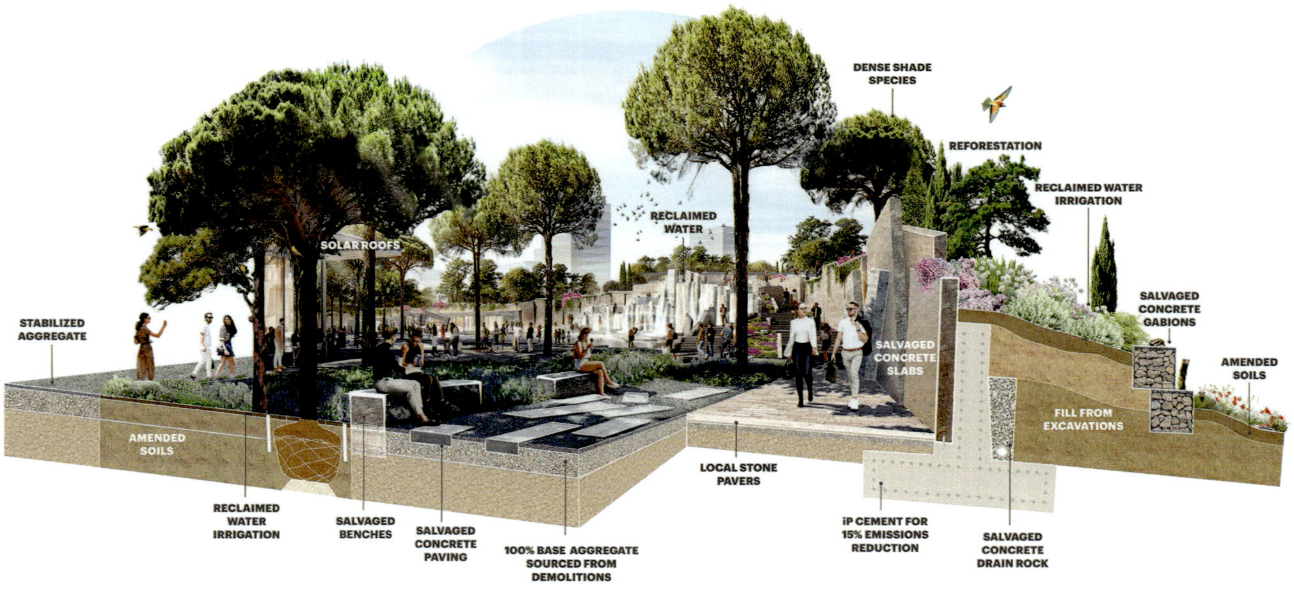

Figure 1.9 Sasaki is transforming the decommissioned Athens International Airport site into a 243 hectare urban coastal park called The Ellinikon Park. The team will repurpose and celebrate the site's heritage through reuse of thousands of tons of on-site structures and materials. Image credit: Sasaki Associates, Inc.

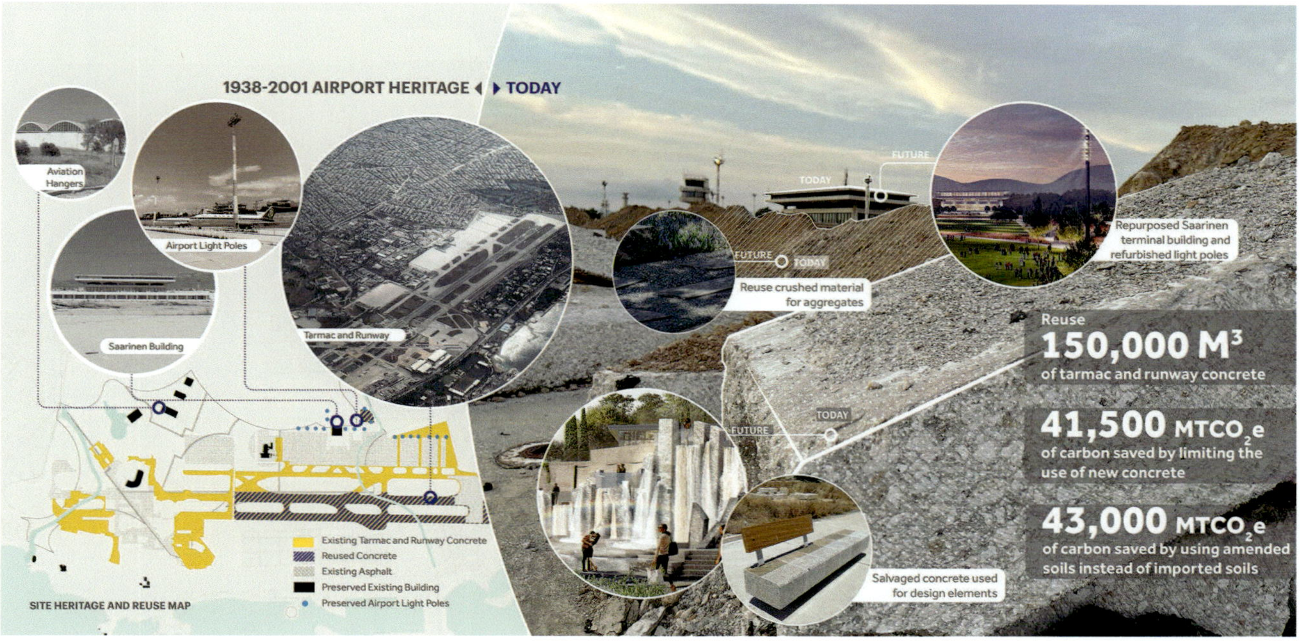

Figure 1.10 Sasaki will repurpose 150,000 cubic meters of tarmac and runway concrete saving virgin resources and substantially reducing the embodied carbon of structures on the site. Sasaki used the Carbon Conscience and Pathfinder tools to reduce embodied carbon and increase sequestered and stored carbon. Image credit: Sasaki Associates, Inc.

Net-Zero Carbon for Resilient Site Structures and Materials

Reducing the embodied carbon emissions of site construction materials and structures is a key priority if we hope to achieve climate positive sites. As more than 80% of the embodied carbon across the life cycle of a typical site is from the materials and structures used to build it, reducing their embodied carbon is critical (Climate Positive Design 2020).

This book demonstrates that it is possible to use standard construction materials such as concrete, asphalt, steel, granite, and brick, but with efficient, reduced emission manufacturing processes, local materials, and material mixes that replace high-carbon constituents.

But we must look beyond materials to reduce the embodied carbon of site structures and even whole sites. The embodied carbon comparison tables in each chapter demonstrate that reductions can be achieved with simple tradeoffs such as designing structures for the strength required, not more; optimizing material use in the structures; and specifying alternative structures. Flexible, repairable structures should be prioritized over rigid walls and pavements that must be replaced if compromised.

The embodied carbon figures used in this book will change rapidly in the coming years as the manufacturing sector improves their practices and transitions to renewable energy sources. The actual figures are less important than the carbon reduction strategies and tradeoffs they illustrate (Figures 1.11–1.13).

Nature-Based, Regenerative, and Biophilic Structures and Materials

The landscape architecture profession is at the forefront of the nature-based solutions approach to mitigate climate change and associated intensive storms. Nature-based landscapes provide ecosystem services of flood mitigation, water quality improvements, carbon sequestration, air pollution reduction, and human health and well-being. **Nature-based solutions must also be employed for site construction materials and structures.**

Living structures such as bioengineered embankments and living fences offer nature-based solutions for retaining slopes, erosion control, and shaping and enclosing space. Employing living structures will avoid use of non-renewable, mineral-based hardscape construction materials, and their associated carbon, resource, and waste impacts.

Where we can't use living structures, we can shift to ethically sourced, renewable biobased materials and structures. It is estimated that this shift to responsibly managed biobased materials for the built environment, such as timber, bamboo, and other biomass-based products could lead to emissions savings of up to 40% by 2050 (UNEP 2023). And if grown and harvested sustainably, these materials can contribute to climate positive sites as they store carbon while in use.

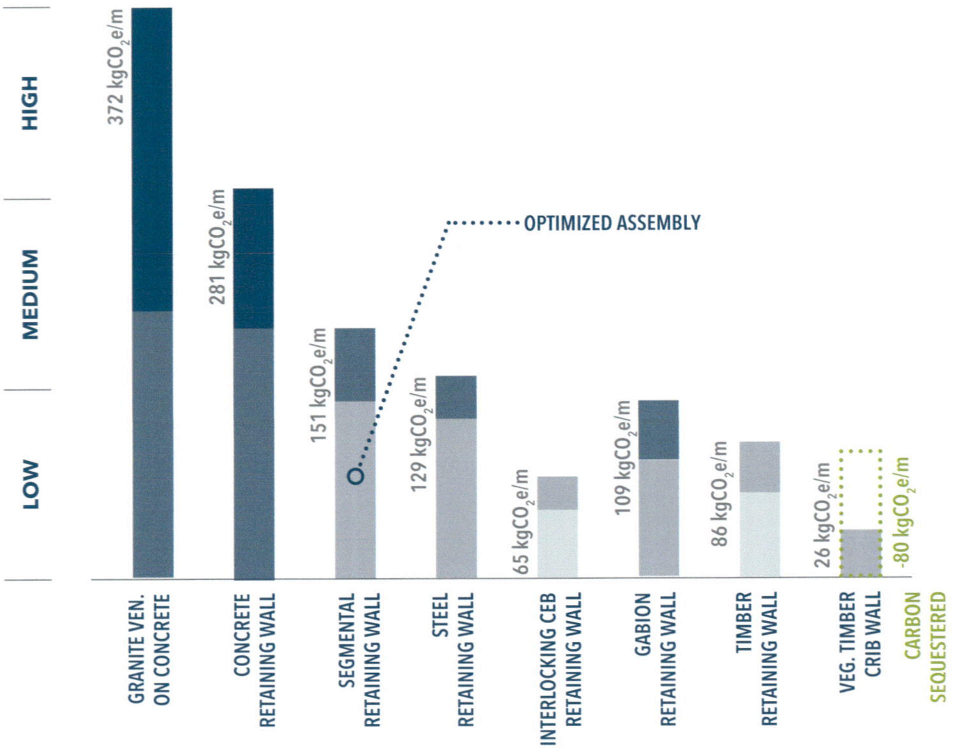

Figure 1.11 Embodied carbon comparisons of typical and alternative site structures such as walls, pavements, decks, and fences are provided throughout the book. These site assembly comparisons demonstrate how embodied carbon can be reduced, sometimes substantially, through material mixes, alternative materials, and smaller, thinner structures. Image credit: Meg Calkins

Figure 1.12 Dan Kiley and Edward Larrabee Barnes collaborated on the design of the Dallas Museum of Art Garden, a great example of modernist landscape architecture. Cast stone and limestone veneer form the clean lines of the pavement and walls. Image credit: Nate Hovee, Alamy, Inc.

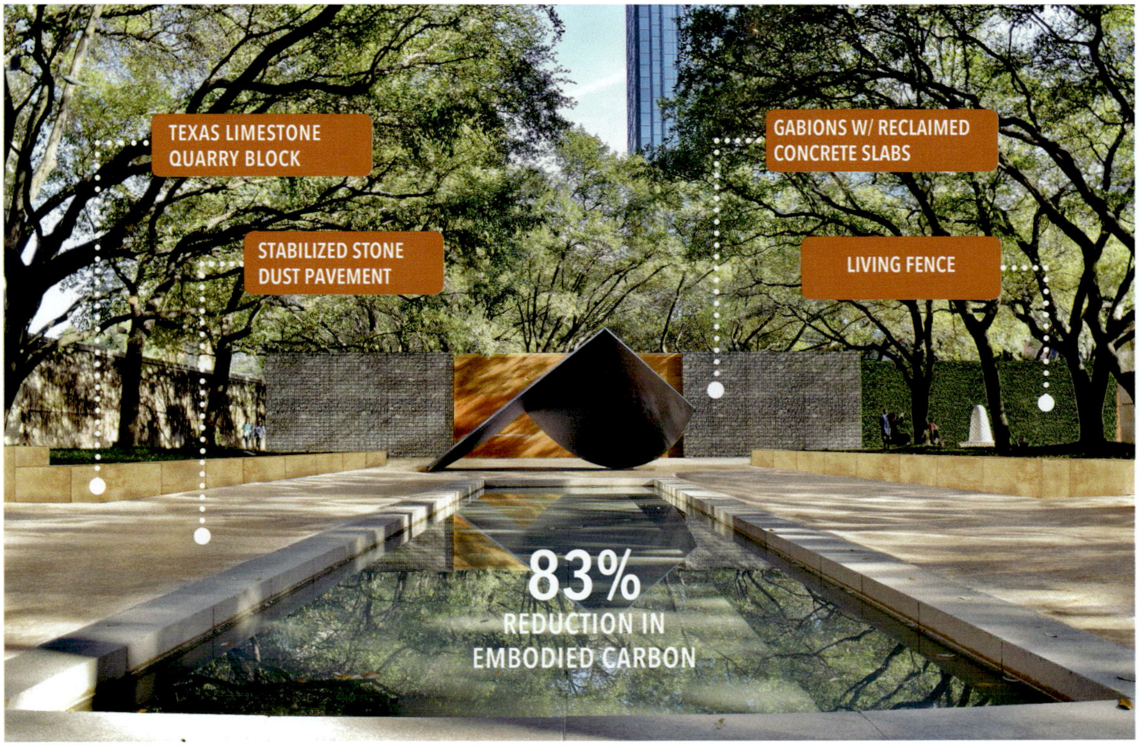

Figure 1.13 This reimagining of Kiley's and Barnes' iconic Dallas Museum of Art Garden substitutes stabilized stone dust pavement, Texas limestone quarry blocks for seat walls, and gabions for the fountain wall. Embodied carbon is reduced by 83%, and roughly estimated construction costs are reduced by 66%. Image credit: Kathleen Elliott and Meg Calkins

Where biobased materials or structures are not possible, we can use minimally processed, local materials such as earth-based masonry and rough stone. These materials are not biobased, but they can be low carbon, biophilic materials that reference the natural landscape of a place.

Exposure to weather, moisture, and decay organisms means that most biobased materials must be treated or coated to protect them. There are good wood modification treatments available now that result in durable, lower carbon lumber (Chapter 18 "Wood Materials and Products"). Products containing lignin and tree resins can stabilize our pavements, and algae and other organisms are used to create biocements. Jute, coir, and other biobased materials can be used in erosion control as they will biodegrade by the time that plant roots establish and stabilize slopes.

Figure 1.14 SCAPE used thermally modified hardwood for the boardwalk and the outdoor classroom at Tom Lee Park in Memphis. Image credit: SCAPE and Brad Howe

A shift to nature-based structures and materials will necessitate a different approach to management and maintenance of sites. The current maintenance regime of "mow, blow, and go" will not support nature-based structures or nature-based solutions. Instead, management of sites and infrastructure must be reconceptualized to allow for regular care of these places by the people who use them. Daily interaction with vegetation, natural structures, and biophilic materials will offer its own kind of ecosystem service for the people who tend and enjoy the landscape (Figures 1.14 and 1.15).

Figure 1.15 The Olin Studio used Kebony lumber for decking and structures throughout Pier 26 in New York City. Immersed in the maritime scrub ecosystem, the viewing shed affords a contemplative moment and a view of the Hudson River. Image credit: Olin Studio

A Shift in Landscape Aesthetics for Resilient, Regenerative Sites

A shift in our aesthetic expectations for sites and infrastructure must happen to embrace use of nature-based structures and biobased and biophilic materials. The current cultural expectation for public open spaces and infrastructure calls for tidy planting beds, green lawns, black seal coated parking lots, and sleek concrete and granite hardscapes. Living structures, nature-based solutions, and some low-carbon materials are not always neat and tidy. They offer a very different aesthetic that will require some getting used to.

A similar shift in landscape planting aesthetics happened over the last 40 years with the acceptance of native plantings that appear rougher and weeder looking than the typical horticultural specimens sold by the mainstream nursery industry. For the first couple of decades, use of native plants was an "alternative" approach to planting design that was only accepted by some. Over the years the native plant aesthetic has become the dominant one in landscape architecture thanks to New York's Highline, Chicago's Lurie Garden, and other high-profile projects that are busy providing ecosystem services within public spaces.

The aesthetic shift to embrace living structures, biobased materials, and biophilic materials must happen in our urban landscapes as well as in suburban and rural settings. Bioengineered embankments belong in a public space between office buildings, living fences screen equipment and services, and gabion seat walls with wood benches can offer biophilic seating. Compressed earth block walls can define space, and some urban open space pavements can be made from stone dust with plant-based stabilizers.

The Structure of This Book

This book provides information to support the multi-pronged approach to materials and structures for resilient sites discussed in this chapter. The first five chapters of the book present an overview of strategies for decarbonization, material circularity, healthy and equitable materials, and tools and resources to obtain transparent information. Themes and techniques discussed in these chapters are woven through the rest of the book.

The next several chapters discuss site structures including dense and porous pavements, bioengineered embankments and retaining walls, freestanding and seat walls, guardrails and fences, and decks and boardwalks. These chapters address both standard approaches to their construction, as well as tradeoffs that reduce embodied carbon, use of resources, and other environmental impacts. Alternative assemblies such as living structures, bamboo structures, and earth masonry construction are addressed.

The second half of the book focuses on reducing the environmental and human health impacts of both standard and alternative construction materials. The chapters provide information and strategies along with the tools, databases, and calculators that are designed to support the critical transition to a regenerative, climate positive, and healthy built environment.

References

Alemu, D., Mesfin, T., and Gudetta Deressa, Y. (2022). "Production of mycoblock from the mycelium of the fungus Pleurotus ostreatus for use as sustainable construction materials". *Advances in Materials Science and Engineering* 2: 1–12.

Bourgoin, C., et al. (2024). "Human degradation of tropical moist forests is greater than previously estimated". *Nature* 631.

Churkina, G., et al. (2020). "Buildings as a global carbon sink". *Nature Sustainability* 3: 269–276.

Circle Economy (2023). *The Circularity Gap Report*. Circularity Gap Reporting Initiative.

Climate Positive Design (2020). *Landscape Carbon Calculator/Pathfinder: Methodology, Data Sources, and Metrics Summary*. Climate Positive Design, CMG and Atelier Ten.

The Concrete Centre (2016). *Whole Life Carbon and Buildings*. Mineral Products Association, The Concrete Centre.

EPD International (2024). "Social product declarations: A declaration of the social sustainability performance of a product life cycle". www.environdec.com/spd, accessed May 25, 2024.

Hio, J. (2022). "Renewable biocement made entirely out of waste material". *Industrial Equipment News*.

International Labor Organization (ILO) (2024). "What is forced labor?" www.ilo.org/topics/forced-labour-modern-slavery-and-trafficking-persons/what-forced-labour, accessed March 31, 2024.

International Labor Organization (ILO), Walk Free, and IOM (2022). *Global Estimates of Modern Slavery: Forced Labour and Forced Marriage*. International Labor Organization (ILO), Walk Free and International Organization for Migration (IOM).

Kriegh, J., Magwood, C., and Srubar, W. (2021). *Carbon-Storing Materials: Summary Report*. Carbon Leadership Forum.

McGhee, G., and Olano, M.V. (2023). *These Three Materials Are Essential, Carbon-Intensive and Produced in Mind Bogglingly Large Quantities*. Canary Media.

Osmani, M. (2011). "Chapter 15 – construction waste". In *Waste: A Handbook for Management*, Second Edition. Eds. T.M. Letcher and D.A. Vallero. Academic Press, Elsevier.

Prince, S., deBaca, L.C., and Thatcher, C. (2020). *Design for Freedom Report: A Call to Action for Practicing Professionals Within the Ecosystem of the Built Environment*. Grace Farms Foundation.

Richardson, Katherine, et al. (2023). "Earth beyond six of nine planetary boundaries". *Science Advances* 9.37: eadh2458.

Rockström, J., Steffen, W., Noone, K., Persson, Å., et al. (2009) "A safe operating space for humanity". *Nature* 461: 472–475. https://doi.org/10.1038/461472a

Schulze, Mark, et al. (2008). "Evaluating ipê (Tabebuia, Bignoniaceae) logging in Amazonia: Sustainable management or catalyst for forest degradation?" *Biological Conservation* 141.8: 2071–2085.

Steffen, W., Richardson, K., Rockström, J., Cornell, S.E., et al. (2015). "Planetary boundaries: Guiding human development on a changing planet". *Science* 347: 736, 1259855

Thacker, S., et al. (2021). *Infrastructure for Climate Action*. Copenhagen, Denmark: UNOPS.

UN Climate Change (2021). *Climate Action Pathway Human Settlements 2021: Vision and Summary*. UN Climate Change.

United Nations Environment Programme (UNEP) (2022). *2022 Global Status Report for Buildings and Construction: Towards a Zero-Emission, Efficient and Resilient Buildings and Construction Sector*. Nairobi: UNEP.

United Nations Environment Programme (UNEP) (2023). *Building Materials and the Climate: Constructing a New Future*. Nairobi: UNEP.

US Environmental Protection Agency (US EPA) (2024). "Heat island effect". www.epa.gov/heatislands, accessed February 12, 2024.

Wilson, E.O. (1986). *Biophilia: The Human Bond with Other Species*. Harvard University Press.

CHAPTER 2

Decarbonization of Materials and Structures

Climate change is the most urgent issue of our time, and construction materials are a significant contributor of the greenhouse gases that are causing it. In the next few years, the decisions that designers, specifiers, engineers, policymakers, and manufacturers make about construction materials can mitigate climate change – or not. Substantial change in the construction materials industry must happen quickly as we are nearing the 1.5°C threshold to limit irreversible impacts to vulnerable countries and ecosystems (WMO 2024).

The 1.5°C Global Warming Threshold

While climate change risks were known to scientists since the 1970s, in the 2010s research showed risk of grave damage to vulnerable regions and ecosystems at less than 2°C of global warming. In 2015, all independent nations in the world (195 countries) signed the Paris Agreement pledging to "hold the increase in the global average temperature to well below 2°C above pre-industrial levels and pursue efforts to limit the temperature increase to 1.5°C above pre-industrial levels" (United Nations 2015).

The 2018 Intergovernmental Panel on Climate Change (IPCC) report (and subsequent reports) have stated that limiting global warming to 1.5°C will reduce impacts on human systems and terrestrial, freshwater, and coastal ecosystems. To limit warming to 1.5°C, "global net human caused emissions of carbon dioxide would need to fall by about 45% from 2010 levels by 2030, reaching net zero around 2050" (IPCC 2018). The report emphasizes that every increment above 1.5°C will increase the magnitude and scope of the impacts. The year 2024 was the first in which global temperature averaged more than 1.5°C (Copernicus and Climate Change Service 2024).

To reduce CO_2 emissions by 45% before 2030, we must act immediately to reduce the embodied carbon of the materials we specify and the structures we build. Manufacturers of construction materials and products must work to decarbonize their industries; and policy and standards must be written to support a shift to low-carbon construction.

Unprecedented Global Growth and a Critical Juncture

Our world is experiencing the largest surge of construction in history due to population growth and rapidly urbanizing development. Between 2020 and 2060, global building floor area is expected to double, and three quarters of the infrastructure that will exist in 2050 is still unbuilt (UN Climate Change 2021).

This is a critical juncture in our attempts to mitigate climate change. If we continue to construct the built environment with the same concrete, steel, and plastics that we have been using for decades, rapid future development will compound the unabated release of greenhouse gas emissions. The gap between our emission targets and carbon emitting activity from the production of construction materials will widen and quickly take us beyond the 1.5°C threshold.

Carbon emissions must be decoupled from population growth and development. Low carbon, resource efficient, and biobased materials should be specified for sites, infrastructure, and buildings. This chapter discusses the impacts that construction materials and material assemblies have on climate change and introduces strategies to decarbonize site construction materials and products. Additionally, detailed information to decarbonize materials, material mixes, and site structures is woven throughout each chapter of this book.

DOI: 10.4324/9781003413974-2

The Carbon Cycle

The carbon cycle is critical to the air we breathe and the viability of life on a habitable planet. Carbon is the fourth most abundant element in the universe and the fifteenth in the earth's crust. Accrued over geological epochs, carbon is stored in the earth's atmosphere, its oceans, forests, soils, and rocky crust. Carbon flows biologically, chemically, and geologically around our planet and into our atmosphere. Some flows, such as the conversion of soil carbon into fossil fuel deposits, are very slow, while others, like the storage of carbon in trees, are relatively fast.

Carbon gives life to plants and animals through a synergistic chemical exchange. Plants absorb atmospheric CO_2 in the presence of light and water in a process called photosynthesis, then emit oxygen into the air we breathe sustaining human and animal life on earth.

The carbon portion of the CO_2 is stored in the structure of plants until the plant is consumed, decays, or burns. In addition to being released to the atmosphere, carbon from decaying plant matter accumulates slowly in soils. Over hundreds of millions of years, the soil-based carbon is transformed into fossil hydrocarbons by temperature and geological pressure. These deposits of coal, oil, and natural gas are locked into the earth crust.

Figure 2.1 The global carbon cycle consists of sources that release greenhouse gas emissions, such as burning fossil fuels and forest fires, and carbon sinks that store and sequester carbon such as vegetation, soils, and the ocean. Image credit: Meg Calkins and Drake Bruner

In the earth's atmosphere, carbon is bonded with two oxygen atoms to form CO_2 gas, an essential component of the global carbon cycle. The absorption of CO_2 by plants and CO_2 emissions from decomposition of organic matter, the combustion of fuel hydrocarbons, and animal and plant respiration is a critical cycle if we are to maintain the habitability of our planet.

Within the last 200 years, relatively recently in geologic time, human activity has disrupted the balance of the carbon cycle by extracting and burning fossil fuels, releasing their embedded carbon at unsustainable rates. This anthropogenic disruption has impacted our habitats, species, air, water, and soil to the point that the earth's geological, chemical, and biological composition is fundamentally altered. The resulting climate change threatens human and species health and well-being around the world (Figure 2.1).

The Scope of Global Human Caused Greenhouse Gas Emissions

Greenhouse gas emissions grew by 53% between 1990 and 2019. In 2021, global GHG emissions from energy combustion and industrial processes reached 36.3 billion metric tons (gigatons) of CO_2 equivalents (IEA 2023). This includes energy to power buildings, manufacturing facilities, kilns, power equipment, and transportation.

Emissions from coal combustion in 2021 were 15.3 Gt, about 42% of all energy sector emissions. Around 7% of total GHG emissions result from the cement and concrete industry, while around 8% result from the iron and steel industry. Three percent of total GHG emissions resulted from land use change, conversion of forests to development, and the forest products sector (WRI Climate Watch 2022).

China is the world's largest carbon emitter at over 12 billion metric tons of CO_2 emitted in 2021. This is over double the CO_2 emissions from the United States in second place. However, per capita emissions in the United States are double the emissions of China and eight times those of India, the third largest emitter (WRI Climate Watch 2022).

Climate Justice and Equity

Climate justice is an approach to climate action that considers the unequal impacts of climate change on vulnerable and marginalized populations. Until recently, most of the greenhouse gas emissions stemmed from North America, Australia, and Europe, yet the climate change impacts were most felt by Bangladesh, central and east Africa, Syria, and Afghanistan (IRC 2023).

The greenhouse gas burden of construction materials and products disproportionately impacts some of the most vulnerable populations; therefore, the embodied carbon of construction materials is a climate justice and equity issue. Our use of high-carbon materials impacts populations far from the project site, and very often these populations have fewer advantages and access to resources to protect themselves. Understanding the embodied carbon of materials and products at all phases of the life cycle, then taking steps to reduce it, is an important step in addressing climate justice.

Glossary of Embodied Carbon and Decarbonization Terms

Greenhouse gas (GHG): Greenhouse gases are a class of gases that have the potential to warm the climate by absorbing infrared energy that is trapped in the atmosphere and prevented from escaping into space, causing the greenhouse effect. The most common GHGs are water vapor (H_2O), carbon dioxide (CO_2), nitrous oxide (N_2O), methane (CH3), and ozone (O_3). GHGs have different global warming potentials, and the amount of time they exist in the atmosphere varies by gas. CO_2 emitted today compounds emissions from the past, overwhelming the earth's ability to reabsorb excess CO_2 photosynthetically leading to lasting climate impacts.

Global warming potential (GWP): Global warming potential is an index that is based on the potential of a greenhouse gas to warm the climate and the

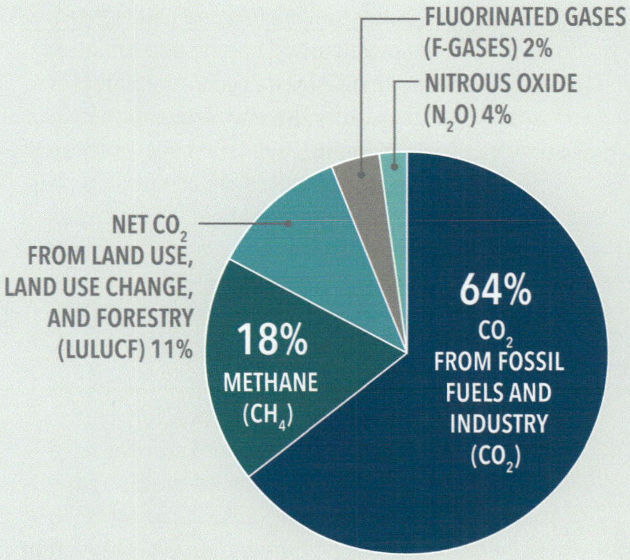

Figure 2.2 Greenhouse gas emissions by type and percentages from the Intergovernmental Panel on Climate Change (2019). Image credit: Drake Bruner and Meg Calkins

amount of time the gas stays in the atmosphere. The GWP index was developed to facilitate comparisons of the GWP of different gases. GWP is the measure of how much energy the emissions of 1 metric ton of gas will absorb over time relative to the emissions of one metric ton of CO_2 over 100 years. CO_2 has a GWP of 1 while methane has a GWP of 27–30 over 100 years. Nitrous oxides have a GWP 273 times CO_2; and CFCs, HFCs, HCFCs, PFCs, and SF6 trap more heat and can exist in the atmosphere for thousands or tens of thousands of years (US EPA 2024) (Figure 2.2).

Carbon dioxide (CO_2): CO_2 gas is a tasteless, odor-free by-product of fossil fuel combustion, burning and decomposition of plant matter, burning of fossil fuels in the chemical industry, and other processes. Aside from water vapor, CO_2 is the most prevalent of greenhouse gases, and its long half-life means it plays a primary role in climate change.

CO_2 equivalents (CO_2e or CO_2 eq.): CO_2 equivalents are the unit of measure used to express global warming potential. The measure is primarily expressed in kilograms of CO_2 equivalents (CO_2 eq. or CO_2e) per functional unit.

Embodied carbon (EC): EC is a measure of the amount of greenhouse gas emissions from a material, product, built site, or building. Embodied carbon is measured by LCAs and expressed in kilograms or metric tons of CO_2 equivalents per unit of measure such as cubic meters, metric tons, or square meters (e.g., 234kg CO_2e/m^3).

Carbon footprint: A carbon footprint is another term for the measure of the amount of greenhouse gas emissions from a material, product, built site, or building. Like embodied carbon, carbon footprints are measured in kilograms or metric tons of CO_2 equivalents per unit of measure such as cubic meters, metric tons, or square meters.

Operational carbon: Operational carbon is the GHGs released from the operations of a site or building during the use phase. These emissions occur from use of electricity or other energy, water, cleaning chemicals, and power maintenance equipment such as mowers and blowers.

Non-renewable energy source: A source of energy that can't be replaced after it is used. These include fossil fuel-based energy and heat sources such as coal, oil, and natural gas. When burned, they release GHGs, chemicals, heavy metals, and particulate matter into the air.

Renewable energy source: Energy or heat that is from a source that can be renewed. Renewable energy sources include wind, sun, water, geothermal deposits, and biomass.

Carbon source: Carbon sources include fossil fuel combustion, burning and decomposition of plant matter, emissions from the chemical conversion of minerals or fossil fuels, and other processes.

Carbon sinks: A carbon sink is a natural or artificial process that removes a greenhouse gas from the atmosphere and stores it. Globally, plants and the ocean are the two largest carbon sinks. These large carbon sinks are sometimes referred to as carbon pools. Artificial carbon sinks include carbon stored in biobased materials and products, and CCUS.

Combustion emissions: GHG emissions result from the combustion of fossil fuels to produce electricity, heat, or energy. Combustion emissions can be reduced by using renewable energy sources.

Process emissions: Process emissions result from the chemical transformation of materials such as fossil fuels or limestone into new materials such as plastics or cement with the use of heat. Process emissions are very hard to reduce so the primary strategy is to reduce use of the material. CCUS is also employed to reduce process emissions.

Net-zero carbon: Net-zero carbon refers to the balance between CO_2e from embodied carbon emissions and/

or operational carbon emissions, and the amount that is removed from the atmosphere through carbon sequestration or storage. Net-zero can be achieved by emissions reduction strategies and emissions removal strategies such as planting more vegetation.

Biogenic carbon: Biogenic carbon is carbon that is stored in biological materials such as plants, trees, and soil. It is removed from the air and distributed in the biomass of a plant through photosynthesis. In biobased products this is referred to as stored carbon, biogenic carbon, or "GWP-biogenic" in carbon accounting. The process that fixes carbon in the biological materials is called *carbon sequestration*, and the resulting carbon stored is referred to as biogenic carbon. Refer to box "Biogenic Carbon and Carbon Stored in Construction Materials" for more information.

Carbon sequestration: Carbon sequestration is the process of plants fixing and storing carbon in their tissues (cellulose, sugars, lignin, etc.) because of photosynthesis, and also in soils from root exudates. This builds soil microfauna and mycorrhizal communities resulting in soil carbon accumulation.

Life cycle assessment (LCA): Life cycle assessment is a method used to quantify the environmental impacts of a product, process, building, or site over its entire life cycle. LCAs measure all inputs such as raw materials, recycled materials, energy, and heat, as well as all outputs including emissions and waste to air, land, and water.

Environmental product declaration (EPD): Environmental product declarations are developed by the manufacturers of construction materials and products to communicate LCA information in an objective, transparent, verified, and standardized format. Theoretically, EPDs allow designers and specifiers to evaluate similar products to make informed choices for their greenhouse gas reduction aims and other environmental priorities (Chapter 5 "Assessing the Impacts of Materials, Products, and Structures").

Embodied Carbon and the Built Environment

The building sector, including buildings, sites, and infrastructure, is responsible for more than 40% of GHG emissions annually. In 2022, 26% of emissions were *operational carbon* from the energy used to heat, cool, and power buildings and infrastructure, while 6% were *embodied carbon* from the materials and products to build buildings, and 7% (2.4GT) are the embodied carbon from the construction of sites and infrastructure (Architecture 2030 2024).

As operating carbon emissions decline with energy efficient buildings and sites and increased use of renewable energy, the embodied carbon of construction materials and products will be responsible for a greater percentage of GHG emissions associated with the construction industry. Therefore, it is critical to focus on the decarbonization of construction materials to meet the IPCC goal of a 45% reduction in carbon emissions by 2030 (from 2010 levels) and achieve net-zero carbon by 2050 (Figure 2.3) (IPCC 2022).

The embodied carbon of the materials and structures used to build site and infrastructure projects are responsible for 80% or more of their total carbon footprint. Operational carbon is only responsible for less than 20% over the life of a typical project (Climate Positive Design 2020). Operational carbon of sites and infrastructures results primarily from lighting, pumps and controllers, and maintenance equipment.

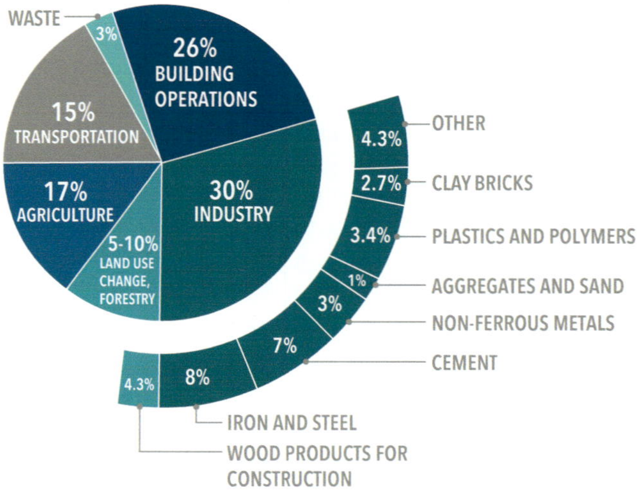

Figure 2.3 World GHG emissions in 2020 by sector from Climate Watch. Total global GHG emissions for 2020 were 47.5 gigatons of C02e (Climate Watch based on data from IEA 2022). Image credit: Meg Calkins and Drake Bruner

Greenhouse Gas Emissions Across the Life Cycle of Construction Materials and Products

Greenhouse gas emissions occur during all life cycle phases of construction materials and products (Figures 2.4 and 2.5). Generally, the greatest carbon emissions occur in

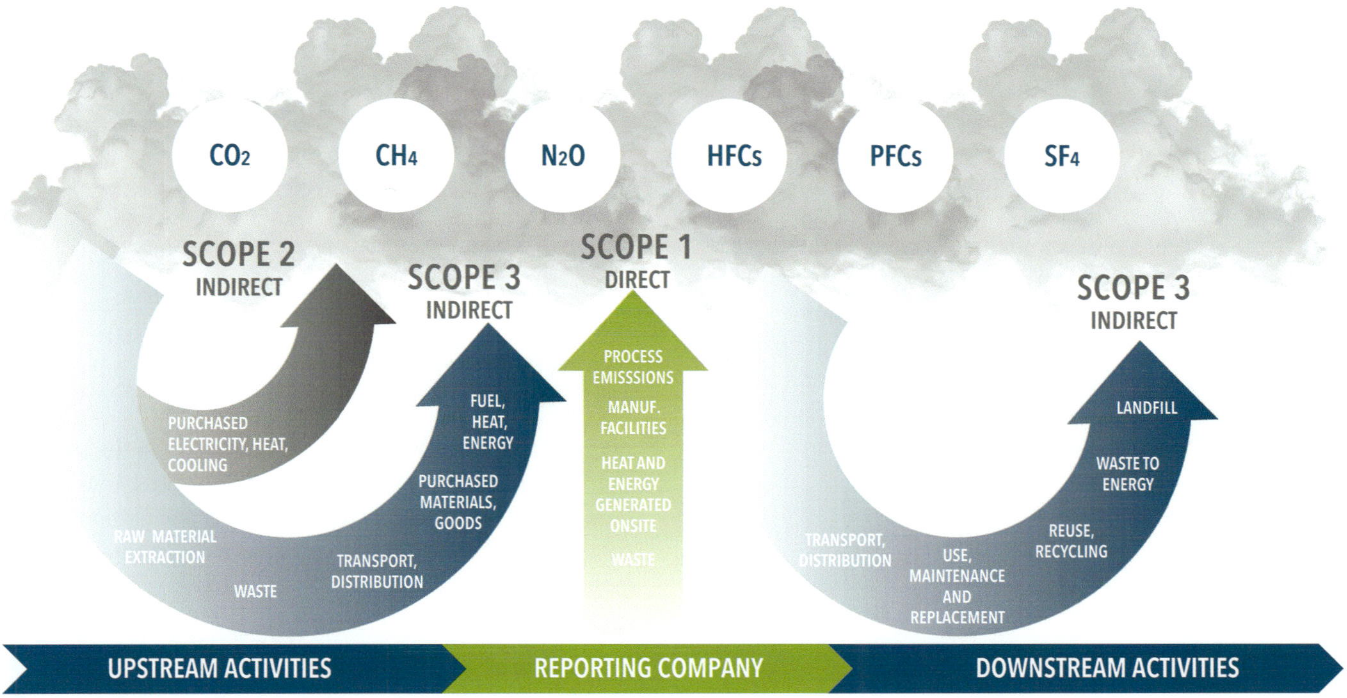

Figure 2.4 Scope 1, 2, and 3 emissions depicting upstream activities, downstream activities, and activities from the reporting company in a life cycle assessment. Adapted from World Resources Institute (WRI) Greenhouse Gas Protocol. Image credit: Meg Calkins and Drake Bruner

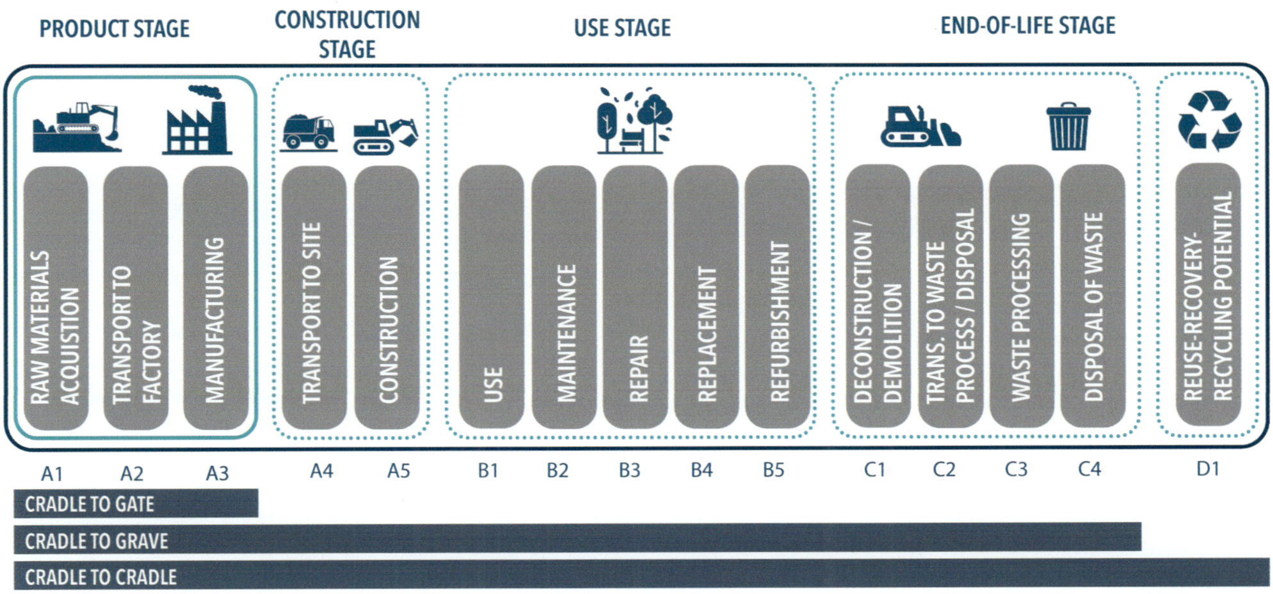

Figure 2.5 Life cycle assessment stages as defined by ISO 14040 and 14044. The typical LCA covers stages A1-A3, Cradle to Gate, although a more comprehensive LCA will examine impacts for all stages. Image credit: Meg Calkins

manufacturing phases, followed by emissions from the raw materials extraction phase. Transportation emissions can have a significant share of the emissions for lower carbon materials or internationally sourced materials.

GHG emissions are classified in three categories:

Scope 1 emissions are from sources that a manufacturer or company controls directly. These include emissions from equipment, buildings, vehicles, and on-site energy and heat.

Scope 2 emissions are indirect emissions that are a result of industrial activities, but they come from sources that the company does not control. Typical scope 2 emissions are purchased electricity, heat, or cooling.

Scope 3 emissions are indirect emissions that occur across the life cycle of a product but are controlled by at least one different company. Scope 3 emissions include resource extraction, manufacturing, fabrication, transportation, distribution, use, and disposal. Scope 3 emissions are often divided into upstream and downstream emissions with downstream emissions coming from use and disposal (WRI 2011).

A1 Raw Material Acquisition

Raw material extraction for some materials, such as dimension stone and metal, is very resource and energy intensive with fossil fuel powered machinery excavating large amounts of overburden material to access high-quality stone or metal ores. Vegetative cover is completely removed to obtain raw materials for nearly all construction materials except for selectively harvested lumber from natural forests or plantations. Once disturbed, most sites are not revegetated, instead they are developed or converted to pastureland. In carbon accounting this is called "land use change".

Carbon dioxide releases from disturbed soil are a major impact of raw material extraction, but these emissions are not accounted for in a typical life-cycle assessment. While this CO_2 release can be substantial, it is nearly impossible to quantify. One study estimates that in the 12,000 years of farming on our planet, soil disturbance has released 110 billion metric tons of CO_2 from the top layer of soil (Sanderman et al. 2017).

Another major carbon impact of raw material extraction is loss of carbon sequestration from removal of vegetation to get at clay, stone, and metal ores. Once vegetation is removed, the carbon sink is lost. This loss is also not usually accounted for in LCAs or life-cycle inventories.

A2, A4, C2 Transportation

Transportation impacts occur between, and sometimes within, each phase of a material or product life cycle, and they can be very carbon intensive particularly for the heavy materials used in site construction. For instance, transporting a metric ton of concrete pavers 250km by truck will release a similar amount of GHG emissions as the production of the pavers (A1–A3) (Table 2.1).

Table 2.1 Greenhouse Gas Emissions Factors of Transportation Types

Mode of Transportation	CO_2e Emissions Factor (kg CO_2e/t-km)	CO_2e Emissions Factors (kg CO_2e/kg-km)
Medium/Heavy Duty Truck	1.07	0.00107
Rail	0.15	0.00015
Waterborne Craft	0.46	0.00046
Aircraft	7.37	0.00737

Source: EPA 2022

Note: Emission factors of transportation will change over time as efficiencies are improved, and more renewable fuel sources are employed.

The 21st century materials industry is global with stone quarried in India, shipped to China for fabrication, then used in Paris. Wood may be plantation grown in Norway, thermally modified in Estonia, then shipped to LA for use. Inexpensive labor outside North America and Europe, combined with low shipping costs, have contributed to the global materials economy.

The best way to reduce emissions during the transportation phases is to specify regionally sourced and produced materials and products. This is not a simple task for two main reasons. First, understanding the path that a material or product travels from raw material extraction to distribution can be very challenging. Often, distributors don't know where the product was manufactured, and manufacturers are not aware of where the raw materials originated.

In the international lumber industry, tropical hardwood goes through so many intermediaries that a chain of custody can be challenging to obtain. Specifying Forest Stewardship Council certified wood with a chain of custody is the only way to know where the wood was sourced. It is not uncommon for "local" dimension stone to be specified within 200 kilometers of a project site only for it to be sent to China to be cut, processed, and fabricated and then sent back to the project site.

Reduced transport emissions can be achieved with on-site reclaimed structures and materials; use of minimally processed materials such as quarry blocks, reclaimed concrete slabs, or living structures that can be procured near the site; or by using lighter weight materials such as wood instead of concrete (Figures 2.6 and 2.7). Other approaches include fuel efficient or alternative fuel trucks, and delivery coordination to ensure full trucks and minimized hauling.

A3 Manufacturing and Fabrication

It is not unusual to have multiple steps in the manufacturing and fabrication of materials and products. Often these steps occur at different locations. The more complex the product, the more complicated it is to quantify the embodied carbon from this phase as well as the transport between manufacturing facilities.

Figure 2.6 08014 architectural used local reclaimed railroad ballast in galvanized gabion wall structures for the landscape regeneration of the Camí de Cabrianes. Image credit: Pol Viladoms

Figure 2.7 On-site reclaimed materials are arranged artfully by hand into a new landscape of ruins called Still Alive at a park in Torino, Italy, by Wagon Landscaping. Image credit Yann Monel

The manufacturing of materials requires *inputs* of raw materials, water, and fuel to power equipment and sometimes heat materials. *Outputs* from manufacturing are the material or product as well as solid, liquid, and gaseous waste.

Manufacturing of materials creates greenhouse gas emissions in three ways. Burning fossil fuels to provide heat releases scope 1 *combustion emissions*. Burning fossil fuels for electricity at a power plant to power factory equipment releases scope 2 combustion emissions. And chemically transforming some materials into new products releases *process emissions*. The most common example of process emissions is the release of CO_2 from calcining calcium carbonate (limestone) to produce cement clinker. These process emissions account for about 60% of the carbon footprint of cement. Other materials that release process emissions are plastics and metals.

Many construction materials such as brick, cement, plastics, and metals require extreme and sometimes prolonged heat for their manufacture. For instance, bricks are fired at temperatures of 900–1250°C for 10–40 hours resulting in an approximate embodied carbon of 0.273kg CO_2e/kg of brick (BIA 2020).

To meet net-zero 2050 targets, materials manufacturing sectors are taking steps to reduce the greenhouse gas releases from their production processes. The simplest way to reduce combustion emissions in manufacturing is to transition from fossil fuel-based heat and energy sources to renewable sources such as solar, wind, hydropower, and biomass.

Manufacturers are using more efficient equipment or modified processes to reduce scope 1 emissions as well. Carbon reductions are already happening with emissions from the manufacturing sector down 17% between 2002 and 2021, even while output increased (CBO 2024). Designers can reward reduced carbon manufacturing practices by specifying materials from producers who have lowered their scope 1 and 2 emissions.

Process emissions are far more difficult to reduce than combustion emissions as the CO_2 emissions result from chemical conversion of the materials. Some manufacturers are installing CCUS equipment to capture process emissions, but this is expensive, and the infrastructure to store the carbon is not yet widespread.

In the concrete sector, process emissions can be captured from cement kilns and stored in cement and concrete. CCUS does not require major modifications to cement kilns and is already being implemented in some plants; however it is a relatively new technology and only stores a fraction of the process emissions (Chapter 14 "Concrete Materials and Products").

The least expensive approach to reducing process emissions from materials such as cement, steel, and plastics is to simply use less of them. Specifying alternative materials, designing smaller structures, and optimizing the use life of these materials will reduce process emissions and could also save money.

A5 Construction and Installation

Measurable CO_2 emissions from the construction phase result primarily from the vehicles, equipment, and power tools used to build a site. Electrified equipment powered by renewable energy sources will make the biggest step toward decarbonizing the construction phase. Other approaches are to require anti-idling, use biofuel blends, avoid generators, and use solar chargers for smaller tools.

Creating a Construction Activity Carbon Reduction Plan (CACRP) outlining requirements for tracking emissions in phases A4 (material transport to site) and A5 can get the contractors, subcontractors, and consultants on the same page with carbon reduction goals. CACRP requirements should be reflected in RFPs, contracts, and subcontractor exhibits.

B1–B5 Use, Maintenance, Repair, Replacement, and Refurbishment

From an embodied carbon standpoint, the use phase of site materials and structures is among the least carbon intensive, although products with shorter use lives such as plastics or pressure-treated softwood will require replacements, adding additional embodied carbon to the life of a site. Asphalt pavement will require resurfacing every 7–10 years, adding to its carbon footprint as well.

Operational emissions from the use phase of materials and products include electricity to power lighting, irrigation, and fountain pumps and controllers. Specifying energy efficient bulbs and equipment and using renewable energy sources will reduce operational emissions.

Operational emissions from fossil fuels to power maintenance equipment such as mowers and blowers was responsible for more than 30 million metric tons of CO_2 emissions in 2020 (EPA 2020). Maintenance or management plans should address the use of power equipment such as power washers, mowers, and blowers. Electrification of power equipment can reduce carbon during this phase.

C1–C4 End-of-Life Stage

While carbon emissions in the end-of-life stage occur from the equipment to deconstruct or demolish the site structures, the largest source of emissions is from the truck transportation to haul high volumes of heavy materials to disposal or recycling facilities. These emissions can be substantially reduced by reusing or refurbishing structures and deconstructed materials on site. This practice will reduce emissions from both debris hauling and from the extraction, manufacture, and transport of new construction materials.

The greenhouse gas emissions from waste processing (C3) and waste disposal (C4) phases vary widely by disposal approach.

Incineration or waste-to-energy is promoted as an alternative to landfilling, but it is very emissions intensive. Each metric ton of incinerated solid waste releases between 0.7 and 1.7 metric tons of CO_2 (ZWE 2020). Additionally, incineration of plastic products can release harmful dioxins, phthalates, and carbon tetrafluoride (CF4), a greenhouse gas that is 6,500 times more potent than CO_2 that is estimated to remain in the atmosphere for 50,000 years. ***Landfilling*** is a common way to dispose of demolition materials, but some states and municipalities are banning disposal of commonly recycled demolition materials such as concrete, asphalt, wood, and stone. Landfills are a major source of greenhouse gas emissions from the release of methane and carbon dioxide as biodegradable materials like food, wood, and yard waste decompose in the presence of moisture. Landfills emit about 400kg of CO_2e per metric ton of organic waste (Nordahl 2020). Methane released from landfills can be captured and used as an energy source for electricity or heat. The benefits are twofold with landfill emissions reduced and reduction of fossil fuel burning for other industrial processes.

When demolition materials are not reclaimed or recycled, there are additional emissions from the production and transportation of new materials. Carbon sequestration is reduced as vegetative cover is removed to mine or quarry new raw materials.

D1 Reuse, Recovery, Recycling

Recycling industries for some site construction materials are robust, and materials like metals, asphalt, concrete, wood, bricks, and stone are easily reprocessed or recycled into new materials. The carbon footprint of phase D1 – reuse, recovery, and recycling – varies by material. If materials are reused on site, their carbon footprint is substantially minimized. If the materials are recycled, their footprint includes transportation to a recycling facility and energy for the recycling process.

Some materials, such as plastics and metals, require energy to heat the recycled material to reconstitute it into a new product. While recycling plastics lowers the demand for new plastics, some, such as PVC, can release harmful toxins when they are heated for recycling. Still, recycling metals can substantially reduce the embodied carbon of new metal products. For example, steel with 90% scrap is just half the embodied carbon of steel with only 30% recycled content.

Construction Materials and Their Embodied Carbon

Three construction material sectors, iron and steel, cement, and fossil fuel derived chemicals (precursors for plastics) were responsible for nearly 19% of all greenhouse gas emissions in 2022. This is more that all the emissions generated by the United States combined (IEA 2024). Billions of metric tons of these emissions-intensive materials are produced globally with a huge impact on our climate.

Most material industries have plans for net-zero emissions by 2050. These plans are detailed in the materials chapters of this book and most depend on a broad array of stakeholders for implementation. Designers can reuse structures and reclaimed materials, design optimized structures, and select lower carbon materials. Designers and engineers can specify lower carbon material mixes for standard materials such as concrete. Material manufacturers can shift to renewable sources of heat and energy, develop process efficiencies, and capture and sequester carbon. Policymakers can set carbon caps and reward producers and specifiers for achieving carbon footprints below baselines.

A very important step that all stakeholders can take is to reduce the transportation of heavy site construction materials in all phases of a material or products life cycle. A transition from global to regionally sourced, manufactured, fabricated, used, and recycled/reclaimed materials will substantially reduce their embodied carbon.

Figure 2.8 The embodied carbon of one typical concrete parking space is just under one ton of carbon equivalents (CO_2e). The actual volume of one metric ton of CO_2 is 10 meters across – larger than the parking space itself. Image credit: Meg Calkins

Cement and Concrete

In 2022, more than 4 billion metric tons of cement were produced for more than 40 billion metric tons of concrete, the most abundantly manufactured material on earth (IEA 2024). Forty-two percent of this was used in infrastructure projects, including landscapes (Figure 2.8). As the binder in concrete, cement makes up only 9–15% of the volume of concrete, but it is responsible for around 90% of the carbon emissions. The carbon emissions from the production of cement were 2.4 gigatons in 2022, over 7% of all CO_2 emissions from energy combustion and industrial processes.

Carbon emissions from cement occur in two ways. First, fossil fuel combustion produces the energy required to heat the limestone to 1480°C (2700°F) and transform it into clinker. Second, as limestone is calcined, it releases CO_2 through the chemical transformation. Accounting for about 60% of the carbon footprint of cement, these are called process emissions.

While the cement industry can transition to renewable energy sources to heat the limestone, the process emissions from portland cement are more challenging to reduce. Some of the process emissions can be captured and stored, but the only way to get to zero is to use cement substitutes for a portion of the portland cement or alternative cementitious materials. These and other decarbonization approaches are discussed in Chapter 14 "Concrete Materials and Products".

Iron and Steel

The iron and steel industry emitted over 2.6 gigatons of CO_2 in 2022 accounting for about 8% of all emissions from energy production and industrial processes globally (IEA 2023). Twenty-one percent of all steel is used in infrastructure projects. In site construction, steel is used for concrete reinforcing, guardrails, fences, handrails, overhead structures, edging, and retaining structures.

Like cement, the production of steel releases two types of emissions. Most emissions stem from energy used to heat the furnaces to 900–1100°C for 15–36 hours while a smaller amount of emissions result from the release of CO_2 during the coking process.

The embodied carbon for a metric ton of steel varies widely by recycled content, the type of furnace producing it, and the production process. Steel produced in China is primarily produced in blast furnaces then basic oxygen furnaces (BF-BOF). This process emits about two metric tons of CO_2 for every metric ton of steel produced. In the United States, over 70% of steel is produced in an electric arc furnace (EAF) with over 90% steel scrap. The emissions from this steel are just under a metric ton of CO_2 per metric ton of steel (Chapter 20 "Metal and metal products").

Hydrogen-based direct reduced iron (DRI) is a very promising emerging DRI technology that reduces iron ore with hydrogen instead of carbon, therefore the waste gas is water, not CO_2. The hydrogen-based DRI process can take place in existing EAF plants and can have a substantially lower carbon footprint at 0.08 t CO_2/t steel produced but it is not yet a widespread technology (Mission Possible Partnership 2022).

Aluminum

Aluminum may have the highest and most varied carbon footprint of any construction material with 3 to 12 metric tons of CO_2 released for every metric ton of aluminum produced. Sixty-five million metric tons of aluminum were produced in 2021 accounting for less than 1% of GHG from energy production and industrial processes (IEA 2023).

Producing aluminum is very energy intensive, so if renewable energy is used, such as hydropower, the carbon footprint can be lower. If the aluminum is made in China, it is likely produced with fossil fuel energy sources, and the footprint is high. Decarbonizing aluminum production will rely most heavily on renewable energy sources and use of higher recycled contents.

Plastics and Chemicals

The chemicals industry emitted 1.3 gigatons of CO_2 in 2022 (IEA 2023). A large portion of the primary raw materials in the chemicals industry are carbon-based fossil fuel feedstocks for polymers and plastic products with process emissions accounting for 24% of the CO_2 released. It is very challenging to decarbonize the chemical industry because carbon is the building block for all chemical-based products such as coatings, plastics, and fertilizers.

To decarbonize, the chemical industry is diversifying their feedstocks away from fossil fuel to biobased materials. Like other sectors, the use of renewable energy sources is also growing. Plastics are discussed in Chapter 21 "Plastics and Plastic Products" and Chapter 19 "Biobased Materials and Products", while coatings are discussed in Chapter 18 "Wood Materials and Products" and Chapter 20 "Metals and Metal Products".

Brick Masonry Materials

Fired bricks are an energy and carbon-intensive material with firing temperatures ranging from 900–1250°C for firing times of 10–40 hours. In addition, 37% of emissions are process emissions from the clay and other elements as it is heated. To decarbonize, the brick industry is transitioning to use of renewable energy sources, incorporation of wastes such as sludges and ashes, and capture and reuse of waste heat (Chapter 17 "Fired Clay Bricks").

Earth-Based Materials

Earthen structures made from compressed earth block, rammed earth, and adobe brick are rare in modern construction despite having been built for millennia. Due to the low-carbon cost and abundantly local raw materials, earth-based materials offer a good low-carbon alternative to highly processed, energy-intensive materials. As a major carbon footprint of site construction is the transport of heavy, bulky building materials, using materials from on, or near the site, will reduce the carbon footprint of site construction substantially.

UN Habitat and other NGOs have identified earth-based construction, primarily compressed earth block (CEB), and rammed earth as local building methods that can address the housing crises in some rapidly developing countries in Asia and Africa.

Earthen materials can also be used to form site structures such as freestanding and retaining walls. Most earth-based masonry is stabilized with higher carbon additives such as cement, lime, or asphalt that, if used in high percentages, can add substantially to the embodied carbon of earthen structures. But cement substitutes can be used as well as a wide variety of waste products (Chapter 16 "Earth-Based Materials").

Stone

Stone can be a very low or a very high-carbon material depending on several factors. Stone that is locally or regionally sourced and fabricated minimally is a durable, low-carbon material that reflects the local geology of a site. Stone quarry waste can be used for quarry block seating, and broken slabs can be used for paving.

Yet dimension stone often has a global supply chain with granite quarried in Brazil, sent to China for fabrication by inexpensive labor, then sent to the United States for use. The economic cost of shipping stone is low enough that it "pays" to send it to a low wage country for fabrication using lower cost labor. This is a tremendous carbon burden adding substantially to the carbon footprint of stone (Chapter 13 "Stone Materials").

Wood Products

Wood is another construction material with a highly variable carbon footprint. As a structural material, it compares favorably to steel, aluminum, and concrete as it is lower embodied carbon when used in a similar capacity for deck structures, fencing, and overhead structures.

But carbon impacts from wood can be substantial if it is harvested from sensitive ecosystems such as tropical rainforests, then shipped thousands of kilometers. Carbon impacts of wood that are not accounted for are loss of carbon sequestration by forests that are clear cut and CO_2 releases from soil disturbance.

Wood grown regionally, harvested sustainably, and treated or thermally modified can be a low-carbon material, and it is renewable. An added benefit is that it stores some carbon until it is burned or decayed (Chapter 18 "Wood Materials and Products").

Biogenic Carbon and Carbon Stored in Construction Materials

Biogenic carbon is carbon that is removed from the air and distributed in the biomass of a plant through photosynthesis. This process, which fixes carbon in the wood, is called *carbon sequestration*, and the resulting carbon stored is referred to as biogenic carbon (CEN 2014a, EN 16449). Wood is roughly 50% carbon measured by its dry weight.

When a tree is cut for wood products or a plant is harvested for a biobased product, the cut biomass continues to store some of this carbon for decades until it

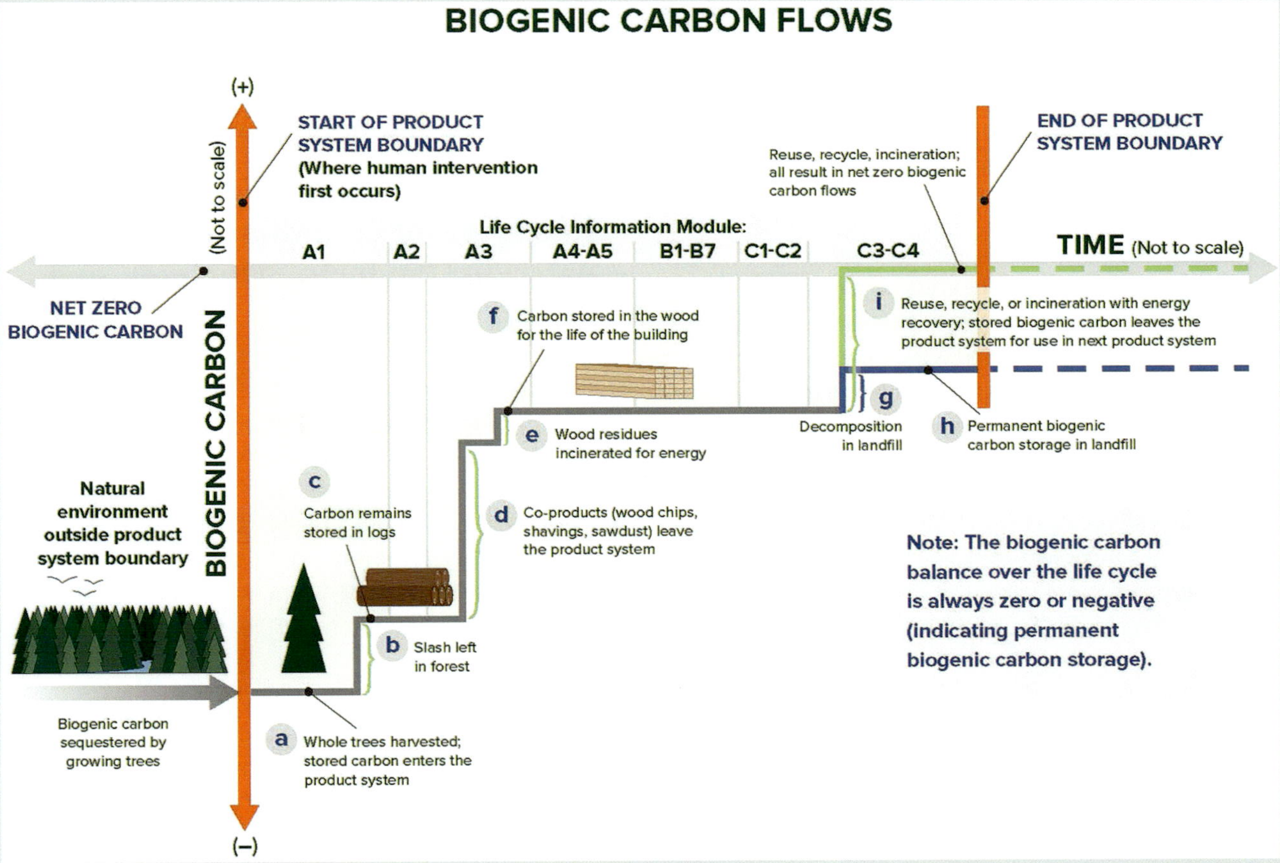

Figure 2.9 Biogenic flows for wood products as per ISO 21930: 2017 will ultimately equal zero over the entire life cycle of the product, however, this does not mean that wood products are carbon neutral as emissions are released from forestry operations, disturbed soil, lumber processing, and transportation. Image credit: Woodworks Innovation Network

decomposes or burns. This is referred to as *stored carbon*, biogenic carbon, or "GWP-biogenic" when referenced in a life cycle assessment for a wood product or other product with biobased content.

Stored/biogenic carbon in wood or plant products is expressed in LCAs and EPDs as a negative value as CO_2 is pulled from the atmosphere and not yet released back. The EN 15804 + A2 Standard endeavors to make information about biogenic carbon at each life-cycle stage clearer, measurable, and traceable. The standard requires accounting of biogenic carbon storage in the product stages, packaging stage, and the end-of-life stages resulting in a zero-carbon balance (EN 15804 + A2 2019). The intent is to encourage information transparency and prevent greenwashing claims of "net-zero" for wood and biobased products.

Accounting for biogenic carbon, particularly in wood products is an ongoing area of research and debate in academia, the building industry, environmental advocacy groups, LCA professionals, standards organizations, and forest product industries (Figure 2.9).

Some argue that wood and biobased products should not be considered carbon neutral as there are too many other sources of GHG emissions resulting from their life cycle that are not accounted for in an LCA. Carbon released from the soil during logging operations or plant harvest is not in the equation. This impact is estimated to be significant but is nearly impossible to quantify.

Additionally, a significant amount of carbon is released from decomposition of roots, slash, and other biomass left in the forest after harvest. This is also challenging to account for. Last, biogenic carbon should only be counted if the wood is from a sustainably managed forest, and this can only be proven if it is certified from the Forest Stewardship Council or the Program for the Endorsement of Forest Certification (PEFC).

In a sustainably managed forest, new trees are planted to replace those that are cut resulting in an offset or even net gain of carbon sequestration. The US Forest Service has found that carbon stores can increase as more and more trees are cut for long lasting wood products and the forests are revegetated (US Forest Service 2024).

Other Biobased Materials

Biobased materials such as bamboo, jute, coir, and agricultural wastes can have relatively low embodied carbon, although the GHG emissions from many bioplastics are just as high as plastics. Most biobased materials are rapidly renewable and store carbon. Biobased materials for exterior applications are not as common as those in building and interior applications due to moisture, UV, and weather exposures that can cause them to biodegrade. An often-overlooked aspect of the embodied carbon of biobased materials are the fossil fuel-based fertilizers, pesticides, and herbicides used to grow biobased materials and bioplastic feedstocks (Chapter 19 "Biobased Materials and Products").

Strategies for Decarbonization of Site Materials, Products, Structures, and Assemblies

Built sites are both sources of greenhouse gases and carbon sinks. The manufacture of materials and products for a site release GHG emissions as do the site's construction, operation, and maintenance. But these landscapes can offset carbon emissions through carbon sequestration by trees and other vegetation on the site. Renewable energy can be produced on a site to power it and any buildings on it (Table 2.2).

Table 2.2 Emerging Approaches for Decarbonization of Materials and Products

Performance-based building codes	Embodied carbon levels for some materials are starting to be incorporated into some federal, state, and local building codes. Most require use of materials with embodied carbon below a baseline.
Low-carbon public procurement policies	Municipal, state, and federal governments are establishing policies for low-carbon procurement. While these policies only pertain to procurement in the public agencies, these policies can have a trickle-down effect on the commercial market through greater information transparency and lower carbon material and product options. The US GSA requires use of concrete, asphalt, steel, and glass products on their projects to have an embodied carbon below a baseline.
Performance-based specifications for low-carbon materials	Use of performance-based specifications for materials and products can result in lower embodied material mixes and products. Performance-based specifications can facilitate reductions in concrete mixes without specifying specific types or quantities of cement substitutes. Specifying a concrete that is 20% below a baseline allows the structural engineer or ready-mix contractor to employ a combination of carbon reduction measures.

(*Continued*)

Table 2.2 (Continued)

Embodied carbon labeling	There is growing support for an international embodied carbon labeling system. Discrepancies exist among countries' LCA standards, and this leads to discrepancies among EPDs reducing the ability to compare products. Additionally, smaller companies are disadvantaged because they can't afford LCAs for their products. Embodied carbon labeling policies are being discussed in North America and in Europe (Chapter 5 "Assessing the Impacts of Materials, Products, and Structures").
Carbon footprint assessments	There are a growing number of tools for carbon footprint assessments for building and site assemblies. Some site and infrastructure specific calculators are Pathfinder, Carbon Conscience, and Athena Pavement LCA. Building specific tools include EC3 Carbon Calculator, Carbon Conscience, Tally, Athena Impact Estimator for Buildings, Open LCA, Glad and WoodWorks Carbon Calculator (Table 2.9).
Industry pledges and carbon reduction plans	Professional organizations for designers and engineers are setting net-zero or climate positive targets for their professionals to achieve in their projects. On the supply side of materials and products, manufacturing sectors are developing detailed plans for carbon neutrality. These plans include a wide variety of measures including using energy and heat from renewable sources, process improvements, and CCUS.
Carbon offsets	Carbon offsets will help projects, municipalities, companies, and organizations meet carbon neutrality deadlines. While carbon offsets can be a good tool, to meet carbon reduction goals, they may allow manufacturers or companies to avoid pushing their organization toward lower carbon practices. Carbon offsets could also allow manufacturers to make claims about a "zero-carbon" products. Carbon offsets should be purchased from companies on the IFLI approved list of carbon offset providers or those listed by Circular Ecology.

Through a variety of approaches, it is possible to design, construct, and achieve a landscape that can be climate positive in a decade or two. Using low carbon, local, and living materials; optimizing resource use through reuse, smaller structures, and recycled content; and employing biobased and living structures will all contribute to carbon reductions.

Tables 2.3–2.7 detail strategies and approaches to reduce the embodied carbon of materials, products, structures, and assemblies on a site. It should be noted that decarbonization will involve a broad group of stakeholders including designers, engineers, manufacturers, policymakers, owners, and regulators.

Symbols in Tables 2.3–2.7 are as follows:

+ High impact strategies for reducing carbon
$ Potentially cost-effective decarbonization strategies
# Easily implemented strategies that are widely available
% Strategies that can be time-consuming
* Strategies that are not yet widely available or they require research and/or testing

Table 2.3 Goals, Targets, Benchmarks, and Baselines to Reduce Embodied Carbon

Planning to reduce embodied carbon in the early stages of projects will help ensure that all team members have time to support project goals. Baselines and benchmarks are key to establishing carbon reduction targets.	
Embodied Carbon Reduction Strategy	**Specific Actions**
Establish whole project embodied carbon baselines +%* Baselining is the first step to reducing embodied carbon on a project. With an established baseline for the project, teams can set goals for reducing embodied carbon from materials and structures and sequestering carbon through planting.	Baselines can be established by gathering data from similar sizes, types, and locations of projects. On previously developed sites, baselines can come from an accounting of the footprint of the previous development. Baselines can also come from the existing carbon sequestration of undeveloped sites. Public agencies, rating systems, and other organizations can establish baselines for specific project types and sizes.
Establish embodied carbon reduction targets for the whole site +%* With established baselines, carbon reduction targets can be set for the entire site early enough in the process that all team members can plan accordingly. As new and existing vegetation can sequester carbon and offset carbon emissions from materials and products, a typical whole site carbon target might be "climate positive within 20 years".	Engage in the Climate Positive Design Challenge by using the Pathfinder calculator to determine the number of years to a climate positive site. Variables that can be adjusted are size of structures, type of materials and material assemblies, and type and quantity of vegetation. The Climate Positive Design Challenge suggested targets are as follows: Five years to positive for parks, residential, on-structure, mixed use, or campus developments Twenty years to positive for streetscapes and plazas (Climate Positive Design 2024).

(Continued)

Table 2.3 (Continued)

Embodied Carbon Reduction Strategy	Specific Actions
Establish material and product embodied carbon baselines +%#* Setting baselines for materials is an approach to reducing embodied carbon that can be used either with or without whole site carbon reduction targets. Some baselines for materials such as concrete, steel, aluminum, asphalt pavement, and some concrete products are widely available. Baselines for plastics, modified lumbers, and other products specific to site construction are not yet available.	Material and product baselines are provided by organizations such as the Carbon Leadership Forum, Embodied Carbon in Construction Calculator (EC3), or Circular Ecology in the United Kingdom. These organizations have different methods for establishing material and product baselines. Some will offer a range of high to low figures. Industry-wide EPDs offer another type of baselines for materials and products. These EPDs are usually an average of data from several manufacturers within a region or a country.
Set embodied carbon targets for materials and products +#%* Embodied carbon reductions below an established baseline are an effective method of reducing embodied carbon. Targets may be "20% below baseline" or without strong baselines, targets might include 20–40% of the materials or products in a category with the lowest embodied carbon. The US EPA and the GSA take this approach.	Setting embodied carbon reduction targets is particularly important for high-carbon materials like concrete, steel, aluminum, or asphalt pavements. This practice can be challenging because embodied carbon data and EPDs are not available for many site construction products.
Monitor progress toward embodied carbon reduction targets +%* If progress toward embodied carbon reduction is monitored on projects, actions may be taken to meet the targets. Even if targets are not met, recordkeeping can inform future projects.	Monitor and track embodied carbon reductions across all phases of the project. Revise action plans and benchmarks as appropriate. Update records to reflect advancements in material production, material baseline updates.
Engage in third-party rating system certifications +% Voluntary rating systems offer credits and prerequisites for reducing the embodied carbon of materials and products and/or whole sites beyond an established baseline.	Rating systems and their embodied carbon reduction standards are discussed in Table 1.1.
Establish embodied carbon inventories +% Collecting, reporting, and archiving material quantities used in projects can help establish future embodied carbon baselines as well as identify potential tradeoffs to reduce embodied carbon on future projects.	Live databases or spreadsheets listing types of materials and material mixes can easily calculate embodied carbon for material assemblies as material baselines change or new data is obtained.
Purchase carbon offsets # Carbon offsets can be a powerful mechanism to reduce "years to climate positive" or to meet other whole site carbon reduction goals. Depending on the type of carbon offset, they can bring benefits to critical and endangered ecosystems. EN 15804 does not allow carbon offsets to be used in the assessment of the carbon footprint indicators. EPD owners can declare their participation in carbon offset programs, however.	Carbon offset purchase should be a last resort when other embodied carbon reduction approaches have been exhausted. There are many carbon offset providers with different methods and standards. IFLI offers an approved list of carbon offset providers as does Circular Ecology.

Table 2.4 Optimization in Material and Product Production to Reduce Embodied Carbon

Material and product manufacturers are key to the reduction of GHG emissions of construction materials. They can optimize their manufacturing processes, recycle waste, and use renewable energy sources. Some provide transparent information on raw material extraction, manufacturing phases, and transportation between all sites for these activities.	
Embodied Carbon Reduction Strategy	**Specific Actions**
Favor materials and products from manufacturers with EPDs +%* or those that provide third-party verified embodied carbon data. This data should be third-party verified using an accepted standard to collect and compile the data (Chapter 5 Assessing the Impacts of Materials, Products, and Structures).	EPDs are available for products such as concrete, asphalt pavement, bricks, wood products, and steel and aluminums shapes. They are less widely available for site construction products such as pavers, interlocking wall blocks, tree grates, fencing and rail systems, lighting, and site furnishings.
Use a whole life-cycle approach when evaluating materials and products +$%* Carbon costs should be considered for all or several phases of a material or product's life cycle. Examining the whole life cycle of a material or product will foster circular, not linear, life cycles where materials are discarded after use.	Most standards for EPDs require that phases A1–A3 (cradle to gate) be considered at a minimum. Including use, maintenance, and end-of-life phases can result in more durable materials and material assemblies. Structures will be designed for deconstruction and reuse of materials and products.

(*Continued*)

Table 2.4 (Continued)

Specify materials from manufacturers with optimized production processes +%* Manufacturers can take steps to reduce the carbon footprint of their raw material extraction practices, transport, and manufacturing processes. They can use renewable energy sources for fuel and heat sources and energy efficient equipment for processing, manufacturing, and fabrication stages. If applicable, they can employ CCUS technologies to capture process emissions. They can also incorporate high quantities of scrap and recycled content. Last, they can locate manufacturing and fabrication facilities close to raw material extraction points.	Look for products with EPDs and/or Corporate Sustainability reports with verified reductions in resource use, energy use, and water. Aluminum products should be manufactured in North America as they are likely to have been produced with renewable energy sources. Steel products should be specified from manufacturers using an electric arc furnace with renewable energy sources.

Table 2.5 Resource Optimization to Reduce Embodied Carbon

Resource efficiency measures are important to decarbonize sites as embodied carbon of new materials and products will be reduced. If materials are reused on the same site, the embodied carbon from transportation will be reduced as well.	
Embodied Carbon Reduction Strategy	**Specific Actions**
Reuse site structures in situ +$% Refurbishment and adaptive reuse of site structures can save resource use and embodied carbon of new material extraction and manufacture. Structures such as old warehouses, factories, or bridge structures can be reused for new landscapes offering cost savings and referencing the history of a place while reducing the carbon footprint of the project (Figures 2.10 and 2.11).	Favor flexible, repairable pavement and wall structures. Some rigid structures are not repairable and must be removed if cracked substantially. Refurbish, repair, and reuse concrete slabs and asphalt pavements in situ. Replace deck boards but maintain structurally sound deck structures. Repair or reapply coatings for metal structures.
Reuse materials on site +#$% Deconstructing, stockpiling, refurbishing, and reusing materials and products on a previously developed site can reduce GHG emissions from both the production of new materials and transport emissions from demolition debris hauling.	Broken concrete slabs can be used in freestanding and retaining walls with or without gabions. They can be used in informal "flagstone" pavements. Building bricks can be used in new walls and pavements. Steel or wood structural members can be reused in new overhead structures.
Reprocess materials on site +$# Deconstructing, crushing, sizing, and replacing concrete and asphalt materials in new pavements or new aggregate bases can save embodied carbon from hauling large quantities of heavy materials away from the site. Bringing crushing and sizing equipment to smaller sites will usually not be cost effective.	Any masonry or asphalt materials can be crushed, sized, and reused on a construction site for use in new pavements or base materials. This is a good option for masonry materials that are mortared together as it is not usually financially feasible to separate and clean the materials for reuse whole.
Deconstruct don't demolish +#% Deconstructing sites and reusing or recycling materials instead of demolishing sites and hauling mixed materials to the landfill will reduce the embodied carbon of new materials used in a site or in other projects.	Reuse or recycle materials such as concrete, asphalt, bricks, stone, aggregates, plastics, metal, and wood. While reuse is preferable to recycling, the embodied carbon of some materials with high recycled content can be 50–95% less than new materials.
Design for deconstruction (DfD) +$# Designing sites for deconstruction will facilitate disassembly and reuse of materials and products after a site's useful life. Prolonging the use life of materials and products will reduce embodied carbon over their full use life. Implementation of DfD strategies can result in 10–50% reductions in GHG emissions (UNEP 2023). ISO 20887:2020 offers guidance on the practice of DfD.	Use materials that are reusable and recyclable. Design reversible attachments that are easily accessed. Avoid using mortar in pavement and wall assemblies. Favor bolted connections over welds and adhesives. Avoid direct burying wood or steel posts in concrete footings. Avoid materials with toxins, such as PVC or foams, that could impact their recyclability.
Design structures for durability and repairability +$# Site structures should be designed for a long use life. Flexible structures can be repaired or resurfaced if settling or frost heave occurs. Rigid structures, particularly concrete walls and pavements, are challenging to repair and are often replaced.	Favor flexible structures such as gabion walls, interlocking retaining walls, stabilized stone dust pavement, and asphalt pavement with reflective or low-albedo surfaces. Use appropriate woods, metals, and metal finishes for the microclimate to prolong the life of wood and metal structures. Avoid plastic products that are exposed to the sun as UV degradation can shorten their life.
Optimize material use and use materials efficiently +$# Dematerializing sites and structures will reduce embodied carbon and likely save material and construction costs. Work to change design standards such as parking quantities and path width requirements. Many standards call for unnecessarily wide paths, thick pavements, or large parking lots.	Use less material in smaller structures, narrower paths, smaller members, smaller parking lots, and drive aisles. Avoid over-sizing structural members and pavement and base thicknesses. Engineers and span tables already build in a margin of error. In areas with deep frost lines (>60cm), consider use of flexible structures to avoid excessive material use in footings. Avoid use of veneers and other unnecessary finishes.

Figure 2.10 Wagon Landscaping created a temporary garden in Kortrijk, Belgium, by saw cutting asphalt pavement, removing some areas, and planting them offering comfortable rooms in a labyrinthine space. This subtractive approach to reusing an existing structure is very low embodied carbon. Image credit: Studio Basta

Figure 2.11 Saw cuts are marked on the existing pavement. Image credit: Wagon Landscaping

Table 2.6 Low-Carbon Materials, Material Mixes, and Products

Favoring low-carbon materials, material mixes, and products can help to transform the construction industry and incentivize manufacturers to improve their practices.	
Embodied Carbon Reduction Strategy	**Specific Actions**
Specify lower carbon material mixes +*% Specifying lower carbon material mixes for concrete, precast concrete products, asphalt, brick, and earth construction can reduce the embodied carbon of built sites. This can be challenging due to industry standards, codes, and entrenched industry practices.	Work with engineers and suppliers to reduce the embodied carbon of concrete by using portland cement substitutes, extending curing times from 28 to 56 days, using the appropriate strength of concrete for the structure, sequestering carbon in the concrete mix, using recycled aggregates and other strategies. Use performance-based specifications for the mix design that specify a percentage carbon reduction. This will allow suppliers to develop custom mixes to meet this requirement.
Specify materials that are lower carbon in their entire life cycle +%* Specify materials and products with lower embodied carbon. It is important to understand carbon impacts for the entire life cycle rather than just the product stage (A1–A3) as some materials and products are not recyclable and pose burdens in disposal.	Brick, stone, and precast concrete products are durable materials that can be reclaimed and reused over and over in multiple structures. While technically feasible, PVC products are not usually recycled because they release toxins when heated. This is not revealed in a cradle to gate EPD. Some metal finishes such as galvanizing can make metal recycling more difficult.
Consider the carbon footprint of entire structures and assemblies +#%* An emphasis on the embodied carbon of materials and products does not tell the whole story. Instead, calculating the embodied carbon for the entire assembly or structure will be more accurate and highlight tradeoffs to reduce carbon burdens. Refer to embodied carbon calculations of assemblies in Chapters 6–11.	Avoid using stone, cast stone, or brick as veneer on concrete structures as it adds unnecessary material and embodied carbon. Instead, use these materials in flexible pavement applications and as solid walls. If mortar is avoided, these materials can be easily reclaimed and reused again and again. Employing flexible structures will use less material for footings extending below frost lines in colder climates.
Specify renewable, biobased materials, and products +# Biobased materials such as wood, bamboo, and jute store carbon until they biodegrade or burn. They are renewable, some rapidly, and are often lower embodied carbon than alternative materials such as steel, aluminum, and concrete for the same type of structure. After their useful life, biobased materials that are free of harmful substances can be composted. Some biobased plastics do not biodegrade, which makes them more durable but not compostable.	Choose biobased materials that are FSC certified to ensure responsible forestry practices, fair labor, and a chain of custody. Avoid tropical hardwoods as they are not usually plantation grown and may be illegally harvested from sensitive ecosystems. Use salvaged wood. Look for locally or regionally harvested and processed biobased materials. Use smaller dimensions of wood to reduce pressure on forests. Beams can be "built up" and glulams can be specified. Consider biobased plastic products that are manufactured from agricultural waste.

(*Continued*)

Table 2.6 (Continued)

Embodied Carbon Reduction Strategy	Specific Actions
Design living structures that store and sequester carbon while in use +%* Living structures are made from live plant material that is trained, pleached, or woven as it grows to form a natural barrier. Living structures that combine living and non-living materials such as vegetated crib walls will continue to retain the slope after the wood crib wall biodegrades due to root stabilization of the slope. The plants in living structures will provide ecosystem services much needed in urban areas such as stormwater attenuation, air pollution removal, and cooler surfaces.	Consider bioengineered embankments such as live staking, live fascines, brush layering, live crib walls, vegetated gabion walls and vegetated rock walls for retaining situations (*Chapter 8 Retaining Wall and Reinforced Embankment Structures*) Build living fences and overhead structures with techniques such as hedge laying, espalier or weaving plants on a built structure, and weaving plants into a plant-only barrier (Chapter 11 "Rails, Fences, and Screens"). Some living structures require seasonal training of the plants over a few years and may not be appropriate for projects with minimal maintenance regimes.
Rewrite standard specifications to reduce carbon +$%* Rewriting standard office specifications to reduce embodied carbon of materials, products, and structures can impact carbon footprints of projects firm wide. This can also influence standard specifications of engineers and impact local concrete and asphalt suppliers. Professionals can lobby local DOTs and other institutions to modify their standard specifications.	Standard specifications should do the following: Employ performance specifications for carbon reductions at 20% or more below state or countrywide benchmarks for concrete, asphalt, steel, and aluminum. Specify EAF produced steel and aluminum from North America. Specify recycled aggregates for base courses such as recycled concrete aggregate (RCA) and recycled asphalt pavement (RAP).
Specify local and regional materials +$%*# Using locally or regionally sourced materials and products can reduce the overall embodied carbon of a built site by reducing GHG emissions of trucks, trains, and ships. It can also stimulate local economies and reduce material costs.	Identify locally produced materials and products during schematic design to ensure local materials. Maintain databases of local and regional material and product manufacturers. Make sure that the local materials are locally processed during all phases. Work with contractors to source and procure local and regional materials and products.
Trace the supply chain of materials and products to ensure reduced carbon in transport +%* It can be very challenging to trace the supply chain for products as components will be sourced in locations sometimes far from the manufacturing plant. Fabrication adds an additional layer of complexity.	Ask material and product manufacturers where raw materials originated and where all stages of manufacturing and fabrication took place. Favor products with transparent supply chains. Require chain-of-custody records where appropriate.

Table 2.7 Reduced Embodied and Operational Carbon in the Construction, Use, and Maintenance Phases

While operational emissions are estimated to be around 20% of the total carbon footprint of a site, energy efficient lighting, controllers, and other equipment can be specified for the use phase. Additionally, operating emissions from construction and maintenance can be reduced as well.	
Embodied Carbon Reduction Strategy	**Specific Actions**
Reduce carbon emissions from construction and installation +*% Emissions from the construction and installation phase result primarily from the vehicles, equipment, and power tools used to build a site. Creating a Construction Activity Carbon Reduction Plan outlining requirements for tracking emissions in phases A4 (material transport to site) and A5 can ensure that the contractors, subcontractors, and consultants understand requirements. CACRP requirements should be reflected in RFPs, contracts, and subcontractor exhibits.	Equipment electrification that is powered by renewable energy will make the biggest step toward decarbonizing the construction phase. Other approaches are to require anti-idling, use biofuel blends, avoid generators, and use solar chargers for smaller tools. Prioritize reduction of truck trips to deliver and haul away materials. Balance cut and fill to minimize hauling. Prefabrication of fence and guardrail panels and other products can use renewable energy sources in factory settings.
Reduce operational emissions +$# Use energy efficient site lighting, controllers, pumps, and other fixtures.	Specifying light fixtures that use LED bulbs will reduce energy loads and fixtures that have built-in solar collectors can make lighting net zero. Minimize light pollution through baffles and other elements that direct the light only where required for safety.
Design sites for renewable energy production +% Some constructed sites offer enough space to accommodate solar panels and smaller wind turbines.	Locate solar panels on structures over parking lots or over seating areas to produce energy and shade paved surfaces. Employ wind turbines in large open spaces. Build geothermal systems under public spaces.
Reduce the urban heat island effect to minimize cooling loads and reduce operational carbon in buildings +$# The urban heat island effect from darker pavements and roofs can increase temperatures in urban and suburban areas.	Employ lighter colored surfaces for pavements and use reflective coatings on darker pavements such as asphalt (Table 6.17). Use permeable pavements, reduce hardscape, and shade pavements with closely spaced trees and overhead structures.

(*Continued*)

Table 2.7 (Continued)

Embodied Carbon Reduction Strategy	Specific Actions
Use low-VOC maintenance materials #% Paints, stains, sealers, and adhesives contain VOCs that can impact worker health in production, fabrication, and maintenance. They can also impact air quality, and some are potent greenhouse gases.	Specify low-VOC products. Refer to the IFLI Red List for substances to avoid and the California SCAQMD VOC limits for paints, stains, sealers, and adhesives.
Reduce emissions from maintenance equipment +# Mowers, blowers, and trimmers are responsible for substantial carbon emissions.	Maintenance or management plans should address the reduced use of power equipment for maintenance. Electrification of power equipment can reduce carbon during this phase. Reducing lawn areas can reduce the use of mowers and blowers.

Embodied Carbon Policies, Codes, and Standards for Construction Materials and Structures

Policies, regulations, and standards offer mandatory or voluntary thresholds for reducing embodied carbon of construction materials, structures, and whole projects. While some thresholds are currently less rigorous, with improvements in material sectors embodied carbon baselines should be lowered over time.

Policy Measures

Policy measures to reduce the embodied carbon of construction materials are in their infancy with a greater number of voluntary reduction initiatives than mandates. European and North American governments have set benchmarks for reducing the operating emissions from buildings, but reductions of emissions from material and products are less common at the time of this writing. This could all change very quickly as several countries have committed to reducing carbon emissions in multiple sectors, including the building and infrastructure sectors, by 2050.

Measuring carbon reductions requires benchmarks for materials, products, or building and site types. The Carbon Leadership Forum (CLF) releases embodied carbon benchmarks biennially for the most common building structural materials, cladding, and some finishes. These benchmarks can be used to specify lower carbon materials and projects; they can also inform state or federal benchmarks for reducing embodied carbon. Whole building or whole project benchmarks can use a comparable existing building or site, or in the case of previously developed sites, the existing site conditions can be calculated for a baseline.

In addition to reducing the operating energy of buildings, federal and state initiatives in North America are currently focused on reducing embodied carbon emissions for high-carbon commonly-used materials such as concrete and steel. Many European nations are taking a slightly different approach by setting targets for reducing whole building embodied carbon emissions.

Whole Project Emissions Reduction Initiatives

Whole building, or whole site, emissions are more challenging to quantify as embodied carbon CO_2e data must be collected on dozens of materials, products, and construction processes. While EPDs are more prevalent in the architecture field than in landscape architecture, this is still a tall order. Other whole building requirements in Europe focus only on the structure, not the finishes.

Buy Clean Policies

Buy Clean policies are gaining traction at the state and federal level in North America and the European Union. Many of these policies require EPDs for reporting the impacts of some building materials for federal and state purchasing, and some require limits on GWP of some materials such as concrete and steel. Given the scale and purchasing power of government procurement, these policies should have an impact on private markets as well.

Material and Product Emission Reduction Initiatives – Federal and State

A growing number of nations and states are implementing policies promoting procurement of materials and products with lower embodied carbon. Use of material baselines such as those provided by EC3 and industry wide EPDs are core sources of information for these initiatives as they establish material baselines for common construction products. Whole building or project reduction initiatives are growing as well with greater progress in the EU.

Voluntary Rating Systems

The ***World Green Building Council's*** Whole Life Carbon Vision calls for all new buildings, infrastructure, sites, and renovations to reduce embodied carbon by 40% by 2030. By 2050, they must have net-zero embodied carbon (WGBC 2024) (Table 2.8).

Over the last decade, voluntary rating systems such as the Sustainable Sites Initiative (SITES), LEED, and the Living Building Challenge (LBC) have added credits or prerequisites that address GWP of construction materials and products and even whole project embodied carbon. Since their inception,

Table 2.8 Embodied Carbon Standards in Voluntary Rating Systems

Rating System	Project Types	Country or Region	Rewards EC Transparency and Accounting (LCAs, EPDs)	Rewards EC Reductions	Rewards Resource Efficiency
Living Building Challenge	Buildings and sites	North America, Europe	–	Required	Required
IFLI Zero Carbon Certification	Buildings and sites	North America, Europe	–	Points	Points
SITES	Sites, infrastructure	North America, Europe, China	Points	Points	Points
LEED BD&C	Buildings, sites, community	International	–	Required	Required/Points
BREEAM Infrastructure	Infrastructure	International	Points	Points	Points
Envision	Infrastructure	International	–	Points	Points
Greenroads	Infrastructure	International	–	Required	Points
INVEST	Infrastructure	US	–	Points	Points

LEED and LBC have had a strong focus on operational carbon reduction.

The credits or prerequisites for these rating systems focus on reductions of 10–40% beyond a baseline for a similar building type. SITES rewards projects for specifying products with transparent information provided by manufacturers such as EPDs or embodied carbon optimization reports. All three systems propose resource efficiency credits and prerequisites that promote structure reuse, reclaiming and reusing materials, resource efficient materials, and recycled content materials. These approaches will reduce embodied as well.

The ***International Living Future Institute*** offers a Zero Carbon Certification to establish a measurable pathway to climate positive buildings and sites. This is a third-party certified standard that includes both operational and embodied carbon emissions of a project. New building projects must demonstrate at least a 20% reduction in the GWP of primary materials and exterior materials beyond a baseline for the building type and size.

Projects must disclose and offset 100% of carbon emissions (tCO_2e) associated with life-cycle phases A1–A5. This can be accomplished with on-site sequestration from vegetation or storage in biobased products or with one-time purchased offsets (IFLI 2024). The Living Building Challenge has similar requirements in their *Net Positive Carbon* imperative.

The ***Sustainable Sites Initiative*** offers a range of points for improving a site's embodied carbon performance beyond a baseline case defined by the project team and approved by SITES. Embodied carbon is measured for a 50-year lifespan, and points are awarded for measurement of GWP. Additional points are awarded for GWP reductions ranging from 10–20% (SITES 2021).

LEED Building Design & Construction (BD&C) offers multiple prerequisites and credits that address GWP assessment and reductions for both embodied and operation carbon. The carbon assessment prerequisite requires projects to develop a 25-year assessment of operational carbon projection, refrigerant management, and embodied carbon (LEED 2024).

Site and Infrastructure Professional Organization Decarbonization Goals

Professional organizations for site design and infrastructure practitioners have developed climate action goals to decarbonize existing and new constructions. Some organizations have detailed plans, while others have policy statements and goals but lack specific objectives and actions.

The ***International Federation of Landscape Architects (IFLA)*** Climate Action Commitment ratified by several landscape architecture organizations worldwide commits to the following (IFLA 2021):

- Advancing the United Nations Sustainable Development Goals
- Attaining global net-zero carbon emissions by 2040
- Enhancing capacity and resilience of livable cities and communities
- Advocating for climate justice and social well-being
- Learning from cultural knowledge systems
- Galvanizing climate leadership.

Table 2.9 Resources for Decarbonizing Sites, Structures, Materials, and Products

Whole Site or Building LCA Tools	
Pathfinder by Climate Positive Design	Free tool for whole site LCA. Estimates "years to (climate) positive" for a site instead of providing total GHG emissions. Types and quantities of assemblies and structures are input along with types and quantities of existing and proposed vegetation. Projects can establish a baseline scenario using standard materials or existing site conditions, then input alternative materials or material mixes for reduced years to positive. Projects can also reduce years to positive through increased use of vegetation.
Carbon Conscience	Carbon Conscience is a free application to assess carbon impacts in the early stages of project planning and design. It does not require information about specific materials and quantities, so it can be used in early concept design for urban design scale projects, buildings, sites, and infrastructure. Users can compare design options, test alternative land uses, structural systems, façade materials, and landscapes with quick sketches.
Early Phase Integrated Carbon (EPIC) assessment	Free whole life carbon tool to support climate positive design decisions early in a project. Primarily building focused.
One Click LCA	Subscription based software, primarily for buildings, that integrates with building information modeling (BIM). Database of EPDs allows for whole project LCA modeling including paving, fencing, rails, walls, and drainage systems.
Tally	Application that allows for whole building LCAs within Revit.
Whole Structure LCA Tools	
Athena Pavement LCA	Free software for pavement assembly life cycle assessment.
LCA Pave tool from the US FHWA Sustainable Pavements Program	Free tool to analyze and quantify environmental impacts for the entire life cycle of a variety of asphalt and concrete pavement mix designs. Data for phases A1–A-3 (cradle to gate) of the tool is provided by industry environmental product declarations while data from other life-cycle phases is based on maintenance activities required to maintain the pavements over a life of 25 years for asphalt pavements and 35 years for concrete pavements.
Databases and Libraries of EPDs	
EC3 by Building Transparency	Open-source database that houses thousands of product-specific and industry-wide EPDs. Multiple search and sort criteria. EPDs can be downloaded and EC3 provides an "Uncertainty Adjusted GWP" for each reported GWP. Focus on A1–A3 impacts and GWP.
EPD International	Searchable library of thousands of product-specific and industry-wide EPDs. Filter by product name, category, PCR, geographical scope, and validity date.
SCS Global Services	Searchable library of thousands of EPDs. Search categories include product types, manufacturers, certifications, and countries.
Sustainable Minds Transparency Catalog	Listing of thousands of North American products by CSI section with links to manufacturers webpages, EPDs, and other sustainability data. EPDs link to the EC3 database.
Spot UL Environment	SPOT UL Environment is a searchable database of construction products and EPDs that is searchable by specification section, manufacturers, product type, certifications, and transparency information type.
EPD Registry	Searchable database of hundreds of EPDs from Europe, Australia, and South America.
Sources of Embodied Carbon Figures for Materials	
Carbon Leadership Forum	The Carbon Leadership Forum publishes baselines for several common construction materials every two years. These are based on industry data.
Carbon Conscience	Carbon Conscience Landscape Data Set Carbon Conscience Architectural Data Set.
Circular Ecology	Inventory of Carbon and Energy Database.
2050 Materials	2050 Materials – Sustainability Data for Construction Materials.
Decarbonization Information Resources	
Carbon Leadership Forum	Information resources, research, partnerships, and initiatives for decarbonization of construction.
EPD International	Resources for information transparency, EPDs, and SPDs.
Greenhouse Gas Protocol	Global standards, guidance, tools, and training to measure GWP emissions. Partnered with World Resources Institute.

(*Continued*)

Table 2.8 (Continued)

Architecture 2030	2030 Carbon Smart Materials Palette, Zero Code, Zero Tool.
Climate Action Resources	
American Society of Landscape Architects	Climate and Biodiversity Action Plan, Field Guide and other supporting publications.
Landscape Institute	Landscape for 2030, Climate and Biodiversity Action Plan.
Australian Institute of Landscape Architects	Climate Positive Design Guidelines and Resources.

The ***American Society of Landscape Architects'*** Climate Action Plan states: "All landscape projects will reach zero emissions by 2040 while providing economic, environmental, social, and cultural co-benefits, including increased sequestration and biodiversity" (ASLA Climate Action Plan 2022).

The ***Institution of Civil Engineers (ICE)*** states, "As an organization, the Institution of Civil Engineers is committed to continuously reduce its negative environmental impact and to enhance its social responsibility and carbon reduction performance". Based in the United Kingdom, they adhere to the UK's Net Zero Strategy to reduce domestic greenhouse gas emissions to net zero by 2050 (ICE 2021).

The ***American Society of Civil Engineers (ASCE)*** Policy Statement 488 states "The American Society of Civil Engineers (ASCE) supports public and private sector strategies and efforts to achieve significant reductions in greenhouse gas (GHG) emissions through the planning, design, construction, renewal, renovation, operation, maintenance, and decommissioning of existing and future infrastructure systems". (ASCE 2022)

The ***American Institute of Architects (AIA) states*** "As a profession, we have the responsibility to prioritize and support effective actions to exponentially decelerate the production of greenhouse gases contributing to climate change. Our goal, as set forward by AIA and partners like Architecture 2030, is net-zero emissions in the building sector by 2040" (AIA 2024).

References

American Institute of Architects (2024). "AIA's commitment to climate action". www.aia.org/about-aia/where-we-stand/climate-action, accessed February 17, 2024.

American Society of Civil Engineers (ASCE) (2022). *ASCE Policy Statement 488*. ASCE, July 22.

American Society of Landscape Architects (ASLA) (2022). *ASLA Climate Action Plan 2022–2025*. ASLA, ASLA Fund.

Architecture 2030 (2024). "Why the built environment: Why buildings, why infrastructure". www.architecture2030.org/old-why-the-built-environment/, accessed August 7, 2024.

Brick Industry Association (BIA) (2020). *U.S.-Canada Industrywide Clay Brick: Environmental Product Declaration*. BIA.

Climate Positive Design (2020). *Landscape Carbon Calculator/Pathfinder: Methodology, Data Sources, and Metrics Summary*. Climate Positive Design, CMG and Atelier Ten.

Comité Européen de Normalisation (CEN) (2014a). *EN 16449 - Wood and Wood-Based Products - Calculation of the Biogenic Carbon Content of Wood and Conversion to Carbon Dioxide*. CEN.

Comité Européen de Normalisation (CEN) (2014b). *EN 16485, Round and Sawn Timber - Environmental Product Declarations - Product Category Rules for Wood and Wood-Based Products for Use in Construction*. CEN.

Congressional Budget Office (2024). "Emissions of greenhouse gases in the manufacturing sector". www.cbo.gov/publication/60030, accessed June 12, 2024.

Copernicus and Climate Change Service (2024). "Copernicus: 2024 is the first year to exceed 1.5°C above pre-industrial level". Global Climate Highlights 2024. https://climate.copernicus.eu/, accessed January 17, 2025.

EN 15804 + A2 (2021). *Sustainability of Construction Works - Environmental Product Declarations - Core Rules for the Product Category of Construction Products*. EN, June 2.

Gajdzik, B., Wolniak, R., and Grebski, W. (2023). "Process of transformation to net zero steelmaking: Decarbonisation scenarios based on the analysis of the Polish steel industry" *Energies* 16.8: 3384. https://doi.org/10.3390/en16083384.

Historic Environment Scotland (2024). www.engineshed.scot/calculate-the-carbon-footprint-of-your-stone-project/, accessed February 24, 2024.

Institution of Civil Engineers (ICE) (2021). *ICE Sustainability Policy Statement*. ICE.

Intergovernmental Panel on Climate Change (IPCC) (2018). *Global Warming of 1.5 °C: An IPCC Special Report on the Impacts of Global Warming of 1.5 °C Above Pre-Industrial Levels and Related Global Greenhouse Gas Emission Pathways, in the Context of Strengthening the Global Response to the Threat of Climate Change, Sustainable Development, and Efforts to Eradicate Poverty*. Cambridge and New York: Cambridge University Press, 616. https://doi.org/10.1017/9781009157940.

Intergovernmental Panel on Climate Change (IPCC) (2022). *Climate Change 2022: Mitigation of Climate Change: Summary for Policymakers*. IPCC, Working Group III.

International Energy Agency (IEA) (2023). "Steel". www.iea.org/reports/breakthrough-agenda-report-2023/steel, accessed February 14, 2024.

International Energy Agency (IEA) (2024). "Cement". www.iea.org/energy-system/industry/cement, accessed February 16, 2024.

International Federation of Landscape Architects (IFLA) (2021). *IFLA Climate Action Commitment*. IFLA.

International Living Future Institute (IFLI) (2024). *Zero Carbon Certification 1.1 Program Manual*. IFLI.

International Organization for Standardization (ISO) (2018). *ISO 14067: Greenhouse Gases – Carbon Footprint of Products – Requirements and Guidelines for Quantification*. ISO.

International Organization for Standardization (ISO) (2020). *ISO 20887: Sustainability in Buildings and Civil Engineering Works – Design for Disassembly and Adaptability*. ISO.

International Rescue Committee (2023). "10 countries at risk of climate disaster". www.rescue.org/article/10-countries-risk-climate-disaster, accessed June 12, 2024.

LEED (2024). "Carbon assessment: Integrative process, planning, and assessments required". www.usgbc.org/credits/new-construction/v5-public-comment, accessed May, 8 2024.

Mission Possible Partnership (MPP) (2022). *Making Net-Zero Steel Possible: An Industry-Backed, 1.5 °C Aligned Transition Strategy*. Mission Possible Partnership, RMI, Energy Transitions Commission, and World Economic Forum.

Nordahl, S., et al. (2020). Life-cycle greenhouse gas emissions and human health trade-offs of organic waste management strategies". *Environmental Science & Technology*.

Sanderman, J., Hengl, T., and Fiske, G.J. (2017). "Soil carbon debt of 12,000 years of human land use". *Proceedings of the National Academy of Sciences* 114.36: 9575–9580. https://doi.org/10.1073/pnas.1706103114.

Sustainable Sites Initiative (SITES) (2021). *SITES Pilot Credit 3: Assess and Improve Site Carbon Performance*. SITES.

United Nations (UN) (2015). *Paris Agreement to the United Nations Framework Convention on Climate Change*. T.I.A.S. No. 16-1104. UN, December 12.

United Nations (UN) (2021). "UN chief promotes 'enormous' benefits of greener cities". https://news.un.org/en/story/2021/10/1101992, accessed April 1, 2024.

United Nations (UN) Climate Change (2021). *Climate Action Pathway Human Settlements 2021. Vision and Summary*. UN.

United Nations Environment Programme (UNEP) (2023). *Building Materials and the Climate: Constructing a New Future*. Nairobi: UN.

US Environmental Protection Agency (US EPA) (2020). "2020 national emissions inventory (NEI) data". www.epa.gov/air-emissions-inventories/2020-national-emissions-inventory-nei-data, accessed June 12, 2024.

US Environmental Protection Agency (US EPA) (2022). *Emission Factors for Greenhouse Gas Inventories*. US EPA, April 1.

US Environmental Protection Agency (US EPA) (2024a). "Understanding global warming potentials". www.epa.gov/ghgemissions/understanding-global-warming-potentials.

US Environmental Protection Agency (US EPA) (2024b). "Inflation reduction act programs to fight climate change by reducing embodied greenhouse gas emissions of construction materials and products". www.epa.gov/inflation-reduction-act/inflation-reduction-act-programs-fight-climate-change-reducing-embodied, accessed June 14, 2024.

US Forest Service (2024). "Carbon". www.fs.usda.gov/managing-land/sustainability-and-climate/carbon, USDA, accessed June 12, 2024.

World Green Building Council (WGBC) (2024). "Whole life carbon vision: Advancing net zero". https://worldgbc.org/advancing-net-zero/whole-life-carbon-vision/, accessed July 13, 2024.

World Meteorological Organization (WMO) (2024). *WMO Global Annual to Decadal Climate Update: Target Years 2024–2028*. WMO.

World Resources Institute (WRI) (2011). *Greenhouse Gas Protocol: Product Life Cycle Accounting and Reporting Standard*. World Resources Institute and World Business Council for Sustainable Development, September.

WRI Climate Watch (2022). "Historical GHG emissions". www.climatewatchdata.org/ghg-emissions.

ZWE (2020). "Understanding the carbon impacts of waste to energy incineration". https://zerowasteeurope.eu/2020/03/understanding-the-carbon-impacts-of-waste-to-energy/.

CHAPTER 3

Resource Efficiency and the Circular Economy

The weight of all human-made structures and products has surpassed the weight of all living biomass on earth (Elhacham et al. 2020). Anthropogenic mass has roughly doubled every 20 years over the past century with 81% used in construction of the built environment (Figure 3.1).

We are living in a time of ecological overshoot where our consumption of resources has outpaced the ability of our planet to support their extraction. In 2023, global use of materials was about 100 billion metric tons. This figure does not include fossil fuel inputs or the substantial waste from raw material extraction and production processes (Circle Economy 2023). We are in a climate emergency with mass biodiversity loss and extinction, pollution, over-use of non-renewable resources, and a global waste crisis.

The take-make-waste use of materials and products is trending in a perilously unsustainable direction with only 7.2% of 2023 waste estimated to be circular, down from 9.1% in 2018. We still rely heavily on virgin materials extracted from the natural environment, and this drives over 90% of global biodiversity loss and water stress (Circle Economy 2023).

The material footprint per capita of higher-income countries in the Global North, Australia, Oceania, and the Gulf is 4.6 times that of lower income countries. These countries are generating more than one third of the world's waste, while only accounting for 16% of the world's population. But while lower income countries create less waste, 93% of it is illegally dumped, leading to unchecked pollution to land, water, and air (Kaza et al. 2018).

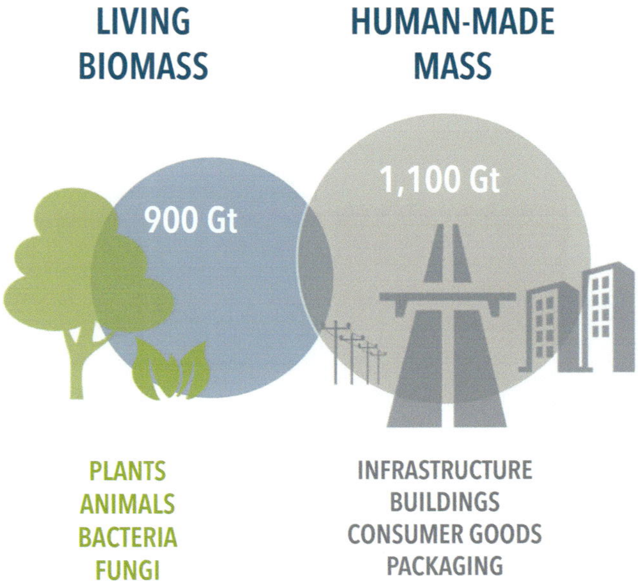

Figure 3.1 Global human-made mass on earth now exceeds the weight of all living biomass.

Adapted from source: Elhacham et al. 2020. Image credit: Drake Bruner and Meg Calkins

Resource Use and Waste Throughout the Life Cycle of Construction Materials

Construction materials contribute significantly to resource use and waste throughout their life cycle. The extraction of minerals to produce construction materials is responsible for a quarter of global land use change and one third of all material consumption (Circle Economy 2023). And demolition activities from the built environment are a significant contributor to the waste stream at 64% of all waste in the United States (US EPA 2020).

To compound the problem, material usage in the construction industry is predicted to exponentially increase in the coming decades with global growth and urbanization. By 2060, the global demand for construction materials and metals is projected to nearly double (OCED 2019). Even with increased resource efficiency and recycling, the production of these materials is estimated to reach 649 billion metric tons of CO_2e by 2100 (Chen et al. 2022).

This chapter addresses resource consumption and waste during all life cycle stages of site construction materials, products, and material assemblies. Strategies to reduce resource consumption, reuse waste, optimize materials and structures, close materials loops, and contribute to the

DOI: 10.4324/9781003413974-3

circular economy are summarized, and they are linked to other book chapters where detailed information is provided.

Construction materials and products are produced through multiple processes with inputs from the natural environment and/or industry. At each phase, solid, liquid, and gaseous waste is generated, sometimes in substantial amounts. It is captured and disposed of in landfills or incinerators, released to the environment, or reused. Resource use and waste of construction materials is discussed by life cycle phase in Table 3.1.

Material flows throughout the life cycle are either direct or indirect. Direct flows are minerals, fuels, and biological materials are used to produce the material or product. Indirect flows, also called hidden flows, are materials that are quarried, excavated, harvested or pumped but do not find their way into the finished product. These flows become waste during the product stage (phases A1–A3). Even though indirect flows are not part of a finished product that enters the economy, they a produced in far greater quantities than direct flows. This is particularly true for metals and dimension stone (Table 3.1).

Table 3.1 Resource Use and Waste in Material and Product Life Cycles

Life Cycle Phase	Resource Use and Waste
A1 Raw Materials	The raw material extraction phase can be very resource and waste intensive for some materials. Mineral-based materials, technical materials, are primarily obtained below the earth's surface; therefore, vegetation and soils are removed at a minimum. Materials like metal ores and stone suitable for dimension stone may be located several meters below the surface; so all the materials excavated above them, called overburden, are indirect flows. Beneficial use of this material in site and infrastructure construction includes aggregate bases, concrete mixes, and asphalt pavements. Harvest of biological materials creates waste with "slash" from tree branches, roots, and leaves resulting from the harvest of trees, bamboo, or other plants. These "waste" materials are often left to biodegrade at the point of extraction becoming food for decay organisms. "Harvest" of raw materials from material banks including the built environment, quarry and mine waste, and landfills is discussed later in this chapter.
A2 Transport to Factory	Transportation distances from raw material extraction locations to a manufacturing facility vary widely and are an important consideration. Some companies will locate their manufacturing facilities near their quarries to save the cost of hauling heavy materials. Others transport large quantities of materials great distances. For instance, bauxite, aluminum's ore, may be mined in South America then shipped to refining plants in Canada or China. The ore-to-waste ratio of bauxite is typically 1:4 with four metric tons of bauxite being shipped these great distances to result in just one metric ton of aluminum.
A3 Manufacturing	Manufacturing phases, including fabrication, can involve multiple stages in different locations. The primary processing phase of manufacturing produces a significant amount of waste as raw materials are refined or transformed. Waste generated is solid, liquid, and gaseous, including greenhouse gas emissions. Manufacturing uses fossil fuel inputs to transform raw materials into new materials (metals, plastics, bricks, and cement) by heating them. Solid waste from manufacturing is increasingly recycled back into processes or sold for use in other products. Toxic waste released from raw material extraction and manufacturing stages must be reported by larger producers in the United States in the Toxics Release Inventory (Table 4.1). The incorporation of recycled materials, also called secondary materials, in manufacturing can reduce resource use and waste from manufacturing.
A4 Transport to site	Transportation of construction materials to the site uses fossil fuel inputs, and the resulting waste is in the form of particulates and gaseous emissions, including GHGs. As site construction materials are heavy and bulky, these impacts can be substantial. Use of locally sourced and produced materials will reduce transportation impacts. It can be challenging to identify all locations in a supply chain due to multiple distributors and intermediates. Chains of custody can reveal the path that the material or product traveled to the site.
A5 Construction	Construction and demolition (C&D) waste reductions are primary considerations during the construction phase. Annual construction wastes are expected to reach 2.2 billion metric tons globally by 2025 (Transparency Market Research 2024). Demolition debris accounts for about 90% of C&D debris while construction waste is just 10% on average. Demolition activities generate both technical and biological wastes. Some states and municipalities regulate C&D waste requiring materials such as concrete, asphalt, gypsum, and wood be reused or recycled. Other regulations prohibit all C&D waste from being landfilled. Recycling markets for these C&D waste materials can be robust where regulations exist. In the United Kingdom approximately 90% of C&D waste is recycled (DEFRA 2018).
B1-B5 Use, Maintenance, Repair, Replacement, Refurbishment	Using durable, repairable materials, and material assemblies that prolong the service life of a structure is the key resource and waste consideration in the use phase. Use of flexible structures will allow for repair and releveling activities without removal and replacement of the entire structure.
C1–C4 Deconstruction/ Demolition, Transport to Waste Processing, Waste processing, Disposal of Waste D1 Reuse, recovery, recycling	The "end-of-life" phase for a project often means that the site will be cleared and rebuilt with demolition debris taken to a landfill. To reduce use of resources and waste for new site development, structures should be reused where feasible, and structures should be deconstructed with materials and products reused on the site to the extent possible (Figures 3.2 and 3.3) (Tables 3.2 and 3.3).

Resource Efficiency Definitions

Dematerialization is a strategy of reducing the quantity of materials in a structure through use of smaller members, thinner sections, or narrower widths. It also includes the concept of structure as finish with avoidance of veneers and extraneous finishes.

Zero waste is the conservation of all resources through responsible production, consumption, reuse, and recycling. There is no burning of waste or discharges to land, water, or air. Zero waste can be a goal in a system, built project, or manufacturing phase.

Circular economy "refers to an industrial economy that is restorative by intention; aims to rely on renewable energy; minimizes tracks and eliminates the use of toxic chemicals; and eradicates waste through careful design" (Ellen Macarthur Foundation 2024).

Reused materials are recovered materials that are reused whole with some processing such as nail removal or refinishing.

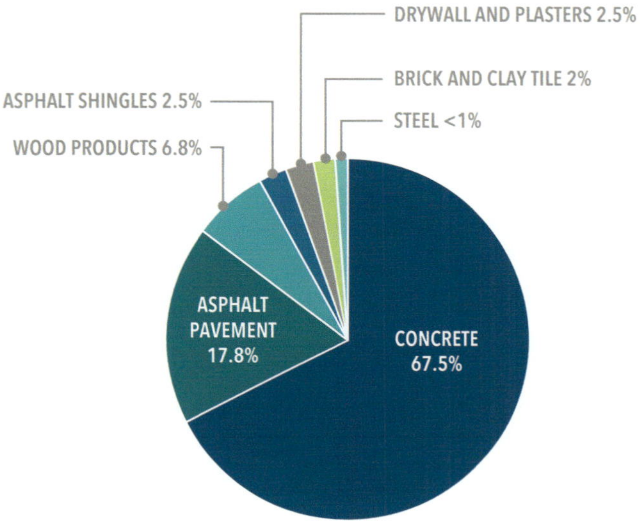

Figure 3.2 United States construction and demolition debris by type in 2018. Adapted from source: US EPA 2020. Image credit Drake Bruner and Meg Calkins

Table 3.2 US Construction and Demolition Debris Generation by Material, 2018

Material	Waste During Construction[1]	Demolition Debris[1]	Total C&D Debris[1]
Concrete	24.2	381.0	405.2
Asphalt concrete	0	107.0	107.0
Wood products	3.4	37.4	40.8
Drywall and plasters	3.9	11.3	15.2
Steel	0	4.7	4.7
Brick and clay tile	0.3	12.0	12.3
Asphalt shingles	1.2	13.9	15.1
TOTAL	**33**	**567.3**	**600.3**

[1]Figures in millions of metric tons

Source: US EPA 2020

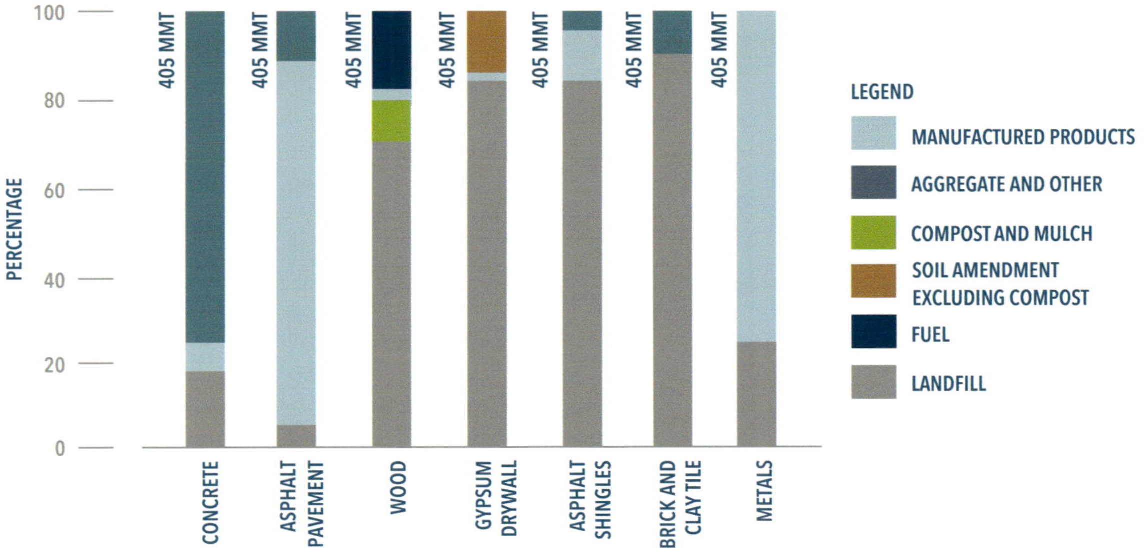

Figure 3.3 United States construction and demolition debris management by destination and type of material in 2018. Adapted from source: US EPA 2020. Image credit Drake Bruner and Meg Calkins

Recycled materials are recovered material that is used to produce a new material or product in manufacturing processes.

Pre-consumer recycled content, also called post-industrial recycled content, is materials diverted from the waste stream during manufacturing. Recycled content is expressed by proportion or mass. Pre-consumer and post-consumer content should be distinguished.

Post-consumer recycled content is material used and discarded by consumers that can no longer be used for its intended purpose. Post-consumer recycled content has a higher value in the circular economy than pre-consumer recycled content.

Refuse is the portion of the waste steam that is disposed of in landfills or waste-to-energy incinerators.

Secondary materials are materials that are recycled.

Demolition is the removal of materials from a building or site by destructive methods. Materials are crushed, fragmented, and comingled reducing their potential for reuse and recycling.

Disassembly is the non-destructive act of taking apart a structure with the intent of salvaging and reusing the constituent parts. Materials are sorted.

Design for disassembly is an approach to the design of a structure that considers the full life cycle of the structure and all the constituent parts. The goal is to facilitate deconstruction, reuse, and recycling at the end of the service life of the structure.

Material banks are repositories or stockpiles of materials that can be recovered. In the circular economy, material banks for construction materials are the existing built environment, mine spoils, and landfills.

Urban mining is the practice of recovering materials and products from a defunct building, site, infrastructure, or other anthropogenic waste. The term is often used in contrast to mining the natural environment for raw materials.

Renewable materials are biological materials that can be regrown within a generation or two.

Rapidly renewable materials are biological materials capable of regrowth within five years.

Table 3.3 US Construction and Demolition Debris Generation by Type of Structure, 2018

Materials	Buildings[1]	Roads and Bridges[1]	Other[1]
Concrete	102.0	168.3	134.9
Asphalt concrete	0	107.0	0
Wood products	39.5	0	1.3
Drywall and plasters	15.2	0	0
Steel	4.7	0	0
Brick and clay tile	12.3	0	0
Asphalt shingles	15.1	0	0
TOTAL	**188.8**	**275.3**	**136.2**

[1]Figures in millions of metric tons.

Source: US EPA 2020

Shifting to a Circular Economy

Shifting the global supply chain to a circular economy offers a comprehensive solution to our unchecked resource consumption and waste production. The circular economy "refers to an industrial economy that is restorative by intention; aims to rely on renewable energy; minimizes tracks and eliminates the use of toxic chemicals; and eradicates waste through careful design" (Ellen Macarthur Foundation 2024).

The Circularity Gap Reporting Initiative estimates that a global circular economy would move human activity back within safe limits of the planet. Human needs could be fulfilled with only 70% of the resources we now extract and use (Circle Economy 2023).

Additionally, transitioning to a circular economy will substantially reduce greenhouse gas emissions (Figures 3.4 and 3.5). By 2050 we could save 10 billion metric tons of GHG emissions – one fifth of total annual emissions – every year if we can close material loops, use resources efficiently, and reduce waste (ILFI 2023). Using reused and recycled materials in buildings and sites could reduce emissions of construction materials by 50–70% by 2050 (IRP 2020).

The circular economy aims to get the most value from the resources we have already extracted. It distinguishes between the technical cycle of minerals and metals, and the biological cycle of plant and animal-based materials. In the technical cycle, products, materials, and structures are kept in use for as long as possible then they are reused or recycled. In the biological cycle, the strategy is to restore nutrients into the biosphere by rebuilding natural capital (McDonough and Braungart 2010).

Figure 3.4 and Figure 3.5 The preserved steel exoskeleton of Mill 19 celebrates the history of the steelmaking site and a regenerative future while providing new landscapes and experiences among post-industrial remnants. An event plaza, stormwater channel, disturbance-adapted gardens, and public gathering spaces use reclaimed materials that celebrate the Mill's industrial legacy. The team from Ten x Ten inventoried materials and remnants of the mill, stockpiled them, and reused them for landscape pavement, walls, and other structures. They carefully refurbished large portions of the mill structure in place for a circulation loggia and other spaces. Image credit: John Rasmussen

Technical Cycles

The most effective technical cycle is to maintain and use materials and products for as long as possible maintaining their value. When a user no longer needs the product, it can be reused by others. When the product can't be used as is, it can be refurbished for reuse. If the product can't be reused or the material must be broken for removal, it can be recycled. When it is recycled, the value of the product is lost but the material's value is preserved.

Biological Cycles

Biological materials such as wood-based products and biobased materials are renewable, but greater value can be achieved by reusing them in whole form and recycling them. The next step would be to anaerobically digest or compost them to extract nitrogen, phosphorous, potassium, and other nutrients. Where this is not possible, they can be burned for energy. Burning and composting releases the carbon stored in these materials.

It is not uncommon for biological nutrients to be mixed with technical nutrients. Engineered wood and bamboo products are examples of this. These products mix wood or bamboo constituents with a fossil fuel-based binder. While these products can be reused in whole form, it is difficult to recycle them because they are mixed materials. Incinerating them can release harmful chemicals from the binders (Figure 3.6).

Core Principles of the Circular Economy for the Construction Sector

The circular economy in the construction sector uses resources efficiently, closes material loops, optimizes the life cycle of structures, and favors local, renewable materials and energy. The core principles of a circular economy for the construction sector are explored next.

Use Less: Reducing the Consumption of Materials and Resources

A major principle of the circular economy is to preserve natural capital. Reducing the consumption of materials and resources involves closing material loops and striving for zero waste. This includes reducing the use of virgin materials, mining the built environment and landfills for raw materials, building smaller more streamlined structures, using local materials, reusing refurbished materials and products, and using new materials and material mixes with high recycled content (Figure 3.7).

Transitioning to a circular economy requires shifting the "end-of-life" perspective to "end-of-use". After the service life of a landscape, structures can be repurposed and products or materials can be reused or recycled. The resources and embodied carbon in the structures and materials are preserved for a new use.

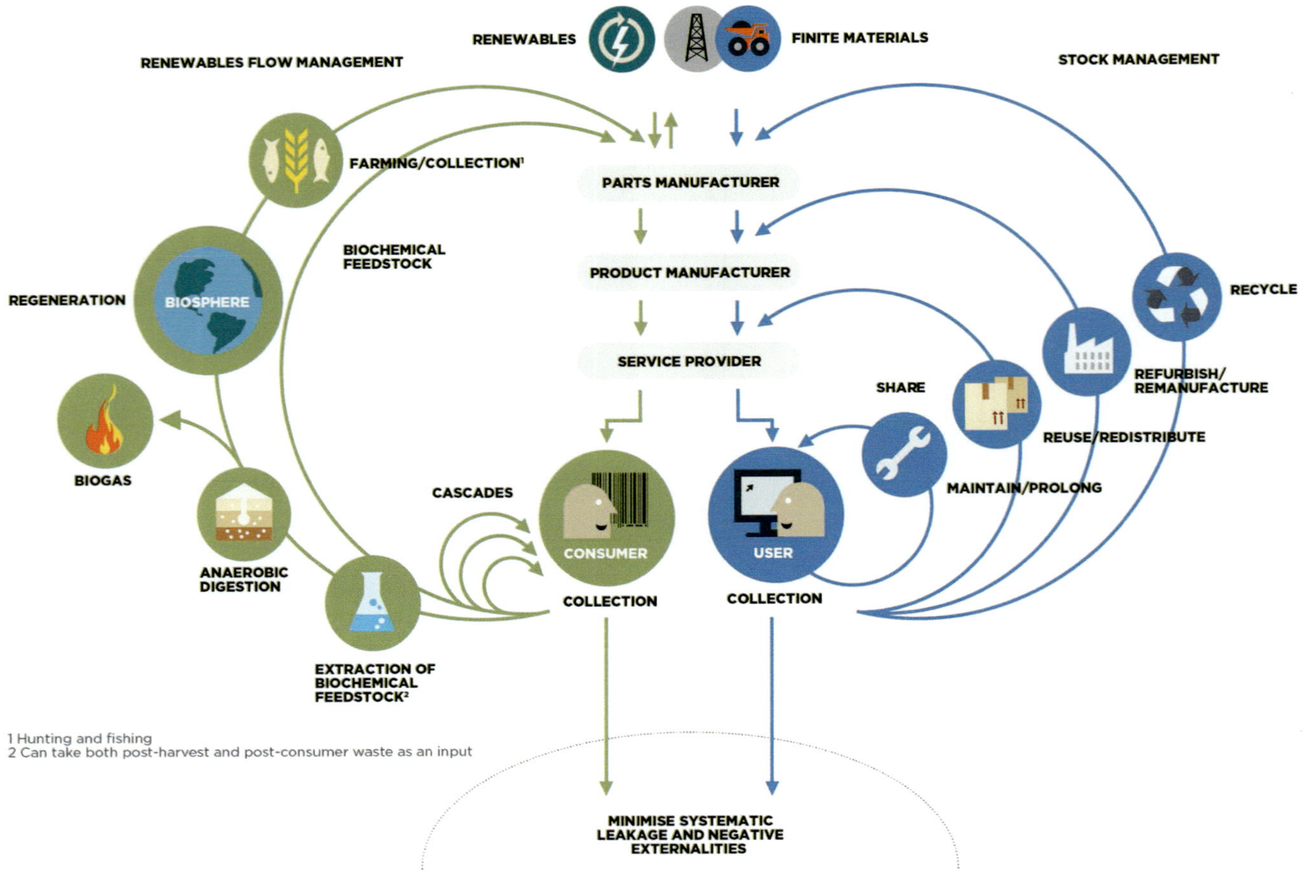

Figure 3.6 The circular economy system diagram of the technical and biological cycles of material resources. Image credit: Ellen MacArthur Foundation, copyright 2019

Figure 3.7 The team from Design Workshop reduced resource use with a two track driveway and a reclaimed stone fountain at a private residence in Aspen, Colorado. Image credit: Brandon Huttenlocher/Design Workshop, Inc

Use Longer: Optimizing the Lifespan of Material and Product Use

Designing sites and buildings to be in use for a long time will reduce the use of new materials and reduce waste. Designing flexible, adaptable, and repairable structures extends use lives. For instance, asphalt pavements can be repaired and resurfaced with recycled materials in a binder to extend the pavement structure's life. Rigid materials, such as concrete are designed for 30-year use lives but are very difficult to repair once cracked and are often removed in just half the time. Flexible retaining walls can be releveled and repaired, whereas rigid walls are challenging to repair if settling or cracking occurs (Figures 3.8 and 3.9).

Use Again: Designing for and Engaging in Practices of Disassembly, Reuse, and Recycling

Designing products and structures for disassembly, reuse, and recycling involves planning for end of use during the design phase. Connections are accessible, standardized, and dismantlable. Technical and biological materials are not mixed to preserve the ability for reuse or recycling. Specifications are written to facilitate deconstruction and reuse. At the end of a project's life, if structures cannot be reused in place, or they are deconstructed, products are refurbished and materials reused or recycled.

Figure 3.8 and Figure 3.9 At the Airport Garden in Callone Ricouart, France, Wagon Landscaping refurbished an existing asphalt roadway into a linear garden by cutting and removing areas that they filled with plants and carefully arranged slabs of the cut asphalt. Wood benches provide seating for airplane viewing. Image credit: Yann Monel

Make Clean and Safe: Regeneration of Nature Through Use of Renewable Materials and Energy

Prioritizing use of healthy, renewable materials and energy sources to produce construction materials will support the circular economy. Cycles of use, waste, and renewal are clean and regenerative. Wood can be used in structural applications and should be favored over concrete and steel where applicable. In exterior settings, use of biobased materials is challenged by moisture and decay organisms, but wood modification technologies can offer good preservation. Another challenge to use of biobased materials can be unsustainable harvesting and agricultural practices (Chapter 18 "Wood Materials and Products" and Chapter 19 "Biobased Materials and Products"). Technical materials and their by-products should be safe and non-toxic.

Lessons From Nature for the Circular Economy

Biomimicry takes inspiration from solutions in nature and applies them to systems in the human world. Ecosystems in nature offer many lessons for the circular economy and goals of net-zero resource use.

Closed, circular loops: In nature, materials are recycled in circular loops that optimize resources. Waste from one process is food for another.

Efficient use of resources: In nature, resources are used over and over. In winter, when the trees shed their leaves, they are food and habitat, then ultimately return nutrients to the soil that feeds the tree. After use, our products can be taken back, disassembled, and reused, or they can provide raw materials for a new product.

Simple materials: Nature uses simple materials that decompose at the end of their use. We currently use complex heterogenous materials that are difficult, and sometimes toxic, to reuse or recycle. We can shift back to use of simple, local materials such as stone, wood, and earth-based materials that can be returned to the soil after use.

Local materials: Nature uses materials that are unique to a specific ecosystem. Ecosystems function with the materials at hand and they adapt to ensure efficient use. Historically, our built environment used primarily locally available materials. Arid areas of the world employed earthen construction using abundant soils to protect inhabitants from hot sun and insulate during cold nights. Wood structures were used in forested areas, and stone from fields was used to enclose the fields to contain animals. Our current global supply chain where materials travel thousands of miles is unsustainable.

Renewable materials: Materials in nature are renewable, and after their useful life they decompose feeding decay organisms and returning to the soil. Currently we extract huge amounts of non-renewable materials for construction, mixing them with other constituents or heating them to transform them into materials and products then shipping them great distances. Wood and some other biobased materials are renewable, some rapidly, and can be returned to the soil after their useful life if they are not treated with toxins or mixed with other materials. Ideally, they will be reused or recycled multiple times prior to decomposition.

Multifunctional solutions: Trees serve multiple functions. They evapotranspirate stormwater, provide oxygen, remove air pollutants, and shade surfaces. Our built landscapes must be designed to provide multiple functions and multi-use spaces. This will ensure their longevity as they can be adapted to accommodate change.

Nature-based structures: We use nature-based solutions in our built landscapes to attenuate storms, cleanse stormwater, reduce urban heat islands, and many more ecosystem services. We can also build nature-based *living structures* to retain embankments (Chapter 8 "Retaining Walls and Bioengineered Embankments"), enclose spaces (Chapter 11 "Rails, Fences, and Screens"), and shade our gathering places.

Mining New Sources of Raw Materials

Within the last four decades, we have depleted a third of the earth's natural resources, and at this rate, we will exhaust them entirely in the next century (OCED 2019). But with a shift to the circular economy there are multiple sources of "waste" materials that can be "mined" for raw materials.

Urban mining: Waste materials from industrial processes, the built environment, disasters, and other anthropogenic products and processes can be recovered and repurposed for use in new materials and structures. In addition to a circular practice, reusing these materials can offer an opportunity to reference the cultural history of a place.

Mining the built environment: After the service life of sites and buildings, current practices demolish them, comingling materials, and disposing of the waste in landfills promoting the linear flow of materials. In a circular economy, anthropogenic materials in the built environment are mined as resources for future projects. Buildings and built sites are the material banks for our future built environment. At the end of the site's life, every brick, bench, concrete slab, light standard, and wood beam has value for another project either on or off the site. With 100 billion metric tons of construction, renovation, and demolition waste generated worldwide annually, there is an abundance of raw materials for new structures and products (Chen et al. 2022). When the built environment is deconstructed then mined for materials and products, pressure on the natural environment for new resources is reduced. Fuel for extraction, transport, and production of new materials is eliminated – along with the associated carbon emissions (Figure 3.10).

Re-mining: Re-mining reprocesses waste materials from mining activities. Tremendous amounts of materials, overburden, are removed from the earth for metals mining and stone quarrying. Re-mining utilizes rubble, gravel, and quarry dust as aggregates for bases, concrete, and asphalt. Quarry blocks that are unsuitable for dimension stone can be used whole as benches and retaining walls. Transporting these heavy materials from the remote locations of mines and quarries can be a challenge to re-mining.

Landfill mining: Landfill mining involves the excavation and processing of waste from closed landfill sites. Mining landfills for recoverable materials is done in North America and Europe but is a costly practice that does not always make financial sense, although it is being reconsidered with increasing focus on reducing GHGs and circular use of materials. The carbon benefits of landfill mining are twofold. The practice could reduce methane emissions from landfills while recovering resources for reuse or energy recovery. The Enhanced Landfill Mining Directive (Hogland et al. 2010) from the European Union has spurred European countries to engage in recovery of materials and energy sources from the 150,000 landfills in Europe. The typical landfill mining process involves excavating the contents of a landfill cell and separating and sorting material into stockpiles. Once soil is removed, ferrous metals can be recovered using a magnetic separator. Plastics and aluminum can also be recovered and recycled. It is estimated that there is more aluminum currently in landfills in the United States than in remaining supplies of bauxite in the natural environment (Davies 2013) (Figures 3.11 to 3.14). Waste-to-energy facilities utilize large amounts of the recovered material, but there are concerns with the release of toxins from some plastics. Other concerns include the release of methane and VOCs from opening the landfill.

Shifting Construction Materials to a Circular Economy

The monumental shift toward a circular, closed loop system will require major structural changes in the construction material supply chain that must be supported by policies, financial incentives, and regulations. A major challenge

Figure 3.10 Assemblage Landscape Architecture used "story mapping" as a strategy to trace the history and geography of the materials used to construct Dolly's Community Park in Brooklyn. They incorporated a variety of recycled and reclaimed materials, which were traced back to their historic point of origin. Image credit: Barrett Doherty and Assemblage Landscape Architecture

Figure 3.11 and Figure 3.12 At Warsaw Uprising Mound Park, topoScape + Archigrest built rubble concrete walls on the 35 meter high mound of debris from the Warsaw city fabric destroyed in World War II. Incorporating fragments of buildings, the slopes of the human-made ravines resemble the geologic layers of an anthropogenic hill. Image credit: Michal Szlaga

Figure 3.13 and Figure 3.14 The Warsaw Uprising Mound Park, built on top of the WWII rubble dumping ground, highlights rubble as a valuable historic resource. Through urban mining the mound, Archigrest and topoScape repurposed the rubble as fill for gabions in the lapidary forming a labyrinth of "ruins". Building fragments are on view for park visitors in the display cases and the gabion walls. Image credit: Michal Szlaga

to the circular economy is the lack of a widespread and standardized infrastructure for exchange of reclaimed and recycled materials. Product and material salvage industries are fragmented and not at a scale that can offer a steady supply to new projects. Recycling industries vary widely by region and country, and, again, steady supplies are not ensured for manufacturers who would like to use the material.

Circular solutions can be supported by databases, online platforms, and other logistical tools to connect manufacturers with recyclers, deconstruction companies, and the designers who will source the material. Modeling tools such as BIM are working to incorporate life-cycle data, life spans, maintenance, and replacement requirements.

Multiple stakeholders will transition manufacturing industries to a circular economy. Policymakers, manufacturers, consumers, designers, engineers, regulators, and others must prioritize and implement circularity principles with long-term vision over the current focus on short-term rewards.

Several cities, states, and countries are setting zero-waste goals with a wide variety of policy approaches. The European Green Deal by the European Union has a Circular Economy Action Plan (CEAP) that funds companies to achieve net-zero waste. Several EU countries have also incentivized extended producer responsibility programs as well.

Policies can heighten the business case and a just transition to a circular economy. They can

- ***Facilitate expansion of the secondary material supply chain*** with financial incentives such as carbon and health related taxes that reflect the true cost and externalities of producing new materials.
- ***Shift jobs*** from extraction of materials from the natural environment to mining the built environment, re-mining, and mining landfills.
- ***Set targets*** for resource reuse, carbon reductions, and other aspects of the circular economy.
- ***Expand the responsibility of producers*** beyond the traditional product stage. This shifts the supply chain away from linear product life cycles.

Extended Produced Responsibility (EPR)

Extended producer responsibility is a policy approach that makes producers responsible for the collection and treatment of their products after consumers use them. This can result in producers considering the whole life cycle of the products they produce, promoting design for durability, reuse, and waste prevention.

Material and Product Takeback Programs

Material and product takeback programs offer a solution for refurbishment, reuse, and recycling at the end of use. These programs incentivize cleaner materials, design for disassembly, and standardization of parts. Packaging takebacks are the most common programs.

Benefits of takeback programs are lower cost of products due to use of secondary materials, stronger customer relationships, safer materials, and reduced resource use and carbon emissions. Materials and product take back programs can be challenging to implement because they can be expensive. Government support can help create the infrastructure for these programs as well as offer incentives to accelerate their implementation.

Product as a Service (PAAS)

Some products can be supplied as a service with specifiers procuring the "use" of a product. After use, the product is taken back and refurbished for reuse. This ongoing relationship with the supplier can facilitate reuse multiple times. Products can be designed for transformation into new products. Lighting, controllers, and site furnishings are potential products for PAAS.

Building Material Passports

Building material passports are a recent tool to support circularity by tracking materials and products in the "material bank" of a built project. Information about products used in a building and site can be mapped and recorded in open-source digital databases to facilitate recovery, reuse and, recycling after the buildings use life. Material passports can be connected to BIM software, digital inventories, and other supply chain databases. All modifications and maintenance of the materials and products are logged with the asset.

Strategies to Promote Circular Construction Material Life Cycles

Some strategies of the circular economy will have greater impact than others. For instance, building nothing new at all will conserve far more resources than crushing and recycling a concrete slab for use in asphalt pavement.

The resource efficiency hierarchy, also called the waste hierarchy, maximizes resource use to preserve value over several lifetimes. Preventing waste by ***refurbishing structures in place*** is the preferred option, followed by reclaiming the material or product through deconstruction and refurbishment for ***reuse***.

Reprocessing and ***recycling*** are less preferred than reuse due to the downcycling of materials. For example, when a concrete slab is removed, it is crushed and reused as aggregate in new concrete, new asphalt pavement, or as

Figure 3.15 A hierarchy for resource efficiency in site construction. Image credit: Drake Bruner and Meg Calkins

aggregate base material. The concrete is not reused in whole form, losing its original value (Figure 3.15).

Other materials such as steel and aluminum can be recycled again and again, but the process requires heat and energy. Some plastics are recyclable, but most plastics recycling technologies are energy intensive. Other plastics give off toxins when heated so, while technically recyclable, they are not often recycled.

Energy recovery is the last option, and only if the incineration processes are efficient with carbon emissions minimized or captured and effective pollution control equipment. About 10% of an incinerated product remains in the form of ash, which is sent to a landfill.

Beneficial use of incineration ash is being explored in the brick, block, cement, and asphalt pavement industries. While energy recovery is not a preferred option, use of the ash in new products could still contribute to a circular economy.

Stakeholder Roles in Implementing Circularity in Construction Materials

Table 3.4 details actions to close material loops and promote the circular economy. Most of the actions will involve communication and collaboration among multiple parties to accomplish. It is important to note that strategies supporting the circular economy are inextricably linked to decarbonization. Reducing use of resources will also reduce the embodied carbon of a built site (Table 2.5).

Table 3.4 Strategies to the Support Circular Economies for Site Construction Materials

Feasibility Planning and Pre-Design for Circularity	
Reuse structures in situ. Do not rebuild, instead refurbish. The first, and perhaps the most important question to ask in the early project planning phase is: *Do we need to build or rebuild?* A major priority of the circular economy is to keep buildings and sites in use as long as possible and to ensure that they are reused instead of being torn down and rebuilt.	Renovate, retrofit, and reuse existing structures and buildings to accommodate new uses. Establish clear goals for reuse of structures early in the design process. Use financial and carbon reduction arguments to convince clients and stakeholders. Reference the cultural history of the place through preservation and adaptive reuse of structures.
Involve project stakeholders in goal setting. Goal setting should occur in the earliest stages of a projects process with as many stakeholders as possible at the table. Goals can be set for waste diversion, material reuse, recycled contents, and embodied carbon. Zero waste is an ambitious goal but one that will become increasingly achievable in the circular economy.	Identify waste reduction baselines then set goals for a percentage below baselines. For example, if the average recycled content of RAP in new asphalt is 25% regionally, set a goal for 40% RAP content. Monitor and document progress toward goals.
Modify standards, codes, and regulations that work against circularity strategies. Standard sizes of structures, material mixes, and recycled content limits in standards, codes, and regulations may work against circular strategies.	Review reference standards in specifications. For example, referencing ACI 150 could prohibit cement substitutes. Instead, ASTM International. *ASTM C595/C595M-23 Standard Specification for Blended Hydraulic Cements* and other applicable standards should be referenced. Work with state and local DOTs and other organizations to modify their standards to reduce resource use and increase allowable recycled contents.
Plan for deconstruction, demolition, and construction waste diversion. Performing reclamation audits prior to design, budgeting for deconstruction and recordkeeping, and getting contractors involved early on will facilitate planning for deconstruction and reuse. Research has demonstrated substantive carbon emissions reductions with deconstruction and reuse as compared with recycling (Di Maria et al. 2018).	Existing assets should be surveyed and inventoried to identify structures, products and materials for reuse. Inventories should include dimensions, quantities, conditions, technical characteristics, and disassembly prospects. Require waste tracking by the contractor during deconstruction and construction. Where possible require contractors to use Recycling Certification Institute (RCI) facilities to help document diversion rates. Build in fees and labor costs for deconstruction, sorting, storage, and refurbishment.
Shift mindsets from linear to circular among clients, designers, contractors, and manufacturers. The product and material supply chain has been linear for the past several decades. Shifting to a circular mindset can get all decisionmakers on board early in the process.	Stakeholders should take a long-term view of structures, products, and materials. Full life-cycle costs should be emphasized instead of first costs, along with full life-cycle carbon accounting. Be prepared to discuss the concept and execution of circularity with project stakeholders. Demonstrate the economic and environmental benefits through use of case study projects and other research.
Engage in third-party rating system certifications. Voluntary rating systems offer standards, credits and prerequisites for reducing resource use, reusing materials and products, and eliminating waste.	Rating systems and their circularity strategies are discussed in Table 2.8.
Design for Circularity	
Optimize material use. Dematerializing site structures will reduce resource use and likely save material and construction costs. Work to change design standards such as parking quantity and size requirements. Many standards call for unnecessarily wide paths, overly thick pavements, or large parking lots.	Use less material in smaller structures, narrower paths, smaller members, and smaller areas. Avoid over-sizing structural members and pavement and base thicknesses. Engineers and span tables build in a margin of error. In areas with deep frost lines (>60cm), consider flexible structures to avoid excessive material use in footings. Avoid use of veneers and other unnecessary finishes. Size structures to standard dimensions of materials to eliminate construction waste.
Design structures for durability and repairability. Site structures should be designed for a long use life. Flexible structures can be repaired or resurfaced if settling or frost heave occurs. Rigid structures, particularly concrete walls and pavements, are challenging and expensive to repair.	Favor flexible structures such as gabion walls, interlocking retaining walls, stabilized stone dust pavement, and asphalt pavement with lighter surfaces. Use appropriate woods, metals, and metal finishes for the microclimate to prolong the life of wood and metal structures. Avoid plastic products that are exposed to the sun as UV degradation could shorten their life.

(*Continued*)

Table 3.4 (Continued)

Design living structures. Living structures are made from live plant material that is trained, pleached, or woven as it grows to form a natural barrier. Some use no additional resources other than the plant material. Living structures are not often used in urban settings or plazas for aesthetic reasons, but we can shift our aesthetic expectations. The plants in living structures will provide ecosystem services much needed in urban areas such as stormwater attenuation, air pollution removal and cooler surfaces.	Consider bioengineered embankments such as live staking, live fascines, brush layering, live crib walls, vegetated gabion walls, and vegetated rock walls for retaining situations (Chapter 8 "Retaining Walls and Bioengineered Embankments"). Build living fences and overhead structures with techniques, such as hedge laying, espalier, or weaving plants on a built structure, and weaving plants into a plant-only barrier (Chapter 11 "Rails, Fences, and Screens"). Some living structures require seasonal training of the plants over a few years and may not be appropriate for projects with minimal maintenance regimes.
Design for deconstruction (DfD). Designing sites for disassembly and deconstruction is a key circularity strategy as it will facilitate disassembly and reuse of materials and products after a site's service life, extending their use life. ISO 20887:2020 offers guidance on this practice.	Specify materials that are durable, modular, and standardized to facilitate their reuse multiple times. Use reversible attachments that are easily accessed. Favor bolted connections over welds and adhesives. Avoid using mortar in pavement and wall assemblies. Avoid direct burying wood or steel posts in concrete footings. Avoid materials with toxic constituents, such as PVC or foams, that could impact their recyclability. Write specifications for DfD. See Model Spec Resources in the resource list at the end of this chapter.
Reuse materials on site or from near the site. Deconstructing, stockpiling, refurbishing, and reusing materials and products on a previously developed site can conserve resources and hauling. Using reclaimed materials from another site will save resource use and energy to produce new materials. Off-site sources of reclaimed materials are other projects by the designers, contractors or owners, materials exchanges, and salvage distributors.	Identify materials for on-site salvage and reuse early in the process with an audit. Inventories from the audit should include type of material, quantities, sizes, condition, and refurbishment steps (Figures 3.16 and 3.17). Let the materials and products inspire the design. Maintain flexibility in the design until materials are found. They should be identified and procured (if off site) during the design process. Reference and celebrate the history of a place with reclaimed materials (Figures 3.18 and 3.19).
Reprocess materials on site. Deconstructing, crushing, sizing, and replacing concrete and asphalt materials in new pavements or new bases can save material and fuel resources from hauling large quantities of heavy materials away from the site. It will also reduce resource use and hauling of new road and base materials.	Many masonry or asphalt materials can be crushed, sized, and reused on a construction site for new pavements or base materials. This is a good option for masonry materials that are mortared together as it is not financially feasible to separate and clean the materials for reuse whole.
Sourcing and Procuring Circular Materials and Products	
Favor materials and products from manufacturers with EPDs. Or those that provide third-party verified data on resource use and raw material extraction practices. This data should be third-party verified using an accepted standard to collect and compile the data.	EPDs will highlight the *Abiotic Depletion of Non-Fossil Resources*, sometimes called *Abiotic depletion potential* (ADPe), for minerals and metals. This includes resource use for the material or product as well as waste and by-products that don't end up in the final product.
Specify local and regional materials and products. Using locally or regionally produced materials and products can reduce resource use by reducing fuel consumption for their transport. Sourcing local and regional materials is particularly critical for heavy materials. It can also stimulate local economies and reduce material costs.	Identify locally produced materials and products during schematic design. Develop a local materials palette as a design strategy. Maintain databases of local and regional material and product manufacturers. Make sure that the local materials are locally processed during all phases. Work with contractors to source and procure local and regional materials and products.
Trace the supply chain of materials and products. It is not unusual for materials such as stone to be sourced locally or regionally, then sent overseas for fabrication.	Ask material and product manufacturers where raw materials originated, and where all stages of manufacturing and fabrication took place. Favor products with transparent supply chains. Require chain-of-custody records where appropriate.
Budget for the additional time it will take to locate circular materials. Responsible material and product sourcing requires a time commitment from the project team.	Responsible material and product sourcing requires a time commitment from the project team. Product research, specification, review, and tracking should be built into the project budget. Databases and in-house records can facilitate these practices.

(*Continued*)

Table 3.4 (Continued)

Purchase from manufacturers who are responsible producers engaging in circular practices. Practices of producers vary widely, and those engaging in optimized production practices, use of local materials, and provision of transparent information should be rewarded for their practices.	Look for producers with takeback programs, product-as-a-service programs, and those who incorporate recycled content and reduce waste. Source products and materials from producers who provide transparent information about their products and practices in chains of custody, EPDs, and HPDs. Favor producers with optimized practices.
Source materials and products from companies that protect and restore ecosystems. Extraction and harvest of raw materials impacts ecosystems far from the project site and even far from manufacturing facilities.	Source materials and products from raw materials suppliers that reduce waste, reduce pollution, and have restorative company practices. Chain-of-custody documentation can reveal raw material suppliers. Certifications for lumber, stone, metals, and concrete will reward good practices in raw materials harvest and extraction.
Specify renewable biological materials and products. Use of renewable materials and energy is an important aspect of circularity. After their useful life, biobased materials that are free of harmful substances can be composted. Some biobased plastics do not biodegrade which make them more durable but not compostable.	Choose biobased materials that are FSC certified to ensure responsible forestry practices, fair labor, and a chain of custody. Or use salvaged wood. Avoid tropical hardwoods as they are not usually plantation grown and may be illegally harvested from sensitive ecosystems. Look for locally or regionally harvested and processed biobased materials. Use smaller dimensions of wood to reduce pressure on forests. Beams can be "built up" and glulams specified. Specify biobased plastic products that are made from agricultural waste.
Specify materials and products are produced with renewable energy. The production of construction materials and products can be energy intensive and polluting. Products such as steel, aluminum, concrete, glass, and brick require fossil fuel resources to heat and transform materials. Plastics use fossil fuel resources as feedstocks. Wood modification technologies use heat to transform the sugars in wood making it decay resistant.	Consider using materials that do not require heat for their production such as natural stone, decay-resistant wood, earth-based materials, and unfired brick. Look for emerging materials that use biotechnologies to produce bricks and blocks instead of heat.
Specify materials, material mixes, and products with high recycled content. Bricks, concrete, concrete products, asphalt, metals, and plastics can all incorporate recycled content in high quantities. Favor post-consumer recycled content over pre-consumer recycled content.	Refer to the EPAs Comprehensive Procurement Guidelines in Chapter 21 for recommended minimum recycled contents by product type. Use performance-based specifications for concrete and asphalt pavement mix designs that specify recycled content for coarse and fine aggregates. This will allow suppliers and engineers to develop mixes and incorporate available recycled materials to meet this requirement.
Specify reusable and recyclable materials and products. Some materials and products can't be reused or recycled because they contain hazardous ingredients including asbestos, polychlorinated biphenyls (PCBs), chlorofluorocarbons (CFCs), heavy metals and other toxins (Tables 4.3, 21.1). Other materials and products can't be recycled because they are mixed assemblies, or the components can't be separated.	Avoid products with mixed biological and technical nutrients. Select materials that can be recycled into an equal value product rather than downcycled. Some types of plastic such as PVC and polystyrene give off toxins when they are heated or burned, so while they are technically recyclable, they are not recycled in most countries.
Rewrite standard specifications to foster circularity. Rewriting standard specifications to reduce resource use and waste can reduce resource use and carbon footprints of projects firm wide. They can also influence standard specifications of engineers and influence local material suppliers.	Standard specifications should do the following: Address construction and demolition practices to deconstruct sites, reclaim materials and products, and reduce waste. Allow resource efficient substitutions. Specify recycled aggregates for base courses such as RCA and RAP (Chapter 6 "Pavements").
Construction and Use Phase	
Reduce construction and demolition (C&D) waste. As C&D waste is a large portion of the waste stream, diverting it through reuse of structures, reuse of products and materials, and recycling will close material loops.	Work with contractors to estimate potential quantities of waste by type. Identify measures to reduce waste including takeback of surplus materials, just-in-time ordering, reducing packaging, reuse of offcuts, and other lean construction strategies. Balancing cut and fill will reduce import or export of soil and associated transportation impacts. Write specifications to allow substitutions of approved reclaimed components and materials.

(Continued)

Table 3.4 (Continued)

Maintain sites and structures for longevity. Regular maintenance of pavements, walls, decks, rails, and fences can prolong their service life and reduce premature replacement and resource use.	Maintenance and management strategies for materials and structures are discussed in most chapters throughout this book.

Figure 3.16 Xuhui Runway Park is a linear park in Shanghai built on a 36-acre site that was China's first civilian airport until its closure in 2011. The runway is reused in place as a major spine of the park. Image credit: Insaw Photography and Sasaki

Figure 3.17 The design of the SITES Gold certified park preserves and repurposes the airport runway's original materials on site using them for special pavement. The concrete slabs were salvaged from the tarmac demolition. Image credit: Insaw Photography

Figure 3.18 Cultural history and sustainability were dual goals for McGregor Coxall's Ballast Point Park in Sydney. The park incorporates several references to prior site uses including indigenous habitation, colonial use as a ballast quarry for ships, and lately an oil terminal from the 1920s until 2002. Artifacts and recycled rubble from the different phases of development fill gabion retaining walls. Image credit: Brett Boardman

Figure 3.19 Ballast Point Park offers opportunities for adventure, exploration, and playfulness. Users of all ages can discover artifacts from the site's richly layered history in the gabion walls. Image credit: Christian Borchert

Figure 3.20 The team from McGregor Coxall optimized resource use and reduced greenhouse gas emissions to build Ballast Point Park by incorporating fly ash and reclaimed aggregates into concrete mixes, reusing almost 20,000m^3 of rubble, constructing benches from reclaimed wood, and repurposing components of oil tanks in new structures. Image credit: McGregor Coxall

Table 3.5 Resources for Circularity in the Built Environment

Information Resources	
Buildings as Material Banks (BAMB)	Buildings as Material Banks is a multi-stakeholder organization that provides resources to facilitate the circular economy in the built environment. Resources include a reversible design checklist, a guide to best practices for materials passports, and a circular building assessment tool.
Common Materials Framework by mindful MATERIALS	The Common Materials Framework by mindful MATERIALS offers a shared language and set of resources for sustainable products. It is an established hierarchical data structure within which a certification or standard can align. The CMF orders standards, ecolabels, and data in categories of circularity, human health, climate health, ecosystem health, and social health and equity.
The Ellen MacArthur Foundation	The Ellen MacArthur Foundation is a network that collaborates with other organizations to facilitate the circular economy. To facilitate the circular economy transition, they run campaigns for industry transformation. Organizations, companies, and firms can join their network.
Recycled Materials Resource Center	The Recycled Materials Resource Center at the University of Wisconsin partners with several state DOTs to provide research and resources for use of recycled materials in road and pavement construction. It offers very detailed information about dozens of recycled materials for use in construction.
Whole Building Design Guide (WBDG)	The WBDG was developed by the National Institute of Building Sciences to foster communication and knowledge sharing to advance high performing facilities. The site offers design recommendations and other resources.
Standards and Product Certifications	
Cradle to Cradle (C2) Products Innovation Institute	Cradle to Cradle (C2C) Products Innovation Institute offers a searchable database of products that meet the C2C standard for material health, product circularity, social fairness, water and soil stewardship, and clean air and climate protection.
US EPA Comprehensive Procurement Guideline (CPG) Program	The US EPA CPG Program promotes use of recycled and recovered materials. It offers minimum recycled content standards for multiple site construction materials. It offers a CPG Product Supplier Directory.
WaterSense	WaterSense is voluntary label for water-efficient products. It is also a water conservation resource.
The Responsible Products Program	The Responsible Products Program from the Green Building Council Australia certifies products and offers Responsible Product Guidelines for responsible, healthy, climate positive, and circular construction products. They also offer the Responsible Products Calculator and the Responsible Products Checklist.

References

Chen, Zhikun, et al. (2022). "Construction, renovation, and demolition waste in landfill: A review of waste characteristics, environmental impacts, and mitigation measures". *Environmental Science and Pollution Research* 29.31: 46509–46526.

Circle Economy (2023). *The Circularity Gap Report 2023*. Circle Economy.

Davies, P. (2013). *Landfill Mining: New Opportunities Ahead?* Macfarlanes LLP, Global Business Publishing, Ltd.

Department for Environment, Food & Rural Affairs (2018). "Resources and Waste Strategy: At Glance". December. www.gov.uk/government/publications/resources-and-waste-strategy-for-england.

Di Maria, A., Eyckmans, J., and Acker, K.V. (2018). "Downcycling versus recycling of construction and demolition waste: Combining LCA and LCC to support sustainable policy making". *Waste Management* 75: 3–21.

Elhacham, E., et al. (2020). "Global human-made mass exceeds all living biomass". *Nature* 588: 442–444. https://doi.org/10.1038/s41586-020-3010-5.

Ellen Macarthur Foundation (2024). "Circular economy introduction". www.ellenmacarthurfoundation.org/topics/circular-economy-introduction/overview, accessed May 8, 2024.

Hogland, W., Hogland, M., and Marques, M. (2010). "Enhanced landfill mining: Material recovery, energy utilisation and economics in the EU (directive) perspective". *Proceedings of the International Academic Symposium on Enhanced Landfill Mining*.

International Living Futures Institute (2023). *The Regenerative Materials Movement: Dispatches from Practitioners, Researchers and Advocates*. Ecotone Publishing.

International Resource Panel (IRP) (2020). *Resource Efficiency and Climate Change: Material Efficiency Strategies for a Low-Carbon Future*. UN Environment Program and International Resource Panel.

International Standards Organization (ISO) (2020). *ISO 20887:2020 Sustainability in Buildings and Civil Engineering Works - Design for Disassembly and Adaptability: Principles, Requirements and Guidance*. ISO.

Kaza, S., et al. (2018). *What a Waste 2.0: A Global Snapshot of Solid Waste Management to 2050*. Washington, DC: World Bank.

Knowledge Hub (2024). "Shift: Circle economy". https://knowledge-hub.circle-economy.com/nations/frameworks/68/434, accessed July 13, 2024.

McDonough, W., and Braungart, M. (2010). *Cradle to Cradle: Remaking the Way We Make Things*. North Point Press.

Organization for Economic Co-operation and Development (OECD) (2015). *Material Resources, Productivity and the Environment*. Paris: OECD Publishing.

Organization for Economic Co-operation and Development (OECD) (2019). *The Economic Consequences of Climate Change*. Paris: OECD Publishing.

Transparency Market Research (2024). "Construction waste market: Global industry analysis, size, share, growth, trends and forecast 2017–2025". www.transparencymarketresearch.com/construction-waste-market.html, accessed July 13, 2024.

United Nations Environment Program (UNEP) (2023). *Building Materials and the Climate: Constructing a New Future*. Paris: UNEP.

United Nations Environment Program (UNEP) (2024). *Global Status Report for Buildings and Construction: Beyond Foundations - Mainstreaming Sustainable Solutions to Cut Emissions from the Buildings Sector*. UNEP.

US Environmental Protection Agency (US EPA) (2020). *Advancing Sustainable Materials Management: 2018 Fact Sheet*. Washington, DC: US EPA.

World GBC (2023). *The Circular Built Environment Playbook*. London: World Green Building Council (World GBC).

CHAPTER 4

Healthy and Ethical Materials and Products

Ethical sourcing is the process of ensuring that all materials and products specified for a project are produced through responsible methods and avoid harmful constituents and by-products. Workers who produce the products are paid fair wages and their human rights are met. All social and environmental health precautions are taken for the workers, the communities surrounding the manufacturing facilities, and the users of the materials.

Manufacturers are not always held accountable for many of the chemical impacts of their products. This, combined with an economy that prioritizes inexpensive, disposable goods, contributes to the use of unsafe chemicals in our construction products.

The burden to identify and avoid toxic chemicals often falls to designers, engineers, and specifiers. Yet it can be challenging and time-consuming to ascertain all the precursors, constituents, and by-products of materials and products.

Identifying ethical issues in material production is also quite challenging. To understand worker health and safety issues, one must trace the supply chain all the way back to raw material extraction or harvesting.

This chapter provides information on typical human and environmental health burdens of materials and products through all stages of their life cycles. Worker health and equity issues are addressed by specific materials and countries; and resources for material health transparency are discussed.

Chemical Pollution and Health Hazards

Chemical pollution is a major cause of human illness and premature death. The World Health Organization estimated that reduction and management of chemicals in the environment could have prevented about 1.6 million deaths and about 45 million disability-adjusted life years in 2016 (UNEP 2019).

Chemical pollution released to air, land, and water contributes to environmental degradation, species loss, and human health impacts. Chemicals are released to the environment through volatilization, chemical degradation, abrasion, leaching, and oxidation. They find their way to humans through multiple pathways such as inhalation or absorption. In the food chain they are absorbed or consumed by smaller species. These species are consumed by larger ones, then sometimes humans.

Health Burdens of Materials and Products

The construction sector is a significant contributor to human and environmental health burdens from chemicals and other health hazards. Exposures can occur across a material's life cycle in production, consumption, and disposal of construction materials. Hazardous chemicals are used in metal finishes, concrete admixtures, plastics, polymer composites, wood preservation, protective coatings, adhesives and sealants, and many other products.

Site construction materials and products, and their by-products, can pose health hazards along the entire supply chain, with certain populations being particularly vulnerable. Workers are exposed to chemicals at all phases of the life cycle as are the fence-line communities adjacent to mines, factories, and recycling facilities. Communities along transportation routes are at risk, and even site occupants and emergency workers can be vulnerable during the use phase of some products. Of particular concern are the millions of workers in forced and illegal labor conditions (Figure 4.1).

Worker health and equity along the supply chain: While forced and child labor are major concerns globally and are discussed later in this chapter, worker health, safety, and well-being are also concerns in most countries. Workers are

DOI: 10.4324/9781003413974-4

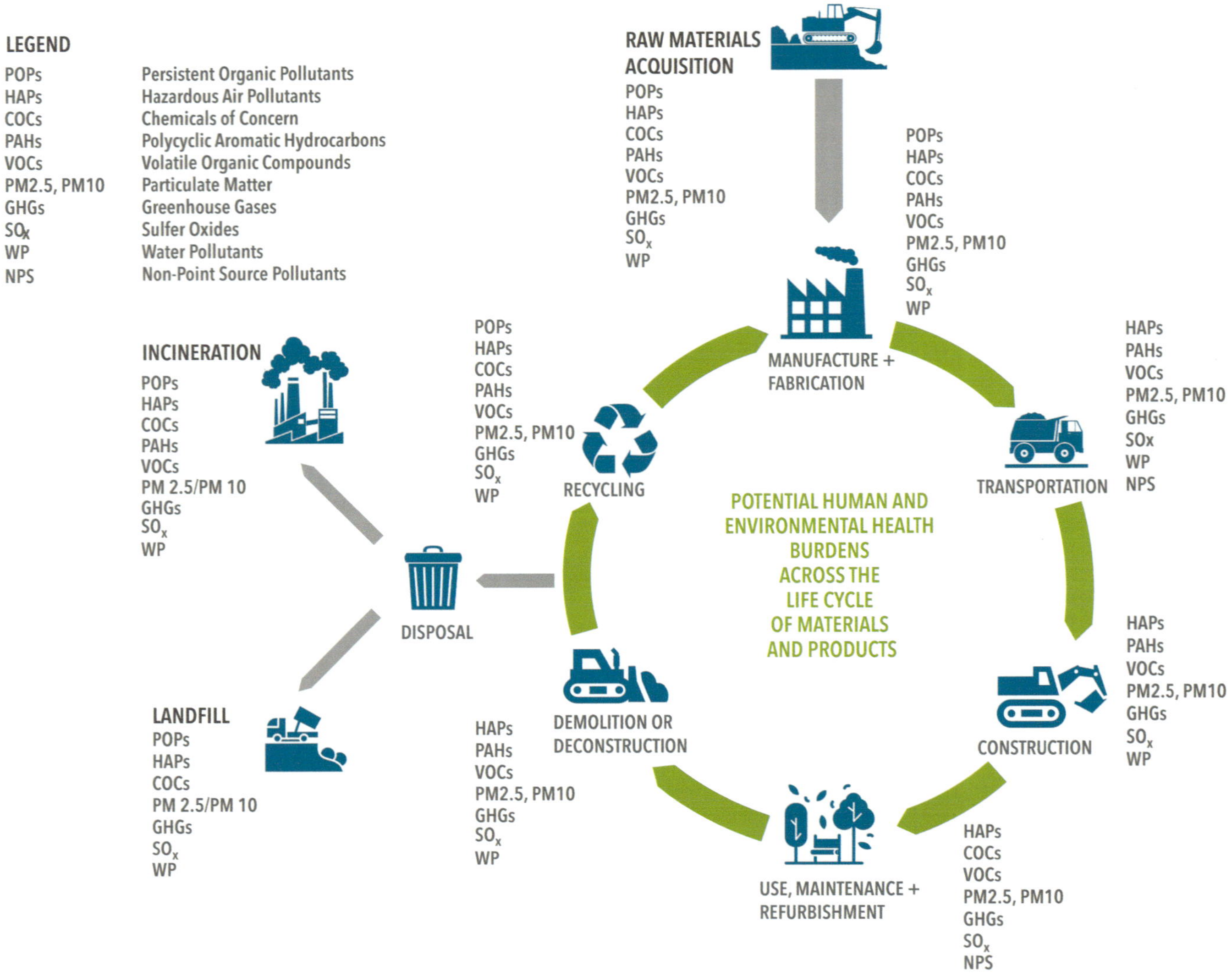

Figure 4.1 Potential environmental and human health burdens across the life cycle of materials and products. It is important to note that life cycle stages of some materials and products pose these burdens, while others do not. The health burdens of some plastics and metals/metal finishing are particularly extensive across their life cycle, while the burdens from stone and wood can be relatively minimal.

not always paid living wages, they are exposed to hazardous chemicals, and illegal laborers are employed at low wages and long hours.

The US Occupational Health and Safety Administration (OSHA) estimates that workers will suffer more than 190,000 illnesses and 50,000 deaths each year that are related to chemical exposures (OSHA 2024). Cancers, lung disease, and other health burdens caused by workplace exposure to chemicals occur slowly and less visibly, therefore exposure trends are very difficult to identify. For example, it took decades to prove that occupational exposure to the vinyl chloride monomer, a precursor to PVC, causes liver angiosarcoma and hepatocellular carcinoma (Fedeli et al. 2019). And still, to a lesser degree, these exposures in the factories and surrounding fence-line communities continue in many countries.

Fence-line community health and equity along the supply chain: Health and environmental hazards don't stop at the factory gate. Communities around mines, factories, and disposal facilities, called fence-line communities, are vulnerable to air and waterborne chemicals. The US Toxic Release Inventory records millions of metric tons of persistent bioaccumulative toxins (PBTs), carcinogens, and hazardous air pollutants (HAPs) released to the air and water each year. These self-reported releases primarily come from the metals and plastics sectors (Table 4.1).

These fence-line communities are often comprised of vulnerable populations that may not be able to move

Table 4.1 US Toxics Release Inventory Self-Reported Releases by Sector 2022 (Metric Tons)

Sector	PBTs and POPs	OSHA Carcinogens	Hazardous Air Pollutants	CERCLA Hazardous Substances
Plastics and Rubber	69	9874	15,080	17,560
Metal Mining	206,291	111,727	386,836	647,016
Primary Metals	11,080	6297	50,891	132,435
Fabricated Metals	623	1914	8703	20,185
Non-Metallic Minerals	183	716	4458	10,739
Wood Products	13	927	4078	5134

Source: US Toxics Release Inventory 2023

Note: CERCLA is Comprehensive Environmental Response, Compensation and Liability Act; OSHA is Occupational Safety and Health Administration.

away from the pollution. Research has shown that low-wealth populations, people of color, and children suffer disproportionate exposures to toxic chemicals and pollution (ATSDR 2023; Woodruff et al. 2023) Systemic racism and racist policies such as redlining over several decades have resulted in a concentration of polluting industries and contaminated soils in BIPOC communities (NMQF 2024).

Starting in the womb and continuing into adolescence, children are particularly sensitive to chemical exposures as their bodies and cells are growing dynamically. Chemical exposures can damage organs, slow development, interrupt hormone systems, and are suspected of causing increased rates of neurological disorders such as ADHD and childhood cancers such as leukemia. Exposure to toxic chemicals in childhood can lead to increased health problems later in life (ATSDR 2023).

Health risks during the use phase: User health risks from site construction materials are far fewer than building materials that can impact indoor air quality, but they still exist. Volatile organic compounds from finishes, adhesives, sealants, and cleaners can impact localized air quality. Polycyclic aromatic hydrocarbons (PAHs) are released from new asphalt sealants and top courses.

Health risks in disposal of materials and products: Environmental and human health burdens during the disposal phase of material and products can be substantial, even when they are recycled. Some types of plastics pose health hazards during disposal when they are incinerated or even just heated for recycling.

Plastics contain toxic additives such as phthalates and PCBs and release pollution such as dioxins that can cause health problems in recycling workers. The sorting, shredding, and melting of these plastics releases the toxins into the local environment impacting the health of workers and communities around the facilities (Human Rights Watch 2022).

Health impacts to other species in the environment: Toxic chemicals can harm animals and organisms at all stages of a material or products life cycle. As toxic substances are released into air, land, and water, they can be carried long distances, impacting terrestrial and aquatic ecosystems far from the chemical release. Some of these substances can bioaccumulate as organisms impacted by the toxic chemicals in the silt of waterways are eaten by small fish, then small

The Relationships Between Toxic Chemicals and Climate Change

Climate change magnifies the effects of some toxic chemicals impacting human and environmental health.

- Warmer temperatures can increase our exposure to toxic chemicals as higher temperatures allow them to vaporize into the air more easily. Additionally, the warmer temperatures can encourage the breakdown of chemicals into toxic by-products (Noyes et al. 2009).
- Extreme heat and forest fires resulting from climate change can release toxic chemicals such as dioxins into the air when buildings burn, or factories are damaged.
- Warmer temperatures exacerbate the smog formation impacts of VOC and pollutants and particulates from fossil fuel combustion (McDonald et al. 2018).
- As severe weather events such as hurricanes and tornados threaten manufacturing and facilities, accidental releases of toxic chemicals could impact fence-line communities.

Table 4.2 Policy Frameworks for PBTs, POPs, and Chemicals of Concern

Region or Country	Regulations and Listings of PBTs, POPs, and CoC
Worldwide	Strategic Approach to International Chemicals Management (SAICM) Stockholm Convention on Persistent Organic Pollutants (POPs) Basel Convention on the Control of Transboundary Movements of Hazardous Wastes and Their Disposal The Rotterdam Convention on the Prior Informed Consent Procedure for Certain Hazardous Chemicals and Pesticides in International Trade The Minamata Convention on Mercury Montreal Protocol on Substances that Deplete the Ozone Layer
Europe	REACH is the Registration, Evaluation, Authorization and Restriction of Chemicals (REACH) regimen requiring that all chemicals produced in volumes greater than one ton per year must be registered and include safety data. The REACH program maintains a database of Substances of Very High Concern (SVHC) that is updated regularly. Chemicals of "very high concern" require authorization from the European Union to be sold.
United States	Toxics Substances Control Act (TSCA) is the main policy for chemical management in the United States. It provides the EPA with the authority to require reporting of toxic releases, recordkeeping, testing requirements, and restrictions of chemical substances or mixtures.
Canada	Canada's Chemicals Management Plan assesses and manages risks to human health and the environment posed by chemical substances.
Latin America	Brazil: National Chemical Law Framework PL 6120/2019 Chile: Decree 57, Regulation on Classification, Labelling, and Notification of Hazardous Chemicals and Mixtures Columbia: Decree 1630/2021
Asia	China: Hazardous Chemical Registration and Inventory, China REACH, Safe Management of Hazardous Chemicals (Decree 591), and the Chemical Import Trade Compliance India: India REACH
Africa	South Africa: Regulation of Hazardous Chemical Agents (HCA)

Source: Ritscher 2021

fish are eaten by large fish who store and concentrate the substance in their fat, impacting mammals who eat the fish.

Impacts to environmental health: Toxic chemicals released to air, land, and water have negative effects on environmental health, sometimes far from the release point. These impacts, defined in Tables 5.2 and 5.3, include climate change, ozone depletion, acidification, eutrophication, smog formation, water consumption, eco-toxicity and abiotic depletion of minerals, metals, and fossil resources.

Chemicals of Concern (CoC) in Construction Materials

Historically, new chemicals have been used without first proving their safety, so there is low availability of chemical impact data for the thousands of chemicals in use today. The speed at which chemicals are developed and placed on the market makes it difficult to keep pace with research documenting their health hazards.

An approach to use of new chemicals called the precautionary principle, employed in the European Union, enables decision makers to adopt precautionary measures when there are threats of serious or irreversible hazards from a chemical. Full scientific certainty is not a requirement to stop using a chemical. It is the job of the chemical company who is producing and selling the product to prove that it is safe.

There are multiple global conventions that list POPs and PBTs to avoid and phase out, but there are many chemicals of concern that are still in use in site construction materials. Table 4.2 lists policy frameworks that address POPs, PBTs, and other chemicals of concern. Table 4.3 lists chemicals and chemical groups of concern found in site construction materials. Table 4.4 contains reference lists for chemical substances of concern.

Persistent organic pollutants are toxic to both humans and wildlife. They are carbon-based chemical substances that when released into the environment remain intact for several years. They become widely distributed throughout the environment due to movement of air, water, and soil. They accumulate in living organisms and concentrate in the food chain becoming even more harmful as they move up the food chain. The Stockholm Convention maintains a widely accepted list of POPs and classifies them in annexes:

Table 4.3 Chemical Groups and Chemicals of Concern in Site Construction Materials

Chemical or Chemical Group	Hazards	Application in Site Construction
Certain phthalates and ortho phthalates	Health burdens include endocrine disruption and reproductive toxicity.	Plasticizers in PVC products, and some adhesives, sealants and varnishes.
Certain halogenated flame retardants	Health burdens include neurotoxicity, potential endocrine disruptor, liver impairments, and indication of immune modulating effects. Bioaccumulation and persistence in the environment.	Expanded polystyrene and some other insulations, engineered lumber, glue, sealants.
Certain solvents, formaldehyde, and other VOCs	Human health burdens include carcinogenic and reprotoxic effects, neurotoxic effects and irritation of eyes and the respiratory system. Environmental health and human health burdens due to the potential of certain VOC for the formation of ground-level ozone and formation of smog.	Some paints, lacquers, varnishes, coatings, adhesives, engineered wood products. VOC can also be released from polymer products.
Per and poly fluorinated substances (PFAS)	Health burdens include liver effects, developmental effects, potential for endocrine modulating effects, and indication for effects on the immune system. Bioaccumulation and high persistence in the environment.	Some engineered lumber products, paint, damp proofing, foams, sealants and adhesives, artificial turf, fabrics.
Some metals and metal compounds such as lead, cadmium, mercury, and chromium VI	Health burdens include carcinogenicity, mutagenicity, toxicity to reproduction, and other systemic toxicity effects. Lead concerns: developmental neurotoxicity, cardiovascular, renal and reproductive effects. Bioaccumulation and high persistence in the environment.	Cadmium stabilizers or pigments in some polymers, such as PVC, PE, and some epoxy resins for fences, decks, pipes. Chromium VI in some wood preservatives, metal finishes, pigment in paints and coatings, and vinyl and cellulose plastics.
Chrysotile asbestos	Human health burdens include asbestosis, mesothelioma and lung cancer.	Some bitumen products, reinforced PVC, other plastic and resin composites.
Dioxins	Health burdens such as carcinogenicity, mutagenicity, reproductive toxicity, and other systemic toxicity effects. Persistent organic pollutants.	Released from combustion or incineration of waste, combustion of PVC products, metal smelting, and manufacture of some herbicides and pesticides.
Dechlorane Plus	Health burdens such as neurotoxicity, potential for endocrine modulating effects, liver impairments, and indication of immune modulating effects. High indications for bioaccumulation and persistence in the environment.	Some flame retardants in polymers, PVC line pipes, expanded polystyrene panels.
Diisocyanates	Human health burdens related to respiratory, dermal and eye irritation, skin and respiratory sensitization, and occupational asthma	Cross-linking agents in polyurethane-based products, such as rigid and flexible foams, and some paints, lacquers, varnishes, adhesives, and glues and sealants
Certain PAHs	Health burdens such as carcinogenicity, respiratory system damage, toxicity to reproduction, and other systemic toxicity effects. Bioaccumulation and high persistence in the environment.	Some asphalt, creosote, bituminous emulsions, primary steel production, primary aluminum production, waste incinerators.
Some organotin compounds	Aquatic toxicity, health burdens include immunotoxicity, reproductive effects, and potential endocrine disruption.	Some paints, coatings, sealants, PVC products, outdoor fabrics.

Sources: DTSC 2024; SAICM 2023; Ritscher 2021

Table 4.4 Reference Lists for Chemical Substances of Concern

Hazard Endpoints	Reference Lists
Carcinogens	International Agency for Research on Cancer as the following: Group 1: carcinogenic to humans Group 2A: probably carcinogenic to humans Globally Harmonized System of Classification and Labeling (GHS) for the carcinogenicity hazard class (Codes H350, H351) Category 1A: Known to have carcinogenic potential for humans Category 1B: Presumed to have carcinogenic potential for humans US National Toxicology Program: Known human carcinogen Reasonably anticipated human carcinogen US Centers for Disease Control, National Institute for Occupational Safety and Health (NIOSH) Occupational Carcinogen List
Mutagens and genotoxins	Globally Harmonized System of Classification and Labeling (GHS) Category 1A: Chemicals known to induce heritable mutations in germ cells of humans Category 1B: Chemicals which should be regarded as if they induce heritable mutations in the germ cells of humans Category 2: Chemicals that cause concern for humans owing to the possibility that they may induce heritable mutations in the germ cells of humans
Reproductive or developmental toxins	State of California Safe Drinking Water and Toxic Enforcement Act (Prop 65) for reproductive or developmental toxicity Globally Harmonized System of Classification and Labeling (GHS) for the Reproductive Toxicity hazard class (codes H360 Categories 1A and 1B, H361, H362)
Respiratory sensitizers	Globally Harmonized System (GHS) Category 1A: high frequency of occurrence or sensitization rate in humans Category 1B: low to moderate frequency of occurrence or sensitization rate in humans
Endocrine disrupters	US National Institute of Environmental Health Sciences List of Endocrine Disruptors Substances identified as endocrine disruptors at EU level
Persistent organic pollutants	Stockholm Convention on Persistent Organic Pollutants
US–Canada Binational Toxics	Chemicals listed in 40 CFR 372.28 due to their PBT characteristics Toxic Substances Control Act (TSCA) Chemical Substance Inventory US EPA Toxics Release Inventory (TRI) Toxic Chemicals
EU Toxics	Registration, Evaluation, Authorization and Restriction of Chemicals (REACH) Restricted Substances List
Volatile organic compounds	California List of Banned Substances State of California, Air Resources Board, VOC Limits

Annex A: POPs to be eliminated
Annex B: POPs subjected to restricted use
Annex C: POPs whose emissions must be reduced

Persistent bioaccumulative toxins are a larger group of bioaccumulate substances that include POPs as well as metals and organo-metal compounds. Metals and their compounds do not break down in the environment, and they accumulate in the food chain.

Chemicals of concern are chemicals that are identified as having potential to cause harm to humans or the environment because of their concentration, toxicity, and distribution. Lists of chemicals of concern vary widely around the world. CoC may be banned or restricted in some countries but not by international treaties (Tables 4.3 and 4.4).

Particulate matter is a sometimes underestimated type of pollution from almost every phase of construction material's life cycles. While they may not be toxic, the small size of PM 2.5 (2.5 microns or smaller) is easily airborne and can be carried hundreds of miles to impact air quality far from the source. PM 2.5 and PM 10 (10 microns or smaller) particles are so small that they lodge in the lungs causing serious health issues. Sources of PM 2.5 and PM 10 include emissions from industrial processes, vehicles, quarrying, mining, and wildfires.

Tools for Understanding Health Impacts of Materials and Products

Globally, there is increasing emphasis on reducing the human and environmental health burdens of construction materials and products; yet it can be challenging to identify all their constituents and by-products. This information is rarely covered in environmental product declarations and life cycle assessments. There are several tools, databases, and ecolabels that can help to identify healthier materials and products.

Health Product Declarations

Health product declarations are standardized documents created by manufacturers to disclose product materials and substances and any associated health hazards. The Health Product Declaration Collaborative (HPDC) offers a standardized method for disclosure of the information as well as a repository for hundreds of product HPDs (HPDC 2024). It is important to note that an HPD does not mean that a product is healthy. It just means that the ingredients are disclosed.

The following information is included on an HPD:

Substances in a product: All substances in a product are listed along with their Chemical Abstracts Service Registry Number (CAS or CASRN). A global database assigns these numbers to all chemicals. The percentage of the finished product is given for each type of substance.

Role of the substance: The role of the substance in the final product is disclosed.

Hazard type and hazard data source: Known hazards and the hazard data sources are disclosed for each substance. Sources are primarily drawn from the GreenScreen List Translator (2018), which includes both known and suspected hazards.

Warnings: Warnings for the substance are identified. Listings on which the substance is included such as the IARC Carcinogen groups or the IFLI Red List are identified.

Level of disclosure: The manufacturer or the product can choose to disclose substances at a variety of thresholds including 100 parts per million (ppm) or 1000ppm. Smaller numbers indicate a higher level of disclosure. Manufacturers do not have to disclose all ingredient names if they are proprietary. Instead, they can reveal their role, percentage, and hazard score.

Additional notes: The HPD allows space for the manufacturer to add any additional information they would like to consumer to know.

In addition to the aforementioned information, the HPD will list a GreenScreen Benchmark score. This offers an important way to understand the disclosures. GreenScreen Benchmark scores are as follows:

BM-1 Benchmark 1 – Avoid chemical of high concern
BM-2 Benchmark 2 – Use but search for safer substitutes
BM-3 Benchmark 3 – Use but still opportunity for improvement
BM-4 Benchmark 4 – Prefer safer chemical
BM-U Benchmark – Unspecified due to insufficient data

Specifying Healthier Materials and Products

- Set material health goals for the project and discuss them with the project team, stakeholders, and contractors. Establish a "do not substitute list".
- Specify materials that are less processed to reduce potential for chemicals of concern. These types of materials include decay resistant or thermally modified lumbers, stone, and earthen materials. Low embodied carbon is a major benefit of these minimally processed materials.
- Consider health and environmental impacts to communities where material resources are harvested, products manufactured, and waste disposed.
- Request transparent and thorough information from product manufacturers on all chemical substances to produce the material or product; mining, manufacturing and fabricating locations; and working conditions at these locations. Ask for chemical inventories to at least 0.1% (1000 ppm).
- Avoid products or materials with chemical precursors, ingredients, or by-products on the Living Building Challenge (LBC) Red Lists or chemicals of concern reference lists (Table 4.3).
- Use products with HPDs that avoid GreenScreen Benchmark 4 substances.
- Look for third-party verified material ingredient documentation on listings such as Declare labels or Cradle to Cradle Certification (Figure 4.2).
- Write specification language to support the material health goals and change firm wide standard specification language. Spec language should include construction product disclosure requirements that encourage the use of products from manufacturers working to minimize environmental and health impacts.
- For paints, coatings, adhesives, and sealants, require compliance with California Banned Solvents (or similarly rigorous listings).
- For plastics and finishes for metal products, require compliance with banned substances from the Stockholm Convention and the LBC Red List.

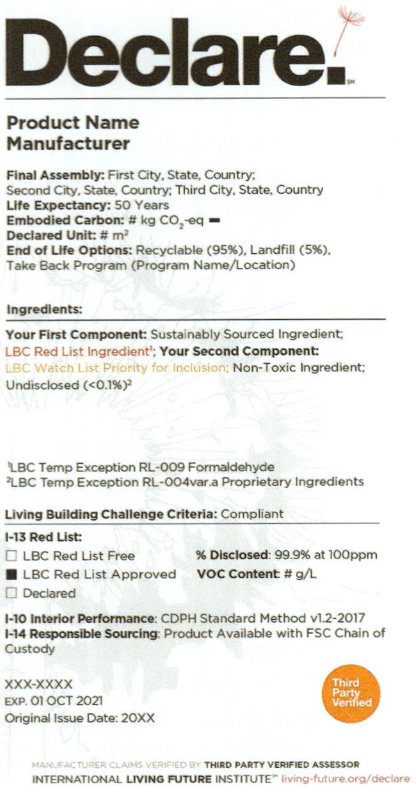

Figure 4.2 Declare by the International Living Future Institute is a platform and database for healthy building products. Declare labels provide third-party verified transparent information about the ingredients of a product. Image credit: Living Future

Social Product Declarations

A social product declaration (SPD) provides information on the social sustainability performance of a material or product at all phases of the life cycle. Structured similarly to its much more common cousin, the EPD, SPDs disclose information about the social and health impacts of a material or product on workers who produce a product or material, communities along the supply and disposal chain, and users of the product or material (EPD International 2024).

Social product declarations are in their infancy with standards developed by the International EPD System. Product category rules will be developed by committees comprised of standards organizations, trade groups, and individual companies within sectors.

SPDs will offer information about the social impacts of materials or product life cycles in three major groups:

Worker subcategories include health and safety, wages, social benefits, working conditions, freedom of association, and collective bargaining.

Local community subcategories include cultural heritage and land rights, poverty, and actions by the company to reduce poverty. This section should also include health and safety impacts to fence-line communities along the supply and disposal chains.

User subcategories include health and safety, feedback mechanisms, consumers privacy, and transparency for consumers.

SPDs are in their infancy, so it will be rare to find two companies with SPDs that allow for comparison of the same product. In the coming decades, SPDs could bring transparency to health impacts of products, equitable working conditions, and social governance of companies.

Forced, Illegal, and Child Labor in Construction Material Supply Chains

The use of forced, illegal, and child laborers subsidizes the production of construction materials around the world. When workers are not paid fair wages to work in safe conditions, producers are able to mine and manufacture products at lower costs. Human rights are violated, and children are exploited.

The International Labor Organization's *Global Estimates of Modern Slavery: Forced Labour and Forced Marriage* (ILO 2022) ranks the manufacturing sector second and the construction sector third for unethically employing a combined 35% of all forced laborers. The site construction materials identified as at the highest risk of embedded slavery are bricks, dimension stone, rubber, steel and iron, timber, and precursors (especially soda ash and limestone) (Grace Farms Foundation 2022). Table 4.5 illustrates the site construction materials produced by forced or child labor. Material-specific information about ethical sourcing is discussed in the individual materials chapters.

Forced Labor

The international legal definition of forced labor is "all work or service which is extracted from any person under the threat of penalty and for which the person has not offered himself or herself voluntarily" (ILO 2024). The International Labour Organization estimates that 28 million people around the world are in forced labor situations with a nearly 10% increase between 2016 and 2021. Illegal profits from

Table 4.5 Site Construction Materials Produced by Forced or Child Labor

Country F = Forced Labor C = Child Labor	**Brick**	**Stone**	**Crushed Stone**	**Cement**	**Timber**	**Jute**	**Bamboo**	**Rubber**	**Plastics**	**Metals**
Afghanistan	CC, F									
Argentina	C	C								
Bangladesh	C					C				
Benin			C							
Bolivia	C									C
Brazil	C				F					
Burkina Faso		CC								
Burma	C, F				C, F		C, F	C, F		
Cambodia	C, F				C			C		
China	C, F								F	
Colombia	C									
Congo, Democratic Republic of										C, F
Ecuador	C									
Egypt	C	C								
Guatemala			C							
India	C, F	C, F								C
Iran	C									
Indonesia								C		C
Kenya			C							
Korea, North	F			F	F					F
Liberia								C		
Madagascar		C								
Nepal	C, F	C, F								
Nicaragua		C	C							
Nigeria		C, F	C, F							
Pakistan	C, F									
Paraguay	C	C								
Peru	C				F					F
Philippines					F			C		
Russia	F									
Sierra Leone		C								
Uganda	C	C	C							
Vietnam	C				C			C		
Zambia		C								

Source: ILAB 2022

forced labor are estimated at 236 billion US dollars per year (ILO 2024).

Forced labor happens around the world, but the biggest segment is in Asia at 15.1 million. Migrant workers are at higher risk for forced labor than other workers as they are not always able to exercise their right to be protected under the laws of the country they have migrated to. Women and girls are more likely to be forced laborers through wage penalties and abuse, while men are coerced through threats of violence and financial penalties.

Child Labor

Child labor is defined as "work that deprives children of their childhood, their potential and their dignity, and that is harmful to physical and mental development" (ILO 2024). Another definition offered by the ILO is "work performed by a child under coercion applied by a third party either to the child or the child's parents, or work performed by the child as a direct consequence of his or her parents being involved in forced labor" (ILO 2022) (Figure 4.3). Close to 160 million children ages 5–17 work in situations that fall under the definition of child labor (Grace Farms Foundation 2022).

Debt bondage is a common type of forced labor where a worker is required to labor to repay a debt and is not allowed to leave the employment. Often wages are so low that the debt is nearly impossible to repay. Just over half of all forced laborers are held in debt bondage. Both children and adults are held in debt-bonded labor situations (ILO 2022).

The US Department of Labor's Bureau of International Labor Affairs (ILAB) maintains a "List of Goods Produced by Child Labor or Forced Labor" by country. The list is based on data from independent research, private sector monitoring, and government research and inspection.

Figure 4.3 Children working in the brick industry in India often work long hours in jobs that prevent critical developmental activities such as school and play. Image credit: Zatletic, Dreamstime.4com

Beginning in June 2022, the United States prohibits the import of any product made in whole or in part in China's Xinjiang Uyghur Autonomous Region (XUAR) region where forced labor is prevalent. China's two largest PVC manufacturers are based in the region producing 10% of the world's PVC (Murphy et al. 2022).

Ethical Sourcing of Construction Materials and Products

- Ask manufacturers about equitable working conditions along the supply chain of their products. Request information about where all components of the product originated. Specify products with supply chain mapping or a chain of custody when possible.
- Employ the resources offered in the Design for Freedom (DFF) Toolkit by the Grace Farms Foundation. DFF offers many useful tools including ethical design specifications, a supplier outreach letter template, supplier questionnaire, and a material tracking schedule.
- Be skeptical about less expensive products from countries listed in Table 4.5 as there is probably a reason for the lower costs. Request information about points of origin and all links in the supply chain. Be sure to examine precursor materials as well.
- Make project team members and clients aware of high-risk building materials such as stone, brick, and timber. Work toward ethical sourcing early in the design process.
- Specify products from third-party certification systems that source responsibly, protect human and worker rights, and track materials along the supply chain. The Natural Stone Sustainability Standard, Responsible Steel Standard, and the Forest Stewardship Council are examples of these systems.
- Refer to the US Department of Labor's Bureau of International Labor Affairs (ILAB) "List of Goods Produced by Child Labor or Forced Labor" to know what products and their country of origin or manufacture to avoid.
- Specify products from companies with fair, just and healthy employment practices (Figure 4.4). Ask about corporate policies and reports on labor practices.

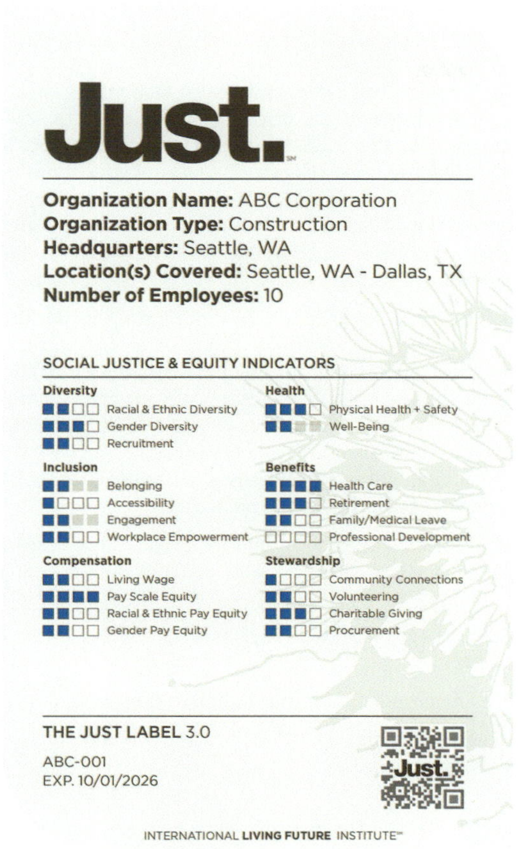

Figures 4.4 The Just label by the International Living Future Institute is a program for companies that focuses on employee experience, health, well-being, and community connections. The Just label provides guidelines for equitable employment policies and is a voluntary disclosure tool rather than a certification program. Image credit: Living Future

Source: Living Future

Table 4.6 Resources for Healthy and Equitably Sourced Products

Resources and Standards for Healthy Materials and Products	
Health Product Declaration Collaborative	The HPDC promotes information transparency of building and site construction products. The group offers a standardized method for HPDs that disclose the chemical composition and potential health impacts of products. HPDC offers a repository for thousands of product HPDs. It is important to note that an HPD does not mean that a product is healthy. It just means that the ingredients are disclosed.
Habitable and Pharos Chemical and Materials Library	Pharos Chemical and Materials Library provides resources to assess the human and environmental health burdens of construction materials and the chemical inputs, ingredients, and by-products. Pharos offers human and environmental health information for over 200,000 chemicals, polymers, metals, and other substances. The database is searchable by CSI format and includes common contents of materials allowing specifiers to know what substances to be on the lookout for. Pharos is administered by Habitable.
Sustainable Procurement of Building Materials: A Progressive Approach to Chemicals of Concern	The Strategic Approach to International Chemicals Management (SAICM) is a policy initiative to promote safer chemicals worldwide. Objectives of SAICM are risk reduction, knowledge and information, governance, technical cooperation, and illegal international traffic. They offer a manual entitled "Sustainable Procurement of Building Materials: A Progressive Approach to Chemicals of Concern".
Living Building Challenge Red List	The International Living Future Institute maintains a Red List of chemicals in the building industry they identify as "worst in class" posing risks to human health and the environment. The list is organized by chemical class and lists individual chemicals by CASRN number. They also maintain a Watch List and LBC Priority List of chemicals that may be added to the Red List.
Declare	Declare is a database of products whose manufacturers have voluntarily disclosed third-party verified product information. Declare is administered by the IFLI. The standardized Declare label includes assembly locations, life expectancy, embodied carbon, and end-of-life options. Listings of at least 99% of the ingredients are included and any chemicals on the IFLI Red List or Watch list are identified. While Declare labels identify substances in the product, they do not identify hazards (Figure 4.2).

(*Continued*)

Table 4.6 (Continued)

Resources and Standards for Healthy Materials and Products	
Safety Data Sheets	Safety data sheets (SDS) are required by the US Occupational Safety and Health Administration and the European Union to be produced by all chemical manufacturers, distributors, and importers. SDS are available for products such as paints or composite lumber when more than one chemical substance is disclosed on an SDS. SDS must be available to all downstream users such as manufacturing workers, construction workers, and users of the product. The SDS includes information on the properties of the chemical(s), the physical health and environmental health hazards, protective measures, and safety precautions for storing handling and transporting the chemical. Some chemicals can be claimed as proprietary, so information is not always disclosed.
LEED	LEED offers a social equity checklist to encourage project teams to focus on integrating social equity into projects. Section four of the checklist focuses on an equitable supply chain asking questions about the following: Communities along the supply chain including impacts on fence-line communities around points of extraction and manufacturing. Communities along transportation routes should also be considered as train derailments and truck accidents can release harmful chemicals and materials. Human rights for workers in the supply chain including human rights abuses, potential "hot spots" for specific materials or precursors, and communication of priorities to manufacturers.
BRE Responsible Sourcing, Construction Products Standard	BES 6001: Responsible Sourcing, Construction Products is a certification from the Building Research Establishment. It addresses stakeholder engagement, labor practices, and management of the supply chains for a product.
Ecologo	UL 2792 Standard for Cleaning and Degreasing Compounds: Biologically based Ecologo/UL 2762 Sustainability Standard for Adhesives.
GreenGuard	UL 2818 GREENGUARD Certification Program for Chemical Emissions for Building Materials.
Green Seal	Green Seal® 34 Standard for Cleaning and Degreasing Agents.
Materials Analytical Services (MAS)	MAS Certified Green.
Resources and Standards for Equitably Produced Materials and Products	
UN Global Compact	UN Global Compact is a corporate sustainability initiative that CEOs and company leaders can join to ensure the adoption of the UN's SDGs into their company values and goals, including Target 8.7, which seeks to eradicate forced labor and child labor.
Design for Freedom	Design for Freedom, part of the Grace Farms Foundation, proffers principles and resources for equity in the product supply chain. They offer a toolkit with sample specifications, a supplier questionnaire and outreach letter, and a materials tracking schedule.
BRE Ethical Labor Sourcing Standard	BES 6002: Ethical Labour Sourcing Standard has been developed by the Building Research Establishment. Companies can become certified and are listed on the website.
Ethical Trading Initiative (ETI)	The Ethical Trading Initiative works with companies, trade unions, and non-profits to provide training on best practices to ensure ethical labor in their supply chains with the ETI base code.
Global GreenTag International Modern Slavery Declaration	The Global GreenTag offers a modern slavery declaration for products. It addresses the need to minimize conditions of modern slavery in global supply chains.
Just	Just is a voluntary disclosure tool and transparency framework that requires reporting on a range of organization and employee related indicators. Just is administered by the IFLI (Figure 4.4). Just offers a database of companies including design and engineering firms, product manufacturers, and organizations who have chosen to disclose information in the following categories: diversity and inclusion, equity, employee health, employee benefits, stewardship, and their purchasing and supply chain.
Blue Green Alliance (BGA)	BGA connects labor unions and environmental organizations to fight climate change, protect worker and community health, and create fair, safe worker environments.
Social Accountability International	The SA8000 Standard and Certification from Social Accountability International ensures fair treatment of workers. Companies can become certified and are listed on their website.
SGE 21 Ethical and Socially Responsible Management System	SGE 21 Ethical and Socially Responsible Management System is a standard for the implementation, auditing, and certification of an ethical and socially responsible management system.
Building Responsibly Worker Welfare Principles	Building Responsibly is a coalition of engineering and construction companies committed to promoting the rights and welfare of workers on construction sites, including those free of forced and child labor.

References

Agency for Toxic Substances & Disease Registry (2023). *Principles of Pediatric Environmental Health: Why Are Children Often Especially Susceptible to the Adverse Effects of Environmental Toxicants?* Agency for Toxic Substances & Disease Registry, May 25.

Association of International Chemical Manufacturers Secretariat (AICM) (2023). *Sustainable Procurement of Building Materials: A Progressive Approach to Chemicals of Concern*. AICM.

California Air Resources Board (ARB) (2021). *Summary of Architectural Coating Rules in California: VOC Limits*. State of California, ARB, September 2.

Department of Toxic Substances Control (DTSC) (2024). "Persistent and bioaccumulative toxic substances". https://dtsc.ca.gov/persistent-and-bioaccumulative-toxic-substances/, accessed June 14, 2024.

EPD International (2024). "Social product declarations". www.environdec.com/spd, accessed July 13, 2024.

Fedeli, U., Girardi, P., and Mastrangelo, G. (2019). "Occupational exposure to vinyl chloride and liver diseases". *World Journal of Gastroenterology* 25.33 (September 7): 4885–4891. https://doi.org/10.3748/wjg.v25.i33.4885.

Grace Farms Foundation (2022). *Design for Freedom Toolkit*. New Canaan, CT: Grace Farms Foundation.

GreenScreen for Safer Chemicals (2018). "GreenScreen method". March 14. www.greenscreenchemicals.org/assess/method.

Health Product Declaration Collaborative (HPDC) (2024). *Making Material Health Evaluation Fast, Easy and Accessible for Every Project*. HPDC, May 8.

Human Rights Watch (2022). *It's as if They're Poisoning Us: The Health Impacts of Plastic Recycling in Turkey*. Human Rights Watch, September 21.

International Labour Organization (ILO) (2022). *Global Estimates of Modern Slavery Forced Labour and Forced Marriage*. Geneva: International Labour Organization (ILO).

International Labour Organization (ILO) (2024). "What is forced labor?". www.ilo.org/topics/forced-labour-modern-slavery-and-trafficking-persons/what-forced-labour, accessed May 10, 2024.

McDonald, B.C., et al. (2018). "Volatile chemical products emerging as largest petrochemical source of urban organic emissions". *Science* 359.6377 (February 16): 760–64. https://doi.org/10.1126/science.aaq0524.

Murphy, L., Vallette, J., and Elima, N. (2022). *Built on Repression: PVC Building Materials Reliance on Labor and Environmental Abuses in the Uyghur Region*. Sheffield Hallam University, Helena Kennedy Centre for International Justice, and Material Research L3C.

National Minority Quality Forum (NMQF) (2024). *People of Color Face Disproportionately Higher Risk of Cancer from Environmental Toxins, New Study Finds*. NMQF, February 22.

Noyes, P.D., et al. (2009). "The toxicology of climate change: Environmental contaminants in a warming world". *Environment International* 35.6 (August 1): 971–86. https://doi.org/10.1016/j.envint.2009.02.006.

OSHA (2024). "Transitioning to safer chemicals: A toolkit for employers and workers". www.osha.gov/safer-chemicals#1.

Ritscher, A. (2021). *Chemicals of Concern in the Building and Construction Sector*. UNEP, SAICM and Global Environment Facility.

Strategic Approach to International Chemicals Management Secretariat (SAICM) (2023). *Sustainable Procurement of Building Materials: A Progressive Approach to Chemicals of Concern*. SAICM, GEF, One Planet.

United Nations Environment Programme (UNEP) (2019). *Global Chemicals Outlook II: From Legacies to Innovative Solutions*. UNEP.

US Bureau of International Labor Affairs (ILAB) (2022). *List of Goods Produced by Child Labor or Forced Labor*. US Department of Labor, ILAB.

US Toxics Release Inventory (TRI) (2023). *TRI Explorer Release Reports 2022*. US Environmental Protection Agency, TRI.

Woodruff, Tracey J., et al. (2023). "A science-based agenda for health-protective chemical assessments and decisions: Overview and consensus statement". *Environmental Health* 21.1 (January 12): 132. https://doi.org/10.1186/s12940-022-00930-3.

CHAPTER 5

Assessing the Impacts of Materials, Products, and Structures

To select materials and products that support GHG reduction goals, ecosystem health, the circular economy, and equitable supply chains, their environmental and human health impacts must be evaluated. And to make these assessments, transparent information about embodied carbon, material ingredients, resource use, waste, pollution, and the chain of custody must be collected and verified.

Techniques such as life cycle assessment (LCA) quantify the inputs and outputs of a material, product, or system; and this information is made accessible to professionals in environmental product declarations (EPD). Human health risks of materials or products are revealed in health product declarations (HPD) (Chapter 4 "Healthy and Ethical Materials and Products"). A wide range of embodied carbon calculators, design integrated LCA tools, material and product certifications, product selection databases, and other tools offer transparent information (Tables 1.1, 2.9, 3.5, 4.6).

While transparent information is key to decarbonization and the circular economy, it can be challenging and time-consuming to gather and may be incomplete. EPDs are not always available for a material or product, and they do not reveal material components or by-products that pose human health impacts or the production locations and working conditions along the supply chain.

This chapter addresses the evolving body of tools, techniques, and resources for evaluating the impacts of site construction materials and products. Evaluation criteria are defined, and questions to ask of manufacturers are offered.

Life Cycle Assessments

An LCA is a tool used to quantify the environmental impacts of a product, process, building, or site over its entire life cycle (Table 5.1). LCAs measure all inputs such as raw materials, secondary materials, water, and fuels, as well as all outputs including greenhouse gas emissions, chemical releases, and solid waste. LCAs measure a variety of metrics such as GWP, ozone depletion potential (ODP), acidification potential (AP), depletion of resources, and water use (Table 5.2).

Data from LCAs is provided in EPDs. Tools for whole project LCAs are discussed later in this chapter.

Product Category Rules

Product category rules (PCRs) govern how LCAs should be performed for a specific product type such as clay bricks or portland cement. They are a set of guidelines that determine what type of data will be collected and how it will be evaluated. PCRs are developed through internationally accepted methods with transparent and participatory stakeholder engagement processes.

PCRs are published by EPD program operators, industry associations, or other organizations and most expire after five years. They are updated regularly or after major changes to the industry.

Environmental Product Declarations

EPDs are published by the manufacturers of construction materials and products to communicate LCA information in an objective, transparent, verified, and standardized format. Theoretically, EPDs allow designers and specifiers to evaluate similar products to make informed choices that support their environmental priorities.

To make legitimate comparisons, functional equivalency of EPDs is crucial. They must have the same LCA scope, be guided by a recognized standard, be third-party verified, and, most importantly, the material or products being examined should all fulfill the same performance needs in the structure or site.

LCAs and EPDs are guided by rules from the International Organization for Standardization (ISO) and European

DOI: 10.4324/9781003413974-5

Table 5.1 International Standards for EPDs, LCAs and PCRs

Standard number[1]	Title
ISO 14025	Environmental labels and declarations – Type III Environmental Declarations – Principles and procedures
ISO 14040	Environmental management – LCA – Principles and framework
ISO 14044	Environmental management – LCA – Requirements and guidelines
ISO 21930	Sustainability in buildings and civil engineering works – Core rules for EPDs of construction products and services
EN 15804:2012+A2:2019	Sustainability of construction works – EPDs – Core rules for the product category of construction products
ISO 14067	Greenhouse gases – Carbon footprint of products – Requirements and guidelines for quantification
EN 15978	Sustainability of construction works – Assessment of environmental performance of buildings – Calculation Method
ASTM E2921	Standard Practice for Minimum Criteria for Comparing Whole Building LCAs for Use with Building Codes, Standards, and Rating Systems
ASTM E2129	Standard Practice for Data Collection for Sustainability Assessment of Building Products

[1]ISO – International Organization for Standardization, Geneva, Switzerland

EN – European Standards

ASTM – American Society for Testing and Materials

Table 5.2 Core Environmental Performance Indicators in an EPD, Mandatory Under ISO 15804+A2

Environmental Performance Indicators	Units of Measure and Discussion of Impact Category
Global Warming Potential The GWP is an index that was developed to allow comparisons of the global warming potential of different gases.	The unit of measure for GWP is kilograms of CO_2 equivalents (CO_2 eq. or CO_2e) per declared unit. 1kg carbon dioxide = 1kg CO_2 eq. 1kg methane = 29.8kg CO_2 eq. 1kg dinitrogen oxide = 273kg CO_2 eq. Four GWP subcategories in EPDs under ISO 15804 +A2 are: ***GWP Fossil*** includes the combustion and process emissions from fossil fuels. ***GWP-Biogenic*** is carbon that is stored in biological materials such as wood fibers. It is considered a negative emission although when the product decays or burns, the carbon is released into the atmosphere, resulting in a zero balance. See "Biogenic Carbon and Carbon Stored in Construction Materials" in Chapter 2 for a greater discussion of biogenic carbon and accounting considerations. ***GWP Land Use and Land Use Change (GWP-LULUC)*** refers to GHG emissions and storage from land use conversions that occurred for the material or product. For example, forested land that is clear-cut then turned to pasture would be accounted for in this category. ***GWP Total*** is the sum of all GWP indicators.
Ozone depletion potential ODP is a measure of the degrading impacts of all substances that contribute to ozone depletion such as CFCs, HFCs, HCFCs, halons, nitrous oxides, chlorine, and bromine.	The unit of measure for ODP is kilograms of trichlorofluoromethane-equivalents (kg CFC-11 eq.) per declared unit. 1kg halon-1211 = 6.9kg CFC 11 eq. 1kg methyl bromide = 0.57kg CFC 11 eq. 1kg CFC 11 = 1kg CFC 11 eq. Ozone depletion is directly linked to GWP because ozone depleting substances are greenhouse gases, and the ozone absorbs solar radiation. Emissions of CFCs, HCFCs, halons, and nitrous oxides result from cooling systems, cleaning methods, and aluminum and steel production.
Acidification potential Acidification occurs primarily from sulfur oxides, nitrogen oxides, and ammonia dissolving in water or adhering to soil particles. It results primarily from air releases from fossil fuel combustion and metal smelting that become acid rain. Land releases occur from acid mine drainage.	Acidification potential is measured in kilograms of sulfur dioxide equivalents (mol H+e) per declared unit. Acidification potential is directly linked to GWP due to fossil fuel combustion and metal smelting. Sulfur dioxide interferes with photosynthesis and plant growth. Acid rain impacts aquatic ecosystems by lowering the pH causing damage to aquatic species. Acid rain can also prematurely age materials such as some metal finishes and porous stones.

(Continued)

Table 5.2 (Continued)

Environmental Performance Indicators	Units of Measure and Discussion of Impact Category
Eutrophication potential (EP) EP occurs by adding nutrients such as nitrogen and phosphorus to soil or water disrupting the natural ecosystem balance.	Three subcategories in EPDs under ISO 15804 +A2 are: Eutrophication Aquatic Freshwater (kg PO4e) Eutrophication Aquatic Marine (kg Ne) Eutrophication Terrestrial (mol Ne) Eutrophication causes overstimulation of plant growth and photosynthetic organisms resulting in algae blooms and other plant growth blocking sunlight in water bodies. Sources of nutrients are manufacturing effluents, fertilizers, waste disposal, and non-point source runoff.
Smog formation potential (SFP), also called **photochemical ozone creation potential (POCP)**, is a measure of the potential impacts from chemicals that contribute to the formation of smog such as nitrous oxides and VOCs.	Smog Formation potential is measured in kilograms of non-methane VOC equivalents (kg NMVOCe) per declared unit. 1kg nitrogen oxides = 1kg NMVOCe 1kg carbon monoxide (fossil) = 0.0456kg NMVOCe 1kg acetic acid = 0.164kg NMVOCe Chemicals that result in smog are released from equipment and vehicular exhausts, and coatings, sealants and adhesives. Smog is a type of air pollution that can pose health burdens such as respiratory disorders, cardiovascular problems and neurological disorders. Ozone in smog can damage plant cells and inhibit their growth, reducing the amount of carbon dioxide they take in during photosynthesis.
Abiotic depletion potential (ADP), minerals and metals sometimes called **abiotic depletion potential elements (ADPe)**, quantifies the use of non-fossil resources used for a material of product. It includes the material waste and by-products that don't end up in the final product as well.	Non-fossil ADP is measured in kilograms of antimony (Sb) equivalents per declared unit: 1kg antimony = 1kg Sbe 1kg aluminum = 1.09 * 10^-9 Sbe 1kg silver = 1.18kg Sbe ADP discloses the magnitude of resources used for a material or product, although it does not reveal what was done with the waste generated. It could have been reused as feedstock for another product, landfilled, or released into air, land, or water. Other optional categories listed in Table 5.3 reveal more information about resource use, waste, and recycled content.
Abiotic depletion potential for fossil resources, sometimes called **abiotic depletion potential (ADPf) of fossil resources**, quantifies the use of non-fossil resources used in a material of product.	Fossil ADPf is measured in megajoules (MJ) per declared unit: 1kg coal hard = 27.91MJ 1kg coal soft, lignite = 13.96MJ Burning fossil fuels such as coal, oil, and natural gas releases GHGs and pollutants including sulfur dioxides, nitrogen oxides, particulates, and heavy metals. These releases contribute to climate change, air pollution acid rain, water pollution, and biodiversity loss.
Water use, sometimes called **water deprivation potential (WDP)** quantifies the water usage and effluent discharges from processing and manufacturing.	WDP is measured in m^3 deprived per declared unit. Use of water in the manufacture of materials and products can pose multiple environmental impacts. Consumption of excessive water and the resulting deprivation from ecosystems can impact their health. Effluent can carry pollutants into receiving waters. With lower water use, pollutants may be concentrated.

Source: EPD International 2024

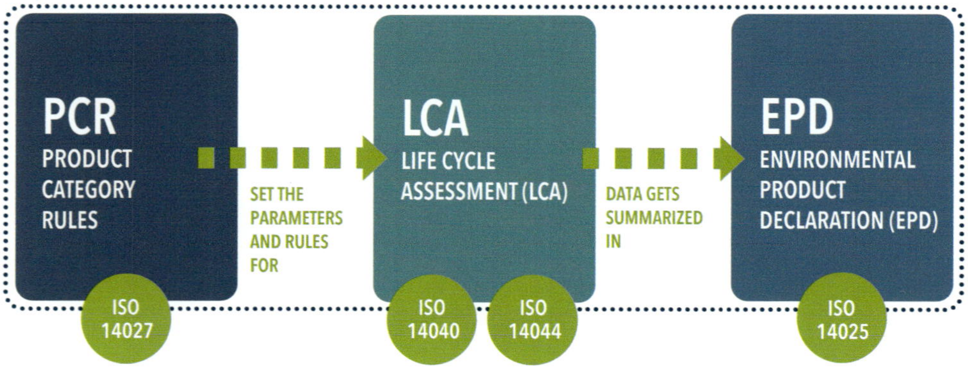

Figure 5.1 Relationships between PCRs, LCAs, and EPDs and their associated standards. Adapted from SCS Global Services: Life-Cycle Assessment, The Future of Sustainability 2024. Image credit: Meg Calkins

Standards (EN). ISO 14040 guides the framework and principles for LCAs, which include goal and scope definition, inventory analysis, impact assessment, and interpretation. ISO 14044 provides detailed methods for conducting an LCA ensuring consistency and rigor.

EN 15804 defines how data for EPDs are collected, reported, verified, and presented, in addition to rules for LCAs and life-cycle inventories. Type III EPDs are compliant with ISO Standard 14025 and must be verified by an unbiased, accredited third-party organization or individual. This will ensure that the data in the EPD is impartial, standardized, and comparable and that the LCA was performed in accordance with the applicable PCRs.

Types of EPDs

EPDs can be published for an entire material or product industry or by individual manufacturers.

Industry-wide EPDs provide baseline data for a particular product industry typically within a country. For instance, a brick trade organization in the United States could use data from several individual brick manufacturing companies to develop an industry-wide LCA and EPD. Data provided in industry-wide EPDs is an average that individual companies can use as a benchmark to improve their practices, or the entire industry can use to track progress. Data from industry-wide EPDs can allow designers and specifiers to understand the environmental impacts and embodied carbon for a material type, as well as serve as a baseline for lower embodied carbon materials and products.

Product-specific EPDs provide data for a specific manufacturer's product or product line. At a minimum, it includes product stage data for phases A1–A3. This means that it includes the manufacturer's Scope 1 emissions released from powering equipment, buildings, vehicles and heat (Figure 2.4).

The EPD will also include Scope 2 emissions generated off site from purchased energy, and indirect Scope 3 emissions from raw materials extraction, transportation and manufacturing processes by other companies. For instance, an EPD for a powder coated steel bench will include Scope 3 raw materials data for coking coal, iron ore, limestone, and steel scrap for the steel as well as the polymers, pigments, and other additives of the powder coating.

Facility specific EPDs are product-specific EPDs where the environmental impacts are quantified from only one manufacturing facility or manufacturer. Facility specific EPDs offer a very high level of resolution to the data of a particular product at a specific manufacturing facility, but they do not communicate impacts across the entire life cycle.

The Typical Format of Environmental Product Declarations

EPDs follow a typical format to present the LCA information as clearly and consistently as possible. Most are based on ISO 14025.

General Information

General information is supplied at the beginning of the EPD. At a minimum, it includes the following:

Program Operator: The company that conducted the LCA and produced the Type III Environmental Declaration Program.

Declaration holder: The company or organization that owns the EPD. This is usually an industry association or an individual manufacturer of a product.

Product: This section states the product(s), material(s) or service under study by the LCA.

Product category: A group of products that can fulfill equivalent functions (e.g., decking, face bricks).

Product Category Rules: One or more PCRs that that set standards for the LCA data collection.

Verifier: The verifier is the independent third-party organization that verified the LCA and EPD.

Verification: The information should be verified by the ISO 14040 series of standards.

Date of issue: The date that the EPD was published.

Period of validity: EPDs are considered valid for five years after their date of issue.

Description of the Product(s) or Assembly

The description of the product(s) includes information about the material composition of the product(s), applicable standards, and technical specifications that the product meets and a brief explanation of applications for the product(s). Technical and performance data such as density, tensile strength, modulus of elasticity, coefficient of thermal expansion, thermal conductivity, melting point, and minimum yield strength are sometimes included.

If an assembly of materials and products is being studied, descriptive information, including dimensions and quantities, will be provided for all components. Information about the product or assembly packaging is also provided.

EPDs will include a ***reference service life*** for the product or assembly if they are cradle-to-cradle in scope. The reference service life is the amount of time the product will last. In an EPD for an assembly or structure, it is the amount of time it will be in use. Some components may need to be replaced at least once in the reference service life of the assembly or structure.

Declared/Functional Units

The terms declared unit and functional unit describe a quantity of a material, product, product system, or process used as a reference unit in an LCA or EPD. Common units of measure for declared and functional units are mass, volume, and area. Other functional or declared units are specific to an individual product size, thickness, or weight.

Declared units are more commonly used in EPDs. They are used when an LCA study does not cover the entire life cycle of a material or product, but only certain modules such as cradle-to-gate.

Functional units define the impact results for a certain service level or performance that the product provides. An example of a functional unit is $1 m^3$ of deck boards that will span a 40cm joist spacing and will be in use for 30 years. Another example from a whole site embodied carbon assessment might be a 10-acre urban park in use for 50 years. The results would include emissions from embodied carbon materials to construct the park, operational emissions, and sequestration of vegetation in the park over 50 years.

Scope and System Boundaries

System boundaries define the scope of data collection for an LCA and EPD (Figure 5.2 and 5.3).

Typical system boundaries are as follows:

Cradle-to-gate, the product stage (A1–A3), is required for LCAs that adhere to the ISO standards. Cradle-to-gate begins with extraction of the raw materials from the earth and continues through all stages to the last factory or fabrication "gate". Transportation between raw materials extraction/harvest and all manufacturing and fabrication phases is included. This is the most common scope for EPDs.

Cradle-to-site includes phases A1–A4 and assumes a hypothetical travel distance from the last fabrication step or the distributor to the construction site.

Cradle-to-grave includes (A1–A3) raw material extraction, transport to and between manufacturing phases, transport to the site (A4), construction (B1), use and maintenance (B2), and the end-of-life phases of demolition/deconstruction (C1), transport to waste processing then disposal (C2–C4). Cradle-to-grave is a linear life cycle of materials and resources as disposal does not reclaim, reuse, reprocess, or recycle materials.

Cradle-to-cradle is the preferred scope for an EPD as it supports the circular economy (Chapter 3 "Resource Efficiency and the Circular Economy"). Products are designed for refurbishment and reuse, and materials are reused or recycled. Site and building structures are designed for deconstruction. End-of-life emissions are increasingly included as a way for manufacturers to promote the reuse and recyclability of their products. EN 15804 requires reporting for end-of-life phases C1–C4 and D (EPD International 2024).

Some products may look good in the product stage (A1–A3) but not be durable or recyclable. This will not be evident

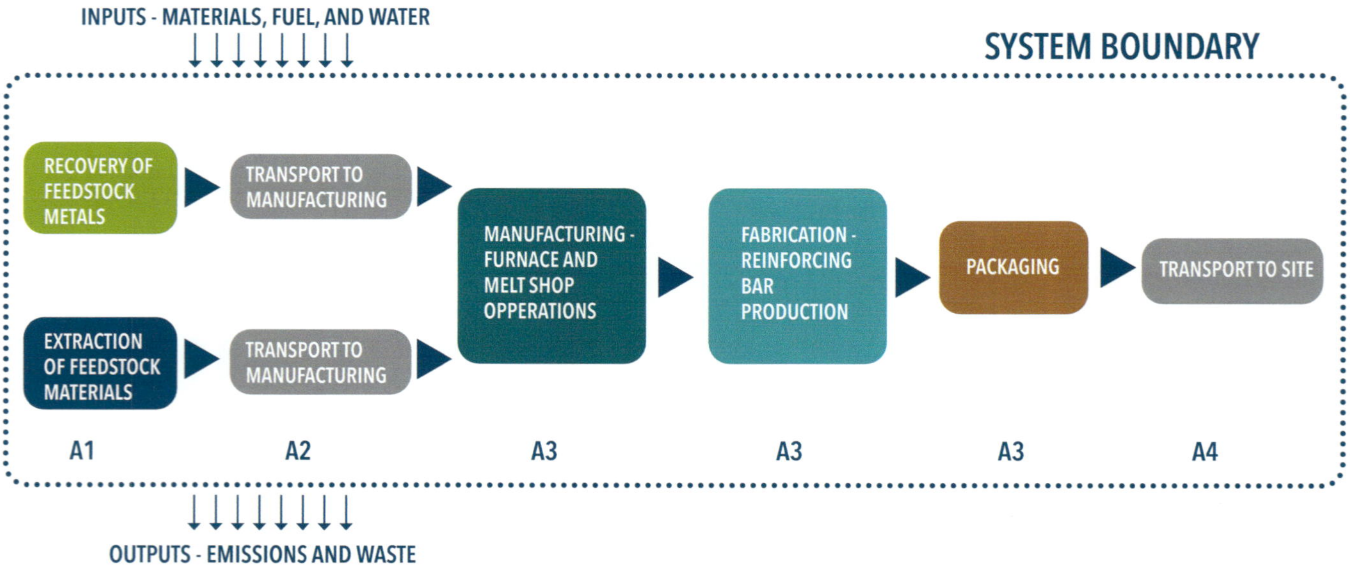

Figure 5.2 An example LCA and EPD system boundary. This one is showing the life cycle stages for a steel reinforcing bar product. Image credit: Meg Calkins

	PRODUCT STAGE			CONSTRUCTION PROCESS STAGE		USE STAGE							END OF LIFE STAGE				RESOURCE RECOVERY STAGE
	Raw Material Supply	Transport	Manufacturing	Transport	Construction installation	Use	Maintenance	Repair	Replacement	Refurbishment	Operational energy use	Operational water use	Deconstruction/Demo.	Transport	Waste Processing	Disposal	Reuse - Recovery - Recycling potential
MODULE	A1	A2	A3	A4	A5	B1	B2	B3	B4	B5	B6	B7	C1	C2	C3	C4	D
DECLARED MODULES	X	X	X	X	MNE	MNE	MNE	MNE	MNE	MNE	MNE	MNE	X	X	X	X	X

x = MODULE INCLUDED
MNE = MODULE NOT EVALUATED

Figure 5.3 An example of life cycle modules included in an LCA and EPD based on ISO 14040 and ISO 14044. At a minimum, the product stage modules (A1–A3) must be included, but a more thorough LCA will include all of the of modules. Image credit: Meg Calkins

in an EPD that just focuses on the product stage. EN 15804 requires EPDs adhering to the standard to show data for life cycle stages C1–C4. The end-of-life stage can demonstrate the durability of the product, as well as possibilities for reuse and recycling.

The system boundary section of an EPD may also show a diagram of the production process for a product. While not required, this is useful information to understand all the steps in the product stage and sometimes the end-of-life stage.

Life-Cycle Assessment Results

LCA results are presented for the declared unit of the product in a table or series of tables. Measures of environmental performance are listed by life cycle phase (e.g., A3) and grouped by life-cycle stage (e.g., product stage). Quantities are listed in scientific notation format (e.g., 2.60E+2 is 260; 7.04E-1 is 0.704) (Figure 5.4).

Environmental performance indicators required by EN 15804 are listed in Table 5.2 and other impacts categories are listed in Table 5.3. There are differences in requirements and terminology between North American and European standards, and this can create a challenge when comparing EPDs.

Other Information in EPDs

EPDs may provide the following information as well:

Cut off criteria: LCAs will usually establish cut off criteria that exclude flows under a certain level from the system boundaries of the LCA. The most common cut off level is 1%.

Assumptions, allocations, and estimates: EPDs will discuss assumptions, uncertainties, exclusions, and other factors in the LCA.

Background data and data quality: Sources of data are disclosed and their limitations are discussed.

Comparability: There is usually a statement on comparability of the EPD with other EPDs for a product, assembly, or system that provides the same function.

Other information: While EN 15804 and other standards (Table 5.1) mandate indicators, measurement methods, and information for LCAs and EPDs, other information such as charts and graphs, product literature, and performance information are added. This is particularly true of EPDs developed for designers, engineers, and specifiers as manufacturers want to market their products.

Limitations and Challenges of EPDs

Challenges to adoption and standardization of EPDs include the following:

- Performing an LCA and compiling an EPD is an expensive process that can be cost prohibitive for some smaller businesses, particularly those with multiple products.
- EPDs are valid for five years, but if a PCR changes after an EPD is completed, it could affect the validity of the EPD.
- Data can be inconsistent among EPDs even when they follow the same PCR. They may be using different background datasets. For instance, one company may use industry-average data, while another may develop facility specific data.

STEEL REINFORCING PRODUCT FOR CONCRETE Declared Unit: 1 metric ton								
INDICATOR	**UNIT**	**A1-A3**	**A4**	**C1**	**C2**	**C3**	**C4**	**D**
Global Warming Potential - Total	kg CO_2eq.	5.05E+02	1.60E+01	5.86E+01	8.73E+00	3.76E+01	2.60E-01	4.85E+00
Global Warming Potential - Fossil Fuels	kg CO_2eq.	4.98E+02	1.60E+01	5.85E+01	8.73E+00	2.36E+01	2.60E-01	3.04E+00
Global Warming Potential - Biogenic	kg CO_2eq.	5.91E+00	1.03E-03	3.85E-03	5.62E-04	1.41E+01	7.04E-05	1.82E+00
Global Warming Potential - Land use and land use change (luluc)	kg CO_2eq.	1.15E+00	3.08E-04	2.40E-03	1.69E-04	8.69E-03	1.65E-05	1.13E-03
Ozone Depletion Potential (ODP)	kg CFC11eq.	6.24E-05	3.39E-07	9.23E-07	1.87E-07	2.71E-07	4.13E-09	3.50E-08
Acidification Potential (AP)	mol H+ eq.	1.72E+00	4.08E-02	5.61E-01	2.20E-02	7.45E-02	2.44E-03	9.61E-03
Eutrophication Potential (EP) - freshwater	kg P eq.	1.45E-02	1.24E-05	5.01E-05	6.76E-06	5.15E-04	3.09E-07	6.65E-05
Eutrophication Potential (EP) - marine	kg N eq.	3.79E-01	1.58E-02	2.64E-01	8.53E-03	4.53E-02	1.15E-03	5.84E-03
Eutrophication Potential (EP) - terrestrial	mol N eq.	4.18E+00	1.67E-01	2.87E+00	9.00E-02	3.08E-01	1.25E-02	3.97E-02
Photochemical Ozone Creation Potential (POCP)	kg NMVOC eq.	1.36E+00	6.53E-02	8.42E-01	3.56E-02	9.96E-02	3.67E-03	1.29E-02
Abiotic depletion potential, minerals and metals (ADPm)	kg Sb eq.	1.99E-03	5.42E-07	2.47E-06	2.98E-07	7.82E-07	1.09E-08	1.02E-07
Abiotic depletion potential, fossil resources (ADPf)	MJ	1.05E+04	2.08E+02	7.68E+02	1.15E+02	1.65E+02	3.41E+00	2.13E+01
Water deprivation potential (WDP)	m^3	1.82E+02	1.90E-01	9.86E-01	1.05E-01	1.46E+00	4.39E-03	1.88E-01

Figure 5.4 An example of how LCA results are shown in a typical EPD. It includes results for environmental indicators by life-cycle stages. Numbers are expressed in scientific notation format. For example, 5.05E+02 = 505 and 1.99E-01 = 0.199.

Table 5.3 Other Indicators in some EPDs

Energy Consumption
Renewable Energy Use (MJ)
Non-renewable energy use (MJ)
Resource Consumption
Use of secondary material (kg)
Renewable material use (kg)
Components for reuse (kg)
Materials for recycling (kg)
Materials for energy recovery (kg)
Waste Generation
Non-hazardous waste generated (kg)
Hazardous waste generated (kg)
Human and Environmental Health
Particulate Matter Emissions (µg/m^3)
Ionizing Radiation, Human Health (electronvolt eV)
Eco-Toxicity (comparative toxic units, CTUe)
Human Toxicity, Cancer Effects (Comparative Toxic Unit for humans, CTUh)
Human Toxicity, Non-Cancer Effects (Comparative Toxic Unit for humans, CTUh)
Land Use Related Impacts/Soil Quality (soil quality index, SQI or normalization – based soil quality index, NSQI)

Source: EPD International 2024

- Some industries do not have industry-wide EPDs, so it is difficult to set a baseline GWP for a particular product.
- Data for raw material extraction is not available in all raw material categories as PCRs have not yet been developed for them.

EPDs are a very good tool, but their focus is primarily on embodied carbon and other environmental impacts. The information listed next is not usually addressed in EPDs.

- They do not reveal where and how raw materials are obtained or where manufacturing takes place. While distances of transport are revealed in life cycle stage A2, it is not clear how far the product will travel from the product stage to the project site.
- They do not give information on the social or cultural impacts of materials at all phases. The social working conditions of the product stage are not made clear. Manufacture in certain countries may use forced or child labor, but an EPD does not tell you where the product was made. They also don't reveal health impacts to fence-line communities around extraction and manufacturing facilities.

- EPDs do not require an accounting of human health impacts from resulting from the life cycle of a material or product. In 2019, EN 15804 added optional categories related to human health impacts including particulate matter emissions, ionizing radiation, human toxicity, cancer, and non-cancer effects.
- LCAs and EPDs do not reveal external and difficult-to-measure impacts such as changes in the biogenic carbon content of the soil from logging for lumber, transportation of labor to factories, production of manufacturing machines, road building in quarries, and transport impacts of recyclable collection for secondary materials.

Using LCAs and EPDs

Information from LCAs and EPDs can inform project design, the detailing of material assemblies, and the selection of products for a project. LCA information for materials and products is used in whole project LCAs, calculation of embodied carbon for material assemblies, and comparisons of products. Multiple tools are available to assist with this evaluation. The last section of this chapter lists questions to ask of raw material suppliers, product manufacturers, contractors, designers, owners, and facility managers to supplement EPDs or when they are not available. See Table 2.9 for a list of resources, standards, certifications, and tools for decarbonization.

Material and Product Selection and Comparison Using EPDs

EPDs are intended to provide information in a format that allows for comparison. Product category rules facilitate this by dictating types of data collection and format. But different countries have different PCRs for the same products, and standards for LCAs and EPDs vary geographically as well. Additionally, it is currently challenging to find multiple comparable EPDs for some site construction products.

For EPDs to be comparable, all the following criteria listed must be met:

- The EPDs must have been created using the same PCR.
- The products must be functionally equivalent. For instance, a 20.7MPa concrete with Type IL cement can't be compared with a 27.6MPa concrete with Type IS cement.
- The EPDs must include the same life cycle stages, although only the product stage (A1–A3) could be considered in the comparison.

Whole Project, Whole Building, and Whole Site Life Cycle Assessments

LCAs for whole projects are rare but increasingly employed. Whole building LCAs (wbLCA) are most commonly performed while whole site LCAs (wsLCA) are less common. Whole project LCAs consider both upstream and downstream burdens for an entire building and site. This includes impacts for most materials and products used in a building and site over the full life cycle. All embodied and operational GHG emissions are totaled and expressed in carbon dioxide equivalents (CO_2e) per unit of measure (e.g., m2 or ft2). For sites, GHG emissions can be offset by the sequestration of trees, shrubs, and grasses, and these offsets are included in the calculations.

System boundaries for whole building and/or whole site LCAs define the following:

- Spatial project limits
- Temporal limits – this is commonly 60 years, but sometimes 30 years
- Systems under consideration – wbLCAs may only include the structural system and the enclosure system
- Delimitations – exclusions may include contractors commuting emissions to the site or temporary structures such as formwork and scaffolding.

In wbLCA and wsLCA, the embodied carbon results are often compared against a baseline scenario. In wbLCA, baselines are identified by building type and area (e.g., 500kg CO_2e/m2 for an office building). Whole site LCA baselines could be expressed in a similar manner. The reference study period, the time period for the site or building operation in the study, is important as it will allow for quantification of operational carbon and energy and water use.

Whole project LCAs provide a more comprehensive picture of the environmental impacts of a project than an LCA for a product or material. But they are more challenging to execute as there are dozens of materials and products in a site or building, and some data may not be available. It is common for wbLCAs to focus on the structure and enclosure materials of a building but not on the interior or finishes. Whole building LCA tools help reduce the complexity of calculations, and LEED has incorporated an option for projects that use one of the approved tools.

Pathfinder

Pathfinder by Climate Positive Design is a free tool to evaluate the GHG emissions and carbon sequestration of a site. Pathfinder estimates "years to (climate) positive" for a site instead of providing total GHG emissions. Types and quantities of material assemblies are input along with

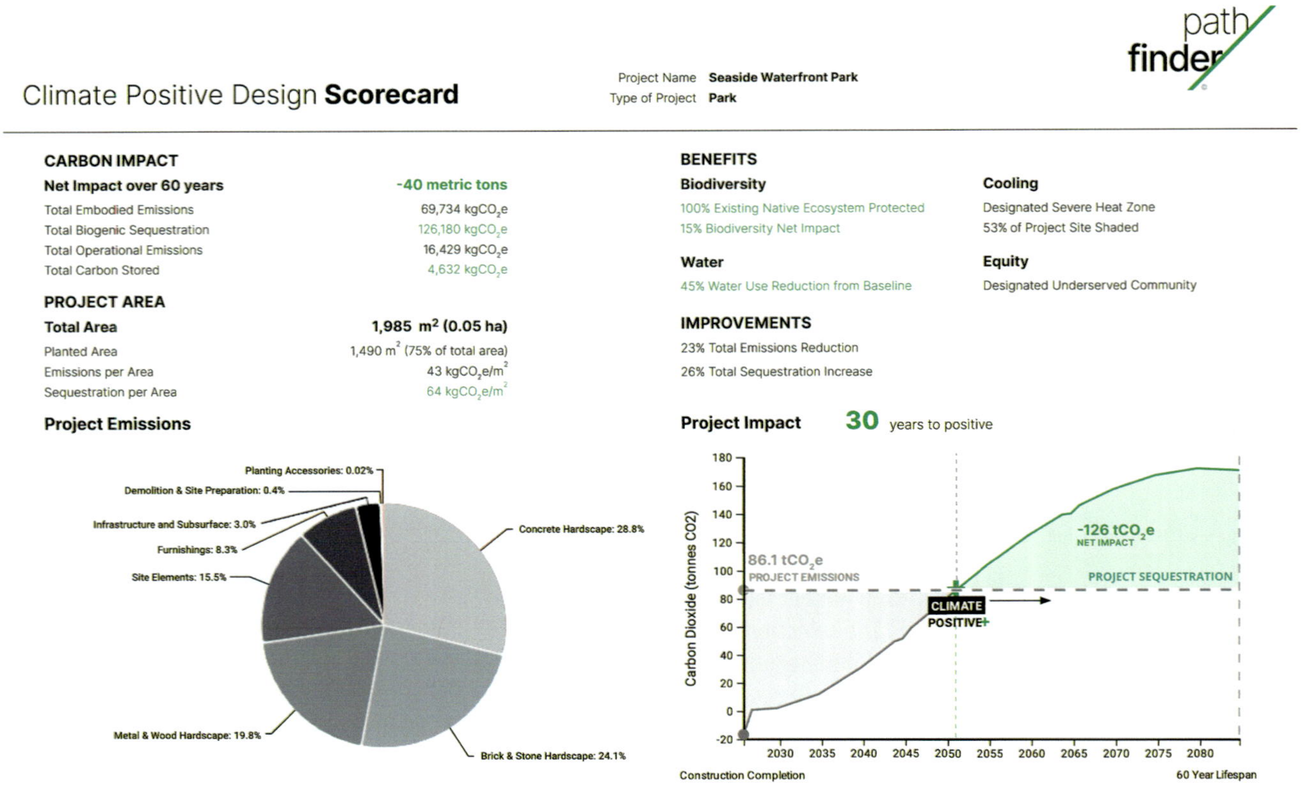

Figure 5.5 A sample Climate Positive Design scorecard from the Pathfinder Tool. Image credit: Climate Positive Design

existing and proposed vegetation. Projects can establish a baseline scenario using existing site conditions or a site design with standard materials and vegetation, then input proposed structures and vegetation to understand the number of years it will take for the site to become climate positive. Years to climate positive can be reduced through use of low-carbon materials and material mixes and increased vegetation (Climate Positive Design 2020) (Figure 5.5). The Climate Positive Design Challenge asks projects such as parks, campuses and residential to become climate positive in five years. Plaza and streetscape projects are challenged to become climate positive in 20 years with infrastructure projects in 25 years.

Carbon Conscience

Carbon Conscience (CC) is a free application to assist designers assess carbon impacts in the early stages of project planning and design. It does not require information about specific materials and quantities, so it can be used in early concept design for urban design scale projects, buildings, sites, and infrastructure. Users can compare design options, test alternative land uses, structural systems, façade materials, and landscapes with quick sketches. Users input site plans of buildings, land uses, number of floors, basic materials decisions, and/or ecosystem typologies. Results are given in ranges and are estimated but are useful for comparing tradeoffs in the conceptual design phase. Users can edit their materials assumptions by land uses from SCMs to recycled content, or types of binders, transportation distances. Future work on Carbon Conscience aims to allow for an export from CC to Pathfinder as the project moves into the schematic design phase (Hardy 2024) (Figure 5.6).

Athena Pavement LCA

Athena Pavement LCA is a free LCA based software for calculating the environmental and embodied carbon impacts of pavements. It considers regional differences in North American roadway materials, construction, and maintenance across the life cycle. Calculations include pavement materials, curbs, shoulder, roundings, base materials, and subbases. The results are given for the US EPA TRACI indicators: global warming potential, acidification potential, human health respiratory effects potential, ozone depletion potential, smog potential, and eutrophication potential.

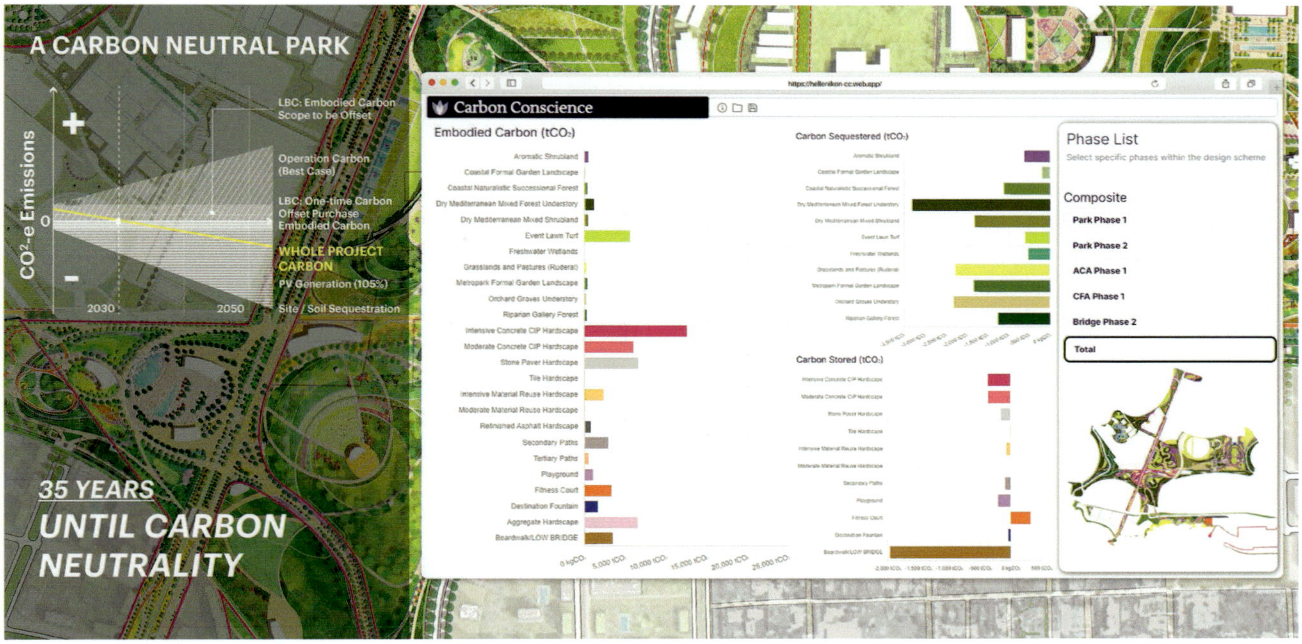

Figure 5.6 Carbon Conscience results from The Ellinikon Park project in Athens by Sasaki. Image credit: Sasaki Associates, Inc.

Calculating the Embodied Carbon of Structures and Material Assemblies

Just comparing the embodied carbon of different types of materials (e.g., concrete vs. asphalt) does not give an accurate accounting of the true impacts of a built structure. Calculating the embodied carbon of material assemblies for a structure will provide far greater accuracy of the true carbon footprint. This type of calculation can also show tradeoffs for optimized sizes, recycled contents, or different mixes.

The embodied carbon of site structures such as retaining walls, pavements, decks, or guardrails can be calculated with embodied carbon data and quantitates for each of the material components. Quantities of materials are totaled for a unit of the assembly such as one linear meter of a wall or rail or one square meter of pavement. For a retaining wall, this would include the wall stem, footing, reinforcing in the stem and footing, and the perforated pipe, drain rock and filter fabric in the drainage assembly behind the wall. The material quantities of each component are multiplied by the embodied carbon of the material for that component.

Embodied carbon figures are derived from product or industry-wide EPDs, material baselines, materials chapters in this book, or other listings of embodied carbon numbers. They are multiplied by the total weight of each material in the assembly.

This method takes time, but it is a helpful approach for understanding the finer tradeoffs of material mix adjustments, alternative materials, optimized structures, and alternative products. A baseline could be established by calculating a standard material assembly for a structure. Then optimized component sizes and thicknesses, lower carbon materials, or alternative structures could be used. The full method is shown in Tables 5.4 and 5.5.

Important considerations when comparing material assemblies are the reference service life of the assembly, and the anticipated service life of the components of an assembly. In Chapter 10 "Decks and Boardwalks", an embodied carbon comparison of decking materials is made for a 30-year service life for a deck. Some of the decking materials will not last the entire service life, so a replacement factor multiplier is employed. As ACQ-treated southern yellow pine deck boards are expected to last 10–15 years, a replacement factor of 2.5 is used, and the total EC of decking is multiplied by 2.5 for the EC over the entire 30 years.

Tables 5.4 and 5.5 show example calculations for a 90cm high retaining wall with granite veneer. The ***typical scenario*** uses imported granite veneer on a 27.6 MPa concrete wall with 20% cement substitutes (SCMs). The total width of the wall is 30cm. The ***optimized scenario*** uses domestically quarried and fabricated granite veneer that is slightly thinner on a thinner 20.7

Table 5.4 Typical Scenario – GWP for One Meter Length of Retaining Wall
30cm thick retaining wall is 90cm above grade with a 60cm frostline

Material	Density of Material (kg/m^3 unless noted)	Volume in One Linear Meter of Wall (m^3 unless noted)	Weight in One Linear Meter of Wall (kg)	Emissions Factor (kg CO_2e/kg)	Emissions From One Linear Meter of Wall (kg CO_2e)
Imported granite veneer, 26mm thickness	2674	0.034	90.916	1.079	98.098
Concrete stem wall and footing, 27.6 MPa with 20% SCM[1]	2300	0.694	1596.201	0.143	228.257
Rebar, 16mm	0.89kg/m	27.82m	24.742	0.854	21.130
Mortar, 10mm	2003	0.016	31.848	0.200	6.370
SS stone anchors, qty. 8	7850	–	1.362	0.696	0.947
Drain rock	1600	0.200	320.040	0.014	4.480
Filter fabric	0.210kg/m^2	2.26m^2	0.475	3.264	1.720
Perforated HDPE drainpipe, 10cm dia., 20% recycled content	0.941kg/m	1.00m	0.941	2.178	2.049
Total GWP from one linear meter of retaining wall					**363.050**

[1] Supplementary cementitious material (SCM), also called cement substitute.

Table 5.5 Optimized Scenario – GWP for One Meter Length of Retaining Wall
21cm thick retaining wall is 90cm above grade with a 60cm frostline

Material	Density of Material (kg/m^3 unless noted)	Volume in One Linear Meter of Wall (m^3 unless noted)	Weight in One Linear Meter of Wall (kg)	Emissions Factor (kg CO_2e/kg)	Emissions From One Linear Meter of Wall (kg CO_2e)
Imported granite veneer, 26mm thickness	2674	0.023	61.235	0.182	11.145
Concrete stem wall and footing, 27.6 MPa with 20% SCM[1]	2300	0.585	1344.352	0.086	142.501
Rebar, 16mm	0.89kg/m	27.82m	24.742	0.854	21.130
Mortar, 10mm	2003	0.016	31.848	0.200	6.370
SS stone anchors, qty. 8	7850	–	1.362	0.696	0.947
Drain rock	1600	0.200	320.040	0.014	4.480
Filter fabric	0.210kg/m^2	2.26m^2	0.475	3.264	1.720
Perforated HDPE drainpipe, 10cm dia., 20% recycled content	0.941kg/m	1.00m	0.941	2.178	2.049
Total GWP from one linear meter of retaining wall					**190.341**

[1] Supplementary cementitious material, also called cement substitute.

MPa concrete wall with 50% cement substitutes (likely fly ash and/or blast furnace slag). The total width of the wall is 21cm.

The optimized scenario has just 52% of the embodied carbon of the typical scenario. There are more of these comparisons in the freestanding walls, retaining walls, pavements, fences and rails, and deck chapters.

Information Transparency Questions for Construction Materials and Products

The following questions can be asked of raw material suppliers, product manufacturers, contractors, designers, owners, and facility managers to supplement EPDs or when they are not available.

A1 Raw Material Supply

a. Have steps been taken to minimize the environmental impacts of raw material extraction for the product (e.g., site restoration, control of toxic releases, and hazardous air pollutants)?
b. Are the raw materials sourced from a company that has fair and safe working conditions? Do they employ children under 18? If so, how many hours do they work per week? Are any raw materials sourced from a country with documented labor abuses (Table 4.5)?
c. Is the product certified for responsible mining or harvesting practices by a standard such as the Stone Sustainability Standard, the Forest Stewardship Council, or the IRMA?
d. Does extraction of this product release any substances that are tracked by the US EPA TRI? Has the company taken steps to reduce or capture these emissions?
e. Does the product contain recycled content? If so, what percentage is post-consumer? Pre-consumer? Does the percentage and type of recycled content meet the US EPA Comprehensive Procurement Guidelines?
f. Is all or part of the product biobased? If so, what is the percentage of biobased material in the product? Is the biobased content sustainably harvested? Or an agricultural or forestry by-product?
g. Is the product made from a renewable resource? Is it rapidly renewable (within five years)?

A2 Transport to Factory

a. How far from the point of extraction or harvest is the manufacturing facility?
b. What is the mode of transport? If trucking is used, is it fuel efficient?

A3 Manufacturing

Note: Some products will have multiple manufacturing and fabrication steps so these questions should be asked of all manufacturers and fabricators.

a. Is the material or product covered by an EPD that meets an accepted standard (e.g., ISO 14025, ISO 14040, ISO 14044, and ISO 21930 or EN 15804)?
b. If the manufacturer has published other information about the environmental or health impacts of their product, is it third-party verified?
c. How much raw material becomes waste during the manufacturing process? What is done with the waste? Is it reused in another process? Or transported off site for reuse elsewhere? Landfilled?
d. What other wastes (including GHG emissions) are produced during the manufacturing stages?
e. Is the product covered under a health product declaration that meets the HPD standard?
f. Does manufacture of this product release any substances that are tracked by the US EPA Toxics Release Inventory or the US National Toxicology Program's (NTP) Report on Carcinogens? If so, what is the substance and how much is released per ton of product? Has the manufacturer taken steps to reduce or capture these releases?
g. Are any material components or by-products on the Living Building Challenge's Red List?
h. What are the energy sources for all manufacturing stages by percentage? Has the manufacturer taken steps to reduce the energy use or use renewables in manufacturing?
i. Are process emissions released in manufacturing? If so, what percentage are captured through CCUS technology? Has the manufacturer taken steps to reduce GHG emissions?
j. How much water is used in the manufacturing process? Have water conservation or recycling measures been put in place?
k. Are there are multiple stages of manufacture/fabrication at different facilities? If so, what is the distance between them? What mode(s) of transport are employed?
l. What is the distance from the last point of manufacture to the distribution facility? What mode(s) of transport are employed?
m. Is the manufacturing facility a safe work environment with fair and equitable treatment of the workers? Is the product produced in a country with documented labor abuses (Table 4.5)?
n. Does the manufacturer have a corporate social responsibility report?
o. Does the manufacturer have a published environmental policy or a corporate sustainability report? Have they taken steps to reduce environmental impacts and improve their processes?
p. Has the manufacturer taken steps to reduce or eliminate packaging of the product?
q. Does the manufacturer have a takeback program or another program to facilitate reuse or recycling of the product?
r. Is the material or product reusable or recyclable?

A4 Transport to Site

a. What is the distance from the last point of manufacture or distribution to the project site? What mode(s) of transport will be employed?

b. Is it possible to combine delivery of the product with other deliveries to make sure only full trucks come to the site?

A5 Construction–Installation

a. Does the product contain or release any substances that are reported in safety data sheets?
b. Does the product's VOC level comply with the California Air Resources Board VOC limits?
c. Is the construction site a safe work environment with fair and equitable treatment of the workers?

B1 Use

a. Is the material or product durable? Is it warranted for the expected service life of the site? Or if not, have replacement factors been used in embodied carbon calculations of assemblies and structures?
b. Does the product meet any efficiency standards such as the EPA Energy Star or the WaterSense label?

B2 Maintenance

a. What are routine maintenance requirements for the material, product, or structure? What is the maintenance schedule?
b. Are non-toxic, low-VOC cleaners, and other maintenance products available for the product or assembly?

B3 Repair

a. Is the product, assembly, or structure repairable? If so, what are the repair procedures?
b. Are the repair materials non-toxic and low VOC?
c. What are the resource and embodied carbon impacts of the repair material?

B4 Replacement

a. If the product, assembly, or structure is not repairable? Is it possible to remove a component or a small area for replacement?
b. If the product, assembly, or structure will not last for the planned service life of the site, is it possible to replace it?

B5 Refurbishment

a. If the product, assembly, or structure must be removed from the site, can it be refurbished and reused elsewhere?

C1–C4, D Deconstruction/Demolition and Reuse, Recovery, and Recycling Potential

a. Does the assembly or structure allow for deconstruction with accessible and dismantlable connections?
b. Is the product reusable in whole form?
c. Is the product recyclable? Can the product be recycled without heating and melting it? Does the material or product give off toxins during recycling?
d. Are there recycling facilities in the region of the site for the product or material?

References

Climate Positive Design (2020). *Landscape Carbon Calculator/Pathfinder: Methodology, Data Sources, and Metrics Summary*. Climate Positive Design, CMG and Atelier Ten.

Environmental Product Declarations International (EPD International). "Environmental performance indicators". www.environdec.com/resources/indicators, accessed May 10, 2024.

Hardy, C. (2024). Personal communication, August 19.

International Standards Organization (ISO) (2012). *ISO 9223 Corrosion of Metals and Alloys - Corrosivity of Atmospheres - Classification, Determination and Estimation*. ISO.

World Resources Institute (WRI) (2011). *Greenhouse Gas Protocol: Product Life Cycle Accounting and Reporting Standard*. World Resources Institute and World Business Council for Sustainable Development, September.

Structures and Assemblies for Resilient Sites

CHAPTER 6

Pavements

Pavements are often the largest hardscape structures in a built landscape. They offer level or uniformly sloped surfaces for pedestrian and vehicular circulation, they provide spaces for site users to gather, and they contribute to the visual qualities of a place. A wide variety of pavement materials are used in site construction for sidewalks, parking lots, roads, plazas, paths, and terraces. These structures have very different performance needs; therefore, their materials, design, and detailing can be widely varied.

Because of their size and weight, pavements are usually the largest contributor to the embodied carbon of a site. Yet they can offer the greatest opportunity to achieve carbon reduction goals, particularly if local, recycled materials are used. While the focus of this chapter is on detailing pavements to extend their service life and reduce their environmental impacts, the most effective approach to reducing embodied carbon will always be to use less pavement.

Pavement materials should be selected for expected vehicular and pedestrian traffic, climate, maintenance, aesthetics, accessibility, cost, reflectivity, and embodied carbon. To extend a pavement's service life, flexible, repairable pavements should be favored over rigid pavements where possible (Table 6.1).

Environmental Impacts of Pavements and Methods to Reduce Them

The environmental impacts of pavements are substantial with millions of metric tons of embodied carbon emissions released and non-renewable resources used globally each year for their construction. In use, pavements can concentrate stormwater quantities, contribute to the urban heat island, disrupt wildlife habitats, and degrade adjacent ecosystems.

Greenhouse Gas Emissions

Greenhouse gas (GHG) emissions from the material extraction, production, transportation, construction, and maintenance of pavements is a major environmental impact. Construction materials such as concrete, concrete pavers, and asphalt are more carbon intensive than gravel and stone dust pavements. For example, just under one ton of CO_2 is released for the construction of a typical concrete parking space.

Even if pavement materials are less carbon intensive, the transport of these heavy materials has a substantial carbon footprint. Transporting 20 metric tons of aggregate base materials 20 kilometers results in emissions of 428kgCO_2e, while producing the aggregate only releases 280kgCO_2e.

Embodied carbon (EC) of pavement assemblies can be reduced by specifying low-carbon materials such as stone dust and gravel or developing low-carbon mixes of standard materials such as concrete with high percentages of portland cement substitutes or warm mix asphalt with RAP aggregate. Embodied carbon can be reduced by specifying locally produced materials and by using less material in pavements through alternative detailing and construction approaches.

The embodied carbon of pavement assemblies is quantified and illustrated later in this chapter. The embodied carbon of individual materials for pavements is listed in the materials chapters of this book. Decarbonization of site materials and structures is discussed in Chapter 2, and methods of assessing embodied carbon are discussed in Chapter 5.

Resource Use

Pavement assemblies use large amounts of non-renewable mineral resources and fossil fuel inputs for their construction. The amounts for a given pavement assembly will vary by standards, climate, and traffic. A typical 60 space concrete parking lot designed to US standards uses 451 cubic meters of mineral resources plus fossil fuel inputs in materials manufacturing. Designing a parking lot to European standards will reduce material use, and embodied carbon, by 26% (BPA 2016; DIN 2005).

DOI: 10.4324/9781003413974-7

Table 6.1 Reducing the Environmental Impacts of Pavements

Reduce the Extent of Pavements
Minimize the size of paved areas. Use alternative materials, shared driveways, smaller parking spaces and drive aisles, and narrower paths and sidewalks.
Design complete, multi-use streets and pavements to reduce auto dominance and allow for multiple users.
Limit pavements to less than 30% of site area where possible. Aim for 70% of sites with vegetation[1].
Reduce Resource Use of Pavements
Extend the service life of pavements. Fix small cracks, ruts, and other deformations before they become large problems. Consider resurfacing options rather than replacement of the entire pavement assembly.
Design durable pavements by improving the load bearing capacity of native subgrade through stabilization where necessary. Design pavements and pavement mixes to accommodate anticipated loads but don't overdesign beyond the recommendations from structural or geotechnical engineers.
Use pavements that are flexible and repairable to extend their use life. Specify perpetual pavements for asphalt pavement where appropriate. Consider unbound pavements for low traffic parking lots and the parking spaces in higher traffic parking lots. Use stabilized crushed aggregate, crushed oyster shells, dense graded RCA or other recycled aggregates.
Refurbish and reuse pavements in place.
Reuse pavement slabs in new pavements.
Avoid mortar to allow for reuse of unit pavers, bricks, and stone.
Crush and grade demolished concrete and asphalt for reuse on site.
Use recycled aggregates in concrete and asphalt pavements, and in base courses.
Reduce the Embodied Carbon of Pavements
Specify low-carbon materials for pavements where appropriate.
Specify warm mix asphalt with high recycled content such as recycled asphalt pavement.
Specify reduced carbon concrete mixes using cement substitutes or alternative cements. Some cement substitutes such as fly and blast furnace slag are industrial wastes.
Sequester carbon in concrete pavements.
Use carbon aggregates for base materials and in pavements.
Source local materials.
Reduce Ecosystem Impacts From Pavements
Use recycled materials in and under pavements to reduce ecosystem impacts from raw material extraction and manufacture of new materials.
Drain pavements to adjacent bioretention areas to attenuate stormwater. Break up large expanses of pavements with bioretention measures.
Use porous pavements to reduce quantities of run off from pavements.
Specify pavement materials from manufacturers who employ effective pollution control equipment to catch particulates and air pollutants during manufacturing.
Minimize the extent of site disturbance during pavement construction. Revegetate as soon as possible.
Advocate for electric-powered construction equipment.
Employ erosion control and drainage measures during construction and use to avoid water eroding pavement layers.
Reduce Urban Heat Island Impacts of Pavements
Use pavements with a minimum initial solar reflectance of 0.33. Refer to the Urban Heat Island discussion later in this chapter.
Reduce the quantity of paved surfaces.
Shade 50% of pavements within 15 years of construction[2].
Use an open grid pavement system with at least 50% unbound aggregate[3].

[1] Conrad 2023

[2] CALGreen 2019

[3] USGBC 2024

Climate positive design recommends that a pavement area should occupy no more than 30% of a site with vegetation occupying the remaining 70% of the area outside of a building footprint (Conrad 2023). Pavement area can be reduced through alternative materials, shared driveways, smaller parking spaces, smaller plazas, and narrower sidewalks, roads, and parking drive aisles. It's also beneficial to incorporate planted areas in parking lots, though they must be scaled to function well for stormwater collection and habitat health.

Resource use for pavements can be reduced by using reclaimed and recycled materials from on-site demolition or from sources nearby. Pavement assemblies can incorporate a very wide range of post- and pre-consumer waste materials (Tables 12.5 and 12.6)

Last, using the minimum thickness of pavement and base materials from the most efficient design dictated by structural or geotechnical engineers will save resources. For instance, increasing a concrete pavement slab thickness from 15cm to 20cm will use 33% more material resulting in 33% higher embodied carbon.

Ecosystem Impacts

Ecosystem impacts of pavements occur from extraction of raw materials, which destroys habitats and disturbs ecosystems around extraction sites. Particulates and emissions from fossil fuel combustion are released during all phases and transport of the materials impacting habitat, air, and water quality in several areas along the journey. Releases from manufacturing of pavement materials may also impact air, land, and water.

In use, pavements can interrupt the flow of wildlife populations and their natural processes. Roads, in particular, can lead to the reduction of connected habitats with an increase in wildlife and vegetation mortality, disease, and fragmented populations. Paved roadways can be vectors for the spread of invasive species. Fences along roads can disrupt the movement of wildlife. Ecosystems adjacent to pavements may be negatively impacted in a zone that is 10–15 times the width of the pavement due to factors such as construction equipment disturbance, soil compaction, removal of native vegetation, stormwater runoff, and erosion (Forman et al. 2003).

Ecosystems are affected by a pavements impervious surface. Soil health under pavements is degraded without water and air. Impervious surfaces concentrate stormwater runoff leading to erosion and flooding, thermal and chemical pollution, siltation, and depletion of groundwater. Areas of large contiguous pavements such as parking lots amplify this problem.

Where appropriate, porous pavements should be employed (Chapter 7 "Porous pavement and green roofs"). This will allow for some infiltration of stormwater as well as maintain soil health under the pavement. Additionally, large contiguous areas of pavement should be broken up with green infrastructure measures such as bioswales, bioretention areas, and stormwater planters to further aid the health of soils under and adjacent to pavements.

Heat Island Impacts

Pavements impact urban and suburban microclimates and contribute to urban heat islands. Darker colored pavements absorb heat from solar radiation and release it into the surrounding atmosphere, raising air temperatures in urban areas by 0.56–3.9°C (2–8°F) and altering weather conditions. Urban heat islands and pavements and the solar reflectance of common pavement types are discussed near the end of this chapter.

Components of Pavements

Pavements are designed to bear pedestrian and vehicular loads with a consistent, well-draining surface. They are classified by their intended use within a design context and their application loading (heavy, medium, or light duty). Other factors that influence pavement material and design are subsoil characteristics, climate, cost, maintenance, pavement porosity, and environmental considerations such as embodied carbon and stormwater performance.

Pavements are composite structures comprised of multiple layers including the surface layer, an aggregate base, and the subgrade. In cases of heavy duty pavements or weaker subgrades, a subbase is included under the aggregate base. Underdrains and geotextiles are employed as required.

Subgrade

The subgrade soil bears the loads on the pavement and the weight of the pavement materials themselves. Subgrade bearing capacity and permeability determine the thicknesses of the pavement layers.

Subgrade soil is subjected to groundwater with upward capillary migration and frost with seasonal fluctuations. It is compacted native soil unless it is not capable of supporting the pavement or if fill soil must be used. The subgrade is "prepared" to ensure the slope, depth, compaction, and uniformity required for the paving design.

Moisture content of the subgrade should be uniform to reduce deformation and swelling of clay or freeze/

Table 6.2 Common CBR Values for Soil Types

Unified Soil Classification Type	California Bearing Capacity Estimated Value
Clay	+/−2%
Sand	7% (poorly graded) to 10% (well graded)
Well-graded sandy gravel	15%
Clayey sand	5–20%
Silty gravel	20–60%
Gravel, poorly graded	30–60%
Gravel, well-graded	40–80%

Source: Woods Ballard 2015

thaw movement. Use of perforated subdrains will help maintain a uniform moisture content of the subgrade. Geotextile separators are employed to prevent fine colloidal soil from migrating up into the aggregate base or subbase and to bridge between stable and unstable subsoil (CA DOT 2013).

A common measure of the strength of a subgrade is the California Bearing Ratio (CBR) test. This test compares the bearing capacity of a soil/granular material with that of well-graded crushed stone. The CBR is expressed as a percentage reflecting the ratio of the bearing capacity of the material being tested and the crushed stone material at 100%. The CBR is primarily intended for the evaluation of cohesive materials having maximum particle sizes less than 19mm. Common CBR values for United Soil Classification types are listed in Table 6.2.

Most codes and standards require that a subgrade have a minimum CBR of 2.5% to create an adequate pavement foundation. Subgrades with a 2.5–5% CBR will require thicker bases, subbases, and/or pavement layers; and those with a CBR lower than 2.5% will require measures to improve the subgrade through the addition of lime, cement, asphalt, or other soil stabilizers (Table 6.2). Excavation to a depth of soil with adequate bearing capacity is another approach.

The slope of the subgrade usually matches the slope of the pavement surface, however for adequate drainage, a subgrade adjacent to structures should have a minimum slope of 2.5%. For soil types with high clay contents, geotextiles or fine sand choker courses can be used to prevent subgrade soil from migrating up into the aggregate base or subbase.

Aggregate Base and Subbase

The graded aggregate foundation that is the base layer transfers the pavement's load to the subgrade and prevents upward migration of water from the soil below. Base thicknesses range from 100–300mm or greater depending on the design load, pavement type, soil bearing capacity, and porosity of the pavement surface layer. Heavy duty pavements or weak subgrade conditions require another layer of coarser graded base material called a subbase.

Both the base and subbase extend horizontally beyond the pavement layer to ensure strong support for the pavement edge. It is important for long-term stability that the base or subbase material does not migrate into the subgrade, so dense graded aggregate and/or a geotextile separator is used.

Aggregate sizes and gradations for pavement bases and subbases are specified based on standards such as

> *ASTM D448 Standard Classification for Sizes of Aggregate for Road and Bridge Construction M43 Standard Specification for Sizes of Aggregate for Road and Bridge Construction*
>
> *AASHTO M147 Standard Specification for Materials for Aggregate and Soil - Aggregate Subbase, Base, and Surface Courses.*

Important characteristics of aggregates for pavement bases are the ability to interlock and form a structurally solid foundation for the pavement.

Crushed stone aggregate is a manufactured product that is produced from crushing larger material. The aggregate will have an irregular, fractured, angular, and rough face. This material has good interlocking properties for bases.

Gravel particles are naturally occurring and have a rounded, smooth surface created by water and natural erosion of the particles surface over time. Gravel does not typically interlock like manufactured aggregate, so it is not recommended for pavement bases. It can be used in concrete, asphalt pavement, or in a pavement surface course.

There are two types of pavement bases:

Unbound bases are compacted crushed stone that is open or dense graded. The base depends on the interlock of the stone particles for stability.

Bound bases are crushed stone or gravel that is mixed with cement, lime, or asphalt then compacted to form a more rigid base. These bases are used in higher traffic applications or with weaker subgrades.

Terms Describing Aggregate Gradations

Dense graded aggregate is a combination of different aggregate sizes with the small particles filling the voids between the larger particles (Figure 6.1).

Open graded aggregate is a narrow range of aggregate sizes that produce open voids between particles (Figure 6.1).

Coarse graded aggregate is dense graded aggregate with predominantly large particles.

Fine graded aggregate is dense graded aggregate with predominantly small particles.

Well-graded aggregate has a balanced distribution of stone sizes.

Gap graded aggregate is two distinct sizes of aggregates in a mix.

Stone dust is fine screenings that remain after screening for size of aggregates.

Screenings are either the larger or smaller aggregate materials that are separated from a mass by a sieve. The term is commonly used to refer to very small, crushed stone particles or 5mm or smaller.

Coarse aggregates range in size from 9.5mm–37.5mm.

Fine aggregates pass through a 9.5mm sieve and are retained on 0.075mm sieve. Sands are considered fine aggregates. Coarse sand is 0.5mm to 2mm.

Fines are different than fine aggregates. They are any material that falls through a 0.075mm sieve.

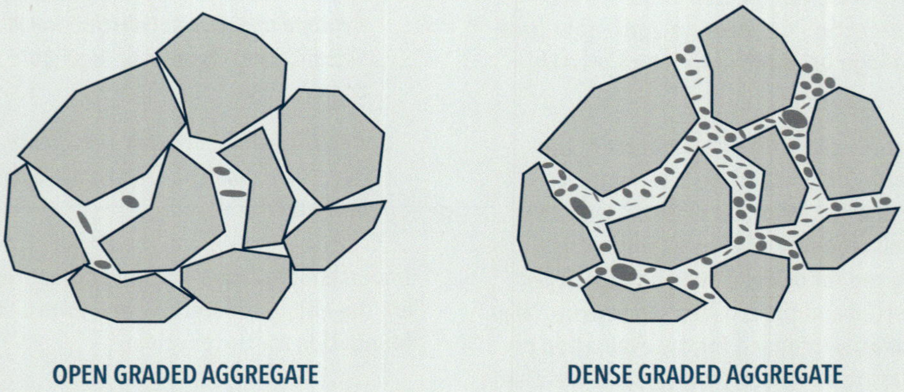

Figure 6.1 Open graded aggregate has a narrow range of aggregate sizes that result in open voids between particles while dense graded aggregate has a range of aggregate sizes with smaller aggregates filling the voids between larger aggregates. Image credit: Drake Bruner and Meg Calkins

Aggregates make up the largest share of the mass and volume of a pavement structure whether used as a constituent in a pavement mixture such as asphalt or concrete, the pavement base or subbase, or as the pavement surface itself. Use of recycled materials for aggregates will reduce the embodied carbon and resource use of pavements, avoid the habitat impacts of excavating virgin aggregates, and reduce landfill waste.

It is important to note that recycled materials should be sourced locally or regionally, or their environmental benefits could be outweighed by their transportation impacts. Chapter 12 discusses criteria for evaluating recycled aggregates and their application in pavement bases and subbases.

Sand Setting Beds

Unit pavements require a coarse sand setting bed, also called a sand leveling bed, bedding layer, or choker course to mediate between the aggregate base and the necessity for a level, consistent pavement surface. The sand setting bed is usually 20–50mm thick and moderates slight variations in unit pavement thicknesses and the aggregate base.

Sand can be produced through mechanical crushing or it can be naturally occurring and extracted. Some naturally occurring sand is rounded and not recommended for sand setting beds. Angular natural sand and mechanically crushed fine aggregates will form a stable and interlocking bed for the unit pavers. Geotextile separators are employed to prevent sand from migrating down into the aggregate base.

Pavement Surface Course

The pavement surface course transfers loads on the pavement to the aggregate base as well as receiving the wear of traffic. Types of pavement surfaces range from rigid concrete to unbound stone dust pavements. Pavement surfaces are determined by expected traffic, climate, aesthetics, cost, maintenance, reflectivity, and porosity.

Pavement Structures

Pavements are either structurally rigid or flexible, and they are monolithic materials or comprised of units. They can be impervious or pervious. Porous pavement structures are discussed in Chapter 7.

Flexible monolithic pavements, such as asphalt pavement or stabilized stone dust pavement, derive their strength and stability from the friction, interlock, and cohesion of the aggregate base layer. They have a relatively thin wearing surface and relatively thicker aggregate bases as it is the base that distributes the design loads over the subgrade. The firmness and stability of flexible monolithic pavements will depend on the binder composition (Figure 6.2).

Flexible unit pavements consist of dry laid, butt jointed paving units (precast concrete, stone, or brick) that are placed on a sand setting bed and an aggregate or asphalt pavement base. Paving units generally range in thickness from 30mm to 150mm (Figure 6.3). Flexible unit pavements can be porous, although very little water will seep through sand swept butt joints. For true porous unit pavement, spacers or precast lugs on the pavers must hold the units apart by 4–7mm or more.

Flexible pavements require stable edges to contain the units or aggregate. Edges can be steel, aluminum, poured in place concrete, brick in a grout base, wood, plastic, or composite lumber (Table 6.3 and Figures 6.10 to 6.13). Raised edges such as curbs will direct water flow, but they can concentrate runoff quantities. Some flexible monolithic pavements such as bike or walking trails without heavy traffic don't require formal edge containment.

Flexible pavements are generally easier to repair or relevel than rigid pavements, but some will require more maintenance activities than a rigid pavement. Flexible pavements are a repairable solution in cold climates where regular expansion and contraction of soils occurs.

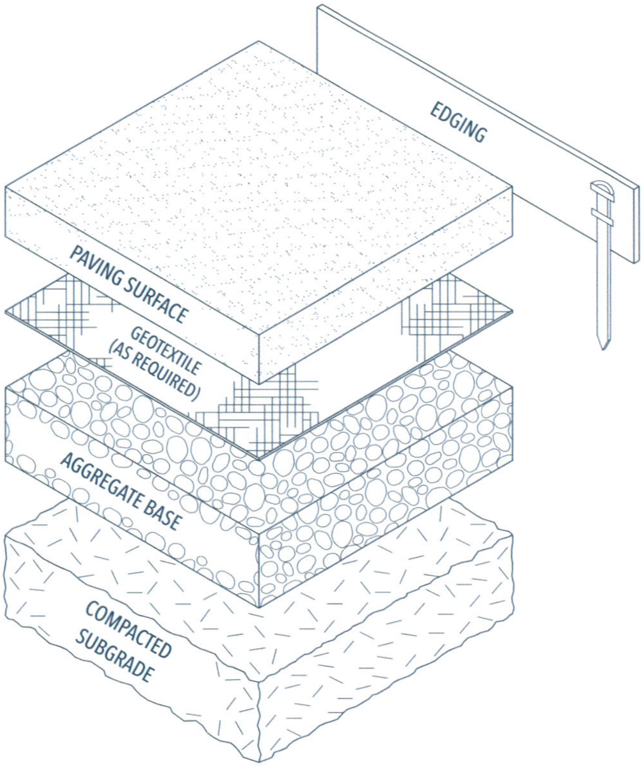

Figure 6.2 Flexible monolithic paving assembly with edge. Underdrain may be required at pavement edge. Flexible pavements can be repaired and releveled for potentially long service lives. Image credit: Kathleen Elliott and Meg Calkins

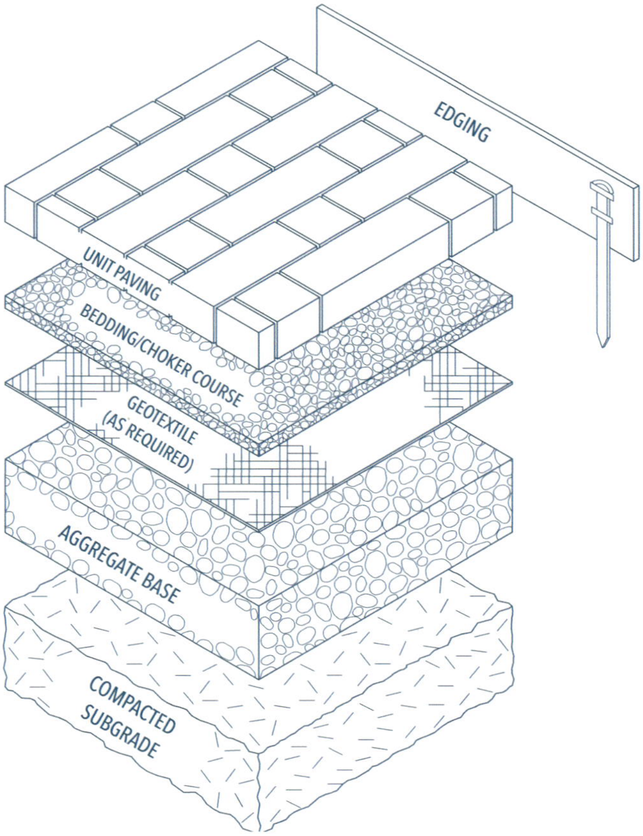

Figure 6.3 Flexible unit paving assembly with edge. Underdrain may be required at pavement edge. Flexible pavements are can be repaired and releveled for potentially long service lives. Image credit: Kathleen Elliott and Meg Calkins

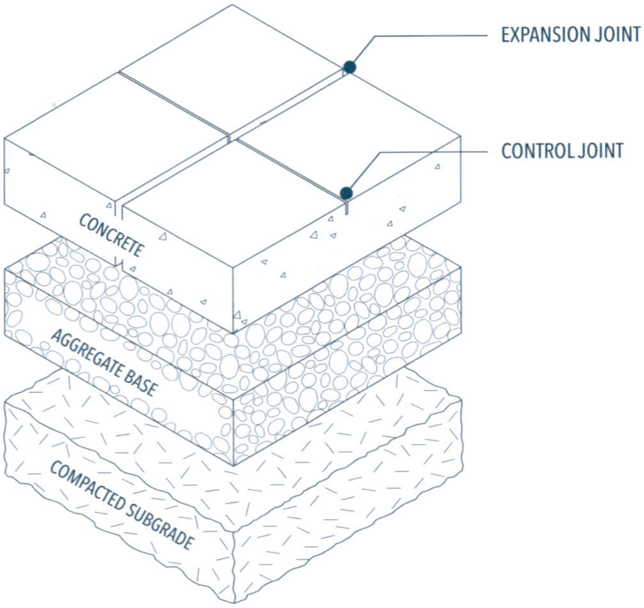

Figure 6.4 Rigid monolithic paving assembly. Underdrain may be required at pavement edge. Rigid pavements are durable, but if cracked through the pavement section, may require premature full replacement. Image credit: Kathleen Elliott and Meg Calkins

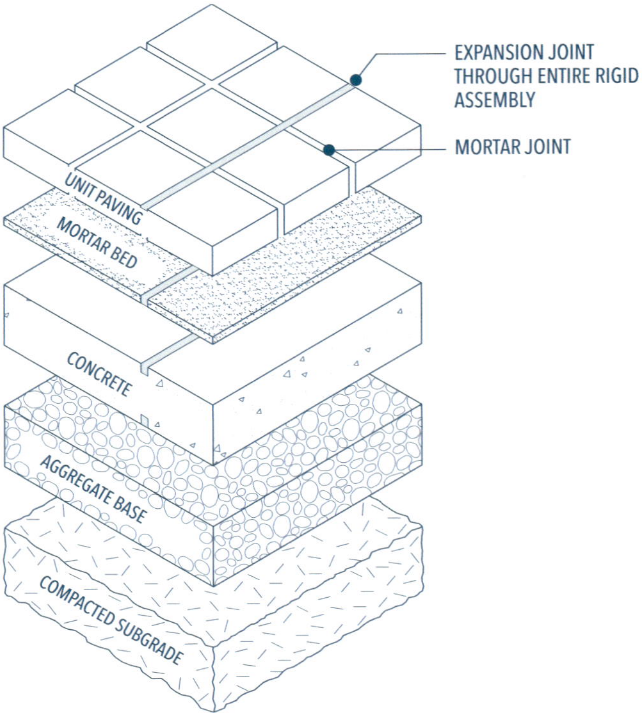

Figure 6.5 Rigid unit paving assembly. Underdrain may be required at pavement edge. Rigid bases for unit pavements are relatively high embodied carbon. and use of mortar makes the pavers very difficult to recycle. Image credit: Kathleen Elliott and Meg Calkins

Figure 6.6 Flexible unit paving (stone) on an aggregate base with a concrete edge. Image credit: Meg Calkins

Figure 6.7 Flexible monolithic pavement at Ballast Point Park in Sydney is stabilized stone dust. Image credit: McGregor Coxall

Rigid monolithic pavements, primarily poured in place (PIP) concrete, perform differently than flexible pavements from a structural standpoint as the loads are distributed internally within the pavement layer. The aggregate base for rigid pavements can be thinner than for flexible pavements as it is primarily used for to insure subgrade uniformity (Figure 6.4).

Rigid monolithic pavements may be initially more expensive than flexible pavements, however, flexible pavements may incur higher maintenance costs over their life. Rigid pavements are suitable for steeper slopes, pavements that will be subjected to higher traffic and intensity of use, and pavements that must be impenetrable to chemicals.

Rigid unit pavements are composite pavements of precast concrete pavers, stone, or brick units on a rigid concrete slab. The paving units can be set on a mortar bed with mortar between the units to form a rigid surface on a rigid slab. This type of pavement is not recommended for cold climates as the mortar joints are vulnerable to cycles of freezing and thawing and deicing materials. The resulting cracks between the mortar and units can let water into the pavement structure (Figure 6.4).

Setting units on an asphaltic binder or a sand bed on a concrete base is a more durable and repairable surface for rigid unit pavement. This type of construction will require

Figure 6.8 Rigid stone unit pavement set on a sand leveling bed on a concrete slab with weep holes for drainage is a rigid pavement structure with a flexible surface that could be deconstructed for reuse. Image credit: Meg Calkins

Figure 6.9 Exposed aggregate concrete bands alternating with brushed finish gray concrete is a rigid monolithic pavement. The change in finishes articulates the pavement pattern without an extra layer of unit pavement. Image credit: Meg Calkins

Table 6.3 Edges for Flexible Pavements and Their Embodied Carbon

Edging Material	Considerations for Use	Embodied Carbon (EC) for 1m and 100m of Edging ($kgCO_2e$ and tCO_2e)[1]
Steel edging is made from steel sheet or plate typically ranging from 14 gauge to 6mm thickness and 102–204mm depths. Steel edges are secured with steel stakes driven into the subgrade, or steel angles are bolted to a concrete base below finished grade.	Very durable if coated, galvanized, weathering steel (WS), or stainless steel (SS). Wide variety of colors and finishes. More expensive than plastic, aluminum, or wood edging but also more durable. Thin section can minimize the edge's presence. Tight curves can be challenging to form in the field.	EC for Type 1: 127mm tall, 5 gauge weathering steel edging, 40cm deep stakes every 50cm **7.4kgCO_2e/meter** **0.74tCO_2e/100 meters** EC for Type 2: 100x100mm WS angle bolted to 150mmx200mm 20.7MPa concrete base with 20% SCMs **16.8kgCO_2e/meter** **1.68tCO_2e/100 meters**
Aluminum edging products are bent or formed aluminum sheet typically ranging from 3 to 6mm in thicknesses and 102–204mm depths secured with aluminum stakes. Available in mill finish, anodized, or powder coated aluminum. Appropriate for residential and light commercial applications.	Does not corrode in most ground contact settings. Lightweight, but top edge can be crushed or bent by vehicles or equipment. Less resistant to material and mechanical wear. Curves are easier to form than with steel edging. Less expensive than steel edging but also less durable.	EC for 127mm tall aluminum edge with 4.76mm thickness at top edge, 30cm deep stakes every 75cm. A replacement factor of two was used as one full replacement is likely during the 30 year service life of the pavement. **30.4kgCO_2e/meter** **3.0tCO_2e/100 meters**

(*Continued*)

Table 6.3 (continued)

Edging Material	Considerations for Use	Embodied Carbon (EC) for 1m and 100m of Edging (kgCO_2e and tCO_2e)[1]
Wood edges are constructed from dimensional lumber or timbers that are placed flush with or raised slightly above the pavement surface. They are stabilized by wood stakes driven 30–60cm into the soil. Stake spacing depends on the form of the edge. Tight curves will require more closely spaced stakes. Wood edging is appropriate for residential and light commercial applications.	Wood edges should use wood products rated for ground contact (Use Class 4.1 or 4.2). Flush wood edges will be less vulnerable to maintenance equipment such as edgers and mowers. Inexpensive first cost, but replacements over the life of the paving will add cost. Curves can be challenging with 38mm thick lumber.	EC for 38x140mm wood edge with 19x38x300mm stakes every 75cm. A replacement factor of three was used as two replacements are likely during the 30 year service life of the pavement. **4.6kgCO_2e/meter** **0.46tCO_2e/100 meters**
Concrete edges or curbs can be flush with the pavement surface or raised to direct stormwater and vehicular movement. Their long thin shape makes them vulnerable to cracking, so control and expansion joints should be employed as required. Steel reinforcing may also be necessary.	Concrete edges are among the strongest, most durable edge types, but they may be the most expensive to construct. They offer a visual band to the edge of a pavement unlike steel or aluminum edges that are less visible. Curves can be achieved.	EC for a 150x300mm rectangular flush concrete edge with rebar. **15.1kgCO_2e/meter** **1.5tCO_2e/100 meters** With 50% SCMs and fiber reinforcing, EC is reduced by 38% to **9.4kgCO_2e/meter.**
Bricks can be embedded in or mortared to concrete to form an edge for flexible pavements. They can be placed upright in a soldier or sailor position and embedded in a concrete footer, or they can be laid in a header or rowlock and mortared to a concrete base.	Brick edges in concrete form a durable edge that offers a visual boundary to a pavement. Curves are limited to radii that work with the brick orientation and size. Relatively expensive type of edge treatment.	EC for brick Soldier course embedded in concrete footer. Footer is 250mm wide by 300mm deep. **31.6kgCO_2e/meter** **3.2tCO_2e/100 meters** If bricks are used in a Sailor position and 50% SCMs are used, the EC is reduced by 30% to **22.2kgCO_2e/meter.**
Plastic edging is made from plastics such as polyethylene (PE) or polyvinyl chloride (PVC) plastic. Appropriate for residential and very light commercial applications.	Curves are easy to form, and it is inexpensive. It can fade and become brittle over time with UV exposure and thermal expansion and contraction. It is less resistant to material and mechanical wear.	EC for 5mm thick 127mm tall plastic edging with (2) 30cm steel stakes per meter. A replacement factor of three was used as two replacements are likely during the 30 year service life of the pavement. **10.97kgCO_2e/meter** **1.1tCO_2e/100 meters**

[1] EC figures are cradle-to-gate and include the product stage life cycle phases A1: Material Extraction, A2: Transport to Manufacturing, A3: Manufacturing and Fabrication. They do not include transportation beyond the factory gate. These figures could be reduced in the future with production efficiencies, increased recycled contents, and renewable energy sources.

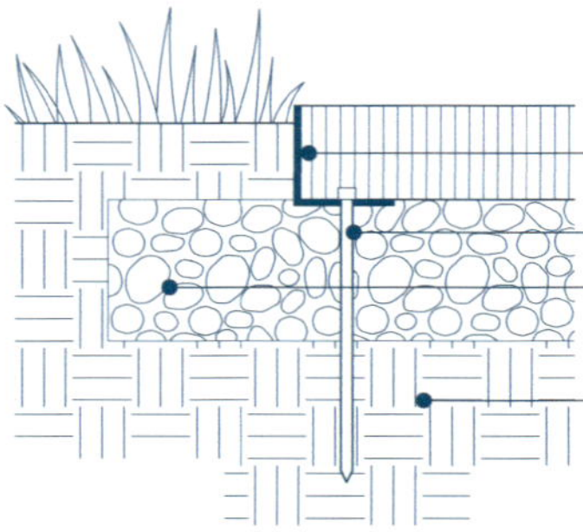

Figure 6.10 Steel edging can be stainless, weathering, or powder coated. It is a very durable type of edging. Image credit: Kathleen Elliott and Meg Calkins

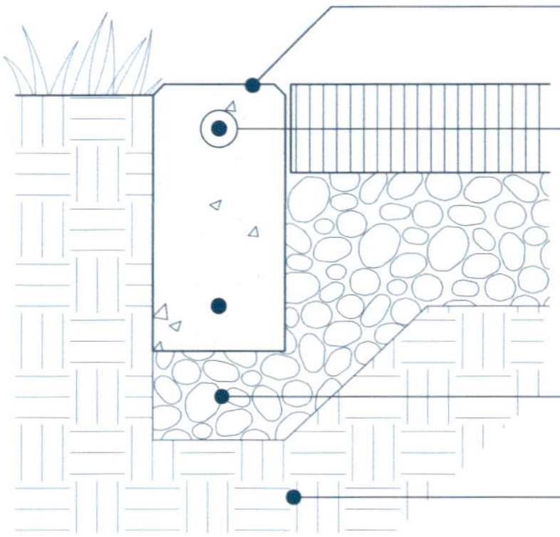

Figure 6.11 Poured in place concrete edges are more visible than steel edging. Image credit: Kathleen Elliott and Meg Calkins

a rigid edge as the unit pavers are not attached to the slab with mortar. It will also require periodic weep holes in the concrete slab to allow water that seeps between the paving joints to drain through (Figure 6.8).

Rigid pavements are more vulnerable to expansion and contraction of the materials than are flexible pavements, which are designed to move slightly. Both rigid monolithic pavements and rigid unit pavements require expansion

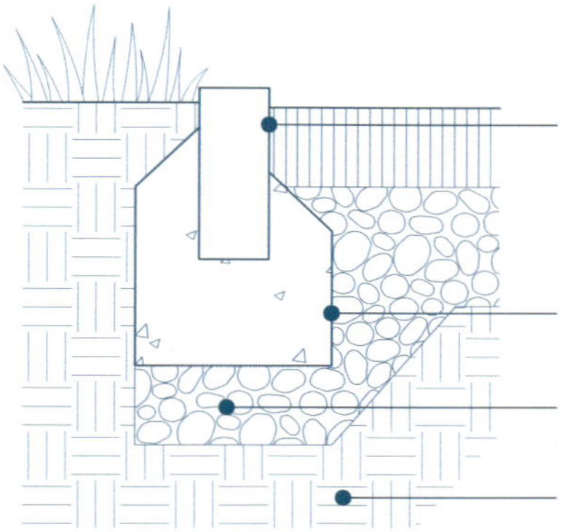

Figure 6.12 Brick edge embedded in a concrete footing can be relatively high embodied carbon but they are a durable edge. Image credit: Kathleen Elliott and Meg Calkins

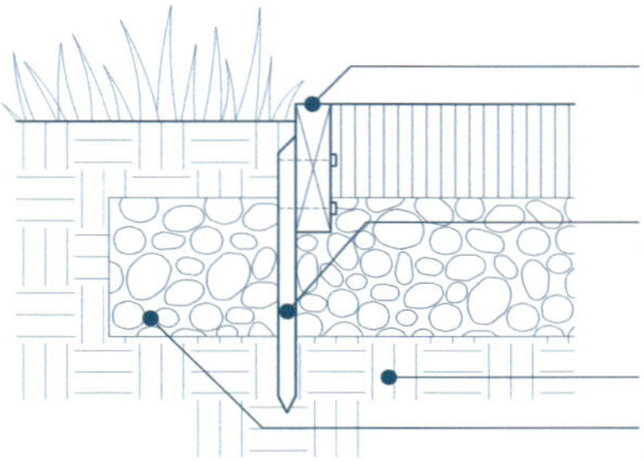

Figure 6.13 Wood edges are low embodied carbon but may require replacement in ten years. Even with multiple replacements, their cost and embodied carbon is relatively low. Image credit: Kathleen Elliott and Meg Calkins

joints typically every 5–8 meters or higher depending on slab reinforcement. Control joints can control cracking and should be placed every 3–4 meters spaced evenly. Where paving units are mortared, expansion joints in the unit paver surface must align with expansion joints in the concrete slab below (Figure 6.5).

Rigid pavements generally require less maintenance than flexible pavements. While they are designed for a use life of 30–35 years, if they crack or heave they are far more difficult to repair than flexible pavements. This could shorten their service life and increase the carbon footprint of the pavement.

Aggregate Pavements

Crushed stone and gravel pavements are less prevalent than asphalt and concrete pavements, yet they can offer a stable, lower impact surfacing option that is appropriate for some low traffic and pedestrian applications. Aggregate pavements are designated as "unbound" as opposed to concrete and asphalt pavements, but they are often stabilized with organic, mineral, or fossil fuel-based stabilizers. Aggregate pavements usually do not meet ADA accessibility requirements unless a stabilizer is used.

In the last century, there has been an emphasis on constructing rigid, low maintenance, and impervious surfaces for all pavement applications without consideration for traffic levels or environmental impacts. As priorities shift to low-carbon site design and green infrastructure, lower impact pavement materials may offer better performance in appropriate applications.

Aggregate pavements are generally less expensive, simple to construct, easily repaired and rejuvenated, and can use locally available and recycled materials. These lower carbon pavements do not require a concrete base and offer a less rigid surface for pedestrians and low traffic vehicular applications. Aggregate pavements are very appropriate for environmental conditions that can deform pavement, such as extremely frost prone areas, swelling subgrade soil, tree roots, and uncompacted subgrade (Table 6.4 and Figure 6.14).

Open graded aggregate pavements make well-draining porous pavements if properly maintained and the aggregates are angular. Porous aggregate pavements are discussed in greater detail in Chapter 7 "Porous pavement and green roofs". Most aggregate pavements will offer good surface reflectance to mitigate urban heat islands, but their specific reflectance is directly related to the color of the aggregate.

Gravel and crushed stone for aggregate pavements are low-carbon materials that can be obtained locally or regionally. Recycled or waste materials such as recycled concrete aggregates, waste oyster shells (WOS), or crushed bricks make good aggregate pavement materials. For greater discussion of the environmental considerations, types, gradations, and recycled materials for aggregates, refer to Chapter 12 "Aggregates and Sand Materials".

Drawbacks of aggregate pavements include the potential for loss/migration of gravel and rutting and lack of suitability for sloped surfaces and disabled access. Increased maintenance of aggregate pavements is a common reason that they are not more widely employed. Seasonal or annual maintenance is required for replacement or redistribution of aggregates, binders need to be rejuvenated every few years, and releveling activities can be required.

Table 6.4 Design and Specification Considerations for Aggregate Pavements

Component	Considerations
Aggregate types	Crushed, angular aggregates will provide a more stable surface as river gravel and pea gravel are too rounded to support loads without displacement.
Recycled aggregates	Consider use of locally available recycled aggregates such as crushed RCA, oyster or clam shells, or crushed bricks for aggregate pavements.
Aggregate sizes	Coarser grade aggregate pavements drain more easily, while finer grades (<14mm) are easier to walk on. Depth of aggregate surface courses should be 75–100mm for greatest interlock of angular materials.
Aggregate gradations	Dense graded aggregates can be compacted to form a stable, impermeable surface. Some dense graded aggregates are sold as "self-binding gravels" with up to 30% dust. Open graded aggregates can be porous and improve drainage through the pavement.
Accessibility	Finer grade stabilized crushed stone pavements can meet accessibility regulations. Refer to stabilizer product information for ADA accessibility performance.
Subgrade preparation	Subgrade does not always need to be compacted as unbound aggregate pavements can accommodate movement. A geotextile separator may be required between the aggregate and the subgrade.
Dust reduction	Open graded aggregates will reduce dust as will stabilizers. Refer to Table 6.5.
Maximum slopes	Max. slope on an unstabilized "self-binding" gravel pavement is 1:12. Refer to stabilizer product literature for max slopes.
Stabilization and dust control	Wide variety of stabilizers and dust control products available. Refer to Table 6.5. Specify stabilizers that are non-toxic, low VOC, and low carbon.
Containment	Aggregate pavements require edges to contain the particles (Figures 6.2 and 6.7). Intermediate containment such as cellular confinement systems can further stabilize aggregate surfaces.
Edge restraints	Aggregate pavements will require edge restraints to prevent lateral migration of the materials. Figures 6.10–6.13 illustrate some potential edges for flexible pavements. Edges should be a minimum of 10mm above the pavement surface to prevent migration of particles.
Construction	Placing the aggregate materials in lifts then compacting will shift particles into place to help resist rutting and displacement under traffic.
Maintenance	Maintenance activities include raking to re-distribute particles, weeding, replacement of aggregates, and releveling. Binder rejuvenation may be required every few years.

Sources: Van Dam et al. 2015; Maher et al. 2005

There are a wide variety of stabilizer products that can inhibit particle displacement and lend rigidity to unbound aggregate and soil pavements. Some stabilizers are biobased materials such as tree resins, plant lignans, or psyllium, while others are polymer or petroleum based. Aggregates can also be stabilized by cellular confinement systems. Table 6.5 details stabilizers for unbound aggregate pavements. Some of these stabilizers are used to stabilize subgrades as well.

Stabilized Stone Dust Pavements (10mm to Dust)

Comprised of small angular particles of 10mm or smaller, stabilized stone dust pavements are called by many names. Decomposed granite is the most common term, but they are also called crusher fines pavement, crushed granite pavement, chat pavement, and many regional terms (Figures 6.15 and 6.16). These pavements offer a low carbon, accessible alternative to concrete and asphalt pavements for pedestrian and low vehicle traffic, particularly if a stabilizer is used.

Figure 6.14 Locally sourced aggregate pavement at Yellowhorn Farm Park in Inner Mongolia by the China State Construction Engineering Corporation. Image credit: Wenzhong Gao

Table 6.5 Stabilizers and Dust Suppressants for Aggregate, Stone Dust, and Soil Pavements and Subgrades

Stabilizers and Dust Suppressants	Life Span, Traffic Volume, and Strength When Used in Surface Applications	Environmental and Health Considerations	Performance Considerations
Tree resin emulsions are derived from pine, fir, or spruce resins and combined with other additives *Dust suppressant* *Soil stabilizer* *Stone dust pavements* *Aggregate pavements* *Base stabilizer*	Very low to low traffic. 5–10 year life expectancy with maintenance applications every 2–3 years.	Tree resins are a by-product of the pulp and paper industry. Materials stabilized with tree resin emulsions can be recycled or returned to the soil.	Form a very hard surface that some manufacturers claim to be stronger than asphalt pavement. Avoid use on steep slopes. Work best on silty sands with fines contents up to 30% and plasticity index less than 8.
Lignosulfonates are derived from plant lignans. These stabilizers draw moisture from the air to suppress pavement dust and at higher application rates, stabilize pavements. *Dust suppressant* *Soil stabilizer* *Stone dust pavements* *Base stabilizer*	Very low to low traffic. 3–10 year life expectancy varies by product. Maintenance involves soaking and raking. New binder can be applied, soaked, and rolled every few years.	By-product of the pulp and paper industry. Self-healing with moisture, yet water soluble so can leach from pavements. In small amounts they are not toxic, but in large amounts they have been shown to reduce the biochemical oxygen demand (BOD) of water bodies.	Best for soils with 8–30% fines and a plasticity index below 8. Can increase the dry strength of soils by a factor of 2 or 3. Max slopes 20%. Chipseal and other asphalt surfaces can be placed on top of lignosulfonate stabilized soils.
Synthetic polymer emulsions are comprised of acetate or acrylic polymers. They cause a chemical bond to form between particles resulting in a dense, water-resistant surface. *Dust suppressant* *Soil stabilizer* *Stone dust pavements* *Aggregate pavements*	Very low to low traffic. 5–10 year life expectancy with maintenance applications every 2–3 years.	Products are relatively high embodied carbon. Some are by-products of the paint and adhesive industries.	Applicable for a wide variety of aggregates and soils. Work best for silty sand materials with fines content between 5 and 20% and plasticity index below 8.
Organic petroleum emulsions (also called emulsified asphalt, cold mix asphalt, cutback asphalt, and bituminous binder) bind soil or aggregate particles together with an asphalt adhesive. *Dust suppressant* *Soil stabilizer* *Aggregate pavements* *Base stabilizer*	Very low to low traffic. 5–9 years life expectancy, varies by product. Periodic spraying of emulsion every 2–3 years.	Cutback asphalt emulsions contain hydrocarbons and solvents with VOCs that can be harmful to installers and the surrounding environment. Reclaimed treated soil can be used as construction material, but not for planting soils.	Performance can vary widely by product. Work on a variety of aggregates and soil types with up to 30% clay fines and a plasticity index of less than 10. Can be sprayed on or mixed into soils or aggregates.
Cementitious materials such as portland cement (PC), hydraulic lime, fly ash, GGBFS, ground limestone, and calcined clay are used to stabilize soil primarily for cement treated pavement bases (CTB), but they can be used in low traffic surface applications. *Soil stabilizer* *Stone dust pavements* *Base stabilizer*	Very low traffic for surface applications. High traffic for pavement base applications, 20–45 year life in base applications.	Can be mixed in place using local soils, reducing carbon footprint of material transport. PC is a high embodied carbon material and is used in percentages of 3–10% to stabilize soil. Reduced amounts can be used with soil testing, and/or other cementitious materials such as fly ash or hydraulic lime.	As a surface material, soil cement can become brittle and crack under heavy loads. Very effective subgrade stabilizer but renders the soil unrecyclable for planting soils.
Cellular confinement of aggregates, also called geocells, are polymer-based honeycomb-like systems that evenly distribute the loads of granular materials. *Stone dust pavements* *Aggregate pavements*	Very low to high traffic volumes, varies by product. 15–20 year life expectancy, surface raking every six months, additional aggregate every 1–2 years.	Can allow for the use of locally available marginal aggregate materials. Can be permeable with appropriate aggregate gradations. Higher EC than some stabilizers.	Can reduce aggregate pavement and base thicknesses. Usually covered with surface of loose aggregate to hide the cells.

(*Continued*)

Table 6.5 (continued)

Stabilizers and Dust Suppressants	Life Span, Traffic Volume, and Strength When Used in Surface Applications	Environmental and Health Considerations	Performance Considerations
Chlorides, liquid or solid compounds, contain chloride salts. They draw moisture from the air to keep the pavement surface relatively moist. They also facilitate compaction. *Dust suppressant* *Soil stabilizer*	3–12 month life expectancy – varies with climate. Low traffic applications. Does not add to pavement strength.	Can cause corrosion damage to vehicles and skin and eye irritation during construction. Can impact adjacent vegetation and water quality. Use buffer zone to protect aquatic species. Relatively impermeable surface.	Proven stabilizer and dust suppressant used for decades. Not effective in very arid or very humid climates. Work best on engineered aggregate surfaces rather than native or uncontrolled surfaces.
Enzymatic emulsions contain enzymes from natural materials or agricultural by-products that react with soil molecules to stabilize the soil's structure and form a cementing bond. *Dust suppressant* *Soil stabilizer*	Very low to low traffic. 5–7 years life with periodic grading may be required every year. Life expectancy varies by product.	Relatively impermeable surface. No leachate concerns. Should not impact water quality. Once diluted, typically biodegradable.	Best for soils with greater than 10% clay content and plasticity indexes greater than 8. Increase CBR from 30–300%. Grade pavement to drain well with no standing water.
Electrolyte emulsions affect the electrochemical bonding characteristics of soils. They replace water molecules within the soil structure and bond particles together. *Dust suppressant* *Soil stabilizer*	Very low to low traffic. 3–5 years life with periodic grading every 6–12 months, expectancy varies by product.	Highly acidic. Check product SDS for health risks, employ proper worker safety precautions. Can impact water quality. Relatively impermeable surface.	Work on soils with greater than 10% clay content and plasticity indexes greater than 10. Increase CBR from 30–50%. Grade pavement to drain well with no standing water.

Sources: Van Dam et al. 2015; Maher et al. 2005

Figure 6.15 Ten Eyck Landscape Architects used stabilized stone dust pavement at The Capri Lounge and Event Space in Marfa, Texas. Image credit: Christy Ten Eyck

Table 6.6 offers a typical aggregate gradation for stabilized stone dust pavements, but these gradations will vary by region, stone type, and stabilizer. Table 6.7 discusses design and specification considerations, and Table 6.5 discusses stabilizers for fine crushed stone pavements.

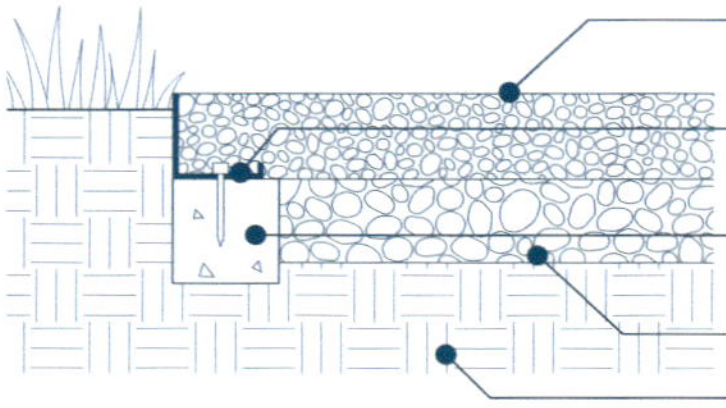

Figure 6.16 Stone dust pavement with edge. Image credit: Kathleen Elliott and Meg Calkins

Stabilized Soil Pavements

Stabilized soil pavements are a variation of stone dust pavements using natural soil. They are often mixed with soil stabilizers used to stabilize subgrade (Table 6.5) but are used in a surface course of low traffic pedestrian pathways. Stabilized soil pavements can be mixed in place with native soils. The stabilizer is mixed in or sprayed on, then mixed with water. It is then rolled and compacted in place. Many of the considerations for stabilized stone dust pavements are applicable to stabilized soil pavements (Table 6.7).

Soil cement pavements are a type of stabilized soil pavement. They are most commonly used as a pavement subgrade stabilizer, but soil cement can also be a wearing course for pedestrian applications. Soil cement incorporates between 3% and 10% cement by weight. It is often mixed in place then rolled for compaction. If soil cement is used, low percentages of portland cement (PC) should be used and cement substitutes should be explored given PC's high embodied carbon. Well-graded sandy and gravely soils will use the least amount of portland cement, and silty and clayey soils will require more cement (Maher et al. 2005).

Table 6.6 Recommended Aggregate Gradations for Stabilized Stone Dust Pavements (10mm to Dust)

Particle Size	Sieve Size (mm)	Percent Passing a 9.25mm Sieve
9.25mm (3/8″)		100%
#4	4.75	90–100%
#8	2.36	55–80%
#16	1.18	40–70%
#30	0.6	25–50%
#200	0.075	6–15%

Source: Bachensky 2024

Concrete Pavements

Concrete pavements are rigid, monolithic pavement types that are strong, durable, and can have long service lives. They can be colored, patterned, and textured for a wide variety of appearances, and they require minimal maintenance. Concrete pavements can be a surface material, or they can be a base material for rigid unit pavements such as brick, stone, or precast concrete pavers.

Concrete pavements designed with current typical mixes are among the most carbon-intensive pavement types. With over 90% of the embodied carbon of concrete resulting from the production of portland cement, cement substitutes should

Table 6.7 Design and Specification Considerations for Stabilized Stone Dust Pavements (10mm to Dust)

Component	Considerations
Aggregate types	Crushed, angular aggregates from limestone, dolomites, and granites. Recycled concrete aggregates can work if properly graded.
Aggregate gradations	Some mixes will specify a 9.5mm (3/8″) minus mix while others will use a 6.35mm (1/4″) minus mix. Refer to Table 6.6 for a typical gradation.
Mix ratio	Work with stabilizer product manufacturer on the mix ratio of stabilizer to aggregate to find the best mix for the location, stone, and application.
Accessibility	Most stabilized crushed gravel pavements are accessible. Refer to stabilizer product information for ADA accessibility performance.
Aggregate base	Crushed stone pavements should be supported on a dense graded aggregate base. Thickness of base will depend on design loads, stabilizer, and soil bearing capacity.
Containment	Edges for crushed stone pavements will contain the particles; however the edge should be at or just below the surface of the pavement so that water is not trapped. Edges can be as simple as grass or vegetation that forms a rougher edge. Consider alternative pavements at building entries so crushed stone is not tracked into buildings.
Drainage	Minimum cross slopes of 1.5–2%. Erosion can occur at path edges unless adjacent surfaces are drained away from pavement. Some products claim to be permeable at just under 1″ per hour at the beginning of a rain event, but heavy rainfall will runoff.
Maximum slopes	Max. slopes vary by product with some allowing slopes up to 20%. Slopes over 10% should use a runoff diverter, such as a concrete curb, railroad tie, or strip of unit paving, to direct concentrated drainage off of pavement. Refer to product literature for max slopes.
Construction	Contractor should discuss installation with the manufacturer's representative. Placing the crushed stone materials in 50mm lifts with compaction will ensure distribution of the stabilizer. The crushed stone mix should be pre-hydrated to a moisture content comparable with ambient conditions (commonly around 14%). Mixture should be turned over as it is wetted.
Maintenance	Make owners and management teams aware of maintenance requirements for crushed stone pavements. Crushed stone should be stockpiled to match installed material for replacement and refurbishment activities. Clay court brooms and brushes can be used to smooth pavements and redistribute particles. Adjacent hard surfaces should be periodically swept. Snowplows should be lifted 12mm above pavement surface.

Sources: Budzinski and Girard 2016; Organic Lock 2024

Table 6.8 Design and Specification Considerations for Stabilized Soil Pavements

Component	Considerations
Soil types	Well-graded sandy and gravely soils will produce stronger soil pavements but may be more easily displaced than clayey soils. Soil testing should be performed and discussed with stabilizer product suppliers.
Accessibility	Soil cement can be accessible.
Grading	Adjacent surfaces should drain away from stabilized soil pavements. Stabilized soil pavements should have minimal cross slopes to prevent erosion.
Design	Thicknesses will vary by soil, stabilizer, subgrade, and expected traffic.
Subgrade preparation	Subgrade should be prepared as required.
Dust reduction	A small amount of moisture should be maintained in stabilized soil pavements. Some dust suppressants will draw moisture from the air.
Edge restraints	Edge restraints are only required when edges are exposed above surrounding grade.
Construction	Can be mixed, compacted, and rolled in place.
Maintenance	Maintenance activities include raking to re-distribute surface particles, weeding, replacement of aggregates, binder rejuvenation every few years, and patching and releveling.

Sources: Van Dam et al. 2015; Maher et al. 2005; Ferguson 2005

Table 6.9 Rigid Pavement Design Loads and Typical Dimensions[1]

Design Loads	Pavement Slab	Aggregate Base
Light Duty	100mm (4")	100mm (4")
Medium Duty	125–150mm (5–6")	125–150mm (5–6")
Heavy Duty	150–200mm (6–8")	150–200mm (6–8")

Source: Harris and Dines 1997

[1] Slab thicknesses are shown for reference only. Geotechnical or structural engineers should determine sizing of concrete pavement assemblies. Actual thicknesses will vary by traffic loading, modulus of subgrade reaction, modulus of rupture for the concrete, and load transfer between joints.

be employed to the extent possible (Table 14.2). Other steps to reduce the embodied carbon of concrete structures include material optimization, carbon sequestration in concrete mixes, recycled aggregates, and many other new technologies and strategies. Detailed discussion of these strategies is provided in Chapter 14 "Concrete Materials and Products".

Most concrete pavements are impervious and will result in concentrated stormwater runoff. Concrete can be made porous by eliminating fine aggregates in the mix and placing the pavement on an open graded aggregate base. Porous concrete is discussed in Chapter 7 with design and specifications considerations detailed in Table 7.6.

Design for durability of concrete pavements is a key consideration to maximize their service life (Tables 6.9 and 6.10). If they are designed with an appropriate mix for the design loads, climate, and subgrade conditions, and placed well, they can be expected to last for 30–40 years with minimal maintenance. However, given their rigid structure, they are vulnerable to cracking from freeze/thaw cycles, subgrade swelling, and tree roots. Once cracked, they are difficult to repair and are sometimes replaced, partially or completely, prior to their full design life (Table 6.11).

Structural Performance of Concrete Pavements

From a structural standpoint, rigid concrete pavements perform differently than flexible pavements. Loads are distributed internally within the concrete slab therefore the aggregate base under the slab can be thinner as it does not serve critical support function. It is a leveling layer between the subgrade and the paving slab and serves to drain soil moisture away from the slab (Figure 6.17 and Table 6.10).

Typical thicknesses of concrete slabs by intensity of use are shown for reference in Table 6.9, but actual thicknesses should be determined by traffic loading, modulus of subgrade reaction, modulus of rupture for the concrete, and load transfer between joints. Typical compressive strengths of concrete pavements range from 17.2 to 50MPa (2500–7250 psi). Required cement quantities increase substantially with compressive strength, therefore the lowest compressive strength for an expected design load should be specified. For instance, the embodied carbon of 20.7 MPa concrete is just 81% of the EC of 27.6 MPa concrete. In some cases, there will be different compressive strengths specified for different pavements on a site.

Joints and reinforcing in concrete pavements are critical to their durability. Concrete has high compressive strength, but the tensile strength is only 8–12% of the compressive strength. Therefore, concrete is vulnerable to cracking from tensile forces. Reinforcing within the slab and joints are the two methods for controlling cracking and accommodating thermal expansion and contraction in concrete pavements.

Table 6.10 Considerations for Design and Specification of Durable Concrete Pavements

Aspect	Considerations
Service life and durability	Lowering the water/cement ratio through admixture use. A 0.40 to 0.45 water/cement ratio can reduce permeability and increase the strength of a concrete pavement. Use of cement substitutes such as fly ash and blast furnace slag can reduce permeability and mitigate ASR and sulfate attack. Air void systems comprised of closely spaced, spherical microscopic sized bubbles can relieve pressure generated as water freezes in concrete pores.
Concrete efficiency measures	Use minimum required thicknesses for concrete slabs knowing that a margin of error is already built into the engineer's calculations. Reduce the extent of concrete pavement. Consider alternative pavements that are lower carbon such as decomposed granite or asphalt pavements with high reflectance surfaces. Reduce or reuse concrete waste.
Cement	Use cement substitutes and cement alternatives in place of part of all portland cement in a mix (refer to Chapter 14). Match the design strength with actual performance needs. Do not over-design concrete structures.
Other carbon reduction measures	Sequester carbon in concrete mixes (refer to Chapter 14). Use cement produced with alternative energy or energy efficient processes.
Cost	Concrete pavements are generally more expensive than asphalt or other flexible pavements, but their life-cycle cost may be lower as flexible pavements will require greater maintenance and refurbishment over their entire life cycle.
Surface finishes	Surface finishes for concrete pavements include textures such as brush or broom finishes; colors such as concrete stains, integral colors and thin tinted cementitious surface overlays; or seeded or exposed aggregates. Concrete stains and applied surfaces can extend the life of concrete pavements (Table 6.11).
Control joints	Control joints are ¼ of the depth of a slab and formed in roughly square shapes not more than 3–4 meters apart. To minimize random cracking, joints should be cut/formed prior to the stress of the concrete exceeding the strength.
Expansion joints	Expansion joints should be placed in concrete slabs where the slab meets another rigid structure such as a building or wall and in large pavement expanses, they should be placed a maximum of 8–10 meters apart.
Drainage	Consider use of porous concrete for low volume roads and other pavements. Break up large expanses of concrete pavement with biological stormwater control measures.
Aggregates	Aggregates should be both physically and chemically stable. Where alkali silica reaction is a risk, reactive aggregates should be avoided. Recycled aggregates can be used for both coarse and fine aggregates in concrete mixes. Refer to Chapter 12 for more information.
Admixtures	Admixtures are chemical or mineral constituents that impart a wide variety of properties to uncured and cured concrete. They can allow for higher percentages of cement substitutes, control curing rates, and increase the performance of concrete. Some may pose risks to worker and/or ecosystem health around concrete pavements. Request information and safety data sheets from product manufacturers.
Heat island performance	Standard gray portland cement concrete will offer enough surface reflectance to minimize the urban heat island effect and achieve applicable LEED and SITES credits. Concrete pavement with white portland cement can cause excessive glare and may stain easily reducing the pavement's service life.
Maintenance	Concrete pavements require minimal preservative maintenance beyond periodic expansion joint resealing every 5–10 years. Crack sealing may be required.
Reuse and recycling	After its useful life, concrete pavement can be excavated and slabs reused for new pedestrian pavement applications (Figure 6.25). It can be crushed for use in concrete, asphalt or an unbound base material. It can also be rubblized in place and covered with an overlay.

Sources: Van Dam et al. 2015; Maher et al. 2005

Reinforcing in Concrete Pavements

Steel reinforcing counteracts small and large tensile forces in a concrete slab. Steel reinforcing bars (called rebar) are used in concrete pavements with larger design loads, such as roadways and highways, while welded wire mesh (WWM) is used to control surface cracking. It is important to note that WWM will not completely control cracking, but it can keep hairline cracks from growing and spreading. All steel reinforcing should be held a minimum of 50mm clear from all concrete surfaces including the top and bottom of the slab.

Fiber reinforcing is increasingly employed to control surface cracking, add durability, and lend additional strength to concrete slabs. Plastic or steel fibers are added to fresh concrete during the batching and mixing process. Fiber reinforcing can eliminate all or some of the metal reinforcing, making slabs easier to replace and repair

Table 6.11 Rehabilitation Methods for Concrete Pavements With Asphalt Materials[1]

Type	Considerations
Thin overlays	Thin overlays (38mm or less) are hot-mix asphalt (HMA) or warm-mix asphalt (WMA) surfaces that are placed on a well-prepared concrete surface as part of a pavement preservation or management program. They are well suited to existing pavements with low to medium levels of surface distress and can offer 6–10 years of service on an existing concrete pavement. They can also offer ten or more additional years of service on an asphalt pavement. Thin overlays may be employed earlier in the deterioration cycle of a pavement to avoid more severe distresses that might result in the need for a thicker structural overlay.
Crack, seat, and overlay	Crack, seat, and overlay is a technique used to restore deteriorated concrete pavement. The concrete pavement is cracked into smaller slabs then overlaid with asphalt pavement. Smaller 120–150cm concrete slabs are less likely to cause reflective cracking in the asphalt pavement overlay. This method will require multiple asphalt pavement layers and should only be used where there is good subgrade support, good existing drainage, and unreinforced concrete.
Rubblization	Deteriorated concrete pavement can be fractured into rubble then overlaid with asphalt pavement. The rubble that is left in place functions as a high-quality base for the asphalt pavement overlay. Rubblization is best used when the existing subgrade provides adequate strength, support, and drainage.

[1] Additional surfaces for asphalt pavements are discussed in Chapter 15 "Asphalt Pavement Materials".

Sources: Muench and Hand 2019; VanDam et al. 2015

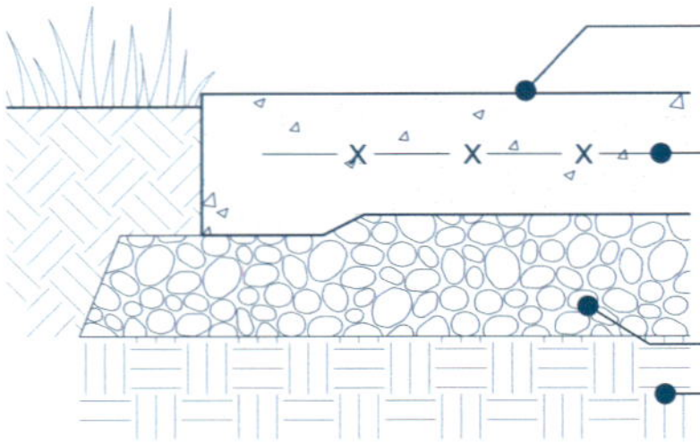

Figure 6.17 Concrete pavement detail with thickened edge. Thickening the concrete pavement slab at the edge adds rigidity where it can be vulnerable to movement, increased forces and moisture. Image credit: Kathleen Elliott and Meg Calkins

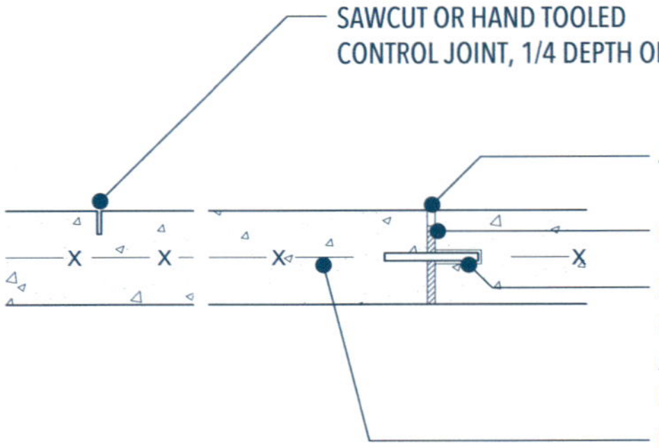

Figure 6.18 Control and expansion joints for rigid pavement. Image credit: Kathleen Elliott and Meg Calkins

and, without rebar, it is easier to reuse and recycle. Fiber reinforcing is often used in precast concrete (Solano 2024).

Other types of fiber reinforcing include cellulose fibers, glass fibers, natural fibers, and composite synthetic fibers. Some recycled plastics and agricultural by-products can be used in fiber reinforcing applications.

Joints in Concrete Pavement

Joints are required in rigid concrete pavements to control cracking, handle expansion and contraction of the material, and counteract minor subgrade movement (Figures 6.18 and 6.19).

Control joints, also called dummy, or fake joints, encourage cracking within the joint to avoid random cracking in the pavement surface. The layout of control joints should be spaced evenly and are most successful when forming roughly square shapes at distances not greater than 25 times the thickness of the slab with maximum spacing at four meters.

Control joints are either hand tooled into the surface prior to the concrete cure, or they are sawcut as soon as the concrete has cured enough to withstand the impact of sawing. The typical depth of control joints is approximately one-fourth of the slab thickness.

Expansion joints, also called isolation joints or spatial joints, are full depth joints employed to relieve flexural stresses of slabs and allow for thermal expansion and contraction of the concrete. Expansion joints are made of asphaltic board, cellulose fiberboard, or closed cell foam and need to be sealed at the concrete surface with a flexible caulk. Expansion joints should be placed in concrete slabs where the slab meets another rigid structure such as a building or wall and in large pavement expanses, they should be placed a maximum of 8–10 meters apart. In some pavements, dowels span across expansion joints to keep the slabs level. Dowels are usually coated

Figure 6.19 Saw cut control joints and an expansion joint with sealant. Image credit: Meg Calkins

rebar or stainless steel. One end of the rebar is either greased or sleeved to allow expansion and contraction (Figure 6.18).

Construction joints allow for large expanses of pavement to be poured and cured in multiple phases. They also allow for two or more different integral colors of concrete to be used. They are butt joints and are not filled with any material.

Synthetic Binder/Resin Bound Pavements

Resin bound pavements, also called synthetic binder, bonded, or synthetic binder concrete pavements, can meet or exceed the performance of asphalt pavements. These types of pavements are a relatively new approach, and products vary in performance and constituents (Pasetto et al. 2023). Resin bound pavements can rest directly on an aggregate base, or they can be a surface layer for an asphalt or concrete pavement.

Resin bound pavements are a type of flexible pavement unless they are used as a surface treatment on concrete.

Table 6.12 Design and Specification Considerations for Resin Bound Pavements

Component	Considerations
Thickness	Minimum thickness should be two to three times the nominal maximum aggregate size. Use as a 6–25mm surface course on asphalt pavement as the embodied carbon of synthetic binders is relatively high.
Appearance	Binders are clear or amber colored. Resulting pavement color is based on color of aggregates. Some can be tinted.
Surface reflectance	As most binders are clear or amber colored, they can offer a greater surface reflectance than standard asphalt pavement. Their surface reflectance is based on the color of the aggregates.
Porosity	Some products can be used for porous pavements.
Cost	Estimated to be 2–4 times the cost of HMA.
Construction	Synthetic binders will require either separate machinery at mixing plants and trucks, or the machinery will need to be cleaned to maintain the binder clarity and avoid black asphalt binder contamination.
Maintenance	Crack sealing and localized patching every 7–10 years. Color matching of patches can be challenging. Surface overlays can extend the life of the pavement application.
Recyclability	Can be crushed and used as an unbound base material, although strength will vary.

Sources: EVIzero 2024; Resin Bound 2024; De Pascale 2023; Pasetto et al. 2023; Van Dam et al. 2015

As compared with asphalt pavement, they have lower susceptibility to temperature extremes, higher fatigue resistance, and reduced rutting potential (Pasetto et al. 2023). They can also have higher surface reflectivity than asphalt pavement as the binder is somewhat transparent.

Some are heated then placed at temperatures ranging from 120 to 140°C, while others are cold mixed and placed. These pavements are a blend of coarse and fine aggregates, mineral filler, and synthetic binders. Synthetic binders are produced from polymers, resins, and oils. Recent products are made from polyolefins, a family of thermoplastics including polyethylene and polypropylene. Some incorporate recycled plastics and can be used in porous applications (De Pascale 2023) (Table 6.12 and Figures 6.20–6.22).

The embodied carbon of pavements stabilized with synthetic binders is typically higher than that of asphalt and even concrete, so synthetically bound pavements should be used as a thin surface course (6mm to 25mm thick) on an asphalt base. Safety data sheets and, where available, HPDs, should

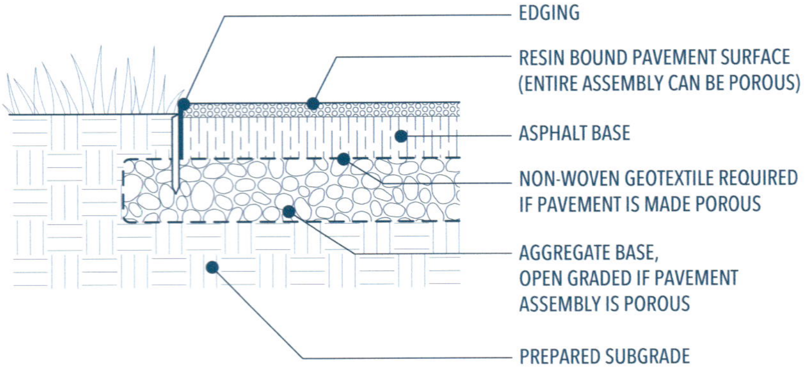

Figure 6.20 Resin bound surface course on an asphalt pavement base. This type of pavement can be made porous. Image credit: Kathleen Elliott and Meg Calkins

Figure 6.21 Tan resin bound pavement at the Serpentine North Gallery in London. Image credit: Meg Calkins

Figure 6.22 Resin bound pavement on stairs at Wolfgang Buttress' Hive at Kew Gardens. Image credit: Meg Calkins

be consulted for human and environmental health risks and VOC levels of synthetic binders.

Asphalt Pavement

Asphalt pavement or bituminous pavement, formally called asphaltic concrete pavement, is the most used type of pavement for vehicular applications. It is a flexible pavement that is relatively low cost and locally available almost anywhere. Asphalt pavement is easily repaired and resurfaced without the removal of the entire structure (Figures 6.23 and 6.24).

The primary drawbacks to asphalt pavements are its contribution to urban heat islands, the energy use and emissions from heating and placement, and its virgin resource use. Impervious asphalt pavement, like other dense pavements, concentrates runoff and contributes to non-point source pollution.

A wide variety of surface overlays can extend asphalt pavement's life cycle and improve it's aesthetic appearance and surface reflectivity. They are discussed in Table 6.11 and Tables 15.1 and 15.6. Asphalt pavement can be made porous to improve stormwater performance, and large quantities of recycled materials, primarily reclaimed asphalt pavement, can be incorporated into new asphalt mixes.

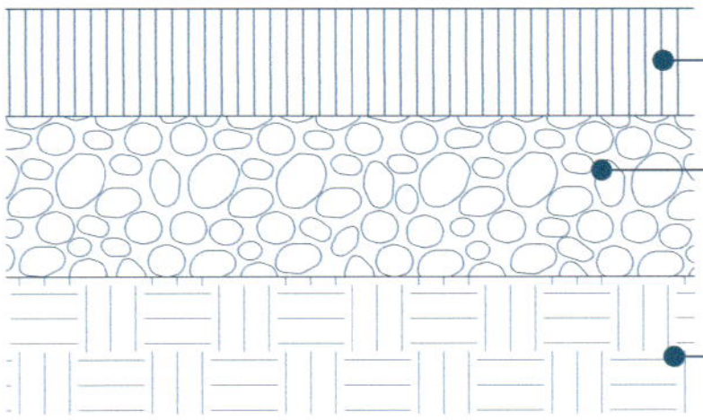

Figure 6.23 Light duty asphalt concrete pavement is used for pedestrian paths and very light vehicular traffic. High volumes of recycled asphalt pavement (RAP) and cooler mixes will reduce resource use and embodied carbon. Image credit: Kathleen Elliott and Meg Calkins

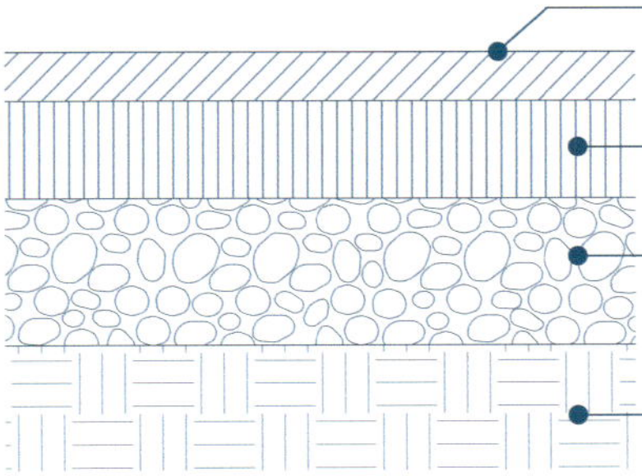

Figure 6.24 Medium and heavy duty asphalt concrete pavement with surface wearing course is used for vehicular traffic. The surface course can be repaired and replaced multiple times extending the service life of the pavement structure. Image credit: Kathleen Elliott and Meg Calkins

Asphalt pavement materials, surfaces, additives, and performance and sustainability considerations are discussed in greater detail in Chapter 15 "Asphalt Pavement Materials". An embodied carbon comparison of a wide variety of asphalt pavement mixes and surfaces is also discussed in Chapter 15.

Figure 6.25 Ten x Ten used large sawcut reclaimed concrete slabs for stairs from the loggia to the Rubble Garden at Mill 19 in Pittsburgh. The slabs were reclaimed on-site. Image credit: Maura Rockcastle

Table 6.13 Embodied Carbon (EC) Comparison of Pavement Assemblies

Pavement Assembly, Components, and Dimensions (Figures 6.26 and 6.27)	**(EC)[1] 1m² ($kgCO_2e$) 200m² (tCO_2e)**	**Alternatives to Reduce EC of Pavement Assembly**
Imported granite on concrete[2] (50mm thick) Sand setting bed (20mm thick) Concrete slab (150mm thick), 27.6MPa, 20% SCM WWM, 152x152mm Aggregate base (150mm thick)	*1m²:* 200$kgCO_2e$ *200m²:* 40.0tCO_2e	Domestic granite – EC reduced by 61% to 77$kgCO_2/m^2$ 20.7 MPA – EC reduced by 12% to 68$kgCO_2/m^2$ 50% SCMs and 20.7 MPA – total reduction of 71% to 57$kgCO_2/m^2$
Poured in place concrete Concrete slab (150mm thick), 27.6MPa, 20% SCM 152x152mm WWM Aggregate base (150mm thick)	*1m²:* 55$kgCO_2e$ *200m²:* 11.0tCO_2e	20.7 MPA – EC reduced by 16% to 46$kgCO_2/m^2$ 50% SCMs and 20.7 MPA – total reduction of 35% to 36$kgCO_2/m^2$

(*Continued*)

Table 6.13 (continued)

Pavement Assembly, Components, and Dimensions (Figures 6.26 and 6.27)	(EC)[1] 1m² ($kgCO_2e$) 200m² (tCO_2e)	Alternatives to Reduce EC of Pavement Assembly
Porous concrete pavement Concrete slab (150mm thick), 27.6MPa, 20% SCM PP fiber reinforcing, 0.9kg/m³ Open graded aggregate base (460mm thick) Filter fabric	*1m²:* 51$kgCO_2e$ *200m²:* 10.2tCO_2e	20.7 MPA – EC reduced by 16% to 43$kgCO_2/m^2$ 50% SCMs and 20.7 MPa – total reduction of 35% to 33$kgCO_2/m^2$
Precast permeable concrete pavers (75mm) Choker course (38mm) Open graded base reservoir (460mm) Filter fabric	*1m²:* 60$kgCO_2e$ *200m²:* 12.0tCO_2e	Thinner pavers (50mm) with 100% recycled aggregate content, EC reduced BY 38% to 37$kgCO_2/m^2$
HMA pavement (100mm thick), 0% RAP Aggregate base (200mm thick)	*1m²:* 21$kgCO_2e$ *200m²:* 4.0tCO_2e	WMA, 0% RAP – EC reduced by 5% to 20$kgCO_2/m^2$ 40% RAP WMA – total reduction of 33% to 14$kgCO_2/m^2$
Porous synthetic binder surface course (25mm thick) on Porous HMA pavement (75mm thick) Open graded aggregate base (460mm thick) Filter fabric	*1m²:* 28$kgCO_2e$ *200m²:* 5.6tCO_2e	40% RAP WMA – EC reduced by 18% to 23$kgCO_2/m^2$
Clay brick pavers[2] (57mm thick) Sand setting bed (25mm thick) Filter fabric HMA base (100mm) Aggregate base (150mm thick)	*1m²:* 51$kgCO_2e$ *200m²:* 10.2tCO_2e	40% RAP WMA base – EC reduced by 16% to 43$kgCO_2/m^2$
Stabilized stone dust pavement[2] (100m thick) with psyllium based stabilizer (3.46kg per m²) Aggregate base (200mm thick)	*1m²:* 5$kgCO_2e$ *200m²:* 1.0tCO_2e	Unstabilized crushed stone pavement 10mm minus – EC reduced by 56% to 2$kgCO_2/m^2$ (not recommended for most applications)
Imported flagstone on aggregate base[2] (50mm thick) Sand setting bed (25mm thick) Filter fabric Aggregate base (300mm thick)	*1m²:* 114$kgCO_2e$ *200m²:* 22.8tCO_2e	Domestic flagstone – EC reduced by 86% to 16$kgCO_2/m^2$

[1] EC figures are cradle-to-gate and include product stage life cycle phases A1: Material Extraction, A2: Transport to Manufacturing, A3: Processing and Fabrication. Figures do not include transportation beyond the factory gate. These figures could be reduced in the future with production efficiencies, increased recycled contents, and greater use of renewable energy sources.

[2] Does not include required edging material.

Unit Pavements

Unit pavement assemblies are comprised of precast concrete units, asphalt pavers, bricks, flagstone, dimension stone, or setts (Figures 6.28–6.33). They can be flexible pavements set on asphalt pavements or compacted aggregate bases, or they can be rigid assemblies that are sand, asphalt, or mortar set on concrete slabs (Figures 6.34–6.37). Refer to Table 6.14 for benefits, drawbacks, and potential performance improvements of unit pavement types and assemblies. Table 6.15 illustrates embodied carbon and tradeoffs to reduce it for various medium duty concrete unit pavement assemblies (Figure 6.38).

Flexible unit pavements are supported by an asphalt base or a compacted, graded aggregate base that distributes loads evenly over the subgrade. Smaller units such as brick, concrete pavers, or granite setts can adapt to slight deformations in pavement bases, whereas larger units are more vulnerable to base movement.

Units are typically laid on a 20–38mm sand setting bed that helps to create a level wearing surface and accounts for minor variations in the aggregate base and paver depths. Joint sand is swept over the pavers after installation.

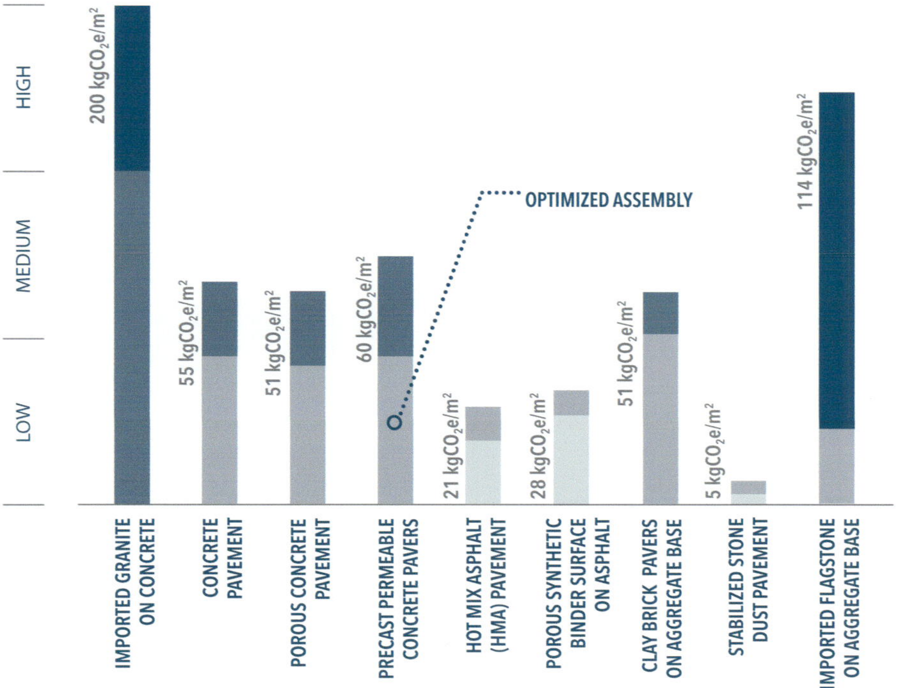

Figure 6.26 Embodied carbon for one square meter of pavement assemblies with a 60 year service life. The top part of each bar and the number shows EC results for original assemblies, and the lower part of the bar shows optimized assemblies. Image credit: Meg Calkins

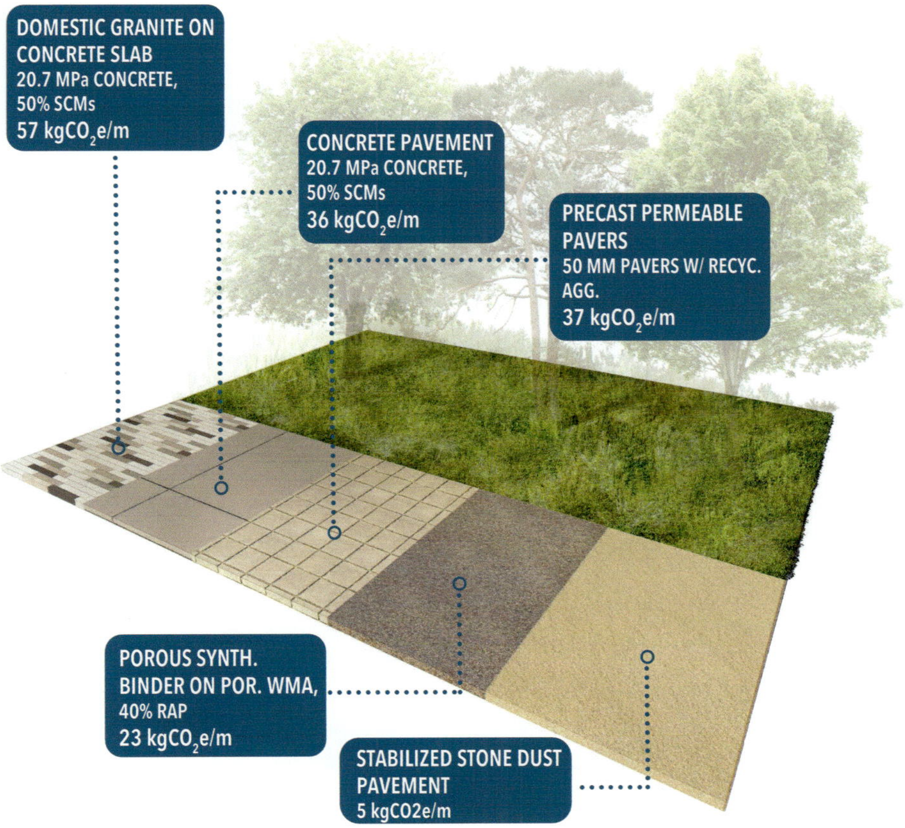

Figure 6.27 Embodied carbon of selected optimized pavement assemblies. The embodied carbon of granite pavement on a concrete base can be reduced by 71% by using domestic granite, 50% SCMs, and 20.7MPa concrete. Image credit: Lizandro Marcial-Armas, Kathleen Elliott, and Meg Calkins

Figure 6.28 Brick herringbone parking lot pavement. Tan bricks demarcate parking lot markings. Image credit: Meg Calkins

Figure 6.29 Asphalt unit pavers at the Miller Garden parking court by Dan Kiley. Image credit: Meg Calkins

Asphalt bases for flexible unit pavements will require weep holes every 1–2 meters to drain any water that seeps between the pavement joints. Weep holes can be lined with HDPE pipe set on end. This is not a permeable pavement as the joints will not admit large amounts of runoff. This assembly can be made porous with use of permeable paver products, larger joints, and a porous asphalt base.

Flexible unit pavements on compacted aggregate bases may require periodic releveling. Settling and distortion can happen within the first year, and after that, it may be required every 8–10 years. Addition or replacement of joint sand will be required every 2–3 years. Polymeric joint sand is a mixture of sand and a polymeric bonding agent that solidifies in the presence of water. It is better retained in joints but will not allow for infiltration.

Rigid unit pavements are supported by a concrete slab and are designed to resist all movement from forces on or below the pavement. Units are either attached to the concrete slab with a mortar setting bed and mortar between the units, or they are placed on a sand or bituminous bed with sand swept joints. In the second scenario, pavers are flexibly placed but supported by a rigid base. Weepholes are required every 1–2 meters in the concrete slab to drain water that seeps between pavers. This is not a permeable pavement as the joints will not admit large amounts of runoff.

Units that are mortared in place will rest on a thin mortar bed (10–20mm) with mortar in the joints between pavers. Expansion joints in the slab should extend through the entire pavement assembly.

Unit pavers used in applications with mortar are challenging to repair and while cracks can be repointed, the entire installation is often removed and replaced. Units with mortar attached are less likely to be reused due to increased labor costs to clean them.

Mortared unit pavements are not recommended in areas with freeze/thaw cycles as the mortar joints are vulnerable to moisture penetration and cracking. Flexible unit installations are far more easily repaired if frost heave occurs.

Figure 6.30 Granite cobbles with aggregate swept joints. If open graded aggregate is used and it is periodically maintained, it can be a pervious pavement. Image credit: Diana

Figure 6.32 Stone pavers in serpentine waves at the Copacabana promenade by Roberto Burle Marx. Image credit: Dan

Figure 6.31 Concrete unit pavers can be a relatively low embodied carbon assembly if set directly on an aggregate base. (Table 6.15). Image credit: Pixabay

Figure 6.33 Black locust pavers at Plitvice Lakes, Croatia. Image credit: Meg Calkins

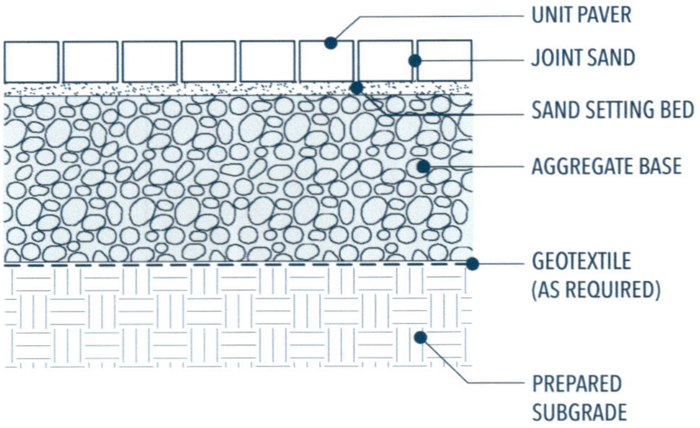

Figure 6.34 Unit pavers on an aggregate base. Image credit: Kathleen Elliott and Meg Calkins

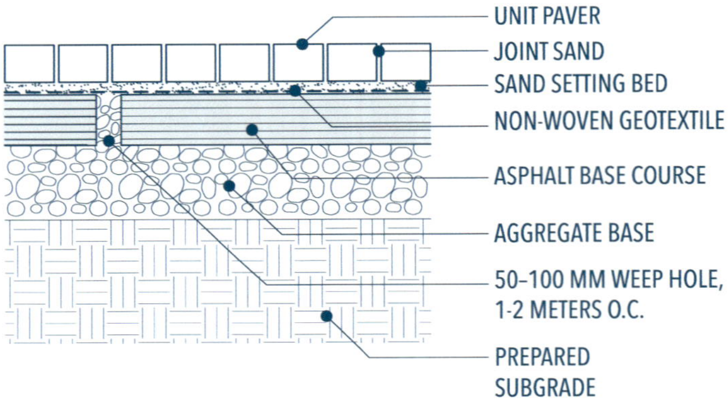

Figure 6.36 Unit pavers on a bituminous asphalt base with weep holes for drainage. Image credit: Kathleen Elliott and Meg Calkins

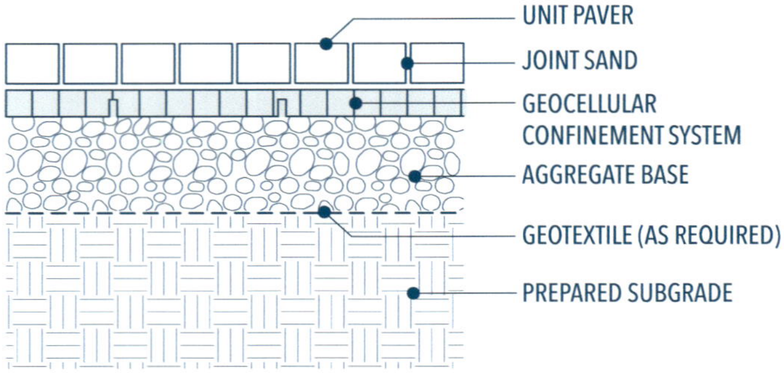

Figure 6.35 Unit pavers on a geocellular confinement system and aggregate base. Image credit: Kathleen Elliott and Meg Calkins

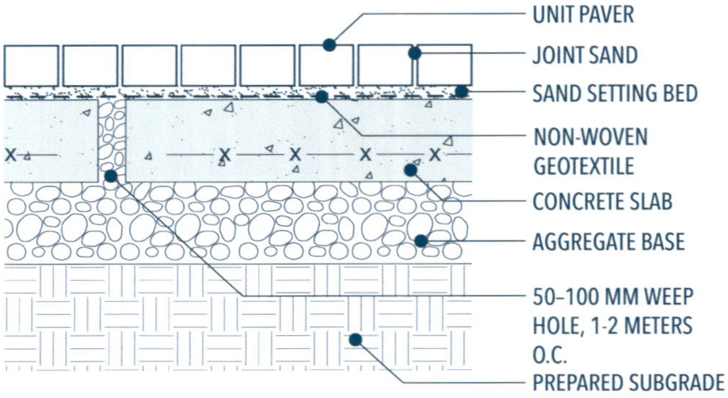

Figure 6.37 Unit pavers on a concrete base with weep holes for drainage. Image credit: Kathleen Elliott and Meg Calkins

Table 6.14 Unit Paving Types

Type	Specification Considerations	Environmental and Human Health Considerations
Brick pavers are kiln fired clay bricks designed for heavier loads than building bricks (Figure 6.28). They come in a very wide range of colors and can be arranged in many patterns including large curvilinear forms. Herringbone pattern is most stable, especially for vehicles. Clay is abundant and brick manufacture is widespread, however, brick pavers may be less locally available. Low speed applications.	Use appropriate brick paver types for weather, traffic, and strength requirements. Grades up to 20% for pathways, 10% for vehicular applications. Polymeric joint sand should be used on slopes over 10%. Releveling may be required when set on an aggregate base. Not a permeable pavement unless spacers are used between bricks and an open graded aggregate base is used.	Pavement service life expectancy is 20–25 years, but bricks can be reused again if not mortared. Low ride quality for fast moving cars so can be used as a traffic calming tool. Brick manufacture is carbon intensive. Refer to Chapter 17 for discussion of strategies to reduce the environmental and human health impacts of bricks.

(Continued)

Table 6.14 (continued)

Type	Specification Considerations	Environmental and Human Health Considerations
Asphalt unit pavers, also called asphalt block, are asphalt based pavers available in a wide variety of shapes and colors (Figure 6.29). Thicknesses range from 32mm for pedestrian applications to 76mm for vehicular applications.	Pavers make a comfortable walking surface as they are just slightly softer than concrete pavers. Absorption rate is less than 1%. Asphalt pavers have much higher resistance to indentations in hot weather than standard asphalt pavement.	Some contain recycled contents up to 87%. Types of recycled content incorporated are asphalt road grindings, fly ash, clam shells, and PVC scrap.
Stone setts or cobbles range in size from 75–200mm and are most commonly granite. Setts are rectangular cuboids and cobbles are rounded. They are available in a variety of colors and finishes (Figures 6.30 and 6.31).	Grades up to 15% for pathways, 10% for vehicular applications. Polymeric joint sand should be used on slopes over 10%. Relatively high material and installation cost. Rough surface and potential tripping hazard with leveling issues Larger joints can result in permeable pavement if base is open graded.	Long life expectancy. Setts and cobbles can be easily reused if not mortared. Relatively low carbon if local stone is used. Light-colored pavers will increase surface reflectivity. Refer to Chapter 13 for greater discussion of environmental and human health impacts of stone.
Concrete unit pavers, also called precast pavers, segmental pavers, or interlocking concrete pavers, are dimensioned, dense precast concrete products that fit together to form a pavement surface. Releveling may be required when set on an aggregate base (Figure 6.32).	They are resistant to point loads and very heavy design loads with thicker pavers. Wide variety of colors, textures, and patterns available. Grades up to 15% for pathways and slow vehicular applications. Polymeric joint sand will help with sand washout on any slope and is strongly recommended on slopes over 10%. Minimum 1.5% cross slopes.	Life expectancy is 20–40 years. Pavers can be reclaimed and reused. Some products form permeable surfaces with large joints or open areas. Low ride quality for fast moving cars so can be used as a traffic calming tool. Light-colored pavers increase surface reflectivity.
Flagstones are flat stone paving slabs from a wide variety of stone types that are rectangular or irregularly shaped. Sizes of flagstones and joint sizes vary widely. A wide variety of colors and patterns are possible.	Rough, irregular surfaces. Larger flagstones are more vulnerable to cracking if less than 50mm thick. Larger joint sizes of irregularly shaped stones can be filled with pea gravel, crushed stone dust, or soil and groundcovers. Water absorption of stone should not be greater than 2.5%. Sealers will extend the life of stone because it reduces absorption rates.	Flagstones can be easily reused if not mortared. Relatively low carbon if local stone is used. Larger joints can result in permeable pavement if base is open graded. Look for non-toxic, low-VOC sealers. Light-colored pavers will increase surface reflectivity.
Dimension stone is cut to specific sizes, shapes, and finishes. Dimension stone for paving is primarily granite, sandstone, limestone, bluestone, quartz, and travertine (Chapter 13). A wide variety of colors and patterns are possible.	Relatively high material and installation cost. Joint sizes can be very small with precisely cut dimension stone. Larger stone slabs are more vulnerable to cracking. Water absorption of stone should not be greater than 2.5%. Sealers will extend the life of stone.	Long life expectancy. Dimension stone can be easily reused if not mortared. Relatively low carbon if local stone is used. Larger joints can result in permeable pavement if base is open graded. Look for non-toxic, low-VOC sealers.
Rubber pavements are synthetic rubber units, sheet material, or monolithic material. Some incorporate recycled content from reclaimed tires. Used for sidewalks, playground surfaces, and other pavement applications.	They can be placed on aggregate, asphalt, or concrete bases. Costs and performance vary by product. Some installations can be porous.	They are a good use for recycled tires, but there are health concerns with recycled tire rubber. It should be tested prior to use as it can contain heavy metals, VOCs, ortho phthalates, and PAHs (US EPA 2024; Habitable 2020, 2024a, 2024b; USGBC 2019). These chemicals are associated with cancer and other illnesses.

Sources: Hanover 2023; Van Dam et al. 2015; Maher 2005

Table 6.15 Embodied Carbon (EC) of Unit Paver Assemblies[1]

Concrete Paver Assembly 30 Year Service Life	EC, 1m² (kgCO_2e) 200m² (tCO_2e)	Tradeoffs to Reduce EC
Precast concrete pavers, (75mm thick) Sand leveling bed (25mm) Filter fabric PIP concrete slab (150mm) 27.6MP, 20% SCM with weep holes Aggregate base (150mm)	*1m²:* 105kgCO_2e *200m²:* 21.0tCO_2e	– 20.7MPa concrete, EC is reduced by 9% to 96kgCO_2/m² – 20.7 MPa concrete with 50% SCMs, EC is reduced by 11% to 87kgCO_2/m² – Thinner pavers (50mm) with 100% recycled aggregate content, 20.7 MPa concrete with 50% SCMs is 47% total EC reduction to 56kgCO_2/m²
Precast concrete pavers (75mm thick) Sand leveling bed (25mm) Filter fabric HMA base (100mm) with 0% RAP with weep holes Aggregate base (150mm)	*1m²:* 71kgCO_2e *200m²:* 14.2tCO_2e	– 40% RAP WMA, EC is reduced by 8% to 65.76kg kgCO_2/m². – Thinner pavers (50mm) with 100% recycled aggregate content, and 40% RAP WMA is 52% total EC reduction to 34kgCO_2/m²
Precast concrete pavers (75mm thick) LDPE geocellular confinement system (50mm) Aggregate fill and base (230mm) Filter fabric	*1m²:* 68kgCO_2e *200m²:* 13.6tCO_2e	– Thinner pavers (50mm) with 100% recycled aggregate content, EC is reduced by 24% to 52kgCO_2/m²
Precast concrete pavers (75mm thick) Sand (25mm) Aggregate base, 30cm Filter fabric	*1m²:* 52kgCO_2e *200m²:* 10.4tCO_2e	– Thinner pavers (50mm) with 100% recycled aggregate content, EC is reduced by 58% to 22kgCO_2/m²
Precast permeable concrete pavers (75mm) Choker course (38mm) Open graded base reservoir (460mm) Filter fabric	*1m²:* 60kgCO_2e *200m²:* 12.0tCO_2e	– Thinner pavers (50mm) with 100% recycled aggregate content, EC is reduced by 38% to 37kgCO_2/m²

[1] Embodied carbon figures are cradle-to-gate and include product stage life cycle phases A1: Material Extraction, A2: Transport to Manufacturing, A3: Processing and Fabrication. They do not include transportation beyond the factory gate. These figures could be reduced in the future with production efficiencies, increased recycled contents, and greater use of renewable energy sources.

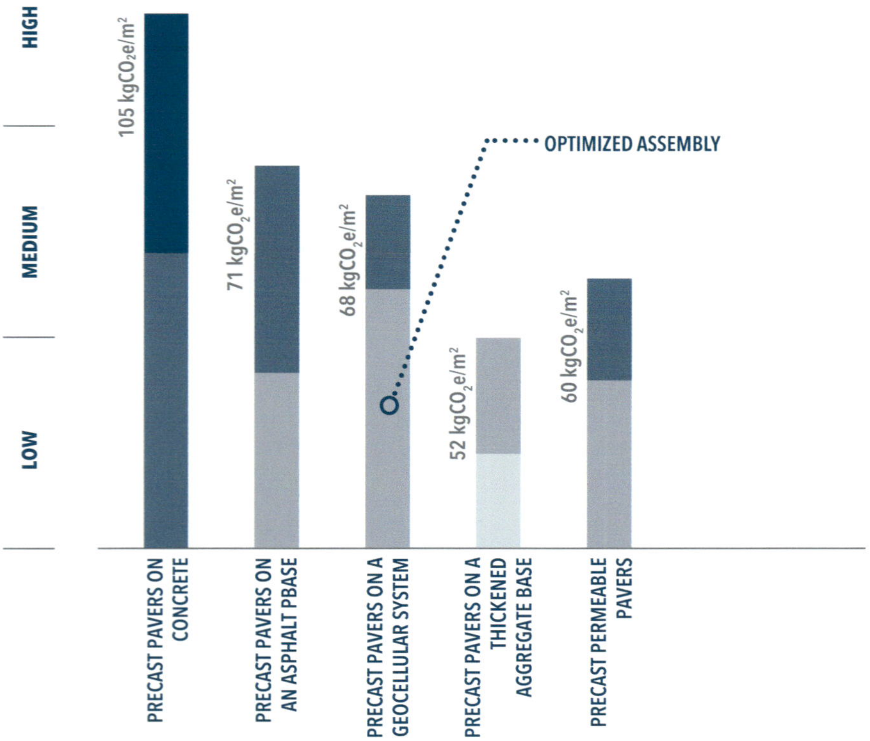

Figure 6.38 Embodied carbon for one square meter of unit pavement assemblies. The top part of each bar and the number shows EC results for original assemblies, and the lower part of the bar shows optimized assemblies. Image credit: Meg Calkins

Pavements and Urban Heat Islands

With pavements comprising 30–45% of land in urban areas, their surface reflectance is an important factor to address the reduction of urban heat islands. Pavements and roofs absorb and re-emit the sun's heat. In urban areas this is particularly concerning as there are large amounts of dark, impervious surfaces with comparatively little vegetation to cool and clean the air. Other causes of urban heat islands are the arrangement of buildings that don't easily release trapped heat and the generation of heat from vehicles, equipment, and industrial processes.

The US EPA defines an urban heat island as "a measurable increase in ambient air temperatures resulting primarily from the replacement of vegetation with buildings, roads and other heat absorbing infrastructure" (US EPA 2008). Dark colored pavement and roof surfaces can reach summertime temperatures as high as 67 °C (150 °F), transferring heat to the air around them. This can elevate temperatures as much as 8% above those of adjacent rural areas resulting in daytime air temperature increases of 0.56–3.9 °C. and nighttime increases of 1.11–2.78 °C (US EPA 2008) (Figure 6.39).

Higher air temperatures cause increased air conditioning loads in buildings and automobiles. Air-quality research in Los Angeles has demonstrated that for every 1 °C rise in summer temperatures, smog formation can increase by 3% (NOAA 2021).

Urban heat islands intensify and lengthen heat waves, increasing the risk of heat exhaustion and heat stroke for vulnerable populations without access to air conditioning. Hotter air traps sulfur dioxide (SO_2), VOCs, and other pollutants forming ground-level ozone and contributing to human respiratory issues. Hot pavements contribute to thermal pollution as stormwater runs across the pavement and eventually into receiving waters.

Increased heat islands can even amplify other weather conditions such as increasing clouds and fog, shifting wind patterns, and more rain or thunderstorms. Regions downwind of urban heat islands experience 48–116% greater rainfall quantities than upwind areas (Shepherd 2002).

Pavement Strategies to Reduce Urban Heat Islands

Cool pavement strategies to reduce heat islands include reducing the quantity of paved surfaces, using lighter colored pavement materials with greater reflectance,

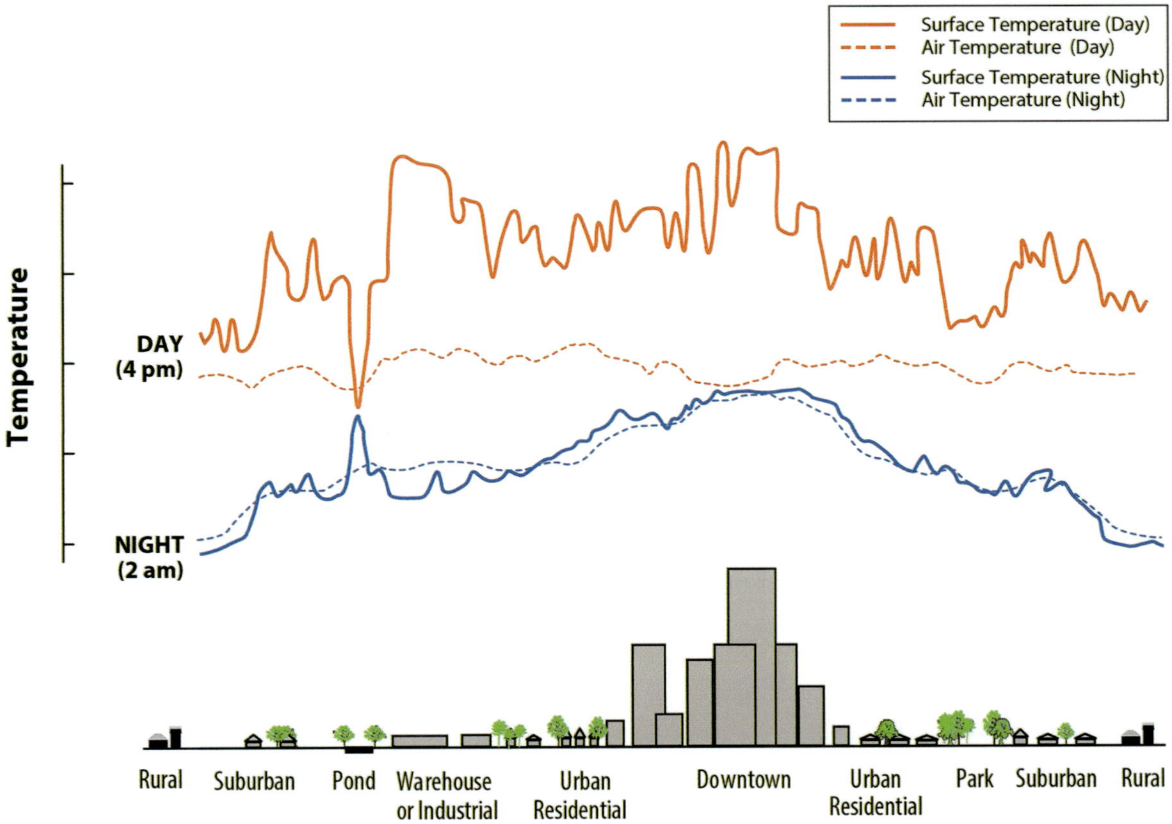

Figure 6.39 Urban heat island temperatures vary by land use contexts with hotter daytime temperatures in urban and industrial areas due to the prevalence of heat absorbing surfaces like rooftops and pavements. Image credit: US EPA 2008

Table 6.16 Pavement Types: Benefits, Drawbacks, and Potential Performance Improvements

Rigid Monolithic Pavements	Embodied Carbon	Initial Cost	Maint. Level	Relevel Possible	Reuse Potential	Edges Req'd	Accessible
Unstabilized aggregate pavements	Low	Low	High	Yes	High	Yes	No
Stabilized aggregate pavements	Low	Low	High	Yes	High	Yes	Varies
Aggregates in cellular confinement pavement systems	Med low	Med	Med high	Yes	High	Yes	Varies
Stabilized crushed stone paving (10mm and less)	Low	Med Low	Med high	Yes	High	Yes	Varies
Asphalt pavement	Med	Med low	Med low	Yes	High	Optional	Yes
Porous asphalt pavement	Med	Med	High	Min.	High	Optional	Yes
Resin bound surface (20–40mm) on asphalt base	High	Med high	Med low	Min.	Med	Optional	Yes
Concrete unit pavers on an aggregate base	Med	Med high	Med	Yes	High	Yes	Yes
Asphalt unit pavers on an aggregate base	Med	Med	Med	Yes	High	Yes	Yes
Brick pavers on an aggregate base	Med	Med high	Med	Yes	High	Yes	Yes
Stone setts or cobblestone pavers on an aggregate base	Med	Med high	Med	Yes	High	Yes	Varies
Dimension stone or flagstone on an aggregate base	Med	Med high	Med	Yes	High	Yes	Varies
Concrete pavers on an aggregate base with cellular confinement	Med high	Med high	Med	Yes	High	Yes	Yes
Porous concrete pavers on an aggregate base	Med	Med high	High	Yes	High	Yes	Yes
Concrete pavers on an asphalt pavement base	High	Med high	Low	Yes	High	Yes	Yes
Asphalt unit pavers on an asphalt base	Med high	Med high	Low	Yes	High	Yes	Yes
Brick pavers on an asphalt pavement base	High	Med high	Low	Yes	High	Yes	Yes
Setts or cobblestone pavers on an asphalt pavement base	Med high	High	Low	Yes	High	Yes	Yes
Stone mortared on a concrete base	High	High	Med low	No	Low	No	Yes
Concrete pavers mortared on a concrete base	High	High	Med low	No	Low	No	Yes
Brick pavers mortared on a concrete base	High	High	Med low	No	Low	No	Yes
Setts or cobblestone pavers mortared on a concrete base	High	High	Med low	No	Low	No	Varies
Concrete pavers on a concrete base with a sand leveling bed and sand swept joints	High	High	Low	Min.	Med	Yes	Yes
Asphalt unit pavers on a concrete base with a sand leveling bed	High	Med	Low	Min.	Med	Yes	Yes
Brick pavers, on a concrete base with a sand leveling bed and sand swept joints	High	High	Low	Min.	Med	Yes	Yes
Setts or cobblestone pavers on a concrete base with a sand leveling bed and sand swept joints	Med high	High	Low	Min.	Med	Yes	Yes
Concrete pavement	Med high	Med high	Low	No	Med	No	Yes
Porous concrete pavement	Med high	High	High	No	Low	No	Yes
Resin bound surface (20–40mm) on a concrete slab	High	High	Med low	No	Low	No	Yes

shading pavements, and using porous pavements. Some of these strategies will offer multiple environmental benefits.

Shading pavements with vegetation will reduce pavement temperatures and provide multiple ecosystem services such as carbon sequestration, improved air quality and habitat. Shading paving with solar panel structures will produce energy. Porous pavements will cool paving materials while infiltrating stormwater, improving water quality, and replenishing groundwater.

Table 6.17 Thermal Performance of Pavement Types

Pavement Type	Solar Reflectance, New[1]	Solar Reflectance, Aged[1]	Comments
Gray portland cement concrete	0.35–0.40	0.20–0.35	SR is affected by the color of both the cement and aggregate.
White portland cement concrete	0.70–0.80	0.40–0.60	Can cause glare when new but can be slightly moderated with the addition of a darker component like lamp black.
Porous concrete	0.35–0.40	0.25–0.35	SR is affected by the color of both the cement and aggregate. SR decreases with porosity increase (Zhang et al. 2015).
Precast concrete pavers	0.30–0.49	n/a[2]	Wide range of paver colors and finishes.
Stone dust pavement	0.35–0.50	n/a[2]	Depends on SR of stone dust's parent material.
Aggregate with synthetic binder	0.35–0.50	n/a[2]	SR is affected by the color of the aggregate.
Asphalt pavement	0.05–0.10	0.10–0.25	Aged SR is affected by the color of the aggregate.
Porous asphalt	0.05–0.20	0.10–0.35	Aged SR is affected by the color of the aggregate.
Chip seal topping on asphalt	0.20–0.50	0.20–0.50	SR is affected by the color of the stone chips.
Asphalt pavement with reflective coating	0.30–0.46	0.20–0.28	Reflectance reduces within a year.
Slurry seal	0.10	0.10–0.25	Similar to unsurfaced asphalt numbers.
Asphalt pavement surface treatments for thermal performance	0.20–0.40	0.25–0.45	See Table 15.6 for specific methods for improving the thermal performance of asphalt pavements.

[1] 0.0 = Absorbs all incident radiation; 1.0 = reflects all incident radiation

[2] n/a = not available

Sources: EVIzero 2024; Resin Bound 2024; Unilock 2018; Stempihar et al. 2012; Tran and Powell 2009

Improving the Thermal Performance of Pavements

The thermal performance of pavements is defined as the change in its temperature over time as influenced by properties of the pavement materials and the ambient environmental conditions (US EPA 2008). Material properties affecting thermal performance include solar reflectance, thermal emittance, thermal conductivity, specific heat, and surface convection.

Solar reflectance (SR), also called albedo, is generally correlated to the color of the surface with lighter colors having a higher reflectance and darker colors lower. Solar reflectance can be expressed as a decimal between 0 and 1 or as a percentage with zero indicating total absorption of solar radiation and 100% indicating total reflection. Solar reflectance of paved surfaces varies by material (Table 6.17).

Pavement SR changes as the surface weathers with asphalt SR increasing as the black binder wears away and concrete SR decreasing as the pavement ages. Asphalt with light-colored aggregate has been shown to lighten substantially with age. One study found that asphalt and concrete pavement have similar SRI values after seven years (Tran and Powell 2009).

The solar reflectance index (SRI) is a composite value based on a surface's SR and it's emittance. ASTM E1980 *Standard Practice for Calculating Solar Reflectance Index of Horizontal and Low-Sloped Opaque Surfaces* provides equations for establishing SRI.

Many rating systems reward points for use of heat island reduction strategies. LEED and SITES require a new SR for pavements of at least 0.33 (Table 6.17). An open grid pavement system with at least 50% unbound will also meet the credits.

Increasing the SR of a pavement can have an impact on surrounding temperatures. One study estimated that increasing the SR by 0.20 of all paved surfaces in a city could reduce summertime air temperatures by about 0.1–0.5°C and save 1kWh per year per m^2 of pavement (Pomerantz 2018).

While higher solar reflectance (high albedo) is desirable for heat island reduction, it must be balanced with the user experience, increased costs, light pollution, and the environmental impacts of producing the pavements. Very light-colored pavement surfaces can create an uncomfortable glare for pedestrians and an additional risk for dangerous vehicular–pedestrian interactions. When combined with lighting, very light-colored pavements can exacerbate light pollution in the night sky.

Thermal Performance Definitions

Solar reflectance, also called albedo, is a measure of the surface's ability to reflect solar radiation. Values range from 0 (no reflectance) to 1 (all sunlight is reflected). This is also expressed as a percentage from 0–100%.

Emittance is the efficiency with which a surface emits radiant energy. It is defined as the ratio of energy radiated by the surface to the energy radiated by a black body at the same temperature. The range is 0 for no emission to 1 for perfect emission. Thermal emittances of dense graded asphalt and concrete are similar (in the range of 0.90 to 0.95).

Thermal conductivity measures the ability of a material to transmit or conduct heat. Materials with high thermal conductivity will transfer heat at a higher rate than materials with low thermal conductivity. The thermal conductivity of pavement materials varies widely from 0.8W/m-K (Watts per meter-Kelvin) to 2.0W/m-K. Dense graded asphalt and concrete have similar thermal conductivity.

Specific heat is the energy required to raise a unit mass of a substance by one unit of temperature. This is typically expressed in J/kg-K (joules per kilogram per kelvin). Again, dense graded asphalt and concrete are similar at 900J/kg-K.

References

American Association of State Highway and Transportation Officials (AASHTO) (2017). *AASHTO M147 Standard Specification for Materials for Aggregate and Soil-Aggregate Subbase, Base, and Surface Courses*. AASHTO.

Bachensky, L. (2024). *What You Need to Know About Building Trails with Crusher Fines*. American Trails. www.americantrails.org/resources/building-crusher-fines-trails, accessed August 27, 2024.

British Parking Association (BPA) (2016). *Parking Know How Bay Sizes*. BPA, July.

Budzinski, N., and Girard, M. (2016). "The right path: A forensic approach found the best decomposed granite solution for Kenyon College". *Landscape Architecture Magazine*, November 8.

CalGreen (2019). *CalGreen Code Section 301 General*. State of California.

California Department of Transportation (CA DOT) (2013). *Subgrade Enhancement Geosynthetic Design and Construction Guide*. CA DOT Pavement Program, Office of Concrete Pavements and Foundations.

Conrad, P. (2023). *Climate Positive Design Toolkit*. Climate Positive Design.

De Pascale, B., et al. (2023). "Innovative light-coloured porous asphalt for low-impact pavements: A laboratory investigation". *Construction and Building Materials* 368: 130482. https://doi.org/10.1016/j.conbuildmat.2023.130482.

Deutsches Institut für Normung e. V. (DIN) (2005). *AS/NZS 2890/1 AMD 1 Parking Facilities Off-Street Car Parking*. DIN.

EVIzero (2024). "EVIzero: Eco-friendly synthetic binder. No bitumen, just natural colors. CORECOM s.r.l". www.evizero.com/en/#gref, accessed September 17, 2024.

Forman, R., et al. (2003). *Road Ecology: Science and Solutions*. University of Chicago Press.

Habitable (2020). *New LEED Circularity Pilot Credit Incorporates HBN Optimize Recycling Research*. Habitable, January 30.

Habitable (2024a). *Flooring Product Guidance*. Habitable, May 8.

Habitable (2024b). *Turf Product Guidance*. Habitable, May 8.

Hanover (2023). *Hanover Asphalt Block*. Hanover Architectural Products: ASPH 2023.

Harris, C.W., and Dines, N.T. (1997). *Time-Saver Standards for Landscape Architecture*, Second Edition. McGraw Hill.

Levinson, R., et al. (2014). *Life Cycle Assessment and Co-Benefits of Cool Pavements*. Final Report for Contract #12-314. Sacramento, CA: California Air Resources Board.

Maher, M., et al. (2005). *Context Sensitive Roadway Surfacing Selection Guide*. FHWA-CFL/TD-05-004. Central Federal Lands Highway Division, Federal Highway Administration, August.

Muench, S.T., and Hand, A.J.T. (2019). *Sustainable Asphalt Pavements: A Practical Guide*. National Asphalt Pavement Association.

Organic Lock (2024). "Design, installation and maintenance". www.organic-lock.com/, accessed January 7, 2024.

Pasetto, M., et al. (2023). "Advances in the rheology of synthetic binder for sustainable road pavements: An improved protocol for DSR testing". *Sustainability* 15.6: 5146. https://doi.org/10.3390/su15065146.

Pomerantz, M. (2018). "Are cooler surfaces a cost-effect mitigation of urban heat islands?". *Urban Climate* 24: 393–397.

Resin Bound (2024). "What is resin bound?". https://resinbondedaggregates.com/, accessed September 17, 2024.

Shepherd, J.M. (2002). *NASA Satellite Confirms Urban Heat Islands Increase Rainfall Around Cities*. NASA Goddard Space Flight Center. https://svs.gsfc.nasa.gov/stories/urban_rain_20020618/, accessed June 18, 2002.

Solano, L. (2024). Personal communication, September 13.

Stempihar, J.J., et al. (2012). "Porous asphalt pavement temperature effects for urban heat island analysis". *Transportation Research Record* 1: 123–130.

Tran, N., and Powell, B. (2009). *Strategies for Design and Construction of High Reflectance Asphalt Pavements*. Report 09-02. National Center for Asphalt Technology (NCAT).

Unilock (2018). *Advanced Tech Guide: Paving Stones, Retaining Walls*. Hengestone Holdings.

US Environmental Protection Agency (US EPA) (2008). *Reducing Urban Heat Islands: Compendium of Strategies*. US EPA.

US Environmental Protection Agency (US EPA) (2024). *Federal Research on Recycled Tire Crumb Used on Playing Fields and Playgrounds*. US EPA Safer Chemicals Research.

US Federal Highway Administration (FHWA). "LCA pave tool". www.fhwa.dot.gov/pavement/lcatool/, accessed August 27, 2024.

US Green Building Council (2024). *Heat Island Reduction Credit, LEED v5 Public Comment*. USGBC.

US National Oceanic and Atmospheric Administration (NOAA) (2021). *Air Quality in the Los Angeles Basin Increasingly Dependent on Temperature*. NOAA Climate Program Office, March.

Van Dam, T.J., et al. (2015). *Towards Sustainable Pavement Systems: A Reference Document*. FHWA-HIF-15-002. Federal Highway Administration.

Woods Ballard, B., et al. (2015). *The SuDS Manual (C753)*. CIRIA, December.

Zhang, R., Guisheng, J., and Liang, J. (2015). "The albedo of pervious cement concrete linearly decreases with porosity". *Advances in Materials Science and Engineering* 1.

CHAPTER 7

Porous Pavements and Green Roofs

Impervious pavements and rooftops concentrate stormwater quantities and impact quality through thermal and chemical pollution. Globally, the area of impervious surfaces is estimated to be 1.29 million km^2 (Sun et al. 2022), with roads occupying roughly 20% of global land area excluding Antarctica and Greenland (Ibisch et al. 2016). In the United States, there are nearly 300m^2 of impervious surface per person with 129m^2 in per person in the United Kingdom (Elvidge et al. 2007).

The green infrastructure approaches of porous pavements, structural soil, suspended pavements, and green roofs are critical to reducing the environmental impacts of impervious surfaces. While not all pavements can be made porous and green roofs are not appropriate on all buildings, these measures are increasingly employed.

These green infrastructure strategies can reduce the extent, size, and area required for engineered stormwater control measures (SCMs). When water is allowed to percolate and recharge soil systems, fewer collection, conveyance, and retention measures are required.

The permeable pavement market, including porous concrete, porous asphalt, permeable pavers, and plastic grid systems is estimated to reach $27 billion by 2031 and is growing at a rate of 5.8% annually. In 2023, the largest market share for permeable paving was North America at 36% with Europe at 27% and Asia Pacific at 23%. In 2022, the global area of permeable asphalt increased by 20% (Industry ARC 2023).

The global green roof market size reached $2.1 billion in 2023 with a projected annual growth rate of 13.3% between 2024 and 2032 (IMARC Group 2024). North America and Europe have the largest market shares supported by state and countrywide legislation.

This chapter addresses porous pavements, structural soils, suspended pavement systems, and green roofs. It does not address stormwater volume calculations or soils and infiltration rates. Additional resources for these green infrastructure measures are listed at the end of the chapter (Table 7.13).

Porous Pavements

Making pavements porous allows for stable pedestrian and vehicular surfaces, while managing stormwater where it falls. Porous pavements, also called pervious pavements, allow for percolation of water through a pavement section, the base and subbase, and into the subgrade. This reduces stormwater runoff quantities and recharges the subgrade under and around the porous pavement. Bruce Ferguson, in his seminal book, *Porous Pavements*, discusses multiple research studies and concludes that about half of all pavements could be made porous (Ferguson 2005). Concrete, asphalt, unit pavers, aggregate pavements, reinforced turf, and boardwalks can all be made porous to allow for movement of water through the pavement assembly to the subgrade below (Figures 7.1–7.4).

Benefits of Porous Pavements

Porous pavements mitigate stormwater impacts from large areas of impervious surfaces in urban and suburban areas by reducing runoff quantities and filtering pollutants and nutrients from stormwater. They manage stormwater at the source, rather than allowing it to concentrate downstream in the watershed or in a stormwater system.

Porous pavements slow or eliminate runoff, allowing stormwater to infiltrate into the soil under or just adjacent to the pavement. This replenishes soil water storage, percolates to the water table, and slows base flow to receiving waters. The downstream flooding and streambank erosion resulting from large areas of dense pavements is decreased, and the size of downstream storm systems can be reduced as well. Porous pavements can also reduce

DOI: 10.4324/9781003413974-8

Figure 7.1 Reclaimed pervious brick pavement in Tampa, Florida. Image credit: Icononiac

Figure 7.2 Elizabeth Gourley used bluestone pavers with planting strips between the units. Plants and soil take up rainwater falling on the pavement. Image credit: Elizabeth Gourley

Figure 7.3 Permeable pavers with lugs on sides to maintain 10mm open joints. Image credit: Sandy Close

Figure 7.4 Permeable concrete. Image credit: Meg Calkins

the need for other stormwater control measures and gray infrastructure.

Stormwater quality is improved as it passes through the pavement layers as it is treated by filtration and the biochemical activity of microorganisms. It is further cleaned and filtered in the subgrade below the pavement.

Porous pavements are beneficial for urban forests as they direct air and water into tree rooting media in densely developed areas. The porous pavement is a stable surface, but the base reservoir provides voids for tree roots to access air and water encourages more root growth than a typical urban tree pit. This is particularly beneficial for tree root establishment. Porous pavements can be combined with

Table 7.1 Benefits, Challenges and Limitations of Porous Pavements

Benefits of Porous Pavements
Stormwater quantity mitigation and flood control
Water quality treatment
Groundwater recharge
Reductions in gray stormwater infrastructure (pipes, catch basins, ponds, curbs)
Suitable for cold climate areas, maintains recharge capacity when frozen
No standing water or black ice development in winter
Good traction in wet weather
Reduced surface temperatures due to greater air circulation and less density within the pavement
Extended pavement life due to well drained base and reduced freeze thaw action
Can reduce stormwater fees
Challenges and Limitations of Porous Pavements
Requires regular routine maintenance of vacuum sweeping or power washing.
Higher installation costs.
Most appropriate for use in pedestrian pavements, sidewalks, low traffic roadways and parking lots. Pervious concrete can bear heavier loads of medium traffic roadways.
Porous pavements are not appropriate for slopes over 5%.
Owners may resist unknown technology.
Requires periodic testing to ensure function.
Subgrade must have an infiltration rate of 6mm to 200mm per hour and be relatively uncompacted.
Construction operations and sequence must be carefully considered to protect subgrade under porous pavements from over compaction and to protect the pavement surface from dust, sediment and debris.
Skilled contractors required for installation, and more construction administration oversight to make sure the pavement is being installed according to specifications and to test the results for compaction.
Soil stabilization and erosion control are key next to porous pavements.
Cellular porous surfaces may not meet ADA, and loose open graded aggregate surfaces will not.

structural soils to promote even richer environments for tree health (Table 7.1).

Limitations and Challenges to Porous Pavements

There are some situations where porous pavements are not appropriate or where they can be used, but water should not be infiltrated directly into the soil. On sites with harmful chemicals or metals in the runoff, other strategies such as bioretention and/or bioremediation should be employed. To avoid drippings from cars, parking lots can be hybrid designs with some areas being dense pavement and other areas porous.

On sites where the water table is within one meter of the bottom of the pavement base, a liner and perforated pipe should be employed to direct water to stormwater control measures that will clean and filter the water more thoroughly prior to infiltration. Pavement with slopes over 5% should not be made porous, and vehicular surfaces with heavy traffic loading or high design speeds are inappropriate for porous pavements.

Higher installation costs, increased pavement function oversight, and inexperienced contractors can all be challenges or perceived challenges to porous pavements. And porous pavements may not be available in every area. However, as they become more common, research shows that the increased cost of porous pavement installation and maintenance can be offset by the reduced need for catch basin and pipe systems, land consumption for detention basins, and, in some areas, stormwater fees (US EPA 2008). Management and maintenance of porous pavements is discussed later in this chapter.

Porous Pavement Components

Like dense pavements, porous pavements are composite structures comprised of multiple layers, and while the layers perform structurally, they must also be constructed to allow for stormwater capture, storage, and infiltration into the subgrade below the pavement. Like conventional paving,

Table 7.2 General Considerations for Porous Pavements

Topic	Design, Specification, and Maintenance Considerations
Soil and hydrology analysis	Soil and hydrological analysis should be performed prior to design. Soils must have adequate infiltration rates (6mm to 75mm per hour) or the base reservoir should incorporate a lateral pipe to another stormwater control measure (SCM).
Water table	Minimum depth to bedrock or seasonal high water from the bottom of the base reservoir is 90cm.
Context of pavement	The context of porous pavements should be considered in their design. Airborne dust, leaf litter, and other particulates will quickly fill voids. In cold climates, water in the base reservoir and subbase can freeze, so surface drainage should be designed accordingly.
Drainage time	Porous pavement assemblies should be designed to completely drain within 48 hours.
Subdrains	Subdrains should be used as required for soil and climate conditions.
Pollutants	Porous pavements should not be used where there is a risk of toxic spills or where there are toxins in the soils that may be carried into the groundwater.
Slopes	Most porous pavements should not be used on slopes greater than 5%. Flatter slopes (1–2%) will allow water to linger on the pavement longer and increase percolation through pores or joints to the base reservoir below.
Edges	Edge restraints are recommended for all porous pavements and are required for aggregate and unit pavements.
Choker course	A 25mm choker course of single-size clean crushed stone will stabilize the paving layer. No. 8 sized stone is recommended.
Aggregate base and reservoir	Uniformly graded, clean crushed stone should be used for the base and reservoir.
Geotextiles	A geotextile should be used between the base reservoir and the subgrade to eliminate migration of soil into the reservoir.
Subgrade	The subgrade should be uncompacted to facilitate stormwater infiltration under the pavement.
Construction	During the construction process, care should be taken to avoid clogging of the surface with dust or sediment laden runoff. The pavement should be constructed as late in the construction schedule as possible.
Maintenance	Clients and site managers should be made aware of the maintenance requirements of porous pavements asphalt, concrete, and unit pavements (Table 7.7). These pavements should be vacuum swept every 3–6 months.

Sources: FHWA 2016; University of New Hampshire Stormwater Center 2015a, 2015b, 2015c; Ferguson 2005

porous pavements will have surface and base courses, but they also need a reservoir, filter layer, and geotextiles or liners. Other components may be necessary for specific types of porous pavement (Table 7.2).

Pavement Surface Course

The surface course receives the pedestrian and vehicular traffic load, collects the stormwater, reflects or absorbs solar radiation, determines accessibility, and establishes the appearance of the pavement. The surface course can be rigid or flexible, and monolithic or comprised of paving units. If it is a monolithic surface like porous concrete or porous asphalt, there will be voids in the pavement to allow for infiltration of stormwater. If it is a unit pavement, the voids for water percolation usually occur in the joints between the units.

Choker Course

Choker courses, also called bedding courses or stabilizing courses, are fine, crushed stone layers that mediate between the surface course and the base course. They provide a smooth platform on top of the open graded base layer on which to place the surface layer. They filter water moving through the pavement layer. And, if the base layer aggregate size is substantially larger than aggregate in the porous surface, it keeps the smaller aggregates from migrating down into the base.

Base Course

The base course under the pavement surface transfers and distributes the traffic loads over the subgrade. In porous pavements, the base course must be permeable, so it is usually open graded aggregate. Open graded aggregate has a narrow range of aggregate sizes that produce open voids between particles. In some instances, a subbase will be added to thicken the pavement structure for structural reasons or to store more water as it infiltrates into the soil or discharges into a pipe.

Aggregate bases will be compacted to ensure structural stability. As they are open graded, compaction should not inhibit their porosity.

Reservoir

A reservoir is the layer of the pavement profile that stores or conveys water while it infiltrates into the soil or a perforated pipe. Very often the reservoir is the base or subbase that is also serving a structural function. This is called the ***base reservoir***. The voids between the aggregate store the water, and the storage volume is based on the estimated void space for the size and gradation of stone. Sometimes the reservoir is called the drainage layer.

Water in the reservoir will discharge relatively slowly into the soil or a pipe, so the base reservoir depth must be sized to accommodate the expected speed of infiltration. This is a factor of subgrade infiltration rate (it is the least permeable layer), soil characteristics (texture, structure, and compaction), climate, and design storm. A base reservoir as thick as one meter could be required, although a typical thickness range is 30–60cm. Base reservoirs should be sized to completely drain within 48 hours.

Lateral Outlet

The lateral outlet, most commonly a pipe, allows for discharge of excess water from all of parts of the base reservoir. Sometimes the discharge pipe is placed at a higher elevation in the base reservoir to allow some water to remain and infiltrate into the subgrade very slowly. A lateral outlet could also be an extension of the base reservoir beyond the surface of the pavement that allows water to flow downhill into another stormwater structure such as a bioswale or bioretention basin.

Filter Layers

Filter layers are used in porous pavements to maintain the porosity and structural integrity of the materials. These layers could be made from aggregate, such as a choker course, or from a permeable geotextile, also called filter fabric. Geotextiles can inhibit the movement of fine particles between layers, and they add tensile strength to the pavement.

Liners

Polyethylene or synthetic rubber liners are occasionally used between the reservoir base and the subgrade to prevent water from moving into the subgrade. High water tables, toxic chemicals in the soils, or swelling soils are some of the reasons a liner would be used to prevent water infiltration. Instead, water in the reservoir is detained then drained in a perforated pipe to adjacent stormwater control measures or to a storm system. While porous pavements with liners do not offer the benefits of infiltration into the soil under the pavement, they still detain stormwater and provide some water quality treatment.

Subgrade

The subgrade soil bears the pavement design load and receives and infiltrates stormwater. In some porous pavements, the only route for stormwater to take is into the subgrade soil. In most porous pavements, there is a lateral pipe that will discharge excess water that does not infiltrate into the subgrade. Geotextile separators are employed to prevent fine colloidal soil from migrating up from the subgrade into the reservoir base.

Subgrade compaction for porous pavements should not be greater than 92% as this could reduce or eliminate infiltration. Care should be taken to avoid inadvertent subgrade compaction during construction activities. Ferguson (2005) recommends assuming a safety factor of 0.5 of the preconstruction infiltration rate for the post construction infiltration rate due to light compaction. Compaction percentage of the subgrade can be verified with standard compaction tests that are typically performed for pavement structures (Solano 2024).

Pavement Edges

The edges of porous pavements can be vulnerable as resistance to lateral spreading ends at the base and surface courses. Unreinforced or unsupported edges can crack, compact, or collapse. There are multiple ways to address the vulnerability of porous pavement edges. Like other pavement assemblies, the base and base reservoir can extend beyond the edge of the pavement surface by at least the thickness of the layer.

Edge thickening by a few inches at the bottom edge of porous concrete pavements can reinforce their edges. Flexible pavements will require edge restraints against lateral deformation (Table 6.3). These include the following:

PIP concrete bands are a common edge restraint producing a clean, finished band at the edge of a pavement. Raised or rolled concrete curbs will contain vehicular wheels, but they will also inhibit the lateral flow of excess water and debris from the pavement. Flush curbs will allow for water and debris to flow into adjacent stormwater control measures.

Edging such as steel, timber, plastic lumber, and aluminum edges are staked in place and provide viable edge restraints for porous pavements. If edging is used for vehicular porous pavements, it should be steel.

Barrier elements such as wheel stops, bumper fences, and bollards will protect edges of porous pavements as well.

Filter Strips

Slopes adjacent to porous pavements should drain away the pavement or flow into bioswales at the edge of the pavement. Where small slopes must drain toward porous pavements and there is no room for a swale, filter strips at the edge of the pavement can catch sediments and debris. Filter strips are vegetated bands of grass or groundcover with complete vegetative cover. Mulch should not be used in filter strips as it can clog porous pavement or prevent the flow of water.

Types of Porous Pavements

Like dense pavements, porous pavements can be made from a wide variety of materials and assemblies. They can be rigid structures, such as porous concrete, and they can be flexible structures such as porous unit pavers on an open graded aggregate base or porous asphalt pavement.

Porous Aggregate Pavement

Porous aggregate pavements can be used for very low traffic applications such as residential driveways, overflow areas of parking lots, or pedestrian paths. Dense graded aggregates are not very porous because the dust and small particles tend to lodge between the larger stones allowing minimal water through. Open graded aggregates create a stone mass with 30–40% void spaces. This allows water to percolate through to the base reservoir and subgrade.

Recommended ASTM D448 gradation numbers for porous aggregate pavements are 57, 67, 78, 89, and 10. Sizes 57, 67, and 78 are larger stones and will drain more quickly, while the smaller sizes 89 and 10 produce smoother surfaces and are easier to walk on (Ferguson 2005; ASTM 2022).

Aggregate surfaces are particularly appropriate for sites with frost prone or expansive soils as deflections are less noticeable. When required, the surface can be easily releveled and lost material replaced. Aggregate is the least expensive surfacing material, is readily available almost everywhere, and can be derived from recycled materials (Figures 7.5–7.7).

Stabilized Porous Aggregate and Turf Pavements

Porous aggregate or turf pavements can be stabilized with structures such as plastic geocells or precast concrete grids. When using aggregates in these applications, displacement of the stones can uncover the cells/grids, and periodic raking will be required. Use of gap graded aggregate can help reduce displacement.

Plastic geocells are three-dimensional lattice products that hold aggregates or topsoil and grass in their cells but still allow full infiltration. These flexible porous paving systems allow for heavier vehicular loads by inhibiting granular displacement and compaction. The visual appearance of the plastic geocells is minimal as aggregate or grass covers the product after installation.

Figure 7.5 Surface 678 designed the aggregate pavement at the North Carlina Museum of Art sculpture gardens to be permeable around the trees allowing water to the tree roots. The aggregate is stabilized in other locations. Image credit: Karen Malinofski, North Carolina Museum of Art

Figure 7.6 HDPE pavement grid with filter fabric backing provides structure for open graded porous pavement. Gravel will entirely cover grid. Image credit: Invisible Structures, Inc

Figure 7.7 Open precast concrete grid with grass. Image credit: Hanohiki

Plastic geocells are often produced with some recycled content, and they are flexible enough to be used with frost prone or expansive soils.

Plastic geocell aggregate pavements offer compressive strengths up to 100Mpa (Invisible Structures 2023). The solar reflectance for these systems is based on the aggregate color, and they can be used on slopes up to 8% for light car and truck traffic.

Plastic geocells can be filled with a specialized soil mixture and seeded or sodded for grass growth in the plastic geocells. The soil mixture is usually planting soil with a higher content of medium and coarse sand to allow for drainage. This application is used for fire lanes and overflow parking. This application will have a similar runoff coefficient as mown grass.

Plastic geocell products rest on a compacted, open graded aggregate base of varying depths from 5–30cm depending on the design load. A base reservoir may be necessary.

Precast or PIP concrete grids are used to contain aggregates or turf and increase the compressive strength of these porous pavements. The concrete lattice is a more visible element of the pavement surface with only around 40% open grids for aggregate or grass growth. Applications, construction, and compressive strengths for this type of assembly are similar to that of the plastic geocells.

Open Jointed Unit Pavements

Open jointed paving blocks are shaped to create widened joints that are filled with fine crushed stone to allow water to flow through the surface layer. Precast concrete pavers are the most common type of open jointed paving blocks, but bricks with lugs, granite cobbles, and flagstone can be used in this porous pavement application. As this is a flexible unit pavement assembly, units can be easily reclaimed and reused after the life of the pavement. Typical precast concrete paver products provide a compressive strength of 55MPa with a 50–75mm thick blocks.

Providing adequate spacing between units and maintaining flow through the joints are key to the porosity of this assembly. Precast concrete units for porous pavements are produced with lugs on the vertical sides of the pavers that maintain a consistent distance from other units to ensure joint space. But flagstone, cobbles, bricks, or other units without lugs can be installed with spacers between them to ensure a 7–12mm space between units.

Joints are filled with ASTM No.8 crushed stone or a similar size and gradation crushed aggregate. Pavers are set on a 20–30mm thick ASTM No. 8 choker course for leveling, or a plastic grid structure may be used for additional structural support. A 15–30cm base course of open graded, crushed stone supports the traffic loads while allowing for percolation of water. A reservoir base of larger crushed, open graded stone provides voids for water storage while water seeps into the subgrade. A geotextile filter fabric is recommended under the subbase to prevent upward migration of subgrade soil into the reservoir (Table 7.3 and Figures 7.8–7.10).

Porous Asphalt Pavement

Porous asphalt pavement is open graded aggregate bound together by a bituminous asphalt binder (asphalt cement) to form a permeable surface. This is a flexible pavement that is structurally similar to standard asphalt pavement, but it is missing sand and mineral filler in the mix. Thicknesses of porous asphalt pavement layers will vary by expected traffic. Pedestrian applications will be thinner.

Table 7.3 Considerations for Open Jointed Permeable Concrete Pavement Units

Topic	Considerations
Strength	Capable of bearing very heavy loads.
Embodied carbon	Embodied carbon of concrete units is higher than concrete and asphalt pavement, but the volume of material used is less, making up for the difference (Table 6.15).
Standards	Refer to ASTM C936 Standard Specification for Solid Concrete Interlocking Paving Units.
Cost	Total project cost can be 25–35% more than dense concrete.
Joints	Joints between pavers must be 7mm minimum. Joints should be filled by small crushed stone or gravel, No. 8 is recommended. Sand should not be used. Joints can be filled with vegetation in sandy soil, but infiltration will be reduced.
Snow removal	Use snowplows with plastic blades or rotational snow brushes. In extreme cold, refer to the manufacturer's recommendations for deicing products and practices.
Maintenance	Joint material will need to be refilled once between 3–6 months after installation then approximately every 5–10 years. Maintenance costs may be offset by a reduction in deicing costs. For preventive and restorative maintenance refer to Table 7.7.

Sources: Unilock 2019; University of New Hampshire Stormwater Center 2015b; Ferguson 2005

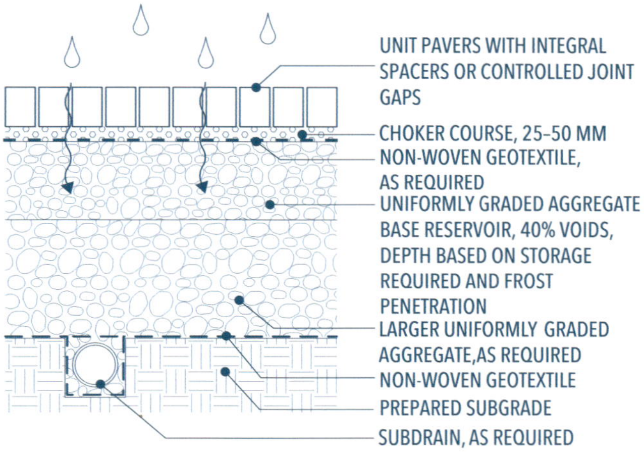

Figure 7.8 Permeable unit paving with underdrain as required. Image credit: Kathleen Elliott and Meg Calkins

Bituminous asphalt binder is a by-product of oil production. When heated, it softens and coats the aggregates holding them together. Reheating of the binder on hot days can cause the binder to drain down into the voids of porous asphalt pavements; however, binder technologies such as polymer-modified binders will reduce or eliminate this issue. Relatively hard, low penetration binders are used in very hot climates to inhibit bleeding of the binder. High penetration binders are used in cold climates to avoid brittleness.

Like dense aggregate asphalt pavements, recycled rubber can be a component of the asphalt binder for porous asphalt. Refer to Chapter 15 "Asphalt Pavement Materials" for more information.

Aggregates in porous asphalt mixes should meet the same standards for angularity, durability, and stability as those of

Figure 7.9 Porous pavement at the North Carolina Botanic Garden parking lot. Spaces and drive aisles are differentiated by paver shape. Image credit: Annie Elvin

Figure 7.10 Porous pavement at the North Carolina Botanic Garden parking lot. Image credit: Stephen Faber

Table 7.4 Considerations for Porous Asphalt Pavements

Topic	Considerations
Specifications	Many states DOTs have specifications for open graded friction courses and asphalt treated permeable bases that can be used when specifying mixes for porous asphalt pavements.
Embodied carbon	The initial embodied carbon of porous asphalt is lower than concrete, but over a 30 year life cycle it is similar.
Cost	Materials cost is 25% greater than standard asphalt pavement, and placement cost will be higher as well.
Surface infiltration rate	The initial surface infiltration rate should be a minimum of 508 centimeters per hour.
Surface compaction	Voids in the asphalt surface can be impacted by turning of heavy vehicles, particularly on hot days.
Construction	The pavement will include one or more layers of open graded stone mixed with asphalt binder. The mixture should achieve 16–25% voids. Porous asphalt should be placed in 25–100mm lifts following state guidelines and compacted with 2–4 passes of a 10 ton static roller. Traffic should be restricted for 24 hours after construction.
Binder drainage	Polymer-modified binders and/or polymer fibers can be added to the mix to reduce binder drain down.
Maintenance	Use of deicing salts can be reduced as porous asphalt will melt snow and ice more quickly than dense pavements. For preventive and restorative maintenance refer to Table 7.7.

Sources: FHWA 2015; University of New Hampshire Stormwater Center 2015c; Ferguson 2005

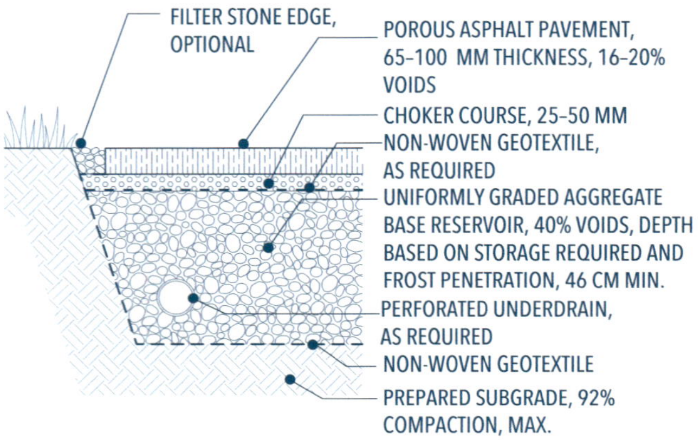

Figure 7.11 Porous asphalt pavement. Image credit: Kathleen Elliott and Meg Calkins

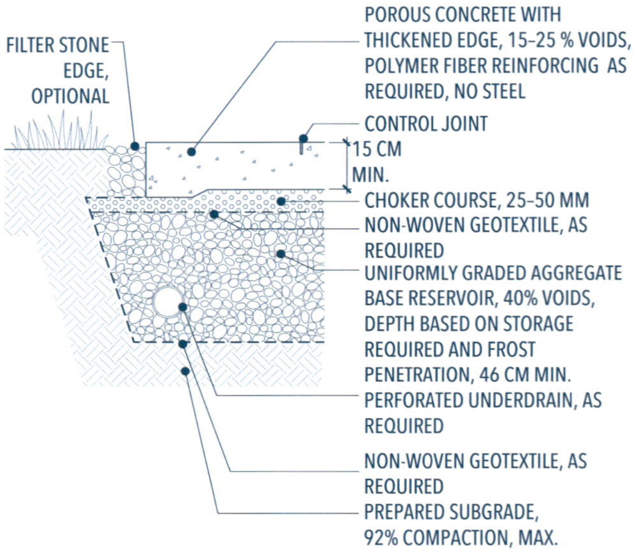

Figure 7.12 Porous concrete pavement. Image credit: Kathleen Elliott and Meg Calkins

standard asphalt pavements. Open graded aggregate mixed should be used to provide voids of 16–25%. Sometimes a small amount of fine aggregate is used to form a matrix with the asphalt binder to control cohesiveness of the mix (Table 7.4 and Figure 7.11).

Porous Concrete Pavement

Porous concrete is comprised of open graded coarse aggregate bound with cementitious materials to form a rigid, porous pavement. Fine aggregate content is limited, and coarse aggregate is kept to a narrow gradation resulting in pavement voids of 12–25% for movement of water through the slab. As the void content decreases, the permeability decreases but the pavement strength increases (Figures 7.6 and 7.7; Figure 7.12).

Pervious concrete slabs are the principal load bearing component of the composite pavement and are typically 15cm thick with edges thickened to 20cm. Slabs of 20cm are used for pavements that bear high volumes of truck (FHWA 2016).

Steel reinforcing, even if it is coated, is not usually recommended for pervious concrete as it will be continuously exposed to moisture. Instead, polymer fiber reinforcing mixed into the concrete will resist tensile forces and help control surface cracking. It will also facilitate crushing and reuse of the pavement after its service life.

Table 7.5 Typical Ranges of Materials in Pervious Concrete[1]

	Proportions (kg/m³)	Proportions (lb/yd³)
Cementitious materials	267–326	450–550
Total Aggregate	1190–1480	200–2500
Water: cement ratio (by mass)	0.27–0.36	0.27–0.36
Fine aggregate	0–297	0–500
Admixtures	As required	As required

Source: Adapted from NRMCA 2024

[1] These proportions are provided for information only. Actual mixes will vary with regional standards, design load, and climate.

Slabs are set on a base of open graded crushed stone that is 10–15cm thick. To strengthen the slab's edges, the base should extend beyond the pavement edge a minimum distance of the slab's thickness. A reservoir base is located below the base and will be 20-100cm with the thickness based on the required water storage.

The proportion of cementitious materials in pervious concrete is higher than in comparable strength dense concrete. The average amount of cement in a standard concrete pavement mix is 9%, and it is around 16% in a pervious mix. Cement substitutes such as fly ash, natural pozzolans, and blast furnace slag can replace portland cement in pervious concrete mixes at similar rates to standard concrete. Fly ash can be used at rates of 5–65%, blast furnace slag at 20–70%, and natural pozzolans will vary by type (NRMCA perviouspavements.org).

The installed cost of pervious concrete pavement is estimated to be 25% higher than standard concrete, and maintenance costs will be higher as well (Table 7.7). However, if pervious concrete is part of a stormwater treatment train, the additional costs can be offset by the reduction in gray infrastructure measures such as curbing, pipes, catch basins, and detention ponds.

The National Ready Mix Concrete Association (NRCMA) has developed a pervious concrete mixture proportioning guideline and spreadsheet software to assist in design of pervious concrete pavements. The software will develop trial batch mixture proportions using volumetric consideration and make the necessary calculations for production batches when mixture proportions are finalized (Table 7.5).

Structural Soils and Suspended Pavements

Trees in urbanized areas provide ecosystem services such as stormwater control, improved air quality, shade, habitats,

Table 7.6 Considerations for Porous Concrete Pavements

Topic	Considerations
Embodied carbon	The initial embodied carbon of porous concrete is higher than asphalt, but over a 30 year life cycle it is similar.
Cost	The materials cost is around 25% greater than dense concrete, and the installation cost can be higher in some regions.
Surface infiltration rate	The initial surface infiltration rate can be 140cm per hour or higher as the entire surface is permeable.
Contractor experience	Porous concrete should be placed by a contractor experienced in the construction process as it differs from that of dense concrete pavements. NRMCA offers a pervious concrete contractor certification program.
Construction	Consolidation is accomplished by rolling over the concrete with a steel roller within 15 minutes of placement. Porous concrete mixtures cannot be pumped, so site access is important to plan. The subgrade must be moist but not saturated when pervious concrete is placed so the concrete does not dry out too quickly.
Joints	Control joints should be placed every six meters maximum. Expansion joints are not usually required.
Maintenance	Repairs can be made with standard concrete. Sealants should never be applied to porous concrete as it will reduce porosity of the pavement. For preventive and restorative maintenance refer to Table 7.7.

Sources: FHWA 2016; University of New Hampshire Stormwater Center 2015a; Ferguson 2005; Solano 2024

and thermal comfort, but pavements and trees often have competing needs. Pavements require a stable base on compacted soil for structural performance, yet these requirements impact the health and vigor of trees by limiting the tree roots access to air, water, and soil nutrients. When the roots of trees find their way under pavements, they can cause heaving and cracking in the pavement surface.

Structural soils and suspended pavements provide appropriate conditions for tree health and stormwater control measures, while offering a stable structural base for porous or non-porous pavements. There are multiple types and proprietary systems available.

These systems offer stormwater performance benefits in two ways. Healthy trees can intercept and evapotranspirate substantial amounts of stormwater. The EPA estimates that tree with a 7.6m diameter canopy and associated soil can manage the 25mm rainfall from 223 square meters of impervious surface (US EPA 2016). The voids in aggregates or sand bases, or cells under these pavement systems, act

Table 7.7 Preventive and Restorative Maintenance for Porous Pavements

Type of Porous Pavement	Considerations
General maintenance for all porous pavements	Detailed maintenance plans should be developed, discussed, and shared with site owners and caretakers. Inspect surface infiltration rates semi-annually during rain events or with a hose. Clogs should be dealt with as soon as they are identified. Porous pavements should not be sanded as a deicing measure as sand will clog pores. Snowplows should be equipped with shoes or high-density plastic blades to reduce the risk of damage to the pavement surface. Snow brooms will remove snow safely as well. Pores should be vacuumed between April and October as it can be less effective in cold weather.
Porous gravel pavements	Generally low maintenance requirements compared to other porous pavements. Particle displacement occurs when vehicles make sharp turns. Particles can be releveled easily with rakes. Keep snowplow blades 12mm above the surface. Permeability can be restored by excavating the surface layer and replacing it with new open graded aggregate. Herbicides should only be spot applied.
Permeable unit pavements	Joint materials should be refilled once 3–6 months after initial installation and again as needed (approx. every 5–10 years). The pavement surface should be swept, blown, or brushed regularly. Care should be taken to remove leaf litter and grass clippings after storms and maintenance activities. If rotary brushes are used, some joint material will need to be replaced. Restorative maintenance activities include use of vacuums and/or power washers. This should occur as required. Maintain a stockpile of joint material to replace as required. Deicing chemicals should be swept to remove excess as the products can damage concrete units. In spring, pavers should be washed to remove any remaining deicing chemicals. Refer to manufacturers literature for deicing chemicals to avoid.
Porous asphalt	Vacuum sweeping and/or power washing should be performed 2–4 times per year. Repairs can be made with standard asphalt. Porous asphalt pavements should never be seal coated or crack sealed. Snowplow blades should be lifted at least 6mm above the pavement and be plastic. Never use fine sand for deicing. Completely clogged porous asphalt can be rehabilitated by milling off the top 25–35mm.
Pervious concrete	Vacuum sweeping and/or power washing should be performed 2–4 times per year. Snowplow blades should be lifted at least 12mm above the pavement and be plastic. Avoid deicers with magnesium chloride, calcium magnesium acetate, and potassium acetate. Never use fine sand for deicing. Repairs can be made with standard concrete. Moss growth can be pressure washed away. Baking soda will control moss growth. Avoid sealants.

Sources: NRMCA 2024; Unilock 2019; FHWA 2016; FHWA 2015; University of New Hampshire Stormwater Center 2015a; Ferguson 2005; Solano 2024

as base reservoirs to hold stormwater as it infiltrates into native soil. Like other stormwater control measures, most systems will require an underdrain to accommodate very large storm events.

When trees are planted in pavement openings, the opening should be expandable to accommodate growth of the tree trunk and root flare. Removable pavers, expandable tree grates, gravel, or mulch can be used.

Suspended Pavement Systems

Suspended pavement systems use a modular structure to support the weight of a pavement while providing void space for soil, tree roots, and stormwater underneath (Figures 7.13 and 7.14). Modular cell systems, primarily plastic materials, support the pavement and some are filled with open graded aggregates. Void space for soil is greater than in structural soils, with some products claiming up to 90% void space. Each suspended pavement system is different, but general design and specification considerations are discussed in Table 7.8.

Structural Soils

Structural soils are two-part systems that provide a load bearing lattice of angular aggregate or coarse sand with stone-to-stone contact to provide a stable pavement base. Voids in the base are filled with soil to accommodate tree root growth. Structural soils encourage tree roots to spread over a larger area, which helps prevent sidewalk heaving during root expansion. Frost heave is also minimized with pavement

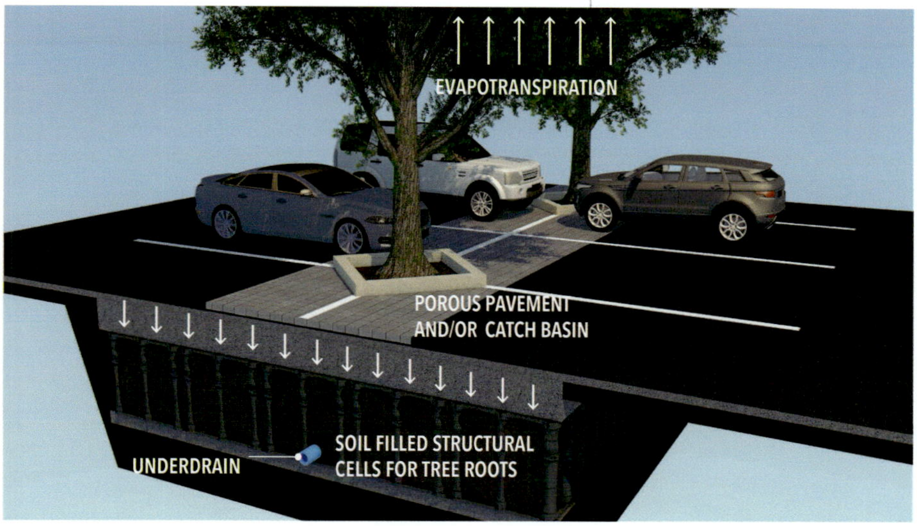

Figure 7.13 Suspended pavement system allowing for tree roots to grow underneath pavement. Image credit: DeepRoot Green Infrastructure and Meg Calkins

Figure 7.14 Installation of a suspended pavement system along a road in Edmonton, Canada. Image credit: DeepRoot Green Infrastructure

over structural soils. Structural soils can link openings in pavements allowing tree roots far more room for expansion than with a typical pavement base (Table 7.9 and Figures 7.15 and 7.16).

There are two main types of structural soils: sand-based structural soils (SBSS) and crushed stone-based structural soils (CSBSS). Both incorporate soil and tackifying agents.

Aggregates or sands for structural soils are uniformly sized angular stone. The angular shape will provide good interlock to support the pavement above, while the uniform size will result in pores for soil, tree roots, air, and water. For CSBSS, crushed stone ranging from 19–38mm is used without fine materials. For SBSS, crushed stone ranging from ½–2mm is used instead. After compaction of the aggregate or sand materials to meet load bearing specifications, about 25–40% void space will remain (River Sand Incorporated 2024; Key 2019; Bassuk et al. 2015) (Figures 7.17 and 7.18).

Soil, the second component of the system, fills the voids between the aggregates and remains uncompacted and root penetrable. A heavy clay loam or loam with a minimum of 20% clay and an organic matter content of 2–5% will ensure nutrient and water holding and adequate cation exchange capacity. While the soil fills the voids, care must be taken to

Table 7.8 Design and Specification Considerations for Suspended Pavement Systems

Topic	Design and Specification Considerations
Volume and depth of soil cell systems	For an estimation of minimum required soil volume, multiply the area of the mature tree canopy and by 0.6m. Stormwater quantity calculations may necessitate a greater volume for the soil cell structure than the tree needs. The system should be located above seasonal high groundwater table.
Extent of soil cells under pavement	Consider use of structural soil cells in a continuous trench under a row of trees and under the adjacent pavement to allow for plenty of root space.
Slopes	Slopes up to 5% when full load bearing capacity if needed. Ten percent is allowable for locations with less load bearing capacity.
Stormwater pre-treatment	Stormwater does not need to be pretreated, but energy dissipation, flow distribution measures will distribute the stormwater throughout the system and avoid erosion of soil in cells.
Soil mix	The soil mix should be consistent with local or regional bioretention soil mixes and properties. A choker course should be provided between the bioretention soil and the drainage layer to avoid migration of fines.
Drainage aggregate	Many systems use a 10–15cm drainage layer of aggregate at the bottom and within the modules. Aggregate is open graded if infiltration is planned.
Subgrade	Most systems are installed on top of a 95% compacted native soil subgrade unless infiltration is planned and a geotechnical engineer identifies that the native soil will support the structure without full compaction. Compaction should extend at least 30cm beyond the system. A geotextile liner recommended between the system and the subgrade.
Irrigation	Use low volume drip irrigation to establish trees.
Drainage	A perforated, wrapped underdrain connected to the stormwater system should be employed when needed. Location of pipe will vary by product.
Specifications and details	Most soil cell brands offer details and specification language for use of their proprietary product.
Maintenance	Inlet/outlet structures, distribution pipes, and underdrain pipes require periodic cleanout. Trees require annual maintenance and sometime irrigation.

Sources: CityGreen 2024; DeepRoot2018

Table 7.9 Design and Specification Considerations for Structural Soils

Topic	Design and Specification Considerations
Volume and depth of structural soil	Two cubic feet of structural soil is recommended for every square foot of tree crown protection. Minimum soil depth of 60cm (24”), but 92cm (36”) is preferred.
Extent of structural soil under pavement	Structural soil should extend under the entire length and width of pavement areas to ensure uniform engineering characteristics.
Aggregate	For CSBSS, uniformly sized angular crushed stone ranging from 19 to 38mm without fine materials. For SBSS, 1/2mm–2mm uniformly sized angular crushed stone should be used. California Bearing Ratio should be greater than or equal to 50.
Soil mix	Heavy clay loam or loam with a minimum of 20% clay, and an organic matter content of 2–5%. It is recommended that structural soils be used in tree pits under the root ball, but if the tree pit is greater than 2.4x2.4m (8’x8’), then sandy loam can be used around the root ball. Soil should be 18–22% of the soil, aggregate, tackifier mix by weight.
Irrigation	Use low volume drip irrigation to establish trees and possibly longer if necessary.
Tackifier	Mix tackifier with aggregate prior to mixing with soil.
Drainage	Positive drainage below the tree root system is recommended. A perforated, wrapped drainpipe connected to the stormwater system should be placed between the structural soils and compacted subgrade as required.
Existing trees	Structural soil can be used under and adjacent to existing trees where pavement retrofits are implemented. Roots should be excavated with an air excavation tool then kept covered and moist until structural soil is backfilled.
Stockpiling	It is best to avoid stockpiling structural soil, but if short-term stockpiling must happen, it should be protected from rain and containments.

Sources: River Sand Incorporated 2024; Key 2019; Bassuk et al. 2015; Lindsey and Bassuk 1992

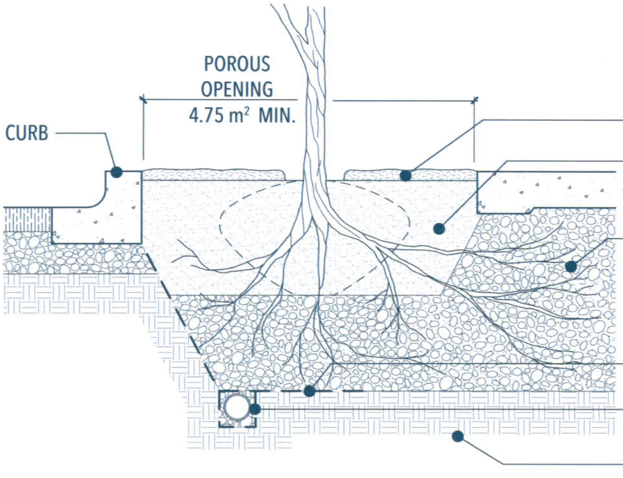

Figure 7.15 Structural soil detail for crushed stone-based and sand-based structural soils at a tree opening. Actual details may vary slightly by type of structural soil. Refer to manufacturer recommendations. Adapted from the Urban Horticulture Institute, Cornell University, Ithaca, NY. Image credit: Kathleen Elliott and Meg Calkins

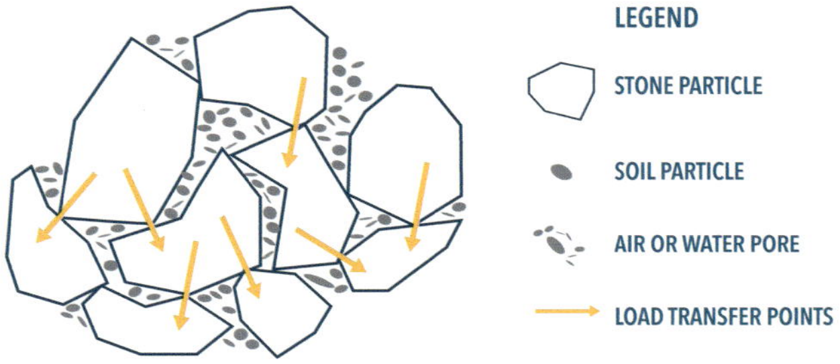

Figure 7.16 Stone-to-stone connection is critical for the structural performance of structural soil. Soil fills the voids between the aggregates and remains uncompacted and root penetrable. Image credit: Drake Bruner and Meg Calkins

Figure 7.18 Local stone boulders define the parking spaces and drive aisles of the permeable parking lot at the Glenstone Museum. Image credit: PWP Landscape Architecture

Figure 7.17 PWP Landscape Architecture used permeable gravel pavement over sand-based structural soil for the parking lot at the Glenstone museum in Potomac, Maryland. The accessible parking spots at the edge of the gravel parking area are ADA compliant chipseal paving. Image credit: PWP Landscape Architecture

not over-fill them so that aeration and bearing capacity are compromised (Bassuk et al. 2015).

Proprietary ***tackifying agents*** allow for the stones and soil to mix uniformly and prevent separation of the materials from vibration during transport and placement.

Porous Pavements and Structural Soils

Structural soils can be used with porous pavements; however, their design and specifications differ from their use under impervious pavements. Structural soil can be used as a base reservoir for porous pavements in place of the typical open graded gravel reservoir as it is designed to be compacted, and engineered to support healthier tree growth. Table 7.10 details reservoir depths and corresponding levels of mitigated rain events for a

Table 7.10 CU-Structural Soils™ Depths for Rain Events

Size of Rain Event (mm)[1]	Depth of CU-Structural Soil™ Reservoir Needed to Mitigate Rain Event (mm)
40mm	152mm
79mm	305mm
119mm	457mm
159mm	610mm
198mm	762mm
239mm	914mm

Source: Bassuk et al. 2015

[1]For reference only, civil engineers should determine actual depths.

CU-Structural Soils™ mix. It is important to keep in mind that infiltration rates for native soils under structural soils vary widely, therefore civil engineers should determine actual depths of base reservoirs.

When using permeable pavers over structural soil, a uniformly graded choker course (ASTM No. 8 recommended) should be used at a depth specified by the paver manufacturer. A geotextile should be placed between the setting bed and the structural soil. Structural soil can also be used under porous asphalt and porous concrete pavements.

Green Roofs

Green roof systems are planted assemblies on the roof of buildings or other structures. They offer many benefits, particularly in urban areas where there are large amounts of impervious surfaces. Stormwater benefits are primary, but they also moderate the urban heat island effect, improve air quality, offer health and well-being benefits, and contribute usable landscape spaces (Table 7.11).

The drainage of green roofs is an important consideration in their design. Rainwater must be retained long enough to sustain plants and allow for transpiration and evaporation. But excess water cannot be allowed to oversaturate soil and/or collect on the roof adding weight and increasing potential for leaks. The roof deck and waterproofing must be sloped toward roof drains tied into the building's roof plumbing system. Collaboration with structural engineers, MEP engineers and architects is crucial for the success of green roofs.

Types of Green Roofs

Green roof types are defined by the thickness of the soil media, weight, and types of plants and activities they can support. Most green roofs are used on roofs with only minimal slope, but some products allow for their use on roof slopes up to 35°. Lower slopes will mitigate greater quantities of stormwater. Multiple types of green roofs discussed in the following sections can be combined on one roof.

Extensive Green Roofs

Extensive green roofs are the lightest weight and thinnest type of green roof with a growing medium depth of 50–150mm. With fully saturated weights ranging from 49 to 170kg/m^2, they can be installed on existing buildings with minimal alternation as long structural capability allows (MNPCA 2008). Most extensive green roofs are not designed for public access except for maintenance activities (Figure 7.19).

The minimal depth of extensive green roofs limits many types of plants, but this means that the maintenance requirements may be lower. Characteristics of plants that will thrive on extensive roofs are relatively fast establishment, spreading, density, drought and cold tolerance, and disease and pest resistance. Types of plants that can meet these criteria are sedums, delospermas, euphorbias, and sempervivums (Getter and Rowe 2008).

Extensive green roofs are the least expensive type of green roof to install and maintain. They can be modular, with trays placed next to each other, or they can be monolithic with green roof layers placed individually. There are several proprietary extensive green roof systems available.

Semi-Intensive Green Roofs

Semi-intensive green roofs have growing media depths of 125–200mm and fully saturated weights of 170–224kg/m^2 (MNPCA 2008). Costs of semi-intensive green roofs are higher than extensive, but lower than intensive green roofs. They can be designed with pathways and deck areas for gathering, but since trees are not possible, shade must be provided by other structures (Figures 7.20 and 7.21).

The increased depth of soil medium for semi-intensive green roofs allows for a wider plant palette that can include perennials, small shrubs, and ornamental grasses. Semi-intensive green roofs will have maintenance and watering needs in some climates. They can offer more habitat for urban wildlife than extensive green roofs (Getter and Rowe 2008).

Intensive Green Roofs

Intensive green roofs, also called roof gardens, have soil depths from 200mm to 1200mm, allowing for trees, shrubs, grasses, and perennials. Most green roofs that are designed for occupation are intensive green roofs. The initial and

Table 7.11 Benefits and Challenges of Green Roofs

Benefits of Green Roofs	
Stormwater benefits	Stormwater is stored, then taken up by plants and returned to the atmosphere through transpiration and evaporation. Green roofs can retain 70–90% of warm weather precipitation and 25–40% of winter precipitation depending on the type of roof and climate. Stormwater peak flow times are slowed. Stormwater is cooled as compared with most standard roof types. Stormwater quality is improved by filtration through soil media and plant roots.
Urban heat island moderation	Plants on green roofs cool air through daily dew and evaporation. Plants cover dark roofing surfaces that could contribute to UHI. Dust and particulate matter in air is reduced as well as the production of smog.
Air quality improvements	Plants on green roofs can filter air pollutants and gases. Plants and soils on green roofs sequester carbon.
Energy efficiency	The increased insulation and cooler surfaces of green roofs can reduce air conditioning and heating loads, reducing demand for fossil fuel based energy.
Habitat provision and biodiversity	Green roofs can provide habitat for insects and birds. Plants on green roofs can add to urban biodiversity.
Waste mitigation	Green roofs reduce the UV degradation of roofing materials and waterproofing membranes, giving them longer service lives. Lightweight recycled materials such as rice hulls, wood residues, almond shells, pine bark, crushed brick, polystyrene flakes, and coconut coir can be used in the growing medium. The service life of HVAC equipment is prolonged with reduced loads due to lower heating and cooling needs.
Aesthetic, health and well-being benefits	Improved views from surrounding buildings to the green roof. Green roofs can offer amenities such as gathering spaces, play and recreation spaces, dining areas, and community gardens. Plants on green roofs can improve human health and well-being of occupants or viewers. Green roofs can foster a sense of community among building occupants.
Challenges of Green Roofs	
Cost	Green roofs have higher first and maintenance costs than standard roofing; however over their life cycle, there may be cost savings due to the longer service life of green roofs.
Weight	Fully saturated, green roofs can increase the dead load on roofs by 50–1400kg/m^2.
Maintenance and monitoring	Green roofs require regular maintenance including weeding, feeding, and watering. Monitoring activities include checking on the continued efficacy of the drainage system. Amount of runoff through storm drains could be measured to monitor stormwater performance of the green roof.
Access	Occupied green roofs must be accessible for all so elevator access and ADA compliant level changes are required. Extensive green roofs require roof access for maintenance activities.
Monitoring	Green roofs must be regularly monitored to ensure hydrologic performance.

Sources: GRHC 2024; Solano 2024; American Hydrotech 2023; MNPCA 2008

maintenance costs for intensive green roofs are higher than other types, but they can be amenities adding considerable value to buildings (Figure 7.22).

Intensive green roofs can be very heavy, particularly if they are supporting trees and associated soil media. The fully saturated weights vary widely from 200 to 1500kg/m^2 (MNPCA 2008). Trees and other heavy elements are often placed directly over building columns, and lightweight fill is employed to create topographic variations. Look for lightweight fill with lower human health risks (Chapter 12).

Components of Green Roofs

Green roofs are assemblies comprised of multiple layers with specific performance expectations. Layers are discussed

Figure 7.19 An extensive green roof covers a mixed use building in Germany. Image credit: JF123

Figure 7.20 This semi-intensive modular green roof at the Talley Student Union on the NC State University Campus was designed and built by Professor Anne Spafford and volunteers. Image credit: Anne Spafford

Figure 7.21 This semi-intensive green roof at Chicago's City Hall was designed by Conservation Design Forum. Image credit: Tony the Tiger

Figure 7.22 These intensive green roofs have a soil profile thick enough to support trees, shrubs, groundcover, and decks. Image credit: Piranka

next in general order from top to bottom of a green roof assembly. Sometimes the order of the insulation layer, root barrier and waterproofing membrane will vary due to climatic needs.

Vegetation

Appropriate vegetation for green roofs is directly related to the depth of the growing medium. As green roofs are subject to excesses of climate such as temperature, humidity, and desiccation, vegetation on green roofs must be able to tolerate these extremes. Vegetation should be drought and wind resistant, tolerant of the extremes of the maps for both the US Department of Agriculture (USDA) cold hardy zones and the American Horticultural Society (AHS) Plant Heat Zones, and able to endure periods of wet and dry soil. Shallow rooted trees, shrubs, grasses, and groundcover will thrive in the limited depth of growing media for all types of green roofs. Vegetation that does not need intensive maintenance such as fertilizers or pruning is best for extensive and semi-intensive green roofs (Figure 7.23).

Growing Medium

Growing media for green roofs is a combination of materials that provide a stable structure for plant roots and their required nutrients. Organic matter is used in growing media,

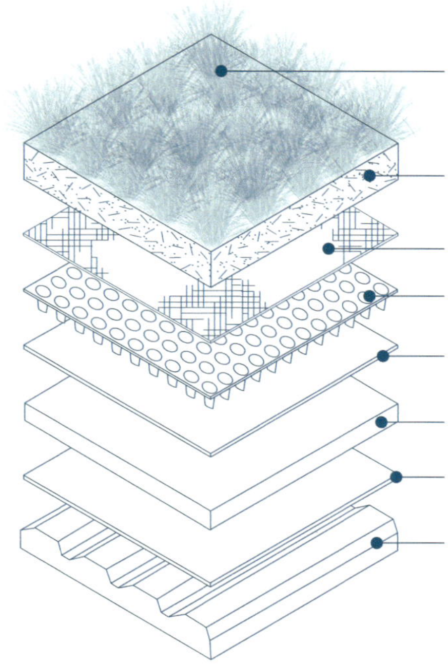

Figure 7.23 Typical layers of green roofs. The growing medium and drainage layer thicknesses will vary by type of roof. The vapor control layer and insulation layers may vary by climate. Image credit: Kathleen Elliott and Meg Calkins

along with lightweight aggregates such as pumice, perlite, expanded clay, sand, or crushed brick that help to aerate the roots, retain water, and minimize compaction. Growing media should be lightweight but not be so light that wind erosion occurs. Degree of porosity and permeability is also critical as it should not drain too quickly. Some green roof systems offer proprietary blends of growing media (SSSA 2024).

Irrigation

Irrigation, either temporary or permanent, may be required for green roofs in some climates. Spray, rotors, sprinklers, and drip irrigation can all work on green roofs. Hoses and sprinklers may be preferable for temporary irrigation to establish plants. Pop-up sprinklers are commonly used for lawns on intensive green roofs, while drip lines can be used for groundcovers and sedums on any type of roof. The best time to water green roofs is within three hours after dawn or three hours before sunset (GeoPard Agriculture 2024).

While green roofs can offer an amenity in arid climates, unoccupied green roofs that must be irrigated are unsustainable in areas with water supply shortages. In these areas, cool roofs (highly reflective) may result in lower energy impacts and costs (Wong et al. 2003; Muga et al. 2008).

Filter Fabric

The filter fabric layer, also called the filter sheet, sits between the growing media and the drainage layer to keep particles of the growing media from clogging the drainage layer. It is a non-woven geotextile that can be its own layer or it can be part of a drainage layer product.

Drainage Layer

The drainage layer's function is to retain stormwater for long enough to allow its evaporation and transpiration and to remove excess water. Drainage layers are often a cellular system that retains water so it can evaporate up into the growing media. The layer allows excess water to drain away through channels or other routes, but thicker drainage layers will detain more water. Drainage board has a thin profile and can be used where space is limited. It keeps water available but can also move water hydrostatically minimizing the slope required for drainage. There are multiple proprietary products for drainage layers and thicknesses vary. Sometimes they are filled with open graded gravel or aggregate (Solano 2024; American Hydrotech 2023).

Some green roofs incorporate a water retention layer with the drainage layer to allow for greater quantities of retention and slow peak runoff rates. These layers can be lightweight aggregate, mineral wool, foam, or heavy duty fabrics (EcoGardens 2024).

Insulation Layer

The insulation layer is placed either between the drainage layer and the root barrier, or between the root barrier and the waterproofing membrane depending on climate. Insulation layers are primarily rigid insulation made from expanded polystyrene (also called geofoam), extruded polystyrene, polyurethane foam, or polyisocyanurate board. Many of these foams contain chlorinated or halogenated flame retardants that pose health risks such as cancer, endocrine disruption, and reproductive toxicity (GSPI 2024). Look for halogen-free polyisocyanurate boards or fiberglass board insulation (Informed 2024).

Root Barrier

Root barriers protect waterproofing membranes from plant roots that find their way into the drainage layer. For extensive green roofs with sedums, this root barrier could be 10mil plastic sheeting; root barriers will need to be thicker (30mil) and tougher for intensive green roofs with trees. Root barriers are polyethylene, polypropylene, or PVC. PVC products should be avoided due to health impacts in manufacture and disposal (Chapter 21 "Plastics and Plastic Products").

Some root barriers are incorporated as a layer in the waterproofing membrane. They may incorporate substances such as copper hydroxide that deter aggressive roots.

Root barriers should adhere to ASTM Standard *D6241 Standard Test Method for Static Puncture Strength of Geotextiles and Geotextile-Related Products Using a 50-mm Probe*. They should be overlapped when installed.

Waterproofing Membrane

Waterproofing membranes are critical to maintaining a dry building. Water leaks into building walls or roof structures is the single largest concern of building owners considering green roofs (Hendricks and Calkins 2006). Waterproofing membrane products can be pre-formed sheets or fluid applied.

Leak tests are carried out after the waterproofing membrane is installed prior to installation of the other layers. It is very difficult to fix leaks after the layers are installed.

Structural Roof Deck

The structural roof deck of the building must be designed to support the weight of a fully saturated green roof, along with other dead and live loads.

Achieving Topographic Variation on Intensive Green Roofs

Lightweight fills are used on intensive green to achieve topographic variation under planted areas. There are multiple types of lightweight fills used with varying weights discussed here and in Table 12.8.

EPS geofoam blocks and sheets are very lightweight and easily stacked and shaped to create topographic forms. Geofoam can be shaped off-site to achieve precise forms. It is strong and resistant to moisture. The bulk density of EPS ranges from 11–32kg/m^3. Geofoam poses human and environmental health impacts discussed in the "Insulation Layer" section, and it's structure may allow rodents to burrow into it (Green Roof Technology 2024).

Foamed glass gravel is produced from ground recycled glass powder mixed with a foaming agent then heated in a kiln. It is a lightweight fill that can be compacted into angles up to 45° with a bulk density is around 160kg/m^3 (Glavel 2024). It is heavier than geofoam, but only around 10% of the weight of soil. The embodied carbon of foamed glass gravel is similar to geofoam but with fewer health impacts.

Lightweight expanded clay aggregate (LECA) is produced from plastic clay with little or no lime that is dried, heated, and burned in rotary kilns at 1100–1300°C. The result is hollow clay pellets. It is mixed with soil for lightweight fill and soil mixes. It is lightweight, but heavier than geofoam and foamed glass at 250–710kg/m^3 (Rashad 2018).

Expanded shale, also called vitrified shale, is lightweight, gray gravel made from shale that has been crushed and fired in a rotary kiln at temperatures over 1100°C. As it cools, gases escape leaving pockets and pores. It is mixed with soil for lightweight fill and soil mixes. It is lightweight, but heavier than geofoam and foamed glass at around 480kg/m^3.

Hardscape Elements on Green Roofs

Decks, elevated pathways, low walls and ramps can be incorporated into green roofs if structural capacities are considered, lightweight materials are prioritized, and detailing can withstand the harsher conditions on roof gardens. For green roofs on new buildings, there may be opportunity to connect hardscape elements to the roof, but on existing building retrofits, foundations for hardscape elements may be gravity based, weighted, or buttressed (Solano 2024).

Elevated Pavements and Decks

Pavements and decks on green roofs are elevated above the roofing membrane/protection layers on pedestal systems as conventional foundations are not possible. Pedestal systems are proprietary products that are primarily made from plastics such as polyethylene, polypropylene, and PVC. Look for products with recycled content.

Pedestals rest directly on the protection board often without fasteners that might puncture the waterproofing layer. Most pedestals have a screw adjustment that allows for leveling to achieve level surfaces.

Pedestals can support masonry pavers, wood deck tiles, or even dimension stone with attachments that hold the units in place (Figure 7.24). Stormwater that falls on the surface of elevated pavements and decks can flow between the units to the roof deck below, then to roof drains. Ideally water is directed to the planted areas of the green roof. Refer to Chapter 10 "Decks and Boardwalks" for more information about pedestals and deck structures.

Edges

Edges define and retain planting areas and can be made from concrete, stainless steel, recycled plastic, or aluminum. They

Figure 7.24 Adjustable deck and pavement pedestals for elevated pavements on green roofs. Image credit: OceanProd

Table 7.12 Standards for Green Roofs

EU Standards	European Union Urban Green Infrastructure Strategy (UGIS)
	FLL Guidelines for the Planning, Installation and Maintenance of Green Roofs
UK Standard	Green Roof Organization (GRO) Green Roof Code, UK
ASTM Standards	E2777 Standard Guide for Vegetative (Green) Roof Systems
	E2400 Standard Guide for Selection, Installation, and Maintenance of Plants for Vegetative (Green) Roof Systems
	E2399 Standard Test Method for Maximum Media Density for Dead Load Analysis of Vegetative (Green) Roof Systems
	E2788 Specification for Use of Expanded Shale, Clay and Slate (ESCS) as a Mineral Component in the Growing Media and the Drainage Layer for Vegetative (Green) Roof Systems

are often "L" shaped and are installed above the filter fabric. Many edges will have perforations to allow water to flow through, or they will be elevated on pedestals with screens for water to move below the edge restraints through the filter fabric and into the drainage base course toward the roof drains.

Drains and Gutters

The roof deck and waterproofing must be sloped toward roof drains tied into the building's roof plumbing system. All drains on green roofs must be accessible for maintenance and protected from leaf litter and other elements that could clog them. Drains are often covered by filters and strainers and contain inspection chambers. Drains should be inspected every three months and after major storms.

Table 7.13 Resources for Additional Information

Porous Pavement, Structural Soil, and Suspended Pavement Resources
Porous Pavements, book by Bruce K. Ferguson
National Ready Mix Concrete Association (NRMCA) Pervious Concrete Pavement Resources
National Asphalt Pavement Association Porous Asphalt Resources
CIRIA SuDs Manual
US Environmental Protection Agency Stormwater Resources
Minnesota Stormwater Manual
North Carolina Stormwater Manual
University of New Hampshire Stormwater Center
Cornell University Urban Horticulture Institute
Green Roof Resources
Green Roofs for Healthy Cities
FLL Guidelines for the Planning, Installation and Maintenance of Green Roofs, Germany
US Environmental Protection Agency Green Roof Resources
European Federation of Green Roof and Living Wall Associations (EFB)
Growing Green Guide: A Guide to Green Roofs, Walls and Facades in Melbourne and Victoria Australia
Green roof systems: A guide to the planning, design, and construction of landscapes over structure, book by Weiler and Scholz-Barth
The Green Roof Manual: A Professional Guide to Design, Installation and Maintenance, book by Snodgrass and McIntyre
Planting Green Roofs and Living Walls, book by Dunnett and Kingsbury
Selecting Plants for Extensive Green Roofs in the United States. Extension Bulletin E-3047. Michigan State Extension.

References

American Hydrotech (2023). *Garden Roof Planning Guide: From Concept to Completion*. American Hydrotech Inc.

ASTM (2022). *ASTM D448-12 Standard Classification for Sizes of Aggregate for Road and Bridge Construction*. ASTM International.

Bassuk, N., et al. (2015). *C-U Structural Soil: A Comprehensive Guide*. C-U Structural Soil.

CityGreen (2024). *Specifier Update, Issue 20*. City Green Urban Landscape Solutions. https://citygreen.com/what-is-structural-soil/, accessed August 19, 2024.

DeepRoot (2018). *Silva Cell Fact Sheet*. DeepRoot Green Infrastructure.

EcoGardens (2024). "What are the layers of a green roof?". https://info.ecogardens.com/blog/, accessed August 29, 2024.

Elvidge, C., et al. (2007). "Global distribution and density of constructed impervious surfaces". *Sensors (Basel)* 7: 1962–1979.

Ferguson, B. (2005). *Porous Pavements*. Integrative Studies in Water Management and Land Development, CRC Press.

GeoPard Agriculture (2024). "What are rooftop irrigation systems?". https://geopard.tech/blog, accessed August 29, 2024.

Getter, K.L., and Rowe, D.B. (2008). *Selecting Plants for Extensive Green Roofs in the United States*. Extension Bulletin E-3047. Michigan State Extension.

Glavel (2024). *Green Roof Fill: High Friction Angle. Lightweight. Free Draining*. Glavel Foam Glass Gravel. www.glavel.com/applications/greenroof/, accessed May 8, 2024.

Green Roofs for Healthy Cities (GRHC)(2024). "About green roofs". https://greenroofs.org/about-green-roofs, accessed August 29, 2024.

Green Roof Technology (2024). "Intensive green roofs". https://greenrooftechnology.com/green-roof-finder/intensive-green-roof/, accessed May 8, 2024.

Green Science Policy Institute (GSPI) (2024). "Some flame retardants of concern". https://greensciencepolicy.org/harmful-chemicals/, accessed August 29, 2024.

Hendricks, J.S., and Calkins, M. (2006). "The adoption of an innovation: Barriers to use of green roofs experienced by Midwest architects and building owners". *Journal of Green Building* 1.3: 148–168.

Ibisch, Pierre L., et al. (2016). "A global map of roadless areas and their conservation status". *Science* 354.6318: 1423–1427.

Informed (2024) *Insulation Product Guidance*. Habitable Future. https://informed.habitablefuture.org/product-guidance/7-insulation, accessed August 29, 2024.

Industry Analytics, Research and Consulting (Industry ARC) (2024). *Permeable Pavement Market Overview*. Industry Arc.

International Market Analysis Research and Consulting Group (IMARC Group) (2024). *Green Roof Market Report by Type (Extensive, Intensive), Distribution Channel (Online, Offline), Application (Residential, Commercial, Industrial), and Region 2024-2032*. Report ID SR112024A4540. IMARC Group.

Invisible Structures (2023). *Gravel Pave2*. Invisible Structures.

Key, A. (2019). "Compaction and sand based structural soil". www.linkedin.com/pulse/compaction-sand-based-structural-soil-al-key/, accessed June 17, 2019.

Lindsey, P., and Bassuk, N. (1992). "Redesigning the urban forest from the ground below: A new approach to specifying adequate soil volumes for street trees". *Arboricultural Journal* 16: 25–39.

Minnesota Pollution Control Agency (MNPCA) (2008). *Minnesota Stormwater Manual, Version 2*. State of Minnesota.

Muga, H., Mukherjee, A., and Mihelcic, J. (2008). "An integrated assessment of the sustainability of green and built-up roofs". *Journal of Green Building* 3.2: 106–127.

National Ready Mix Concrete Association (NRMCA) (2024). "Pervious concrete pavement maintenance and operations guide". www.perviouspavement.org/downloads/pervious_maintenance_operations_guide.pdf, accessed August 29, 2024.

Rashad, A.M. (2018). "Lightweight expanded clay aggregate as a building material – an overview". *Construction and Building Materials* 170: 757–775. https://doi.org/10.1016/j.conbuildmat.2018.03.009.

River Sand Incorporated (2024). "Structural soil". https://riversandinc.com/soil/structural-soil/, accessed September 16, 2024.

Soil Science Society of America (SSSA) (2024). "Green roofs". www.soils.org/about-soils/green-roofs/, accessed August 29, 2024.

Solano, L. (2024). Personal communication, September 13.

Sun, Z., et al. (2022). "Global 10-m impervious surface area mapping: A big earth data based extraction and updating approach". *International Journal of Applied Earth Observation and Geoinformation* 109.

Unilock (2019). *Permeable Paver Maintenance Guide*. Hengestone Holdings.

University of New Hampshire (UNH) Stormwater Center (2015a). *Pervious Concrete for Stormwater Management*. UNH Stormwater Center.

University of New Hampshire (UNH) Stormwater Center (2015b). *Pervious Interlocking Concrete Pavement (PICP) for Stormwater Management*. UNH Stormwater Center.

University of New Hampshire (UNH) Stormwater Center (2015c). *Porous Asphalt Pavement for Stormwater Management*. UNH Stormwater Center.

US Environmental Protection Agency (US EPA) (2008). *Urban Stormwater Management in the United States*. US EPA, Committee on Reducing Stormwater Discharge Contributions to Water Pollution.

US Environmental Protection Agency (US EPA) (2016). *Stormwater Trees: Technical Memorandum*. EP-BPA-13-R5-0001. Great Lakes Restoration.

US Federal Highway Association (FHWA) (2015). *Tech Brief: Porous Asphalt Pavements with Stone Reservoirs*. FHWA-HIF-15-009. FHWA.

US Federal Highway Association (FHWA) (2016). *Tech Brief: Permeable Concrete Pavements*. FHWA-HIF-16-004. FHWA.

Wong, N.H., et al. (2003). "The effects of rooftop garden on energy consumption of a commercial building in Singapore". *Energy and Buildings* 35: 353–364.

Woods Ballard, B., et al. (2015). *The SuDS Manual (C753)*. CIRIA, December.

CHAPTER 8

Retaining Walls and Bioengineered Embankments

Retaining wall structures or bioengineered embankments are required when a grade differential is greater than the angle of repose of the soil or when slope conditions require stabilization. Low retaining walls can serve as seat walls in areas where seating is needed, and they can define space. Retaining walls and embankments facilitate level outdoor spaces adjacent to buildings and level changes at stairs.

Retaining walls and bioengineered embankments can also protect and restore riparian edges. In these situations, bioengineered embankments may be preferable to retaining walls as plant roots in the structures will reinforce the soil behind the structure and take up any additional soil moisture, in addition to providing other environmental benefits.

The primary consideration with retaining structures is stability as the structure must be designed to resist forces above and behind the wall. These forces can be lateral soil pressure, surcharge above the wall, expansion of soils from freeze/thaw cycles, soil surcharge, earthquake forces, or vehicles and structures above the wall. Soil type and bearing capacity under the footing or base of the wall are other important factors for stability.

This chapter addresses a range of retaining structures from live fascines to rigid retaining walls. It begins with bioengineered embankments (also called bio-reinforced embankments), which are generally the lowest impact structures from an embodied carbon, resource use, and cost standpoint. Flexible retaining walls structures are discussed and the chapter concludes with a discussion of rigid retaining walls including poured in place concrete walls and veneer retaining walls with brick, stone, or precast concrete. Embodied carbon is calculated for a range of rigid and flexible retaining wall assemblies.

Bioengineered Embankments and Living Walls

Bioengineered embankments and retaining structures offer a low impact alternative to masonry retaining walls in many settings. Traditionally these structures have been used in rural and natural settings, but they can offer much needed ecosystem services in urban settings as well. As urban landscapes become more environmentally driven, these bioengineering solutions will be important design options that could be cost effective and aesthetically enhancing. If reducing the embodied carbon of a site design is a primary goal, bioengineered embankments and natural field stone walls should be employed instead of masonry retaining wall structures wherever they are feasible.

Bioengineered structures combine living vegetation with non-living materials to stabilize slopes and stream banks. Plants and live branch cuttings are integrated with logs, geotextiles (both biodegradable and non-biodegradable), stone, and soil to form living structures that reinforce slopes. The structures discussed in this chapter result in strong living structures that can stabilize and revegetate vulnerable slopes and streambanks. The embodied carbon of bioengineered structures is far less than for conventional retaining walls and will ultimately result in carbon positive structures due to their vegetation.

Rigid retaining walls can be 5–6 times as expensive to construct as bioengineered embankments and retaining structures; however, bioengineered structures will require periodic maintenance and in some instances refurbishment. Some structures may not be successful in arid climates where the soil is very dry and not enough water is available to support the living materials. In these climates, stone

DOI: 10.4324/9781003413974-9

embankments and rubble or recycled concrete filled gabions may be more appropriate.

Bioengineered structures require larger areas of construction than retaining walls as many are shallow slopes instead of vertical walls. They may not be appropriate in tight urban spaces. However, the extent of soil disturbance and excavation could be less than is required for the deep footings and backfill of a rigid retaining wall, which could save costs and limit the carbon released when soil is disturbed.

Live Staking

Live staking, also called live cutting or pole planting, creates a living root mat that stabilizes soil by both reinforcing it and extracting excess moisture from the soil. Live cuttings of willow or other rapidly rooting species are inserted and tamped into the ground. Willow stakes can be used to hold other erosion control measures such as fascines, fiber rolls or biodegradable erosion control matting in place. Rip rap may be used at the toe of the slope, and stakes can be inserted in this area as well (Figure 8.1).

Live staking is useful in areas of consistent moisture such as humid and temperate climates and streambanks. Stakes should be 12–38mm in diameter with side branches removed but bark intact, and they should be 60–90cm long. Place live stakes in a triangular spacing at 60–90cm (NRCS 2021b).

Live Fascines

Live fascines are living branches bundled into 16–22cm tubular shapes and staked with dead stout stakes into a slope. They are placed into shallow contour trenches perpendicular to dry slopes and angled into trenches on wet slopes. The fascines provide immediate slope protection, then long-term protection as the live branches root into the slope. Fascines are held in place with live stakes at 75–100cm on center that will also root into the slope. Dead stout stakes further stabilize the fascines (Table 8.1; Figure 8.2).

Live fascines can trap and hold soil on the face of the slope, creating small terraces. Cuttings for live fascines should be 18–50mm in diameter and 1.5–2.5 meters long to form 2.5m length bundles. Side branches should be removed from the cuttings, but the bark should remain. Species with long straight branches that quickly root such as young willows or shrub dogwoods should be used (LaRiMiT Toolbox 2023).

Brush Layering

Brush layering alternates layers of brush and benches of wrapped earth lifts stacked on a slope between 2:1 and 4:1. The cuttings of quickly rooting plants, such as willows or poplars, are layered between wrapped earth lifts and set perpendicular to the slope with the ends protruding from the face of the slope. The protruding brush ends provide frictional resistance to minor soil slides and runoff.

Table 8.1 Live Fascine Installation Guidelines

Slope	Slope Distance Between Trenches (m)	Maximum Slope Length (m)
1:1 to 1.5:1	0.90–1.22	4.57
1.5:1 to 2:1	1.22–1.52	6.10
2:1 to 2.5:1	1.52–1.84	9.14
2.5:1 to 3:1	1.84–2.44	12.19
3:1 to 4:1	2.44–2.74	15.24
4:1 to 5:1	2.74–3.04	18.29

Source: Adapted from NRCS 2021b; Norwegian Geotechnical Institute 2023

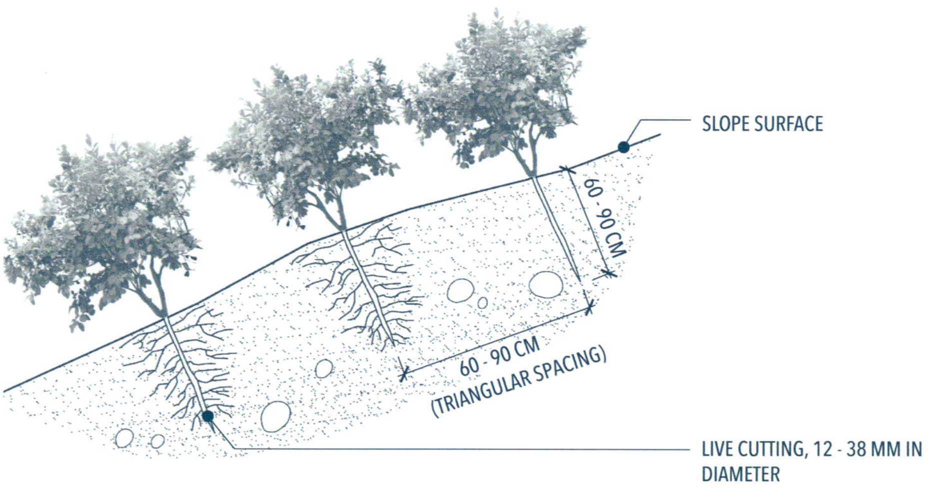

Figure 8.1 Live staking detail for an embankment. Adapted from NRCS 2021b. Image credit: Kathleen Elliott and Meg Calkins

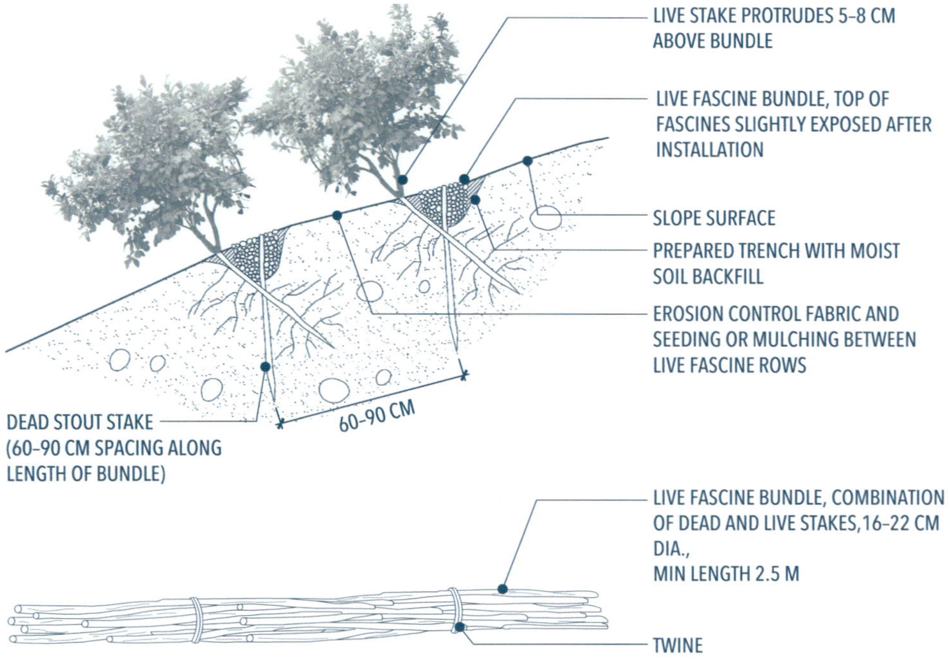

Figure 8.2 Live fascines detail for an embankment. Adapted from NRCS 2021b. Image credit: Kathleen Elliott and Meg Calkins

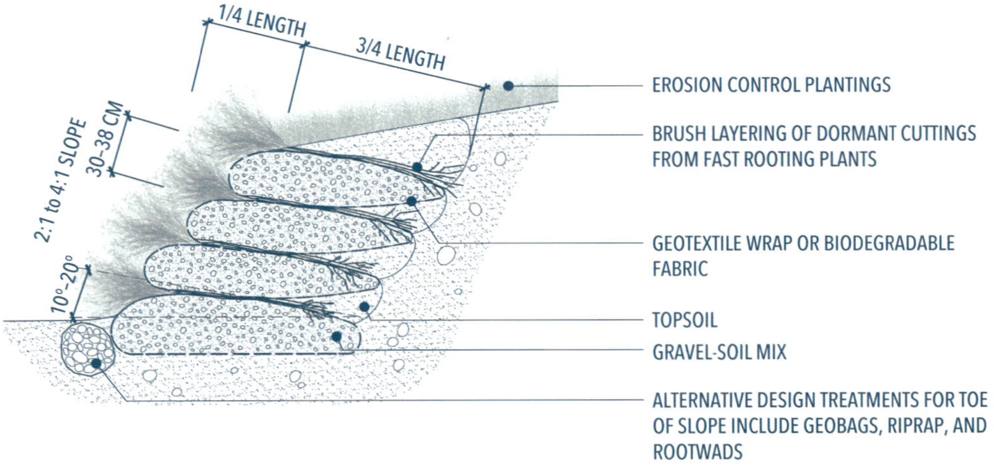

Figure 8.3 Brush layering detail. Adapted from Norwegian Geotechnical Institute 2023. Image credit: Kathleen Elliott and Meg Calkins

Brush layering breaks up the slope into multiple shorter benches separated by layers of brush and offers multiple additional functions including reinforcing the soil once roots develop, adding plant material to a slope, trapping debris, aiding infiltration on dry sites, and drying excessively wet sites (Figure 8.3).

Live branches should be 12–38mm in diameter, and their length will depend on how far they are inserted into the slope. The growing end should protrude 150–600mm from the slope face (Table 8.2).

Live Crib Wall

A live crib wall is a box shaped arrangement of untreated logs or timbers forming a container filled with backfill and layers of live branch cuttings that will root inside the structure and extend back into the slope. This type of structure is appropriate to the base of a slope where a low wall may be required to stabilize the toe.

The crib structure, tilted back into the slope, is comprised of 10–30cm logs and is 120–150cm wide. It extends 60–90cm below grade with a rubble or aggregate base. The length parallel to the slope can vary widely, but the maximum total height is 2 meters. Taller live crib walls can be constructed in a stepping form with higher layers stepping back toward the slope.

This is a useful structure where space is limited as it is comparatively thinner than other bioretention measures and can retain steeper slopes. The slope is immediately

Table 8.2 Brush Layer Installation Guidelines

Slope	Slope Distance Between Benches for Wet Slope (m)	Slope Distance Between Benches for Dry Slope (m)	Maximum Slope Length (m)
2:1 to 2.5:1	0.92	0.92	4.6
2.5:1 to 3.5:1	0.92	1.22	4.6
3.5:1 to 4:1	1.22	1.52	6.0

Source: Adapted from NRCS 2021b; Norwegian Geotechnical Institute 2023

reinforced by the crib wall, and as the wood biodegrades, the cuttings will have rooted enough to stabilize the slope (Figures 8.4 and 8.5).

Figure 8.5 Live crib wall with logs. Image credit: Norwegian Geotechnical Institute 2023

Vegetated Gabion Retaining Walls

Stone or masonry filled gabion baskets and mattresses can be supplemented with soil and live cuttings to encourage plant growth. Rock to rock contact in the gabion must be maintained for structural stability, but voids between the rocks can be filled with soil then seeded and/or staked with live cuttings. Cuttings between gabion baskets should extend into the undisturbed soil behind the gabion basket. A biodegradable erosion control blanket such as a coconut coir net or mat is secured inside the gabion face with lacing wire to prevent soil from washing out of the gabions until the plant roots are established (Figure 8.6).

Woven Walls

Woven walls are constructed from live or non-living branches woven around vertical stakes. If live cuttings are used, they can sprout and root to form a more permanent wall structure. This type of low wall is also called wicker work and willow spiling and can be used to retain slopes and small stream banks (Figures 8.7–8.9).

Woven walls are low structures with heights not exceeding 75cm, although taller slopes can be reinforced with terracing of lower walls. The woven branches should be 8–20mm in diameter. Smaller diameter branches can be woven in groups while larger and drier branches are more difficult to weave. Vertical wood stakes should be a minimum of 5cm in diameter and be driven into undisturbed soil at a depth equal to the wall height above grade. The spacing of the stakes will vary by their size and the diameter of the woven branches. A geotextile can be installed on the soil side of the wall to allow for moisture drainage while soil is retained. This is required if pavement is adjacent to the downhill side of the woven wall (Steele 2004; Polster 2003).

Log Terraces

Log terraces are similar to woven walls as they are supported by vertical stakes or posts, but the horizontal logs or

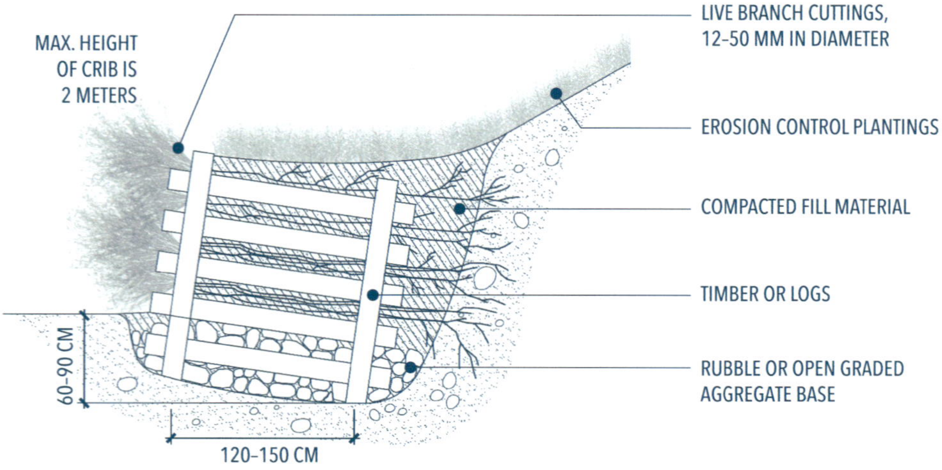

Figure 8.4 Live crib retaining wall detail. Adapted from NRCS 2021b. Image credit: Kathleen Elliott and Meg Calkins

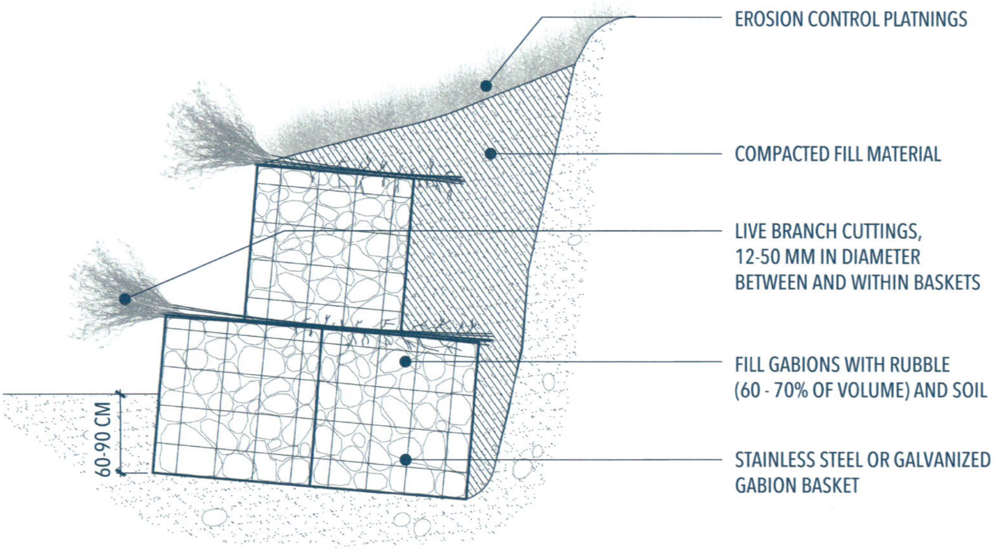

Figure 8.6 Vegetated gabion retaining wall detail. Adapted from NRCS 2021b. Image credit: Kathleen Elliott and Meg Calkins

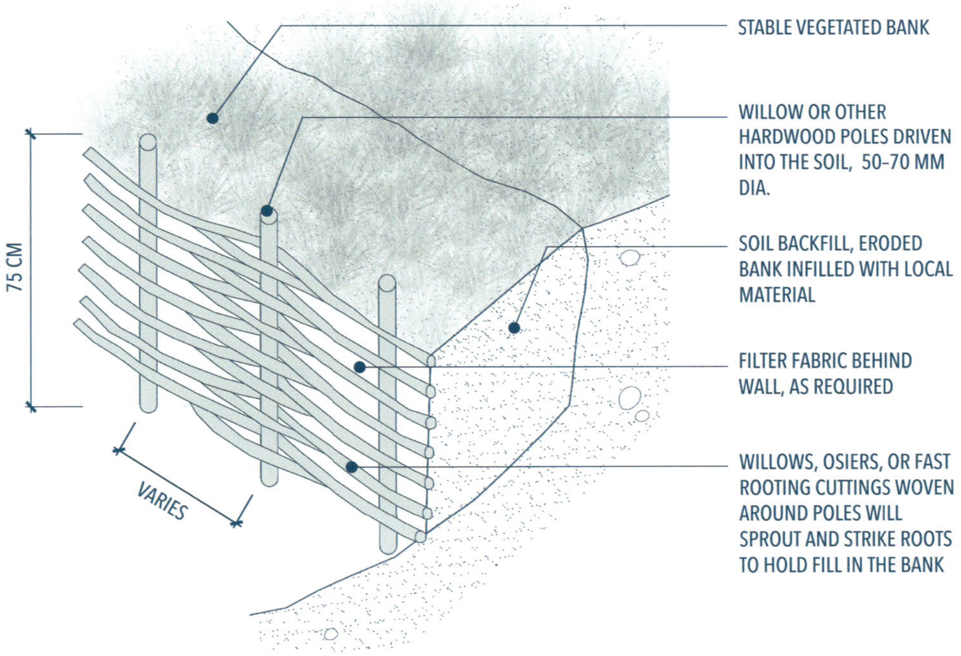

Figure 8.7 Woven retaining wall detail. Adapted from Steele 2004. Image credit: Kathleen Elliott and Meg Calkins

branches are held in the space between the stakes and the soil backfill. The logs or branches are often non-living woody material, but they could be quick rooting living branches. Logs and branches with 3–10cm diameter will work best, and posts should be sized to sufficiently retain the logs behind. Posts should be embedded in undisturbed soil a minimum of 60cm or more for higher log walls.

A variation of a log terraces embeds larger logs in soil, or sometimes concrete, vertically. This can be a good use for small trees removed from a site that are not appropriate for lumber processing. Wood is generally most decay resistant when the bark is removed.

Anytime non-living wood is in continuous contact with damp soil, it will biodegrade. To maximize the structure's life span, end cuts of logs or pressure-treated lumber should be treated with a brush-on preservative. Open graded aggregate bases and backfill should be used under and behind wood retaining structures.

Figure 8.8 Low woven walls at the Tower of London. Image credit: Meg Calkins

Figure 8.9 Cuttings are woven between and around poles. Image credit: Meg Calkins

Vegetated Rock Wall

A vegetated rock wall is a combination of large stones and live cuttings that can be used to stabilize the toe of a slope. Vegetated rock walls are useful where natural stone is abundant, and space is limited. Ideal stones are 20–60cm with a rectangular shape.

These walls differ from a traditional dry stack retaining wall as they are placed against undisturbed soil and are not designed to resist substantial lateral soil pressures. These vegetated rock wall structures are toe walls that are 90–150cm high with a 60–90cm rock footing below grade. They should have a minimum batter of 1:6.

Live branch cuttings with a diameter of 12–25mm are inserted between the stones in the wall with brush extending beyond the face of the wall. The cuttings should be long enough to extend back into the undisturbed soil behind the wall (Figure 8.10).

Stone and Boulder Embankments

Stone and boulder embankments can be durable, low-carbon structures particularly if stones are excavated or collected on-site. They are constructed with field or rubble stone varying in size from large boulders to 8–13cm rubble from quarry waste, also called rip rap. Local stone can offer an aesthetic reference to the natural landscape, and it is a low embodied carbon material for grade changes and reinforced slopes.

Stone embankments consist of larger rip rap stone, field stone or rubble placed on an aggregate base with a geotextile filter fabric between the base and soil underneath. If heavy materials such as large stones are used on the embankment, the toe of the slope could require reinforcing with an aggregate base or a concrete edge.

As vegetation is not part of the structure, stone embankments will require periodic maintenance to remove volunteer vegetation and debris. Unlike bioengineered embankments, stone embankments don't offer ecosystem services such as carbon sequestration, improved air quality, or habitat, however they are relatively low-carbon retaining structures (Figure 8.11).

Boulder embankments employ larger stones to retain taller and steeper slopes than stone embankments using smaller rubble or field stone. Boulder embankments should have an overall batter of 20° minimum, and taller structures can use geogrid tiebacks into soils behind the boulders (Figures 8.12 and 8.13).

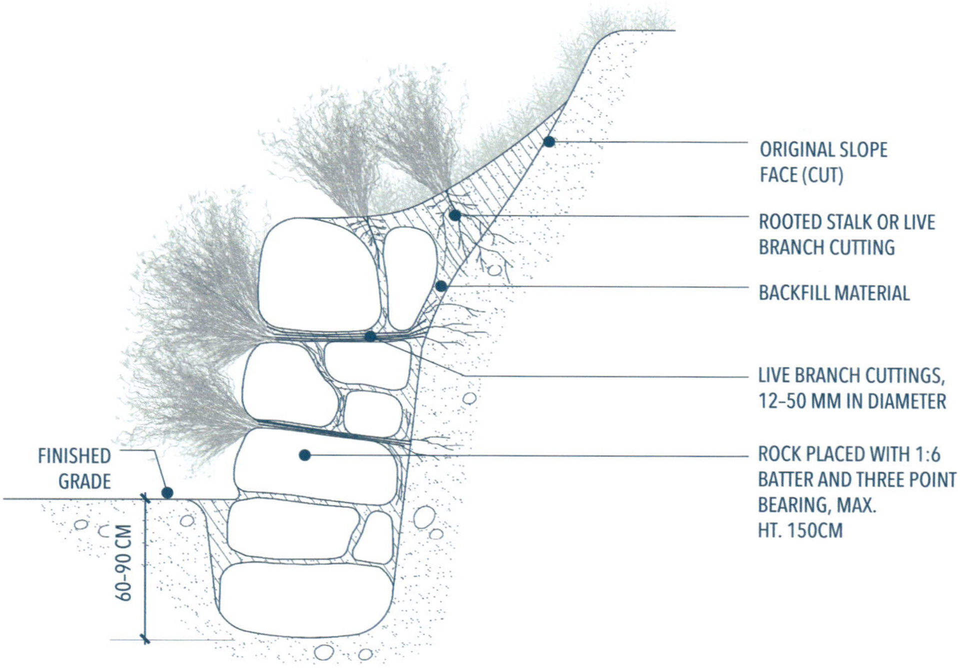

Figure 8.10 Vegetated rock wall. Adapted from NRCS 2021b. Image credit: Kathleen Elliott and Meg Calkins

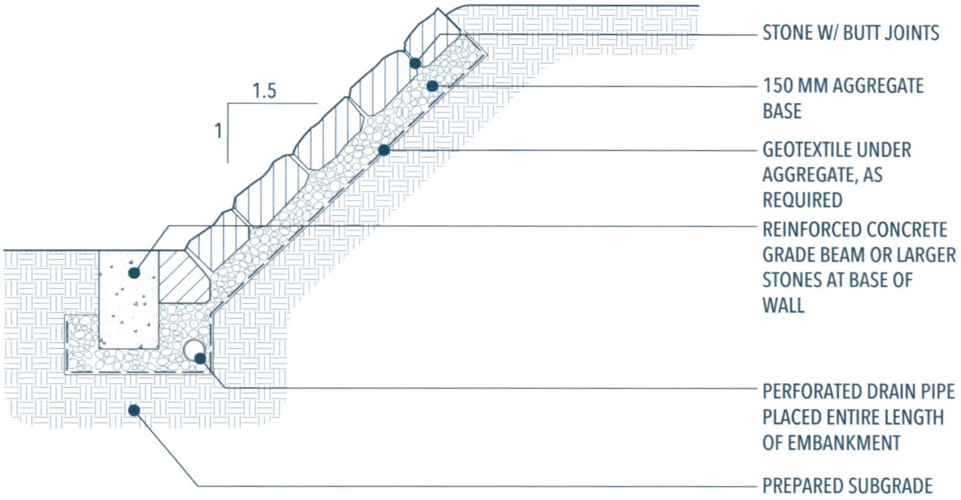

Figure 8.11 Stone embankment detail with concrete grade beam or larger boulders at base. Image credit: Kathleen Elliott and Meg Calkins

Retaining Wall Structures

The primary consideration for the design of retaining wall assemblies is mechanical strength as it ensures stability, therefore performance and durability. All retaining walls must resist or accommodate forces from expansion and contraction of the soil, hydrostatic pressure, and, where applicable, earthquake forces. Other forces on retaining walls are those from vehicles and structures above the wall.

While most retaining walls are constructed from masonry materials (e.g., concrete, precast concrete, brick, stone), other materials such as wood or steel are also employed. Design of retaining walls varies widely depending on expected forces, height of grade retained, soil moisture and bearing capacity, aesthetic intentions, locally available materials, budget and maintenance capacities.

Applicable building codes should be consulted for retaining wall specifications such as height, width, and reinforcing requirements. Structural engineers should be consulted for retaining walls above one meter, or those in challenging conditions. All footing sizes, reinforcing, and other details in the illustrations following are for illustrative purposes only and should not be directly used for construction details.

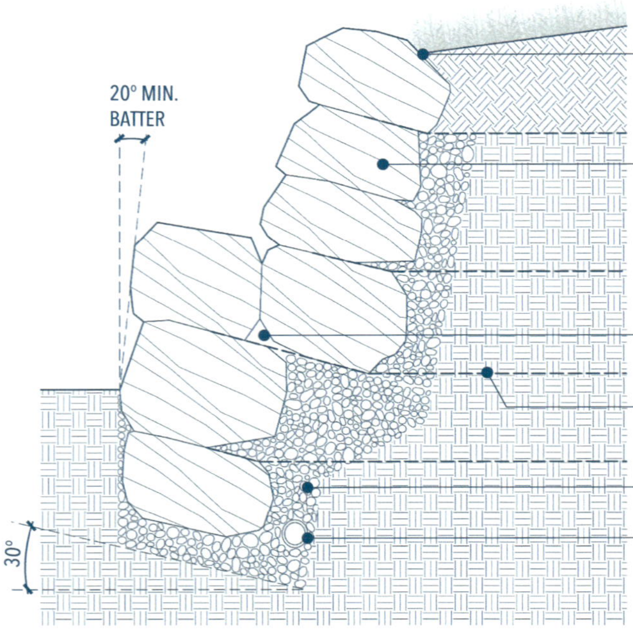

Figure 8.12 Boulder embankment detail. Image credit: Kathleen Elliott and Meg Calkins

Figure 8.13 Boulders excavated for streets on Yerba Buena Island were repurposed by CMG Landscape Architecture for embankments at a dog park. Kids and dogs enjoy climbing on them. Image credit: CMG Landscape Architecture

Forces on and Failures of Retaining Walls

Basic forces on retaining walls are discussed in this section, as are the five main types of retaining wall failure due to these forces. Retaining walls must be designed to counteract the forces and maintain a state of static equilibrium at all times.

Statics and mechanics of retaining structures are outside the scope of this book. Landphair's and Klatt's *Landscape Architecture Construction* and Zimmerman's *Constructing Landscape: Materials, Techniques and Structural Components* offer good information for calculating the strength, statics, and mechanics of site structures (Zimmermann 2015; Landphair and Klatt 1999).

Forces on retaining walls include the following:

Lateral soil pressure: Pressure from the soil behind the retaining wall is a lateral load. The weight of the soil and resulting pressure will vary by soil type and moisture content. Only soil above the natural angle of repose of the soil is bearing on the wall.

Surcharge loads: Forces from the slope above the top of the wall. They can be sloping backfill, pavements, or other structures.

Other forces: Include seismic forces, impact forces, and many other elements that affect the stability of retaining walls.

Retaining walls fail in the following ways (Figure 8.14):

Overturning: Walls can fail through overturning when lateral forces behind the wall exceed the dead load of the wall.

Sliding: The retaining wall will slide if the lateral thrust exceeds the frictional resistance of the footing on the subgrade or aggregate base. A shear key on the bottom of the wall footing can be employed to resist lateral thrust as required.

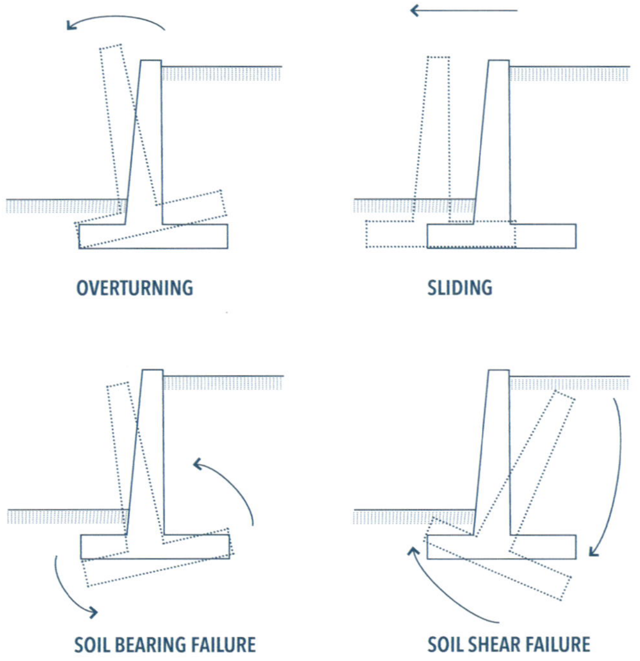

Figure 8.14 Types of retaining wall failures are overturning, sliding, soil bearing failure, and soil shear failure. Image credit: Meg Calkins

Bearing failure: Retaining walls can fail when the normal pressure between the footing or the base of gravity walls exceeds the bearing capacity of the subgrade. The pressure distribution across the base/footing is not consistent as there is greater pressure at the front toe of the base or footing.

Rotational slip: Failure of the wall and a large amount of soil behind the wall can occur when the shear resistance of soil along a rotational arc behind and under the wall is exceeded.

Material failure: Retaining walls can fail if their materials or connections are not capable of withstanding forces on the wall.

Rigid Retaining Walls

Rigid walls are designed to resist all movement with a structural system of reinforced concrete or masonry units held in place with mortar to form a composite structure. Stone, brick, or precast concrete veneer can be attached to these wall structures.

Steel reinforcing is utilized in the wall structure to counteract tensile forces, and tie stems to footings. Footings of rigid walls must rest a minimum of 15cm below the frostline, also known as frost depth, to avoid frost heave. In cold climates, the frostline can be one to two meters deep, while in temperate climates it is 0.5–1.0 meter.

Generally rigid retaining wall structures will use greater quantities of material and have higher embodied carbon, particularly if built in temperate or cold climates. In some instances, more of the structure will be underground than above. For instance, 60% of a one meter high concrete cantilever retaining wall with a spread footing in Ohio will be below grade.

Wall thicknesses and footings sizes of rigid walls will increase with greater loads. Rigid walls can have a very long life cycle; yet they are vulnerable to differential settlement and other forces that may cause cracking or even failure. It is difficult to repair rigid walls with substantial cracking, and this can result in replacement before their designed service life.

Flexible Retaining Walls

Flexible walls can accommodate small amounts of differential settlement or movement and remain structurally sound. Flexible wall types include precast concrete segmental retaining walls (without mortar), dry stack stone walls, gabion walls, timber walls, geocellular plastic stack wall systems, and others. These flexible unit walls do not use concrete or mortar for rigidity, instead they rely on the weight of the assembly and interlocking units or pins for stability.

Flexible retaining wall bases will extend 20–50cm below grade, but as they are designed for minor movement, they do not need to extend below the frost line. This can result in substantially less material and lower embodied carbon in cold and temperate climates.

Types of Retaining Walls

There are five major types of retaining wall structures.

Gravity retaining walls depend on their substantial mass and volume for stability. They can be structurally rigid (e.g., PIP concrete, stone with mortar) or flexible (e.g., segmental retaining walls, gabion cages with stone, dry stack stone walls). Gravity walls are the simplest to construct because they don't have a spread footing, therefore, they can be cost efficient, although some will use more material resources than other wall types.

The sectional shape of gravity walls is often trapezoidal with a wider base and narrower top. For lower gravity walls, the batter can be on the front or back of the wall, but it will be on the front of the wall leaning back into the sloped backfill for taller walls. Base thicknesses depend on height of wall, soil bearing capacity, and forces behind the wall. The ratio of base to height ranges from .35 to .75 depending on these factors.

The base of rigid gravity walls should be placed below the frost line on well-compacted subgrade or an aggregate base as required. In areas with average to deep frost lines, this will necessitate a large amount of wall material below grade, particularly since the base of gravity walls gets wider with the overall height of the wall. Flexible gravity walls usually extend only 15–60cm below grade and rest on a compacted aggregate base.

Cantilever retaining walls consist of a base spread footing and a stem that are tied together by steel hooks or reinforcing bars extending from the base up into the stem. The weight of the backfill on the footing behind the wall counteracts forward pressure of the retained height and resists overturning of the wall. Structurally, the stem and base perform as one unit.

Both the base and the stem are relatively thin as compared with gravity walls. Therefore, reinforcing bars run both vertically and horizontally through the stem providing resistance to tensile forces on the wall. Generally, the stem sits one third of the way back on the base footing measured from the toe although this may vary as required. Dimensions of a cantilever retaining wall will vary by soil conditions and expected forces on the wall.

Counterfort walls are very tall retaining walls (over 4 meters) used primarily in highway walls, railroad cuts, and water channels. These tall walls require additional support against soil pressure and have buttresses bracing the front of the wall or counterforts on the back of the wall. These additional elements provide stability against increased compressive and tensile stress. These walls are typically constructed from reinforced concrete and/or steel. More economical and potentially lower carbon alternatives for very tall retaining walls are soil reinforced segmental retaining walls, stacked gabion walls, or earth tieback walls with precast face panels.

Piling retaining walls are constructed from sheet piling spanning between vertical structural members that are driven into the ground. I-beams and other steel shapes are often used for the vertical piers and interlocking metal shapes that form a wall between the vertical piles.

Anchored retaining walls employ the weight and stability of soil or embedment into rock to hold the wall in place. These walls have a relatively thin wall face that is tied back into undisturbed or minimally disturbed soil or rock with cables, anchors, timbers, or geogrids.

Moisture Considerations for Retaining Walls

Adequate drainage behind retaining walls is very important as hydrostatic pressure from buildup of water in the soil behind retaining walls can compromise their stability. Water in the soil can create greater compressive loads against the back of the wall, and freeze/thaw action of water in the soil can negatively impact the wall and footing as well.

Multiple measures to drain water from around walls are necessary. Surface water on slopes behind walls can be caught in an impervious clay lined swale behind the wall and carried away from the structure. This will prevent water from seeping into the soil and building up behind the wall.

Behind the wall, granular backfill can catch water and a perforated pipe within the gravel can convey it to a storm system. The granular backfill should be wrapped in filter fabric to prevent penetration of fine soil particles. If a perforated pipe is employed, it should be placed in the drain rock near the base of the wall with perforations pointed down.

An alternative or supplemental drainage measure for rigid walls is a series of weep holes through the wall located just a few centimeters above finished grade on the downhill side of the wall. Weep holes are generally lined with a non-corrosive pipe and are spaced at regular intervals (60–130cm) depending on the expected moisture behind the wall. Weep holes let water escape continuously so that water does not build up in the soil behind the wall.

Last, waterproofing on the soil side of rigid retaining walls is recommended for concrete, concrete masonry unit (CMU), and solid brick walls in temperate and humid climates. Waterproofing can be bituminous coated felt, a sheet of elastomer, a waterproofing coating, and/or drainage board applied on the back of the wall stem. Some waterproofing coatings and sheets contain hazardous chemicals that can be released in use or disposal. Look for products with low VOCs that avoid Living Building Challenge Red List ingredients (ILFI 2024).

Foundations and Footings for Retaining Walls

Foundations for retaining walls are based on the wall type, soil bearing capacity, the coefficient of friction values of the soil type, the wall height, and the required design strength of the wall. For rigid walls, the depth of the wall footing is based on the frost depth and the depth of the soils with adequate bearing capacity. The placement of the bottom of the footing should be a minimum 15cm below the local code requirements for frost depth to avoid movement from frost action. It is recommended that soils be tested for bearing capacity and coefficient of friction or a structural or geotechnical engineer be engaged in the design of the wall and foundation.

As foundations for flexible retaining walls do not need to extend below the frost line, they should rest on soils with adequate bearing capacity and drainage. Where bearing capacity is low, or soils are poorly drained, flexible wall structures can be placed on a compacted aggregate base at least 20cm below finished grade.

Durable Retaining Walls

A long service life is a key consideration for retaining wall structures. Quality construction is key to the long service life of walls, yet they should not be "overbuilt" beyond the recommendations of engineers. If a wall is in use for a long period, the impacts of the embodied carbon and resource consumption of its materials are somewhat offset. For instance, if two walls are constructed and they have the similar embodied carbon, but one wall must be replaced twice in 60 years, their embodied carbon is no longer equal. Therefore, detailing and construction of durable structures is critically important to reducing the carbon impacts of site construction (Table 8.3).

Table 8.3 Strategies for Durable Wall Construction

Control Moisture Behind and Under Retaining Walls
Use drainage assemblies and/or weep holes to avoid moisture build-up behind retaining walls. Typical drainage assemblies are perforated drainpipes in drain rock wrapped in filter fabric. Weep holes penetrate the wall with plastic pipe lined holes that are placed a few centimeters above finished grade on the downhill side of the wall. Weep hole spacing varies with the wall material and expected water behind the wall, but they are generally placed 1–2 meters on center. Use waterproofing on the back of rigid retaining walls. Waterproofing can be a liquid agent or sheet material.
Design Rigid Walls to Resist Moisture Penetration
Use capstones and flashing on rigid cavity walls, veneer walls, and multiple wythe brick walls. Use joint sealants and silicone water repellant surface treatments where required. Consider use of clean cavities with airspace in combination with weepholes or other openings for drainage of moisture out of walls in humid climates. In hurricane or high moisture regions, favor masonry headers over metal wall ties to tie solid brick wall wythes or brick cavity walls together as metal ties can corrode with moisture. Hold all steel reinforcing in joints back from the mortar face by at least 20mm. Specify appropriate brick, wall block, and mortar for climatic conditions. Examine walls periodically for moisture penetration. If it is occurring, address the source of moisture immediately.
Design Walls to Resist Wind, Freezing Soil, Earthquakes, and Other Forces
Consult a structural engineer when snow, earthquakes, and high winds loads are expected and/or when walls are retaining soil more than one meter in height. Place the bottom of rigid wall footings a minimum of 15cm below the frost line. Place the bottom of flexible walls at least one course and/or a minimum of 20cm below finished grade on 95% compacted aggregate and/or compacted subgrades where applicable. Strengthen structural bonds between veneers and wall stems with use of corrosion-resistant unit wall ties, masonry ties, and/or continuous joint reinforcements. Strengthen structural bonds between solid or cavity brick wall wythes with masonry headers, masonry ties, and/or continuous joint reinforcements.
Design Rigid Walls to Control Cracking From Expansion and Contraction
Employ control joints every 2–3 meters in concrete walls. Place expansion joints every 8–10 meters in rigid walls structures.
Design Flexible Wall Structures to Be Stable, Easily Repaired, and Releveled
Avoid over-use of hidden mortar in "dry stack" walls. Purchase extra segmental blocks and caps, or gabions and gabion stone, in case of damage during releveling activities. Specify 95% compacted aggregate and/or compacted subgrades where applicable.

Reducing the Environmental Impacts of Retaining Walls

General strategies for reducing embodied carbon and resource use of retaining walls include reducing wall thicknesses and heights to the extent possible, using flexible and repairable wall structures and systems, and using locally produced materials. Where possible, alternative structures such as bioengineered embankments or living walls can be used; and where concrete must be used, cement substitutes, recycled aggregates, and optimized mixes can be employed (Table 8.4) (Chapters 13 and 14).

Flexible Retaining Walls

Flexible retaining walls can accommodate small movements and minor differential settlement while remaining structurally sound. These flexible unit walls do not use mortar or concrete for rigidity, instead, interlock of units, pins, and/or the weight of the assembly results in a stable retaining wall structure. Drainage measures behind the walls should be employed in most situations.

Gabion Retaining Walls

Gabion retaining walls are steel wire mesh cages filled with heavy stone or masonry materials. They are widely used in streambank restoration and highway cuts as they are relatively inexpensive to construct, durable, and can use local stone, saving truck transport costs to bring in heavy materials. Gabion retaining walls can be low embodied carbon structures, particularly if the steel wire mesh gabion cages have a high recycled content.

Stainless steel, decorative, and custom gabion cages are increasingly used for retaining wall structures in public landscapes. Gabion seat walls will require a wood or stone bench seat affixed to the top of the gabion cage.

A wide variety of stone or reclaimed masonry materials can be used in gabion cages. Unlike stone walls that depend on an appropriate shape of stone for interlock, just about any shape of stone can be used in gabions as the cage holds the masonry material in place. It is important that stone-to-stone contact be achieved to maintain the shape of the gabion cage as steel wire is flexible (Figures 8.15 to 8.18).

Welded gabion cages are made from dimensional panels of wire mesh that are fabricated by welding transverse and longitudinal wires at each intersection to form a grid. These panels are wired together with interwoven wires to hold the panels in place and form a rectangular cube. Less common is a trapezoidal prism shaped gabion basket.

Woven gabion cages are formed by weaving strands of wire and twisting them at intervals. This hexagonal mesh is wrapped around a heavier transverse edge wire to form the panels of the cage. This type of gabion basket is much more easily deformed than welded gabions and is more

Table 8.4 Strategies to Reduce the Environmental Impacts of Retaining Walls

Strategy	Considerations
Reduce wall and footing sizes	Design thicknesses of retaining walls to the thinnest section required by structural engineers for performance and stability.
	Comfortable seat wall thicknesses are 30cm, with a minimum of 23cm.
Use flexible, repairable wall structures and systems	In areas with deep frost lines, the use of flexible walls will reduce the overall amount of wall and footing material used. Interlocking segmental block walls, rubble filled gabion cages, geocellular stack walls, and timber walls do not need to extend below the frost line.
	Wall structures that can be releveled and refurbished will have a longer service life.
Locally sourced materials	Retaining walls use heavy materials, therefore the carbon footprint of transport of these materials can be substantial. Wall materials should be excavated, quarried, processed, and sourced locally. Aggregates and concrete should be sourced within 80km of the project site. Dimension stone, brick, and precast concrete should be sourced within 500km. Using stone in its natural form has a low-carbon footprint since fabrication and waste are minimal. Gabion retaining walls may be appropriate for sites with large amounts of deconstructed concrete, and geocellular stack walls could be used on sites with excess fill material.
Alternative structures and strategies	The most effective approach to reducing the environmental and economic impacts of retaining walls is to avoid their construction. Vegetated slopes and bioengineered embankments can serve the same function as a retaining wall but at lower economic and environmental costs. While they are not possible in some very tight situations, they can introduce additional vegetation into urban and suburban environments. As the vegetation on these structures will sequester carbon, they can contribute to reducing the carbon footprint of a built site.
Low-carbon material mixes	Use of portland cement alternatives (SCMs, alternative cements), recycled content materials, and reprocessed and reclaimed materials will reduce the embodied carbon and resource consumption of retaining walls.

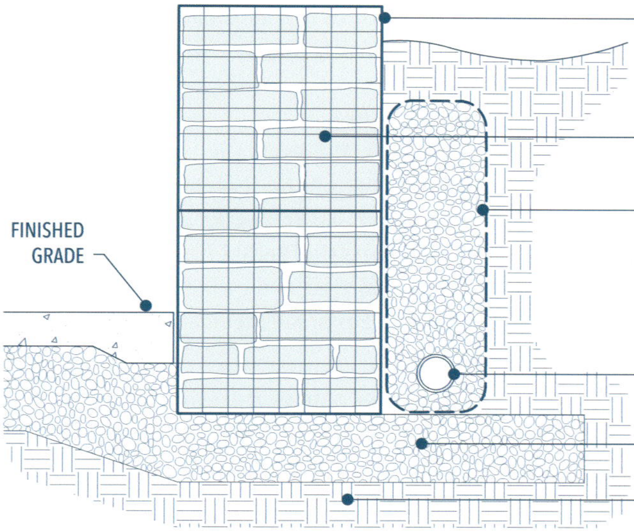

Figure 8.15 Gabion retaining wall detail. Image credit: Kathleen Elliott and Meg Calkins

Figure 8.16 At Ballast Point Park in Sydney, McGregor Coxall diverted 22,200 tons of demolition waste from landfills by repurposing it for use in gabion retaining walls on the site. The rubble in the gabion walls stands in contrast to the native sandstone cliffs of the site. Image credit: Brett Boardman

Figure 8.17 Park users circulate among the gabions containing rubble and artifacts reclaimed from the site's richly layered history. Image credit: Brett Boardman

appropriate in informal retaining walls where maintaining the rectangular shape of the gabion is less important.

Custom gabion cages can be formed from punched sheet metal panels that are wired or welded along the edges. Other custom cages will be formed from steel angles at the intersection of the panels. Increased use of metal materials will substantially increase the embodied carbon of the structure.

Gabion wall construction will vary by the height of the wall, type of masonry filling the baskets, the bearing capacity of the soil under the wall, and the type of backfill. Welded gabion cages for retaining walls under 150cm in height can be stacked without a batter. Taller gabion walls should have a minimum 6° batter, or the cages should be stepped back toward the slope.

Gabion retaining walls are placed on an aggregate base leveling course, although some will be placed on a concrete base if differential settling of the retaining wall structure cannot be tolerated. If retained soil migrating into the gabions is a concern, filter fabric behind the gabions will prevent migration of soil to the wall face while allowing drainage. Some conditions will require a complete drainage assembly of gravel wrapped in filter fabric and a perforated pipe to a storm system.

Dry Stack Retaining Walls

Dry stack retaining walls, flexible gravity walls, can be constructed from simply dressed quarry stones, field stones, or dimension stones. They should have a minimum 1:10 batter with the base of the wall being ⅓ to ½ the overall height. Greater batters should be used for rounder stones.

Stone is laid to achieve as much interlock as possible with very small cavities between the stones. Precisely fitting stone chips wedge between the larger stones providing

Figure 8.18 The tall retaining walls at Ballast Point Park are constructed from reinforced earth and faced with gabions. Image credit: McGregor Coxall

additional interlock. Soil can be packed into the cavities in the wall and be planted for a vegetated dry stack wall (Figure 8.10).

Dry stack stone walls are more commonly used for freestanding walls, and if used in retaining situations, they must be kept below one meter unless mortar will be used in the joints. Refer to the discussion of freestanding dry stack walls in Chapter 9 "Freestanding Walls" for design standards.

Reclaimed concrete slabs and curbs with their consistent thickness are ideal materials for dry stack retaining walls. In some ways, the one flat side of concrete slabs makes stacking them to achieve a level wall easier; however it does not result in the interlock necessary for a taller retaining wall. Laying the slabs or curbs in a batter back toward the slope assists with wall stability (Figures 8.19 and 8.20).

Segmental Retaining Walls

Segmental retaining walls (SRWs) are precast concrete gravity retaining walls that rely on their mass and interlocking elements to resist forces from retained soil and surcharges above the wall. Precast concrete units are dry stacked without mortar and connected through interlock such as shear keys or mechanical connectors (Table 8.5).

SRWs are flexible retaining walls as they accommodate small amounts of movement and differential settlement

Figure 8.20 At Raleigh Iron Works, Stewart used Jersey barriers reclaimed on-site from the former Peden Steel warehouses for a tall retaining wall. Image credit: Steve Badger

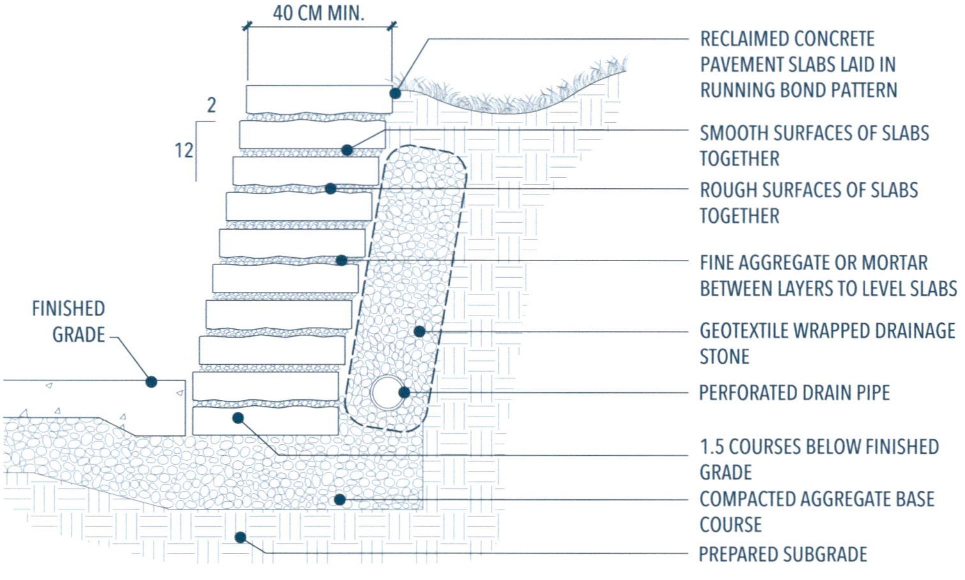

Figure 8.19 Reclaimed concrete dry stack gravity retaining wall. Image credit: Kathleen Elliott and Meg Calkins

Table 8.5 Components of Segmental Retaining Wall (SRW) Assemblies

Component	
Foundation soil	The foundation soil must provide adequate bearing capacity to support the leveling pad, the SRW blocks, retained soil influences, and external surcharges on the system.
Leveling pad	A level surface consisting of either well-compacted gravel or an unreinforced concrete pad used to distribute the weight of the SRW column over a wider area and to create a working surface during construction. The leveling pad typically extends 15cm from the toe and heel of the lowest SRW unit and is a minimum of 15cm deep.
Wall units	The units create the stability of conventional SRWs and the face of the structure for soil reinforced SRWs. They determine the visual appearance of the wall. Standards for the units are in ASTM C1373, Standard Specification for Dry-Cast Segmental Retaining Walls Units.
Gravel fill	Clean, free-draining aggregate placed in the open cores of hollow SRW units to increase system weight and strength. Gravel fill is extended at least 30cm behind the units as well. This fill acts as a compaction aid to reduce horizontal compaction stresses on the back of units and to facilitate drainage behind the wall.
Drainpipe	A perforated pipe placed behind the lowest SRW course in the gravel fill to catch and convey water from behind the wall.
Reinforced soil	Compacted structural fill placed behind soil reinforced SRW units. This low plasticity reinforced backfill should be free of debris and consist of inorganic gravel or sand. Ideally it is native soil, although imported soil may be required.
Geogrid reinforcing	Synthetic geogrid made from polypropylene, polyester, or polyethylene. Layers extend into fill soil for a soil reinforced SRW.
Planting voids	Some segmental retaining wall systems are designed with voids and a stepped wall face offering space for soil and plants. The soil adds mass to the system and the plants can spill over the wall. Some block products are designed to accommodate irrigation for the planted cells as well.

and can be repaired or releveled if required. They can be disassembled and reused after the life of a site extending their life cycle. SRWs are effective retaining structures in tight spaces and/or for very tall walls. They are less expensive than rigid masonry retaining walls.

By weight, the embodied carbon of SRW blocks can be higher than poured in place concrete since these dry-cast concrete products generally use more portland cement and fewer cement substitutes such as fly ash or blast furnace slag. SRW companies are developing new concrete mixes to reduce their embodied carbon.

Yet, while the EC of SRW blocks is high, SRW retaining wall assemblies are often lower in EC because the total amount of masonry material required to construct a comparable height wall is lower. SRW structures only require 1–2 courses of blocks below finished grade that rest on a compacted aggregate base, whereas rigid wall structures must set footings 15cm below the frost line.

There are two distinct types of segmental retaining walls, conventional SRWs and soil reinforced SRWs.

Conventional SRWs depend solely on their mass for stability. The blocks must have sufficient mass to prevent sliding at the base or overturning at the top or base of structure. The blocks are shaped to interlock and/or are pinned together to resist shear forces. Blocks may be solid or hollow and filled with aggregate. Conventional SRWs are usually lower or the blocks are much larger and heavier to achieve stability. The wall often has a batter that is achieved through set back of the units. Increasing the wall batter will generally increase the stability of the wall (Figure 8.21).

Soil reinforced SRWs are used when the maximum height for conventional SRWs is exceeded, when the wall has a surcharge such as sloping backfill, or when soil conditions are poor. Soil reinforced SRWs are stabilized by horizontal layers of geogrids placed between the blocks that extend into compacted backfill behind the wall. These layers increase the effective width and weight of the gravity mass. This system will allow for greater wall height with less use of carbon-intensive precast block material as it relies on the weight of existing soil for stability. Soil reinforced SRWs allow for the use of smaller or hollow blocks (Figures 8.22 and 8.23).

Timber Retaining Walls

Timber retaining walls use renewable, biobased materials for relatively low embodied carbon structures compared with masonry walls. Wood stores some carbon until it decomposes or burns. If the wood was harvested from a sustainably managed forest that is revegetated, this could result in a net increase in carbon storage.

The two main drawbacks to timber retaining walls are their relatively short use life and height restrictions. When wood is in ground contact and exposed to soil moisture, it will

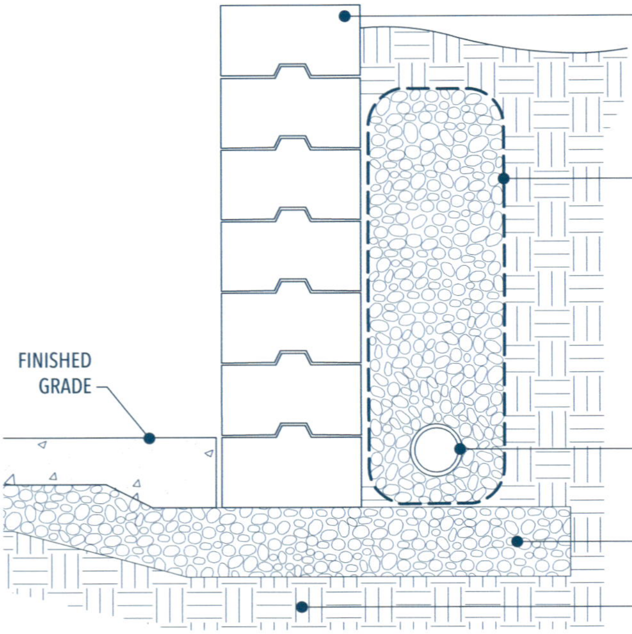

Figure 8.21 Interlocking block SRW detail. Taller SRWs will incorporate a batter and/or blocks that step back. Hollow blocks filled with aggregate will be substantially lower embodied carbon. Image credit: Kathleen Elliott and Meg Calkins

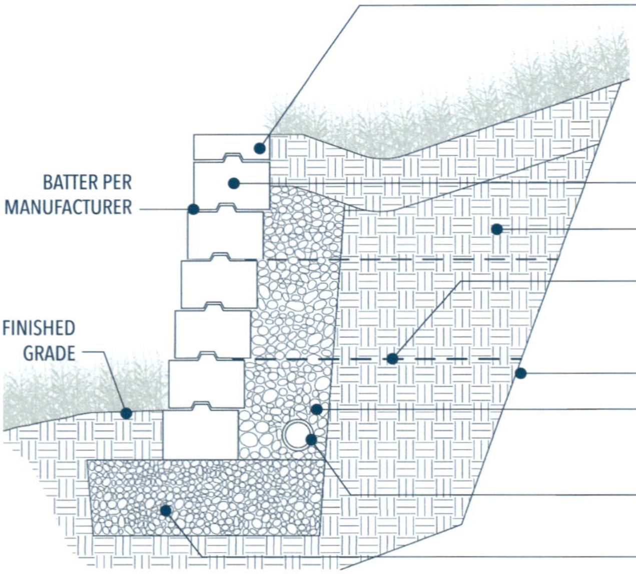

Figure 8.22 Soil reinforced segmental retaining wall. Geogrid layers tie into the backfill and stabilize the wall. Image credit: Kathleen Elliott and Meg Calkins

decompose. Pressure-treated wood timbers in a retaining wall with aggregate behind and under the wall can be expected to last from 15 to 30 years before they will require replacement. This service life will be extended in drier climates and reduced in humid climates.

Timber crib walls are stacked and interlocked horizontally in a crib form. The maximum height of timber crib walls is three meters (Figure 8.4) (Harris and Dines 1997).

Horizontal timber walls stack large wood timbers, usually 140x140mm or 190x190m and fasten them together with steel rods or reinforcing. The stacking of timbers could include a batter or stepping back of the timbers. They are not recommended for retaining heights above 1.2 meters. These walls either require vertical stakes at one meter on center or deadmen every 2 meters. "Deadmen" offer lateral stabilization for timber walls. These are timber or concrete anchors placed into the undisturbed soil behind the wall in an orientation parallel to the face of the wall. Deadmen are tied to the wall with a steel rod or timber (Figure 8.24).

Horizontal plank and pile walls are recommended for retaining heights up to 1.2 meters. The piles should be

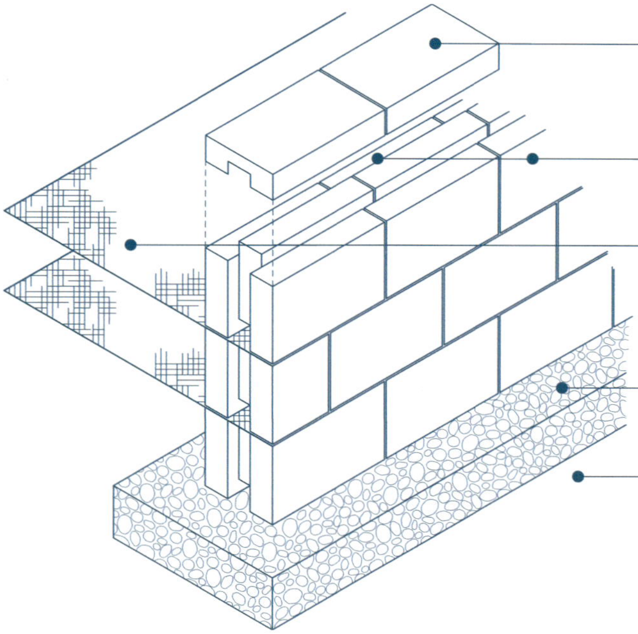

Figure 8.23 Interlocking CEB soil reinforced retaining wall. Image credit: Kathleen Elliott and Meg Calkins

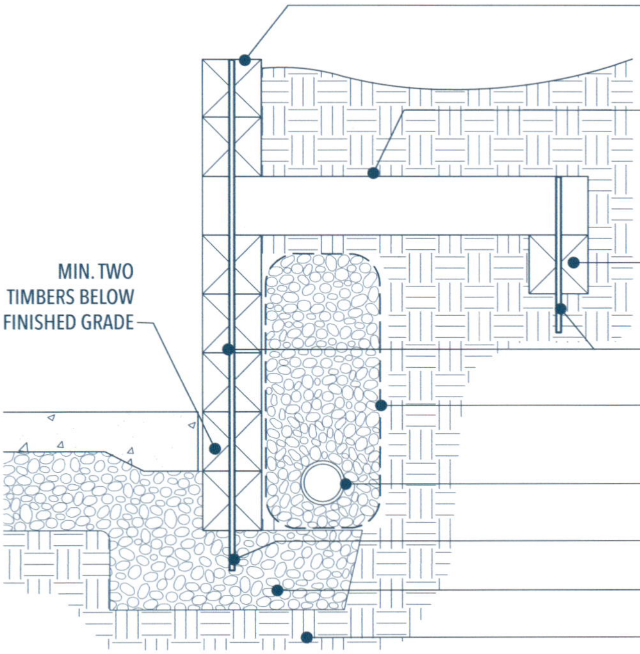

Figure 8.24 Timber retaining wall with horizontal timbers and timber tieback/deadmen assemblies. Image credit: Kathleen Elliott and Meg Calkins

embedded in the subgrade at least 75cm and spaced at 1 meter on center. Tiebacks or deadmen are recommended (Figure 8.25).

Geocellular and Modular Confinement Systems for Walls and Embankments

A wide range of geocellular and modular bag systems are available for reinforced embankments and retaining walls. Some systems allow for very high retaining walls that rely on the weight of soil and stone in the cells or bags for strength and stability. These are flexible systems that allow for some deformation to occur without failure. Most of these systems rely on vegetation in the cells to cover and protect the plastic containment from UV degradation. These plastic geocellular confinement products are primarily made from polyethylene or polypropylene plastics and some incorporate substantial recycled content (Figures 8.26 and 8.27).

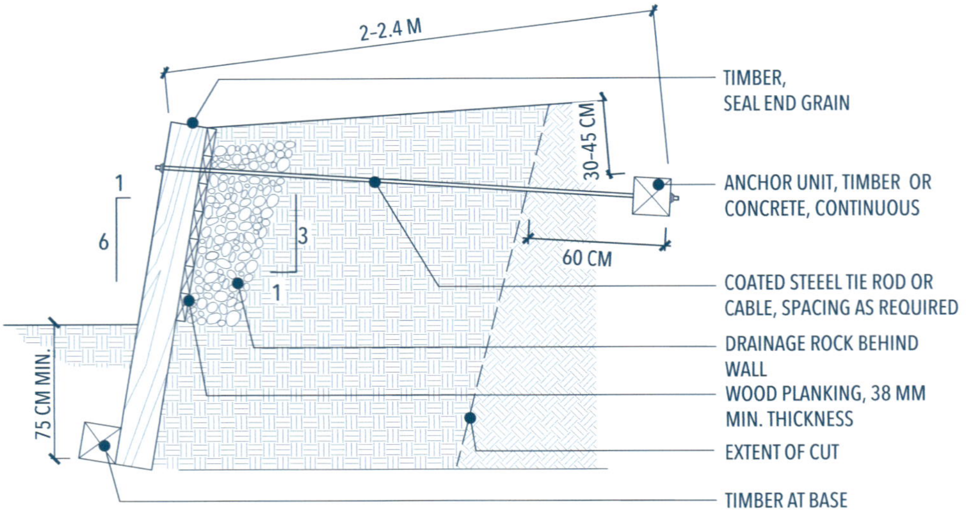

Figure 8.25 Horizontal wood plank and pile retaining wall with timber of concrete anchor and tie rod assemblies. Image credit: Kathleen Elliott and Meg Calkins

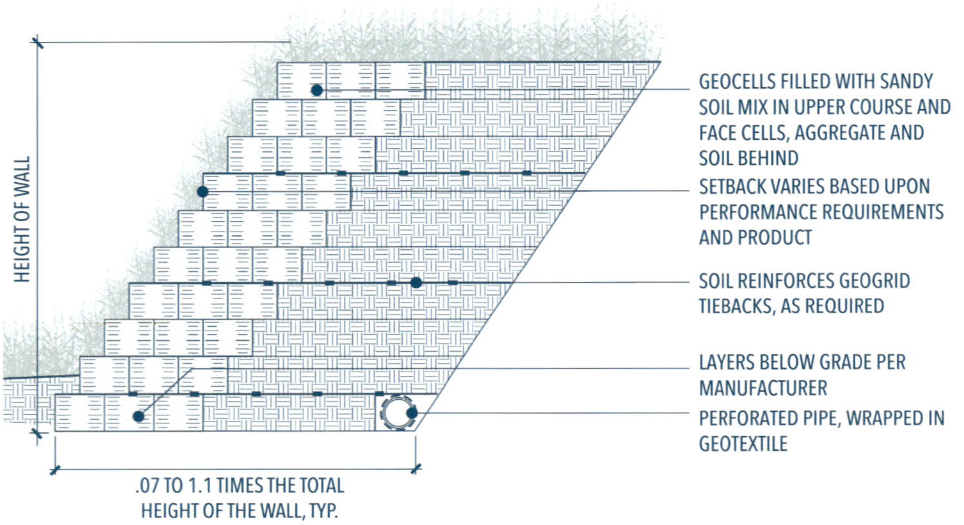

Figure 8.26 Geocellular retaining wall detail example. Product systems vary. Image credit: Kathleen Elliott and Meg Calkins

As retaining structures, the cost and embodied carbon of geocellular structures is relatively low as compared with masonry based retaining structures because the greatest mass of material in the structures is soil that can often be obtained on the project site. If the system uses large amounts of aggregate, local sources should be used. On a previously developed site, aggregate could be obtained from crushed reclaimed concrete or asphalt.

Most geocellular systems allow for water to flow through the wall and the plant roots take up excess water. They will also absorb a small amount of rainfall and surface runoff on and immediately adjacent to the wall.

Geocellular walls and embankments are designed to house plants and ultimately should have a completely vegetated surface. Durability of the polymer materials for these green

Figure 8.27 Geocellular retaining wall with mature plants. Wall cells should be fully covered with plantings. Image credit: Flex MSE

walls depends on plant cover shielding the material from the sun. Most of these products incorporate UV inhibitors, but if exposed to intense sunlight they may be vulnerable over time.

Geocellular walls and embankments are pinned together and into the slope with fiberglass or metal connectors. Some systems expand to create cells that are filled with soils and compacted. For taller walls, this is executed in layers that are pinned together and stepped back in the direction of the hillside. To create even taller walls, the cellular systems can be tied back into the backfill with geogrid layers.

Rigid Retaining Walls

Rigid retaining walls are designed to resist all forces without movement. Structural materials for rigid retaining walls are PIP concrete or masonry units held in place with mortar forming a composite structure. Steel reinforcing is utilized in rigid retaining walls to resist tensile forces and tie wall components together. Rigid walls will be cantilever, gravity, and counterfort wall structures.

Poured in Place (PIP) Concrete Cantilever Retaining Walls

Concrete cantilever retaining walls are the most common type of rigid retaining wall, although low gravity retaining walls are common where frost lines are shallow. PIP cantilever walls are reinforced with one or two layers of vertical and horizontal rebar depending on the width and height of the stem, and the expected forces on the wall. The size and extent of rebar used is determined by a structural engineer based on the calculated loads the wall is supporting.

Cantilevered wall stems are tied to the base by steel dowels or hooks that bend into the width of the base. They extend up into the wall stem at least 30 bar diameters and are tied to the vertical rebar in the stem. Where all or part of the wall stem is PIP concrete, a 50x100mm keyway will also secure the stem and footing (Figures 8.28 and 8.29).

A shear key on the bottom of the footing is sometimes employed for taller walls or soils with less frictional resistance. This will prevent the footing from sliding forward due to the compressive soil forces behind the wall. In addition to vertical expansion joints, control joints should be placed every one to two meters to control cracking of the wall face.

While durable, PIP concrete is a relatively carbon-intensive material. Refer to Chapter 14 "Concrete Materials and Products" for information about reducing the environmental impacts of concrete.

Concrete Masonry Unit Retaining Walls

CMU retaining walls stack precast concrete blocks in a common or stack bond pattern with mortar. Cells of blocks

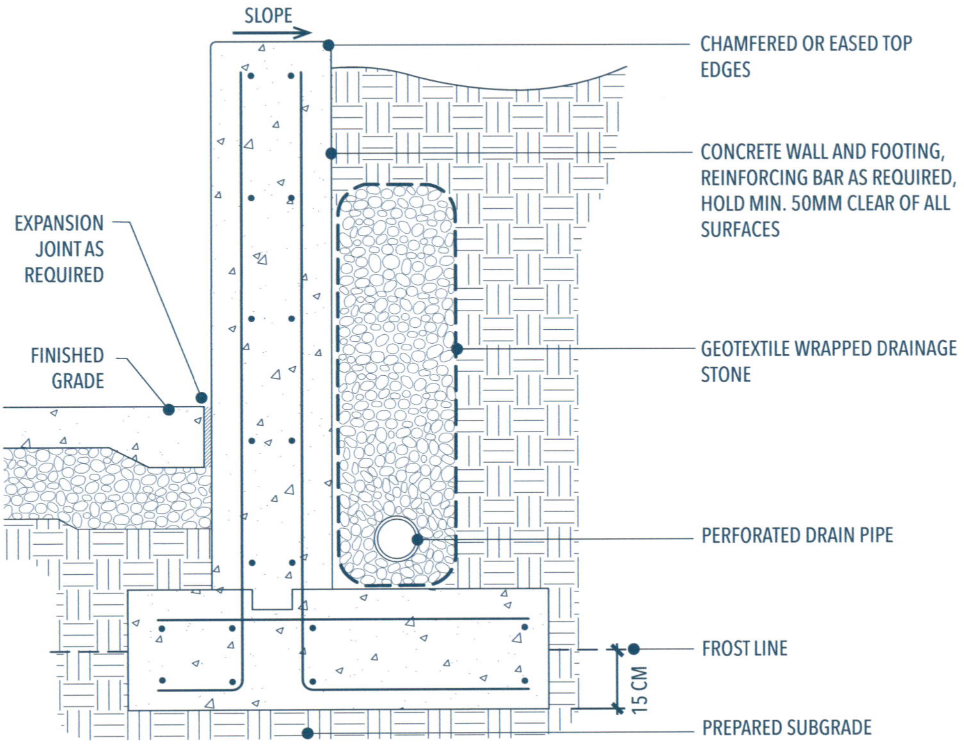

Figure 8.28 Concrete cantilever retaining wall detail. Thickness of wall stem should be thinnest recommended by structural engineers to reduce EC and resources. Image credit: Kathleen Elliott and Meg Calkins

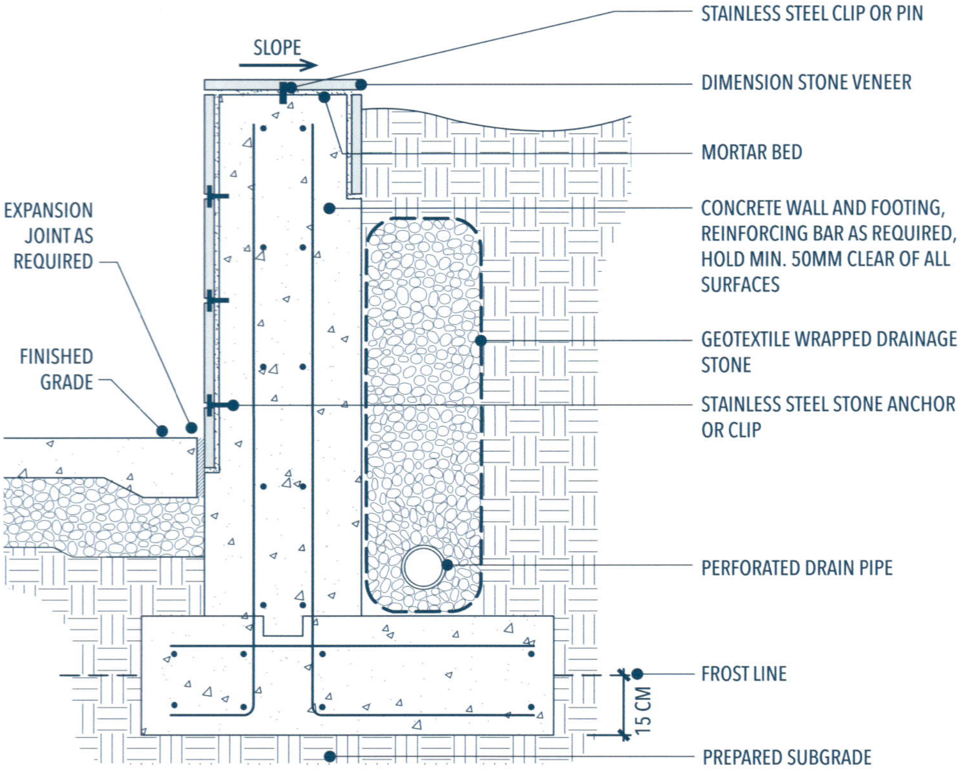

Figure 8.29 Granite veneer on concrete retaining wall detail. Use of domestic granite and reduced wall thickness will lower the embodied carbon of the structure. Image credit: Kathleen Elliott and Meg Calkins

are aligned for vertical reinforcing and/or dowels that tie the wall stem into the footing. Grout is placed around the vertical reinforcing, and sometimes in other open cells, to form a rigid composite wall stem. Horizontal steel reinforcing, called ladders, are used as required in horizontal joints (NCMA 2007).

Wall caps are required to keep the wall cavity dry. The cap is held in place by mortar and weather resistant steel dowels, and if the cap is a unit material such as brick or stone, flashing should be employed under the cap to further protect the wall stem (Figure 8.30).

Veneer on Rigid Retaining Walls

Stone, brick, or cast stone (precast concrete) veneer can be attached to concrete or CMU wall stems with mortar and stainless steel masonry ties or clips. Thickness of the veneer will vary by type of material from 15–150mm. Veneer plays no structural role except to support its own weight. Wall caps are required to keep the wall cavity and bed joints between the veneer and wall structure dry (Figures 8.29 and 8.30).

Brick Retaining Wall Structures

Brick retaining wall structures are the least common type of rigid retaining wall. When a brick wall is required, it is usually constructed as a veneer on a concrete or CMU wall (Figure 8.30). However, double wythe and grouted cavity brick wall structures can be employed as retaining walls.

Climatic conditions are a primary consideration for durable brick walls. Brick is rated by weathering capability with respect to temperature and water absorption. As with all rigid unit retaining walls, waterproofing measures such as flashing, drainage assemblies behind the wall, and waterproofing on the back of the wall are also important (Haseltine et al. 1991).

Bricks are a durable, yet relatively high embodied carbon construction material. They are easily reused from sand set pavement applications but reuse of wall bricks is often compromised by the mortar that is attached when they are reclaimed. Refer to Chapter 17 "Fired Clay Bricks" for more information about reducing the embodied carbon of brick masonry structures.

Steel Retaining Walls

Steel retaining walls are comprised of sheet or plate steel that is stiffened by posts or gussets anchored to concrete. The spacing of the support elements is directly related to the gauge of steel and anticipated forces behind the wall. If some movement will be tolerated

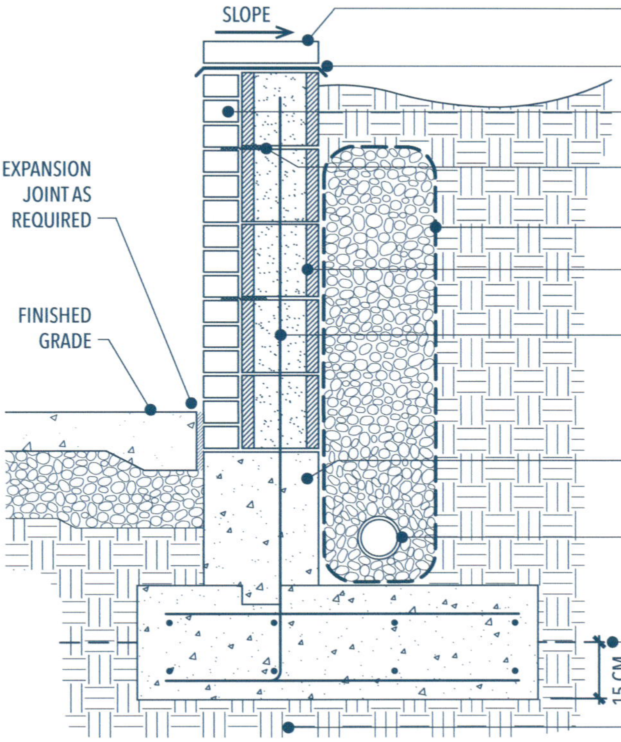

Figure 8.30 Brick veneer on CMU retaining wall detail. If structurally recommended, filling every other core with grout will reduce embodied carbon and resource use. Image credit: Kathleen Elliott and Meg Calkins

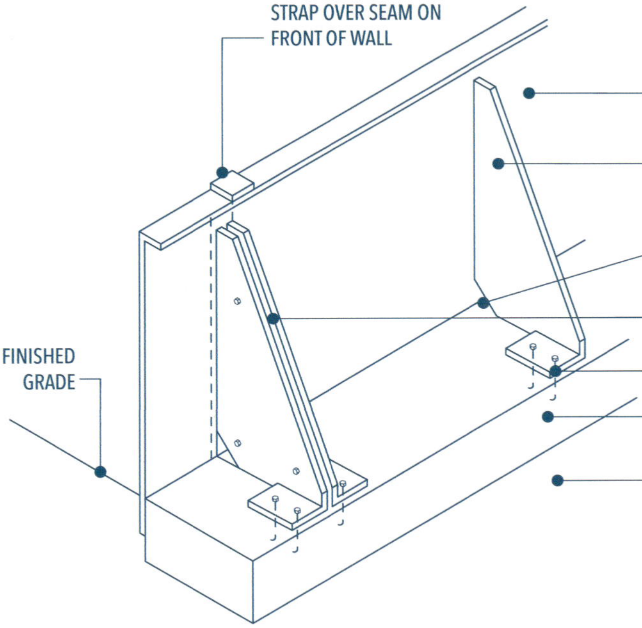

Figure 8.31 Steel plate or sheet retaining wall on a concrete base. Image credit: Kathleen Elliott and Meg Calkins

and the wall is low, it could be attached to a concrete base that is 15–30cm thick but does not extend down to the frost line. Concrete footings for taller steel retaining walls should rest 15cm below the frost line (Figures 8.31–8.34).

Sheet pile is interlocking sheets of steel that form a sheet pile retaining wall. These walls may be supported by vertical posts. Prefabricated sections are driven into the ground in sequence, and the joints are connected with fasteners.

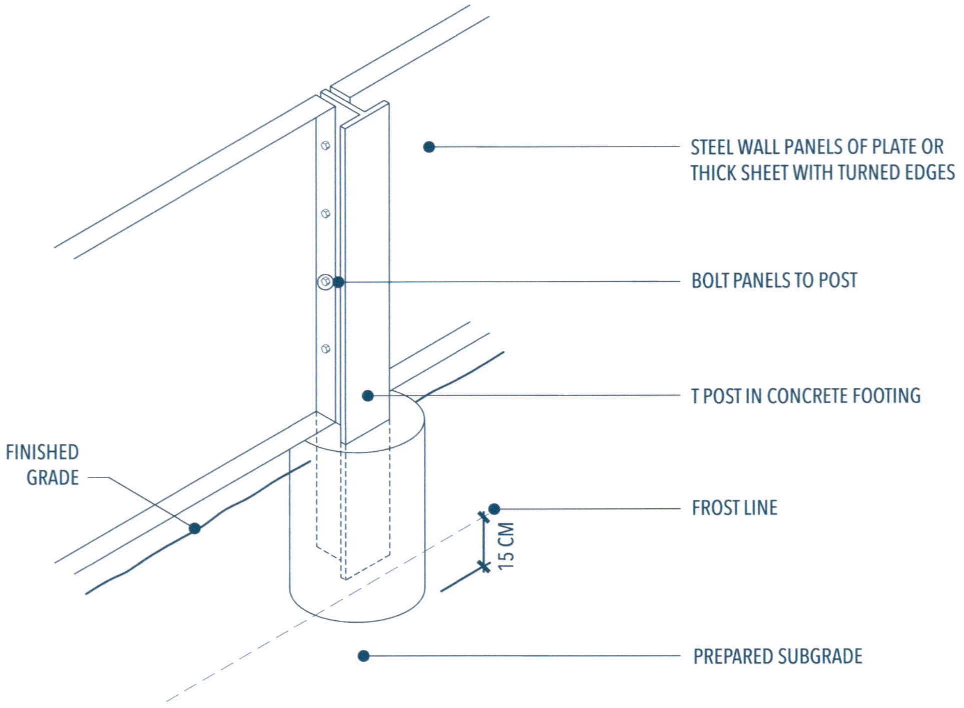

Figure 8.32 Steel plate or sheet retaining wall with concrete pier footings. Image credit: Kathleen Elliott and Meg Calkins

Figure 8.33 Design Workshop used steel retaining walls engraved with interpretive information at the Natural History Museum of Utah. Image credit: D. A. Horchner/Design Workshop, Inc.

Figure 8.34 At the Dillon in Raleigh, Stewart used weathering steel plate for a retaining wall. Perforations allow for lighting just under the top of the wall. Image credit: Tzu Chen

Steel used in exterior settings must be stainless steel, weathering steel, or coated steel to protect the structure from corrosion. Bolted connections will make steel structures easier to deconstruct than welded connections. To prevent galvanic corrosion, metal fasteners should be specified to be compatible with the steel or steel finish of the wall. Dissimilar metals should be isolated with neoprene or other plastic washers and separators (Chapter 20).

Steel sheet or plate thickness used in retaining walls will increase with the height of the wall and the soil retained. Steel can warp and bend due to soil pressure and heat from weathering, even while it remains structurally sound. To maintain the appearance of stability, a 6mm minimum thickness should be used (Faber 2024).

Retaining Wall Embodied Carbon Comparison

A variety of typical retaining wall assemblies are evaluated for their performance, embodied carbon, and environmental impacts. The retaining wall assemblies are designed for a

60 year service life and the same intensity of use in a public landscape (Table 8.7 and Figures 8.35 and 8.36).

The retaining walls have a height above grade of 90cm. The frost line depth is 60cm, therefore the bottom of the spread footing for the two rigid walls is 75cm below finished grade. For rigid walls, the footing thickness is 30cm with a 90cm width. The semi-flexible wall stems end 15–30cm below finished grade and rest on compacted aggregate bases.

A drainage assembly consisting of a 10cm HDPE perforated pipe in drainage stone wrapped in non-woven geotextile filter fabric is located behind each of the retaining walls. The embodied carbon figures are provided for both one linear meter of retaining wall assembly and a ten-meter length of wall.

Table 8.6 Retaining Walls: Benefits, Drawbacks, and Potential Performance

Segmental Retaining Walls		
Benefits Relatively long use life Very low maintenance Can accommodate differential settlement. If settling occurs, releveling can be done without replacing the wall In cold climates, SRWs are lower EC than rigid retaining walls since the base of the wall doesn't extend below the frost line Lower construction costs Ease of construction (Figure 8.21)	**Drawbacks** Aesthetic qualities may not be as desirable as other materials in commercial and institutional landscapes. Transport distances could be long as blocks are not as widely produced as PIP concrete. Precast concrete uses higher volumes of cement and often no cement substitutes.	**Potential Performance Improvements** Blocks with cement substitutes or alternatives. EC can be reduced with use of hollow blocks filled with recycled aggregates. Recycled aggregates can reduce use of virgin aggregates in the base. Incorporation of recycled materials in the concrete block mixture.
Segmental Retaining Wall With Soil Reinforcing		
Benefits Relatively long use life Very tall wall heights can be achieved Very low maintenance Can accommodate differential settlement. If settling occurs, releveling can be done without replacing the wall In cold climates, lower EC than rigid retaining walls since the base of the wall doesn't extend below the frost line Lower construction costs Ease of construction (Figure 8.22)	**Drawbacks** Aesthetic qualities may not be as desirable as other materials in commercial and institutional landscapes. Transport distances could be long as blocks are not as widely produced as PIP concrete. Precast concrete uses higher volumes of cement and often no SCMs.	**Potential Performance Improvements** Soil reinforcing allows for smaller and lighter weight blocks to be used in the wall. Blocks with cement substitutes or alternatives. EC can be reduced with use of hollow blocks filled with recycled aggregates. Recycled aggregates can reduce use of virgin aggregates in the base. Incorporation of recycled materials in the concrete block mixture.
Gabion Retaining Walls		
Benefits Long use life, repairable Can accommodate minor differential settlement Relatively low embodied carbon Local stone and recycled masonry materials can be used Lower cost than granite veneer and PIP concrete walls (Figure 8.15)	**Drawbacks** Gabions with rubble or reclaimed concrete slabs can be perceived as rough or inappropriate in a formal public landscape. Maintenance is required to make sure soil does not accumulate in the gabions allowing weeds to take root, unless they are intended to be planted gabions.	**Potential Performance Improvements** 100% recycled content stainless steel gabions will reduce EC. Any reclaimed masonry can fill the gabions. Recycled materials can reduce use of virgin aggregates in base.
Geocellular Confinement Retaining Walls		
Benefits Relatively low EC compared to masonry walls Tall retaining walls can be achieved Wall systems can accommodate minor differential settlement Relatively low cost Ecosystem services are provided by vegetation on walls (Figure 8.26)	**Drawbacks** Plastic is vulnerable to UV degradation if not covered by plants. Can be unattractive if plant cover is patchy. Not suitable for dry climates where plants will not easily grow on the wall without irrigation.	**Potential Performance Improvements** Products with higher recycled content. Use of on-site soil reduces the embodied carbon from transportation. Use of vegetation that requires minimal inputs.

(*Continued*)

Table 8.6 (continued)

Timber Retaining Walls		
Benefits Low embodied carbon Stores some carbon until it decays Can accommodate minor differential settlement Lower cost than granite veneer and PIP concrete walls. (Figures 8.24 and 8.25)	**Drawbacks** The expected use life for a timber retaining wall is only 15–30 years. Timber walls can be perceived as rough or inappropriate in a formal public landscape.	**Potential Performance Improvements** Use of smaller, local timbers will reduce EC. 100% recycled content galvanized steel rods will reduce EC. Sustainably harvested timbers will minimize environmental impacts from harvesting.
Poured in Place Concrete Retaining Wall		
Benefits Very low maintenance Relatively long use life yet wall structure could be compromised by settling of subgrade or unexpected forces behind wall Widely available materials Relatively tall walls are possible. (Figure 8.28)	**Drawbacks** Cracks are very difficult and costly to repair. High embodied carbon. Relatively high construction costs but could be offset with a long use life.	**Potential Performance Improvements** High volumes of SCMs or cement alternatives. 100% recycled content steel for rebar. Recycled materials can reduce use of virgin aggregates in the concrete mix. Wall thickness could be reduced if structurally possible.
Masonry Veneers on CMU Retaining Walls		
Benefits Very low maintenance Relatively long use life, yet wall structure could be compromised by settling of subgrade or unexpected forces on wall (Figure 8.29)	**Drawbacks** High embodied carbon for masonry materials. Highest construction costs but could be offset with a long use life. Transport distances could be relatively long for brick, cast stone, or dimension stone veneer.	**Potential Performance Improvements** High volumes of SCMs and cement alternatives. 100% recycled content steel for rebar, dowels, and masonry ties. Wall thickness could be reduced if structurally possible. Reclaimed bricks or stone.
Steel Retaining Walls		
Benefits Steel retaining walls can be durable if steel is protected from corrosion Stainless steel is relatively high embodied energy. Weathering steel is lower (Figures 8.31 and 8.32)	**Drawbacks** At risk for corrosion. Can be expensive, particularly if piles must be drive into bedrock.	**Potential Performance Improvements** 100% recycled content steel. Optimized concrete footings or slabs.

Table 8.7 EC for Retaining Wall Assemblies

Retaining Wall Type 90cm Retaining Height, 60cm Frost Depth, 60 Year Service Life	**Retaining Wall Components and Sizes**	**Retaining Wall EC, 1 Linear Meter ($kgCO_2e$), 10 meters (tCO_2e)[1]**	**Optimized Specifications and Sizes to Reduce EC**
Imported granite Veneer on PIP concrete retaining wall and footing	Imported granite veneer is 26mm thick and attached to a 30cm 27.6 MPa concrete stem with mortar and galvanized stone anchors. Footing is 30cmx105cm. Vertical and horizontal rebar form a cage in the stem and tie into footing. Footing is reinforced both ways.	*1m:* 372$kgCO_2e$ *10m:* 3.72tCO_2e	Domestic granite is 23% reduction to 288$kgCO_2e$/m 50% SCMs and 20.7MPa is an additional 31% reduction to 198$kgCO_2e$/m 23cm wall thickness, 20.7MPa, 50% SCMs is a total 55% reduction to 166$kgCO_2e$/m
PIP concrete retaining wall	27.6MPa concrete stem is 30cm wide x 135cm and footing is 30cmx105cm. Two layers of vertical and horizontal rebar form a cage in the stem and tie into footing. Footing is reinforced both ways.	*1m:* 281$kgCO_2e$ *10m:* 2.80tCO_2e	20.7MPa concrete is 16% reduction to 237$kgCO_2e$/m 50% SCMs is an additional 22% reduction to 185$kgCO_2e$/m 20cm thickness, 20.7MPa, 50% SCMs is a total 47% reduction to 148$kgCO_2e$/m

(*Continued*)

Table 8.6 (continued)

Retaining Wall Type 90cm Retaining Height, 60cm Frost Depth, 60 Year Service Life	Retaining Wall Components and Sizes	Retaining Wall EC, 1 Linear Meter ($kgCO_2e$), 10 meters (tCO_2e)[1]	Optimized Specifications and Sizes to Reduce EC
Segmental retaining wall, solid blocks	Solid interlocking precast concrete blocks are 30cm wide. One course of blocks is below grade on a compacted aggregate base.	*1m:* 151kgCO_2e *10m:* 1.5tCO_2e	Hollow blocks with 40% open cells and three layers of geogrid soil reinforcing at 1.75m deep is a 27% reduction to 110kgCO_2e/m
Steel retaining wall	7mm thick weathering steel plate (80% RC) retaining wall with T angle posts embedded in 30cm dia. 27.6 MPa, 20% SCM concrete footings every 1.5m.	*1m:* 129kgCO_2e *10m:* 1.29tCO_2e	98% RC WS and 50% SCMs is a 19% reduction to 104kgCO_2e/m
Interlocking CEB wall	Interlocking CEB are 30cm wide. One course of blocks is below grade on an aggregate base. Wall is stabilized with three layers of geogrid tiebacks, 1.75m deep.	*1m:* 65kgCO_2e *10m:* 0.65tCO_2e	50% SCMs in CEB is a 34% reduction to 43kgCO_2e/m
Gabion retaining wall filled with recycled concrete slabs	Stainless steel (SS) gabion cages are 4 gauge welded wire mesh and are 60cm wide x 60cm high x 122cm long and stacked two gabions high. SS is 50% recycled content. The gabion baskets are set 30cm below finished grade on a compacted aggregate base.	*1m:* 109kgCO_2e *10m:* 1.09tCO_2e	85% recycled content SS is 32% reduction to 74kgCO_2e/m
Timber retaining wall	ACQ-treated southern yellow pine (SYP) timbers are 150mm square. Two courses of timbers are below finished grade resting on an aggregate base. Galvanized steel rods hold the timbers in alignment. A 1m long horizontal timber tie extends into the backfill with a 1m timber "deadman" parallel to the wall face every 2 meters. Calculations include one full replacement of wall assembly within 60 year service life.	*1m:* 86kgCO_2e *Biogenic carbon stored:* 223kgCO_2e[2] *10m:* 0.86tCO_2e *Biogenic carbon stored is:* 2229kgCO_2e	100x100mm timbers is a 30% reduction to 60kgCO_2e/m Biogenic carbon stored is reduced to 143kgCO_2e/m
Vegetated timber crib wall	Untreated timber crib wall of 89x89mm SYP timbers on compacted aggregate base. Crib wall is filled with soil and planted with red osier dogwood cuttings. Crib will biodegrade and cuttings will root to stabilize slope.	*1m:* 26kgCO_2e *Biogenic carbon sequestered by veg:* 80kgCO_2e *10m:* 0.26tCO_2e *Biogenic carbon sequestered by veg. is:* 799kgCO_2e	

[1] EC figures are cradle-to-gate and include product stage life cycle phases A1: Material Extraction, A2: Transport to Manufacturing, A3: Processing and Fabrication. They do not include transportation beyond the factory gate. These figures could be reduced in the future with production efficiencies, increased recycled contents, and greater use of renewable energy sources.

[2] Biogenic carbon stored in the timber wall for the first wood used (30 years). Replacement wood is not included in the biogenic carbon calculations.

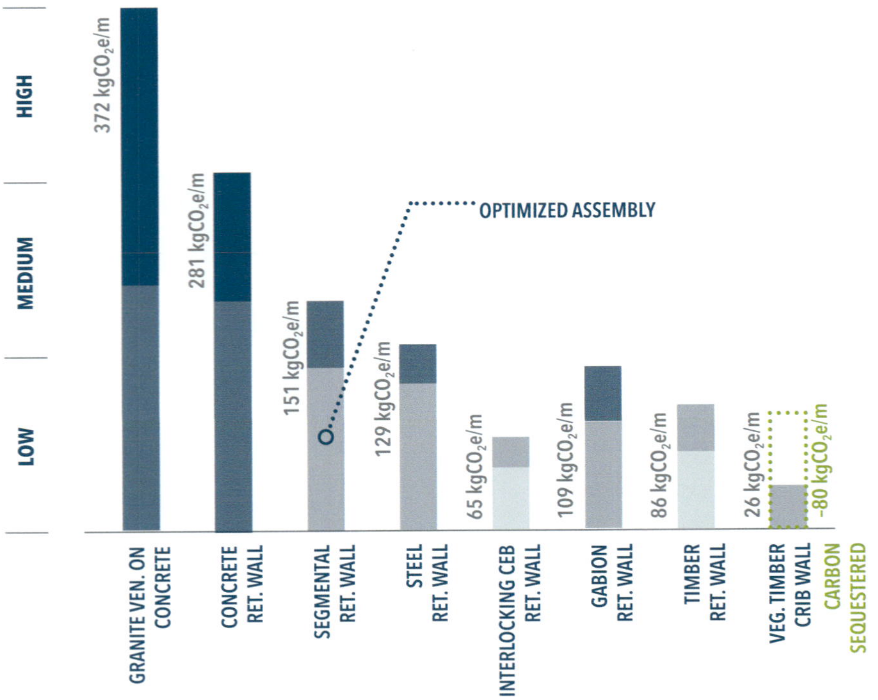

Figure 8.35 EC results for one linear meter of a retaining wall assembly with a 60 year service life. The retaining height is 90cm, and the frost depth is 60cm. Footings for rigid walls extend 15cm below the frost depth. The top part of each bar and the number shows EC results for original assemblies, and the lower part of the bar shows optimized assemblies. Image credit: Meg Calkins

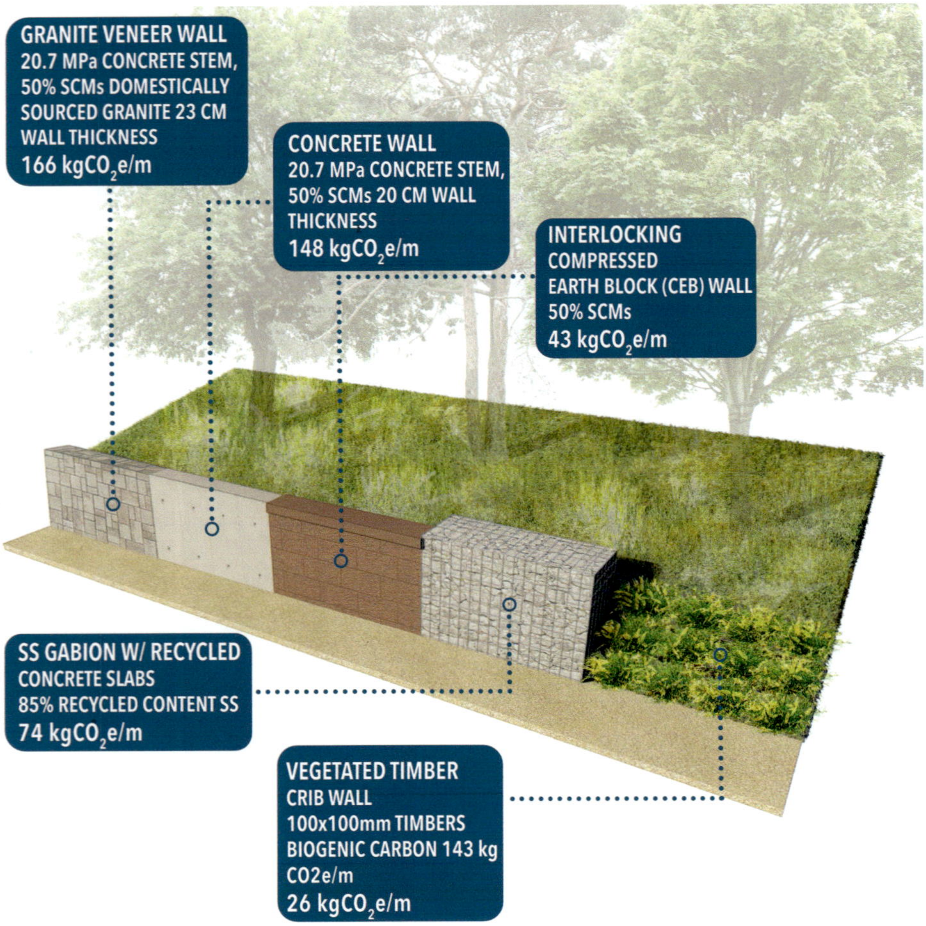

Figure 8.36 Embodied carbon of selected optimized retaining assemblies. Image credit: Lizandro Marcial-Armas, Kathleen Elliott and Meg Calkins

Table 8.8 Resources for Retaining Walls and Bioengineered Embankments

"Chapter 16 Soil Bioengineering for Streambank and Shoreline Protection" "Chapter 18 Soil Bioengineering for Upland Slope Protection and Erosion Reduction" *Title 210 National Engineering Handbook Part 650 Hydrology*. US Natural Resources Conservation Service (NRCS)
LaRiMiT Toolbox. "Mitigation Measures". Norwegian Geotechnical Institute
Landscape Architecture Construction, Third Edition. Landphair and Klatt 1999
Segmental Retaining Walls Best Practices Guide, National Concrete Masonry Association (NCMA)
TEK Guides, National Concrete Masonry Association
Constructing Landscape: Materials, Techniques, Structural Components. Astrid Zimmermann, ed.
Time-Saver Standards for Landscape Architecture, Second Edition. Harris and Dines 1997
The Brick Development Association Guides
The Brick Industry Association Technical Notes

References

Faber, S. (2024). Personal communication, September 14.

Harris, C.W., and Dines, N.T. (1997). *Time-Saver Standards for Landscape Architecture*, Second Edition. McGraw Hill.

Haseltine, B.A., et al. (1991). *Design Guide No. 2: The Design of Brick Retaining Walls*. Brick Development Association (BDA).

International Living Future Institute (2024). *Living Building Challenge Red List*. IFLI.

Landphair, H.C., and Klatt, F. (1999). *Landscape Architecture Construction*, Third Edition. Prentice-Hall, Inc.

National Concrete Masonry Association (NCMA) (2007). *Concrete Masonry Cantilever Retaining Walls*. TEK 15-07B. NCMA.

Natural Resources Conservation Service (NRCS) (2021a). "Chapter 16 Soil Bioengineering for Streambank and Shoreline Protection". In *Title 210 National Engineering Handbook Part 650 Hydrology*, Second Edition, NRCS, 210-650-H. United States Department of Agriculture, February.

Natural Resources Conservation Service (NRCS) (2021b). "Chapter 18 Soil Bioengineering for Upland Slope Protection and Erosion Reduction". In *Title 210 National Engineering Handbook Part 650 Hydrology*, Second Edition, NRCS, 210-650-H. United States Department of Agriculture, February.

Norwegian Geotechnical Institute (2023). *LaRiMiT (Landscape Risk Mitigation Toolbox)*. www.larimit.com/, accessed November 11, 2023.

Polster, David F. (2003). "Soil bioengineering for slope stabilization and site restoration". *Mining and the Environment* 3: 25–28.

Solano, L. (2024). Personal communication, September 13.

Steele, D.P., et al. (2004). *The Use of Live Willow Poles for Stabilizing Highway Slopes*. Transportation Research Laboratory.

Zimmerman, A. (Ed.). (2015). *Constructing Landscape: Materials, Techniques, Structural Components*, Third Edition. Birkhäuser.

CHAPTER 9

Freestanding Walls

Freestanding walls play an integral role in built landscapes by defining and enclosing spaces, offering privacy, providing seating, creating microclimates, blocking sounds, and establishing the visual qualities of a place. While most freestanding walls are masonry (e.g., concrete, precast concrete, brick, stone, adobe, rammed earth), wood or metal fences could serve the same function of boundary and privacy walls at a potentially lower economic and environmental cost. Fences are discussed in Chapter 11 "Rails, Fences, and Screens".

The performance, aesthetics, durability, cost, and carbon footprint of freestanding walls varies widely. Some types of walls can be carbon and resource intensive structures, while others use locally reclaimed or earth-based materials. When selecting a freestanding wall type, the focus should be on maximizing the life cycle of the structure and reducing environmental impacts while meeting aesthetic, cost, and performance needs (Table 9.1).

Seat walls can be freestanding walls or low retaining walls. Comfortable seating heights are between 38 and 52 centimeters (15–18") and comfortable seating widths can be as little as 23cm (9") (Cooper Marcus 1990). A width of 60cm (23.6") or greater will allow for back-to-back seating, although this will use substantially more material resources in the wall.

This chapter addresses the full range of freestanding wall types from flexible to rigid structures. Construction considerations are discussed and illustrated, and strategies to reduce environmental impacts are offered. Embodied carbon comparisons are performed for selected types of seat walls.

The two major structural approaches to walls, rigid walls and flexible walls, are discussed in Chapter 8 "Retaining Walls and Bioengineered Embankments". Strategies for durable wall construction are discussed in Table 8.3, and strategies to reduce the embodied carbon and resource use of freestanding walls are listed in Table 9.1 and discussed in more detail in Table 8.4.

The primary forces on free standing walls are the "dead loads" from the weight of their materials, as well as lateral forces from wind and, where applicable, earthquakes. Expansion and contraction of soil due to freezing and thawing will impact wall bases and footings. The structural footings of rigid freestanding walls should be placed on undisturbed soil a minimum of 15cm below the frost line.

Applicable building codes for wall height, width, footing, reinforcing, and other requirements should be followed. Structural engineers should be engaged for tall freestanding walls, or wall footings in challenging soil conditions. All footing sizes, reinforcing, and other details in this chapter are for illustrative purposes only.

Flexible Freestanding Wall Structures

Flexible freestanding wall structures are designed to accommodate slight movement and remain structurally sound. Because they do not employ mortar or concrete, repair and releveling activities are possible. Bases for flexible walls do not need to extend below the frost line, therefore they will conserve resources in areas with substantial frost depths. Flexible freestanding wall types include stacked gabions, dry stack reclaimed concrete slab walls, dry stack stone walls, monolithic quarry block walls, and concrete segmental block walls.

Freestanding Gabion Walls

Gabion walls are a form of gravity wall that rely on the mass of the masonry materials within a steel gabion cage for stability. As the cage serves to contain the materials, no mortar or other stabilizing material is required. Gabion cages can be filled with rubble stone, reclaimed concrete slabs, reclaimed bricks, and other heavy masonry materials. Unlike a stone or brick wall, they do not require skilled labor to construct, and they can take less time to build (Figure 9.1).

DOI: 10.4324/9781003413974-10

Table 9.1 Strategies to Reduce the Embodied Carbon and Resource Use of Freestanding Walls

Strategy	Considerations
Optimize materials	Reduce the extent of the wall (height and length). Reduce the thickness of the wall stem. Adhere to building codes and structural design by structural engineer. Use pier footings rather than continuous spread footings where applicable.
Alternative structures	Consider alternatives to freestanding walls such as fences, bio barriers, living fences, or hedges.
Reducing embodied carbon	Use locally sourced and produced materials. Specify cement substitutes or cement alternatives in concrete structures. Specify 100% recycled content steel reinforcing. Sequester carbon in concrete.
Durability	Use flexible wall structures that do not need to extend below the frost line and are more easily repaired.
Resource efficiency	Fill gabions with reclaimed masonry material from on-site or near site demolition debris. Use reclaimed bricks and stone. Use stone from quarry waste. Use earth-based materials. Specify recycled aggregates for wall bases. Specify recycled aggregates in concrete mixes.

If a gabion wall will be used as a seat wall, a bench should be installed on the gabion for comfortable seating. This bench could be constructed from wood, stone, precast concrete, or steel. It can be attached to the gabion basket or could be supported independently of the gabion. Since gabions are flexible structures, the bench must also accommodate some movement if it is attached to the gabion. Gabion retaining wall structures are discussed in Chapter 8.

Dry Stack Stone Walls

Dry stack stone walls can be constructed from simply dressed quarry stones, field stone, rubble, or dimension stone. Stone is laid to achieve as much interlock as possible with very small cavities between the stones. Precisely fitting stone chips wedge between the larger stones providing additional stability.

The pattern of stone placement will depend on the type and shape of the stone. Stones with a rectangular shape are laid in a common bond coursing pattern, while rounded or irregularly shaped stones are laid in a more random pattern. Larger stones are placed at the base of the wall, and stones get smaller higher in the wall. About 25% of the stones are laid in a header orientation to tie the wall together with some extending all the way through the thickness of the wall in the higher courses.

Dry stack stone walls should have a minimum batter of 1:10 for rectangular stones and 1.5:10 for rounded stone. The base thickness of the wall should be at least ⅓ the overall height of the wall. The top of the wall is 30cm minimum and sloped to drain. The wall extends 30cm below grade, and

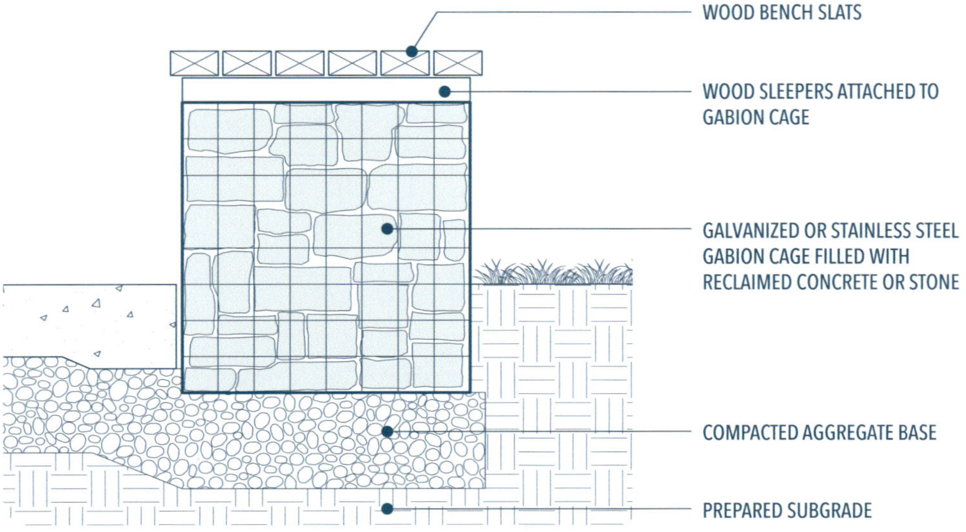

Figure 9.1 Gabion seat wall with wood bench detail. Image credit: Kathleen Elliott and Meg Calkins

Table 9.2 Design Considerations for Freestanding Dry Stack Stone Walls

Topic	Considerations
Type of stone	Quarry stone, field stone, rubble, reclaimed concrete slabs, or dimension stone can be used.
Stone interlock	Rectangular stone should be placed in a common bond coursing pattern with 25% of stone laid as "headers" extending through the wall section periodically. Wedge precisely fitting stone chips between larger stones in the wall interior for additional stability.
Distribution of stone color	If stone has a wide color range, colors should be evenly distributed across the wall face.
Stone sizes	Larger stones should be placed toward the bottom of the wall. The exception is the wall cap course. This should be larger stones that span the width of the wall section at a minimum of 30cm.
Wall batter	Minimum batter of 1:10 for rectangular stone walls and 1.5:10 for rounded stone.
Wall base	The base thickness of the should be at least ⅓ the overall height of the wall. Bottom of wall extends 30cm below grade and rests on undisturbed soil or a compacted aggregate base.
Mortar	Hidden mortar should be used sparingly as this flexible wall structure should be allowed to move slightly. Hold mortar back from face of wall by a minimum of 30mm.

rests on undisturbed soil or a compacted aggregate base (Table 9.2 and Figures 9.2 and 9.3).

Dry stack stone walls can be relatively low embodied carbon if locally quarried stone is used. As stone is a heavy material, the carbon footprint of stone that is transported long distances will substantially increase.

It is not uncommon for mortar to be added to interior joints of the wall for ease of construction and additional stone stability. If a large amount of mortar is used, this will increase the embodied carbon of the wall assembly. More critically, it could make the flexible wall too rigid and subject it to failure in a freeze/thaw climate.

Reclaimed concrete slabs and curbs can be used to create dry stack retaining walls. In some ways, the flat surface of concrete slabs makes stacking them to achieve a level wall easier, however sliding could be an issue as there is no interlock (Figure 8.19). Use of small amounts of mortar will contribute to their stability.

Monumental Stone and Cast Stone Walls

Low walls and seat walls can be constructed from solid quarry stone or 'cast stone' (precast concrete) blocks. These blocks

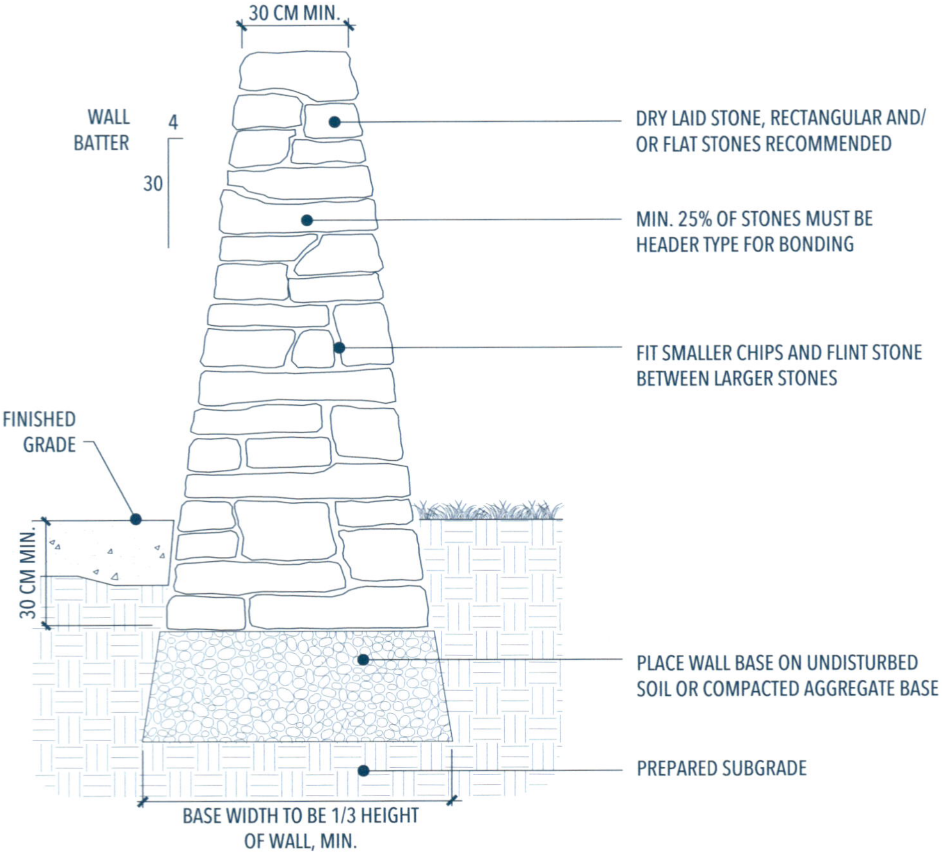

Figure 9.2 Dry stack stone wall detail. Stone interlock is key to durability. Image credit: Kathleen Elliott and Meg Calkins

can be placed on concrete footings for extra stability or on aggregate bases if some minor movement will be tolerated.

Quarry stone are large stone blocks found abandoned at many stone quarries because they are unusable for cutting

Figure 9.3 Freestanding dry stack wall in the Cotswolds. Stones were carefully fit together with larger stone at the bottom. Image credit: Meg Calkins

into dimension stone. These quarry blocks are technically stone waste because the dimension stone industry can't use them, but they can make useful benches, walls, and embankments. Quarry blocks vary substantially by type of stone, method of stone quarrying, and location. The embodied carbon of local, unfinished stone is very low, and if it is a waste product from stone production, it is even lower. From a design standpoint, the color and texture of local quarry blocks can reference the surrounding natural landscape (Figures 9.4 and 9.5).

The embodied carbon of solid cast stone blocks will be substantially higher than natural stone as it is a precast concrete material produced in a factory environment then transported to a project site (Figure 9.6). Precast production allows for finer concrete mixes, greater control of environmental factors, and more sculptural or intricate forms than those of PIP concrete. Most precast concrete does not incorporate supplementary cementitious materials (SCMs) like fly ash or slag as these materials slow down set times and production. Additionally, precast concrete uses higher percentages of cement. Ultra-high performance concrete (UHPC) uses as much as 30% cement; however many of these structures are hollow using less material than solid cast concrete.

Segmental Block Freestanding Walls

Precast concrete segmental block walls rely on their weight and interlock for stability. Mortar is not used. While segmental block walls are more commonly used as retaining wall systems, they are also employed for low freestanding walls. The wide variety of blocks products are designed to interlock with shear keys and/or mechanical connectors

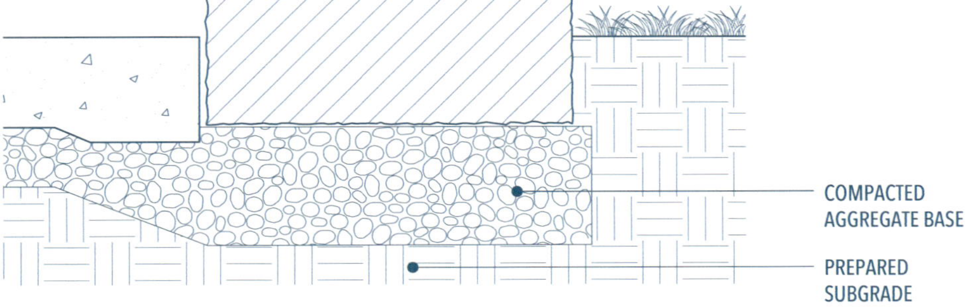

Figure 9.4 Quarry block seat wall detail. Image credit: Kathleen Elliott and Meg Calkins

such as pins. Segmental retaining wall systems are discussed in greater detail in Chapter 8.

Segmental block wall systems accommodate small amounts of movement and settlement and remain structurally sound. Because no mortar is used, they can be repaired or releveled if necessary. They can also be disassembled and reused after the life of a site extending their life cycle. These flexible segmental block walls usually have one course of blocks below finished grade that rests on a compacted aggregate base.

Figure 9.5 At Kingsbury commons in Austin, Texas, Ten Eyck Landscape Architects used locally sourced limestone quarry blocks set on a compacted aggregate base for seat walls. Image credit: Casey Dunn

By weight, the embodied carbon of precast concrete blocks can be higher than PIP concrete as precast blocks don't commonly incorporate concrete substitutes such as fly ash or blast furnace slag. However, the EC of these structures may be reduced without the need for spread footings below the frost line and hollow blocks filled with aggregate. There are multiple blocks that incorporate alternative cements, and recycled content making their way to market at the time of this publication.

Rigid Freestanding Walls

Rigid freestanding walls are designed to resist all movement therefore wall bases extend a minimum of 15cm below the frost line, and rest on undisturbed subgrade or a compacted aggregate base if soils are expansive or wet. Rigid freestanding walls require expansion joints every

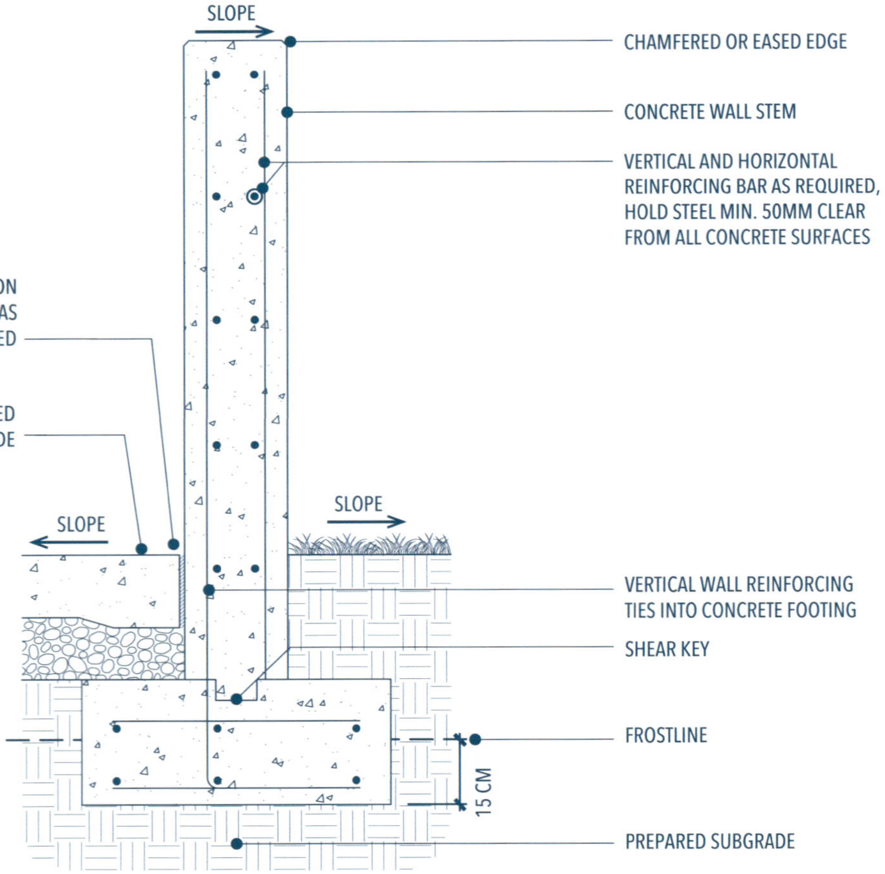

Figure 9.6 Freestanding poured in place concrete wall detail. A thinner wall stem can substantially reduce embodied carbon and resource use if allowed by the structural engineer. Image credit: Kathleen Elliott and Meg Calkins

8–10 meters and near wall corners. Expansion joints in wall veneers should be aligned with expansion joints in wall structures.

Poured in Place (PIP) Concrete Freestanding Walls

PIP concrete freestanding walls are rigid structures that do not tolerate differential settlement without cracking; therefore they are designed for stability. Concrete is a rigid material in compression and steel reinforcing, also called rebar, is used to counteract tensile forces on the wall. Freestanding concrete walls are usually cantilever walls with a vertical wall stem tied into a spread footing that rests a minimum of 15cm below the frost line on undisturbed soil (Figures 9.7 and 9.8). If the soil bearing capacity is low or the soils are wet, the footing may be placed on a compacted aggregate base. Lower PIP concrete walls might be solid gravity walls with a wider base and batter on at least one side that results in a trapezoidal section.

Reinforcing in concrete cantilever walls runs both vertically and horizontally with vertical reinforcing or bent steel rods tying the stem to the footing. Size and spacing of reinforcing will vary based on wall height and width and expected loads and forces. Reinforcing should always be held a minimum of 5cm clear of all concrete surfaces to prevent corrosion. Control joints may be required in the wall face to control cracking, but horizontal reinforcing may suffice.

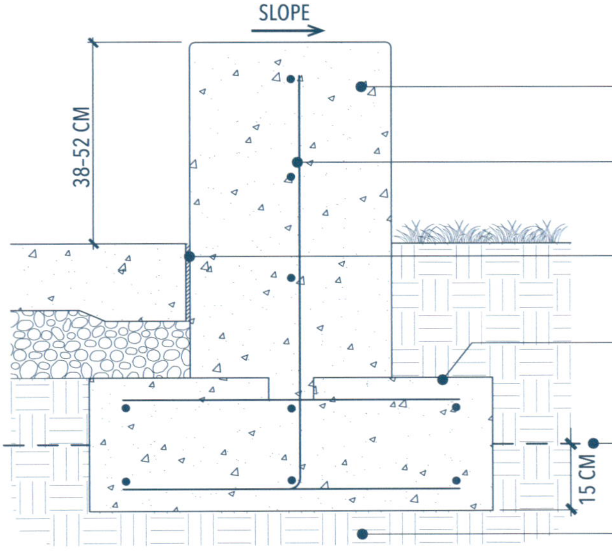

Figure 9.7 Poured in place concrete seat wall detail. A thinner seat wall can substantially reduce embodied carbon and resource use. Image credit: Kathleen Elliott and Meg Calkins

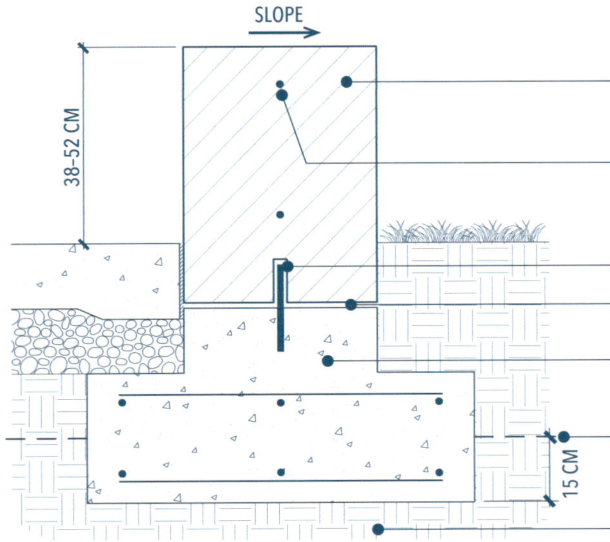

Figure 9.8 Cast stone seat wall on concrete footing detail. Cast stone can be very carbon intensive, so reducing the thickness of the wall can reduce embodied carbon substantially. Image credit: Kathleen Elliott and Meg Calkins

Concrete Masonry Unit Walls

Freestanding CMU walls are a rigid cantilever wall that rests on a PIP concrete footing below the frost line. Where frost is not an issue, the minimum depth to the top of the footing is 460mm for walls 1.8 meters or less and 610mm for walls that are 2.4mm or less (NCMA 2016b, 2003b) (Figure 9.9).

Concrete masonry units are primarily laid in a common bond pattern with vertical steel reinforcing and grout placed in some or all open cells as required. Steel dowels with a standard hook tie the wall stem to the footing and are wired to vertical rebar. The direction of the hook should be alternated in the footing and should extend to within about 80mm of the bottom. Dowel spacing should be at least equal to that of the vertical reinforcing. The length of lap for the dowels will vary by wall height and expected stresses.

Horizontal reinforcement is provided by steel bed joint reinforcement and sometimes a bond beam or joint reinforcement at the top of the wall. Where CMU block will be exposed, wall caps or copings are used to prevent moisture from penetrating the wall stem. Copings should project beyond the wall stem and provide a drip edge. Flashing may also be used.

Unless decorative CMU are used, it is common to use an exterior grade stucco plaster material or masonry veneer to provide aesthetic opportunities and protect the wall from moisture penetration. Veneer walls are discussed later in this chapter as are stucco plasters.

Rammed Earth Walls

Rammed earth freestanding walls are rigid earth-based wall structures stabilized with cement or lime. Rammed earth is placed in 150–200mm lifts in a temporary formwork and compacted either pneumatically or manually. While still not a common modern construction method, rammed earth structures are gaining traction in some regions as testing data has informed recently developed standards and codes (Minke 2022) (Figures 9.10 and 9.11).

Stabilizers for modern rammed earth structures are usually 5–10% portland cement or lime, although other stabilizers such as lignite ash and ground blast furnace slag can be used. PIP concrete incorporates an average of 10% portland cement, so if the impetus for using rammed earth is to reduce embodied carbon beyond standard concrete materials, cement reductions or substitutes should be explored.

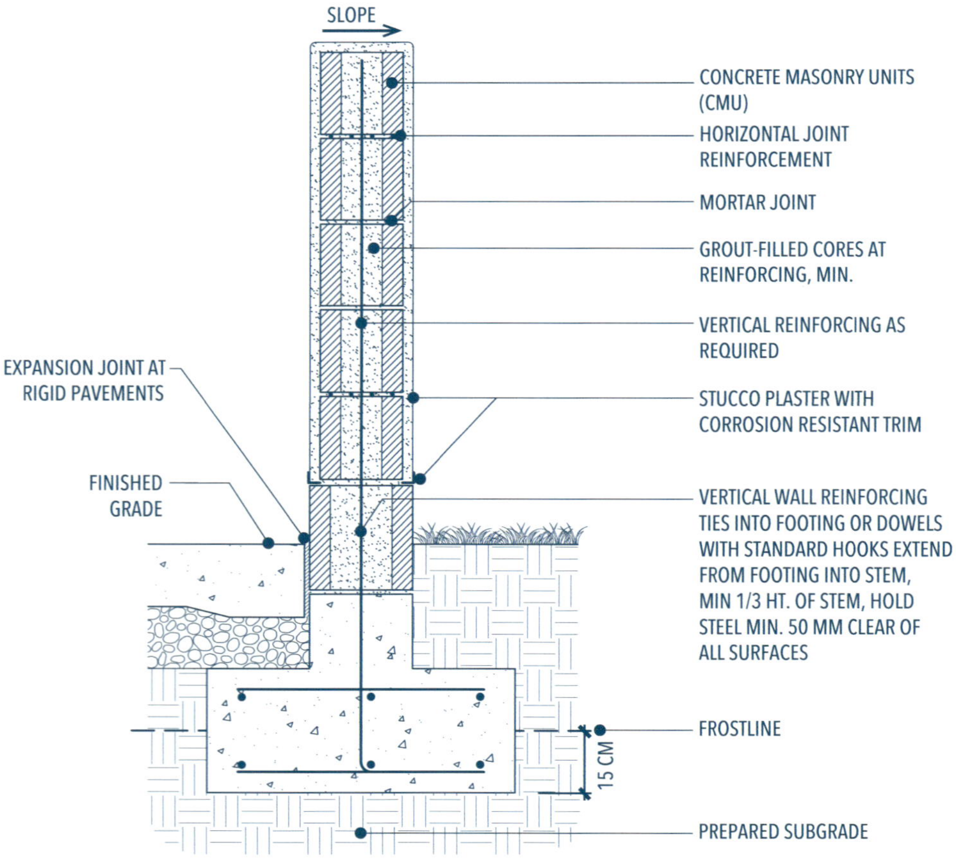

Figure 9.9 Freestanding CMU wall with stucco plaster detail. Plaster wall finishes can be repainted or even replaced without compromising the wall structure leading to a relatively long service life for the wall. Image credit: Kathleen Elliott and Meg Calkins

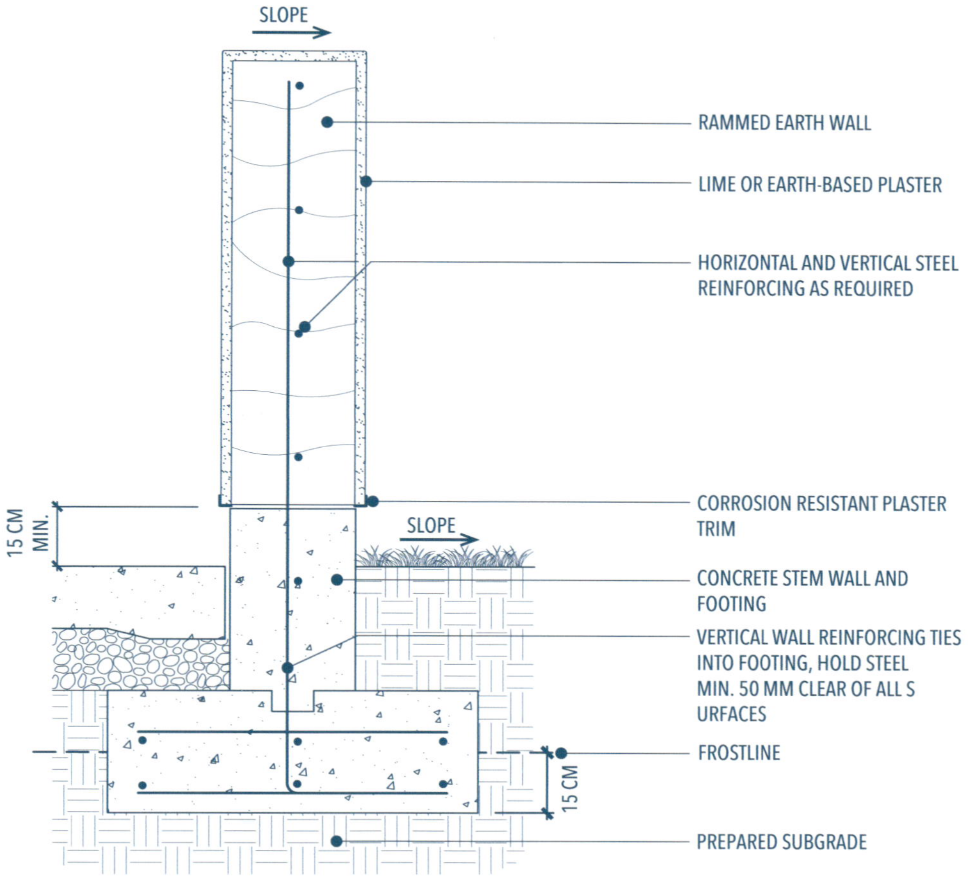

Figure 9.10 Rammed earth wall with plaster detail. In temperate and humid climates, rammed earth should be finished with plaster or a clear waterproofing coat. Image credit: Kathleen Elliott and Meg Calkins

Figure 9.11 Rammed earth walls support the steel roof structures in an open space at the Ling Vai School in Minh Tan, Vietnam. The color of the rammed earth reflects colors in the surrounding landscape. Image credit: Trieu Chien, 1 + 1 > 2 Architects

If stabilized, rammed earth structures can be used in any climate, particularly if protected with a natural stucco finish. Rammed earth materials and their construction considerations are discussed in detail in Chapter 16 "Earth-Based Materials".

In most climates, a freestanding rammed earth wall stem will rest on a low concrete stem wall and spread footing on compacted subgrade a minimum of 15cm below the frost line. In climates where frost is not a concern, lower rammed earth walls can rest on a compacted aggregate foundation. Rammed earth should be held a minimum of 15cm above finished grade to protect it from rain splash.

Adobe Brick Walls

Adobe wall construction is similar to that of brick walls in that the modular blocks are bonded together with mortar to form a rigid composite wall, although the materials and mortar types differ. Adobe, also called unfired clay bricks, sundried earth blocks, or mud bricks, are made from soil, water, and organic materials such as straw or other fibers. They are air dried and often stabilized with bitumen emulsion. Adobe materials and construction considerations are discussed in detail in Chapter 16 "Earth-Based Materials".

Modern adobe brick walls are usually 26–40cm thick with wider blocks used for earthquake prone areas and taller walls. Adobe walls should not be higher than eight times their width. In most locations, blocks are placed on a PIP concrete foundation and low stem wall per local codes and conditions. Unstabilized and semi-stabilized adobe bricks should be held a minimum of 15cm above finished grade, and the bottom few courses should be stabilized with asphalt.

For lower walls in areas where frost heave is not a concern, adobe can be laid directly on a well-compacted gravel trench foundation. This could act as a form of base isolation in earthquake prone areas. Other foundations for freestanding adobe brick walls are constructed from stone and mud mortar to hold the adobe away from moist soil.

Vertical reinforcing between the adobe blocks will tie into the stem wall and footing. Horizontal reinforcing in mortar joints can be steel wire reinforcing. This should be placed every three courses.

Mortar for adobe block walls should be a similar soil mix as the adobe with the addition of cement or asphalt emulsion for stabilization. Use of typical portland cement mortar is not recommended as it is too rigid for the relatively soft adobe blocks. Instead, mud or lime mortar is recommended (Table 9.12).

Unless adobe blocks are fully stabilized with 4–6% asphalt emulsion, walls should be protected with a mud or lime plaster coat that is reapplied every 1–3 years. Alternately, cement stucco on wire mesh can be used as a longer lasting, but higher embodied carbon plaster. Synthetic stucco should not be used as it can trap moisture in the adobe block.

Compressed Earth Block Walls

Compressed earth block (CEB) walls are a modular block like adobe or fired bricks. CEB are made from soil, water, sand and sometimes cementitious stabilizers that are compressed under very high pressure lending them compressive strength and stability. A recent innovation is interlocking CEBs also called interlocking stabilized soil blocks (ISSB). These blocks have deformations that interlock, eliminating the need for mortar and increasing their resistance to out-of-plane displacement. ISSB freestanding walls are flexible structures that can be placed on a compacted aggregate base (Figure 8.23).

CEB can be produced with portable mixing and compressing equipment on or near a construction site with local soil and portland cement or lime added for stability. CEB structures will be laid with lime or mud mortar as it has similar strength, coefficient of expansion, breathability, and moisture content as the CEB. Materials and construction

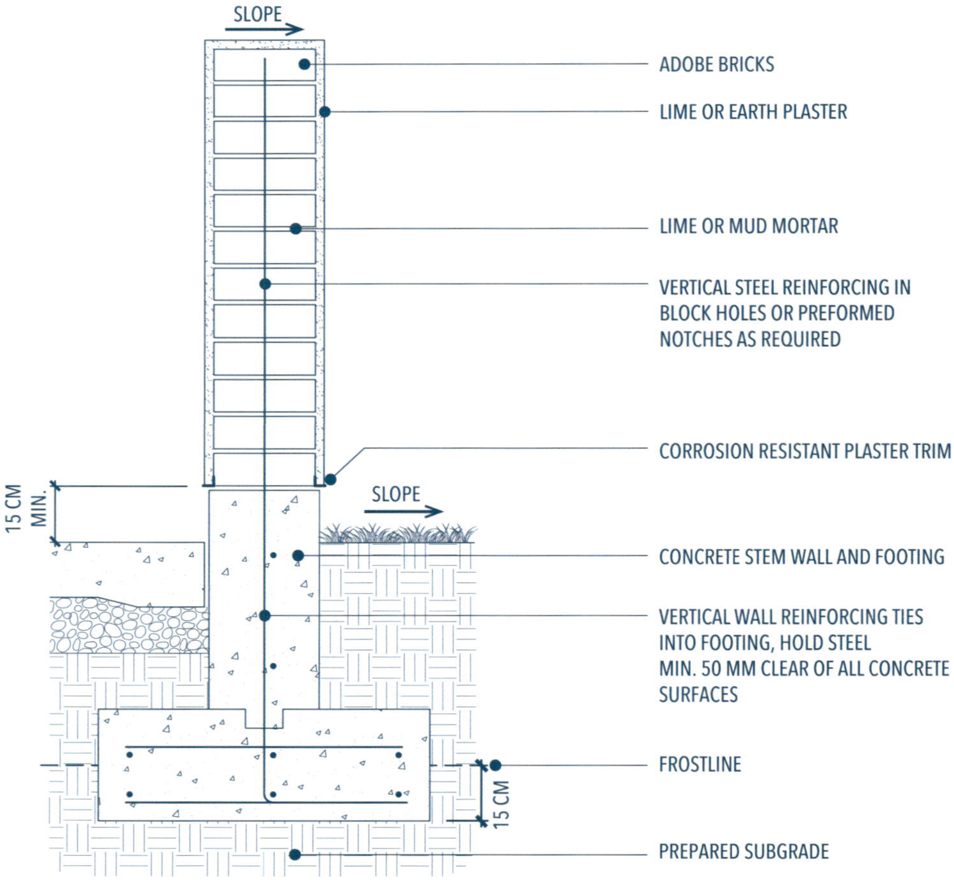

Figure 9.12 Adobe block wall with plaster detail. In temperate and humid climates, rammed earth should be finished with plaster or a clear waterproofing coat. Image credit: Kathleen Elliott and Meg Calkins

considerations for CEB structures are discussed in detail in Chapter 16 "Earth-Based Materials" (Table 16.10).

A compressed earth block wall will typically rest on a low concrete stem wall and spread footing like adobe brick freestanding walls. CEB with voids will receive vertical reinforcing tying the wall stem and footing together. These voids can also be filled with mud or lime grout. Freestanding interlocking compressed earth block walls can rest on a compacted aggregate base as they are technically flexible structures since mortar is not used.

Double Wythe Brick Walls

Freestanding solid brick wall structures were historically very common but are used less frequently in modern construction due to comparatively inexpensive CMU wall structures with brick veneer. A double wythe freestanding brick wall can be as thin as the length of one brick (about 215mm). Other types of solid brick walls include a triple wythe wall, a double wythe wall with a grouted core (Figure 9.13), or a double wythe wall with an open core. The double wythe wall with an open core is a cavity wall and is discussed in the "Cavity Walls" section of this chapter.

The traditional construction method for a double wythe wall lays the bricks in a common bond pattern with periodic header bricks or header rows every few courses to bond the wythes. The length of the header brick (perpendicular to the wall face) influences the overall width of the wall. Both the US Brick Industry Association and the UK's Brick Development Association now recommend use of horizontal masonry ties or horizontal joint reinforcing located in bed joints to tie the wythes together (BDA 2023a, 1999).

Most solid brick walls are cantilever walls with a brick stem and a PIP concrete footing. Vertical steel reinforcing runs through the bricks or between the brick wythes connecting the wall stem to the concrete footing. Steel dowels with a standard hook tie the wall stem to the footing and are tied to vertical rebar. The direction of the hook should be alternated in the footing and should extend to 8cm from the bottom. Dowel spacing should be at least equal to that of the vertical reinforcing. Length of lap for the dowels will vary by wall height and expected stresses (BDA 2023a).

As bricks are relatively high in embodied carbon, it is essential to build durable brick structures. Specifying bricks and mortar rated for the climatic conditions and preventing

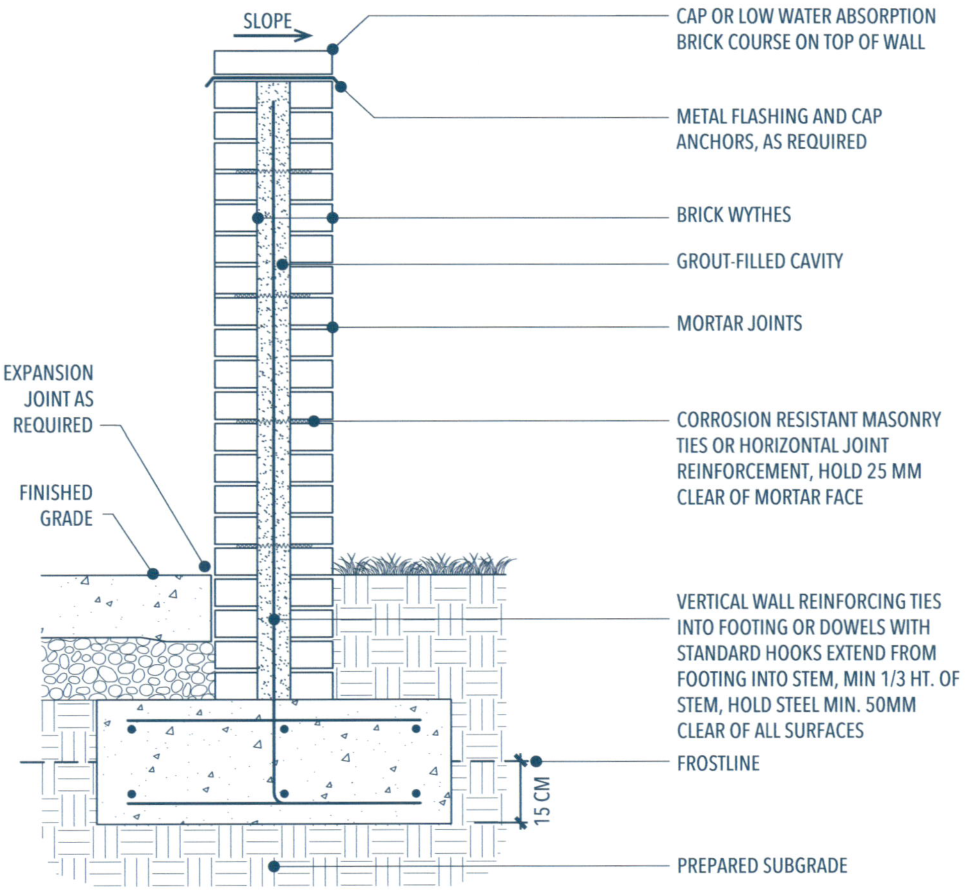

Figure 9.13 Double wythe brick wall with grout filled cavity detail. Locally produced and sourced bricks will reduce greenhouse gas emissions in transportation of heavy bricks. Image credit: Kathleen Elliott and Meg Calkins

moisture penetration into the wall cavity with caps, flashing, and other measures will produce durable walls. Innovations are being made in brick production by using waste heat, renewable fuels, and recycled content (Chapter 17).

Cavity Walls

Cavity walls are comprised of two masonry wythes separated by a continuous 5–10cm vertical airspace. Other cavity walls are a brick wythe veneer attached to a PIP or CMU structural wall. In both assemblies, the cavity allows moisture inside the wall to drain down to flashing near the bottom of the wall where it is diverted out of the wall through weep holes. This type of wall is recommended in areas of high moisture.

Wythes are bonded together with corrosion-resistant metal ties embedded in the horizontal mortar joints allowing the wall to perform structurally as one unit. Metal ties should be held back from the outer face of the mortar by at least 20mm. Codes usually dictate minimum spacing and frequency of ties.

Weep holes should be located just above the flashing near the bottom of the wall. Simple weep holes could be a header brick left out every 2–4 brick lengths. Cavity walls require caps to keep water out of the cavity.

Freestanding Walls With Stone, Brick, or Cast Stone Veneer

Veneer walls are a reinforced concrete or concrete masonry unit wall structure with a veneer of stone, cast stone, or brick attached to at least one face of the wall. As the veneer is not performing a structural function other than supporting itself, veneer can be very thin to reduce costs, embodied carbon, and resource use. Minimum thicknesses

Figure 9.14 and Figure 9.15 HOOD Design Studio used perforated brick walls to weave along the boundary of the sweetgrass field at the African Ancestors Memorial Garden in Charleston, South Carolina. Brick materials were chosen as a testament to the historical labor of enslaved Africans as brickmaking and laying were tasks commonly assigned to them. This undulating wall is a poignant reference to the robust, curved brick structures built by enslaved Africans under Thomas Jefferson's direction at the University of Virginia. From a sustainability point of view, perforated brick walls use fewer resources than solid and the openings allow for ventilation. Image Credit: Fernando Guerra

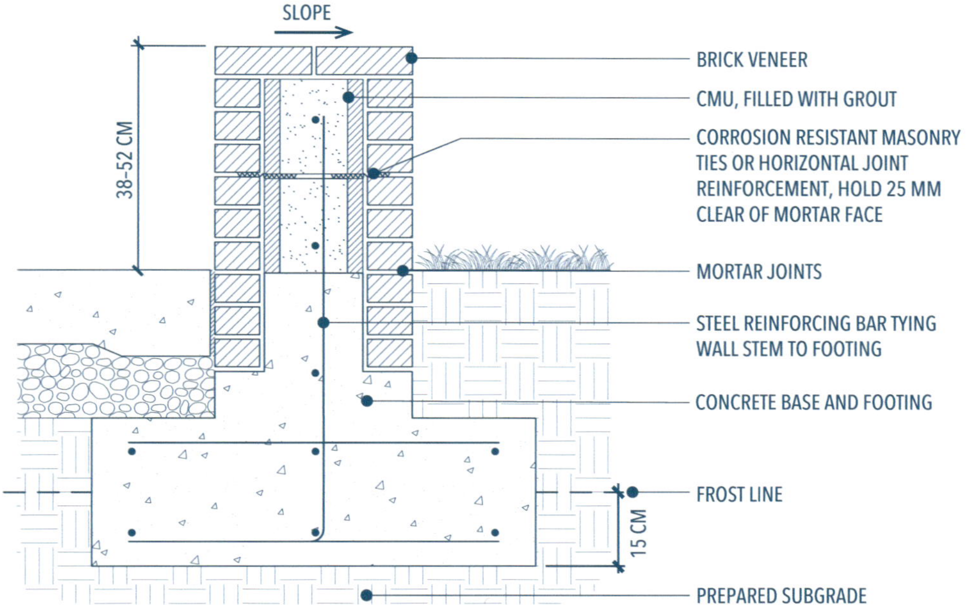

Figure 9.16 Brick veneer on CMU seat wall detail. Image credit: Kathleen Elliott and Meg Calkins

of stone veneer will vary by type, shape, and strength of stone (Figure 9.16).

The veneer is attached to the stem with corrosion-resistant mortar and/or masonry ties, veneer ties or anchors. Codes usually dictate minimum spacing and frequency of ties. Metal ties should be held back from the outer face of the mortar by at least 20mm. In humid climates, there may be a 2–3cm airspace between the veneer and the wall stem to conduct moisture as in a cavity wall.

Veneer walls require caps that shed water and prevent moisture from penetrating the bed joint or cavity between the veneer and wall structure. Caps may match the wall veneer material or be a contrasting material such as a cast stone cap on a stone veneer wall. Overhangs are recommended for wall caps so water does not drain down the face of the wall. Drip kerfs, a small slot that breaks water adhesion, are located on the underside of the cap overhang. If there is no cap overhang, metal flashing between the cap and the wall stem can prevent moisture penetration.

Stone or precast caps are held in place with corrosion-resistant dowels embedded or attached to the stem wall. Rebar should not be used for this purpose as it is vulnerable to corrosion.

Pier and Panel Walls

Pier and panel walls consist of panels braced and supported by structural piers. The piers or pilasters are the primary structural support transferring the wall loads to the soil. As these are rigid structures, the piers must extend below the frost line. The panels rest on or are held above finished grade. Sufficient bonding between the panel and piers is critical for resisting wind and other forces (Figure 9.17).

Pier and panel walls can be easily adapted to slopes as they can step down at the piers. They may be more economical than a solid freestanding wall due to ease of construction, lack of spread footings, and reduced wall panel thicknesses. Embodied carbon will vary widely based on materials, but in general less material is used with pier footings than with the spread footings of cantilever walls.

Brick Pier and Panel Walls

One or two wythe brick panels are supported by horizontal steel joint reinforcing in the mortar joints of the panel and/or a steel bar or angle at the bottom of the panel. A single wythe panel is most common for brick pier and panel walls, but it may not be appropriate in high wind conditions or in areas with expansive soils.

Finished dimensions of piers should at least be twice the width of the panel, and they can be brick, concrete or steel supported by concrete footings that extend a minimum of 15cm below the frost line. For masonry piers such as PIP concrete or bricks, reinforcing in the piers will connect to the concrete footing (BIA 1999).

CMU Pier and Panel Walls

CMU pier and panel walls are very similar to brick pier and panel walls with CMU panels supported by steel joint reinforcing in the panel's horizontal mortar joints and/or a steel plate at the bottom of the panel. Finished dimensions of

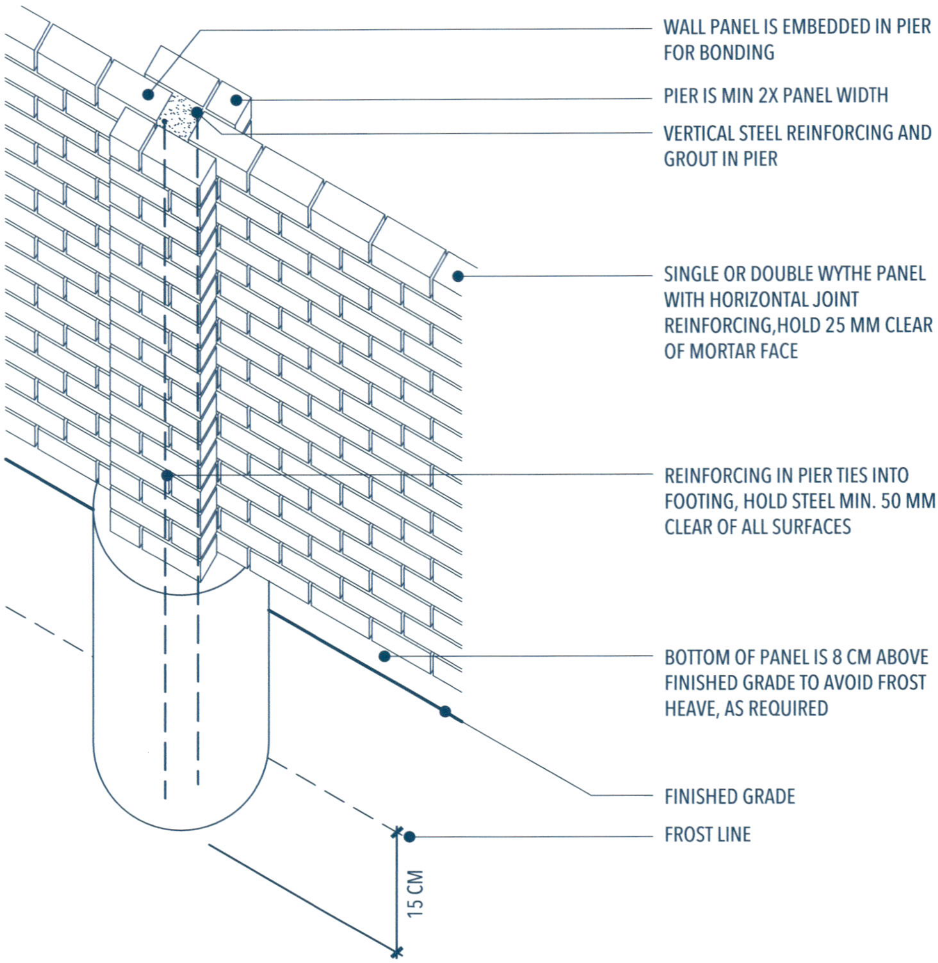

Figure 9.17 Brick pier and panel wall axonometric detail. This type of brick wall allows for a single wythe of bricks which could result in lower embodied carbon for the structure. Image credit: Kathleen Elliott and Meg Calkins

masonry piers (CMU or PIP concrete) should be at least twice the width of the panel, and they are supported by concrete footings that extend a minimum of 15cm below the frost line. Reinforcing in these piers will extend into the concrete footing. Corrosion-resistant steel piers such as I-beams, angles, or bar stock can also be used for piers and will be attached to or embedded in concrete footings (NCMA 2003b).

CMU Screen Walls

CMU screen walls are another type of pier and panel wall that provides some enclosure but has decorative openings for visibility and cooling breezes. Embodied carbon for CMU screen walls will be less than for a solid CMU wall as there is less CMU material, no grout filled cells, and the panels don't extend below grade.

CMU screen walls require lateral support from piers and sometimes bond beams because the openings in blocks prohibit the vertical steel reinforcing and grouted cells of a typical CMU cantilever wall. Piers can be constructed from CMUs with vertical reinforcing and grout filled cells, PIP concrete, or corrosion-resistant steel such as I-beams, angles, or bar stock. All piers will connect to PIP concrete footings (NCMA 2016a).

Plaster for Freestanding Walls

Wall plaster can offer aesthetic opportunities through color and texture while protecting freestanding walls from moisture penetration. Stucco plaster is the most common type of plaster for site walls, although mud plaster can be used on earth-based structures as it breathes similarly to the earth materials.

There are two types of stucco plaster for exterior walls – traditional stucco and synthetic stucco, also called acrylic stucco. Traditional stucco, made from portland cement, lime, and coarse sand, is a rigid plaster finish for rigid walls. Synthetic stucco contains acrylic resins with sand to give the appearance of traditional stucco. Traditional and synthetic stuccos are compared in Table 9.3, while earthen plasters are discussed in Chapter 16 "Earth-Based Materials".

Mortar and Grout for Masonry Walls

Mortar is used in rigid unit walls to bonds masonry units together into a composite assembly. It seals joints against water and air penetration, accommodates very small movements within the wall, and allows for slight size differences between units.

Masonry grout is a cementitious mixture used to fill cores or cavities in masonry construction. It is primarily used for structural stability to bond wall elements into a rigid structure acting as one system.

As most conventional mortar and grout contains relatively high proportions of portland cement (20–50%), cement substitutes, alternative cements, or lime-based mortars should be considered (Table 9.4). Flexible structures such as interlocking blocks or gabions will avoid mortar and grout altogether and allow for easier deconstruction and reuse of the materials.

Table 9.3 Traditional Stucco Plaster vs. Synthetic Stucco for Freestanding Walls

	Traditional Stucco Plaster	Synthetic Stucco Plaster
Life span	Up to 50 years with repairs	Up to 50 years with repairs
Embodied carbon	Can be high due to portland cement and lime; explore cement alternatives	Higher EC than traditional stucco as it incorporates both acrylic resins and cement
Porosity	More porous, dries out quickly	Waterproof, but water can be trapped between stucco and wall stem
Elasticity	More rigid than synthetic stucco.	Synthetic stucco has greater elasticity and will adapt to expansion and contraction from temperature extremes more easily
Fire resistance	Greater	Less
Colors and textures	Available in more textures, integral colors are muted, can be painted	Brighter colors, fewer textures, can be painted
Cost	Lower	Higher
Application	Applied in three coats – a rough scratch coat, a second base coat and a finish coat. Wire mesh is sometimes attached as a base if the wall surface is not rough enough	Varies by product, but generally thinner application
Maintenance	Cleaning, patching and painting	Periodic cleaning. Minimal painting unless integral color product is used or color change desired

Sources: BDA 2023b; NCMA 2003a

Table 9.4 Mortar and Grout Types

Mortar Types	Mix Details and Considerations for Use
Portland cement mortar[1] Types I, IA, II, IIA, III or IIIA *ASTM C150* *ASTM C270*	This hydraulic cement mortar is comprised of portland cement and sand mixed with water. It sets and hardens through a chemical reaction with water in the same way as portland cement concrete. A typical ratio is one part cement and two parts sand.
Masonry cement mortar[1] Types N, S, M *ASTM C91* *ASTM C270*	The binder for this mortar is a mix of portland cement or blended hydraulic cements and plasticizing materials such as limestone, hydrated or hydraulic lime. Sand and water are blended with the binder to form masonry cement mortar. Embodied carbon is reduced by 35% or more below portland cement mortar. Type N is recommended for exterior walls subject to severe exposure. Type S is recommended for use in masonry structures where flexural strength is required, and Type M is recommended for masonry below grade and in contact with the soil.
Mortar cement *ASTM C1329*	Mortar cement is like masonry cement, but it has a minimum bond strength requirement.
Blended hydraulic cement mortar[1] Types IS, IS-A, IP, IP-A, I(PM), I(PM)-A *ASTM C595*	The binder for this mortar is a mix of portland cement and blast furnace slag or pozzolans such as fly ash or calcined clay. Sand and water are blended with the binder. Embodied carbon reductions vary by cement substitutions, but they can be substantial.

(*Continued*)

Table 9.4 (Continued)

Mortar Types	Mix Details and Considerations for Use
Hydrated lime mortar[1] and natural hydraulic lime (NHL) mortar[2] Type S, SA *ASTM C207*	Hydrated lime and hydraulic lime are produced by burning different types of limestone to quicklime powders then slaking them. These mortars are a mix of hydrated or hydraulic lime, sand, and water. Both set initially when mixed with water and then have a second set through absorption of CO_2 as they harden. Hydraulic lime sets through a chemical reaction with water rather than just absorbing CO_2 like non-hydraulic hydrated lime. NHL mortar has greater compressive strengths than hydrated lime mortar and it cures more quickly (Lime Planet 2024). The compressive strengths of standard types of NHL are NHL2 (2.98MPa), NHL3.5 (6.00MPa), and NHL5 (8.80MPa). NHL mortars take 24 months to reach full strength. Embodied carbon of these lime mortars is only 13–20% of the EC of portland cement mortar.
Quicklime mortar[1] *ASTM C5, ref. 4a*	This mortar is calcined limestone that is combined with water prior to use. The resulting lime putty must hydrate for 24 hours before use. This type of mortar is no longer in widespread use.
Earthen mortar[3] *ASTM E2392*	Earthen or mud mortar is a mix of clay soil, water, straw, sand, and sometimes lime. Mud mortars are used in adobe and compressed earth block construction as they breathe like lime mortars. Cement is added for weatherproofing, but it may not be necessary if the mud mortared structure will be covered with plaster. Studies have shown that use of cement does not increase compressive strength (Erdogmus and Garcia 2015). Embodied carbon will vary by lime or cement content.
Grout Types	
Masonry grout[4] *ASTM Standard C476 Standard Specification for Grout Masonry*	Masonry grout is composed of portland cement, sand, aggregate, lime (sometimes), and water. Fine grout contains sand smaller than 9.5mm while coarse grout allows for pea gravel and sand smaller than 13mm.
Lime grout[2] *ASTM Standard C476 Standard Specification for Grout Masonry*	Lime grout is composed of hydraulic lime, sand, aggregate, and water. Fine grout contains sand smaller than 9.5mm while coarse grout allows for pea gravel and sand smaller than 13mm. Like lime mortar, lime grout is more breathable than cement-based masonry grout.

[1] NCMA 2001

[2] DGLW 2022a, 2022b, 2022c

[3] Minke 2022

[4] NCMA 2004

Hydrated and hydraulic lime-based mortars and grouts have been used in historic restoration projects to match existing historic mortars, but they are gaining use as an environmentally preferable alternative to cement-based mortars and grouts. Like cement mortars, lime-based mortars and grouts undergo the process of carbonation, absorbing CO_2 as they cure. Hydraulic lime mortars can absorb carbon up to 75% of the weight of the hydraulic lime in a mortar or grout mix. Their low production energy combined with carbonation result in an embodied carbon that is only 13–30% of the embodied carbon of portland cement mortars (DGLW 2022c).

Lime mortars are more flexible than cement-based mortars and can accommodate the very slight movements of rigid structures, reducing cracking and structural damage. Lime mortars and grouts can absorb and release moisture better than cement-based mortars. This ability to "breathe" makes them more appropriate for cavity walls and adobe and compressed earth block construction (Minke 2022).

Seat Wall Embodied Carbon Comparison

A variety of typical seat wall assemblies are evaluated in Table 9.5 for their embodied carbon, along with optimization measures to reduce the embodied carbon. The seat wall assemblies are designed for their applicability to the same intensity of use in a public landscape and for a 60-year life cycle. The seat walls have a height above finished grade of 46cm and a width of 46cm.

The frost line depth is assumed to be 60cm, therefore the bottom of the spread footing for the rigid wall alternatives is 75cm below finished grade. Embodied carbon figures are provided for 1 and 10 linear meters of each seat wall assembly (Table 9.5 and Figures 9.18 and 9.19).

Table 9.5 Embodied Carbon (EC) for Seat Wall Assemblies

Seat Wall Assembly Type	Components and Rough Dimensions	EC for One Linear Meter ($kgCO_2e$), 10 Meters (tCO_2e)[1]	Optimized Seat Wall Assemblies to Reduce EC
Precast concrete seat wall on PIP concrete footing	Precast concrete seat wall on 20.7MPa concrete footing with 20% SCMs 46x60cm precast wall stem with horizontal steel rebar SS dowel and sleeve assembly ties precast to short 46x30cm concrete wall stem, 30x60cm footing with horizontal rebar	*1m:* 263kgCO_2e *10m:* 2.63tCO_2e	50% SCMs in PIP concrete reduces EC by 9% to 240kgCO_2e/m Reduced seat wall thickness (30cm) and 50% SCM reduces EC by 34% to 171kgCO_2e/m.
Brick veneer on CMU wall with PIP concrete footing	Brick veneer attached to CMU wall with galvanized masonry ties. Concrete footing and stem wall, 20.7MPa, 20% SCMs Vertical reinforcing and grout filled cores 30x60cm footing with horizontal rebar	*1m:* 255kgCO_2e *10m:* 2.55tCO_2e	50% SCMs in PIP concrete reduces EC by 10% to 229kgCO_2e/m Reduced seat wall thickness (30cm) to a double wythe mortar filled cavity wall, eliminating the CMU and using 50% SCM reduces EC by a total of 19% to 206kgCO_2e/m.
Poured in place concrete seat wall	20.7MPa concrete with 20% SCMs 46x90cm wall stem 30x90cm footing Vertical and horizontal rebar in stem and footing	*1m:* 168kgCO_2e *10m:* 1.68tCO_2e	50% SCMs in PIP concrete reduces EC by 11% to 150kgCO_2e/m Reduced seat wall thickness (30cm) and 50% SCM reduces EC by 42% to 97kgCO_2e/m.
Wood timber seat wall on concrete base	(4) ACQ-treated softwood timbers, 200x230mm each Concrete base, 20.7MPa, 20% SCMs, 15x38cm Aggregate base 15x38cm Two full replacements of the timber are assumed for a 60 year service life	*1m:* 67kgCO_2e Biogenic carbon stored 151kgCO_2e[2] *10m:* 0.67tCO_2e	50% SCMs in PIP concrete base reduces EC by 6% to 63kgCO_2e/m Reduced seat wall thickness (30cm) and using 50% SCM reduces EC by 39% to 41kgCO_2e/m
Gabion and wood bench seat wall	The stainless steel gabion cage is 60x60x122cm It is filled with rubble stone and is supported by a 20cm compacted aggregate base The bench is thermally modified (TM) lumber on TM sleepers wired to the gabion cage Two full replacements of the lumber are assumed for a 60 year service life	*1m:* 58kgCO_2e Biogenic carbon stored 20kgCO_2e[2] *10m:* 0.58tCO_2e	Reduced seat wall thickness (40cm) reduces EC by 33% to 39kgCO_2e/m
Quarry block seat wall	The unfinished quarry block is 46x70cm. The block is set 15cm below finished grade on a 46x17cm aggregate base	*1m:* 19kgCO_2e *10m:* 0.19tCO_2e	Reduced seat wall thickness (30cm) reduces EC by 32% to 13kgCO_2e per meter.

[1] EC figures are cradle-to-gate and include product stage life cycle phases A1: Material Extraction, A2: Transport to Manufacturing, A3: Processing and Fabrication. They do not include transportation beyond the factory gate. These figures could be reduced in the future with production efficiencies, increased recycled contents, and greater use of renewable energy sources.

[2] Wood products store some CO_2 until the product biodegrades or is burned (Chapters 2 and 18).

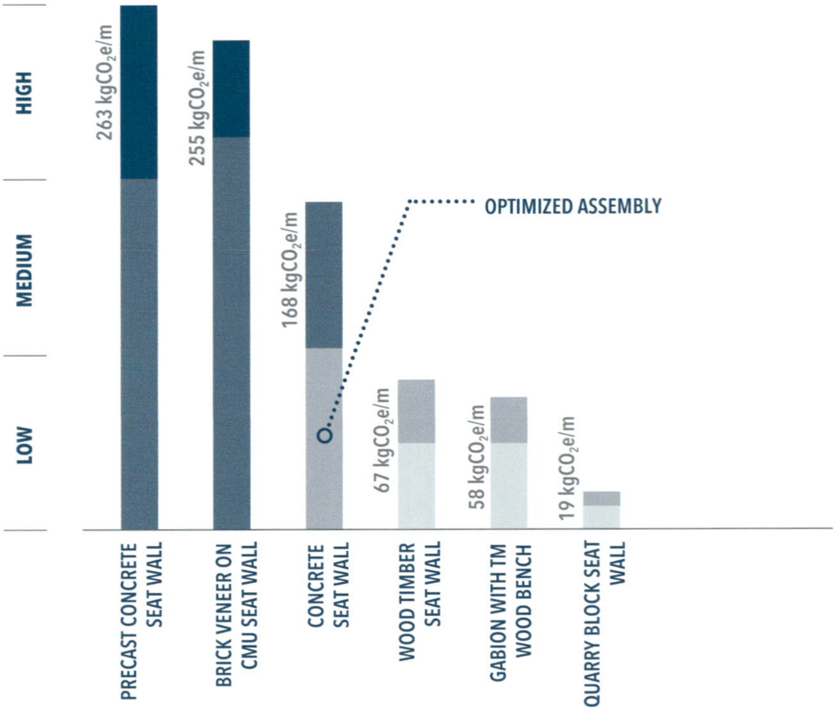

Figure 9.18 EC results for one linear meter of a seat wall assembly with a 60 year service life. The seat wall height is 46cm, and the frost depth is 60cm. Footings for rigid seat walls extend 15cm below the frost depth. The top part of each bar and the number shows EC results for original assemblies, and the lower part of the bar shows optimized assemblies. Image credit: Meg Calkins

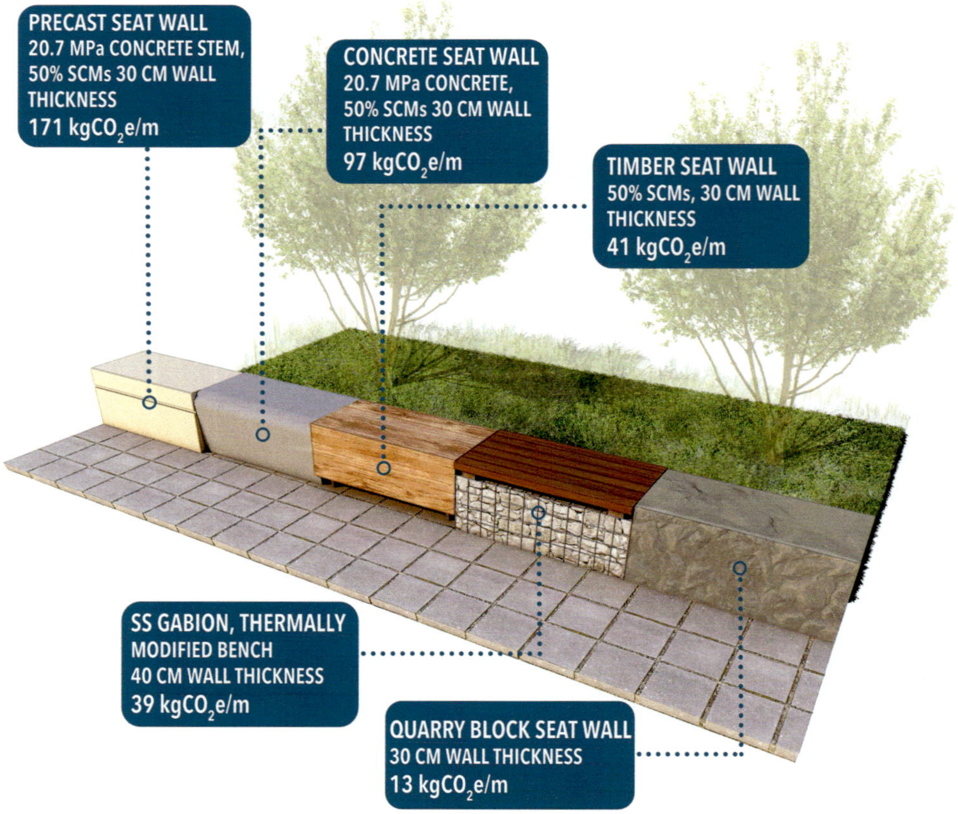

Figure 9.19 Embodied carbon results for selected optimized seat wall assemblies. Image credit: Lizandro Marcial-Armas, Kathleen Elliott and Meg Calkins

References

Brick Industry Association (BIA) (1999). *Technical Notes 29A Brick in Landscape Architecture Garden Walls*. Reston, VA: BIA.

Brick Development Association (BDA) (2023a). *Good Site Practice and Workmanship*. London: BDA.

Brick Development Association (BDA) (2023b). *Resisting Rainwater Penetration*. London: BDA.

Cooper Marcus, Clare. (1990). *People Places Design Guidelines for Urban Open Spaces*. Van Nostrand Reinhold Company.

deGruchy's Lime Works (DGLW) (2022a). *Saint-Astier Natural Hydraulic Lime 2 (NHL2)*. DGLW. www.limeworks.us.

deGruchy's Lime Works (DGLW) (2022b). *Saint-Astier Natural Hydraulic Lime 3.5 (NHL3.5)*. DGLW. www.limeworks.us.

deGruchy's Lime Works (DGLW) (2022c). *Saint-Astier Natural Hydraulic Lime 5 (NHL5)*. DGLW. www.limeworks.us.

Erdogmus, E., and Garcia, E. (2015). "Influence of stabilizers on the compressive strength of compressed stabilized earth block masonry". *Proceedings of the 12th North American Masonry Conference* 12.

Houben, H., and Guillard, H. (1994). *Earth Construction: A Comprehensive Guide*. London: Intermediate Technology Publications.

Landphair, H.C., and Klatt, F. (1999). *Landscape Architecture Construction*, Third Edition. Prentice-Hall, Inc.

Lime Planet (2024). "The difference between hydrated lime and hydraulic lime". https://limeplanet.co.uk/, accessed September 1, 2024.

Minke, Gernot. (2022). *Building With Earth: Design and Technology of a Sustainable Architecture*, Fourth Edition. Birkhäuser.

National Concrete Masonry Association (NCMA) (2001). *Mortars for Concrete Masonry*. TEK 09-01A. NCMA.

National Concrete Masonry Association (NCMA) (2003a). *Plaster and Stucco for Concrete Masonry*. TEK 09-03A. NCMA.

National Concrete Masonry Association (NCMA) (2003b). *Annotated Design and Construction Details for Concrete Masonry*. NCMA.

National Concrete Masonry Association (NCMA) (2004). *Grout for Concrete Masonry*. TEK 09-04A. NCMA.

National Concrete Masonry Association (NCMA) (2016a). *Concrete Masonry Screen Walls*. TEK 03-16A. NCMA.

National Concrete Masonry Association (NCMA) (2016b). *Concrete Masonry Fence Design*. TEK 14-16B. NCMA.

Zimmerman, A. (Ed.). (2015). *Constructing Landscape: Materials, Techniques, Structural Components*, Third Edition. Birkhäuser.

CHAPTER 10

Decks and Boardwalks

Decks and boardwalks allow for site access and gathering spaces above grade and can offer a softer, more biomimetic surface than masonry pavement. Some decks and boardwalks can be lower embodied carbon than a concrete pavement, but not always as the embodied carbon of these structures varies just as widely as the choice of deck boards.

Structures for decks and boardwalks can be constructed of wood, composite lumber, coated steel, concrete, and even aluminum (Figures 10.1–10.5). Footings are primarily poured in place concrete, but helical piers and pin foundations are increasingly used in areas with deep frost lines, sensitive ecosystems, or challenging soils. Plastic pedestals support decks on rooftops or on other hard surfaces.

The economic and environmental costs of deck boards, structures, and footings vary. While some types of deck boards have much higher upfront costs, maintenance and replacement costs must be factored into decisions about materials for decks. As with any landscape structure, durability and a long service life are key considerations. The longer the deck or a boardwalk is in use, the lower the ultimate economic and environmental costs of its construction. This chapter provides information for decks and boardwalks; however, given their similarities, the term deck will be used to apply to both decks and boardwalks.

Optimizing Materials for Decks

Optimizing the use of materials in a deck structure can conserve resources, reduce embodied carbon, and alleviate pressure on forests. It may also offer economic benefits. The strategies discussed in this section focus on using less material, designing durable decks, and reusing salvageable components after the service life of the deck.

Efficient Use of Materials

Avoiding ***oversized structural members*** will reduce the amount of wood or steel used in a deck structure. It is relatively common for wood deck members to be oversized when it may not be necessary for structural purposes. Structural engineers and span tables build tolerances into their calculations, so building in an additional margin of is unnecessary. Structural member sizes should *never* be reduced below those recommended by structural engineers or in span tables, but member sizes don't need to be increased either.

Often designers specify larger members for aesthetic reasons, but ***cultivating a leaner aesthetic*** can result in a more efficient use of materials. For instance, 140x140mm (6x6) posts are commonly specified for a deck when 89x89mm (4x4) posts may be all that is required for structural purposes. This practice uses over twice as much wood and requires larger trees to yield the 140x140mm posts.

When large beams are required for a structure, specifying ***exterior grade glue laminated timber*** (glulams) can make efficient use of wood resources by using smaller members glued into a larger composite member. For instance, when an 89x286mm (4x12) beam is required for a structure, building up a glue laminated beam from six or seven 38x89mm (2x4s) members will reduce the harvesting of large trees and make better use of wood resources.

Specifying ***lower lumber appearance grades*** for structural members that are not visible will allow for a more efficient use of wood resources. It is not necessary to specify high-quality appearance or "select" grades for beams and joists of a deck that will not be visible. General purpose "common" grades can be specified if their structural grade is appropriate.

Designing deck structures to work with ***standard lumber sizes*** can reduce wood waste. Joists, beams, and deck

DOI: 10.4324/9781003413974-11

Figure 10.1 Metal boardwalks at Warsaw Uprising Mound Park by Archigrest and topoScape hover above the 35-meter high pile of WWII rubble on which the park is built. Park visitors can view historic fragments of destroyed buildings as they walk through the new park. Image credit: Michal Szlaga

boards should be sized to the standard lumber lengths. For instance, a deck width of 3.1m will use 3.3m joist lengths resulting in .2m cut offs for each joist. Designing either a 2.9m or 3.2m deck width will use wood joist lengths more efficiently. Structure sizes should be planned for a 50mm cut off at each end to reduce any check or cracks that might develop during the drying process. Cutting any more than 150mm total off the length is an inefficient use of wood lumber.

Circular Materials and Deck Design

Wood and steel deck structures can be ***designed to facilitate their deconstruction*** at the end of their service life. Designing connections that are accessible and easy to dismantle is important to preserve the integrity of structural members during the deconstruction process, and it all also allows for replacement of members without replacing the full deck.

Accessible bolted and screwed connections will be the easiest to separate. Nailed, welded, and glued connections are difficult to dismantle and can result in tearing, breaking, or cutting of the structural member. This can reduce the amount of material reclaimed, or it could discourage deconstruction altogether due to higher costs or perceived difficulty.

Partial reuse of deck structures can eliminate waste and resource use. Very often posts, beams, and joists are still structurally viable when deck boards need to be replaced. But it is not uncommon to for an owner to decide to replace the entire deck structure. Use of durable deck detailing strategies such as wood joist caps, holding wood 15cm above grade, and regular maintenance can increase the service life of a deck (Table 10.1). And detailing a deck for ease of deconstruction can facilitate partial replacement of deck members.

Use of ***reclaimed decking and structural members*** for decks will facilitate the circular economy and can inspire the deck design. Reclaimed wood used in structural applications may require regrading. If reclaimed wood is not treated or decay resistant, it should be coated with a wood preservative such as copper naphthenate and not used in ground contact. Reclaimed wood is discussed in detail in Chapter 18 "Wood Materials and Products", and considerations for using reclaimed wood are detailed in Table 18.3.

Designing and Detailing Durable Deck Structures

Designing and maintaining durable wood structures will reduce both the environmental and economic life-cycle costs of a deck structure. The primary considerations for extending the service life of a deck include matching the wood treatment to the decay hazards and setting, detailing for durability, and engaging in appropriate and timely maintenance techniques.

Deck Structures

Platform framing and plank and beam framing are the two major approaches to framing decks and boardwalks. Platform framing is far more prevalent except on roof decks and/or pedestal decking systems where closely spaced pedestal supports allow for the elimination of joists.

Platform Framing

Platform framing employs joists to transfer loads to beams then posts, footings, and subgrade. Deck boards span on top of joists, and the joist spacing is a direct function of

Figure 10.2 Metal boardwalks at Warsaw Uprising Mound Park by Archigrest and topoScape hover above the 35-meter high pile of WWII rubble on which the park is built. Park visitors can view historic fragments of destroyed buildings as they walk through the new park. Image credit: Michal Szlaga

Figure 10.3 Organic debris lodged between deck boards and a wall or fascia can trap moisture against the end grain of wood decking. Deck boards should be spaced a minimum of 10mm from walls and adjacent structures. Image credit: USDA 2021

the spanning ability of the decking. Joists bear on beams either directly or in a flush application employing steel "joist hangers". Beams bear on posts and posts are supported by footings. Less commonly, beams are supported directly on helical pier footings or pedestals. Fewer beams are required because joists have the ability to carry loads over longer spans. Platform framing can result in a thicker deck profile if the joists bear on top of the beams. If they are flush with the beams, the profile will be thinner.

Plank and Beam Framing

Plank and beam framing does not use joists; instead deck boards span between closely spaced beams. This requires the deck boards or prefabricated decking panels to span at least 60cm and more ideally 75–120cm. Plank and beam decks have shallower profiles, but they have a greater number of footings, so if the footings will be concrete, this can be cost prohibitive and platform framing will offer a more economical option.

Plank and beam framing is a common method of framing for roof decks that are supported on plastic pedestals resting directly on flat roofs. Pedestals are also employed on top of concrete slabs. They allow for very low decks that meet, or are just above, surrounding grades.

Sizes and types of structural members for decks can be determined by span tables in codes if the use, size of deck,

Figure 10.4 The furfurylated lumber deck at the end of the pier hovers over the river, providing extraordinary vistas, but then spirals around embracing the rocky tidal zone, an ecotone teeming with life in the pools and grasses. A "get down" to the rocks is accessible for guided educational tours. Image credit: Olin Studio

Figure 10.5 The 800′ long folded furfurylated lumber furnishing climbs and winds its way through an interpretation of the pre-colonial ecosystem of the Hudson River estuary: the upland woodland (pictured here), the coastal grassland, the maritime scrub, the rocky tidal zone, and, in a final celebratory gesture, the deck reaches out over and embraces the Hudson River. Image credit: Olin Studio

and loads are allowable by code, or if not, they should be determined by a structural engineer. The publication *Deck Construction Based on the International Residential Code* contains span tables and detailed code diagrams for decks (ICC 2021) (Table 10.4).

Decking

Decking is the walking surface of the deck and is primarily comprised of deck boards but can be prefabricated wood, steel, or aluminum panels. The spanning ability of the decking dictates the spacing of the joists in a plank and beam deck, or the beams in a platform framed deck. Most wood decking laid flat spans 30–40cm (12–16") with typical board thicknesses of 19–38mm (3/4"-1-1/2").

Wood decking is usually 89mm or 140mm in width because deck boards any wider than this could warp. Deck boards should be laid with a 3.5mm (1/8") space between them,

Table 10.1 Strategies for a Durable Wood Deck

Category	Consideration
Wood durability	Match wood preservation level to decay hazards, setting, and climate. Wood durability classes, wood use categories, and naturally durable lumber, also called decay-resistant lumber, are discussed in Chapter 18 "Wood Materials and Products". Naturally durable lumber should be primarily heartwood with a maximum of 10% of a board face being sapwood (Figure 10.2).
Wood in ground contact	Wood rated for ground contact should be specified unless it is held a minimum of 15cm above finished grade. Poor grading or an overabundance of mulch adjacent to a wood structure can inadvertently eliminate the 15cm clearance over time. Maintaining this 15cm requirement should be written into maintenance plans.
Deck design	Maintain airflow under decks with at least two open sides. Grade under decks for water drainage. Consider covering the ground underneath decks with an impermeable membrane that will facilitate water runoff and suppress the growth of vegetation. Trapping moisture against board ends with a fascia can increase the chance of decay as wood absorbs moisture more quickly through its end grain than through side grain. Another area where end grain is vulnerable is the butt ends of deck boards over joists as they can block drainage below the boards. A solution to this is to double up and slightly space joists where deck boards meet (Figure 10.3).
Shedding water from decks and rails	Wood members that are laid flat such as deck boards and fence rails should be sloped and spaced to shed water. Decks should be sloped at 1–2% away from adjacent structures such as buildings or walls.
Deck board spacing	Space deck boards a minimum of 3.5mm (1/8") apart to allow for water drainage, air circulation, and free movement of dirt and debris through the space. Wider boards should have larger spacing. A measurement of 6.5mm is recommended if there is poor air circulation under the deck or it is close to grade.
Deck board orientation	Deck boards should be placed with the pith side down and the bark side up as they will tend to warp slightly in this direction and shed water, rather than cup and hold water.
Field cut treatment	Treat field cuts, holes, and notches with copper naphthenate containing 1% or 2% elemental copper or an oil borne solution containing at least 0.675% oxine-copper.
Screw depth	Avoid sinking screws into deck boards as this creates a tiny reservoir that can hold water. Screws should be just flush with the deck board surface. Driving screws from below works better on thicker deck boards of 38mm or more.
Deck fasteners	Use high quality, weather resistant metal fasteners that are recommended for the type of wood preservative treatment. Consider use of undermount or proprietary deck fasteners. This eliminates the risk of water puddling at screws and gives a clean look to the deck. Access underneath the deck is necessary for undermount fasteners.
Joist covers	Cover joists with joist tape, caps, or another material that keeps water from ponding directly on joists between deck boards. Note that joist covers can be visible in spaces between deck boards.
End joints between deck boards	Doubling up and spacing joists where deck board ends meet will allow for free drainage of water that could otherwise be trapped between the boards on top of a joist. This will use a few more joists but can prolong the life of the deck boards.
Deck to wall connections	Hold deck boards a minimum of 10mm from walls and adjacent structures. Use joist tape or joist caps on the leger beam at the building wall. Employ elastomeric spacers between the leger beam and the building wall to allow for moisture and debris to move through. Consider building a freestanding deck structure that is not attached to a building.
Footings	Avoid burying wood deck posts in soil even if they are rated for ground contact. Instead, use a steel post holder in a concrete footing. Ensure airflow between posts and footings by holding the post 6mm above the concrete. Use of helical piers can eliminate the need for a post and hold wood members above grade.
Deck maintenance	Decks should be cleaned annually with a hose and mild detergent. Power washing should be avoided as it drives water into the side grain. Decks should be oiled or stained as needed. Oils and stains should be penetrating. Maintenance of specific types of deck boards is discussed later in the chapter.

although some manufacturers of modified wood advise spaces of 5mm (3/16″) for 140mm wide deck boards. Composite lumber manufacturers advise 5mm spaces between deck boards give their relatively high expansion and contraction rates.

Prefabricated wood, steel, and aluminum panels can span greater distances than wood or composite deck boards. Prefabricated wood panels usually incorporate deck boards in a frame. Steel and aluminum grating panels use bars, grating, or expanded sheet metal in a frame. This type of decking is not commonly used for decks that will have chairs or tables; instead it is used for boardwalks and simple bridges over swales or small waterways.

Decking materials are subject to the greatest moisture and UV exposure of any deck members, so choosing deck boards for their durability in a particular climate and/or specific application is key. While pressure-treated wood deck boards can be replaced multiple times over the life of a deck, often the whole structure is removed after the decking wears out in 10–15 years. If properly cleaned and maintained, wood and composite deck boards can last from 15–30 years or more. Deck maintenance activities and their schedule should be incorporated into site maintenance or management plans. Types of wood and composite decking are discussed in Tables 10.6–10.8. Cleaning and maintenance considerations for each type of decking are discussed in Tables 10.9–10.11.

Joists

Joists bear the loads that occur on the deck boards. Joists span between and transfer these loads to the beams. Joists are structural dimension lumber, composite lumber, or a similar shape of steel that orients the wider part of the member vertically.

Joists bear on beams in one of three ways:

1. They ***bear directly on top of the beam*** and are stabilized with steel joist straps or plates. Blocking at least 60% of the height of the joist is placed between the joists over the beam to prevent their overturning (Figures 10.6 and 10.7).
2. They are ***notched and bear directly on the beam*** and are toenailed or toe screwed to the beam. Blocking is employed between the joists over the beam.
3. They are ***flush with the beam and supported by mechanical connections,*** often joist hangers, that are attached to the beam. While this method results in a lower profile deck, it may be the least durable as the joist hangers are solely responsible for supporting the joists. If the hangers rust or otherwise fail, the deck structure will be compromised (Figure 10.8).

Joists are commonly pressure-treated lumber even when the decking is modified wood, a naturally durable wood, or a composite material. Wood joists should be strength graded lumber. They can also be composite lumber or steel.

While appearance is important for deck boards, structural integrity is most important for joists, so lower appearance grades could be specified, reducing pressure on forests and saving some cost. Joists are typically spaced 40cm on center, but sometimes 30cm o.c. if the spans are long, or the live loads are high. Joists can be spaced 60cm o.c. if deck board and joist spans allow.

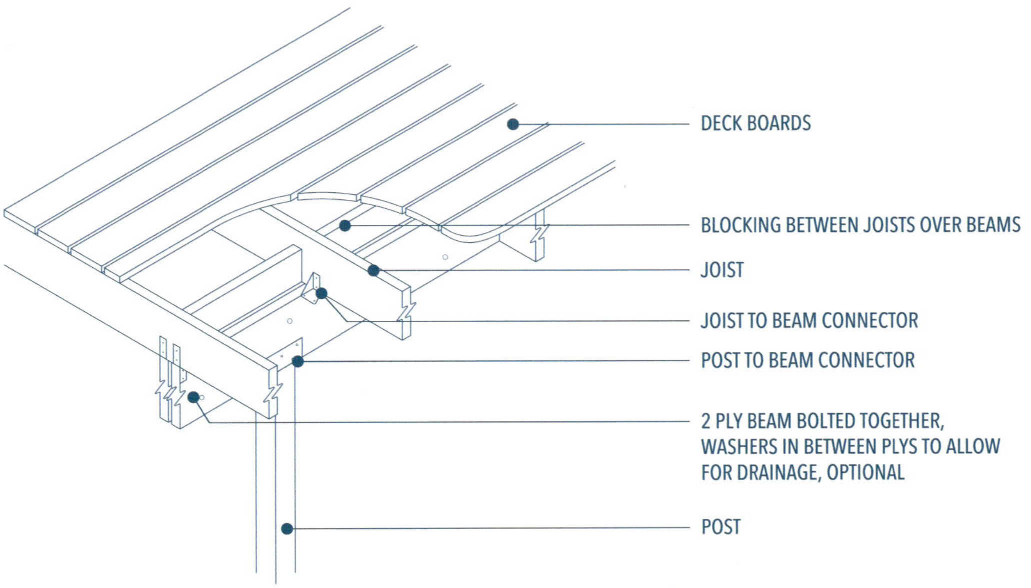

Figure 10.6 This deck assembly employs a two-ply beam of two pieces of dimension lumber bolted together but spaced 4mm apart with washers for drainage. Joists bear directly on the two-ply beam that is bearing on a post. Structural members are stabilized and attached with metal connectors. Image credit: Kathleen Elliott and Meg Calkins

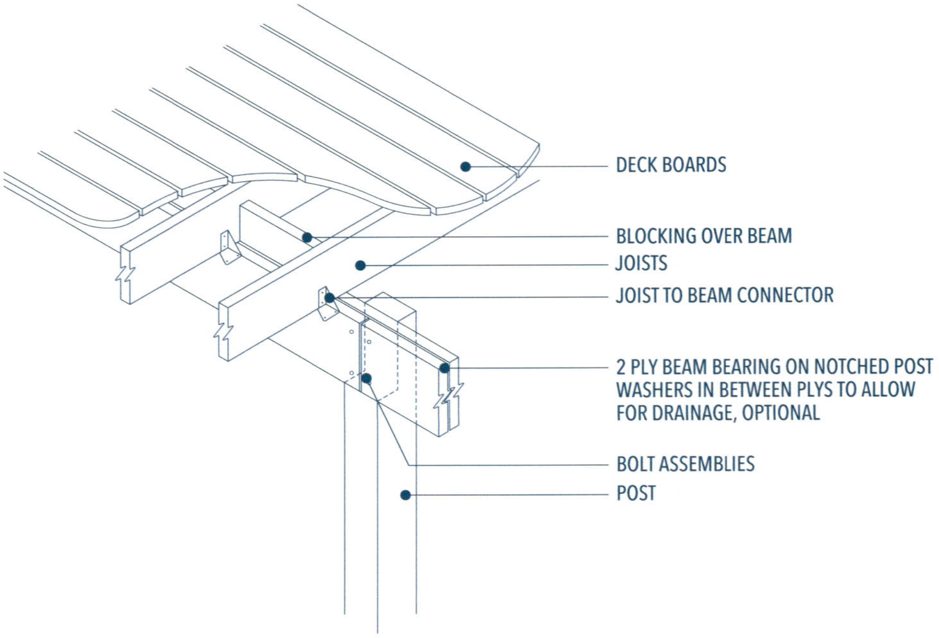

Figure 10.7 This deck assembly employs a two-ply beam of two pieces of dimension lumber bolted together resting on a notched post. Joists are bearing directly on the two-ply beam. Structural members are stabilized and attached with metal connectors. Splices in multi-ply beams should not all be in the same bearing location, and no beam splices should be located between bearing locations. Source: IRC 2021. Image credit: Kathleen Elliott and Meg Calkins

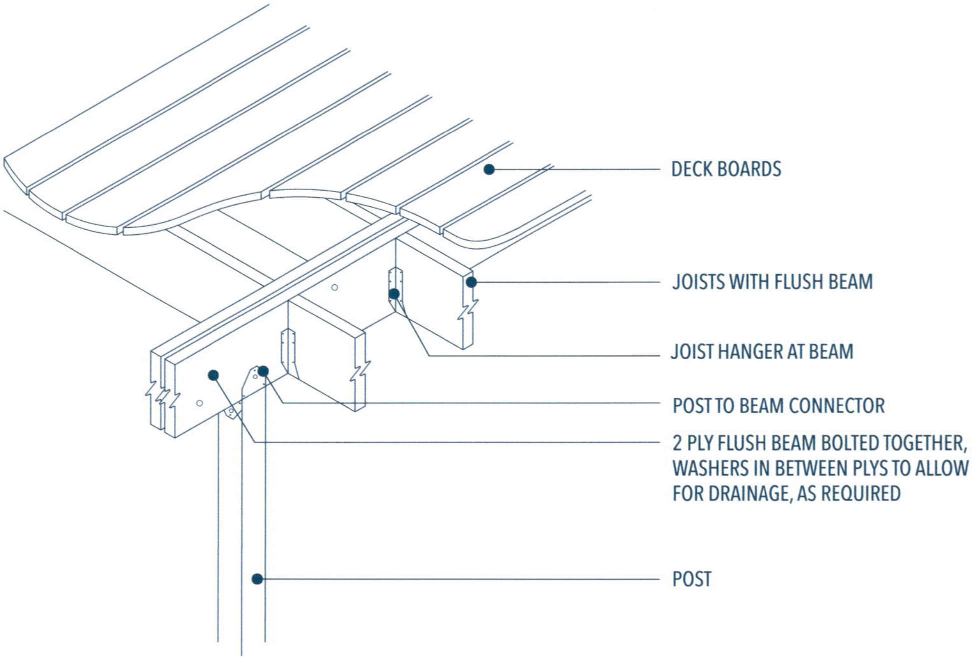

Figure 10.8 The thinner section of this deck assembly places the joists flush with the top of the beam. Joists are bearing on a two-ply beam supported by steel joist hangers. Beams bear directly on posts. Image credit: Kathleen Elliott and Meg Calkins

Beams

Beams support the weight of joists, decking, and live loads and transfer forces to the posts or sometimes directly to the footings or a foundation wall. The spacing of beams is determined by the joist spans, or in the case of plank and beam framing, the decking spans. The cross-sectional dimension of beams is determined by their span and loads and, if wood, should be graded lumber. Beams orient the longitudinal axis of the member in the vertical dimension (Figures 10.6–10.8).

Wood beams comprised of multiple plys of dimensional lumber are common as they are less expensive, and it is easier to obtain two 38x286s instead of one 89x286mm member. Current codes dictate that the two or three plys of a beam must be bolted together with a maximum space of only a few centimeters between them. They can no longer be "sandwiched" around a post (ICC 2021).

Where larger wood beams are required, pressure-treated glue laminated (glulam) wood can be a sustainable alternative. Glulams are built up members of glued boards. Use of glulams can reduce pressure on forests and even sometimes costs.

Beams can also be made of steel or concrete, both of which will span greater distances than dimensional lumber, although glulams can cover large spans. Concrete deck beams are less common but are employed when long spans or exceptionally large loads are expected.

Ledger beams are supported continuously along a masonry wall with bolts. Because the bolts and the wall, usually concrete, support the beam, the "spans" are quite small and so is the thickness of the beam. It is usually 38mm wide (Figure 10.9).

Blocking and Bracing

Blocking and bracing help to stabilize the deck by limiting lateral movement of the structure and structural members. Blocking is placed between joists on top of a beam to keep them upright and to resist lateral forces. Current code requires that the height of blocking must be at least 60% of the height of joists (ICC 2021).

Diagonal bracing is recommended and may be required for decks over 1.5 meters above grade. Bracing will prevent wracking of the deck posts and beams. It is particularly recommended at the corners of decks. Bracing can employ wood members in compression or cables in tension.

Posts and Columns

Posts carry the weight of the deck to the footings, and their spacing is directly related to the allowable span of the deck's beams. Posts are not required if the beams are directly supported by foundation walls, helical piers, pin foundations, or pedestal foundations. In rare cases, the beam will rest directly on grade or a slab. Wood posts are most common for decks, but steel posts or concrete posts are used to support steel or concrete beams, and larger glulams.

The height of the deck above finished grade and the deck loads determine the size of the posts. Square 89x89mm or 140x140mm posts are common sizes for wood posts. Taller posts or heavier loads will require 135x135mm posts, but if smaller posts can be used, they should be as this will conserve wood materials.

Footings

Footings transfer deck loads to the subgrade, and they anchor the structure. Footings must extend below the frost line in cold and temperate climates to prevent movement from freeze/thaw action of the soil. The weight of the deck and the bearing capacity of the soil determine the footing sizes. Poured in place concrete footings are most used to support decks; however, helical piers and pin foundations are alternative systems (see Table 10.2 for a comparison). Other footing systems are described in Table 10.3.

Design of footings supporting wood posts are an important consideration for the durability of a wood deck structure. Some footings bury the wood post in soil, which could compromise the structure over time. CCA treated lumber,

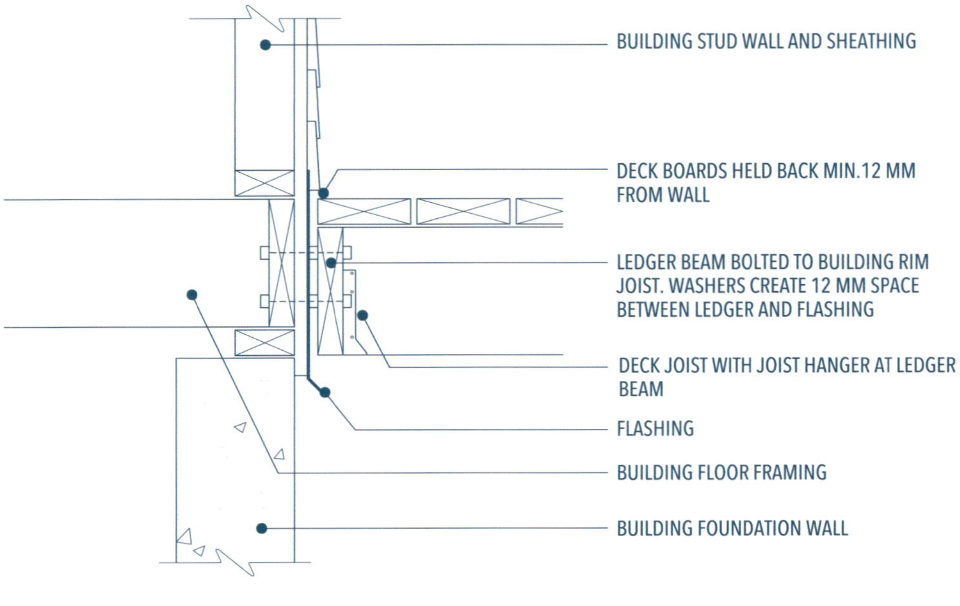

Figure 10.9 The deck's ledger beam is bolted to the building rim joist. Washers create a space between the ledger beam and the building wall for air flow and cleaning of debris. Flashing is continuous from under the siding to below the deck frame. Ledger beams can also be bolted to concrete walls with anchor bolts. Image credit: Kathleen Elliott and Meg Calkins

voluntarily phased out in 2003, was very durable in ground contact, but newer copper treatments are not quite as durable. The best approach for wood posts is to hold them 15cm above grade, treat end cuts with a brush-on preservative, and specify lumber rated for ground contact for posts. Tops of concrete footings should be sloped to shed water, and in humid climates, steel post bases should allow for a space below the post and base for air flow and water drainage.

Concrete Footings

Concrete footings can be a cylindrical form with a wider cylindrical base, a square prism form with a wider square prism base, or just a wide cylinder or a wide square prism.

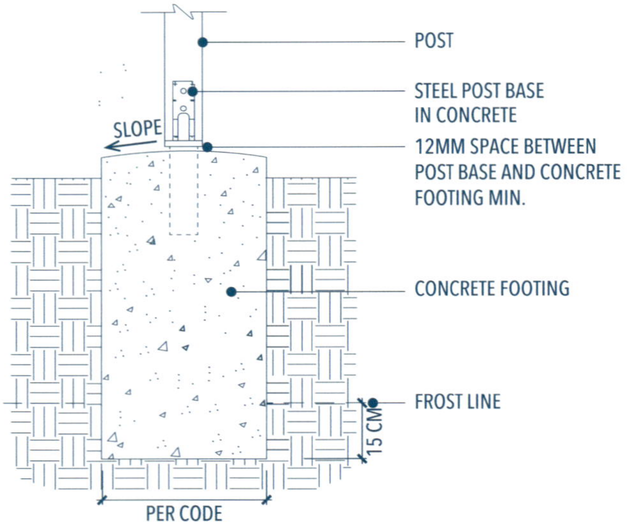

Figure 10.10 The post is attached to a concrete footing with a steel post base. The steel post base base is embedded in uncured concrete. The top of the concrete footing slopes to drain away from the post. This is the most durable wood post and footing detail as the post is held away from soil moisture. Image credit: Kathleen Elliott and Meg Calkins

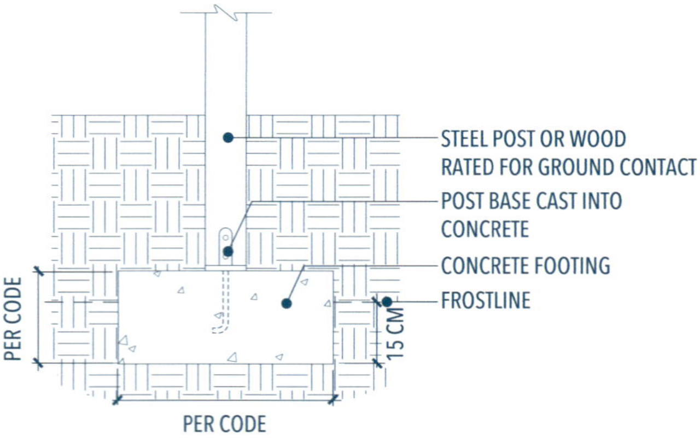

Figure 10.12 The post is attached to the concrete footing with a post base below grade. This approach is not advisable for wood posts as they will be in contact with moist soil. Image credit: Kathleen Elliott and Meg Calkins

Precast concrete pier blocks are no longer approved by the IRC to support beams. If they are used, they must support each joist.

There are three ways that deck posts meet and attach to concrete footings:

Post attached with mechanical connector to top of footing. A steel post base is embedded in the concrete footing and receives the post once the concrete has cured. There are a wide variety of steel post base products on the market. Most have deformations to resist pulling out of the concrete from deck uplift. The most durable post base

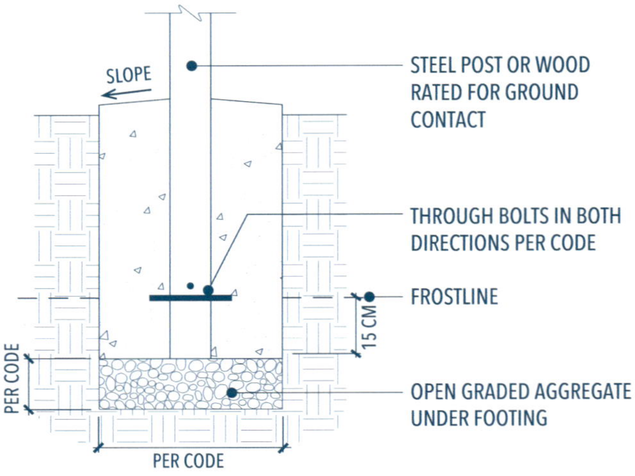

Figure 10.11 The post is embedded through the depth of the concrete footing. The top of the concrete footing slopes to drain away from the post. Image credit: Kathleen Elliott and Meg Calkins

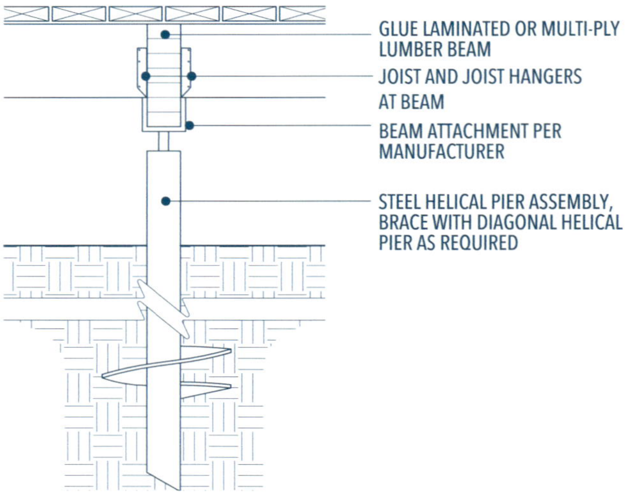

Figure 10.13 Helical pier footing detail. A helical pier footing can reduce site disturbance as there is no excavation of soil required. Image credit: Kathleen Elliott and Meg Calkins

connection will allow for water drainage where the post meets the concrete to not hold moisture against the end grain of the wood. Steel posts are welded to plates that are bolted to a concrete footing with anchor bolts placed in uncured concrete (Figure 10.10).

Post embedded through the depth of the concrete footing. Wood or steel posts can extend through the entire depth of the concrete footing with a compacted gravel base below the footing for drainage. Concrete is poured around the posts. Current code requires bolts through the posts in two directions to resist uplift. As wood posts embedded in concrete will have contact with moisture, they must be rated for ground contact (Figure 10.11).

Post attached to footing below grade. Wood or steel posts can be attached to cylindrical or square prism concrete footings below grade. This method directly buries part of the post, so it is not recommended for wood posts. If wood is used, it must be rated for ground contact, and steel posts must be galvanized, powder coated, stainless, or weathering steel (Figure 10.12).

Helical Piers

Helical piers, also called screw piers, screw piles, or helical piles, are an in-ground anchoring system for decks, boardwalks, buildings, roads, and other infrastructure. The piers consist of steel tubes with a helical flight near the end that anchors into the soil. Helical piers are installed using hydraulic torque equipment to drive the piers into the ground (Figures 10.13–10.15).

Site disturbance can be minimized with use of helical piers as there is no excavation required as there is for concrete footings. Helical piers allow for anchoring into greater depths in areas of poor soils, and they can support substantial loads. Table 10.2 compares concrete footings, helical piers, and pin foundations. The embodied carbon of deck structures with concrete footings and helical pier footings is compared later in this chapter.

Figure 10.14 Helical pier footings for a deck can be used for anchoring structures to greater depths in areas of poor soils or when greater loads are expected. Image credit: Dancing Man

Pin Foundations

Pin foundation systems combine steel bearing pins and a high strength precast concrete head that angles the pins into the soil, spreading out loads. The precast concrete head has an embedded anchor bolt in the post bracket seat that allows for a secure connection to the deck or boardwalk structure. The bearing pins are driven into intact penetrable soils by handheld equipment at angles requiring no excavation. The pins provide bearing, resistance to uplift, and lateral capacity (Figures 10.16 and 10.17; Table 10.2)

Rails

Decks are required to employ railings when they are more than 75.2cm above grade. However, use of deck railings on lower grade changes can create a safer deck and enclose the space. Deck posts can extend vertically to act as structural support for the deck rails.

Rails for decks can be wood, composite lumber, glass, or metal. Design, detailing, materiality, and embodied carbon considerations are discussed in Chapter 11 "Rails, Fences, and Screens".

Connections and Fasteners for Decks

The connections between structural members are critically important to the durability of decks as they can limit strength and are often where structural failures occur. Fasteners and hardware can corrode due to moisture and galvanic reactions with wood preservation treatments. Therefore, it is very important to use appropriate metal fasteners for the wood treatment and detail connections so that structural members bear directly on beams and posts to the extent possible. This transfers loads primarily through the structural members instead of through the fasteners.

Use of connections that allow for easy replacement of individual wood members can increase the life of a structure. Bolted connections offer the possibility of replacing just a few members instead of the entire structure. For instance, if a single post is compromised and bolted connections for the post are used at the footing and beam, it can be replaced without removing the entire structure. If deck boards are screwed to joists, a single rotted board can be unscrewed and replaced.

Figure 10.15 PWP Landscape Architecture used helical piers for a boardwalk through a natural area at the Glenstone Museum in Potomac, Maryland. Image credit: PWP Landscape Architecture

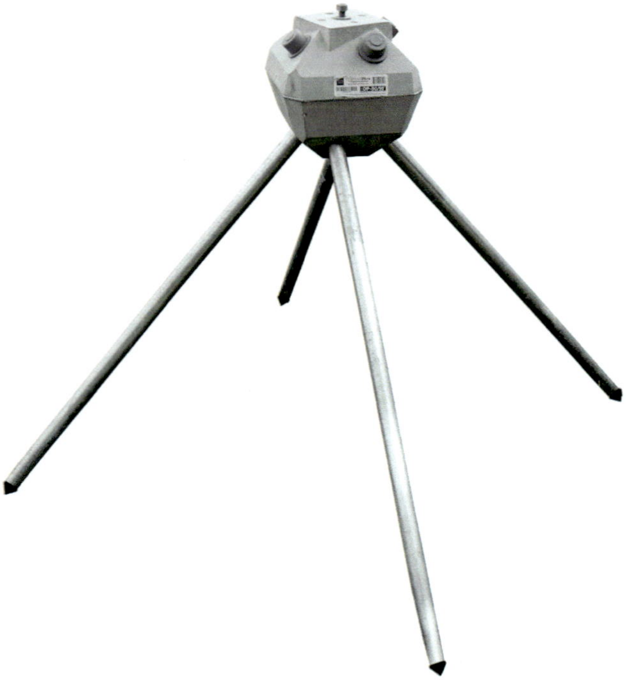

Figure 10.16 A pin foundation with four steel bearing arms and a precast concrete head to receive a deck or boardwalk post. Image credit: Courtesy of Pin Foundations, Inc. (www.diamondpiers.com)

Figure 10.17 Installation of pin foundations can be done with handheld equipment and is a good footing type for rocky soils. Image credit: Courtesy of Pin Foundations, Inc. (www.diamondpiers.com)

Table 10.2 Advantages and Disadvantages of Selected Deck Footing Types

	Concrete Footings	Helical Piers	Pin Foundations
Site disturbance	Greater site disturbance from excavation and excess soil stockpiling.	Minimal site disturbance as there is virtually no soil excavation, although the installation equipment could compact soil.	Minimal site disturbance as there is virtually no soil excavation. Installation equipment is worn by workers.
Load capacity	Load capacities are limited by the soil bearing capacity a few feet below grade.	Very good load capacities even for smaller piers. Can be installed to greater depths.	Loads are distributed along the four bearing pins and limited by bearing capacity of subgrade.
Cost	Lower material cost, higher installation costs.	Higher material cost, higher installation cost.	Higher material cost, potentially lower installation cost.
Depth constraints	Deep footings are limited except for concrete pilings.	Easier to achieve deeper footings than with concrete.	Max. pin lengths are around 3.2m.
Soils	Footing bases are widened for poor bearing capacity of soils.	Can go deeper to avoid poor soils, but not viable with bedrock near surface.	Soils must have min. 72kPa bearing capacity; not as viable in stony soils.
Installation process	Slower with excavation then concrete placement and curing.	Quick but requires specialized equipment. The height of the piers can be easily adjusted as can the head length at the top of the pier.	Quickly installed with handheld equipment.
Recognized by the ICC's International Residential Code	Yes	Yes	Yes

Sources: Diamond Pier Foundations 2023; Goliath Tech 2023; ICC 2021

Mechanical Fasteners and Connectors

Mechanical fasteners and connector products must be compatible with the wood preservation treatment of the lumber members they connect. The copper in pressure-treated wood can cause a galvanic reaction with some metal screws resulting in corrosion of the lesser metal. This can cause black stains around fasteners on wood or the fastener itself can become corroded. The acids in some modified woods can cause corrosion of fasteners as well. High-grade stainless steel fasteners should be used in marine environments (Table 10.5). Current building codes require use of stainless steel, hot dip galvanized, bronze, or copper fasteners for most wood decks (ICC 2021).

Bolted Connections

Bolted connections can be very durable. Use of washers is critically important to protect wood from damage due to over tightening of bolts. Use of bolts allows for ease of deconstruction and reuse of viable wood lumber. Bolts should be long enough to allow for use of a washer under both the head and the nut with at least 6mm protruding beyond the nut.

Metal Connector and Fastener Products

There is a wide array of metal connector and fastener products for a myriad of wood connections scenarios. Manufacturers of connectors will recommend appropriate bolts, screws, or nails for use with their product.

Decking Clip Systems

Manufacturers of decking offer proprietary hidden clip systems to attach deck boards to joists or beams. Many of these clip systems hold deck boards up off the joists, creating a separation for air flow and drainage.

Design, Construction, and Maintenance Considerations for Decks

While most deck structural members will be pressure-treated lumber, steel, or composite lumber, there is a wide variety of materials used for deck surfaces. Criteria for selection of decking materials include appearance, durability, service life, cost, ease of construction, maintenance requirements and embodied carbon.

Table 10.3 Other Deck Footing Types

Footing Type	Considerations
Plastic pedestals	Plastic deck pedestals are designed for use on flat, stable surfaces such as roof decks or concrete slabs, although some products claim to be useable on well compacted soils (Figure 10.18). These products are adjustable in height accommodating surface inconsistencies or slight roof slopes. They can be easily adjusted even after the deck is constructed. Products have a variety of attachments to connect to beams, joists, or posts. They are most often used in plank and beam framed decks. Many have high recycled content (Bison Innovative Products 2024; TileTech 2024). PVC pedestals should be avoided.
Footing pads	Footing pads are fiber reinforced composite products that are placed at the bottom of a post hole as a base for a deck post. The pad spreads the loads out over the soil and can be used with wood, composite, precast, or PIP concrete posts. Some products are accepted by the ICC. They do not provide uplift resistance, so if required, posts will need to have that built in. Not ideal for wood posts in damp soils as the wood will be in constant ground contact (FootingPad 2024).
Stackable concrete footings	These footings stack precast concrete tubes or pads in excavated holes to form a concrete footing. Product manufacturers market these foundations as no mess, quick installation, and strong footings. Precast pieces are very heavy and EC of precast is relatively high as compared with PIP concrete (EZ-Tube 2024).
Steel anchored footings	Steel anchored footings are a smaller version of helical piers that are driven into the ground with an electric impact wrench or a hand auger tool. They are for freestanding smaller decks and consist of a steel anchor with a steel load plate on top and the option for a post base attachment.
Precast concrete pier blocks	Precast concrete pier blocks rest on grade and are produced with deformations or attachments to receive deck framing lumber. They are allowed by the IRC if the joists bear directly on the pier blocks, the area of the deck is less than 18.6m^2 and the surface of the deck is not more than 508mm above grade. They should not be used with expansive soils or in areas prone to frost heave (ICC 2021).

Sources: Bison Innovative Products 2024; DecksGo 2024; EZ-Tube 2024; FootingPad 2024; TileTech 2024

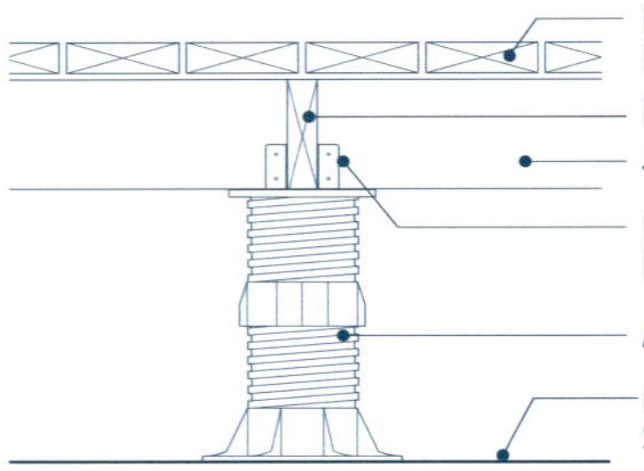

Figure 10.18 Pedestal footing detail footings can reduce the need for joists, and are adjustable for leveling of surfaces. Image credit: Kathleen Elliott and Meg Calkins

The service life of deck boards varies from 10–30 years or more, and service life is related to cost and embodied carbon. For instance, ACQ-treated decking will last from 10–15 years depending on climate and regular maintenance and preservation activities. It can be the least expensive decking option with the lowest embodied carbon. Composite decking types are the most expensive option and the highest in embodied carbon, but they can be expected to last over 30 years.

Tables 10.6–10.8 provide design and construction considerations for modified wood decking types, black locust decking, copper pressure-treated decking, carbon-based pressure-treated decking, fused bamboo decking, rice hull composite, and wood composite decking. These treatments and wood and composite lumbers are discussed in Chapter 18 "Wood Materials and Products".

Maintaining Deck Structures

Even when wood is treated for decay and insect resistance, it is still vulnerable to UV degradation, mold, and mildew. Regular maintenance and finishing will extend the life of wood structures. Decks should be cleaned annually and many types of decking will require applied finishes periodically protect their surfaces (Tables 10.9–10.11).

There are two categories of wood finishes: film forming finishes and penetrating finishes. ***Film forming finishes***, including paints, solid color stains, and varnishes, are not recommended for deck surfaces as they don't always tolerate the moisture cycling of the wood. Once the film is cracked or chipped, water can get under it and cause the film to peel or blister. Film forming finishes on deck surfaces are also vulnerable to foot traffic.

Penetrating finishes penetrate the surface of the wood. These include oil or water-based semitransparent stains, oil or water-based clear finishes, and oil or water-based water repellents and preservatives. Penetrating wood finishes with a UV inhibitor and mildewcide are recommended to maintain deck surfaces. Some products claim to only require

Table 10.4 Selected Items From the International Residential Code (IRC) for Decks

Item	Code Intent
Wood grade	Wood materials are No. 2 grade or better.
Wood preservatives	Wood is preserved in accordance with Section R317.
Naturally durable lumber	Approved naturally durable lumber and termite protected in accordance with Section R318. Lumber should be primarily heartwood with a maximum of 10% of a broad board face being sapwood. This does not apply to narrow edges.
Field treatment	All cuts, notches, and holes of preservative treated lumber are field treated per Section R317.1.1.
Wet use factor	Where applicable all deck designs must use a wet use factor. The wet use factor, also called wet service factor, is applied when wood moisture content will be above 19% for prolonged periods and its strength could be affected.
Ground contact lumber requirements	IRC does not require ground contact rated treatment for structural wood in decks unless it is in actual ground contact. Industry standards do recommend ground contact treatment for structural lumber when it is used in a wet climate or the structure is difficult to access and maintain (AWPA 2024; Lebow et al. 2019).
Plastic composite	Plastic composite deck boards, stair treads, guards, and handrail must comply with Section R507.2.2. Must adhere to *ASTM D7032 Standard Specification for Establishing Performance Ratings for Wood-Plastic Composite and Plastic Lumber Deck Boards, Stair Treads, Guards, and Handrails* (ASTM 2019).
Fasteners and connectors	Metal fasteners and connectors must comply with Section R317.3 and Table R507.2.3. Fasteners for preservative treated wood must be hot-dipped, zinc coated galvanized steel, stainless steel, silicon bronze, or copper. ASTM A653 type G185 zinc coatings or equal must be used (ASTM 2023b).
Deck joist bearing	The ends of joists must have 38mm or greater bearing on the wood or metal joists and not less than 76mm of bearing on concrete or masonry.
Deck joist lateral restraint	Lateral restraint must be provided by joist hangers or blocking. Both must be at least 60% of the height of the joist.
Deck beams	Beam plies must be fastened together with two rows of 10d nails at a minimum of 406mm o.c. along each edge. Beams can cantilever at each end up to ¼ of the beam span. Multi-ply beams cannot have the plies separated around a post. Spacers (12.7mm max.) can be used for better drying and thickening the beam to a standard 89x89mm post size.
Deck beam bearing	The ends of beams must have 38mm or greater bearing on wood or metal posts and not less than 76mm of bearing on concrete or masonry.
Notches	Notches in solid lumber joists, rafters, and beams should not exceed 1/6 of the depth of the member and not longer than ⅓ of the depth of the member. They should not be in the middle third of the span. Notches at the end of the member should not be greater than ¼ the depth of the member.
Footings	Decks must be supported on concrete footings or other approved structural systems and be designed to accommodate all loads. Helical piers, pin foundations and some other systems are approved by the ICC.

Source: ICC 2021

Table 10.5 Recommended Fastener Type by Wood Preservation Treatment

Wood Preservation Treatment	Recommended Faster Type
Thermally modified lumber	High-quality stainless steel screws as lumber is slightly acidic. Some manufacturers offer propriety clip systems.
Acetylated lumber	High-quality stainless steel screws as lumber is acidic.
Furfurylated lumber	High-quality stainless steel screws as lumber is slightly acidic. Some manufacturers offer propriety clip systems.
ACQ-treated lumber	Corrosion-resistant fasteners and connectors. Use hot-dipped galvanized fasteners (meeting ASTM A153) and connectors (ASTM A653) or stainless steel fasteners. For coastal installations, use code approved stainless steel.
Copper azole treated lumber	Corrosion-resistant fasteners and connectors. Use hot-dipped galvanized fasteners (meeting ASTM A153) and connectors (ASTM A653) or stainless steel fasteners. For coastal installations, use code approved stainless steel.
EL2 and PTI treated lumber	Corrosion-resistant fasteners and connectors. Use hot-dipped galvanized fasteners (meeting ASTM A153) and connectors (ASTM A653) or stainless steel fasteners. For coastal installations, use code approved stainless steel.

Sources: Accoya 2024; Kebony 2024; Thermory 2024; ASTM 2023a, 2023b; Lebow 2019

Table 10.6 Design and Construction Considerations for Modified Wood Decking

	Thermally Modified Wood[1,2]	Acetylated Wood[3,4]	Furfurylated Wood[5,6]
General construction considerations	Gloves and a dust mask should be worn during construction.	Gloves and a dust mask should be worn during construction.	Gloves and a dust mask should be worn during construction.
Deck board thicknesses available	20–26mm thick decking	19–25mm thick decking	22–38mm thick decking
Joist spacing per deck board thickness (actual joist spacing may vary by design load)	40cm o.c.	25mm thick, 60cm o.c. 19mm thick, 40cm o.c.	22mm thick "clear", 50cm o.c. 28mm thick "character", 60cm o.c. 34mm thick "character", 90cm o.c. 38mm thick "clear", 110cm o.c.
Decking joint end matching	yes	no	No Double up joists under decking butt joints to allow for water drainage. Use 3–5mm gap between butt joints
Deck board orientation	Install spruce pith side down.	Pith side down	Pith side down
Gaps between deck boards	4mm gap for 89mm boards 6mm gap for greater than 89mm.	$\geq$ 3.2mm gap If poor air circulation under deck or close to grade, $\geq$ 6mm is rec.	7.5mm for clips 6mm for SS screwed boards 15mm to building wall
Ground contact	Durability class 1 Min. 50mm above grade	Durability class 1	Durability class 1 Min 20mm above hard surface Min. 50mm above soil
Deck design	Design deck for airflow underneath. Joist tape recommended. Use spacers to elevate deck boards 4–8mm above joists when using screws. Deck should slope 1–2% to drain deck boards.	Copper from pressure-treated deck structural members can leach and stain deck boards. Isolation techniques such as joist tape, strips, or spacers should be used. Deck should slope 1–2% to drain deck boards.	Design deck for airflow underneath. Use moisture barriers between decking and joists (spacers or joist tape). Deck should slope 1–2% to drain deck boards.
Used in structural applications	no	no	no
Other lumber sizes available	Thicknesses up to 75mm.	Thicknesses up to 75mm.	Thicknesses up to 38mm.
Cutting, planning, and sanding	Similar to standard structural wood.	Similar to standard structural wood. Tungsten carbide blade with 60 or more teeth. Ripping to create thinner boards is not recommended.	Very resistant to warping, cupping, and splintering. Very hard but easier to work with than tropical hardwoods. Carbide tools should be used. "Clear" can be ripped and crosscut. "Character" should not be ripped.
Predrilling recommendations	Predrill when using SS screws.	The need for predrilling, countersinking, and keeping sufficient distance from the edge of wood is similar to other softwoods.	Predrill when using SS screws.
Fasteners	High-quality SS screws as wood is slightly acidic. Use self-tapping SS composite deck crews or predrill. Clips can be used. Some products offer propriety clip systems.	A2 type SS screws as wood are acidic and corrosive to steel and galvanized steel. Use A4 type 316 SS in coastal environments. Proprietary hidden fastener system or toenailed screws can be used.	Proprietary clip system is recommended but high-quality SS screws can be used.
End sealing	Oil or wax seal end cuts.	Does not require end grain sealer.	Finish ends for color. Seal "character" with anti-parasite/fungal agent.
Factory applied finishes	Factory oiled finishes available.	Factory applied integral color is available. Color extends through the entire section.	Factory oiled finishes available.

[1] Thermory 2022 [2] Arbor Wood Co. 2024 [3] Accoya 2022 [4] CAMO & Accoya 2022 [5] Kebony 2016 [6] Kebony 2024

Table 10.7 Design and Construction Considerations for Black Locust and Pressure Treated Wood Decking

	Black Locust Lumber[2]	Copper-Based Pressure-Treated Lumber (ACQ and Copper Azole)[3]	EL2[4] and PTI[5] Treated Lumber (Carbon-Based Pressure Treatments)
General construction considerations	Boards should be acclimated at the jobsite for 5 days prior to installation. Wear a dust mask when working with wood. Deck should slope 1–2% to drain deck boards.	Wear gloves and a dust mask when working with treated wood. Deck should slope 1–2% to drain deck boards.	Wear gloves and a dust mask when working with treated wood. Deck should slope 1–2% to drain deck boards.
Deck board thicknesses available	20–38mm thicknesses.	20–38mm thicknesses.	20–38mm thicknesses.
Joist spacing per deck board thickness (actual joist spacing may vary by design load)	38mm thick, 80cm o.c. 25mm thick, 60cm o.c. 20mm thick, 40cm o.c.	38mm thick, 60cm o.c. 25mm thick, 40cm o.c. 25mm thick, 30cm o.c.[1]	38mm thick, 60cm o.c. 25mm thick, 40cm o.c. 25mm thick, 30cm o.c.[1]
Decking joint end matching	Butt joint over joists. Butt damp boards tightly together during installation as they will shrink slightly in width and length as they dry. Consider doubling up joists with spacer at butt joints to allow for water drainage.	Butt joint over joists. Butt damp boards tightly together during installation as they will shrink slightly in width and length as they dry. Consider doubling up joists with spacer at butt joints to allow for water drainage.	Butt joint over joists. Butt damp boards tightly together during installation as they will shrink slightly in width and length as they dry. Consider doubling up joists with spacer at butt joints to allow for water drainage.
Deck board orientation	Pith side down. Avoid very long lengths and screw decking to every joist.	Pith side down.	Pith side down.
Gaps between deck boards	6mm spacing	3mm spacing	3mm spacing
Ground contact	Very durable in ground contact.	Min 15cm above finished grade unless rated for ground contact.	Min 15cm above finished grade unless rated for ground contact.
Used in structural applications	no	yes	yes
Deck design	Allow 60cm of unobstructed airflow under decks. Joist tape is recommended.	Allow 60cm of unobstructed airflow under decks. Joist tape is recommended.	Allow 60cm of unobstructed airflow under decks. Joist tape is recommended.
Cutting, planning, and sanding	Boards can be worked with standard tools. Carbide blades are recommended. Lightly sand imperfections prior to installation.	Standard tools and techniques.	Standard tools and techniques.
Predrilling recommendations	Yes	No	No
Fasteners	Boards requiring straightening should be clamped into position until fastened. High-quality SS screws should be used. For coastal installations, use code approved stainless steel.	Use corrosion-resistant fasteners and connectors. Use hot-dipped galvanized fasteners (meeting ASTM A153) and connectors (ASTM A653 Class G185 sheet) or stainless steel fasteners. For coastal installations, use code approved stainless steel.	Use corrosion-resistant fasteners and connectors. Use hot-dipped galvanized fasteners (meeting ASTM A153) and connectors (ASTM A653 Class G185 sheet) or stainless steel fasteners. For coastal installations, use code approved stainless steel.
End sealing	Seal with protective oil or wax immediately after cuts.	Coat all cut ends, holes, or other intrusions into the wood with a suitable wood preservative product containing a minimum of 0.675% copper as oxine-copper (copper-8 or copper-8-quinolinolate) or 1% copper as copper naphthenate.	Coat all cut ends, holes, or other intrusions into the wood with a suitable wood preservative product containing a minimum of 0.675% copper as oxine-copper (copper-8 or copper-8-quinolinolate) or 1% copper as copper naphthenate.
Factory applied finishes	No It is recommended to oil the boards on all six sides prior to installation.	Yes. Some decking is stained to cover the greenish tint from the copper treatment.	No

[1] ICC 2021 [2] Robi decking 2024 [3] Lebow et al. 2019 [4] Viance EcoLife 2024 [5] Lonza 2017

Table 10.8 Design and Construction Considerations for Composite, Fused Bamboo and Rice Hull Composite Lumber Decking

	Wood Plastic Composite Lumber[1,2]	Fused Bamboo Lumber[3,4]	Rice Hull Composite Lumber[5,6]
General construction considerations	Gloves and a dust mask should be worn during construction.	Gloves and a dust mask should be worn during construction.	Gloves and a dust mask should be worn during construction.
Deck board thicknesses available	20–51mm thicknesses, solid and hollow.	20–40mm thicknesses.	25–38mm thicknesses, hollow.
Joist spacing by deck board thicknesses (actual joist spacing may vary by design load)	22–25mm thick, 40cm o.c. 51mm, thick 60cm o.c. in residential and light commercial apps. 20–25mm thick, 30cm o.c. 51mm thick, 40cm o.c. in heavier commercial loads.	20mm thick, 40cm o.c. 40mm thick x140mm width, up to 91cm o.c.	25mm thick, 30–40cm o.c. 30cm o.c. is recommended by one manufacturer for hot climates and commercial projects.
Decking joint end matching	No. Butt joint ends of deck boards over joists. Gaps of butt joints are determined by climate and product.	Deck boards are grooved for structural end interlock with other boards. They do not need to meet over joists.	No. Butt joint ends of decking over double joists. Gaps of butt joints are determined by climate. Some products use a breaker board system at decking ends.
Gaps between deck boards	Boards will expand and contract with temperature changes. Gaps should be 5mm in areas where temperatures reach less than 4.5°C and 3–5mm in areas above 4.5°C.	Determined by deck board clip fasteners or 5mm spacing recommended.	Boards will expand and contract with temperature changes. Manufacturers dictate spacing based on temperature at installation and overall climate.
Ground contact	Use min. 38mm sleepers to hold composite decking above grade.	Hold 10cm above permeable surfaces. Ensure good drainage and airflow under deck on impermeable surfaces. Vapor barriers may be necessary if very close to finished grade.	15cm min. ventilation under deck boards.
Used in structural applications	No. Decking, siding and fencing only.	Decking, siding, and some lumber.	Decking, siding, and hollow shapes for fencing and trellises.
Deck design	Low-emissivity (Low-E) glass can potentially harm decking as it can cause heat to build up on the deck surface. May cause decking to melt, sag, warp, discolor, or accelerate weathering.	Design deck for airflow underneath. Deck should slope 1–2% to drain deck boards.	Low-emissivity (Low-E) glass can potentially harm decking as it can cause heat to build up on the deck surface. May cause decking to melt, sag, warp, discolor, or accelerate weathering.
Cutting, planning, and sanding	Use carbide tipped saws. Notched areas should be fully supported. Cannot be routed.	Worked similar to tropical hardwoods. Cut with carbide tipped saw blades. Can be sanded (with the grain only), drilled, and planed.	Use carbide tipped saws. Can be ripped near a vertical rib.
Predrilling recommendations	Yes	Predrill holes if using screws, countersinking is recommended.	Predrill holes if using screws, countersinking is recommended.
Fasteners	Proprietary hidden fastening and clip systems. Use screws made for composite decking. Stainless steel screws should be used in marine environments.	Proprietary hidden fastening and clip systems or stainless steel screws. Deck plugs available if SS screws are countersunk. Use SS 316 screws in marine environments.	Proprietary hidden fastening and clip systems. Use screws made for composite decking. Stainless steel screws should be used in marine environments.
Factory applied finishes	Some products are "capped" with a surface material that could be dulled with certain cleaners or abrasives.	Pre-primed penetrating oil but wears off a few months after installation. Coat 3–4 months after installation with an oil finish to maintain color.	Some products are "capped" with a surface material that could be dulled with certain cleaners or abrasives. Some manufacturers offer pre-finished products and can be re-stained.

[1] Trex 2024a [2] Azek 2023 [3] Dasso 2020 [4] Moso® Bamboo 2021 [5] TruNorth Decking 2023 [6] Resysta 2023

re-application every six years, but most finishes will need to be reapplied every 1–2 years. Semitransparent stains last longer than clear finishes because the pigment protects both the wood and the stain from UV damage.

It is recommended that a finish be applied to structures built with pressure-treated lumber within two to six months after the structure's installation. Kiln-dried pressure-treated wood structures can be finished immediately.

Table 10.9 Maintenance Considerations Modified Wood Decking

	Thermally Modified Wood	Acetylated Wood[2]	Furfurylated Wood[5]
Weathering	Wood left unfinished will weather to silver/gray.	Uncolored wood will bleach then weather to light silver/gray Light color of uncoated wood may cause early molds and blemishes to stand out more than on other wood types, prior to it weathering to gray. "Colored" products will weather more slowly in 3–4 years.	Wood left unfinished will weather to silver/gray.
Expansion and contraction over time	Small surface checks will appear and disappear with expansion and contraction of wood.	Small surface checks will appear and disappear with expansion and contraction of wood. Surface fibers can appear but are easily removed.	Very small surface checks will appear and disappear with expansion and contraction of wood.
Cleaning	Clean annually to remove surface molds and stains once per year. Avoid harsh cleansers and heavy power washing. Clean minor stains with dish soap. Lightly sand major stains.	Clean annually to remove surface molds and stains once per year. Avoid harsh chemical cleansers and heavy power washing. Avoid any power washing on coated surfaces. Lightly sand major stains. Avoid high pH cleaners such as bleach.	Clean annually to remove surface molds and stains once per year. Avoid harsh chemical cleansers and heavy power washing. Avoid any power washing on coated surfaces. Lightly sand major stains. For heavily contaminated surfaces, alkaline detergents can be used.
Maintenance finishes	To maintain color and prevent small surface checks, use oil-based finish with UV protectant. Some products recommend specific finishing products.	Acetylated wood is more porous than hardwoods and some softwoods. Oils should contain a mildewcide. Non-film forming stains and oils can be used. Water-based stain may not penetrate as quickly as the modified wood is hydrophobic. Darker or more highly pigmented coatings typically have longer maintenance intervals.	Use UV inhibiting hardwood oil. Can use stain or paint.
Finishing intervals	Will vary with climate, use, and traffic. 1–2 years will maintain color.	Will vary with climate, use and traffic. 1–2 years will maintain color.	Will vary with climate, use, and traffic. 1–2 years will maintain color.
To restore original color	If grayed, restore original color by sanding then oiling.	If grayed, restore original color by sanding then oiling.	If grayed, restore original color by sanding then oiling.
Warranty	20 years	25 years	15 years for "character" 25 years for "clear"
Reuse	Yes	Yes	Yes
Recycling	Yes	Yes	Yes
Incineration for energy production	Yes	Yes, in commercial incinerators.	Yes, in commercial incinerators.

[1] ArborWoodCo Decking Installation Guide

[2] Thermory Maintenance and Care

[3] Accoya decking care

[4] Accoya Warranty

[5] m.Kebony Use and Maintenance

[6] Kebony Warranty

Table 10.10 Maintenance Considerations for Black Locust and Pressure Treated Wood Decking

	Black Locust Lumber[1,2]	Copper-Based Pressure-Treated Lumber (ACQ and Copper Azole)	EL2[3] and PTI[5] Treated Lumber (Carbon-Based Pressure Treatments)
Weathering	Wood left unfinished will weather to silver/gray.	Wood left unfinished will weather to silver/gray.	Wood left unfinished will weather to silver/gray.
Expansion and contraction over time	Surface checks and cracks will appear and disappear with expansion and contraction of wood.	Small surface checks will appear and disappear with expansion and contraction of wood.	Very small surface checks will appear and disappear with expansion and contraction of wood.
Cleaning	Clean to remove surface molds and stains once per year. Avoid harsh cleansers and heavy power washing. Clean minor stains with dish soap. Lightly sand major stains. Liquid detergents, water, and a stiff bristle brush will remove most mildew and dirt.	Clean to remove surface molds and stains once per year. Avoid harsh chemical cleansers and heavy power washing. Lightly sand major stains. Liquid detergents, water, and a stiff bristle brush will remove most mildew and dirt. For hard to clean surfaces, use a deck brightener containing oxalic acid and follow the manufacturer's instructions.	Clean to remove surface molds and stains once per year. Avoid harsh chemical cleansers and heavy power washing. Liquid detergents, water and a stiff bristle brush will remove most mildew and dirt. For hard to clean surfaces, use a deck brightener containing oxalic acid and follow the manufacturer's instructions.
Maintenance finishes	The density and tight grain of black locust requires special finishes as standard products may not penetrate fully and leave a sticky feel. To maintain color, use oil or water-based finishes with UV protectant that are specifically recommended for black locust lumber.	After installation, allow the wood to dry for 2–6 months prior to applying any finish. Use high-quality oil or water-based finish with UV protection. Apply a water repellent sealer every two years after initial staining and sealant are applied.	After installation, allow the wood to dry prior to applying any finish. Apply a high-quality oil or water-based finish with UV protection to slow down the process of wood turning gray from exposure to the sun.
Finishing intervals	Will vary with climate, finish type, and traffic. To maintain natural wood color, apply finishes every 1–5 years.	Will vary with climate, finish type, and traffic. Some oils and stains are rated to last 6 years.	Water repellency of EL2 means than sealing can be delayed for up to 3 years. Apply a water repellent sealer every 2 years after initial staining and sealant are applied.
To restore original color	If grayed, restore original color by lightly sanding then oiling.	If grayed, restore original color by lightly sanding then oiling.	If grayed, restore original color by lightly sanding then oiling.
Warranty	Estimated to last 30–50 years. Warranties not found.	Limited 15 year warranties are provided by some companies.	"Lifetime limited warranty".
Reuse	Yes	Yes	Yes
Recycling	Yes	Yes, but do not mulch.	Yes, but do not mulch.
Incineration for energy production	Yes	Yes, in commercial incinerators.	Yes, in commercial incinerators.

[1] Robi Decking 2024

[2] Black Locust Wood 2024

[3] Viance EcoLife 2024

[4] Viance EcoLife 2021

[5] Lonza 2017

Table 10.11 Maintenance Considerations for Composite, Fused Bamboo and Rice Hull Composite Lumber Decking

	Composite Lumber[1,2]	Fused Bamboo Lumber[3,4]	Rice Hull Composite Lumber[5,6]
Weathering	Colors will fade over time, particularly with intense UV exposure.	Changes from dark brown to brown to gray if not oiled.	Colors will fade over time, particularly with intense UV exposure.
Expansion and contraction over time	Composite lumber will expand and contract substantially with changes in temperature.	Very small fissures can appear and disappear with expansion and contraction.	Rice hull composite lumber will expand and contract substantially with changes in temperature.
Cleaning	Clean the deck annually. Use a medium plastic bristle brush and mild liquid soap and water. Manufacturers list approved cleaners for heavier dirt and stains. Power washers can be used, look for recommended max. pressure. Avoid use of metal snow shovels. Calcium chloride deicers can be used but should be cleaned off after the snow melts.	Annual cleaning with warm water, a mild deck cleaner, and a non-metallic stiff bristle brush. Rinse thoroughly with clean water. Dry for 24–48 hours before applying a finish. Some manufacturers do not recommend power washing. Sodium percarbonate or potassium hydroxide can be used for heavily soiled areas. Use a broom, soft bristle brush, or plastic shovel along the length of boards for snow removal. Avoid rock salt and deicers.	Clean the deck 1–2 times per year with mild soap and water. Clean spills immediately. Light pressure washing with a fan tip nozzle. Plastic shovels or leaf blowers should be used to remove snow. Shovel along length of deck boards. Deicing salts can be used but should be cleaned off after use.
Maintenance finishes	N/A	Deck oil is recommended 3–4 months after installation then once per year. Material is not porous and will not absorb oil like wood. Wipe off excess oil within 15 minutes. Deck brighteners and stains can be used.	Can be stained.
Finishing intervals	N/A	Annually	N/A
To restore original color	Cleaning	Lightly sand then stain or oil.	Cleaning and possibly staining.
Warranty	Warranties vary widely by product and use.	Products have 25–30 year limited warranties.	Products have 25 year limited warranties.
Reuse	Yes	Yes	Yes
Recycling	No	Yes, but do not mulch.	No
Incineration for energy production	No	No	No. At least one product contains PVC which releases harmful dioxins when incinerated.

[1] Azek 2023

[2] Trex 2024b

[3] DassoXTR 2018

[4] Moso® Bamboo 2022

[5] Resysta 2024

[6] TruNorth Decking 2023

Embodied Carbon of Deck Structures

The embodied carbon (EC) of decks is impacted by the wide variety of structural materials, footings, and decking types. Replacement factors of deck boards are a key consideration when evaluating the EC of decks. In a typical EC comparison, a service life of decks is assumed to be 30 years, therefore, some decking materials will need to be replaced at least once during this time. The replacement factor is a multiplier that is applied when the deck boards will need to be replaced. For instance, pressure-treated SYP deck boards have a replacement factor of 2.5 because they will last an average of 12.5 years (Table 10.12; Figures 10.19 and 10.20).

Embodied Carbon Calculations for Typical Deck Structures

The EC of multiple deck structures is evaluated for a freestanding deck that is 9 meters by 12 meters. The structures vary by joist spacing, post spacing, footing type, and structural materials. Deck boards were not included in these calculations because the wide variety of deck boards would add too many variables. The EC of decking types is shown in Table 10.12 and Figures 10.19 and 10.20.

Four distinct deck structures were evaluated, and both 40cm o.c. and 60cm o.c. joist spacing were calculated for each. EC was calculated for the deck structures with both concrete and helical pier footings (Table 10.13).

Table 10.12 Embodied Carbon of Deck Boards, 30 Year Service Life

Deck Board Material and Density (kg/m^3)	Embodied Carbon $kgCO_2e/kg$	Carbon Stored $kgCO_2e/kg$	Replacement Factor for 30 Year Deck Life (1 = No Replacement)	Total EC for $108m^2$ Deck, 40cm o.c. Joists[1] (Deck Board Thickness)	Total EC for $108m^2$ Deck, 60cm o.c. Joists[1] (Deck Board Thickness)
Pressure-treated lumber (copper based) $512kg/m^3$	0.180[2]	1.59	2.5	25mm **622kgCO$_2$e**	38mm **946kgCO$_2$e**
Pressure-treated lumber (LOSP) $468kg/m^3$	0.241[2]	1.64	2.5	25mm **761kgCO$_2$e**	38mm **1157kgCO$_2$e**
Thermally modified lumber, softwood $447kg/m^3$	0.582[2]	1.76	1.5	20mm **843kgCO$_2$e**	25mm **1054kgCO$_2$e**
Thermally modified lumber, hardwood $563kg/m^3$	0.535[2]	1.50	1.2	20mm **781kgCO$_2$e**	25mm **976kgCO$_2$e**
Acetylated lumber, softwood $515kg/m^3$	0.633[4]	1.56	1	20mm **704kgCO$_2$e**	25mm **880kgCO$_2$e**
Furfurylated lumber, softwood $570kg/m^3$	0.588[5]	1.38	1	20mm **724kgCO$_2$e**	25mm **905kgCO$_2$e**
Tropical hardwood deck boards $1050kg/m^3$	0.49[3]	1.65	1	20mm **1111kgCO$_2$e**	25mm **1389kgCO$_2$e**
Fused bamboo composite $680kg/m^3$	1.060[1]	1.65	1	20mm **1572kgCO$_2$e**	25mm **1965kgCO$_2$e**
Composite 48% recycled PE; 50% fiber $930kg/m^3$	1.49[2]	0.83	1.2	24mm **3592kgCO$_2$e**	Max 40cm joist spacing
Composite PVC and wood fiber $1300kg/m^3$	3.23[3]	Not available	1.2	24mm **10,017kgCO$_2$e**	Max40cm joist spacing
Rice hull composite (some contains PVC) $1200kg/m^3$	1.650[3]	0.92	1.2	24mm **5132kgCO$_2$e**	Max40cm joist spacing

[1] Fasteners not included. These figures could be reduced in the future with production efficiencies and increased use of renewable energy sources.

[2] Average of Product EPDs 2024

[3] Hardy 2024

[4] Agrodom A&D 2022

[5] Norwegian EPD Foundation 2022

[6] Zhejiang Daocheng Bamboo Industry Co., Ltd 2020

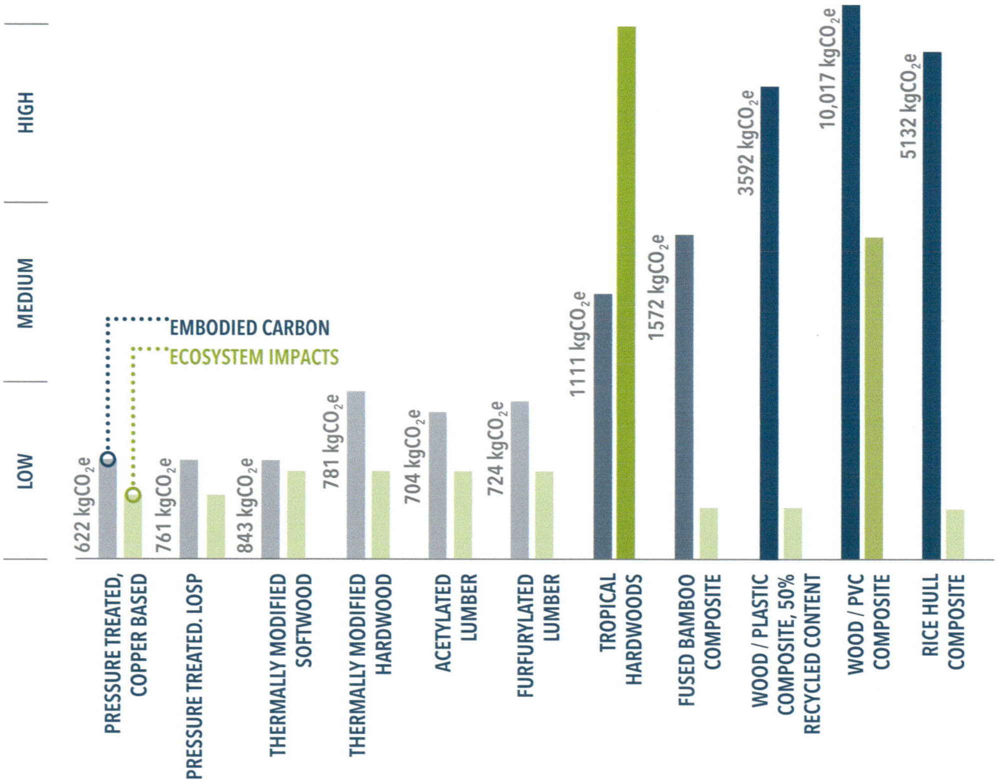

Figure 10.19 Embodied carbon results and estimated ecosystem impacts for 108 square meters of deck boards with a 30 year service life. Replacement factors were included for the deck board types that will not last 30 years. The deck boards are sized to span 40cm joist spacing. Image credit: Meg Calkins

The four main deck structures are as follows:

- Pressure-treated SYP deck structure with 30 posts and relatively short joist and beam spans
- Pressure-treated SYP deck structure with 20 posts and relatively long joist and beam spans
- Pressure-treated SYP deck structure with pressure-treated glue laminated beams and longer joist and beam spans
- Powder coated steel deck structure with long joist and beam spans.

Deck structures were designed to adhere to the International Residential Code for Decks (ICC 2021). Span tables in the code were used to size structural members and concrete footings for the SYP pressure-treated lumber structures. Member sizes for the steel deck structure are based on span tables from a global manufacturer of steel decks, and the sizes of members reflect their proprietary products and allowable spans. Connectors and fasteners are not included in these calculations (Table 10.13).

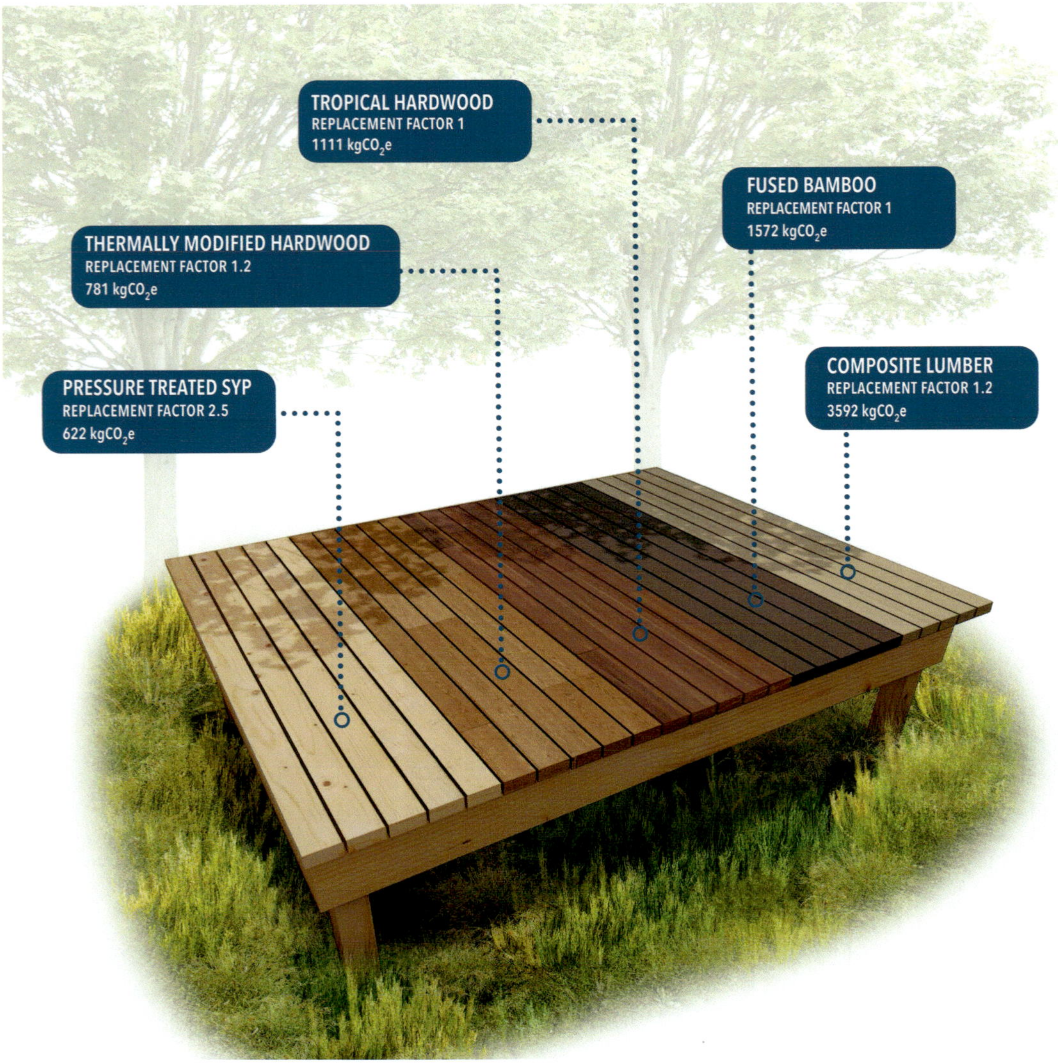

Figure 10.20 Embodied carbon and replacement factors of selected deck boards. Image credit: Kathleen Elliott and Meg Calkins

Table 10.13 Comparison of the Embodied Carbon of Typical Deck Structures

Joists, Beams, Posts, Footings, Blocking, Rim Joists[1] Deck Boards Are Not Included in These Calculations (Table 10.12)	Embodied Carbon (EC) of Deck Framing Only ($kgCO_2e$)	EC Deck Framing With Concrete Footings ($kgCO_2e$)	EC Deck Framing With Helical Pier Footings ($kgCO_2e$)	Key Observations
SYP deck structure 40cm o.c. joists 2.1m joist span 2.28m beam span (30) 89x89mm posts	365	1080	1272	EC of deck framing is lower for smaller wood joist and beam spans. Concrete footings have lower EC than helical piers for smaller tributary areas.
SYP deck structure 60cm o.c. joists 2.1m joist span 2.28m beam span (30) 89x89mm posts	369	1085	1277	Deck framing EC is lower for smaller wood joist and beam spans. Concrete footings have lower EC than helical piers for smaller tributary areas. Wider spacing of joist results in similar wood quantities because joists must be larger.
SYP deck structure 40cm o.c. joists 2.8m joist span 2.85m beam span (20) 89x89mm posts	468	1198	1073	Helical piers are lower EC than concrete footings for larger tributary areas.
SYP deck structure 60cm o.c. joists 2.8m joist span 2.85m beam span (20) 89x89mm posts	447	1177	1052	Helical piers are lower EC than concrete footings for larger tributary areas. Wider spacing of joist results in similar wood quantities because joists must be larger.
Glulam beam and SYP deck structure 40cm o.c. joists 4.0m joist span 5.5m beam span (9) 140x140mm posts	583	1268	946	Decks with larger tributary areas using glulam beams and helical piers have lower EC than most decks in this study. Glulams are lower EC than steel for the same span.
Glulam beam and SYP deck structure 60cm o.c. joists 4.0m joist span 5.5m beam span (9) 140x140mm posts	477	1162	840	This deck with larger tributary areas using glulam beams and helical piers and 60cm joist spacing is the lowest EC of all decks in this study at $840kgCO_2e$. Glulams are lower EC than steel for the same span.
Steel deck structure 40cm o.c. joists 4.0m joist span 3.67m beam span (12) 89x89mm posts	2483	3396	2967	40cm o.c. joist spacing results in much higher EC than 60cm o.c. spacing for steel decks. Helical piers are lower EC for larger tributary areas.
Steel structure A60 60cm o.c. 4.0m joist span 3.67m beam span (12) 89x89mm posts	1906	2820	2390	40cm o.c. joist spacing results in much higher EC than 60cm o.c. spacing for steel decks. Helical piers are lower EC for larger tributary areas.

[1]Decking, connectors, and fasteners are not included in these calculations. Refer to Table 10.12 for the EC of decking for a $108m^2$ deck. These figures could be reduced in the future with production efficiencies and increased use of renewable energy sources.

All decks have blocking over beams. The steel frame deck has steel blocking between beams.

All decks have rim joists.

The wood decks have 30cm joist and beam cantilevers.

The steel deck and glulam beam deck have 50cm joist and beam cantilevers.

EC for materials is found in Chapter 18 "Wood Materials and Products" and Chapter 20 "Metals and Metal Products".

There are no posts with the helical pier footings. Beams rest directly on the pier bracket.

References

Accoya (2020). *Accoya Performance Testing Summary V4.0*. Accoya.

Accoya (2022). *Color Decking Installation Guide*. Accoya.

Accoya (2024). *Accoya Wood Information Guide V3.9.1*. Accoya.

Agrodome Advice & Development (Agrodome A&D) (2022). *Accoya Wood Environmental Product Declaration*. Agrodome A&D.

American Society for Testing and Materials (ASTM) (2021). *ASTM D7032 Standard Specification for Establishing Performance Ratings for Wood-Plastic Composite and Plastic Lumber Deck Boards, Stair Treads, Guards, and Handrails*. ASTM International.

American Society for Testing and Materials (ASTM) (2023a). *ASTM A153 Standard Specification for Zinc Coating (Hot-Dip) on Iron and Steel Hardware*. ASTM International.

American Society for Testing and Materials (ASTM) (2023b). *ASTM A653 Standard Specification for Steel Sheet, Zinc-Coated (Galvanized) or Zinc-Iron Alloy-Coated (Galvannealed) by the Hot-Dip Process*. ASTM International.

American Wood Protection Association (AWPA) (2024). *2024 AWPA Book of Standards*. AWPA.

Arbor Wood Co. (2024). *Decking Installation Guide*. Arbor Wood Co. https://arborwoodco.com/resources, accessed September 6, 2024.

Azek (2023). *Timber Tech Composite and Specialty Decking Installation and Maintenance Guide*. The Azek Company.

Bison Innovative Products (2024). "Deck pedestals". https://bisonip.com/products/pedestals/, accessed April 1, 2024.

Black Locust Wood (2024) "The lumber". blacklocustwood.com/lumber/, accessed November 11, 2024.

Building Transparency (2024). *Embodied Carbon in Construction Calculator (EC3) Tool*. Building Transparency. https://buildingtransparency.org/ec3.

CAMO & Accoya (2022). *Marksman System: Hidden Fastening for Accoya Wood Boards*. CAMO Fasteners.

DassoXTR (2018). *DassoXTR Fused Bamboo® Maintenance. Version 10.23.18*. DassoXTR.

DassoXTR (2020). *DassoXTR Fused Bamboo Decking Installation Guidelines. Version 12.15.20*. DassoXTR.

DecksGo (2024). "Deck foot anchor". https://store.decksgo.com/products/deck-foot-anchor, accessed April 6, 2024.

Diamond Pier Foundations (2023). "Quick guide for the pro-builder". www.diamondpiers.com/quick-guide-pro-builder, accessed November 12, 2023.

EZ-Tube (2024). *EZ-Tube: An Evolution in Footings*. EZ-Tube, April 6.

FootingPad (2024). "Below grade footings". https://footingpad.com/, accessed April 6, 2024.

Goliath Tech (2023). "Helical screw piles for decks". www.goliathtechpiles.com/application/helical-screw-piles-decks, accessed November 11, 2023.

Hardy, C. (2024). *Carbon Dataset Version 3*.

International Code Council (ICC) (2021). *2021 Deck Construction Based on the International Residential Code*. ICC, April.

Kebony (2016). *Use and Maintenance for Kebony Terrace Cladding and Roof Boards*. Kebony Wood.

Kebony (2024). "Product guidelines". https://us.kebony.com/wp-content/uploads/2024/06/Kebony_Product-Guidelines_V9.pdf, accessed May 8, 2024.

Lebow, S., et al. (2019). *Guidelines for the Selection and Use of Pressure Treated Wood*. General Technical Report. FPL-GTR-275. US Department of Agriculture (USDA), Forest Products Laboratory.

Lonza (2017). *Wolmanized Outdoor Wood Product Info Sheet*. Lonza, Wolmanized Outdoor Wood.

Meier, E. (2024). "The wood database". www.wood-database.com/, accessed January 1, 2024.

Moso® Bamboo (2021). *Moso Bamboo X-treme Outdoor Decking Installation Instructions*. EN-NA-2021-11. Moso® Bamboo.

Moso® Bamboo (2022). *Moso Bamboo X-treme Outdoor Decking Care & Cleaning*. EN-NA-2022-06. Moso® Bamboo.

Norwegian EPD Foundation (2022). *Kebony Character Environmental Product Declaration*. Kebony AS.

Resysta (2023). *Resysta Installation Guide: Resysta Gold and Platinum Decking System*. Resysta Building Products.

Resysta (2024). "Resysta care and cleaning recommendations". https://resystausa.com/resysta-owner/care-and-cleaning/, accessed September 6, 2024.

Robi Decking (2024). "Technical information and product data sheets". robidecking.com/resources/technical-data/, accessed August 4, 2024.

Thermory (2022). *THM-533 Decking and Porch Flooring Installation Guide*. Thermory.

Thermory (2024). "Thermory: Decking boards and modular systems". https://thermory.com/technical-area/?filter=gallery_category_taxonomy%3A111, accessed August 24, 2024.

TileTech (2024). "Roof pedestals that elevate". https://tiletechpavers.com/decking/pedestals, accessed April 1, 2024.

Trex (2024a). *Decking Installation Guide*. Trex Company Inc.

Trex (2024b). *Trex Transcend®, Trex Enhance®, and Trex Select® Decking Care and Cleaning Guide*. Trex Company Inc.

TruNorth Decking (2023a). *Installation Guide*. TruNorth Decking.

TruNorth Decking (2023b). *Care & Maintenance Guide*. TruNorth Decking.

Viance EcoLife (2021). "Lifetime limited warranty". www.treatedwood.com

Viance EcoLife (2024). "EcoLife fact sheet". www.treatedwood.com/products/product-information, accessed September 6, 2024.

Zhejiang Daocheng Bamboo Industry Co., Ltd (2020). *Dasso Strand Woven Bamboo Environmental Product Declaration*. Zhejiang Daocheng Bamboo Industry Co., Ltd.

CHAPTER 11

Rails, Fences, and Screens

Rails, fences and screens offer a good opportunity to customize a site design with the variety of materials, finishes, member sizes, and types that comprise these structures. Innovative custom structures can be created, or prefabricated gate, fence, or rail systems can be specified. Either choice influences the aesthetic qualities of a place. The service life, costs, and embodied carbon of these structures vary as widely as the materials.

The primary performance consideration for guardrails and handrails is safety between grade changes in a landscape, but guardrails can also provide a sense of enclosure to a space (Table 11.1; Figures 11.1–11.4). Fences and screens provide privacy, physical enclosure and exclusion, spatial definition, environmental control, and security (Table 11.2 and Figure 11.5).

Guardrail, Fence, and Screen Structural Design

A variety of loads act on fences and rails influencing the footings, connections, and member sizing. Primary loads on guardrails and fences are lateral from people leaning against the structure or bumping into it. Wind loads also play a role in the design of solid fences and rails. Vertical loads consist of people climbing on fences and rails (Figure 11.6).

Fences, screens, and guardrails are cantilever structures that are connected at their base to a slab, footing, or other foundation. Forces on the rail or fence and the height of the structure will determine the depth of embedment of the posts or the load value of the bolts connecting the post and plate to a footing. The International Building Code requires that guardrails be able to withstand 75kg/meter (50lbs/ft) (ICC 2021). If greater loads are expected, a structural engineer should be involved in the guardrail and footing design. If guardrail or fencing product systems are specified, the manufacturers should provide installation recommendations for anchoring and footings.

Posts

Guardrail and fence posts resist lateral and horizontal forces and tie the structure into footings or slabs. Posts are primarily steel or wood but can also be aluminum, plastic, or composite lumber. Wood posts are vulnerable to moisture, so attaching the post to a footing above grade is the most durable approach. Wood posts should be pressure treated, modified, or naturally durable lumber that is Durability Class 1. Steel posts in ground contact should be heavy galvanized or powder coated.

The size of posts will vary by material with metal posts being smaller than wood or plastic. Steel or aluminum fence posts can be 38–76mm square or in diameter. Typical wood fence posts are 89x89mm, although 140x140mm posts are sometimes used for aesthetic reasons. If embodied carbon or resource conservation is a priority, posts should be the smallest size that is structurally sound. Larger wood posts may be required at heavy gates, the corner of fences, or when they are higher than 1.8m. Post spacing is a factor of the size, number, and orientation of rails spanning between the posts (Figures 11.7–11.10).

Rails or Cross Bars

Rails or cross bars span between posts, and they support fence panels, pickets, or slats. Handrails are often affixed to guardrail structures, but as codes dictate, they are a lower height than the top of the guardrail, so they are not usually part of the structural frame (Figures 11.7–11.10).

Steel or aluminum rails offer multiple shapes, sizes, and color opportunities. To maximize resource efficiency and minimize embodied carbon, steel, or aluminum rails should be the smallest size structurally possible. It is important to note that a steel "T" or angle could have the same structural capability as a similarly sized square steel tube at just 50% of the weight and embodied carbon (Table 20.4). Steel

DOI: 10.4324/9781003413974-12

Table 11.1 Design Considerations for Guardrails and Handrails

	Design Considerations
Dimensions for guardrails	The International Code Council (ICC) and the US Americans with Disabilities Act (ADA) dictate standards for guardrails although individual countries, states, and municipalities may have stricter codes. ICC and ADA require a 106.7cm (42″) guardrail when there is a difference in finished grades of 76.2cm (30″) or more.
Openings for guardrails	ICC and ADA require a maximum opening size of 10.16cm (4″) on a guardrail. There are exceptions that allow for some 11.11cm (4.375″) openings.
Dimensions for handrails	ICC and ADA require handrails for exterior stairs with more than three risers and ramps on accessible routes. The diameter of handrails ranges from 31.8–50.8mm (1.25–2″) and the height above risers is 86.4–91.4cm (34–36″). Handrails extend at the top and bottom of stairs. Refer to codes for extension lengths.
Post spacing	Post spacing of guardrails is a factor of the structural elements spanning between the posts and the expected forces on the guardrail (Figure 11.1).
Materials	Powder coated steel or stainless steel are the most common materials for exterior handrails and guardrails, but they are also constructed from weathering steel, aluminum, wood, glass, and composite materials.

Sources: ICC 2022a; USAB 2016, 2015

Figure 11.1 The stainless steel cable guardrail at Pier 26 by the Olin Studio allows for views through to the water and cityscape beyond while using minimal material. The furfurylated lumber top rail offers a softer surface for viewers matching the decking. Image credit: Meg Calkins

Figure 11.2 The weathering steel guardrail at Little Island Park by MNLA and Heatherwick Studio forms a sculptural edge to the Island referencing wooden piles in the water. Image credit: Meg Calkins

rails are often welded to posts, while aluminum rails are connected with proprietary connection systems.

***Wood fence* rails** are typically 38x89mm. Rails in a flat orientation should not span more than 1.5m or they may sag over time. Oriented on edge, they can span 2.1m (RealCedar 2024; SFPA 2024). Three rails in a wood fence will result in greater stiffening than two rails.

Bottom and middle rails are often oriented on edge and can be bolted and/or notched into posts. Top rails may be

Figure 11.3 A bronze alloy guardrail lines John Pawson's Lake Crossing Bridge at Kew Gardens. Image credit: Meg Calkins

Figure 11.4 A stainless steel guardrail in Glasgow makes minimal use of materials. Image credit: Meg Calkins

Table 11.2 Design Considerations for Fences and Screens

	Design Considerations
Spatial delineation	Fences and screens, even low ones, can delineate spaces. Hedges or allées of trees are a lower carbon way to define space. Lower fences around smaller spaces will help the space feel more connected to areas outside the fence. Taller fences around small spaces will make the space feel smaller and cast more shade in the space.
Environmental control	Fences can block wind and sunlight and mitigate noise. Staggered or spaced fence boards and perforated metal fence panels will allow for airflow while providing sound attenuation and shade.
Property definition	Lower more transparent fence designs will delineate property edges but still be open to sunlight, air flow, and views. Lower fences are perceived to be more friendly to neighbors and many regulations limit height to 1.2 meters and opacity greater than 50%. The "back" of the fence should face inward toward the property.
Privacy	Privacy needs will determine height and opacity of the fence. If complete privacy is required, fence height should be over 1.7m as this is an average eye height. Fence panels can be solid, or visual exclusion can be achieved with staggered boards that allow for penetration of light and air flow but exclude visual access. The closer the fence is to the source of exclusion, the easier it is to block views.
Security	Security fence design will discourage climbing with minimal horizontal elements, and articulation at the top that makes it difficult to climb over the fence.
Safety	Fences can be used to exclude people from dangerous equipment such as electrical transformers or mechanical equipment. The ICC requires a 122cm (48") fence around swimming pools and water bodies deeper than 30cm (ICC 2022b). Fences and barriers should be designed to discourage climbing. Safety fencing around pools should be transparent or semitransparent so adults can see if a child has breached the fence.
Animal exclusion	Fences can be used to exclude wildlife but should be used judiciously as fencing can disrupt habitat and wildlife movement. Fencing to exclude deer should be minimum 2.4 meters high. Welded wire mesh with 15cm openings can be used for deer fencing as it will allow smaller wildlife to move through the fencing.
Animal containment	Fencing to contain a dog will vary by the size of the dog, with a 1-meter height containing smaller dogs and a 1.8m fence height containing larger dogs.
Child containment	Openings in the fence should not be larger than 102mm. Fences and barriers should be designed to discourage climbing.

Figure 11.5 Reclaimed steel from on-site oil tanks was repurposed by McGregor Coxall for screens and signage at Ballast Point Park in Sydney. Image Credit Brett Boardman

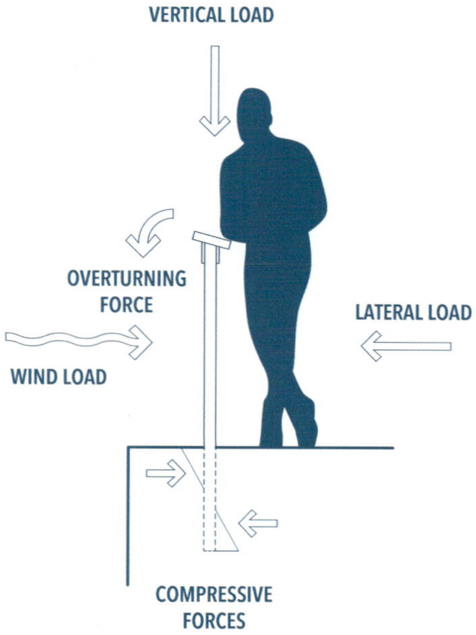

Figure 11.6 Typical forces on fences, rails, and screens. The strongest forces on these cantilever structures are often lateral loads from people leaning against guardrails of wind loads on solid fences. Image credit: Kathleen Elliott and Meg Calkins

Figure 11.7 Steel cable guardrail detail. Steel posts and rails can be stainless, weathering steel, or powder coated. Cables are most often stainless steel. Tightening mechanisms are employed to adjust slack from expansion of cables over time. Image credit: Kathleen Elliott and Meg Calkins

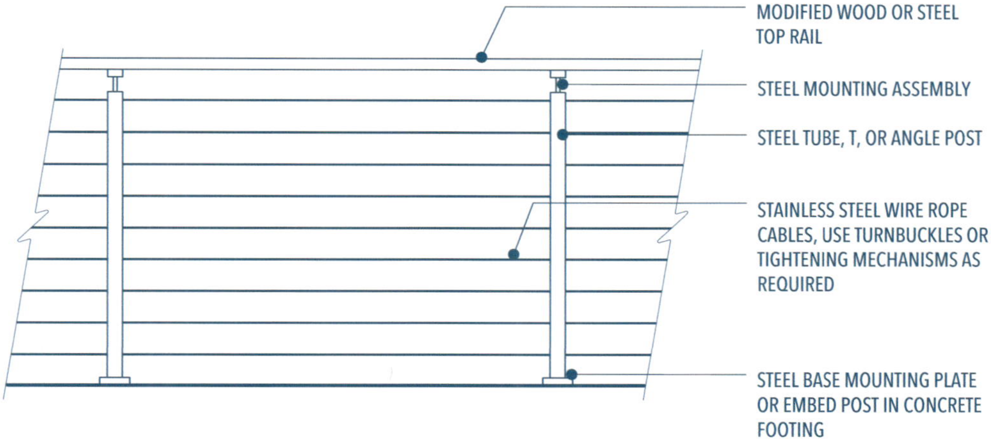

oriented flat if they rest directly on the top of posts. The top rail can protect the end grain of vertical wood pickets. Top rails should be slightly sloped to drain water.

Panels, Slats, Pickets, or Cables

Panels, slats, pickets, or cables determine the opacity of the fence or rail structure for aesthetics and performance. They can be spaced, perforated, or arranged for a feeling of openness, or they can form a solid screen with staggered boards or panels. Rails and/or frames support panels, slats, or pickets. Cables may be supported by posts (Figures 11.7–11.10).

Steel or aluminum shapes and tubes are used in both vertical and horizontal fence pickets, and steel or aluminum sheet is used in panels. Metal sheet used in fence panels is

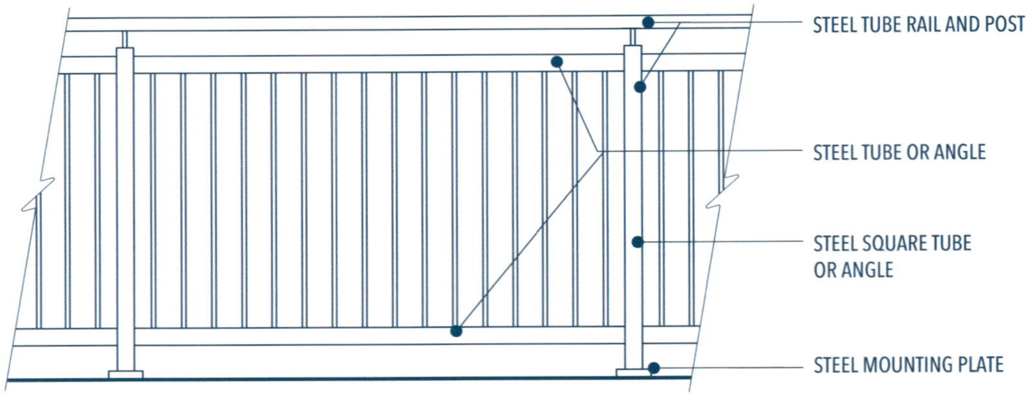

Figure 11.8 Steel guardrail with welded connections. Steel can be stainless, weathering steel, or powder coated. Steel angles will use less material than steel tubing for a similar structural performance. Image credit: Kathleen Elliott and Meg Calkins

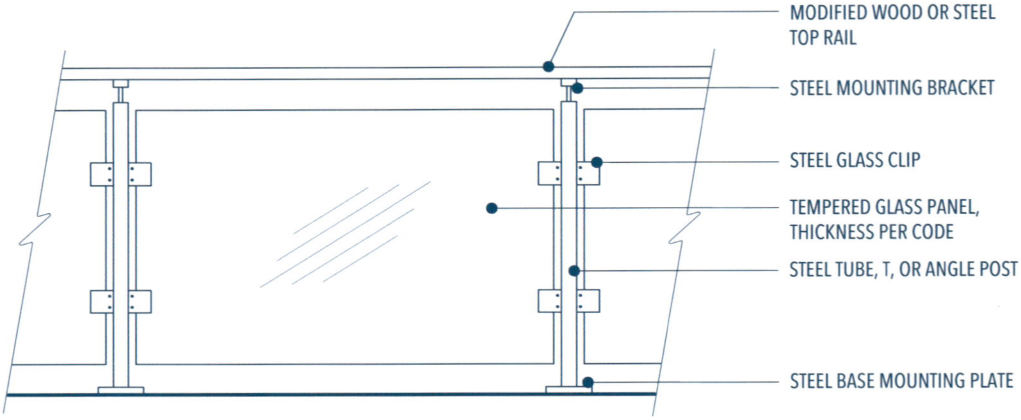

Figure 11.9 Steel and glass guardrail. Steel can be stainless, weathering steel, or powder coated. The embodied carbon of glass is very high, so thicknesses should not be oversized beyond structural requirements. Image credit: Kathleen Elliott and Meg Calkins

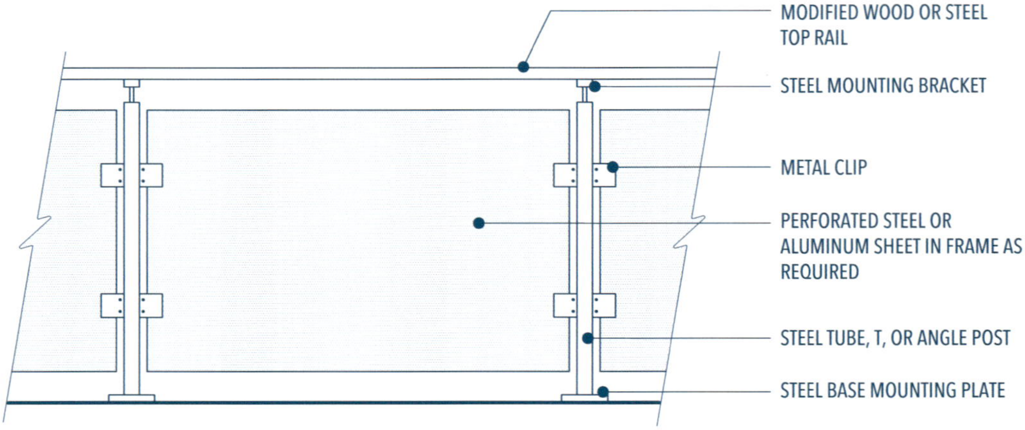

Figure 11.10 Perforated steel or aluminum panel guard rail with steel structure. Higher open area percentages in the perforations will reduce the embodied carbon of the structure. Steel can be stainless, weathering steel, or powder coated. Image credit: Kathleen Elliott and Meg Calkins

stiffened by a metal frame attached to the fence rails or the posts. Perforated or expanded metal sheet materials can allow light and air flow while providing privacy, safety, and shade.

Sheet glass is used in guard rail panels, and it must be laminated, fully tempered, or heat strengthened. It must meet the glazing safety requirements of Category II or CPSC 16 CFR Part 1201, or Class A of ANSI Z97.1. The minimum glass thickness is 6.4mm, and it must be able to resist a single concentrated load of 91kg (200lbs) (ICC 2021).

Wood boards oriented vertically or horizontal form pickets or slats. Fence boards are usually 19mm thick by 89 or 140mm wide. Smaller 19mmx38mm pickets can be used but should be nailed or screwed to rails at least every 60cm to avoid warping, particularly if they are made from wider boards that have been ripped down to size.

Composite lumber is used for pickets, slats, and panels. It is usually supported by a wood or steel fence structure.

Preassembled vinyl (PVC) and aluminum fence panels and fencing systems are prefabricated and designed for ease of installation. While initial costs of preassembled systems can be low, they may be challenging to repair if one picket or rail becomes damaged. This can result in premature replacement.

Preassembled bamboo roll products are available for panels on wood or steel fence structures. They are a rapidly renewable fence panel that is very low in embodied carbon. If sealed or treated and protected from ground contact, these bamboo panels can last 10–15 years. They also may be challenging to repair.

Biobased and living fences are trained or woven into panels. They are discussed in detail later in this chapter.

Footings for Guardrail and Fence Posts

Fence or guardrail posts are anchored with footings, attached to concrete slabs, or connected to other structures (Figures 11.11–11.14). Concrete slabs will provide anchoring for guardrail and handrail posts either by embedding the post in the slab or attaching it to the slab with a steel plate and bolts. Guardrails on decks are attached to the deck beam or may be supported by the extension of the deck's posts.

Fence and screen post footings are primarily poured in place concrete, but helical piers may be used for fence posts in some applications. These posts can be attached to the top of the footing with a plate and bolts or they can be directly buried in the footing. Refer to the discussion of deck footings in Chapter 10 "Decks and Boardwalks" for information about helical pier foundations and deck post foundations. It is important to note that the sizing of deck footings is larger than is required for fence or rail footings as both the dead and live loads on rails and fences are substantially lower than those of decks.

Footing depths for rails and fences will be based on anticipated forces, height of the structure, soil types, local codes, and depth of frost line. Footings should extend to the depth of 60cm plus 10% of the height of the fence or rail structure, or 15cm below the frost line, whichever is deeper. High wind loads with an opaque fence design or poor soil bearing capacity may necessitate even deeper footings (Landphair and Klatt 1999).

Materials for Fences, Rails, and Screens

Steel and wood are the primary materials used to construct guardrails, fences, and screens. Aluminum, vinyl, and composite lumber are also used, while rapidly renewable materials such as bamboo and living fences offer a very low-carbon alternative. The performance, construction, and embodied carbon considerations for rails, fences, and screens are discussed next and in Table 11.5.

Steel

Steel is a very common materials for handrails, guardrails, screens, and fences as its high strength to size ratio allows for relatively small members. Thin posts, rails, and cables or pickets can form a very open railing or fence. The wide variety of steel shapes, grilles, and perforated panels provide limitless design opportunities. Combined steel and wood fences and guardrails employ steel for the load bearing components, and wood pickets or slats for the "panels".

Given the relatively high cost and embodied carbon of steel fences and rails, their durability is a primary consideration. Structures must be detailed to shed water and appropriate metal finishes used for the climate and site conditions (Figures 11.15–11.17 and Table 11.3). Types of metals and their protective metal finishes are discussed in Chapter 20 "Metal Materials and Products".

Steel Connections

Connections join two members together, facilitate the transfer of forces, and lend stability to a structure. Steel

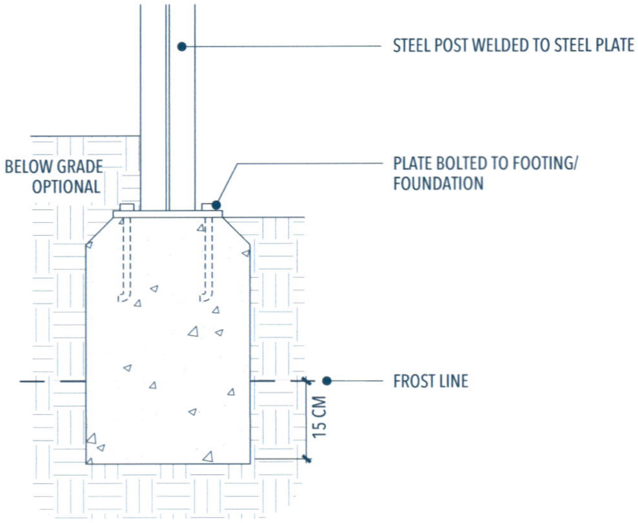

Figure 11.11 Steel post and plate on concrete footing detail. Image credit: Kathleen Elliott and Meg Calkins

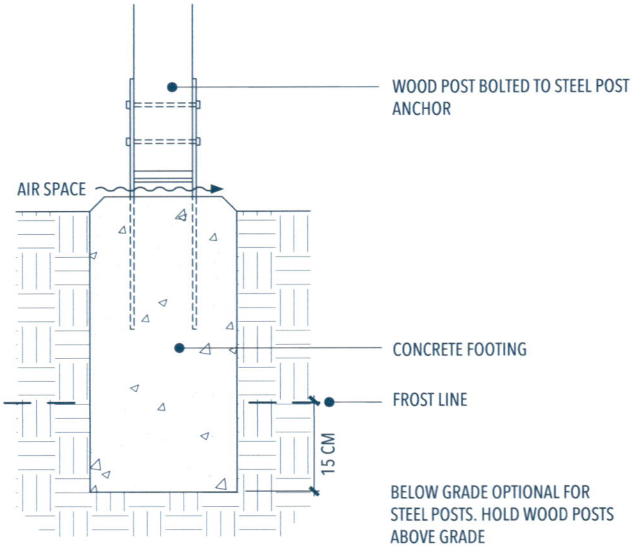

Figure 11.13 Wood post, post anchor, and concrete footing detail. Image credit: Kathleen Elliott and Meg Calkins

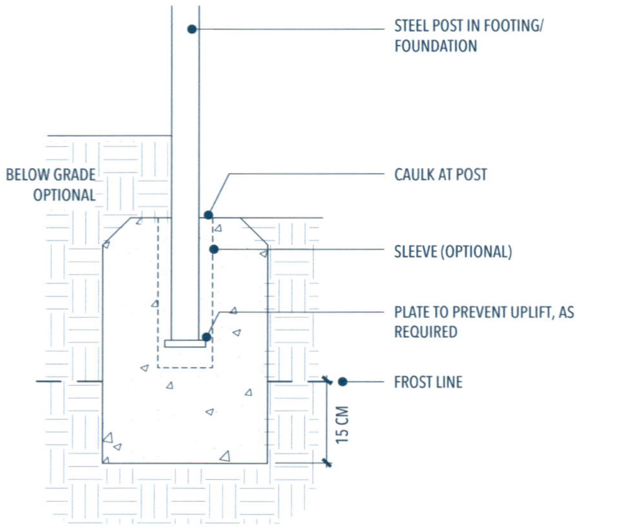

Figure 11.12 Steel post embedded in concrete footing detail. Image credit: Kathleen Elliott and Meg Calkins

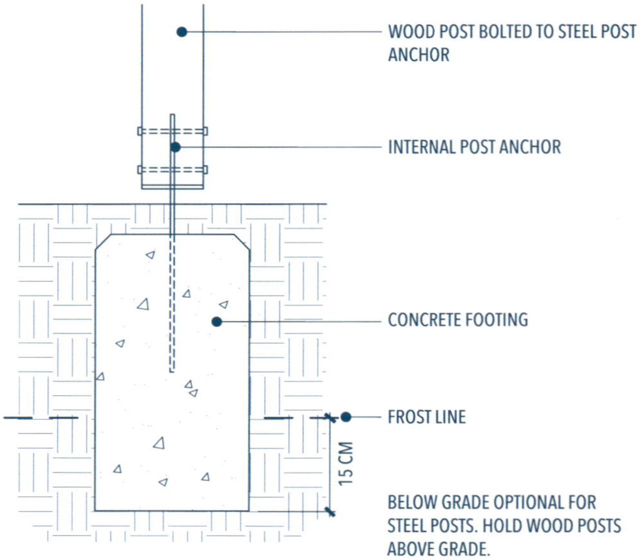

Figure 11.14 Wood post with internal post anchor in concrete footing detail. Image credit: Kathleen Elliott and Meg Calkins

connections for fences and rails are often made in two phases. First, elements of a fence, rail, or screen are prefabricated and joined in a fabrication facility. Panels, bars, and pickets are welded together then finished with a protective coating.

Next, these prefabricated assemblies are brought to a construction site and connected to on-site constructed elements. Field connections may be bolts, welds, or rivets. Connections made in the field can be vulnerable to corrosion as they do not have the same level of protective coating as those elements finished in a factory setting. When on-site welding or bolting occurs, care should be taken not to damage protective finishes.

When designing for deconstruction, bolted connections are far easier to disassemble than welded connections. Members that are welded are usually cut when removed

Table 11.3 Design and Construction Considerations for Durable Steel Fences and Rails

Category	Consideration
Appropriate metals for the environment	In marine or heavy pollution environments use higher grade stainless steels and avoid aluminum and galvanized steel. Powder coated steel is durable in marine environments. Table 20.11 has detailed information about metals and environmental conditions.
Protective coatings	Use factory applied protective coatings for steel structures to the extent possible (Tables 20.12 and 20.13). Stainless steel does not require a coating and weathering steel only needs a clear coating that could be applied on site.
Prefabrication	Prefabricate assemblies then powder coat or galvanize in a factory setting. Minimize field welding and protect finishes when installing prefabricated assemblies.
Connections	Use fasteners that have an equal or greater corrosion resistance than the metal for the structure but not so much that bimetallic corrosion occurs. Isolate dissimilar metals with inert washers or other physical barriers (Figure 20.14). Weld imperfections should be ground away to avoid pockets that may collect water and pollutants. Figures 11.8 and 11.10 illustrate connections to avoid moisture build up.
Design to clean away pollutants	Design the structure to allow rain to rinse away pollutants. Slightly slope rectilinear top rails or use round ones. If rain will not rinse elements, consider specifying a higher grade of stainless steel or thicker protective coatings.
Detail to shed water	Orient steel members to avoid catching water. Place I-beams with the web in the vertical orientation, avoid exposing lapped joints vertically, and avoid using two angles back-to-back, instead use a T shape. Refer to Figure 11.9 for metal member orientations to shed water.
Posts	Posts can be directly buried in concrete footings, but this detail makes the fence or rail challenging to deconstruct and the posts are not easily replaced if necessary.

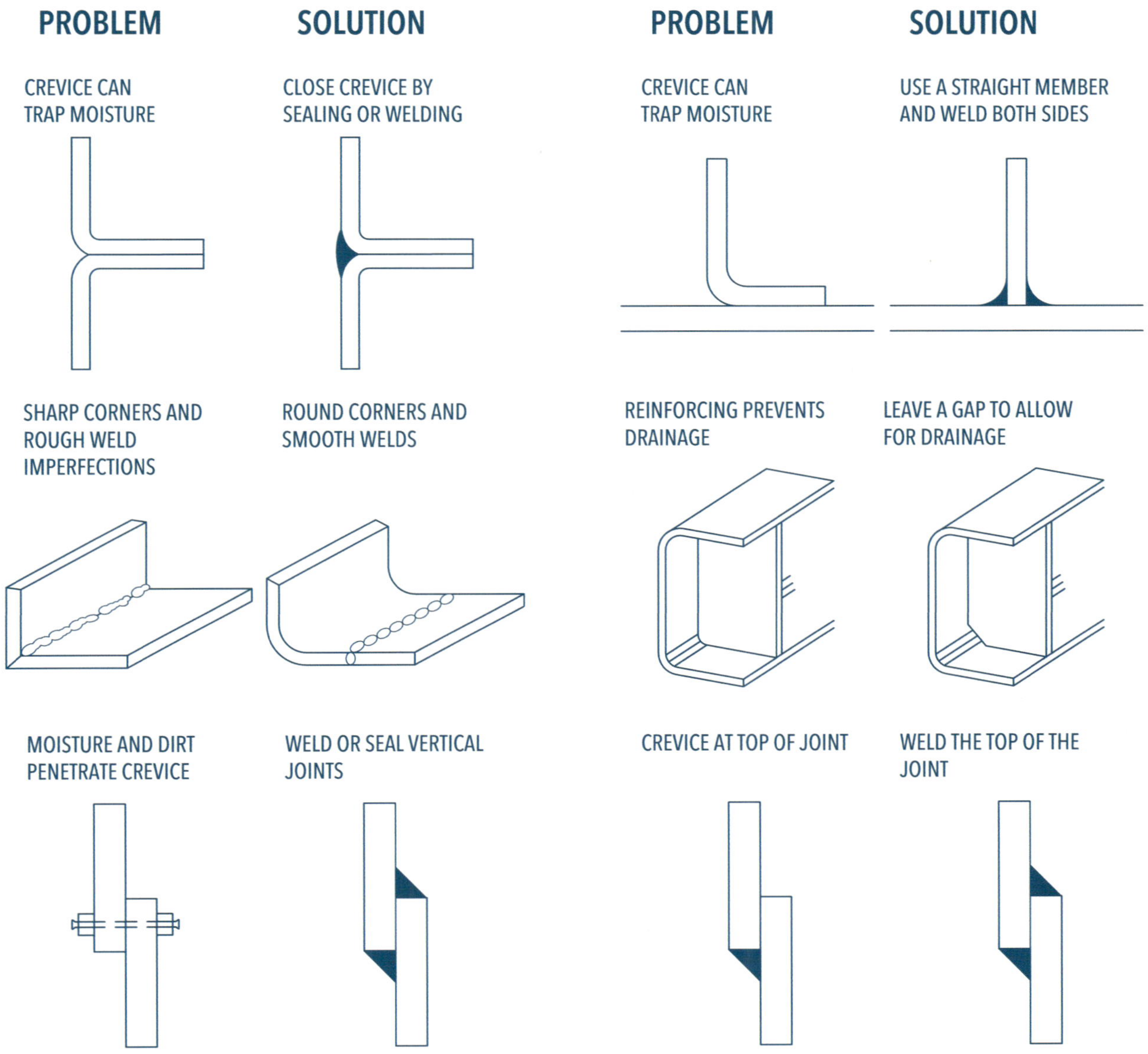

Figure 11.15 Metal joints to avoid standing water, moisture, and accumulation of chemicals that could reduce the service life of the structure. Image credit: Kathleen Elliott and Meg Calkins

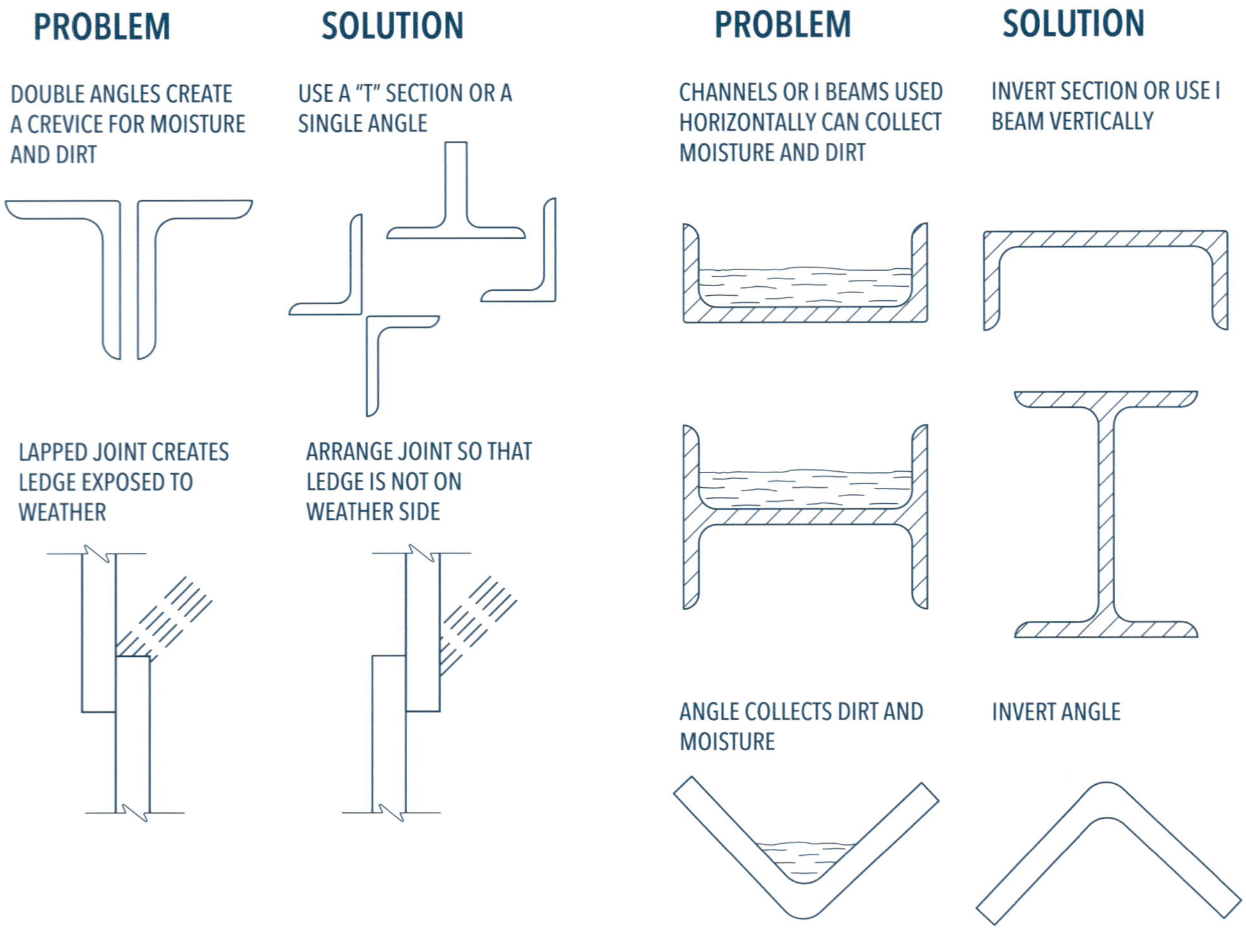

Figure 11.16 Orientations of metal members to avoid standing water, moisture, and accumulation of chemicals that could service reduce the life of the structure. Image credit: Kathleen Elliott and Meg Calkins

from a structure; however, a steel guardrail could have the individual elements welded to form a panel that is bolted or screwed to posts. The entire panel could be removed and reused whole.

Bolted connections are commonly used in steel structures for their ease of installation, flexibility, and strength. Bolted connections employ assemblies comprised of bolts, nuts, and washers. The metal of bolt assemblies should be compatible with the type of metal members being connected to avoid galvanic corrosion.

There are two major types of bolted connections:

Shear/bearing connections transmit loads through one member resting on another, perpendicular to the bolt axis. A single or double plate connected with bolts into each member is used to stabilize them. An example of shear/bearing bolted connection is a post base attached to a concrete footing with anchor bolts embedded in the concrete (Figure 11.11).

Friction grip connections rely on the clamping force from the nut tightened on the bolt to hold the connection together. Types of friction connections are shear tabs, slotted holes, and snug-tightened bolts.

Welded connections involve heat applied to melt and fuse steel members for a permanent joint that is very strong and rigid. The two major types of welding are *fusion welding*, where two metal faces are melted and joined

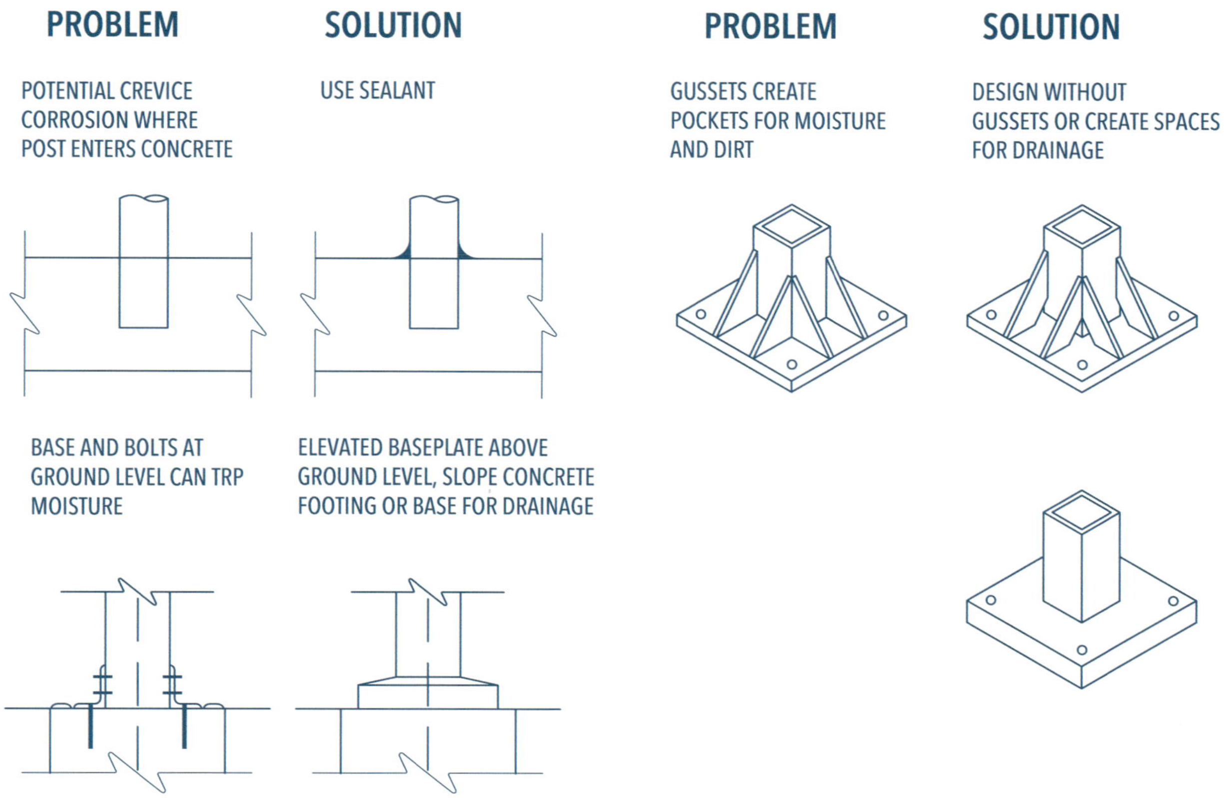

Figure 11.17 Post to footing or slab connections to avoid standing water, moisture, and accumulation of chemicals that could reduce the service life of the structure. Image credit: Kathleen Elliott and Meg Calkins

together in a linear seam, and *pressure welding*, where two metal faces are heated up then joined together through mechanical compression. Pressure welding results in discrete welds.

The two main types of welds are as follows:

Fillet welds are a very common type of weld for steel connections. They are formed in a triangular cross section along the edges of the connected components. They are very effective as they distribute loads over a larger area.

Butt welds are used to join two steel members together that are aligned in the same plane. The ends of the members are prepared to form a groove or a bevel, which ensures a full penetration weld. This type of weld is frequently used in post to rail and beam to column connections.

Embodied Carbon of Metal Rail and Fence Members

While relatively high in embodied carbon, metals can incorporate high recycled contents, and they can be very durable if finished appropriately. Most structural steel in North America contains over 90% recycled content (AISI 2024) (Tables 11.6 and 11.7; Figures 11.20–22).

Additionally, some sizes and shapes of materials have substantially lower EC than others for the same structural applications. The same sizes of square tubes and angles or T shapes can span or support similar loads, but Ts and angles use just half the material, resulting in about half the embodied carbon (Figures 11.18).

Wood Guardrails and Fences

Wood guardrails and fences offer a relatively low-carbon alternative to steel, aluminum, and PVC. Wood lumber is discussed in detail in Chapter 18, and design, construction, and maintenance considerations for wood products are discussed in detail in Tables 10.6–10.11 in Chapter 10. Design and construction considerations for durable wood fences and rails are discussed in Table 11.4.

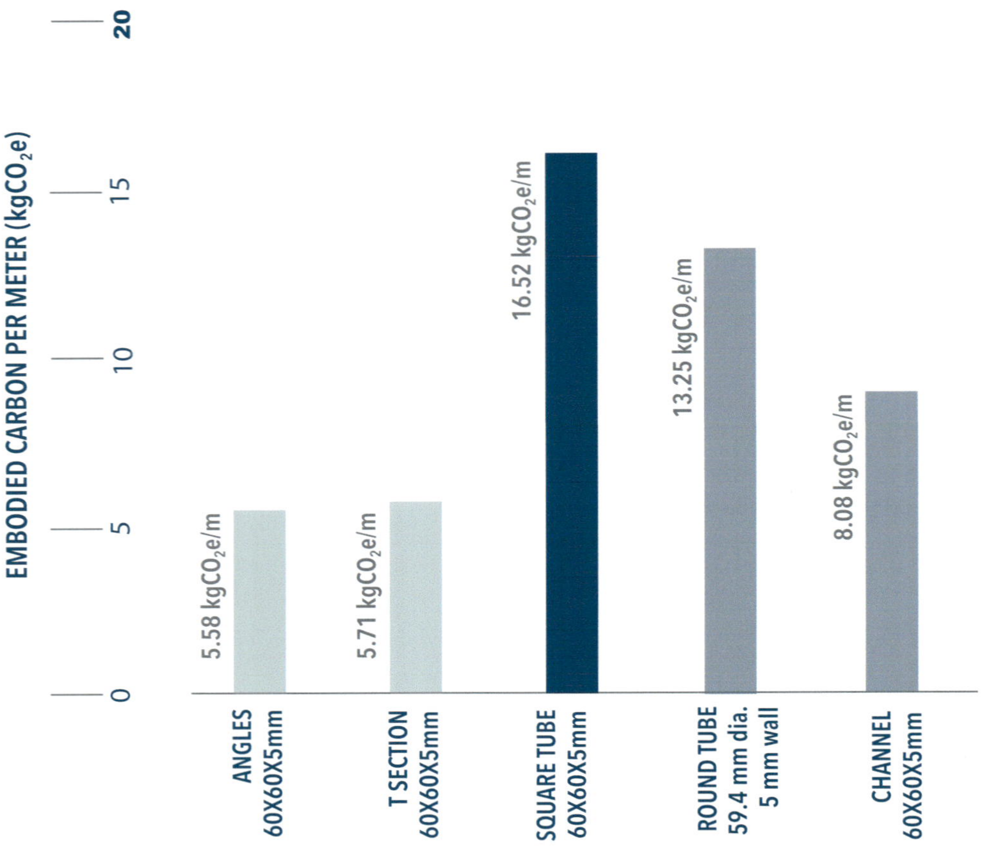

Figure 11.18 This graph is an EC comparison of selected 60mm steel shapes that could fulfill the same structure function in a guardrail or structures. The angle and T section have both a lower weight and a lower embodied carbon, so the difference in the embodied carbon between them and the square and round tubes could be substantial over an entire structure. Image credit: Meg Calkins

Table 11.4 Design and Construction Considerations for Durable Wood Fences

Topic	Consideration
Fence opacity	Follow height and transparency requirements of neighborhood and/or municipality regulations. A solid fence design will provide privacy but will block air flow and cooling breezes. It will also experience greater wind loads. A fence design with staggered boards will allow for some air flow while still obscuring views. Lower open fences can contain children and animals, while allowing for visual connections and air flow.
Adjacent structures	Avoid footings of adjacent structures by placing fence footings at least 15cm away. Locate wood fences away from exhaust outlets to avoid excessive moisture.
Field treatment of end cuts and holes	Field treat all cuts and holes in preservative treated lumber with copper naphthenate, oxine-copper, or inorganic boron.
Posts	Posts should be rated for ground contact and should be bolted to a SS or heavy galvanized post holder embedded in a PIP concrete footing. In wet climates, there should be a minimum 6mm air space between the post holder and the bottom of wood post. The tops of exposed posts should be angled to shed water and field treated. The bottom of the post should be field treated as well.
Footings	Concrete footing depth is a factor of soil type and wind loads. A general rule adds 10% of the fence height to a 60cm depth or 15cm below the frost line – whichever is greater. Avoid directly burying post in concrete or soil.
Rails	Rails in a vertical orientation will span farther between posts than rails in a horizontal orientation. Posts should be spaced at 2 meters o.c. max. when 38mm rails are placed horizontally. Three rails should be used to support fences that are 180cm or taller. The bottom rail should be held a min of 15cm above finished grade.
Boards	Fence boards or slats span horizontally between posts. Nineteen mm thick boards can span 2 meters if they are at least 89mm deep.
Pickets	Pickets are vertical boards attached to the rails. End cuts at the top and bottom of the pickets should be field treated. Pickets greater than 10cm in width should use two fasteners per rail.

(Continued)

Figure 11.19 CMG Landscape Architecture used local, plantation grown redwood, a naturally durable wood, for the Yerba Buena Dog Park fence. Welded wire mesh is the infill for a very low-carbon fence structure. Image credit: CMG Landscape Architecture

Bamboo Fences and Screens

Bamboo is a rapidly renewable material that can offer a low embodied carbon fencing option. As bamboo is technically a grass, it can be harvested on a regular cycle of 1–5 years depending on the species. Bamboo stalks are called *culms*, and the rings along a bamboo culm are called *nodes*. The interior part of the culm at the node is the *nodal diaphragm*, and the bamboo between nodes is the *internode* (Figure 11.23).

The best species of bamboo for fencing and overhead structure applications are *Bambusa balcooa*, *B. nutans*, *B. polymorpha*, *Gigantochloa apus*, *Guadua angustifolia*, *Phyllostachys edulis* (Correal 2020). These species offer the greatest strength and most consistent performance and are used for low buildings and other structures.

Table 11.4 (continued)

Topic	Consideration
Fasteners	Screws and bolts are more durable than nailed connections. Be sure to use metal fasteners that are compatible with the preservative treatment when using wood (Tables 10.6–10.8).
Gate bracing	Gates should have a wood cross brace from the bottom of the gate on the hinged side to the upper corner away from the hinges. Alternatively, a tension cable can extend from the upper corner of the gate on the hinged side to the lower corner away from the hinges. Tension cables should have turnbuckles or a method of tightening the cable as it will stretch over time, particularly in warm climates.
Gate stops	Mount a wooden stop strip on the non-hinged post behind the gate frame to keep it from being pushed beyond flush.

Table 11.5 Material and Sustainability Performance Considerations for Guardrails, Handrails, and Fences

Types of Rails, Fences, and Screens	Material, Design, and Construction Considerations	Sustainability Performance Considerations
Steel guardrails, fences, and screens are comprised of tube stock, shapes, bars, and sheet metal that are SS, weathering steel, powder coated, or galvanized for protection from corrosion.	Detail for deconstruction and reuse of steel members. Specify bolted connections between posts and panels where possible for ease for disassembly. Specify SS or steel coatings that are appropriate for the climate and air quality. Avoid chain link and other steel fencing details that will compromise protective coatings. Refer to Chapter 20 for more information on steel and steel coatings.	Specify high recycled content steel materials. Use smallest members and sections for structural performance. Consider using Ts and angles instead of tube stock as this is less steel for similar structural performance. Avoid use of solid stock unless very small members. Do not embed steel posts in concrete footings, instead, attach with anchor bolts and plates.
Wood guardrails, fences, and screens are constructed from dimensional wood lumber. A wide variety of wood materials and preservative/modification treatments are used in fences.	Can be painted or stained. Detail to avoid ground contact. Detail for deconstruction and reuse of wood members. Refer to Chapters 10 and 18 for more information on types of wood, wood preservative treatment, wood construction, and maintenance. Wood fences can be repaired and, if detailed for durability, compromised wood members can be replaced.	Relatively low-carbon material and continues to store some carbon until it burns or decays. Use certified wood. Avoid tropical hardwood fencing materials. Replacement cycles should be factored into EC accounting for wood fences.

(Continued)

Table 11.5 (continued)

Types of Rails, Fences, and Screens	Material, Design, and Construction Considerations	Sustainability Performance Considerations
Wood and steel guardrails, fences, and screens employ steel for the structural elements and wood for rails, slats, or pickets.	Detail wood to steel connections that are accessible for wood replacement. Detail structures to shed water and not trap moisture against wood, particularly the end grain.	See sustainability performance considerations for wood and steel in this table. Replacement cycles should be factored into EC accounting for wood in wood and steel fences.
Aluminum fences and screens can be prefabricated aluminum fencing systems or custom designed. Many use steel posts.	Specify aluminum finishes that are appropriate for the climate and air quality. Avoid using aluminum in marine environments. Refer to Chapter 20 for more information on aluminum and aluminum treatments and coatings.	Use high recycled content aluminum materials. Use the thinnest and most open aluminum panels and members that are structurally appropriate and meet code. Consider using steel for posts and rails. Do not embed aluminum posts in concrete footings, instead, attach with anchor bolts and plates.
Vinyl (PVC) fences are prefabricated fencing systems. Many are designed to look like white painted wood fences. They are primarily used in residential applications.	Can be used in ground contact. Perform well in humid environments, but vinyl fences can become brittle, crack, or shatter with large temperature and humidity fluctuations. Relatively low maintenance with regular cleaning. Prefabricated panels simplify construction. Refer to Chapter 21 for more information on the health burdens of PVC.	Relatively high embodied carbon. Low initial costs, but these should be weighed against durability as vinyl fences are very difficult to repair. Many product manufacturers state a 20–30-year use life, but do not back this up with substantive warranties. Vinyl is not easily recycled in the United States or Europe, as harmful toxins can be released during the recycling process.
Plastic lumber is used as fence boards, slats, or pickets with wood or steel posts and rails.	Plastic lumber is relatively heavy, so fence structures should be designed accordingly. Long spans of plastic lumber can sag as it has a high coefficient of expansion. Brightly colored plastic lumber will fade over time with UV exposure.	Some plastic lumber has a high post-consumer recycled content. While resource efficient, plastic lumber is relatively high in embodied carbon. Can be recycled.
Composite fences Composite boards, panels, and pickets are used in fences and rails. Usually, the posts and rails are wood or steel.	Most composites are wood and plastic, although plastic and agricultural waste composites are available for fencing as are fused bamboo materials. Refer to Chapters 10 and 18 for more information on composite lumber.	Specify composite lumber with high recycled content. Avoid composite lumber containing PVC. Detail connections for deconstruction and reuse. Not easily recycled.
Bamboo fences Bamboo is used in fencing applications as prefabricated panels, woven wattle fences, and fence structures.	Prefabricated panels are inexpensive. Other bamboo construction can be expensive because it is unusual construction in North America and Europe. Several recent publications could make it more prevalent and supported by codes and regulations. Bamboo fences are discussed in greater detail later in this chapter (Figures 11.23–11.27).	Bamboo is a rapidly renewable material regenerating in 1–5 years. Very low embodied carbon. Durable if sealed or preserved and not used in ground contact. Replacement cycles must be factored into EC accounting.
Wattle fencing weaves green, flexible cut branches between durable vertical stakes or posts to form panels or "hurdles". A variety of species can be woven into wattle fencing.	Posts or uprights for wattle fencing should be decay resistant or treated hardwood or softwood that is 38–50mm in diameter. Weavers should be "green" (recently cut) shoots or small branches that are flexible and easily bent between the posts (Figures 11.28 and 11.29).	Wattle fences are very low embodied carbon, although their service life may be shorter than steel and lumber fences. They will require greater maintenance and some replacement of weavers over their service life.
Living fences Living fences are made from live plant material that is trained, pleached, or woven as it grows to form a natural barrier.	Living fences are time and labor intensive as they require seasonal attention over multiple years to train the plants into a structure. Living fences are discussed in greater detail in this chapter (Figures 11.30–11.33).	They are truly carbon positive structures as they sequester carbon over their lifetime. With care and management, they can last longer than other types of fences.

Table 11.6 Embodied Carbon (EC) of Guardrail Assemblies

Guardrail Components and Rough Dimensions Height 107cm (42") One Bay Width Is 1.5m Center to Center of Posts 30 Year Service Life	EC for One Bay (1.5m) and Ten Bays (15m) of Guardrail ($kgCO_2e$)[1]	Alternatives to Reduce Embodied Carbon
Stainless steel (SS)[2] cable guardrail *Posts and mounting assemblies:* 50x50x3mm square SS tube with SS mounting brackets and plates at top and bottom *Cable assemblies:* 4.76mm SS cable *Top rail:* 38x89mm polymerized lumber (Replacement factor of 1.25)	*One bay:* 22$kgCO_2e$ *Ten bays:* 217$kgCO_2e$	SS angles for posts instead of square tubes reduces EC by 27% to 16$kgCO_2e$
Weathering steel (WS) guard rail *Posts and mounting assemblies:* 50x50x3mm square WS tube post with WS mounting brackets and plates at top and bottom *Panel:* 14 gauge 40% perforated WS panel with u-edging frame mounted with WS clips welded to posts *Top rail:* 38x89mm polymerized lumber (Replacement factor of 1.25)	*One bay:* 28$kgCO_2e$ *Ten bays:* 277$kgCO_2e$	Angles for posts and 16 gauge 58% open perforated WS reduces EC by 28% to 20$kgCO_2e$
Powder coated (PC) steel guardrail *Posts and mounting assemblies:* 50x50x3mm PC square steel tube with PC steel mounting brackets and plates at top and bottom *Vertical pickets:* 19x19x1.6mm PC steel square tube *Horizontal bars:* 25x50x3mm PC steel bars at top and bottom of pickets *Top rail:* 50x50x3 PC steel square tube	*One bay:* 44$kgCO_2e$ *Ten bays:* 443$kgCO_2e$	Angles for posts and pickets, and top rail 50mm diameter reduces EC by 27% to 32$kgCO_2e$
SS[2] and glass panel rail *Posts and mounting assemblies:* 50x50x3mm square SS tube with SS mounting brackets and plates at top and bottom *Panel:* 11.5mm tempered glass panel mounted with SS glass clips *Top rail:* 50x50x3mm SS square tube	*One bay:* 74$kgCO_2e$ *Ten bays:* 740$kgCO_2e$	9mm tempered glass panel and SS angles for posts reduces EC by 22% to 58$kgCO_2e$
SS[2] and perforated aluminum *Posts and mounting assemblies:* 50x50x3mm square SS tube with SS mounting brackets and plates at top and bottom *Panel:* 8 gauge thick 40% perforated aluminum panel with u-edging frame mounted with SS clips welded to posts *Top rail:* 50x50x3mm SS square tube	*One bay:* 60$kgCO_2e$ *Ten bays:* 602$kgCO_2e$	58% open perforations reduces EC by 27% to 44$kgCO_2e$ 58% perforations and 14 gauge perforated aluminum reduce EC by 40% to 36$kgCO_2e$

[1] EC figures are cradle-to-gate and include product stage life cycle phases A1: Material Extraction, A2: Transport to Manufacturing, A3: Processing and Fabrication. They do not include transportation beyond the factory gate. These figures could be reduced in the future with production efficiencies, increased recycled contents, and greater use of renewable energy sources.

[2] 75% recycled content stainless steel is the US average in 2024.

Table 11.7 Embodied Carbon of Fence Assemblies

Fence Components and Rough Dimensions Height 183cm (6'-0") One Bay Width Is 1.83m Center to Center of Posts 30 Year Service Life	EC for One Bay and Ten Bays of Fence ($kgCO_2e$)[1]	Optimized Assemblies to Reduce Embodied Carbon
Wood fence, ACQ-treated southern yellow pine *Posts:* 89x89mm *Top and bottom rail:* 38x89mm *Horizontal slats:* 19x140mm *Post Base:* Stainless steel (SS) *Footings:* 30cm diameter, 65cm deep (Assumes a replacement factor of 2)	*One bay:* 25$kgCO_2e$ *Ten bays:* 250$kgCO_2e$	Woven fence or living fence
Weathering steel (WS)[2] and wood fence *Posts:* 50x50x3mm square WS tube post *Vertical slats:* 19x140mm furfurylated lumber *Top and bottom rails:* 50x25mm WS tube *Footings:* 30cm diameter, 65cm deep (Assumes a replacement factor of 1.25 for furfurylated lumber)	*One bay:* 37$kgCO_2e$ *Ten bays:* 365$kgCO_2e$	T posts instead of square posts All wood fence

(*Continued*)

Table 11.7 (continued)

Fence Components and Rough Dimensions Height 183cm (6′-0″) One Bay Width Is 1.83m Center to Center of Posts 30 Year Service Life	**EC for One Bay and Ten Bays of Fence** ($kgCO_2e$)[1]	**Optimized Assemblies to Reduce Embodied Carbon**
Steel fence, powder coated (PC)[2] *Posts:* 50x50x3mm square PC steel tube post *Panel:* Prefabricated PC steel comprised of square tubes *Footings:* 30cm diameter, 65cm deep	*One bay:* $70kgCO_2e$ *Ten bays:* $704kgCO_2e$	Angles for posts Fewer tubes in panel
Vinyl (PVC) fence *Posts and caps:* 127x127mm PVC *Prefabricated panel and brackets: PVC* *Footings:* 30cm diameter, 65cm deep (Assumes a replacement factor of 1.25)	*One bay:* $74kgCO_2e$ *Ten bays:* $740kgCO_2e$	Steel fence, steel, and wood fence
Aluminum fence, 50% recycled content, powder coated *Posts:* 75x75mm square aluminum tube post *Panel:* Prefabricated aluminum panel comprised of square tubes *Footings:* 30cm diameter, 65cm deep	*One bay:* $86kgCO_2e$ *Ten bays:* $862kgCO_2e$	Steel fence, steel, and wood fence
Woven willow fence *Posts:* Uprights are black locust poles, 40mm dia *Panel:* Weavers are hazel cuttings	*One bay:* $4kgCO_2e$ *Ten bays:* $40kgCO_2e$	
Hedgelayed living fence *Posts:* Uprights are black locust poles, 40mm dia *Panel:* Living hawthorn	*One bay:* $4kgCO_2e$ Carbon sequestered $88kgCO_2e$ *Ten bays:* $40kgCO_2e$	

[1] Embodied carbon figures are cradle-to-gate and include product stage life cycle phases A1: Material Extraction, A2: Transport to Manufacturing, A3: Processing and Fabrication. They do not include transportation beyond the factory gate. These figures could be reduced in the future with production efficiencies, increased recycled contents, and greater use of renewable energy sources.

[2] 75% recycled content stainless steel is US average.

Bamboo can be stained or left to weather to a silver sheen. It is relatively durable in heat, sun, rain, and snow as compared with some softwood lumber.

Drawbacks to bamboo fences are cost, availability, construction, and weathering. Construction with bamboo can be challenging because whole bamboo shoots are not easily connected with standard fasteners such as bolts, nails, and screws.

As bamboo is a natural material, it will absorb water, particularly through the end grain of cuts in a culm. It should be held a minimum of 5cm above finished grade. Pressure treatment of bamboo using boric acid, borax, and boron is the most durable treatment method. End cuts can be sealed with a varnish or sealant or rubbed with linseed oil to increase water resistance (Table 11.8).

Bamboo Fencing Types

There are multiple ways that bamboo is used in fencing and screening applications:

Prefabricated rolls of bamboo attached to a fence frame are the most common and are a widely available type of bamboo fencing. A wood or steel frame is constructed with posts and rails and the bamboo fence rolls are attached to the frame with decking screws or proprietary fasteners. Bamboo should be predrilled to avoid cracking.

Bamboo wattle fences weave culms between closely spaced posts. Posts are treated larger diameter culms, decay-resistant wood, or treated lumber. Bamboo makes a good material for weaving between posts if it is not overly dry. If too dry to bend easily, shoots can be soaked in water to make them more flexible for weaving. Freshly cut bamboo shoots used as posts should be allowed to cure for a few days prior to building the fence. The best species of bamboo for wattling are *Bambusa balcooa*, *B. nutans*, *B. polymorpha*, *Gigantochloa apus*, *Guadua angustifolia*, and *Phyllostachys edulis* (Correal 2020). Wattle fences are discussed in greater detail later in this chapter.

Pole construction bamboo fence structures are often inspired by traditional Japanese fence construction (Figures 11.24 and 11.25). Connections are lashed, bolted, pinned, or doweled. Internodes of some structures are filled with cement mortar. Refer to Figures 11.26–11.27 for typical modern bamboo connections.

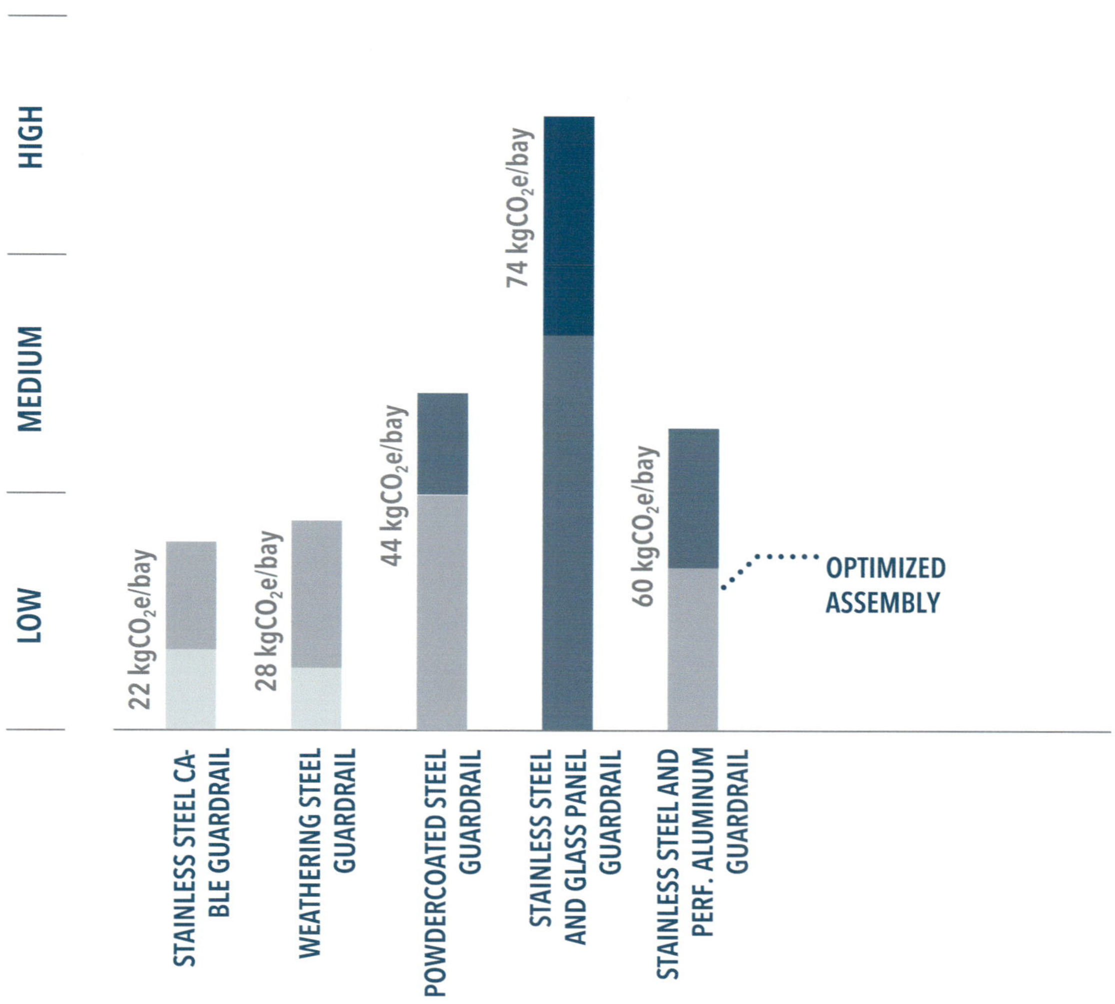

Figure 11.20 Embodied carbon results for 1.5m bays of guardrail assemblies with a 30 year service life. The guardrail height is 107cm (42"). The top part of each bar and the number shows EC results for original assemblies, and the lower part of the bar shows optimized assemblies. Image credit: Meg Calkins.

Maintenance and Repair of Bamboo Fence and Screen Structures

Regular cleaning and sealing every 2–3 years will extend the life of bamboo fence and screen structures (Table 11.8). Sealants should provide resistance to moisture, UV protection, and protection from wood consuming insects such as termites.

Repair of bamboo structures can be challenging as the attachment techniques are different than standard wood joinery. If bamboo stalks are lashed together in a prefabricated panel, it is not easy to replace one or two broken stalks, instead, the panel may need to be replaced.

Wattle Fencing

Wattle fencing weaves green, flexible cut branches between durable vertical stakes or posts to form panels or "hurdles". A variety of species can be woven into wattle fencing (Figures 11.28 and 11.29).

Posts or uprights for wattle fencing should be decay resistant or treated hardwood or softwood that is 38–50mm in diameter. Posts can be finished poles or rough branches with bark remaining or removed. The posts should be pounded into the soil 30–60cm or more for taller structures. Post spacing will vary widely with the type of cuttings that are woven between them. Posts are usually spaced

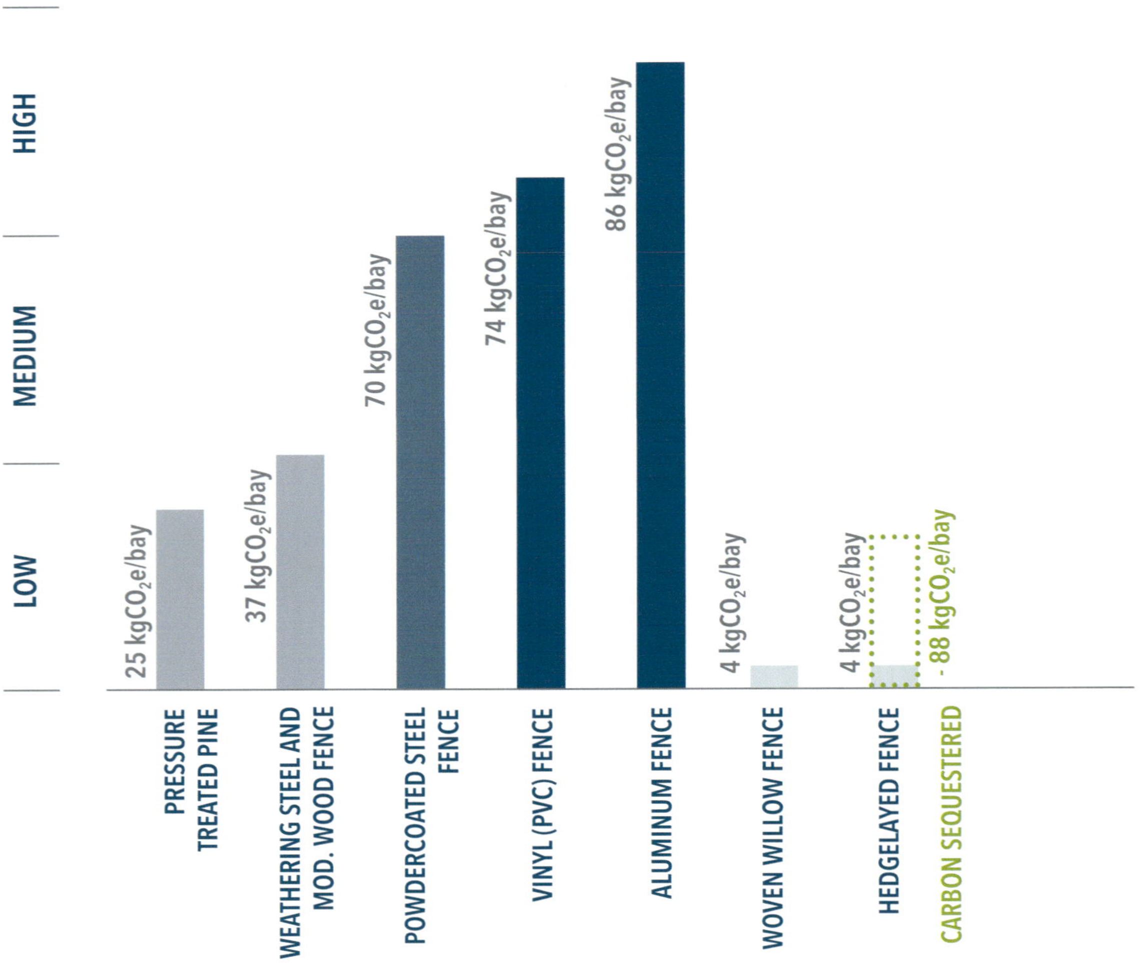

Figure 11.21 Embodied carbon results for 1.83m (6'-0") bays of a fence assemblies with a 30 year service life. The fence height is 1.83m (6'-0"). Replacement factors are included in the calculations where applicable. Concrete footings for fence posts are 65cm deep. Image credit: Meg Calkins.

25–100cm apart, but they can be as much as 2 meters on center if they are larger posts and the woven material is larger as well. Appropriate wood for stakes includes black locust, willow, cedar, alder, hazel, maple, and birch.

Weavers should be "green" (recently cut) shoots or small branches that are flexible and easily bent between the posts. Cuttings for wattle fencing are typically 25mm or less in diameter with a length of 1.8–2.4 meters. Weavers are most commonly willow (*Salix sp.*) or hazelwood cuttings, but bamboo, birch, poplar, ash, fruit tree, and other year old shoots can be used. If a tree is cut near the ground, it will "coppice" by regenerating long straight shoots or whips that make ideal weavers.

Living Fences

Living fences are made from live plant material that is trained, pleached, or woven as it grows to form a natural barrier. Sometimes simple hedges are called living fences, however this section addresses living structures that are trained to form structural barriers or enclosures.

Living fences enclose and define space, confine animals and children, and provide the same ecosystem services as other types of living plants. They are truly carbon positive structures as their plants sequester carbon over their lifetime. With care and management, they can last longer than other types of fences.

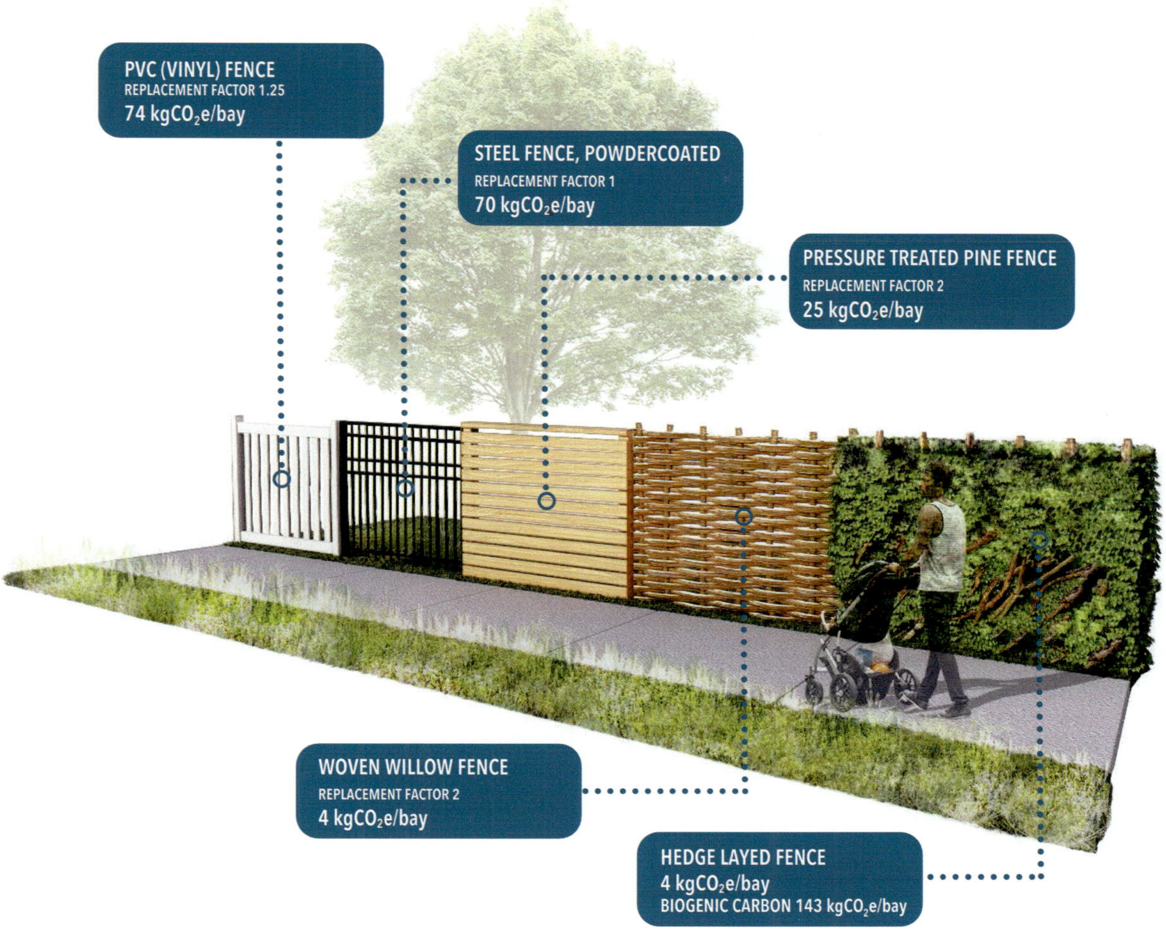

Figure 11.22 Embodied carbon results of selected fence assemblies. Image credit: Lizandro Marcial-Armas, Kathleen Elliott, and Meg Calkins

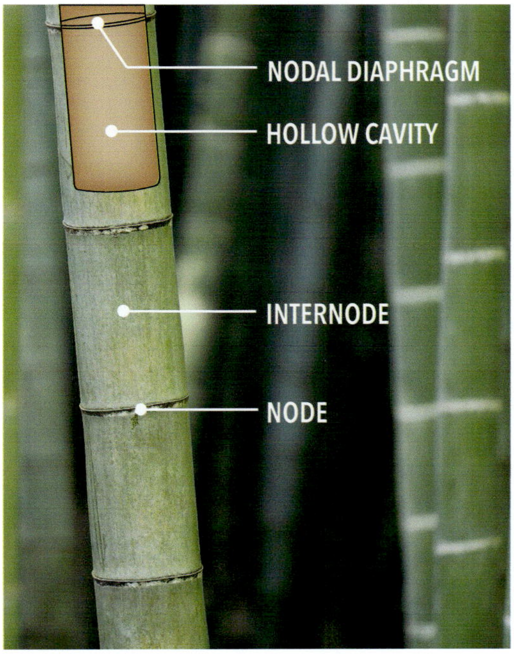

Figure 11.23 The anatomy of a bamboo culm. Image credit: Meg Calkins

Figure 11.24 Bamboo fence with traditional lashed connections. Image credit: Sawitre

Living fences were historically a common landscape structure in agricultural settings in the United Kingdom and Europe, as they were inexpensive, offered food for livestock while containing them, and built quality soil with their leaf litter. Most importantly, they were made with the materials

Figure 11.25 Bamboo pergola with traditional lashed connections. Image credit: Lord Baileys

at hand. Prior to the proliferation of barbed wire, thorny trained hedges such as Osage orange were a simple way to contain livestock. In Europe, living fences were incorporated into landscaped gardens as a romantic interpretation of rural life and as sculptural elements.

Today, living fences are rarely constructed as they require seasonal attention over multiple years to train the plants into a structure. They are far too time and labor intensive for the project delivery settings of commercial construction, and the scale of modern industrial agriculture. Yet while they will likely never find their way into mainstream landscapes, they could be used in residential landscapes or institutional settings where the owner or another stakeholder is willing to engage in training and weaving the plant shoots semi-annually.

The main approaches to living fences can be summarized into three categories: hedge laying, espalier or weaving plants on a built structure, and weaving plants into a plant-only barrier.

Hedge Laying

Hedge laying involves partially cutting through the stems of a line of shrubs or small trees, then bending them down into a diagonal orientation (Figure 11.30 and 11.31). This piece, called a *pleacher*, contains some sapwood and bark. The pleacher will produce new growth near the base of the shrub creating an enclosure. Small shoots off the pleacher are either removed or woven into the structure. While there are several regional styles of hedge laying in the United Kingdom, most

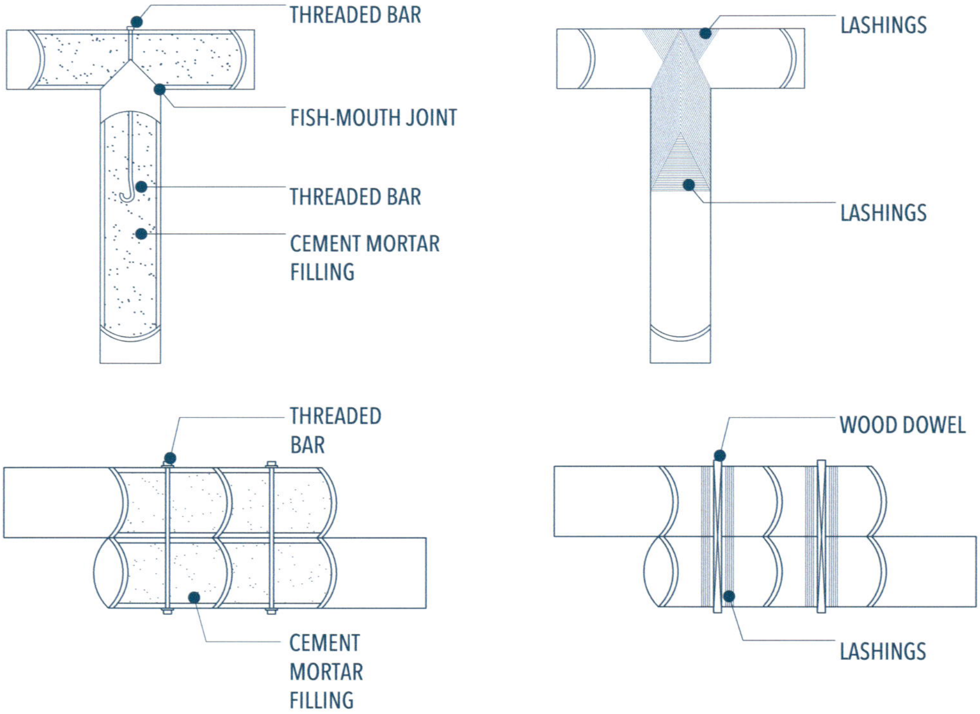

Figure 11.26 Typical orthogonal and longitudinal bamboo structure connections. Adapted from Juan Francisco Correal Daza. Image credit: Kathleen Elliott and Meg Calkins

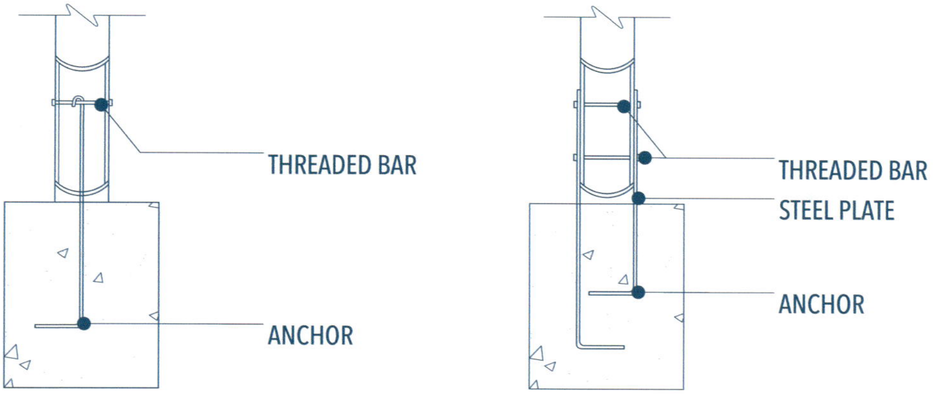

Figure 11.27 Typical post to foundation connections for bamboo structures. Adapted from Juan Francisco Correal Daza. Image credit: Kathleen Elliott and Meg Calkins

Table 11.8 Construction and Maintenance Considerations for Bamboo Structures

Issue	Consideration
Standards for bamboo construction	Refer to *ISO 22157 Bamboo Structures – Determination of physical and mechanical properties of bamboo culms – Test Methods* for bamboo testing standards (ISO 2019) and *ISO 22156 Bamboo Structures – Bamboo culms structural design – Test Design* (ISO 2021) for structural design standards.
Preserving bamboo	Borate emulsion is used to preserve bamboo culms (Chapter 19). Thermal modification treatment is not common but could offer a more durable treatment for culms in exterior use.
Bamboo drying	Bamboo should be dried to less than 20% moisture content prior to use. It can be air dried or kiln dried.
Ground contact	Hold bamboo at least 5cm above grade.
Cutting bamboo	Use a bamboo saw or a fine hack saw. Cut the top of the cane just above a node. The membrane at the node will keep water out of the cane.
Connections	Connections are lashed, bolted, pinned, or doweled. Internodes of some structures are filled with cement mortar. Refer to Figures 11.26 and 11.27 for typical modern bamboo connections.
Lashed connections	Wet twine prior to tying the knot as it will shrink when dry for a tighter connection.
Fasteners	Do not nail directly into bamboo. Instead, create a pilot hole slightly larger than the nail. This will allow for expansion. Nail drivers will prevent crushing of the bamboo.
Construction process	Gloves should be worn when working with bamboo to avoid splinters, except when tying knots.
Maintenance and repair	Regular cleaning and sealing every 2–3 years will extend the life of bamboo fence and screen structures. Ideal sealants will provide resistance to moisture, UV protection, and protection from wood consuming insects such as termites.

Sources: ISO 2021, 2019; Correal 2020

incorporate sawn timber stakes for weaving the pleachers and shoots. (National Hedgelaying Society 2024).

Common plants for hedge laying include hawthorn (*Crataegus sp.*) and blackthorn (*Prunus spinosa*). Other suitable plants are field maple (*Acer campestre*), hazel (*Corylus sp.*), ash (*Fraxinus sp.*), dogwood (*Cornus sp.*), *Vibernum opulus*, spindle (*Euonymus europaeus*), lilac (*Syringa sp.*), and elm (*Ulmus*) (RHS 2024).

Espalier Fences and Screens

Espalier is the horticultural practice of training woody plant growth on a frame or wall in two-dimensional form to produce fruit in a small space. When tree shoots are espaliered horizontally or diagonally on frame or trellis structures, they can perform as fences or spatial enclosures. This is not the type of fence that will contain or exclude animals, instead it is just a perceived barrier (Figure 11.32).

Structures to support espaliered trees can be very simple with posts and wires stretched between them for training the branches. One approach to form a three-tiered espalier embeds two decay-resistant wood posts in soil about 2.25 meters apart. Eyebolts are located at three points up the posts starting at 45cm above grade. Twelve-gauge wire is threaded through the eye bolts and pulled tight with turnbuckles used to take up slack if necessary. The wires form the path that the branches will be trained along (Arnold 2013). This structure could have a top rail for a more finished frame.

Figure 11.28 Wattle fencing weaves cuttings into "hurdles". The cuttings are 12–30mm in diameter and are sourced from coppiced hazel or other trees with vigorous shoots such as *Salix alba* and *Salix viminalis*. Image credit: Meg Calkins

Figure 11.29 Cuttings are woven in bunches of 3–5 around at least three posts, then all the way around a larger end post. Green cuttings should be obtained in the winter or early spring when cuttings are sold in "wads". Image credit: Meg Calkins

Figure 11.30 Hedge laid fence in springtime. Image credit: mtreasure

Figure 11.31 Freshly cut hedge laying plashing. Image credit: daseaford

Figure 11.32 An espaliered apple tree fence in an English kitchen garden. Image credit: Stocknshares

Belgian fences are a style of espalier where small trees planted closely together are trained to form a diamond lattice pattern. The plants are pruned to result in thin profile that is a screen. Each individual tree is trained in a "Y" form with two arms at 45°. With training over time, the arms interlock and inosculate. The result is a dramatic diamond patterned fence with leaves and fruit filling in the open form. Apples and pears are good species for a Belgian fence, although dwarf rootstocks can be closely spaced for a lower fence (Thevenot 1999).

Willow Fences and Screens

Willow fences are comprised of closely spaced living willow "wands" (*Salix sp.*) that are woven and trained into a barrier screen. Some fast-growing willow species will grow 2 meters per year and can be woven quickly into a fence structure. Willows require full sun and are very tolerant of a wide variety of soils. As their roots tend to crawl toward water, they should not be constructed near septic or sewer lines.

One approach to constructing a live willow fence inserts one-year rods of 2-meter length into soil about 25cm apart and 25–45cm deep. The stems are placed at 45° angles crossing each other and forming a diamond pattern like a Belgian fence. Some joints are tied for stability. As roots develop and the willows grow, new growth is trained and woven into the existing form or allowed to grow vertically to make the fence taller (RHS 2024).

Figure 11.33 Living willow fence in a woven diamond pattern. Image credit: Luis Lindwurm

Willow fences are also grown on simple fence frames of posts and rails. Willow shoots are placed upright and attached to the rails. As the shoots sprout and root, they fill in the spaces to make a thick barrier. With proper training, this type of fence can provide full enclosure (Figure 11.33).

After establishment, willow fences should be pruned twice per year to keep the fence structure or pattern clear, otherwise it will take on the appearance of a hedge. The area at the base of the willow fence should be kept clear of plants and weeds that will compete for water, nutrients, and sunlight (RHS 2024).

References

American Iron and Steel Institute (AISI) (2024). "Sustainability: Recycling and recyclability". www.buildusingsteel.org/why-choose-steel/sustainability/, accessed January 1, 2024.

Arnold, M. (2013). "An easy approach to espalier: Train a fruit tree into a beautiful and practical form with just a few simple steps". *Fine Gardening* 152.

Correal, F.F. (2020). "Bamboo design and construction". In *Nonconventional and Vernacular Construction Materials*. Elsevier Ltd.

International Code Council (ICC) (2021). *2021 International Building Code*. ICC, Inc.

International Code Council (ICC) (2022a). *2021 International Property Maintenance Code*. ICC, Inc.

International Code Council (ICC) (2022b). *2021 International Swimming Pool and Spa Code*. ICC, Inc.

International Organization for Standardization (ISO) (2019). *ISO 22157 Bamboo Structures - Determination of Physical and Mechanical Properties of Bamboo Culms - Test Methods*. ISO 22157:19. ISO.

International Organization for Standardization (ISO) (2021). *ISO 22156 Bamboo Structures: Bamboo Culms Structural Design - Structural Design*. ISO 22156:21. ISO.

Landphair, H.C., and Klatt, F. (1999). *Landscape Architecture Construction*, Third Edition. Prentice-Hall, Inc.

National Hedgelaying Society (2024). "Regional styles". www.hedgelaying.org.uk/pg/info/styles.aspx, accessed March 25, 2024.

RealCedar (2024). "Western red cedar fence specifications". www.realcedar.com/outdoor/fence-specifications/, accessed September 7, 2024.

Royal Horticultural Society (2024a). "Hedges renovation". www.rhs.org.uk/plants/types/hedges/renovation, accessed March 25, 2024.

Royal Horticultural Society (2024b). "Living willow". www.rhs.org.uk/garden-features/living-willow, accessed March 25, 2024.

Southern Forest Products Association (SFPA) (2024). "Joists and rafters span tables". www.southernpine.com/resources/span-tables/joists-rafters/, accessed September 7, 2024.

Thevenot, Peter A. (1999). "An espalier for every garden". *Fine Gardening* (December): 54–59.

US Access Board (USAB) (2015). "Chapter 4: Accessible routes". In *Guide to the ADA Accessibility Standards*. Department of Justice, US Access Board.

US Access Board (USAB) (2016). "Chapter 5: General site and building elements". In *Guide to the ADA Accessibility Standards*. Department of Justice, US Access Board.

Zimmerman, A. (Ed.). (2015). *Constructing Landscape: Materials, Techniques, Structural Components*, Third Edition. Birkhäuser.

Materials for Resilient Sites

CHAPTER 12

Aggregates and Sand Materials

Aggregates and sand make up the largest share of the mass and volume of site structures. They are used in composite mixes such as concrete, cast stone, and asphalt pavement; as base and drainage materials for pavements, walls, and other structures; and as reservoirs for porous pavements and other stormwater control measures (SCMs). They are also used in standalone aggregate pavements with or without binders. In site construction, a simple 100 square meter, 150mm thick concrete pavement on a 150mm aggregate base uses a total of 27m^3 (43.2 metric tons) of aggregate. The base is 24 metric tons while the coarse and fine aggregate in the concrete comprise the remaining 19.2 metric tons.

While the embodied carbon of virgin aggregates and sands is relatively low, they are non-renewable materials, and the habitat impacts from their quarrying, extraction or dredging can be significant. Additionally, at 1200–1750kg per cubic meter, carbon impacts from the transportation of heavy aggregates can be substantial.

Aggregate applications offer one of the best opportunities in site construction for reuse of a wide variety of waste materials. This can reduce the embodied carbon of site structures as well as close material loops contributing to the circular economy.

Aggregate Use

The value of the global aggregate market was estimated at $544 billion in 2023 and is forecast to grow at 7.2% each year to around $951 billion by 2031. Just under 50 billion metric tons of aggregate were used worldwide in 2023. Of this figure, crushed stone is the largest market share at 60% with sand and gravel at 23% and 17%, respectively (Verified Market Research 2024).

The global recycled construction aggregates market is estimated to be valued at $42.8 billion in 2022 and expected to exhibit a compound annual growth rate (CAGR) of 9.5% each year through 2030. These figures include materials from construction demolition such as recycled concrete aggregate (RCA) and recycled asphalt pavement (RAP), but they do not include materials such as steel slag, blast furnace slag, mineral processing wastes, waste glass, or recycled tires (CMI 2023).

Classifications and Sources of Aggregates

There are multiple classifications and sources of aggregate that have varying qualities, performance, and environmental impacts. Construction aggregates are mechanically sorted by particle size through sieving and are categorized as either coarse or fine (Table 12.1). Fines, different from fine aggregates, are materials less than 75µm (0.075mm). This material is referred to as dust or filler in asphalt pavement mixes. AASHTO M147 *Standard Specification for Materials for Aggregate and Soil – Aggregate Subbase, Base, and Surface Courses* designates potential gradations for aggregate surface, base, and subbase courses (Table 12.1 and Figure 12.1).

Crushed stone aggregate is a manufactured product that is produced from larger material excavated from hard rock quarries. It is processed through crushing and grading to specific sizes, and the resulting aggregate has an irregular, fractured, and angular form. This material has good interlocking properties for construction bases.

Construction gravels and sands are primarily mined from alluvial sources by scraping or bucketing directly from the deposits. Naturally occurring gravel and sand particles have a rounded, smooth surface created by water and natural erosion of their surfaces over time. These materials do not interlock like crushed stone aggregate unless they are crushed.

Manufactured aggregates are created to achieve a unique quality like a lighter weight or greater porosity. They can be derived from natural materials that are heated, such

DOI: 10.4324/9781003413974-14

Table 12.1 Definitions Related to Aggregates and Sand

Item	Definition
Aggregates	Aggregate is a broad category of coarse to medium-grained particulate material used in construction, including sand, gravel, crushed stone, and recycled materials. The upper grain size limit is 75mm.
Sand	A non-cohesive mineral granular material where at least 50% of the particles are smaller than 4.75mm and less than 15% is less than 75μm.
Gravel	A non-cohesive mineral granular material where 50% of the particles are larger than 4.75mm but smaller than 75mm. Less than 15% of material is smaller than 75μm.
Crushed stone, crushed aggregate	A crushed non-cohesive mineral granular material where 50% of the particles are larger than 4.75mm but smaller than 75mm. Less than 15% of material is smaller than 75μm.
Fine aggregate	Fine aggregates range in size from 150μm to 9.5mm. Sands are considered fine aggregates.
Coarse aggregate	Coarse aggregates are 9.5mm–37.5mm.
Fines	Fines, different from fine aggregates, are materials smaller than 75μm.
Primary aggregates	Crushed stone, gravel, and sand extracted from the natural environment.
Secondary aggregates	Recycled and reused aggregate materials from sources including recycled crushed concrete, recycle asphalt pavement, recycled glass, recycled brick, and recycled foundry sand (Table 12.5).

Sources: ASTM 2023a; UNEP/GRID-Geneva 2022

Figure 12.1 Crushed stone compared to naturally occurring gravel. Crushed stone is more angular providing for more stable aggregate bases. Image credit: Meg Calkins

as expanded shale, clay, and slate (ESCS), or processed in other ways. ESCS aggregates are strong, yet lightweight with air pockets in the material that are formed by hot gases expanding the material to nearly twice its original size. This material is then crushed, resulting in hard, porous particles with rough surfaces.

Recycled aggregates are derived from a wide variety of recycled, co-product, and waste materials (RCWMs) (Table 12.5). Reuse of waste materials keeps them from landfills; eliminates habitat impacts from quarrying, dredging, and excavation; and often reduces embodied carbon.

Environmental and Human Health Impacts of Aggregate Production

While aggregates and sand are abundant around the world, the sheer volume required for the current rapid urbanization, particularly in developing countries, is increasingly characterized as a "shortage". While aggregates are primarily used in construction, sand is also used to make glass, and as technology increasingly relies on screens and photovoltaic panels, sand is at a premium (Table 12.2).

Sands and gravels are extracted from alluvial sources such as lakes or rivers where their removal can impact water quality, stream flow, and aquatic habitat. Other alluvial sources are floodplains and glacial deposits. Some countries engage in the practice of mining offshore deposits of sand and gravel; however, this is not done in the United States except for beach erosion control and replenishment activities.

The United Nations Environment Programme writes, "We now find ourselves in the position where the needs and expectations of our societies cannot be met without improved governance of sand resources" (UNEP 2022). A 2022 plan by UNEP offers recommendations for optimization in the sand industry including recognition that sand delivers critical ecosystem services, yet it is key to expanding infrastructure and development. They emphasize a transition to circular practices and use of secondary sand materials (Table 12.5) and to ecosystem restoration where primary sand is extracted.

Crushed coarse and fine aggregates are often the overburden and excess process material from dimension stone excavation. They may have fewer habitat impacts than sands and gravels as they are co-products of dimension stone production. While dimension stone quarries disturb habitats, they are generally much deeper and less expansive than the dredging or excavation sites for comparable quantities of sands and gravels.

Table 12.2 Environmental and Human Health Impacts of Aggregates, Sand, and Gravel

Life Cycle Phase and Subprocesses	Air Emissions and Solid Waste	Water Impacts	Ecosystem and Human Health Impacts
Raw material extraction Mining, milling, crushing, dredging, bucketing	CO_2 and other emissions from fossil fuel combustion for vehicles and machinery; dust and particulates from drilling, bucketing, and preliminary crushing; sludge from quarrying operations.	Surface runoff containing dust, sediment, and debris impacts receiving waters; groundwater seepage; water consumption for extraction and dust suppression.	Vegetative cover and soil removed; wildlife habitat loss; noise and dust in surrounding communities; dust inhalation risks for workers without proper safety equipment.
Transportation Loading, conveying, unloading	CO_2 and other emissions from fossil fuel combustion for rail and trucking transport.	Water use for truck washout and dust suppression can impact surrounding receiving waters.	Exhaust and emissions can impact habitats around transportation routes; consumption of non-renewable fossil fuel resources.
Processing Crushing, grinding, screening, washing, drying, heating	CO_2 and emissions from fossil fuel combustion to power equipment; dust and particulates from crushing, grinding, and screening operations; sludge is generated from dust suppression activities.	Substantial process water used for washing, dust suppression, and separation; wastewater can impact surrounding receiving waters.	Dust, noise, and wastewater from operations can impact adjacent aquatic and terrestrial habitats; dust inhalation risks for workers and residents of surrounding communities, particularly from silica-based aggregates.

Table 12.3 Carbon Factors by Aggregate Type

Aggregate Type	Carbon Factors ($kgCO_2e/kg$)	Source and Notes
Aggregate, crushed	0.0046	Average of several product EPDs
Gravel, not crushed	0.002–0.004	Hardy, Carbon Data Set, Version 3, 2024
Aggregate base, 100% recycled	0.002–0.004	Hardy, Carbon Data Set, Version 3, 2024
Sand	0.002–0.024	Hardy, Carbon Data Set, Version 3, 2024

Table 12.4 CO_2e Emissions Factors of Transportation Types

Mode of Transportation	CO_2e Emissions Factor (kg/t-km)	CO_2e Emissions Factors (kg/kg-km)
Medium/heavy duty truck	1.07	0.00107
Rail	0.15	0.00015
Waterborne Craft	0.46	0.00046
Aircraft	7.37	0.00737

Source: EPA 2022

Embodied Carbon of Aggregates

The embodied carbon to produce aggregates is quite low as compared with other construction materials; however, at weights of around 1600kg/m^3, the largest embodied carbon impact stems from their transportation (Table 12.3). For example, road base for a 60 space asphalt parking lot will use 361 metric tons of aggregate requiring around 15 truck loads.

If high-quality aggregates are specified, they can be transported over hundreds of kilometers from other states. The resulting carbon impacts of this can be substantial if they are transported by truck rather than trains and barges. Use of high-quality aggregates can ensure durability of structures, but where they are not required, they should not be specified.

Increasingly, quarries are located greater distances from the urbanized areas that consume the aggregates, as land values are high, and communities do not want the habitat disruption, dust, and noise nearby. Waste materials such as RCA and RAP are available from infrastructure improvements and building demolition in urbanized areas, closer to new development jobsites.

The carbon emissions from transporting 10 metric tons of crushed aggregate 50 kilometers by truck is 535kgCO_2e while the carbon emissions from the production of the 10 metric tons of crushed gravel is just 140kgCO_2e. Table 12.4 lists CO_2e emissions factors for modes of transport by ton-miles and ton-kilometers.

Recycled, Co-product, and Waste Materials for Aggregates

Use of recycled, co-product, and waste materials (RCWMs) for aggregate applications offers one of the best opportunities for closing materials loops and supporting the circular economy in site construction. Use of RCWMs reduces use of virgin resources along with the ecosystem impacts from their excavation (Tables 12.5 and 12.6; Figures 12.2–12.7).

Reuse of waste materials reduces pressure on landfills and can offer cost savings over natural aggregates. Embodied carbon of virgin aggregate excavation is reduced, although energy and emissions from processing recycled aggregates can be similar to that of natural aggregate processing. Use of RCWMs strengthens the market for reuse of waste materials.

There are drawbacks and challenges to the use of RCWMs. Supplies may be limited or variable in regions where landfill regulations are less stringent. Quality of RCWMs

Table 12.5 Recycled, Co-Product, and Waste Materials for Aggregate Applications

Recycled Material	Applications	Performance Considerations
Mineral processing wastes are wastes generated during the extraction and beneficiation of ores and minerals such as waste rock, stone screenings, mill tailings, coal refuse, and spent oil shale. Over two billion metric tons of these materials are produced in the United States each year. Compacted density and absorption vary widely by type.	Can be used in every aggregate application, however, many mineral processing wastes have limited potential for use because of their high impurity content, fineness, trace metal leachability, potential for acid generation, or the remote locations of their occurrence. Select mineral processing wastes are used as clinkers in cement production and as cement substitutes.	Mineral processing wastes should be tested for strength, durability, abrasion resistance, porosity, leachate, and other important qualities prior to use.
Recycled concrete aggregate is produced when concrete is crushed for aggregates in base, subbase, or paving applications. RCA is discussed in greater detail in Chapter 14 "Concrete Materials and Products". *Compacted density:* 1400–1750kg/m^3 *Absorption:* 2–8%	When used as a base or subbase both the coarse and fine RCA materials are used. In new concrete mixes, only the coarse fractions are used as the fines create a significant increase in water demand.	RCA can contain unhydrated cement that results in increased stiffness in bases and subbases when mixed with water. RCA may also contain increased chlorides if deicing chemicals were used on the old concrete pavement.
Reclaimed asphalt pavement (RAP) is produced from milled layers of asphalt pavement or from entire asphalt pavement structures that have been removed and processed. When properly crushed and screened, RAP consists of high-quality, well-graded aggregates coated by asphalt cement. *Compacted density:* 1120–1360kg/m^3 *Absorption:* 1–6%	The predominant use for RAP is in new asphalt pavement, but it is also used in aggregate base, embankment, and fill applications. Coarse fractioned RAP is starting to be used in new concrete mixes. Use of RAP is discussed in the Chapter 15 (Table 15.4) and Chapter 14 (Table 14.8). When used as a base material RAP is crushed, screened, and blended with conventional coarse aggregate or reclaimed concrete aggregate.	The quality of RAP is heavily dependent on the quality of the asphalt pavement from which it was derived. When RAP is heated, the asphalt binder will soften then harden with cooling. This lends stability to bases and new asphalt pavement. Blending RAP with other aggregates is necessary to attain some bearing strengths. RAP can be blended with other reagents to attain stabilized bases. Use of RAP in embankments or fill applications is a good use for RAP that is comingled with other materials.
Air-cooled blast furnace slag (ACBFS) is an industrial co-product from iron blast furnaces. It is the remaining molten material after the extraction of pig iron that is directed into cooling pits. Once cooled, it is crushed and graded. *Compacted density:* 1120–1360kg/m^3 *Absorption:* 1–6%	ACBFS can be used in a wide variety of applications such as bases, asphalt pavement, and as coarse aggregate in concrete mixes. It is also ground for use as a cement substitute and raw material in cement production.	ACBFS is available in regions with iron ore processing, although as use of recycled steel grows, supplies of slag will be reduced. When crushed, it is angular and provides good interlock with strength similar to hard igneous rock. Pelletized blast furnace slag can be used as a lightweight aggregate.
Steel furnace slag (SFS) is a co-product of steel manufacture that is produced during the separation of molten steel from impurities in steel making furnaces that solidifies on cooling. *Unit weight:* 1600–1920kg/m^3 *Absorption:* higher than natural aggregates.	The qualities and applications of SFS are controlled by the method of processing. Most SFS can be used in asphalt pavements and base applications, but some SFS is not suitable for use in concrete mixes as it can lead to expansion and deterioration. SFS should be tested prior to use to determine if expansion will be a problem for structures using the material.	Can exceed aggregate requirements for base applications. Provides good frictional properties in asphalt surface courses. The pH leachate from steel slag can exceed 11. This can corrode nearby steel and aluminum as well as impact adjacent soils and water. Drainage from steel slag aggregates can result in the formation of tufa like precipitates, which are powdery deposits that consist primarily of calcium carbonate.

(*Continued*)

Table 12.5 (Continued)

Recycled Material	Applications	Performance Considerations
Foundry sand is high-quality silica sand generated by the ferrous and non-ferrous casting industries. *Dry density:* 1750kg/m^3 *Absorption:* 0.76–6.2%	Foundry sand can be used as a partial replacement of fine aggregate in concrete, in asphalt pavement mixes, in mortar, and as engineered fill material in embankments and flowable fill applications.	Foundry sand is normally of higher quality than typical bank run or natural sands. It displays good resistance to weathering. Foundry sand should be tested for its impact on the mix's performance and for the presence of heavy metals prior to use.
Recycled tires are a ubiquitous waste material that are shredded, chipped, or ground into a variety of sizes for use in a variety of applications. Steel is stripped out of the tires during processing. Tire chips: 13–76mm Ground rubber: 0.15–19mm Crumb rubber: 0.075–4.75mm *Compacted unit weight:* 322–725kg/m^3 *Absorption:* up to 6.7%	Ground rubber and crumb rubber are used as an asphalt binder modifier and a fine aggregate substitute in asphalt pavements. These rubberized asphalt pavements are discussed in Chapter 15 "Asphalt Pavements Materials". Shredded or chipped tires are used as a lightweight fill material in embankments. Crumb rubber use in concrete mixes is in the research stage.	Heating recycled tires can produce PAH emissions that are hazardous to human health. Heavy metals may be present in recycled tires. Accidental combustion could be an issue when shredded or chipped tires are used in embankments. In concrete applications, there can be poor bond strength between the RA and the cement matrix. Research on pre-stressing the RA is ongoing.
Waste glass is primarily post-consumer glass that is hand sorted by color and crushed for size reduction into cullet. *Compacted density:* 1120kg/m^3 and higher *Absorption:* 0.06%	Glass cullet is used in bases, asphalt pavements, and to a far lesser degree in concrete mixes. Glass cullet exhibits properties like that of gravel or sand. The smaller the glass cullet particles, the less angularity they possess. In base applications, cullet should be limited to the replacement of fine aggregate sizes as gravel size glass cullet (greater than 4.75mm) exhibits less strength than typical aggregates.	Consistent supplies of waste glass can be a challenge in smaller communities or those with less robust recycling programs. Alkali silica reaction can occur between certain types of glass cullet and high alkali portland cement. Glass has low thermal conductivity, so a "glassphalt" pavement will take longer to cool, however greater than 15% glass in asphalt surfaces can reflect substantial heat from the sun.
Crushed brick and crushed vitrified clay come from discarded bricks and clay products during manufacturing and construction processes. Chipped, deformed, or unacceptable products are crushed for use as aggregate. *Compacted density:* 1120kg/m^3 *Absorption:* 0.06%	Crushed brick is very strong and can be used in base applications or, if crushed to smaller sizes, as a surface course for unbound aggregate pavements.	Sources of crushed brick are co-located with brick manufacturing facilities, so it is not available everywhere and in large quantities.

Sources: RMRC 2024; Van Dam 2015

Table 12.6 Applications for Recycled Aggregates

Application	Granular Base	Embankment or Fill	Aggregate Pavement Surface	Concrete	Asphalt Pavement
Mineral processing wastes	some	some	some	some	some
Recycled asphalt pavement	yes	yes	no	some	yes
Reclaimed concrete aggregate	yes	yes	yes	yes	yes
Air-cooled blast furnace slag	yes	yes	yes	yes	yes
Steel furnace slag	yes	some	no	no	yes
Foundry sand	no	yes	no	yes	yes
Tire shreds and chips	no	yes	yes	no	no
Crumb rubber	no	no	no	some	yes
Ground rubber	no	no	no	no	yes
Waste glass	yes	no	no	some	yes
Crushed brick and clay products	yes	yes	yes	no	some

Sources: RMRC 2024; Van Dam 2015

Figure 12.2 Typical recycled aggregates from construction and demolition or other recycling sources. Image credit: Meg Calkins

Figure 12.3 and Figure 12.4 Assemblage Landscape Architecture, Field Form, and Sunnyside Masonry used waste masonry materials to create pavers celebrating NYC's legacy building materials. Image credit: Assemblage Landscape Architecture

Figure 12.5 The finished pavers, revealing "Bits of NYC", will be used in local projects. Image credit: Assemblage Landscape Architecture

Figure 12.6 and Figure 12.7 Design Workshop used recycled asphalt pavement as aggregate pavement for the driveway at the Highlands Retreat in Aspen, Colorado. The drive was carefully sited to minimize tree removal. Image credit: Brandon Huttenlocher/Design Workshop, Inc.

can vary widely, and quality assurance and quality control can be difficult to implement without testing the materials. Some RCWMs may contain contaminants such as heavy metals or high lime contents that can impact worker health and/or adjacent ecosystems, soils, and waterways.

RCWMs should be locally obtained otherwise their transportation impacts could outweigh their environmental benefits. When RCRMs are used in concrete or asphalt pavement mixes, RCWMs should be well graded and tested for contaminants, heavy metals, and other components that could impact the integrity of the mix and the performance of the resulting structure. Types, qualities, applications, and performance of recycled aggregates from RCWMs are discussed in Table 12.5.

On-Site Aggregate Recycling

On large sites with existing hardscape or buildings to be removed, it can be economically viable to bring crushers and grading equipment to the site to produce base aggregate, drain rock, and even larger pieces for rip rap. Compact crushers are increasingly available and the cost savings from not hauling large amounts of debris off site to disposal facilities along with savings from purchasing aggregates for the new construction will likely pay for the on-site crushing.

Another option with existing pavement to be demolished is to remove the pavement layer and leave the base intact. New pavement can be applied over the top of the existing base. The existing base should be carefully evaluated by geotechnical and/or civil engineers.

Lightweight Aggregates and Fill

Lightweight aggregates and fill are used in site construction, infrastructure projects, and green roofs to reduce loads on structures and soil. Lightweight fill allows for soil and aggregate use on structures such as green roofs and green spaces over structures. It can be used in situations where subgrade can't support the weight of conventional aggregates or imported soil and it reduces the dead load of assemblies lowering the chances of slope failures and settlement. It is also used to improve the earthquake resistance of structures.

There are many organic, mineral, and fossil fuel based materials used as lightweight aggregates and fills, although some are synthetic. The most commonly used materials are

Table 12.7 Approaches for Improving the Sustainable Performance of Aggregates

Objective	Strategy	Economic Impact	Environmental, Human Health, and Performance Impact
Reduce the amount of virgin aggregate used	Use aggregates derived from RCWM sources (refer to Tables 12.5 and 12.6).	Costs vary. Some recycled aggregates could be lower cost, while others may be higher. Recycling waste materials will save landfill disposal costs.	Reduced energy, emissions, and resource use for virgin aggregates; beneficial reuse of waste material; some RCWMs may pose issues from pH, toxicity, and contaminants; use of RCWM can require additional knowledge and care to ensure performance. Testing is recommended.
	Optimize aggregate bases by using the thinnest section recommended to meet performance needs by the geotechnical and/or civil engineer. Do not over design or build in contingencies.	Using less material will save construction costs.	Potentially reduced energy, emissions, and resource use for production of virgin aggregates; beneficial reuse of waste.
	Stabilizing dense graded bases with asphaltic emulsion, small amounts of cement or other stabilizers may allow them to be thinner.	The cost will be similar to or even higher than a thicker base due to the cost of stabilizers.	The EC of a stabilized base could be similar to cement, and asphalt will add to the carbon footprint.
Reduce the impacts of virgin aggregate acquisition and processing	Specify aggregates from companies that disclose and reduce ecosystem impacts from quarrying, dredging, excavation, and processing activities.	Costs of aggregates could be higher.	Reduced habitat destruction, and pollution impacts.
	Specify aggregates from companies that disclose and reduce energy and water use and emissions from their quarrying, dredging, excavation, and processing practices.	Costs of aggregates could be higher.	Reduced energy use, water use, and emissions.
	Use aggregates derived from RCWM sources on and off site (refer to Tables 12.5 and 12.6).	Costs vary. Some recycled aggregates could be lower cost, while others may be higher. Recycling on-site waste materials will save disposal costs.	Potentially reduced energy and resource use for virgin aggregates; beneficial reuse of waste material; some RCWMs may pose issues from pH, toxicity, and contaminants; use of RCWM can require additional knowledge and care to ensure performance. Testing is recommended.
Reduce the impacts of aggregate transportation	Use locally available materials. Write local sourcing requirements into specifications. Make sure it is not just the distribution center that is local.	Can reduce initial cost, may influence structure service life if local aggregates are of lesser quality. Strengthens local economies.	Reduced embodied carbon and related emissions of transport. Use recycled aggregates in regions where natural aggregate sources are far from site.
	Minimize transportation impacts by maximizing use of barge and rail transport and minimizing truck transport.	Can reduce cost. Strengthens local economies.	Reduced embodied carbon and related emissions of transport.

Sources: RMRC 2024; UNEP-GRID-Geneva 2022; Van Dam 2015

shown in Table 12.8. Other lightweight fill materials are chipped bark, sawdust, dried peat, fly ash, slag, cinders, and pumice.

Embodied Carbon of Lightweight Aggregates and Fills

The embodied carbon of lightweight fills is substantially higher than full weight natural aggregates or recycled material aggregates due to their additional processing and sometimes heating. While lightweight aggregates and fills are required in some conditions, they should not be used unless they are absolutely necessary. At the same time, if their use allows for lower quantities of structural materials, they may likely be worth it (Table 12.7).

When comparing embodied carbon of different types of lightweight fill, it is important to note that the $kgCO_2e$ is meaningless without factoring in the weight of the fill. For instance, cellular concrete has just $0.323kgCO_2e/kg$, but its density is much higher than other lightweight products at

Table 12.8 Lightweight Aggregate and Fill Materials

Material	**Density** kg/m^3	**EC** kgCO_2e/kg	Performance Considerations
Natural aggregate	1600	0.014	Not lightweight. Provided for reference.
Expanded polystyrene (EPS) – EPS, also called geofoam, is a rigid cellular foamed polystyrene board or block material.	45.7	7.543	1–2% of the weight of soil EPS contains chlorinated or halogenated flame retardants that pose health risks such as cancer, endocrine disruption, and reproductive toxicity (GSPI 2024).
Extruded polystyrene (XPS) is a rigid cellular foamed polystyrene board or block material.			1–2% of the weight of soil More waterproof than EPS Higher compressive strength than EPS XPS contains chlorinated or halogenated flame retardants that pose health risks such as cancer, endocrine disruption, and reproductive toxicity (GSPI 2024).
Lightweight cellular concrete (LCC), also called foam concrete, is a type of concrete made from water portland cement and pre-formed foam.	600	0.323	Heavier than other options Very high EC, higher than EPS and XPS Low cost Foam types vary but may pose health risks.
Lightweight foamed glass aggregate is produced from ground recycled glass powder mixed with a foaming agent then heated in a kiln.	180	0.622	10% of the weight of soil but substantially heavier than EPS and XPS Non-toxic, non-combustible. Highly porous
Expanded shale, also called vitrified shale, is lightweight, gray gravel made from shale that has been crushed and fired in a rotary kiln at temperatures over 1100°C. As it cools, gases escape leaving pockets and pores.	480	0.256	10–20% of weight of soil 1/3 the EC of XPS and EPS Used in lightweight soil blends Highly porous Non-toxic
Lightweight expanded clay aggregate is a lightweight aggregate made from plastic clay with little or no lime that is dried, heated, and burned in rotary kilns at 1100–1300°C. As it cools, gases escape leaving pockets and pores.	250–719	Similar to expanded shale	10–20% of weight of soil Used in lightweight soil blends Highly porous Non-toxic
Pelletized or foamed blast furnace slag is produced by cooling and solidifying molten blast furnace slag in a spinning drum with air and water. Rounded pellets are created (Pasetto 2023).	840		About half the weight of natural aggregate or soil New product Limited to steel producing regions Highly porous Non-toxic
Tire derived aggregate (TDA) is made from shredded or chipped recycled tires.	322–725		Widely abundant material Some concern about heavy metals on the tire leaching into the environment Accidental combustion could be an issue when shredded or chipped tires

Sources: RMRC 2024; Hardy 2024; Nicholson 2015; Van Dam 2015

600kg/m^3. Ten cubic meters would have an EC of 1.94tCO_2e. In comparison, the EC of lightweight foamed glass is nearly twice that of cellular concrete, but the total EC for 10m^3 is just 1.12tCO_2e due to its much lighter weight.

References

ASTM Standard D448/D448M. (2022). *Standard Classification for Sizes of Aggregate for Road and Bridge Construction*. West Conshohocken, PA: ASTM International. www.astm.org.

Coherent Market Insights (CMI) (2023). *Recycled Construction Aggregates Market Size to Surpass $88.4 Billion by 2030, Exhibiting a CAGR of 9.5%, Says Coherent Market Insights (CMI)*. CMI.

Green Science Policy Institute (GSPI) (2024). "Some flame retardants of concern". https://greensciencepolicy.org/harmful-chemicals/, accessed August 29, 2024.

Hardy, C. (2024). *Carbon Data Set Landscape – Version 3*, September 15. Sasaki Associates, Inc.

Islam, M.M., et al. (2022). "Design and strength optimization method for the production of structural lightweight concrete: An experimental investigation for the complete replacement of conventional coarse aggregates by waste rubber particles". *Resources, Conservation and Recycling* 184: 106390.

Nicholson, P.G. (2015). "Chapter 17: Lightweight fill materials". In *Soil Improvement and Ground Modification Methods*. Ed. P.G. Nicholson. Butterworth-Heinemann, 413–431.

Pasetto, M., et al. (2023). "The use of steel slags in asphalt pavements: A state of the art review". *Sustainability* 15.11.

Recycled Materials Resource Center (RMRC) (2024). "Materials". https://rmrc.wisc.edu/ug-materials/, accessed March 31, 2024.

UNEP/GRID-Geneva (2022). *What Is Sand – Results from a UNEP/GRID-Geneva Expert Discussion*. Technical Report GSOI-GSA-2022-002. UNEP.

US Environmental Protection Agency (US EPA) (2022). *Emission Factors for Greenhouse Gas Inventories*. US EPA, April 1.

US Geological Survey (USGS). (2023). *Mineral Commodity Summaries 2023*. US Department of the Interior, USGS.

Van Dam, T.J., et al. (2015). *Towards Sustainable Pavement Systems: A Reference Document*. FHWA-HIF-15-002. Federal Highway Administration.

Verified Market Research (2024). *Aggregates Market by Product Type (Crushed Stones, Manufactured Sand, Natural Sand), by Application (Commercial, Residential, Industrial, Infrastructure), and Region for 2024-2031*. Report ID 156909. Verified Market Research.

CHAPTER 13

Stone Materials

Natural stone offers many positive attributes as a construction material for walls, pavements, embankments, and other site features. Its tactility and appearance evoke biophilia and, with locally sourced stone, a deep connection to a place. Its geologic form memory reveals natural processes from across millennia, and, reflecting the local geology, it can seamlessly settle a built landscape into the larger natural context.

Stone can inspire the design of structures as form often follows the type, shape, and qualities of the stone products used. Stone construction materials take many shapes from rough boulders to polished dimensioned granite. Imperfections, veining, and color variations in the material provide natural character.

Stone holds great potential to be a sustainable construction material as it is extremely durable, strong, abundant locally almost everywhere, easily reclaimed and reused, and relatively low embodied carbon when quarried and fabricated regionally. But stone can be more carbon intensive than PIP concrete or cast stone (precast concrete) if it is imported.

In 2023 86% of dimension stone used in the United States was imported, primarily from Brazil, China, and India (USGS 2024). This is due to substantially lower labor costs, fewer worker safety regulations in these countries, and inexpensive shipping.

It is not unusual for granite harvested in Minnesota to be sent to China for cutting and finishing, then installed in Paris – offering no connection to the local landscape and creating a substantial carbon footprint. This international path that a lot of dimension stone takes is often the single largest environmental impact of its use – and the determinant of whether stone can be considered a sustainable material or not.

An additional concern with imported aggregate, rubble, and dimension stone is potential human rights abuses from unsafe working conditions, lack of protective equipment, extremely low pay, forced labor, and, to a lesser extent, child labor. These issues are most prevalent in parts of India, Africa, and Brazil, and even in China, but to a lesser degree. It is very difficult to track stone's path to market through all the processing stages and distributers, so ethically sourced international stone can be difficult to verify.

Another determining factor for the sustainability of stone is how it is used in site structures. When quarry blocks are stacked without mortar for a retaining wall, the embodied carbon can be quite low. When dimension stone is applied as a veneer over a PIP concrete cantilever retaining wall, the carbon footprint grows exponentially. Likewise, dimension stone pavement on a sand leveling bed and compacted aggregate base is far lower than the same dimension stone mortared to a concrete slab.

Use of stone can be challenging in our modern construction industry where mass-produced inexpensive materials in standard sizes and formats are valued. It is easy to spec concrete, cast stone, or bricks from a company website and know exactly what you will get. As stone is a natural material, it is highly variable in appearance, performance, and availability if one is prioritizing regional over imported stone. Designers often need to visit quarries early in the design process to find and tag excess process material slabs, blocks, and boulders that will inform their design. Stone color and texture can vary by location within the same quarry, so specs need to be written to address this, and extra materials for replacements must be procured and stored (Tables 13.1–13.3; Figures 13.1 and 13.2).

Stone Use Statistics

In 2023, approximately 2.3 million metric tons of dimension stone was used in the United States valued at around 3 billion US dollars. Eighty-six percent of this was imported from Brazil (23%), China (19%), Italy (16%), and Türkiye (14%). Eighty-eight percent of granite was imported worth

DOI: 10.4324/9781003413974-15

Table 13.1 Stone Definitions

Dimension stone is natural rock material quarried and fabricated for the purpose of obtaining blocks or slabs that meet specifications as to size (width, length, and thickness) and shape. Dimension stone usually has one or more dressed or finished surface.
Dressed stone, also called cut stone, is cut, shaped, and finished to specific dimensions.
Ashlar stone is finely cut, shaped, or worked so that it has smooth and even faces and square edges. An ashlar laying pattern in pavement or walls offsets the stones by one half the stones width. Stones are used in a variety of modular sizes, e.g., 100x200mm, 200x200mm, 200x400mm, 400x400mm.
Stone slabs have two parallel surfaces produced by splitting or shaving in the first milling or quarrying process.
Stone panels are cut stone with large face dimensions in relation to its thickness for use on a building, wall, or frame assembly.
Cladding is a stone facing material on a wall structure that contains other materials. It is non-load bearing other than its own weight and the vertical dead load of the cladding above. It is 30–80mm thick and is attached to a structural wall with mechanical fasteners, ties, or pins. In site construction, stone cladding is primarily attached to a PIP concrete or concrete masonry unit wall.
Thin stone, also called **stone veneer,** is cladding under 50mm thick.
Monumental stone is rock that is quarried and cut in large, thick pieces for the monument and memorial industry, although it is used as signage and other structures in site construction.
Rubble stone is broken quarry stone that are not dimensioned or finished. Sometimes it is very roughly dressed. Rubble stone is an excess process material from dimension stone production, but it is usable in walls and embankments. Usually 50–300mm in size.
Coursed rubble stone is laid with continuous and approximately level courses with varied heights along a wall.
Uncoursed rubble masonry is laid in unlevel courses. Stones are placed for interlock and stability.
Quarry stone, also called **quarry block**, is a rough piece of stone extracted directly from the quarry, sometimes dressed or sawn on one or more faces to the shape of a regular rectangular prism. It is often excess quarry material as it is not usable for dimension stone for a variety of reasons.
Stone boulders are rough, undressed, rounded stones greater than 250mm.
Flagstone, also called **flagging**, are nominally flat slabs of stone in irregular shapes with broken edges used for paving, walks, and patios.
Shaped stone is dimension stone that is processed by carving, sawing, or grinding into specific non-planar applications.
Aggregate is a broad category of coarse to medium-grained particulate material used in construction, including sand, gravel, crushed stone, and recycled materials. Aggregates are discussed in Chapter 12.
Stone fabrication is any process that changes a raw stone piece into its end use form. This includes cutting, splitting, routing, carving, drilling, grinding, and face-finishing.
Stone grain has two definitions (1) a direction in a body of rock along which it is more easily split, broken, or cut; (2) a rock constituent with a distinct identity such as an oolith, a mineral crystal, or a rock fragment in a sedimentary stone.
Stone rift is a consistent direction in a rock body along which the rock is most easily split or broken; it is a grain orientation in stone that is particularly visible in sedimentary stone showing how the stone was originally bedded.

Source: ASTM 2023a

Table 13.2 Classifications of Stone

Classification by Mineral Composition	
Siliceous stone	Siliceous stone are comprised of silica or silicates accounting for nearly 95% of the earth's crust. Siliceous stone includes quartz, feldspar, and mica. They are usually strong and resistant to acids. Siliceous stone typically used in construction are granite, sandstone, bluestone, quartzite, basalt, and slate.
Calcareous stone	Calcareous stone are comprised of calcium carbonates that tend to be vulnerable to acids. Examples typically used in construction are limestone, marble, and travertine.
Classification by Origin and Formation	
Igneous stone	Igneous stone is formed by the solidification of magma, molten mineral mixtures, below the surface of the earth or when it is forced to the surface and cools. Granite, basalt, and gabbro are common types of igneous stone. Igneous stone makes up about 95% of the volume of the earth's crust. These are commonly used for dimension stone due to their very dense structure and low porosity and water absorption.
Sedimentary stone	Sedimentary stone is produced by the weathering and erosion of older rocks or by the deposit of shells from aquatic species. It makes up about 1% of the volume of the earth's crust. Sedimentary rocks are usually hardened and consolidated enough to be used as structural stone. Typical sedimentary stone used in construction are bluestone, sandstone, dolomite, and limestone. Sedimentary stone varies widely in its performance characteristics (Table 13.3)
Metamorphic stone	Metamorphic stone are formed by recrystallization of older rocks when exposed to intense heat, pressure, or both. The stone is altered in appearance, density, crystalline structure, and mineral composition from the original rock. It makes up about 4% of the volume of the earth's crust. The structure of metamorphic stone is dense, and the porosity and water absorption are low. Compressive strength ranges from medium to high. Slate, marble, and quartzite are typical metamorphic stone used in construction.

Source: Marshak 2019

Table 13.3 Types, Properties, and Performance of Stone

Type and Properties	Characteristics and Performance	Considerations for Use
Granite *Igneous stone* Absorption, max: 0.4% Density, min: 2560kg/m^3 Compressive strength, min: 131MPa Abrasion resistance, min: 25Ha[1] *ASTM C615/C615M*	Main minerals of granite are quartz and feldspar resulting in high compressive strength and good chemical resistance. One of the hardest stones, and very resistant to stains and scratches. Very wide variety of colors available with fine to course-grained textures. Uniform appearance, but some colors have spots and veining. Many types of stone are sold as granite including gabbro, diorite, gneiss, and schist.	Used in dimension stone applications for paving and walls. Primarily used as a veneer material in walls due to its high cost. Granites are relatively easy to split, and their surface takes texturing, polishing, and carving. They have good frost resistance. International sources of granite can be less expensive but will have much higher EC and potential for human rights issues in quarrying and processing.
Basalt *Igneous stone* Absorption: 0.1–1.0% Density: 2,700–3,000kg/m^3 Compressive strength: 110–338MPa	Uniform dark and dense stone that is strong and weather resistant. It has dense pores, which make it good for polishing. Silica content of 45–55%.	Used in dimension stone applications of pavements and walls but also used in rubble and crushed stone applications. Behaves and performs similarly to granite (MatWeb 2024).
Sandstone *Sedimentary stone* Absorption, max: 8% Density, min: 2000kg/m^3 Compressive strength, min: 28MPa Abrasion resistance, min: 2Ha *ASTM C616/C616M*	Layered grain, rough textures, uniform appearance across stone. Sandstone color and characteristics vary widely. Gray, yellow, brown, red, rust colors. Lighter colored sandstones contain quartz while dark brown to dark red sandstones are a mix of quartz and feldspar. Breaks along rift and grain, so cleft surfaces are common.	Paving and wall applications. Full bed veneer and thin veneer, dimension stone, quarry blocks. Some sandstones are more porous than others. Porous sandstones should not be used for pavement in cold climates or in areas with very heavy foot traffic. Some sandstones have a cross bedding pattern with gentle diagonal or u-shaped lines.
Limestone *Sedimentary stone* Absorption: 3–12% Density: 1760–2560kg/m^3 Compressive strength: 12–55MPa Abrasion resistance, min: 10Ha *ASTM C568/C568M*	Fine grained to minimally textured, small fragments of marine species and shells can be seen in some limestones Uniform in color; buff, yellow cream, gray colors, and rare pinkish hues. Primarily calcium carbonate (CaC03) and is easily attacked by acids that can dissolve it. Vulnerable to excessive acid rain and pollution. Porous and susceptible to staining.	Full bed veneer and thin veneer, dimension stone, quarry blocks. Should not be used in pavements as it can be slippery when wet, can wear unevenly, and stain. Very workable, good potential for sculptural forms. Quarry blocks and quarry waste are very low EC if used locally. International sources of limestone can be less expensive but will have much higher embodied carbon and potential for human rights abuses in quarrying and processing.
Bluestone *Sedimentary stone* Absorption: 0.5–3% Density: 2400kg/m^3 Compressive strength, min: 69MPa Abrasion resistance, min: 8Ha *ASTM C616/C616M*	A very hard type of sandstone. Smooth to rough finishes. Uniform in color – blue, gray, purple, and rarely pink. Breaks along rift and grain, so cleft surfaces are possible.	Primarily limited to the US Northeast. Durable, yet easy to work. Dimension stone, full and thin veneer, paving, slabs. Thermal, brushed, honed, flamed, tumbled, and natural cleft finishes.
Quartzite *Metamorphic stone* Absorption, max: 1% Density: 2560kg/m^3 Compressive strength, min: 140MPa Abrasion resistance, min: 8Ha *ASTM C616/C616M*	Hard stone with 85% quartz content. Durable and resistant to staining and abrasion; can be polished. White to gray colors with veining. Beige, pink, and blue are less common colors.	Used in paving and walls. Full bed veneer and thin veneer, dimension stone, rubble. Appearance can be similar to marble, but it is much harder and abrasion resistant. Some dense sandstones are sold as quartzite. Can be expensive as it is hard to work like granite, although stratified structure makes it easy to split.
Marble *Metamorphic stone* Absorption, max: 0.2% Density, min: 2600–2800kg/m^3 Compressive strength, min: 52MPa Abrasion resistance, min: 10Ha *ASTM C503/C503M*	Wide variations of veining and colors. Metamorphosized limestone, but denser and harder. Polishes well and colors become more vibrant. Most are some form of white or gray color; pink, green and deep red are less common.	Full bed veneer and thin veneer, dimension stone. Should not be used in pavements as it will be slippery when wet and wears unevenly. Workable. Fine grained and polished textures. Thin slabs can be translucent if lit from behind. Prone to acids and staining.

(*Continued*)

Table 13.7 (continued)

Type and Properties	Characteristics and Performance	Considerations for Use
Slate *Metamorphic stone* Absorption: 0.25% Modulus of rupture, min. along grain: 62MPa; min across grain: 50MPa Abrasion resistance, min: 8Ha *ASTM C629/C629M*	Smooth to rough textures. Fine-grained stone, but generally rough appearance. Relatively uniform colors including blue, gray, green, reddish.	Splits easily into slabs, but layers can spall with freeze thaw cycles as water penetrates under layers.
Travertine *Sedimentary stone* Absorption, max: 2.5% Density, min: 2300kg/m^3 Compressive strength, min: 52MPa Abrasion resistance, min: 10Ha *ASTM C1527/C1527M*	Appearance is roughened by voids and veining. Earth tones to brown colors. Potential for staining and acid attack. Smooth to semi-rough textures.	Full bed veneer and thin veneer, dimension stone. Contains voids in the surface that can be filled during finishing with a synthetic material or left open. If unfilled, they can become embedded with dirt or collect water and be vulnerable to freezing. Should be sealed to avoid staining and acids.
Schist *Metamorphic stone* Absorption: 0.2–2.3% Density: 2510–2880kg/m^3 Compressive strength: 60–150MPa Abrasion resistance, min: 16.5–26.5mm (Carvalho et al. 2013)	Beige, buff, brown, gold, gray. Contains mica and quartz. Sometimes labeled as granite, slate or phyllite.	Pavement and walls, full bed veneer and thin veneer, dimension stone, rubble, and boulders. Higher mica contents can result in a flaky stone. Does not polish well. Eased edges recommended over saw cut. Avoid large overhangs.

Sources: NSI 2022, 2024a, 2024b; ASTM; Zimmerman 2015

[1] Ha is the Abrasion Index for natural stone from ASTM C1353 (2020a).

Figure 13.1 Ten Eyck Landscape Architects used local limestone quarry blocks for a water feature at Kingsbury Commons in Austin. Image source: Adam Barbe

one billion dollars. Forty-five percent of this was imported from Brazil and 23% from India, while other granite quarried in India was sent to China for processing then imported to the United States (Figure 13.11) (USGS 2024).

Dimension stone produced in the United States, valued at $410 million, was just 14% of the dimension stone used in the United States in 2023. There are 176 companies operating 224 dimension stone quarries in 33 states (Figure 13.12). Leading states by tonnage are Texas, Wisconsin, Indiana, Vermont, and Georgia. Fifty-one percent of dimension stone sold or used was limestone with granite (20%), sandstone (14%) and dolomite, quartzite, marble, traprock, and slate following (USGS 2024).

Like the United States, dimension stone and unfinished stone blocks worth 4.3 billion US dollars were imported into Europe and the United Kingdom in 2021, primarily from India and China, with a lesser amount imported from Brazil (OEC 2021). Sources of natural stone are abundant in Europe, but open space is at a premium, land values are high, environmental regulations are strict, and labor is more expensive.

Environmental and Human Health Impacts of Stone Production

The primary environmental and human health impacts from stone include greenhouse gas (GHG) emissions and others from transportation of heavy stone long distances, worker health impacts from stone dust and unsafe working conditions, and habitat disturbance at the point of extraction. The environmental and human health impacts of stone are discussed in this section by life cycle phase (Figure 13.4).

Figure 13.2 Common types of stone used in site construction. Colors within each type of stone can vary widely. For instance, granite is available in pink, red, tan, black, brown, and even, rarely, blue.

Phase A1: Raw Material Extraction of Stone

Stone is primarily extracted from the earth in quarries although some stone is underground mined. Pit quarries are the most widespread and economical as uniform banks of block can be gradually and evenly excavated. Hillside quarries are shaped like an amphitheater with terraces of stone excavation. Mines are used only when open quarries are not feasible as they are much higher cost to operate.

The stone extraction process begins with removal of vegetation and soil from the surface. These ecosystems are lost, and wildlife is displaced. Adjacent waterways can be impacted by sedimentation from runoff and dust.

Next, stone overburden, also called spoil, is removed and stockpiled to access the quality of stone usable for dimension stone. Deposits of quality stone can be quite deep due to cracks and fissures caused by subsurface water seeping into stone and weathering it. It is not uncommon for 10–20 meters of overburden removal before unweathered bedrock is reached and viable dimension stone can be excavated (Figure 13.3).

Figure 13.3 Stone viable for dimension stone can be several meters below grade. The stone that must be removed to get to the viable stone can be substantial; however, landscape and infrastructure projects can use this excess process material in multiple applications. Photo credit: Vallefrias

Stone excavation activities include dynamic splitting, diamond wire cutting, chain cutting, and quarry block cutting. Powering machinery for these activities consumes non-renewable fossil fuels and releases pollution and particulates. Water is consumed for dust control, although it is increasingly recycled and reused.

Excess Process Material and Waste from Quarrying Operations

Waste from dimension stone quarrying operations can be substantial with a review of multiple studies concluding that an average of 51% of all material extracted from dimension stone quarries becomes waste during the extraction operation (Montani 2020). A majority of this waste is solid stone called "excess process material" because it can be reused in multiple applications. Sludge from the combination of dust and water used for dust control, and airborne particulates are waste that is released to the land or air.

Optimization Measures in Stone Extraction

There are several optimization methods that are gaining use at some stone quarries (Jalalian et al. 2021):

Ground penetrating radar is used to identify quality stone deposits without needless removal of overburden only to find that the stone deposit is not viable.

3D Linear Isometric Projection of Rock Mass (LIP-RM) has been used to estimate discontinuities in a quarry bench and determine its recovery rate, reducing waste. This is an emerging technology.

Computer algorithms are being used to estimate stone qualities, cutting geometries, and extraction efficiencies prior to excavation.

Phase A3: Manufacturing, Processing, and Fabrication of Stone

Manufacture of stone products involves multiple steps to cut, shape, and finish large quarry blocks into stone products. In the first processing step, blocks from the quarry are cut into regular blocks then cut into slabs in a cutting machine. Thicknesses of resulting slabs vary by market demand and/or intended application. After sawing, the surface textures (polishes, flaming, honing, bush hammering) of slabs will be applied then the slab will be cut into specific dimensioned sizes. Sometimes there is a third process that shapes the stone into custom shapes or carves the stone with text, patterns, or images. These processes are relatively high energy and water intensive, although water is increasingly recycled.

Some smaller quarry owners don't engage in processing of their stone, instead they sell the large blocks to fabricators who may be located far from the quarry and even overseas. Where stone must be cut to very precise dimensions for applications with tight tolerances, it is more likely that it will be shipped overseas where labor costs are low for skilled stone workers.

Excess Process Materials and Waste in Stone Processing

Quantities of stone waste are significant in stone processing phases, although not as much as from quarrying activities. On average, 41% of stone material that arrives at dimension stone processing plants becomes waste during fabrication operations (Montani 2020). Waste in processing includes crushed blocks resulting from joints, cracks, and poor decisions in cutting geometry; sawdust and sludge resulting from mixing of water and dust during sawing and polishing/texturing; and crushed slabs resulting from the final cutting process.

Again, a majority of this is excess process material that are reused as smaller dimension stone pieces, irregular flagstone pavements, and aggregate (Figures 13.5–13.8). A small fraction is actual waste from sludge and emissions released to air and land.

Optimization Measures in Stone Processing

New fabrication technologies, production efficiencies, and waste recoveries are reducing stone waste and allowing for fabrication of stone products by relatively unskilled laborers.

Computer numerical control (CNC) routing of stone is used to create sculptural stone forms that only very skilled stone workers could have created at considerable expense. Routers do the stone carving from 3D computer programs. While this is still an expensive approach, growth in the CNC industry could eventually allow for more localized stone fabrication (Figures 13.9 and 13.10) (Baca Systems 2024).

Computer algorithms are being used to identify cutting geometries to minimize waste and maximize cutting efficiencies.

Renewable energy sources are increasingly used by European processing companies, but worldwide stone processing energy sources are still primarily fossil fuel based.

Saw technologies can offer energy and water efficiency, and reduced waste. Diamond multi-wire saws (DMWS) have been shown to use 30% less water and 40% less energy with an 80% less material loss (Mendoza et al. 2014).

Reuse of Excess Process Materials

Site and infrastructure projects are uniquely positioned to use large percentages of excess stone process materials. The main challenge to reuse of these materials is the distance of most stone quarries from points of reuse. The cost to haul these materials may sometimes be prohibitive. In addition, supplies can be irregular.

Aggregates are crushed from quarry waste for use in road base, backfill, and drain rock. They are also used as coarse and fine aggregate in concrete, asphalt pavement, and unbound pavement surfacing. While there is an almost

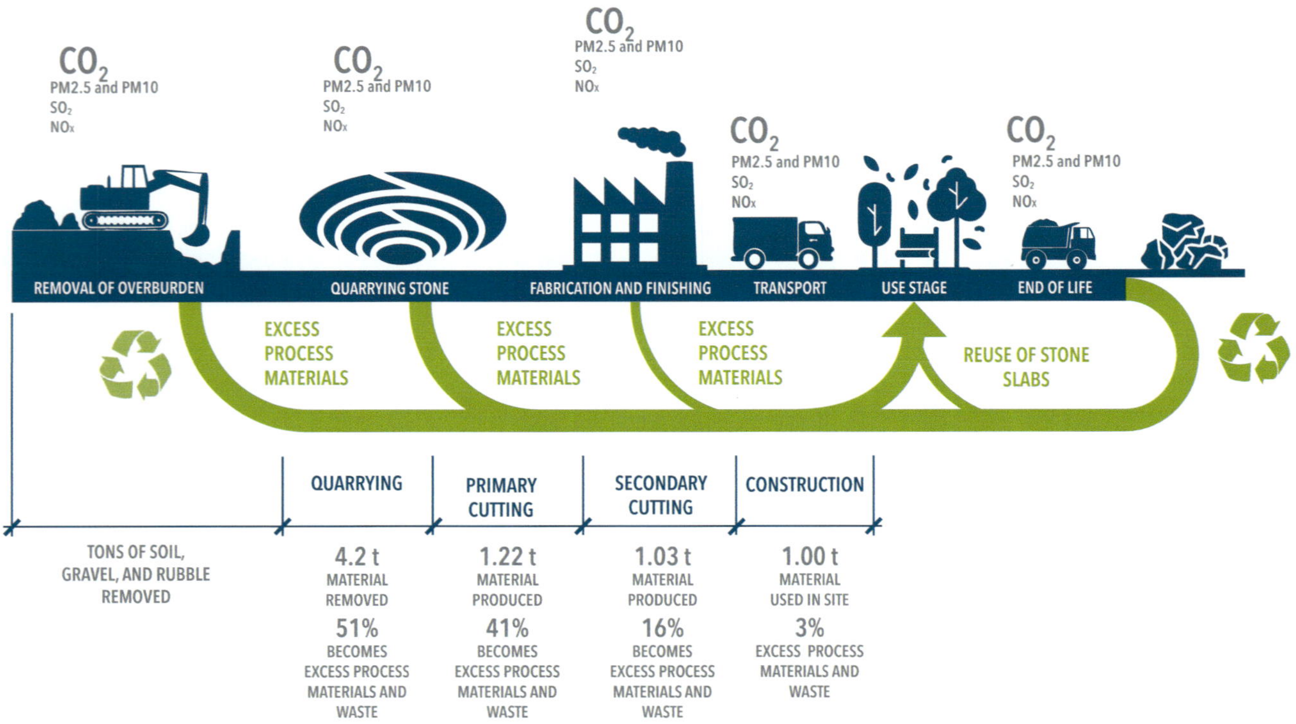

Figure 13.4 The life cycle of stone in site construction. Excess process material quantities and stone waste varies widely by quarry and type of stone. On average it takes 4.2 metric tons of quarried material to yield one metric ton of dimension stone. Sources Montani 2020; Jalalian et al. 2021. Image credit: Meg Calkins

Figure 13.5 08014 architectura used local limestone, known as Sallent stone, from just across the river along with reclaimed railroad ballast in gabion wall structures at the Camí de Cabrianes. This local stone references the color of the surrounding landscape. Image credit: Pol Viladoms

Figure 13.6 A retaining wall along the Canal in Indianapolis is constructed from local Indiana limestone quarry block. Image credit Meg Calkins

Figure 13.7 A variety of stone remnants from other on-site stone work form the base of a pergola at Duke Gardens in Durham, North Carolina. Contrasts in color, texture, and size reference stone types from around this extensive garden. Image credit: Meg Calkins

Figure 13.8 Small stone pieces form a sculptural drainage structure at Duke Gardens in Durham, North Carolina. Image credit: Meg Calkins

Figure 13.9 and Figure 13.10 The team from Gustafson, Porter + Bowman used locally sourced granite from Cornwall for the Princess of Wales Memorial Fountain in London's Hyde Park. The intricate grooves and patterns were cut with CNC milling technology in the United Kingdom. Image credit: Hélène Binet

unlimited supply from quarrying operations, it is still best to use aggregates from locally reclaimed concrete and other post-consumer sources as this keeps these materials out of landfills.

Rubble from quarry waste is sold for use in freestanding and retaining walls, embankments, and erosion control structures. Quarry blocks that are unusable for processing into dimension can be used for retaining walls and seat walls. Quarry waste is often left at the quarry rather than shipped to vendors, so visiting local quarries early in the design process can yield design inspiration and beneficial use of waste materials (Figure 13.5–13.8).

Sludge from quarrying and processing operations is more challenging to reuse at its properties are highly variable. Sludge consists of fine particles, water, and potentially heavy metals and or petroleum hydrocarbons from processing machinery drippings. Sludge has successfully been used in concrete, mortars, fired clay bricks, compressed earth blocks, asphalt pavement, ceramics, glass, rubber, agriculture, stabilized substrates, and engineered fill (ANNIET 2024).

Phases A2 and A4: Transportation of Stone

Transport of stone occurs between quarrying, primary and secondary processing, distribution, then use in the stone structure. During these steps, stone can be transported multiple times, sometimes literally around the world.

It is possible for regional or local stone to be specified resulting in minimal transportation impacts and low embodied carbon. But as 86% of stone used in the United States is imported, it is more likely that stone will travel thousands of kilometers between the quarry, fabrication, and distribution (Table 13.4 and Figure 13.11).

A typical path for some granite involves quarrying in India, processing in China, then distribution in the United States or Europe. The carbon impacts of this path can be around ten times higher than stone that is quarried, processed, and used within the same country (Historic Environment Scotland 2024).

Another challenging path to identify is when stone is quarried within a few hundred miles of the project site but shipped overseas for processing by less expensive skilled stone workers. This practice is not uncommon for North American and European quarries, and unless asked it is not often disclosed to consumers. In North America, the

Table 13.4 Major International Stone Exporters and Importers

	Worldwide Exporters of Crude Blocks and Stone Slabs	Worldwide Exporters of Worked Stone	Worldwide Importers of Stone	North American Sources
Granite	India, China, Norway, Brazil, Italy, Zimbabwe, South Africa	China, Brazil, India, Italy	South Korea, United States, Japan, Germany, Vietnam, France	Vermont, New Hampshire, Minnesota, Texas, North Carolina, Wisconsin, Pennsylvania, South Dakota, Texas, Oklahoma, Georgia, Massachusetts, Indiana, British Columbia, Manitoba, Ontario, and Quebec
Sandstone	India, China, Spain, Italy	India, Spain, Italy	United Kingdom, United States, France, Germany, Romania, Belgium	Nevada, Arizona, Colorado, California, Utah, Wyoming, Montana, New York, Ohio, Ontario, and Quebec
Limestone	UAE, Japan, Malaysia, Vietnam, India	India, Bangladesh,	India, Chinese Taipei, South Korea	Indiana, Texas, Georgia, Ontario, Wisconsin, and Minnesota
Bluestone	N/A	N/A	N/A	Pennsylvania, New York
Quartzite	Brazil, Italy, China, Türkiye, Germany, Canada, India	Brazil, Italy, China	Italy, United States, China, Iceland, Japan, Malaysa	Oregon, Nevada, Utah, and Quebec
Basalt	Norway, India, Italy, China, Belgium, Germany	China, India, Italy	China, Bangladesh, France, Italy, Maldives, Germany, United Kingdom	Western United States
Marble	Türkiye, Italy, Iran, Greece, Portugal, Spain	China, Türkiye, Italy	United States, Vietnam, South Korea, UAE	Tennessee, Vermont, Georgia, Alabama

Source: OEC 2021

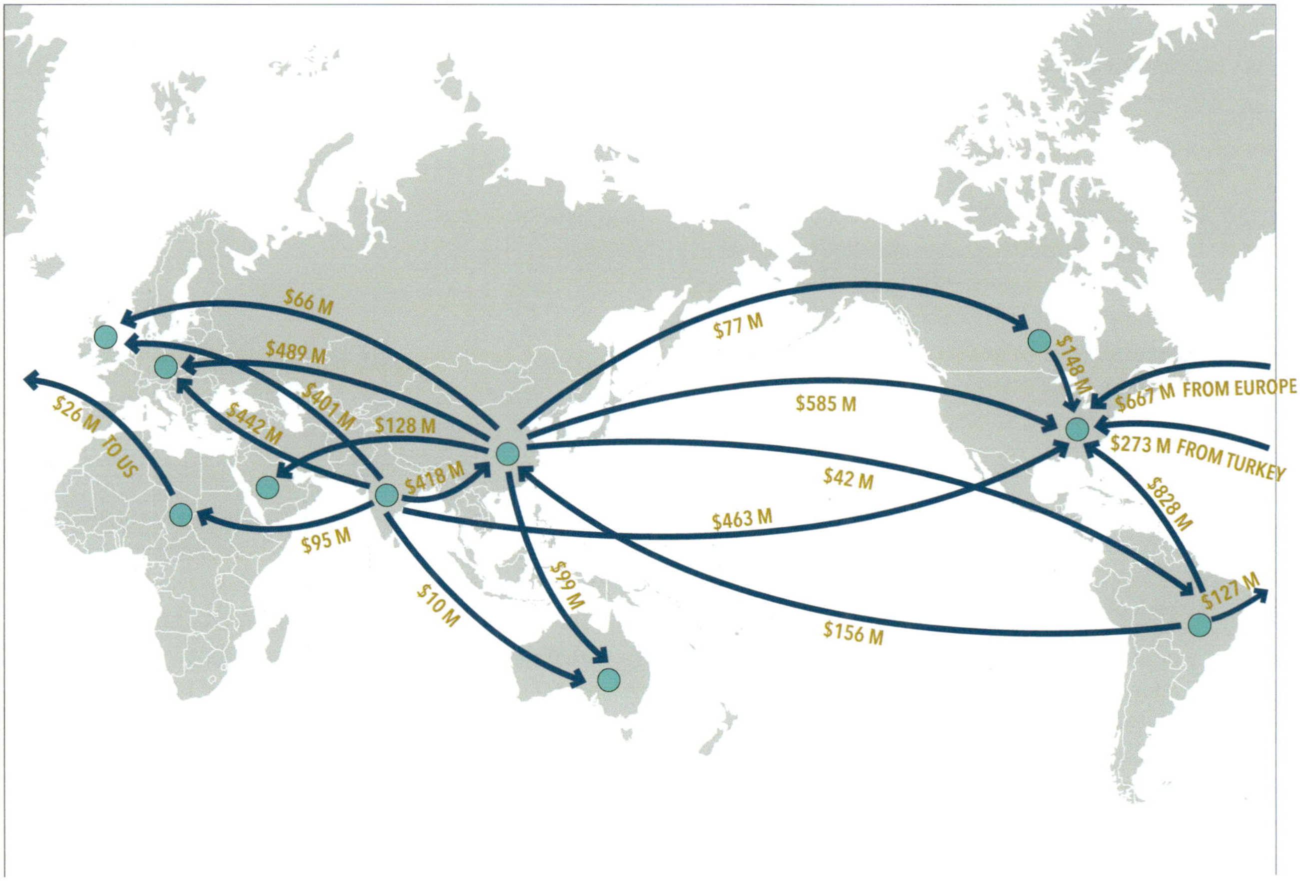

Figure 13.11 The international stone trade is a multibillion dollar business with millions of tons shipped around the world. Source: OEC 2021. Image credit: Meg Calkins

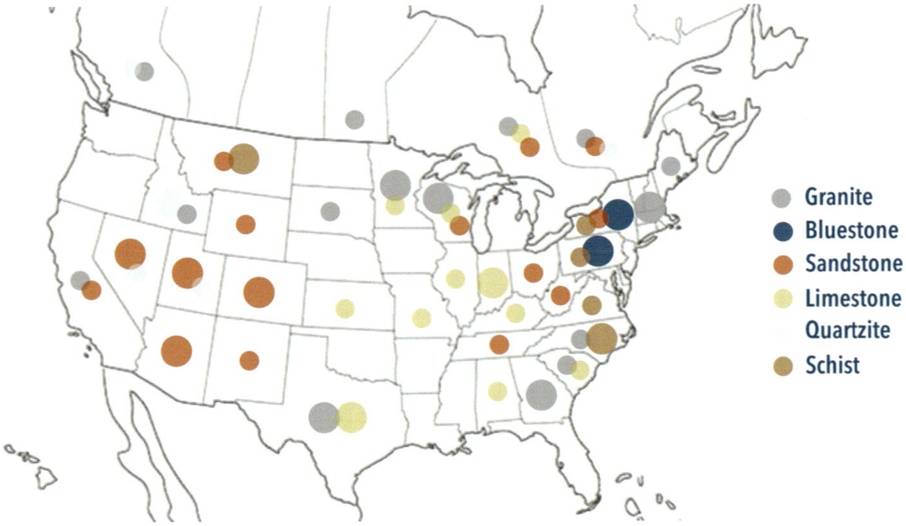

Figure 13.12 North American stone quarrying by type of stone. Sources: NSI 2024a; OEC 2021. Image credit: Meg Calkins

Natural Stone Institute offers a searchable database by type of stone, state, and sustainability criteria such as EPDs, certified quarries, and fabrication, other aspects of information transparency (Figure 13.12) (NSI 2024a).

Embodied Carbon and Stone

Greenhouse gas emissions are another waste product from dimension stone quarrying, processing, and transport (Tables 13.5–13.7). As stone is a very heavy material, transportation distances and modes can impact the embodied carbon of the stone substantially. The embodied carbon of dimension stone that is transported 800 kilometers (around 500 miles) is similar to that of PIP concrete. After around 700 kilometers, the carbon footprint of transport by truck exceeds the embodied carbon of quarrying and fabricating dimension stone (Historic Environment Scotland 2024).

Historic Environment Scotland offers a carbon calculator for both granite and sandstone with locations for each phase of dimension stone processing as variables. Embodied carbon of some typical paths for dimension stone are shown in Table 13.5 and 13.6.

Quarrying, manufacturing, and fabrication activities also contribute to the embodied carbon of stone, and the more the stone is finished, the higher the embodied carbon. For instance, flamed granite dimension stone for a wall must go through multiple steps in fabrication, while rubble stone is used with minimal or no processing directly from the quarry. The embodied carbon of natural stone products along with alternatives to natural stone is shown in Table 13.7. Like most material industries, the stone quarries, and fabricators are taking steps to reduce the embodied carbon of their manufacturing and fabrication processes.

Table 13.5 Embodied Carbon of Stone Transport

Phases A1–A4	Granite $kgCO_2e/ton$	Sandstone $kgCO_2e/ton$
Quarried in India Worked in China Shipped to central Scotland	1349	1176
Quarried in India Worked in India Shipped to central Scotland	1020	906
Quarried in China Worked in China Shipped to central Scotland	1079	952
Within Scotland	156	120

Source: Historic Environment Scotland 2024

The embodied carbon of stone is not the true measure of a stone structure's impact. Instead, all materials in the pavement or wall assembly must be considered. For instance, local, low EC stone veneer on a CMU seat wall with grouted cores and a PIP spread footing will still have a substantial carbon footprint due to the PIP concrete footing and precast concrete stem. A quarry block wall will have a lower embodied carbon, even if the stone was transported 200 kilometers. The embodied carbon of stone structures is shown in Chapter 6 "Pavements", Chapter 8 "Retaining Walls and Bioengineered Embankments", and Chapter 9 "Freestanding Walls".

Table 13.6 Embodied Carbon Examples for Imported and Domestic Stone by Life Cycle Stage

	A1 Quarrying	A2 Transport to Fabrication	A3 Fabrication	A4 Transport to Site	Total EC $kgCO_2e/t$
Imported granite	India	To China	China	To Glasgow	
Embodied carbon for one metric ton of stone ($kgCO_2e$)	52	719	67	512	**1349**
Domestic sandstone	Central Scotland, south	To Scotland, England border region	Scotland, England border region	To Central Scotland, south	
Embodied carbon for one metric ton of stone ($kgCO_2e$)	38	39	54	25	**156**

Source: Historic Environment Scotland 2024

Table 13.7 Embodied Carbon of Stone Products and Stone Alternatives

Stone Product	Embodied Carbon $kgCO_2e/kg$	Embodied Carbon $kgCO_2e/t$	Notes and Sources of Data
Stone, quarried and unfinished	0.02	20	Hardy median carbon factor (MCf) 2024; granite is assumed. Phases A1–A3.
Stone, quarried and dressed	0.076	76	Hardy MCf 2024; granite is assumed. Phases A1–A3.
Stone, quarried, complex cutting (CNC) and dressed	0.182	182	Hardy MCf 2024; granite is assumed. Phases A1–A3.
Granite dimension stone, local	0.156	156	HES Carbon Calculator 2024, quarried and processed within Scotland. Phases A1–A4.
Sandstone dimension stone, local	0.120	120	HES Carbon Calculator 2024, quarried and processed within Scotland. Phases A1–A4.
Granite dimension stone, imported	1.079	1079	HES Carbon Calculator 2024, internationally quarried and fabricated, EC can vary widely. Phases A1–A4.
Sandstone dimension stone, imported	0.952	952	HES Carbon Calculator 2024, internationally quarried and fabricated, EC can vary widely. Includes Phases A1–A4.
Quarry block	0.035	35	Based on Hardy MCf 2024, dressed one side. Phases A1–A3.
Rubble stone/rip rap	0.025	25	Hardy MCf 2024. Phases A1–A3.
Alternatives to Natural Stone	**Embodied Carbon** $kgCO_2e/kg$	**Embodied Carbon** $kgCO_2e/t$	**Notes and Sources of Data**
Cast stone/precast concrete	0.28	280	Hardy MCf 2024. Phases A1–A3.
PIP Concrete, 27.6MPa, 10% or less SCMs	0.16	160	NRMCA Baselines 2022. Phases A1–A3.
Fired clay brick	0.273	273	BIA 2020 Industry-Wide EPD. Phases A1–A3.

Natural Stone Sustainability Standard ANSI/NSI 373: Sustainable Production of Dimension Stone

The *ANSI/NSI 373: Sustainable Production of Natural Dimension Stone* is a multi-attribute, voluntary standard designed to encourage and recognize sustainable practices in the natural dimension stone industry. Developed by a wide range of stakeholders and maintained by the Natural Stone Institute, the standard offers a series of environmental, social, and human health metrics for stone quarries and fabricators to improve their practices.

Quarriers and fabricators can become certified under the standard, and they must periodically recertify to maintain their status. Certification under the standard is third-party verified by NSF International (Table 13.8).

The standard requires certified companies to develop, implement, and maintain a documentation and tracking system that conforms to the standards of NSI's Chain of Custody Standard (NSC 2014). Much like a lumber chain-of-custody, this requires documentation of the stone at every phase of its journey to the project site providing assurance to the consumer that the stone has not been sent overseas or across the country for processing without disclosure (NSI 2023).

ANSI/NSI 373 is recognized by the voluntary rating systems of LEED, SITES, and the Living Building Challenge. Use of stone from certified sources can contribute to their credits.

Ethical Sourcing of Stone Products

Human rights abuses have been well documented in the stone industries in South America, India, China, and Africa. Some quarries in these countries have unsafe working conditions, bonded or forced laborers, and even child labor. With the majority of all US dimension stone imported from Brazil,

Table 13.8 ANSI/NSI 373 Sustainable Production of Natural Dimension Stone Standard

	Strategies That Are Required	Strategies to Earn Points
Water	Inventory water use and create water management plan. Implement systems that enhance discharge water quality.	Points for recycling water, sourcing water from non-potable sources.
Transportation	Provide a chain of custody (COC) using the NSI COC standard. Create program to continuously improve transportation efficiency.	Document and demonstrate processes to improve transportation efficiency.
Site management	Create a site management plan for maintenance and operations of facilities.	Points for establishing ecosystem boundaries, executing environmental impact reports.
Site reclamation and adaptive reuse	Create a post-closure reclamation plan.	Points for community involvement with plan, public access to plans and exemplary plans.
Social governance	Document prohibition on forced and child labor, and prevention of discrimination.	Points for employee participation in community service work, creation of a social accountability plan, and publication of a corporate sustainability report.
Energy	Inventory energy use and type of energy consumed. Create a management plan to reduce energy use and GHG emissions.	Points for energy reductions, carbon footprint analysis, and renewable and alternative energy sourcing.
Excess process materials and waste	Inventory excess process materials and solid waste. Implement a program to manage excess process material and reduce waste.	Points for excess materials and solid waste reduction; for diverting sludge from landfill; for reuse of excess process material.
Solid waste	Inventory solid waste. Create a solid waste management plan.	Reduce solid waste.
Chemical management	Create a chemical inventory. Create a chemical management plan.	Points for safer chemicals management program and elimination of chemicals of concern.
Human health and safety	Create an occupational safety plan. Create a silica exposure control plan. Inventory air emissions and create an air emissions management plan.	Point for documenting improved workplace safety records.
Innovation		Points for documenting and demonstrating innovative processes that lead to improvement in human health and safety; transportation efficiency; and waste reduction or reuse.

Source: NSI 373 2023

India, and China, it would be helpful to know the locations and working conditions of its quarrying and manufacture.

Unfortunately, transparency in the international dimension stone industry is lacking so it is nearly impossible to know where the stone originated and even where it was fabricated. Some of this stone was quarried in one country and processed in another country before being shipped to the United States for distribution and use. Wholesalers and distributors add additional steps to the chain of custody of stone products (Figure 13.13).

India is one of the top stone producers in the world, exporting limestone, sandstone, and granite. Several NGOs have identified issues in Indian quarries including child labor, forced or debt-bonded labor, unsafe working conditions, and a lack of protective gear. While India has laws against child labor and bonded labor, there are many small quarries and some illegal mines, so they are not easily regulated (BHRRC 2016).

Fifty percent of granite used around the world originates in India (Glocal Research et al. 2017). While only around 13% of dimension stone sold the United States is imported from India, large quantities are supplied to China for processing and fabrication then exported to the United States and Europe.

In 2015, nearly 10% of stone quarry laborers were children, and 50–86% were bonded to quarries through debts with high interest. (Glocal Research et al. 2017; Glocal Research and India Committee of the Netherlands 2015). Most of the human and labor rights violations in India are in the South Indian granite and sandstone industries. Problematic regions are Rajasthan, Karnataka, and Tamil Nadu. Unsafe working conditions and excessive working hours are a concern in both India and China (TruStone Initiative et al. 2021).

Labor abuses also happen in the stone industry in Brazil. Between 2008 and 2021, over three hundred enslaved workers were rescued from Brazilian mines (Angelo 2021). While child labor is decreasing in Brazil, there are still incidences of children working in stone excavation and processing in small and/or illegal quarries that are not easily regulated (Angelo 2021; Rizzini 2016).

Under the Trafficking Victims Protection Reauthorization Act (TYPRA), the United States lists *Goods Produced by Child Labor or Forced Labor*. In addition to stone from India and Brazil, it lists dimension stone from Egypt, Madagascar, Nepal, Paraguay, Uganda, and Zambia as potentially resulting from child or forced labor (ILAB 2022) (Table 13.9).

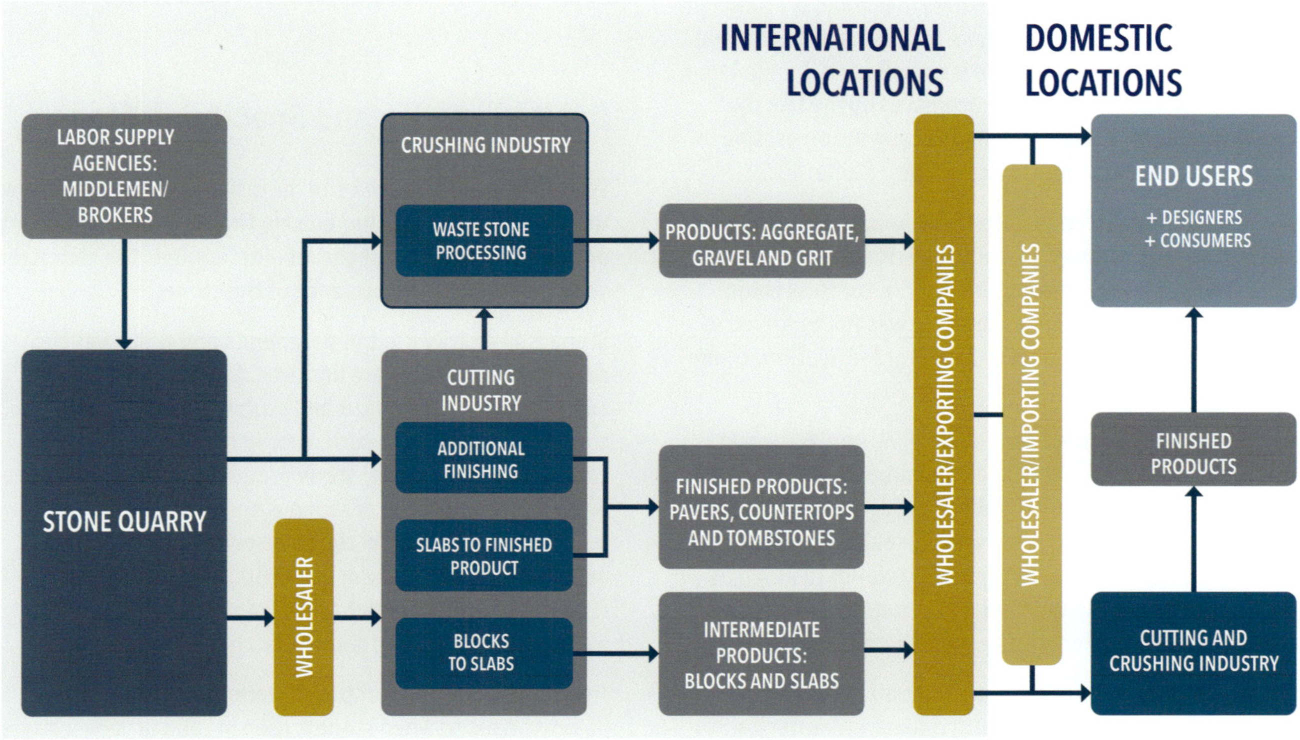

Figure 13.13 A typical chain of custody for imported stone. This illustrates the challenge of tracking a chain of custody for a stone product. Image credit: Meg Calkins and Drake Bruner

Table 13.9 Human Rights Issues and Stone

Stone	Country	Child Labor	Forced Labor
Granite	Burkina Faso	x	
	Nigeria	x	x
	Sierra Leone	x	
Limestone	Egypt	x	
	Paraguay	x	
Sandstone	India	x	x
Gravel/Crushed stone	Benin	x	
	Guatemala	x	
	Nicaragua	x	
	Nigeria	x	x
Sand	Kenya	x	x
	Nigeria	x	
	Uganda	x	
Stone (unclassified)	India	x	x
	Madagascar	x	
	Nepal	x	x
	Uganda	x	
	Zambia	x	

Sources: ILAB 2022; DFF 2022a

Specifying Ethically Sourced Stone

While there are ethical stone registries in Europe, procurement of ethically sourced stone elsewhere can be more challenging. The dimension stone supply chain is complicated, unregulated, and opaque. The complexity is exacerbated by the fact that stone extraction often happens in one country, processing in another and distribution in a third country (Table 13.10).

The simplest approach to using ethically sourced stone in North America is to specify locally quarried and processed stone. This will involve asking the distributor for a chain of custody to make sure that the stone was not quarried in North America, sent to China or India for fabrication, then back to North America.

ANSI/NSI 373 *Sustainable Production of Natural Dimension Stone Standard* requires certified companies to ensure that natural stone products are handled through the supply chain in a manner that is transparent and traceable. The Standard is modeled after other chain of custody standards such as the Forest Stewardship Certification. Producers, fabricators, distributors, and brokers can become certified.

ANSI/NSI 373 requires compliance with all US OSHA and MSHA laws for occupational health and mining safety. Outside the United States, organizations must comply with all applicable international and regional laws addressing occupational health and safety. The standard prohibits child and forced or bonded labor.

Table 13.10 Ethical Stone Labels and Organizations

Organization	Standard or Label
Ethical Stone Register	The Ethical Stone Register is a UK-based listing of ethical sources of stone for granite, limestone, sandstone, slate, and marble.
Fair Stone	Fair Stone offers a standard for ethical sourcing of stone, traces fair stone with a software application, and lists fair stone partners in Europe who are committed to importing ethically sourced stone.
Ethical Trading Initiative	Ethical Trading Initiative is an NGO that influences businesses to act responsibly and promote fair working practices. They offer a guide for Ethical Sourcing of Natural Stone from Rajasthan, India.
Natural Stone Council	ANSI/NSI 373 Sustainable Production of Natural Dimension Stone Standard (Table 13.7).

Design for Freedom offers specification language for ethical sourcing of materials. The specification language is not specific to stone, but it can be applied to sourcing of all construction materials and is written to be a part of the *Special Conditions* specification section. The specification language requires that contractors respond to requests for information from the design team and that they provide *Ethical Supply Chain Documentation Submittals* for designated materials and products (DFF 2022b).

Designing with and Specifying Stone

Stone varies by type, texture, porosity, and strength. How the stone formed – either quickly through volcanic activity or over millennia layer by layer – determines the character, appearance, and performance of the stone.

Stone is primarily used in pavements and walls, yet the design opportunities are limitless. Stone can be rustic and rough or precisely cut and sculpted. Unlike some construction products such as fired brick, integral color concrete, or cast stone, stone products are all unique. Even when a type and color of stone is specified, the actual stone product may differ from the same color specified the year before as variations occur even within the same quarry. It is always best to purchase extra stone for replacements, otherwise, it could be hard to match in the future.

The rift and grain of the stone determine the form of its natural break, so when stone is used in solid or veneer stone walls, some stone will form an easier interlock than other stone. The rift and grain also determine the appearance and fit of stone pavements. Irregular flagstone will result in larger joints between stones than saw cut dimension stone units.

The tables in this chapter discuss types and classifications of stone (Tables 13.2 and 13.3), stone cuts and finishes (Table 13.11), and considerations for designing with and specifying both new and reclaimed stone (Tables 13.12 and 13.13). When specifying stone for a durable structure, performance factors to consider are moisture absorption, density, compressive strength, and abrasion resistance (Table 13.2).

Maintenance, repair, and stain removal for stone structures is discussed in Table 13.15. Detailing considerations for stone walls are considered in Chapter 8 and 9; and detailing for stone pavements is covered in Chapters 6 and 7.

Table 13.11 Stone Cuts and Finishes

Stone Cuts and Finishes	Granite	Sandstone	Limestone	Bluestone
Finishes	**Granite**	**Sandstone**	**Limestone**	**Bluestone**
Saw cut produces crisp, precise edges.	x	x	x	x
Snap cut breaks stone along the rift or a natural breaking point in the stone structure. Stone is often scored for a semi-precise cut. Wider joints are required due to irregular edges.	x	x	x	x
Cross cut, also called **fleuri-cut,** is cut parallel to the natural bedding plane or veining of the stone.	x	x	x	x
Vein cut is cut perpendicular to the natural bedding plane or veining of the stone.	x	x	x	x
Natural cleft finish is an irregularly textured surface that results from slitting stone along its rift or bedding plane. Very rustic appearance.		x	x	x
Split face is a slightly convex or concave finish produced by stone splitters driving wedges into a stone without natural cleft surfaces. Used on igneous and metamorphic stone.	x	x	x	x
Rock face, also called rock pitch is a split surface that has been dressed to create a convex or concave bold projection along the face of the stone. Bolder and larger appearance than split face. Appearance can differ when separated along the bedding, stratification, or rift, or perpendicular to it.	x	x	x	x
Bush hammering creates a rough, but uniformly textured surface with small evenly spaced pits.	x	x	x	x
Water jet finishes are textured surfaces produced by exposure to a high-pressure stream of water. The texture will vary by the stream type, the nozzle speed, the pressure of the water and position of the nozzle with respect to the stone. Fine to coarse textures can be achieved and water jet treatment can be applied to other finishes to enhance them.	x	x	x	x
Thermal finish, also called **flame finish**, is a roughly textured, non-reflective finish that is produced by applying a high temperature flame to the stone. It exfoliates the stone and may change the color of the stone depending on mineral composition, particularly if the stone contains iron.	x			x
Sandblasted surface finishes are an irregular, pitted surface created by impact from sand or metal particles sprayed at high velocity against a stone surface. Effects vary by type of stone and pressure.	x		x	x
Tooled finishes are a linear pattern surface of parallel grooves.	x	x	x	x
Diamond sawn finishes are a low-relief patters of linear or curved grooves produced by diamond saw blades.	x	x	x	x
Shot sawn finishes are random markings and grooves produced by gang sawing with chilled steel shot.	x	x	x	x
Tumbled finishes are worn surfaces produced by rotating stone in a drum sometimes containing sand or aggregates until the edges and faces become eroded.	x	x	x	x
Hand rubbed finishes are non-reflective finishes with a slight stipple patter produced by abrasive pads.		x	x	x
Honed finish is a smooth, non or semi-reflective satin like finish produced from mechanical abrasion.	x		x	x
Polished is a smooth, highly reflective finish that intensifies the natural colors of a stone. Reflective of light.	x		x	

Sources: ASTM 2023a; NSI 2022, 2024a

Table 13.12 Designing With and Specifying Natural Stone

Criteria	Considerations
Stone type selection criteria	Stone selection considerations include cost, performance, durability, embodied carbon, sustainable quarrying and processing, and ethical sourcing. Consider specifying stone from quarries and fabricators certified under *ANSI/NSC 373: Sustainable Production of Natural Dimension Stone*.
Stone performance	Stone used in pavements and walls should meet standards for absorption, compressive strength, density, and abrasion resistance (Table 13.3). Frost resistance is an important criterion in temperate and cold climates.
Regional and local stone	Regionally or locally quarried and fabricated stone can reduce transport emissions and cost of stone. The color can reference the surrounding landscape (Figures 13.17 and 13.18).
Ethical stone	Locations of quarrying and fabrication should be identified to ensure that human rights were not violated in quarrying or production of the stone.
Stone veneer	Stone veneer can give the appearance of a solid stone structure, but at lower economic cost. Embodied carbon of stone veneer walls is usually higher than solid stone walls, as concrete or CMU and grout are used for the wall structure. Stone veneer may not be reusable because cleaning mortar is time intensive.
Quarry waste	Quarry blocks, rubble stone, and boulders are considered quarry waste but can be used in stone walls, embankments, erosion control, and seating applications. Their use can result in very low-carbon site structures.
Stone weathering	Stone in exterior settings will acquire a natural patina and weather over time. Igneous and metaphoric stone will weather less than sedimentary stone. Sandstones acquire a patina more quickly than others. Limestone will darken unless it is sealed.
Mortar joints	Where possible, stone structures should avoid use of mortar as it can compromise reuse of stone. When it is used, mortar joints should be compressed and tooled to force mortar into the joints making them watertight.
Accessibility on stone pavements	Polished and honed finishes do not usually meet ADA standards for textured surfaces.
Detailing stone structures	Detailing considerations for stone structures are discussed in the pavement, retaining wall, and freestanding wall chapters.
Finishes	Finishes vary widely by stone fabricator, so request finish samples early in the design process (Table 13.11).
Availability and procurement of stone products	Stone materials should be located and sourced during design or design development as availability can vary. Visits to local quarries can reveal potential "quarry waste" that can inspire the design of stone structures. Additional quantities of stone should be procured as color and type of stone will be difficult to match a few years after construction. Be aware that stone is a natural product with natural variation. Writing flexibility into specifications, and designing flexibility into project aesthetics, can allow for more latitude for inclusions, imperfections, and stratification, ultimately resulting in fewer wasted stone blocks.
Maintenance	Stone is generally a low maintenance, durable material. Primary maintenance for stone structures involves periodic cleaning and stain removal (Table 13.15).

Table 13.13 Locating, Designing With, and Specifying and Reclaimed Stone

Criteria	Considerations
Sources of reclaimed stone	Most common sources of reclaimed stone are curbing, cobblestones, wall stone, and pavers (Figure 13.16).
Stone history and stories	Reclaimed stone often has an interesting history that can offer meaning to a landscape; however, stone removed from historic towns and landmarks should not be specified.
Appearance	Reclaimed stone often has a patina or weathered appearance that lends a feeling of age to a structure. It may be mossy or have lichen growing on it.
Reclaimed stone types	Granite, limestone, and sandstone are the most common types of reclaimed stone. Granite is commonly reclaimed from curbs and street cobbles.
Carbon considerations	Reclaimed stone is very low carbon as it usually takes less energy to reclaim the stone than it does to quarry and produce new stone products. Reclaimed stone should be reused near where it is reclaimed.
Cost considerations	Reclaimed stone is relatively expensive as there is a very strong market for it.
Designing with reclaimed stone	Reclaimed stone should be sourced during the design or design development processes as it will have unique properties that can influence the design. Additionally, availability of sizes and types of reclaimed stone varies widely (Figure 13.16).

Table 13.14 Natural Stone vs. Cast Stone

	Natural Dimension Stone	Cast Stone (Precast Concrete)
Raw material extraction impacts	Can be similar, although some types of dimension stone result in substantial overburden to get to the usable material for dimension stone.	Limestone for cement is often near the surface as is aggregate.
Manufacturing impacts	Less processing than cast stone.	Limestone for cement is heated to very high temperatures.
Embodied carbon	About ⅓ the EC of cast stone (Vierra 2022). If stone is imported, the EC could be much higher.	Three times the embodied carbon of domestic natural stone.
Durability	Granite, bluestone, and basalt are likely more durable than cast stone.	Not as durable as granite, bluestone, or basalt, but likely more durable than softer stones such as limestone or sandstone.
Cost	Varies by type of stone. Granite could be more expensive.	Varies by type of stone and shapes of cast stone. Complex shapes may be less expensive as they are cast into molds rather than carved like natural stone.
Reusability	Can be easily reused if little or no mortar.	Can be reused but not as common.
Maintenance	Harder stones are easily cleaned. Limestone absorbs air pollution and dirt more than cast stone.	Light colors could stain, but not as porous as limestone.
Certification	Natural Stone Sustainability Standard ANSI/NSI 373	Not available
Human rights impacts	Stone from some countries could be quarried and/or processed by forced or child laborers (Table 13.9).	Minimal risk of human rights abuses.

Sources: Schrenk 2024; NSI 2023; Vierra 2022

Table 13.15 Maintenance for Stone Structures

Maintenance	Considerations
Surface cleaning	Porous stones will grow algae, mildew, fungi, moss, and lichen due to water in their pores. Dirt and air particulates can build up on stone surfaces.
Surface cleaning, pressure washing	Pressure washers can be used on granite walls and pavement but should not be used on softer stones such as limestone and some sandstones. It should be tested prior to use on bluestone and harder sandstones. Stone surfaces should be rinsed thoroughly after cleaning.
Surface cleaning, hose	Softer stones such as limestone and some sandstones should be cleaned with a hose. Rinse thoroughly after cleaning.
Types of cleaners	Non-acidic, non-bleach detergents and stain removers should be used to remove growths, pollution, and dirt from calciferous stone such as limestone, travertine, and marble. Stiff bristle brushes will remove organic materials. Granite and other siliceous stone are not usually vulnerable to acids, but cleaners with acids should still be avoided. Good cleaners are neutral cleaners, stone soap, or mild dishwashing liquid. Ammonia solutions will dull polished surfaces. Scouring powders can scratch polished and honed surfaces.
Stain removal	Pollution, dirt, particulates, and organic materials such as mold and mildew can stain the surface of stone.
Stain removers and degreasers	Mild and diluted bleach, ammonia, and hydrogen peroxide solutions can be used on granite and siliceous stones for stain removal but should be tested on softer and/or calciferous stones.
Oil-based stains	Oil-based stains can be removed with household detergents, mineral spirits, or acetone.
Organic stains	Organic stains will generally fade with exposure to sunlight, but a diluted ammonia solution can be used.
Biological stains	Biological stains from mildew, algae, lichens, moss, and fungi can be removed with a very diluted bleach or ammonia solution.
Metal stains	Metal stains should be removed with a poultice. Do not use rust removers.
Grease, oil and waxes	Water-based degreasers specially formulated for stone should be used.
Stone sealers and enhancers	Stone sealers and enhancers can protect stone from stains and traffic, while enhancing the color of natural stone.

(*Continued*)

Table 13.15 (continued)

Maintenance	Considerations
Stone sealers	Some types of natural stone are porous and are vulnerable to staining, salt damage, acid erosion, and algae growth. Sealing stone will help protect it. Stone surfaces absorb moisture and allow it to evaporate, so stone sealers should be breathable and soak into the stone, so they allow for more than just surface protection. Use sealers made for natural stone. They should not leave any sheen or surface coating behind. Stone should be cleaned and dried fully before application of sealer.
Stone enhancers	Stone fades in color when dry so if a wet look is preferred, stone enhancers are sealers that leave the stone surface darker. They should be specifically made for natural stone and should be penetrating to protect more than just the surface of the stone.
Mortar joint maintenance	If mortar is used in a stone structure, mortar joints should be checked periodically for cracks between the stone and mortar. Cracks can allow water to seep into the wall or pavement, and in freezing climates can expand, damaging the stone or mortar.

Sources: NSI 2024b, 2022

CNC Stone Fabrication

CNC stone milling can achieve sculptural and precise fabrication of stone without skilled stone carvers as it uses CNC technologies (Figures 13.9, 13.10, 13.14, and 13.15). Computer aided manufacturing (CAM) software controls CNC equipment directly from a 3D computer model. CNC machines use diamond bits for milling three-dimensional stone works and waterjets for two-dimensional stone carving.

CNC offers the potential for stone fabrication closer to the point of extraction instead of sending the stone overseas to be worked by skilled stone workers in China or Italy. The cost of CNC technology and equipment is high, but as the technology becomes more prevalent, it is becoming more affordable and is being embraced by stone industries in North America and Europe.

Figure 13.14 Gustafson Guthrie Nichol maximized the accuracy of CNC milling to create sculpted undulating forms for a black granite fountain cascade at the Plaza at City Center in Washington DC. CNC stone milling allows for complex, highly accurate and sculptural forms to be carved in stone. Image credit: GGN Landscape Architecture Ltd.

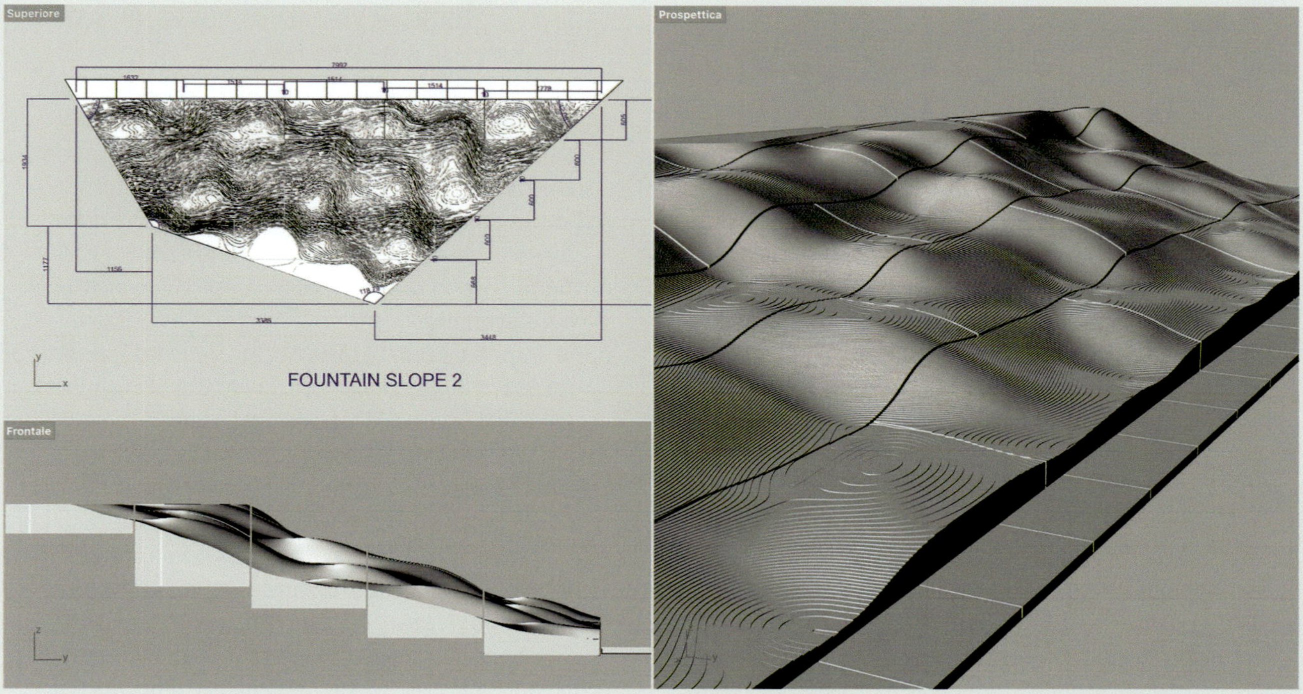

Figure 13.15 Gustafson Guthrie Nichol used computer modeling software to direct the CNC router. While CNC milling is still expensive, the increasing prevalence of CNC equipment may eventually allow for more stone fabrication to happen closer to quarries instead of overseas. Image credit: GGN Landscape Architecture Ltd.

Figure 13.16 MVVA reclaimed slabs of New York bluestone abandoned nearby for the Max Family Garden at St. Ann's Warehouse in Brooklyn. Slabs were used for pavement and stacked for seating. Image credit: Lexi Van Valkenburgh

Figure 13.17 SCAPE's use of locally sourced limestone at Town Branch Commons in Lexington, Kentucky, was inspired by the karst geology of the area. Limestone is used in nearly every aspect of the project in a variety of forms in and around water and along paths. Image credit: SCAPE & Ty Cole

Figure 13.18 The signature stacked stone walls at Town Branch Commons are inspired by the honored traditions and techniques of local stacked stone masons in the region. Some members of SCAPE's design team attended wall building workshops to understand this process. The stone walls simply sit on a mortar setting bed with minimal amounts of mortar between the stone flags. Image credit: SCAPE & Ty Cole

References

Angelo, M. (2021). "333 people rescued from slavery in Brazil mines since 2008, exclusive report shows". *Mongabay*. www.mongabay.org, accessed August 3, 2021.

Associados Nacional da Industria Extractiva E Transformadora (ANNIET) (2024). https://aniet.pt/pt/, accessed February 25, 2024.

ASTM Standard C97/C97M (2018a). *Standard Test Methods for Absorption and Bulk Specific Gravity of Dimension Stone*. West Conshohocken, PA: ASTM International. www.astm.org.

ASTM Standard C99/C99M (2018b). *Standard Test Method for Modulus of Rupture of Dimension Stone*. West Conshohocken, PA: ASTM International. www.astm.org.

ASTM Standard C119 (2023a). *Standard Terminology Relating to Dimension Stone*. West Conshohocken, PA: ASTM International. www.astm.org.

ASTM Standard C170/C170M (2023b). *Standard Test Method for Compressive Strength of Dimension Stone*. West Conshohocken, PA: ASTM International. www.astm.org.

ASTM Standard C241/C241M (2021). *Standard Test Method for Abrasion Resistance of Stone Subjected to Foot Traffic*. West Conshohocken, PA: ASTM International. www.astm.org.

ASTM Standard C616/C616M (2022). *Standard Specification for Quartz-Based Dimension Stone*. West Conshohocken, PA: ASTM International. www.astm.org.

ASTM Standard C880/C880M (2023c). *Standard Test Method for Flexural Strength of Dimension Stone*. West Conshohocken, PA: ASTM International. www.astm.org.

ASTM Standard C1353/C1353M (2020a). *Standard Test Method for Abrasion Resistance of Dimension Stone Subjected to Foot Traffic Using a Rotary Platform Abraser*. West Conshohocken, PA: ASTM International. www.astm.org.

ASTM Standard C1528/C1528M (2020b). *Standard Guide for Selection of Dimension Stone*. West Conshohocken, PA: ASTM International. www.astm.org.

Baca Systems (2024). "Choosing between a stone CNC router & a water jet machine". www.bacasystems.com/choosing-between-a-stone-cnc-router-waterjet-cutting-machine/, accessed September 2, 2024.

Business & Human Rights Resource Center (BHRRC) (2016). *Rajasthan Stone Quarries: Promoting Human Rights Due Diligence and Access to Redress in Complex Supply Chains*. BHRRC.

Carvalho, Cristina, et al. (2013). "'Schist' from trás-os-montes and alto douro (NE of Portugal): Potential use as natural stone". *Key Engineering Materials* 548: 205–211.

Design for Freedom (2022a). *Design for Freedom Report: A Call to Action for Practicing Professionals Within the Ecosystem of the Built Environment*. New Canaan, CT: Grace Farms Foundation.

Design for Freedom (2022b). *Design for Freedom Toolkit*. New Canaan, CT: Grace Farms Foundation.

Ethical Stone Register (2024). https://ethicalstoneregister.co.uk/, accessed February 18, 2024.

Ethical Trading Initiative (2014). *A Guide for the Ethical Sourcing of Natural Stone from Rajasthan, India*. London: Ethical Trading Initiative.

Fair Stone (2024). www.en.fairstone.org/, accessed February 18, 2024.

Glocal Research & India Committee of the Netherlands (2015). *Rock Bottom: Modern Slavery and Child Labour in South Indian Granite Quarries*. Utrecht: India Committee of the Netherlands, 19, May.

Glocal Research, India Committee of the Netherlands and Stop Child Labour (2017). *Dark Sites of Granite: Modern Slavery, Child Labour and Unsafe Work in Indian Granite Quarries*. Utrecht: India Committee of the Netherlands, Stop Child Labour and Kerk in Actie.

Hardy, C. (2024). *Carbon Data Set Landscape - Version 3*. Sasaki Associates, Inc.

Historic Environment Scotland (2024). www.engineshed.scot/calculate-the-carbon-footprint-of-your-stone-project/, accessed February 24, 2024.

Jalalian, M.H., Bagherpour, R., and Khoshouei, M. (2021). "Wastes production in dimension stones industry: Resources, factors, and solutions to reduce them". *Environmental Earth Sciences* 80: 1–13.

Marshak, S. (2019). *Essentials of Geology*, Sixth Edition. W.W. Norton & Company.

MatWeb Material Property Data (2024). "Basalt stone". www.matweb.com/search/datasheet.aspx?matguid=9642fc0c676740659233201627b67b73&ckck=1, accessed February 29, 2024.

Mendoza, J.-M.F., et al. (2014). "Environmental management of granite slab production from an industrial ecology standpoint". *Journal of Cleaner Production* 84: 619–628.

Montani, C. (2020). *Marble and Stones in the World - XXXI Report*. Carrara, Italy: Aldus Casa di Edizioni.

Natural Stone Council (2014). *Chain of Custody Standard for Natural Dimension Stone*. Natural Stone Council.

Natural Stone Institute (NSI) (2022). *Dimension Stone Design Manual*. Oberlin, OH: Natural Stone Institute.

Natural Stone Institute (NSI) (2023). *Natural Stone Sustainability Standard ANSI/NSI 373: Sustainable Production of Dimension Stone*. Ann Arbor, MI: NSF International.

Natural Stone Institute (NSI) (2024a). "Natural stone catalogue". https://stones.naturalstoneinstitute.org/, accessed March 1–March 3, 2024.

Natural Stone Institute (NSI) (2024b). "Get to know your stone". www.naturalstoneinstitute.org/consumers/care/#:~:text=Clean%20stone%20surfaces%20with%20a,a%20film%20and%20cause%20streaks, accessed March 2, 2024.

Observatory of Economic Complexity (OEC) (2021). *Data*. Categories accessed: Building Stone 6802; Granite: Merely cut into blocks 251612; Granite: Crude or Rough Trimmed 251611; Sandstone, merely cut into blocks 251622; Sandstone, whether or not roughly trimmed, cut by sawing, etc. into blocks or slabs 251620; Quartzite, crude or roughly trimmed 250621; Quartzite, merely cut or not into blocks, 250620. https://oec.world/en, accessed February 29, 2024.

Rizzini, Irene. (2016). *Child Labor in Brazil: The Stone Quarry and Stone Finishing Industries*. The Ministry of Health, Equalities, Care and Ageing (MGEPA) of the state of North Rhine Westfalia Rio de Janeiro, Brazil, October.

Schrenk, S. (2024). *Façade Systems: Characteristics and Performance of Cast Stone vs. Natural Stone*. Polycor. https://blog.polycor.com/facades-systems-characteristics-and-performance-of-cast-stone-vs-natural-stone, accessed September 1, 2024.

TruStone Initiative, International Responsible Business Conduct, and Socio-Economic Council (2021). *Social Conditions in the Natural Stone Sector in China*. TruStone Initiative.

US Bureau of International Labor Affairs (ILAB) (2022). *List of Goods Produced by Child Labor or Forced Labor*. US Department of Labor, ILAB.

USGS Summaries, Mineral Commodity (2024). *Mineral Commodity Summaries*. Reston, VA: US Geological Survey.

Vierra, S. (2022). *Manufacturing Impacts: Natural Stone vs. Precast Concrete*. Natural Stone Institute.

Zimmerman, A. (Ed.). (2015). *Constructing Landscape: Materials, Techniques, Structural Components*, Third Edition. Birkhäuser.

CHAPTER 14

Concrete Materials and Products

After water, concrete is the second most used material on the planet with around 30 billion metric tons of concrete produced globally each year (Architecture 2030 2024; Mission Possible Partnership 2023). It is an important construction material for sites and infrastructure projects due to its ability to support heavy loads, withstand extreme weather conditions, become permeable, and be formed into distinctive shapes. Few materials are as versatile, durable, easy to produce, and relatively low cost as concrete. Mix specifications can modify the performance, appearance, and environmental impact of a concrete structure; thus, they hold potential to substantially reduce its carbon footprint.

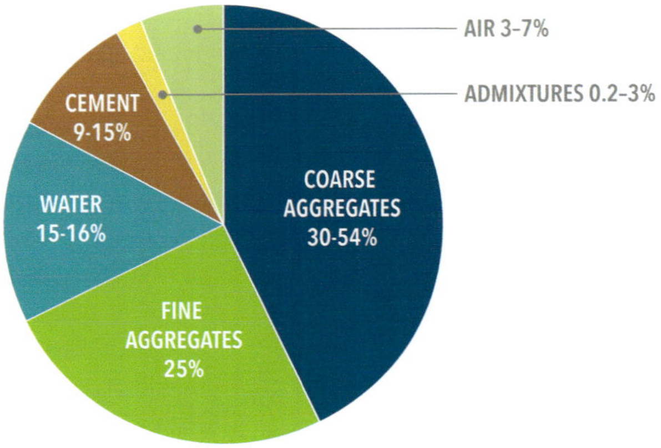

Figure 14.1 Typical concrete constituents by volume. Image credit: Drake Bruner and Meg Calkins

Concrete typically consists of portland cement, coarse and fine aggregates, admixtures, air, and water (Figure 14.1). Portland cement binds the stone matrix together, and 4.1 billion metric tons of it are used globally each year.

While portland cement only comprises 9–15% of a concrete mixture, it is responsible for over 90% of concrete's embodied carbon, resulting in about 7% of global greenhouse gas emissions annually (Architecture 2030 2024; GCCA 2024; Mission Possible Partnership 2023). Therefore, reducing the use of portland cement is key to lowering the carbon footprint of concrete.

In a business-as-usual scenario, concrete consumption is expected to increase with the UN Environmental Program's Sustainable Development Goals of housing and infrastructure development around the world. The UN Global Climate Action estimates that 75% of the infrastructure that will exist in 2050 has not even been built today – and much of it will be made from concrete (UN Climate Change 2021).

As use of concrete increases, the construction industry faces a complex challenge to decarbonize cement and concrete (Figure 14.2). Like many other industries, some industry groups and individual companies within global cement and concrete markets have set the ambitious goal of net-zero carbon by 2050. Industry partnerships such as the Mission Possible Partnership have set goals for each decade and provided detailed strategies to achieve them (Mission Possible Partnership 2023).

Strategies to reduce the embodied carbon of concrete include reducing the clinker-to-cement ratio with alternative materials, electrifying kilns and using renewable energy sources, capturing CO_2 emissions and sequestering them in concrete, optimizing use of concrete in durable structures, and incorporating recycled aggregates into the mix. While some of these strategies are the responsibility of cement manufacturers and concrete producers, others must be adopted by professionals who design with and specify concrete.

Site and infrastructure designers can reduce the amount of concrete they use by specifying alternative materials where applicable or designing smaller concrete structures with longer service lives. Instead of a concrete plaza in an urban setting, the pavement could be primarily stabilized

DOI: 10.4324/9781003413974-16

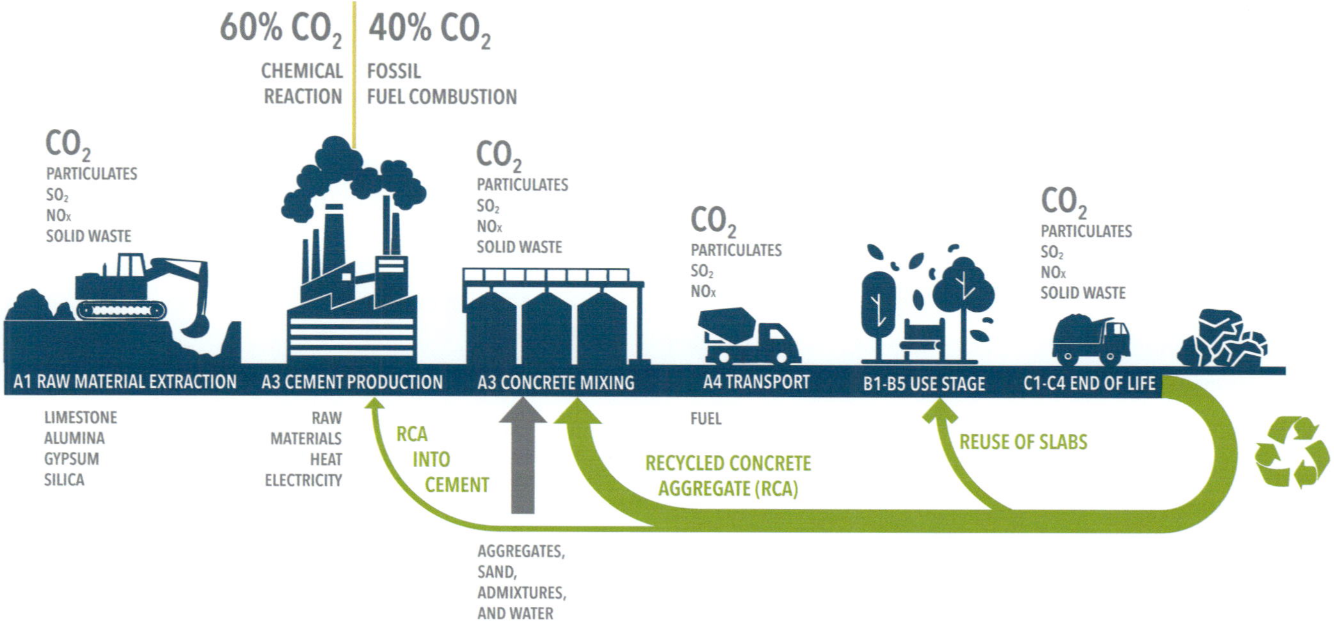

Figure 14.2 Emissions and waste across the life cycle of cement and concrete. Image credit: Meg Calkins

stone dust with the major path(s) of travel being concrete pavement. Instead of concrete, seat walls can built from stone quarry blocks, and bioengineered embankments can retain slopes instead of concrete retaining walls.

Where concrete walls and pavements must be used, durable structures with longer service lives will reduce replacement cycles and the overall embodied carbon of the structure over time. These structures can be the minimum thickness dictated by the structural or geotechnical engineers instead of building in extra thicknesses. A 4-meter-wide path could be reduced to a 2.5-meter width. Flexible precast concrete structures can allow for repair and releveling. Interlocking retaining wall blocks can be used in areas with deep frostlines, and concrete pavers can be set on stable aggregate bases.

Substantive efforts are being made to address the industry's massive carbon footprint, but it is not yet clear if the change will happen quickly enough to meet industry goals of decarbonization by 2050. Early adopters are using relatively new technologies such as carbon capture, utilization, and storage (CCUS) alternative cements, and other innovations, but they have limited availability and a lack of long-term performance data. For widespread adoption, policymakers must drive change with policies that support low-carbon concrete strategies, and standards organizations must adapt or write new standards for these and other recent innovations in the concrete industry.

Cement

Hydraulic cement is a material that sets and hardens when mixed with water. This process, called hydration, is a non-reversible chemical reaction. Hydraulic cements include ordinary portland cement (OPC), calcium aluminate cement, some calcium sulfoaluminate cements, and some fly ash-based cements. Ordinary portland cement is the main binding agent in most concrete.

OPC has a high-carbon footprint, resulting in just under one metric ton of CO_2 released for every metric ton of cement clinker produced. Carbon emissions occur from two major sources in the production of clinker. Process emissions from the calcination of limestone for lime account for around 60% of CO_2 emissions, and about 40% of emissions stem from fossil fuel combustion to heat the kilns to a sintering temperature of 1480°C (2700°F). As cement plants transition to renewable energy sources, the emissions from fossil fuel combustion will be reduced (Czigler et al. 2020; GCCA 2021).

The share of clinker in cement, measured by mass, is the clinker-to-cement ratio. OPC typically contains more than 90% clinker with the remaining constituents being gypsum and fine limestone. This clinker-to-cement ratio, 0.9, is expressed as a decimal.

Other cement constituents, called supplementary cementitious materials, may include calcined clay, volcanic materials, ground limestone, and industrial by-products such as fly ash, silica fume, and ground granulated blast furnace slag (Tables 14.1 and 14.2).

Given the embodied carbon burden of cement, an important way to reduce the carbon footprint of concrete is to reduce the use of portland clinker. The global concrete industry aims to reduce the average clinker factor of cement to 0.6 by 2050 (GCCA 2021). This means that only 60% of the cement in a mix would be portland cement while the remaining material would be alternative cements, fillers, SCMs, or a combination of the three.

As the 0.6 goal is an average clinker-to-cement ratio, some mixes would need to be substantially less than 60% portland cement clinker since others will be greater. The current global clinker factor is around 0.76. In the United States, the average clinker factor is 0.88; in the European Union, India, and China it is between 0.64 and 0.76 (Gangotra et al. 2024).

Table 14.1 Supplementary Cementitious Materials (SCMs) by Function

Active Pozzolanic Addition		Active Non-Pozzolanic Addition	Fillers	
Natural	**Artificial**	**Artificial**	**Natural**	**Artificial**
Pumice	Ground granulated blast furnace slag	Ground granulated blast furnace slag	Limestone	Marble powder
Tuff	Siliceous fly ash	Burnt shale		Concrete powder
Diatomaceous earth	Calcined clay/Metakaolin	Calcareous fly ash		
Opaline rock	Silica fume			
Moler	Agroforestry waste			
Gaize	Masonry waste			
Clays	Mixed C&D waste			
	Ornamental stone waste and sludge			
	Paper sludge			
	Bentonite			
	Recycled glass powder			

Adapted from: Hart 2022

Table 14.2 Clinker Substitutes: Supplementary Cementitious Materials and Fillers

Clinker Substitute and Substitution Quantities	Advantages	Disadvantages	Comments
Ground limestone is added to portland cement in percentages of 5–15% to make portland limestone cement (PLC). This is ASTM C595 Type IL cement **5–15% substitution for OPC** *ASTM C595 Standard Specification for Blended Hydraulic Cements*	Higher compressive strength at one and seven days, and at full strength. Moderate to high sulfate resistance.	Low substitution rate of 15% max. in North America. See "High-Volume Ground Limestone" for more information.	Virtually the same handling and performance as OPC. Other SCMs can be used in conjunction with PLC. Most state DOTs accept use of Type IL cement.
Natural pozzolans are siliceous or siliceous and aluminous materials that possess little cementitious value. When finely ground and in the presence of water, they will react chemically with calcium hydroxide to form compounds that possess cementitious qualities. Examples of natural pozzolans are volcanic ash, rice husk ash, calcined clays, calcined shale, and diatomaceous earth. Performance characteristics vary by type of pozzolan. **5–40% substitution for OPC** *ASTM C618 Standard Specification for Coal Fly Ash and Raw or Calcined Natural Pozzolan for Use in Concrete*	Reduce the heat of hydration 10–40%, lowering thermal cracking. Increased late compressive strengths. Improved resistance to chemically aggressive environments.	Increased water demand. Primary barriers are lack of experience using them, and lack of widespread availability in supply chains despite their natural occurrence worldwide.	Kaolin clay and other natural pozzolans are abundant around the world. As supplies of fly ash and GGBFS diminish, natural pozzolans will increasingly be used.

(Continued)

Table 14.2 (Continued)

Clinker Substitute and Substitution Quantities	Advantages	Disadvantages	Comments
Limestone calcined clay cement (LC³) is a ternary mix that is gaining use around the world. LC³ is comprised of finely ground limestone and calcined clay and can replace up to 50% of portland cement. Calcined clay requires energy to heat and activate but the temperatures are only 600–800°C.	Process CO_2 emissions are minimal as clay contains very little carbon. Finely ground limestone is not heated.	Low early strength, increased water demand. Some cement plants will need to invest in new technology.	Research shows up to a 40% reduction in the carbon footprint of concrete with LC³ cement. Indian building code has a provision for LC³.
Fly ash (FA), a by-product of coal combustion, is removed from flue-gases by pollution control equipment. It is well used in amounts of 5–15% as a standard practice but can replace OPC up to 40%. Class F fly ash has a sum of oxides greater than 70% and is pozzolanic. Class C fly ash has a sum of oxides of 50% or greater and is hydraulic with some pozzolanic properties. **5–40% substitution for OPC** *ASTM C618 Standard Specification for Coal Fly Ash and Raw or Calcined Natural Pozzolan for Use in Concrete*	Produces ultimately higher strength than OPC concrete. Reduces moisture permeability. Resistance to alkali silica reaction is improved.	Slows cure times. Can contain mercury and other heavy metals. Studies have shown that solid concrete can encapsulate hazardous substances and reduce exposure risks, although there are concerns with the handling of FA and recycling FA concrete.	Performance of uncured and cured concrete varies with higher volumes of FA. FA will become limited with transition away from coal. One study showed increased mercury levels in water from porous concrete pavement with 15% FA. Fly ash is on the ILFI Red List.
Ground granulated blast furnace slag is a by-product of the smelting of iron in a blast furnace. It contains silica and calcium and can almost entirely replace OPC. **25–95% substitution for OPC** *ASTM C989 Standard Specification for Slag Cement for Use in Concrete and Mortars*	Ultimately higher strength than OPC concrete. Effective at mitigating ASR. Increased hardened cement paste (HCP) density, resulting in a lower permeability.	Reacts slower than OPC and slows cure times. Can be a positive for thick pours or hot weather. Can contain chromium and is caustic so safe handling is important.	Supply can be limited to iron and steel processing areas of the world. Could become difficult to procure as more steel is produced outside the United States and/or is recycled. Costs of GGBFS are increasing.
Silica fume is produced in arc furnaces with the production of silicon alloys. It is a very fine amorphous silica that is highly pozzolanic. **12% substitution for OPC** *ASTM C1240 Standard Specification for Silica Fume Used in Cementitious Mixtures*	Increased concrete strength and durability. Produces higher density concrete with low permeability. Effective at mitigating ASR.	Increased water demand so high-range water reducers are used. Increased heat of hydration. Typically more expensive than OPC.	Not commonly used in site or road construction unless low permeability, ASR and sulfate attack mitigation, or high strength is required.
Ground glass pozzolans (GGP), a post-consumer waste product, can be processed into a fine amorphous silica powder to act as an SCM. **5–40% substitution for OPC** *ASTM C1866/C1866M Standard Specification for Ground Glass Pozzolan for Use in Concrete.*	Improves strength and durability and reduces permeability. Uses post-consumer waste.	Not a long use history in concrete except in CMUs. GGP is currently not widely found in the concrete supply chain.	Use could increase as fly ash and GGBFS supplies are reduced. Challenges of inconsistent supply and non-glass materials mixed in.
High-volume ground limestone is used in a relatively new technology called a "high-filler cement". These products used sophisticated grinding processes to optimize particle size distribution of limestone. Unlike portland clinker, they require no calcination and avoid the CO_2 emissions associated with this process. Other SCMs can be used with high-filler cement to further reduce the portland clinker.	High-filler, low-clinker cements can reduce the carbon footprint of cement by 75%. Use as little as 20% portland clinker. Develops strength quickly.	They work best with much less water than traditional cement, however, it can be difficult to place without use of plasticizers.	Other SCMs can be used with high-filler cement to further reduce the portland clinker. Relatively new cement type that is not widespread.

Sources: ASTM C595 2023; Natural Pozzolan Association 2024; Taylor et al. 2019; Pillai et al. 2019; John 2018; ILFI 2024; Bernot et al. 2011

Glossary

Cement is a powder made by grinding clinker with various mineral components such as gypsum, limestone, calcined clay, fly ash, blast furnace slag, coal, and natural volcanic material. Cement sets by a hydraulic reaction with water, hardens, and sticks to other materials such as sand, gravel, or crushed stone, binding them together to form concrete. Hydraulic cements include portland cement, calcium aluminate cement, some fly ash-based cements, and calcium sulfoaluminate cements.

Clinker is an intermediate product in the production of cement. Conventional clinker (also called portland clinker) is a gray substance, consisting of marble-sized granules that are formed from heating limestone and other materials in a cement kiln.

Clinker substitutes are materials added to cement to lower the share of conventional clinker. They could be SCMs or fillers.

Supplementary cementitious materials (SCMs) have cementitious properties and react with clinker playing a role in the strength development of concrete. They include natural pozzolans such as calcined clay, volcanic ash, rice husk ash, and diatomaceous earth; ground limestone; fly ash; ground granulated blast furnace slag (GGBFS); silica fume; and ground glass.

Pozzolans are natural or artificial materials that contain reactive silica. They chemically react with calcium hydroxide and water to form compounds with cementitious properties.

Fillers are clinker substitutes such as ground limestone and quartz that are only slightly reactive with conventional clinker.

Alternative clinkers or binders are made from different materials or processes from those associated with conventional portland clinker. They are man-made or natural materials ground to a fine powder that react with water and/or CO_2 to form a hardened mass. Geopolymers and belitic clinkers are examples of alternative clinkers. They are discussed in later in the chapter.

Portland Cement Clinker Reducing Technologies

Reducing the portland clinker-to-cement ratio is one of the most effective means of reducing the embodied carbon of concrete as portland cement is responsible for around 90% of the total embodied carbon of concrete. Portland clinker can be reduced through the substitution of SCMs or fillers and/or use of alternative binders.

Clinker Substitutes: Supplementary Cementitious Materials and Fillers

Substituting supplementary cementitious materials for portland clinker can be an opportunity for cost-effective decarbonization of concrete. SCMs are industrial by-products or naturally occurring pozzolans that are used as a partial replacement of portland clinker. Some have hydraulic properties (Tables 14.1 and 14.2; Figure 14.3).

Fly ash, GGBFS, and silica fume are industrial by-products. Ground glass is a waste product, and ground limestone, calcined clay, and volcanic ash are abundant, naturally occurring pozzolans.

The amount of portland clinker that can be replaced depends on the type of substitute material(s) utilized, the performance requirements of the concrete in a specific application, and codes and standards. Some clinker substitutes can improve strength development, durability, and other performance characteristics of concrete. Clinker substitution can be used by ready mix or precast companies relatively inexpensively as it does not require an investment in new equipment or a change in fuel sources.

Figure 14.3 New concrete structure structures at Ballast Point Park in Sydney incorporate 40% fly ash as a cement substitute and recycled concrete aggregates from on-site rubble. Other existing walls are retained and refurbished with new concrete facing. Image credit: Brett Boardman

Challenges to the use of clinker substitutes include codes and standards, reluctance by specifiers to vary their standard specs, and lack of contractor experience with placement and curing of alternative concrete mixes. SCM

Table 14.3 ASTM C595 Standard Specification for Blended Hydraulic Cements

Cement Type	Alternative Materials and Max. Percentages	Comments
Type IS(X), portland blast-furnace slag cement	Up to 95% ground granulated blast furnace slag.	Over 70% slag can incorporate hydrated lime.
Type IP(X), portland-pozzolan cement	Up to 40% fly ash, calcined clays, volcanic ash, rice husk ash, calcined shale, diatomaceous earth, and other pozzolans.	Lower percentages of fly ash are commonly used.
Type IL(X), portland limestone cement	5–15% interground limestone	Used for general construction. Can be used in combination with other pozzolans.
Type IT(X), ternary blended cement	15% max. limestone. 40% max. pozzolans. 70% max. total content of pozzolan, limestone, and slag.	Ternary cements are blends of two complementary supplementary cementitious.

Note: X refers to the percentage of alternative material in the mix. Cement with 80% portland cement and 20% slag is Type IS(20).

Source: ASTM 2023

Table 14.4 BS 8500 Blended Cement Types

Cement Type	Alternative Materials and Max. Percentages	Embodied Carbon (kg CO_2e/metric ton)
CEM I portland cement	Less than 5%	840
CEM II/A-L portland limestone cement	6–20%	673–791
CEM II/A-M portland composite cement		
CEM II/A-V portland fly ash cement		
CEM II/B-V portland fly ash cement	21–35%	553–693
CEM II/B-M portland composite cement		
CEM II/B-S portland slag cement		
CEM II/C-M portland composite cement	36–50%	452–569
CEM III/A blast-furnace cement	36–65%	362–575
CEM III/B blast-furnace cement	66–80%	252–355
CEM IV/B-V siliceous fly ash cement	36–55%	390–545
CEM V (S-L) composite cement	51–65%	342–459

Source: Adapted from BS 8500–2:2023

availability varies by region with GGBFS available in steel processing regions and volcanic ash available in areas of volcanic activity.

Supplies of fly ash have been abundant but are expected to decrease in Europe and North America over the coming decades with shifts away from coal fired power plants. GGBFS supplies will decline with increased use of secondary steel. China and India, however, are still producing very high volumes of these materials. Volcanic rocks and ash will be used in Italy, Greece and the US west coast. Calcined clays offer a great opportunity to increase clinker substitution around the world as clay is an abundant material.

Blended Hydraulic Cements

Supplementary cementitious materials are combined with OPC to form blended cements and are sold by cement manufacturers. They are defined by *ASTM C595 Standard Specification for Blended Hydraulic Cements*. While blended cements are pre-blended and packaged, the alternative materials in Table 14.2 can be used by concrete producers in custom quantities as well.

Blended cements have lower clinker to cement ratios than standard OPC, therefore they produce concrete with reduced carbon footprints. The reduction can be substantial if high volumes of SCMs are used (Tables 14.3 and 14.4).

Ternary materials can be combined at the batch plant or obtained as a pre-blended product. Ternary mixtures perform in a way that can be predicted by knowing the characteristics of the individual ingredients. Sometimes the negative properties of one alternative material can be reduced by the positive aspects of another.

Embodied Carbon of Concrete Types

The embodied carbon of concrete varies widely by concrete type, performance requirements, and mix constituents and percentages. The amount of portland cement increases substantially as compressive strength requirements increase. Concrete with 20.7MPa (3000psi) strength uses just 80% of the portland cement that 27.6MPa (4000psi) concrete does. Matching the mix design to the expected use conditions and strength requirements is critically important to performance, longevity, and the carbon footprint of a concrete structure (Table 14.5).

The US GSA drafted GWP limits for concrete that are performance-based by mix type and strength class. These limits apply to only some GSA structures, but they are useful baselines for concrete produced in North America (US GSA 2023) (Table 14.6).

Table 14.5 Embodied Carbon of Concrete Types

Concrete Compressive Strength at 28 Days	SCM (%) and Clinker Factor	Global Warming Potential kg CO_2/kg kg CO_2/t	Notes and Source
OPC 17.2MPa, 2500psi			
OPC 2500psi, 17.2MPa	< 10% SCM, <0.9 clinker factor	0.122 122	279.56kg CO_2e/m^3 (Mix name 2500-00-FA/SL) NRMCA 2022
OPC 2500psi, 17.2MPa	20% fly ash, 0.8 clinker factor	0.105 105	241.63kg CO_2e/m^3 (Mix name 2500-30-FA) NRMCA 2022
OPC 2500psi, 17.2MPa	30% fly ash, 0.7 clinker factor	0.096 96	221.18kg CO_2e/m^3 (Mix name 2500-30-FA) NRMCA 2022
OPC 2500psi, 17.2MPa	40% slag, 0.6 clinker factor	0.087 87	199.80kg CO_2e/m^3 (Mix name 2500-40-FA) NRMCA 2022
OPC 2500psi, 17.2MPa	50% slag, 0.5 clinker factor	0.078 78	179.86kg CO_2e/m^3 (Mix name 2500-50-SL) NRMCA 2022
OPC 2500psi, 17.2MPa	>20% fly ash, >30% slag, 0.5 clinker factor	0.078 78	178.72kg CO_2e/m^3 (Mix name 2500-50-FA/SL) NRMCA 2022
OPC 20.7MPa, 3000psi			
OPC 3000psi 20.7MPa	< 10% SCM, <0.9 clinker factor	0.135 135	311.39kg CO_2e/m^3 (Mix name 3000-00-FA/SL) NRMCA 2022
OPC 3000psi 20.7MPa	20% SCM, <0.8 clinker factor	0.117 117	311.39kg CO_2e/m^3 (Mix name 3000-00-FA/SL) NRMCA 2022
OPC 3000psi 20.7MPa	30% fly ash, 0.7 clinker factor	0.107 107	245.16kg CO_2e/m^3 (Mix name 3000-30-FA) NRMCA 2022
OPC 20.7MPa, 3000psi			
OPC 3000psi 20.7MPa	40% slag, 0.6 clinker factor	0.096 96	220.73kg CO_2e/m^3 (Mix name 3000-40-FA) NRMCA 2022
OPC 3000psi 20.7MPa	50% slag, 0.5 clinker factor	0.086 86	198.37kg CO_2e/m^3 (Mix name 3000-50-SL) NRMCA 2022
OPC 3000psi 20.7MPa	>20% fly ash, >30% slag, 0.5 clinker factor	0.086 86	197.09kg CO_2e/m^3 (Mix name 3000-50-FA/SL) NRMCA 2022
OPC 27.6MPa, 4000psi			
OPC 4000psi 27.6MPa	< 10% SCM, <0.9 clinker factor	0.167 167	383.6kg CO_2e/m^3 (Mix name 4000-00-FA/SL) NRMCA 2022
OPC 4000psi 27.6MPa	20% SCM, <0.8 clinker factor	0.143 143	329.16kg CO_2e/m^3 (Mix name 4000-00-FA/SL) NRMCA 2022

(*Continued*)

Table 14.5 (Continued)

Concrete Compressive Strength at 28 Days	SCM (%) and Clinker Factor	Global Warming Potential kg CO_2/kg kg CO_2/t	Notes and Source
OPC 4000psi 27.6MPa	30% fly ash, 0.7 clinker factor	0.130 130	299.71kg CO_2e/m^3 (Mix name 4000–30-FA) NRMCA 2022
OPC 4000psi 27.6MPa	40% slag, 0.6 clinker factor	0.117 117	268.97kg CO_2e/m^3 (Mix name 4000–40-FA) NRMCA 2022
OPC 4000psi 27.6MPa	50% slag, 0.5 clinker factor	0.104 104	240.33kg CO_2e/m^3 (Mix name 4000–50-SL) NRMCA 2022
OPC 4000psi 27.6MPa	>20% fly ash, >30% slag, 0.5 clinker factor	0.104 104	238.71kg CO_2e/m^3 (Mix name 4000–50-FA/SL) NRMCA 2022
Concrete Compressive Strength at 28 Days	**SCM (%) and Clinker Factor**	**Global Warming Potential kg CO_2/kg kg CO_2/mt**	**Notes**
OPC 17.2MPa, 2500psi			
OPC 2500psi, 17.2MPa	< 10% SCM, <0.9 clinker factor	0.146 146	279.56kg CO_2e/m^3 (Mix name 2500–00-FA/SL) NRMCA 2022
OPC 2500psi, 17.2MPa	20% fly ash, 0.8 clinker factor	0.126 126	241.63kg CO_2e/m^3 (Mix name 2500–30-FA) NRMCA 2022
OPC 2500psi, 17.2MPa	30% fly ash, 0.7 clinker factor	0.115 115	221.18kg CO_2e/m^3 (Mix name 2500–30-FA) NRMCA 2022
OPC 2500psi, 17.2MPa	40% slag, 0.6 clinker factor	0.104 104	199.80kg CO_2e/m^3 (Mix name 2500–40-FA) NRMCA 2022
OPC 2500psi, 17.2MPa	50% slag, 0.5 clinker factor	0.094 94	179.86kg CO_2e/m^3 (Mix name 2500–50-SL) NRMCA 2022
OPC 2500psi, 17.2MPa	>20% fly ash, >30% slag, 0.5 clinker factor	0.093 93	178.72kg CO_2e/m^3 (Mix name 2500–50-FA/SL) NRMCA 2022
OPC 20.7MPa, 3000psi			
OPC 3000psi 20.7MPa	< 10% SCM, <0.9 clinker factor	0.162 162	311.39kg CO_2e/m^3 (Mix name 3000–00-FA/SL) NRMCA 2022
OPC 3000psi 20.7MPa	20% SCM, <0.8 clinker factor	0.140 140	311.39kg CO_2e/m^3 (Mix name 3000–00-FA/SL) NRMCA 2022
OPC 3000psi 20.7MPa	30% fly ash, 0.7 clinker factor	0.128 128	245.16kg CO_2e/m^3 (Mix name 3000–30-FA) NRMCA 2022
OPC 3000psi 20.7MPa	40% slag, 0.6 clinker factor	0.115 115	220.73kg CO_2e/m^3 (Mix name 3000–40-FA) NRMCA 2022
OPC 3000psi 20.7MPa	50% slag, 0.5 clinker factor	0.104 104	198.37kg CO_2e/m^3 (Mix name 3000–50-SL) NRMCA 2022
OPC 3000psi 20.7MPa	>20% fly ash, >30% slag, 0.5 clinker factor	0.103 103	197.09kg CO_2e/m^3 (Mix name 3000–50-FA/SL) NRMCA 2022
OPC 27.6MPa, 4000psi			
OPC 4000psi 27.6MPa	< 10% SCM, <0.9 clinker factor	0.200 200	460.4kg CO_2e/m^3 (Mix name 4000–00-FA/SL) NRMCA 2022
OPC 4000psi 27.6MPa	20% SCM, <0.8 clinker factor	0.172 172	395kg CO_2e/m^3 (Mix name 4000–00-FA/SL) NRMCA 2022
OPC 4000psi 27.6MPa	30% fly ash, 0.7 clinker factor	0.156 156	359.7kg CO_2e/m^3 (Mix name 4000–30-FA) NRMCA 2022
OPC 4000psi 27.6MPa	40% slag, 0.6 clinker factor	0.140 140	322.8kg CO_2e/m^3 (Mix name 4000–40-FA) NRMCA 2022
OPC 4000psi 27.6MPa	50% slag, 0.5 clinker factor	0.125 125	288.4kg CO_2e/m^3 (Mix name 4000–50-SL) NRMCA 2022

(*Continued*)

Table 14.5 (Continued)

Concrete Compressive Strength at 28 Days	SCM (%) and Clinker Factor	Global Warming Potential kg CO_2/kg kg CO_2/t	Notes and Source
OPC 4000psi 27.6MPa	>20% fly ash, >30% slag, 0.5 clinker factor	0.125 125	286.5kg CO_2e/m^3 (Mix name 4000-50-FA/SL) NRMCA 2022
Limestone Calcined Clay Concrete (LC^3)			
LC^3 2500psi 17.2MPa	30% calcined clay, 15% ground limestone, 5% gypsum and 50% clinker	0.088–0.111 88–1110.094	201.3–254.98kg CO_2e/m^3 Boanda-Fuchs et al. 2024; Raju and Lalit 2023; IIT Madras, Tara and IIT Delhi 2015
LC^3 3000psi 20.7MPa	30% calcined clay, 15% ground limestone, 5% gypsum, and 50% clinker	0.097–0.123 97–123	224.2–284.01kg CO_2e/m^3 Boanda-Fuchs et al. 2024; Raju and Lalit 2023; IIT Madras, Tara and IIT Delhi 2015
LC^3 4000psi 27.6MPa	30% calcined clay, 15% ground limestone, 5% gypsum, and 50% clinker	0.120–0.152 120–152	276.24–349.90kg CO_2e/m^3 Boanda-Fuchs et al. 2024; Raju and Lalit 2023; IIT Madras, Tara and IIT Delhi 2015
Geopolymer Cement Concrete			
Geopolymer cement concrete	32MPa (4641psi) concrete	0.051 51	117.3kg CO_2e/m^3 Noller 2020

Notes: GWP values represent life cycle stages A1-A3 (Raw material supply, transportation, and manufacturing) unless stated otherwise. Bulk density varies, but CEMEX UK states an average of 2300kg/m^3. Embodied Carbon figures will change over time as processing efficiencies are realized and more renewable energy sources are employed.

Table 14.6 GSA IRA Limits for Low Embodied Carbon Concrete

	GSA IRA Limits for Low Embodied Carbon Concrete (EPD-Reported GWPs, kg CO_2e/m^3)		
Concrete Strength Class (psi and MPa)	**Top 20% Limit**	**Top 40% Limit**	**Better Than Average Limit**
≤2499 ≤17.2	228	261	277
3000 20.7	257	291	318
4000 27.6	284	326	352
5000 34.5	305	357	382
6000 41.2	319	374	407
≥7200 ≥49.6	321	362	402
Add 30% to these numbers for GWP limits where high early strength concrete mixes are required for technical reasons.			

Source: US GSA 2023

Alternative Clinker Technologies

Alternative clinker technologies either fully or partially avoid use of portland clinkers. By using alternative raw materials and substantially reducing or eliminating the use of limestone, these clinkers can reduce the process emissions from limestone calcination that currently account for 60% of the carbon emissions of OPC.

Most of the alternative clinkers discussed in Table 14.7 are not widely used while some are used in niche applications but not yet commercially available. Those that are commercially available are rarely cost competitive with OPC, and they face raw material shortages and/or resistance from specifiers or contractors.

Performance-Oriented Specs

It can be challenging to specify concrete mixes with SCMs, recycled aggregates, CCUS, and technologies to reduce its embodied carbon. Performance-oriented specifications can allow concrete companies to innovate with low-carbon, high performance solutions.

Prescriptive specifications have been the industry norm, particularly for engineers and designers who favor tried-and-true concrete mixes. But this overreliance on traditional blends can result in a high-carbon footprint for concrete. Replacing mix requirements with performance requirements (e.g., strength, embodied carbon, frost resistance, sulfate resistance, etc.) can encourage innovation by those who have the most sophisticated understanding of the performance and science of concrete mixes – the concrete ready-mix company.

The National Ready Mix Concrete Association (NRMCA) supports the transition to performance specifications for concrete and published the *2021 Guide to Improving Specifications for Ready Mixed Concrete* (NRMCA 2021). The Federal Highway Administration and the Port Authority of New York and New Jersey are also deploying performance-oriented specifications, but most public agencies and departments of transportation could be slow to transition to this practice (Yu et al. 2017).

Table 14.7 Alternative Clinker Technologies

Activated Binders	**Performance**	**Comments**
Geopolymer cement/alkali activated binders are made by alkali activation of aluminosilicate source materials (fly ash, GGBFS, calcined clays, zeolite) in a polymer formation.	60–90% lower embodied carbon than OPC. Competitive with portland cement in cost and performance (Provis 2018). Strengths vary but generally higher than OPC. Ready mix, precast and spray applications (Scrivener et al. 2018). Performance standards are minimal for geopolymer cements.	Primarily niche markets in Australia, South Africa, Brazil, India, Ukraine and North America. Raw material supply constraints, customer resistance, and challenges with industry buy-in (Scrivener et al. 2018).
Alternative-Clinker Cements		
Reactive calcium and silicates Non-carbonate materials are fed through an electrolytic reactor to form reactive calcium and silicates. The electrons break the ingredients down at room temperature instead of the 1500°C kilns used to produce OPC.	>90% less than OPC carbon footprint. Production of cement differs but can be "dropped in" to a concrete plant without modifications (Atunes et al. 2021).	Demonstration phase, scaling up in a few years. The crystalline structure is different than PC, but it adheres to the same standards and is used in the same way in concrete (Sublime Systems 2024).
Carbonatable calcium silicate clinkers (CCSC) are low-lime calcium silicates that are made for carbonation curing instead of the typical hydration curing.	Around 30% less embodied carbon than OPC3. High material availability. Lower energy demand. Lower water demand. CCSC require a constant source of CO_2. The cement cures with exposure to a 60–90% CO_2 gaseous environment.	Primarily limited to precast products. A lower burning temperature is required. Captures up to 300kg CO_2 per ton of cement binder. 43% reduction in process emissions from those of OPC (Atunes et al. 2021).
Low carbonate clinkers with pre-hydrated calcium silicates have a low calcium share. Raw materials include marl, limestone, natural sand, slags, glass, and fly ash.	>50% less embodied carbon than OPC. High material availability. Similar costs. Lower energy demand. Lower water demand.	Demonstration phase, and niche projects. Can be used in ready mix or precast applications.
Belitic clinkers containing Ye'elimite (CSA) are belite-based clinkers that contain Ye'elimite or calcium sulphoaluminate. They are produced with the same process as OPC, but they use less limestone and more aluminum as raw materials. Sintering temperatures are lower as is the energy required for grinding.	Embodied carbon is around 50% less than OPC. Can match OPC performance. Gain strength more slowly than OPC. Higher costs than OPC. Lower grinding energy (Gartner and Sui 2018).	Niche markets in China and Europe; demonstration phase elsewhere. Limited by bauxite supplies. Covered by Chinese standards but not currently by North American or European standards.
Belite Ye'elemite-ferrite clinkers (BYF or BCSA) are clinkers based on belite, Ye'elimite and ferrite. They are produced with the same process as OPC. Sintering temperatures are lower as is the energy required for grinding.	Around 20% lower embodied carbon than OPC due to reduced amounts of limestone. Can be produced in conventional cement plants. Can match OPC performance. Higher costs than OPC. Lower grinding energy. Concrete sets more rapidly than OPC.	Demonstration phase, more research and development needed. 23–33% reduction in process emissions from those of OPC.
Reactive Belite-rich PC (BPC) clinkers are based on belite rather than alite and produced with the same process as OPC. They use a lower limestone content and lower calcination temperature. Less fuel for heating is required, and CO_2 emissions from calcination are reduced.	Higher early strength by lower later-age strength gain than OPC. Greater durability than OPC – better resistance to sulfates and chlorides. Can be produced in conventional cement plants. Still considered a portland cement so governed by those standards.	Commercialized. Niche applications. Used in China. Meets Chinese standards. 7% reduction in process emissions from those of OPC.
Magnesium-based clinkers (MOMS) are based on magnesium oxide clinkers produced by calcinating natural magnesite.	Currently high embodied carbon from calcination of the magnesium oxides, but theoretically substantial reductions could be achieved if ultramafic rocks (basic magnesium silicates not magnesium oxides) are used. Industrial processes for this are in the research phase.	Research phase. 100% reduction in process emissions from those of OPC.

Source: Except where noted Lehne and Preston 2018

Standards and testing protocols exist for some alternative cements but not all of them. This makes it challenging to experiment with and scale up the use of these products.

Aggregates for Concrete

Coarse and fine aggregates make up the largest mass and volume of concrete mix at 65–80% of the mix. Fine aggregates are typically natural sand or crushed stone with particles passing through a 9mm sieve. Coarse aggregates range from over 9mm to 38mm.

Although they have relatively low costs and low embodied carbon, they are resource intensive, and extraction of virgin aggregates can impact sensitive ecosystems. Nearly half of all aggregates used in concrete in North America are natural gravels dug or dredged from a pit, river, lake, or seabed. Excavation of these materials removes vegetative cover, impacts water quality, and harms ecosystem functions and services.

Given the abundance of Recycled, co-products, and waste materials (RCWMs), there is an opportunity to improve supply chains to facilitate beneficial reuse of these materials as both coarse and fine aggregates in concrete structures. The most common RCWM used in new concrete is RCA.

On large previously developed sites, recycled aggregates can be deconstructed, crushed, and graded on site. If they are obtained from an outside source, they should have originated within 80km of the construction site to curtail the carbon impacts of their transport.

Recycled Concrete Aggregate

The World Business Council for Sustainable Development's (WBCSD) Cement Sustainability Initiative has a 'zero-landfill' goal for concrete worldwide. Yet globally, only around 5% of concrete removed from use in 2019 was recycled (Islam et al. 2019). This is skewed by countries with minimal construction and demolition recycling infrastructure. Europe and North America have far higher concrete recycling rates.

In the United States, an estimated 140 million metric tons of concrete are recycled each year primarily for aggregate (EPA 2022). This is about 13% of the total natural aggregate produced for construction uses. In the United Kingdom, recycled concrete aggregates made up 28% of annual demand for construction aggregate (Moore 2020).

Compared to other wastes, concrete waste is relatively inert. The one exception is reclaimed concrete with high volumes of fly ash as it can contain heavy metals such as mercury, lead, nickel, zinc, and manganese. The extent of the heavy metals is dependent on both the minerology, the particle size of the original coal burned to produce the fly ash, and the combustion temperature.

Recent studies have shown that the heavy metals in fly ash are bound into concrete as it cures and do not pose a threat to indoor air quality and human health. However, when the concrete is crushed and used as a base course, heavy metals from the RCA could be released into the environment (Fan et al. 2021; Kurda et al. 2018). More research is required in this area.

As reclaimed concrete can be comingled with other C&D debris, care should be taken to prevent contamination by other materials such as asphalt, soil, chlorides, glass, gypsum board, wood, and roofing materials. Also reinforcing steel, welded wire mesh, and other embedded items must be removed during processing and cleaning of the RCA.

While the carbon reduction benefits of using RCA vary widely depending on transportation and the extent of reprocessing, use of RCA instead of natural aggregates can offer up to a 50% cost savings. When RCA is removed, processed and recycled on large project site, a direct cost savings of 60–80% can be achieved due to reduced hauling costs and landfill tipping fees (Ohemeng and Ekolu 2020; Tam 2008) (Figures 14.4 and 14.5).

Recycled Concrete Aggregate for Use in Unbound Applications

Unstabilized, granular base applications are the primary use for recycled concrete aggregates. Its use in unbound pavement bases and sub-bases is widespread and quite common in the North America and Europe. It is also used for backfilling pipe excavations, and wall and building foundations.

Use of RCA as base aggregate has been promoted by the US Federal Highway Administration and is used by numerous state departments of transportation. *AASHTO M 319 Standard Specification for Reclaimed Concrete Aggregate for Unbound Soil - Aggregate Base Course* sets forth standards for use of RCA in unbound base courses. *ASTM D8038-16 Standard Practice for Reclamation of Recycled Aggregate Base (RAB) Material* sets forth guidelines for reclamation of recycled aggregate base material derived from concrete and asphalt pavement related to materials quality, testing, and operator qualifications.

Recycled Concrete Aggregate for Use in New Concrete

Recycled concrete can be used as coarse or fine aggregate in concrete mixes. It is generally accepted that about 30% of aggregate content in new concrete can be replaced by RCA for structural applications without significantly affecting the mechanical properties of the new concrete. Only 10–20%

Considerations for Use of RCA in Base Applications

- The angular, rough textured nature of the RCA particles can provide a very stable construction platform and offers good interlocking of particles.
- RCA can be produced to provide excellent free-draining base material that is both permeable and highly stable with appropriate crushing and screening processes.
- Concrete pavement design thicknesses could be reduced with the increased levels of support from an RCA base that has re-cemented.
- RCA (and blends of RCA and natural aggregate) should be placed close to the optimum moisture content to ensure that compaction efforts are efficient. Optimum moisture content for RCA is typically significantly higher than for natural aggregate because of the higher absorption capacity of typical RCA.
- Residual concrete dust can be a minor issue. Washing RCA prior to use can reduce residual concrete dust that could impact filter fabrics and drainage systems.
- Drainage systems should be designed to allow residual crusher dust to settle in a granular filter layer at the bottom of the trench rather than allowing direct entry to the pipe.
- Rainwater runoff from RCA stockpiles on site can have a high pH.
- Water percolating through RCA bases can result in effluent that is initially higher alkaline, but this effect diminishes with time in service. The FHWA does not consider this to be an environmental hazard because it is diluted within short distances.
- Frost or moisture heave can happen in more densely graded RCA bases. Slightly more open gradations should be used in frost prone areas.
- Soil and clays from the demolition and removal processes can contaminate RCA. Screening can remove these particles.

Figure 14.4 Four different sizes of on-site reclaimed concrete were used by Ten x Ten at Mill 19. Crushed concrete was used for some pathway surfaces, concrete slabs with flat faces were stacked for walls, larger slabs were used for pavements and stairs, and 45cm concrete "rock" was used in the rubble gardens pictured here. Image credit: Gaffer Photography

Figure 14.5 Linear arrangements of reclaimed bunker blocks form stairs and walls while salvaged I-beams from the Mill building are used as low retaining walls and bench supports. Image credit: Gaffer Photography

RCA should be used for fine aggregates in concrete (PCA 2024). Some states and countries allow higher percentages.

Characteristics of RCA are slightly different from virgin aggregates. They are comprised of both the natural aggregate used to produce the original concrete as well as the adhered mortar. The characteristics of the original recycled concrete will influence how the RCA performs in new concrete (FHWA 2022) (Figures 14.6–14.9).

Figure 14.6 and Figure 14.7 Rubble concrete walls incorporating building fragments from World War II bombing form human-made "ravines" at Warsaw Uprising Mound Park. The concrete mix incorporating recycled aggregate was based on the rubble concrete that was used for reconstruction in the years following WWII. Rubble such as stove tiles, bricks, and broken balustrades "mined" from the mound were hand placed in layers with sand at the outside edge of the formwork. These form the distinctive striations that reference layers of debris and sedimentation. Image credit: Michal Szlaga

Figure 14.8 and Figure 14.9 In an effort to increase biodiversity at the park, moss spores obtained from the site were mixed into a buttermilk and water suspension and sprayed on the wall surfaces with a suitable microclimate. Striations of the 40–60° sloped rubble concrete walls offer a bio-receptive habitat for moss and self-seeding plants. Image credit: Archigrest + topoScape

Considerations for Use of RCA in New Concrete Mixes

- As RCA replacement amounts increase, drying shrinkage will increase, and tensile strength and modulus of elasticity will decrease. Compressive strength and freeze thaw resistance are not significantly affected.
- Use of RCA may reduce concrete set times.
- The quantity and quality of adhered mortar has been shown to play an important role in the strength of RCA concrete. Smaller sized fractions of RCA usually comprise higher volumetric percentages of adhered mortar with some particles consisting of only mortar (FHWA 2022).
- Because RCA contains adhered mortar, which can reduce performance of concrete, more cement may be required, offsetting any gains in carbon performance from the RCA. Using smaller amounts of RCA may not require additional cement.
- On a mass basis, RCA has less than half the carbon footprint of virgin aggregates, but these gains could be more than offset by the need for additional cement in a concrete mix with high amounts of RCA.
- Blending RCA with virgin aggregates may reduce the quantity of adhered mortar and mitigate the impact of RCA on strength.
- Unwashed, dusty RCA will have reduced bond strength and, subsequently, lower concrete strength. Washing and air-blowing of RCA to remove dust can mitigate this issue (FHWA 2022).
- RCA mixtures exhibit increased water demand and reduced workability due to RCAs' higher absorption, angular shape, and rougher surface. Use of smaller sized RCA fractions will increase these impacts. Prewetting of RCA before batching can help offset strength reductions (FHWA 2022).
- The water demand of RCA concrete can be reduced by use of fly ash (at 20–30% substitution rates) and a superplasticizer.
- RCA should only be used if it is from a local source. Trucking it over 80km could cause a greater carbon footprint than is saved by its use.

Cement SCMs and Fillers from Recycled Concrete

The hardened cement fraction of recycled concrete can be recycled through the cement manufacturing process in controlled amounts in two ways. As an alternative raw material to produce clinker, or as an additional component when grinding clinker, gypsum, and other additives to cement. In these methods, RCA is considered an SCM.

Currently, these techniques are inhibited by lack of widespread and efficient separation and treatment techniques. Techniques that are cost competitive are mechanical abrasion, heat treatment, ultrasonic water cleaning, and carbonation for fortification. Carbonation is seen as a promising technology as it could enhance the RCA properties as well as act as a permanent mineral store of CO_2. This area of innovation is largely still in research and testing.

Other Recycled Materials for Coarse and Fine Aggregates in Concrete

While less used than RCA, other pre-and post-consumer waste can be used for a percentage of aggregate in concrete. Table 14.8 lists recycled materials that can be substituted for coarse and fine natural aggregates in concrete.

Precast Concrete

Precast concrete is created in a factory or fabrication setting, then transported to a site for use in construction. In site construction, it is used for unit pavers, wall blocks, pipes, site furnishings, fountain coping, and other structures. Precast has some environmental advantages over poured in place concrete as it is produced in a controlled environment. Use of portland cement and other cementitious materials can be precisely controlled eliminating over-use due to the variable site conditions of poured in place concrete. Waste and water can be more easily recycled.

Controlled climate conditions in a precast fabrication facility and precise mixes result in durable products that can be reclaimed and reused multiple times. Many precast concrete products can be installed in such a way that their removal and reuse is facilitated. Precast concrete pavers are sand set on an aggregate base, a concrete slab, or asphalt slab. Precast wall blocks are interlocking and/or stabilized with pins that allow for their removal after their useful life.

Some precast products use higher volumes of portland cement causing them to have a higher carbon footprint by weight than PIP concrete. Supplementary cementitious materials such as fly ash and GGBFS have not been used

Table 14.8 Recycled Materials for Coarse and Fine Aggregates in Concrete

Recycled Material	Performance	Comments
Recycled concrete aggregate	Refer to the discussion in previous sections of this chapter.	
Air-cooled blast furnace slag is a non-metallic co-product of iron production. It is used as an aggregate in concrete mixes and, more commonly, as a base material.	Aggregates have a rougher texture with a greater surface area and bond well with cement. Low water absorption. Lighter weight.	Availability is limited to iron and steel processing regions. Blast furnace slag is primarily ground, granulated, and used as an SCM.
Reclaimed asphalt pavement (RAP) is produced from the millings of an existing asphalt pavement. RAP is most used in new asphalt pavement and pavement bases, but coarse fractioned RAP (FRAP) is occasionally used as coarse aggregate in concrete mixes.[1]	Workability of concrete with RAP is slightly increased. Toughness related properties are increased. Slight reduction in compressive, flexural, and tensile strength. Greater porosity and permeability. Good abrasion resistance.[3]	Recommended for pavement applications rather than wall or building structure applications. Surface treatment of RAP aggregates will increase bonding with cement paste.[3] RAP is abundant.
Crushed glass, called cullet, can be used in concrete in both aggregate and, when crushed to a powder, in SCM applications.	When using cullet for coarse aggregates alkali silica reaction (ASR) can be an issue and concrete strength can be reduced.[4] Cullet for fine aggregates can replace up to 20% of fine aggregates.[4] Lighter weight Decreased slump.	Crushed glass is a post-consumer recycled material. Supplies of cullet are highly variable.
Crushed brick can be used as a coarse aggregate in concrete mixes.	Brick aggregates have a higher porosity and absorption than natural aggregates.[5] Increased workability Improved compressive strength. Decreased chloride resistance.	Not recommended if reinforcing steel is used in the concrete. Crushed brick supplies are variable. Color of bricks can add visual interest to concrete.
Recycled plastics can be used for coarse aggregate in concrete. Number 1–7 plastics are shredded and sometimes melted down to form an inert hybrid mineral-polymer then graded as per stone aggregate standards.	Some plastic aggregates demonstrate an increase in compressive strength, flexibility, fire resistance, thermal resistance, and acoustic properties. Allows for increased deformation of concrete.[2]	Can be used in precast and PIP applications. Some products contain more than 95% post-consumer plastics. Lighter weight than natural aggregates.[2]
Recycled crumb rubber from tires can be used for coarse aggregate in concrete.[1]	There is a weak bond between the cement paste and the crumb rubber so it must be treated with a coating. Pores in the rubber aggregate cause higher water demand. When in compression, tensile stress on the surface can occur with the soft surface of rubber.	Research is ongoing for surface treatments of the rubber particles to improve adherence to the cement paste. Methods to compress the rubber aggregate to reduce its softness in cured concrete are being investigated.
Waste foundry sand is a fine, high-quality silica sand by-product of metal casting industries. It is used as fine aggregate in concrete.	It can be used for 10–30% of fine aggregate.[6] Increased compressive and tensile strength.	The use of foundry sand in cement is starting to be explored. Foundry sand is black and could impact the color of finished concrete.

[1] Taylor et al. 2019

[2] CRDC Global 2024

[3] Debbarma et al. 2020

[4] Harrison et al. 2020

[5] Alsadey 2019

[6] Manjunatha and Rakshith 2022

widely in precast applications because they slow set times and consequently slow production. Sometimes set accelerator admixtures are used to allow incorporation of SCMs.

There are multiple precast concrete products on the market that incorporate plastic, agricultural, and mineral wastes. In addition, precast pavers and wall blocks incorporating alternative cements are increasingly available.

Table 14.9 Reinforcing for Concrete Structures

Type	Considerations
Steel reinforcing	Steel dowels, tie bars, and reinforcing are used in concrete structures to help the concrete carry tensile stresses and transfer loads across joints. Their use can maintain stability and lengthen the service life of a pavement or wall by keeping it rigid and reducing cracking. But steel reinforcing can also shorten the life of a concrete structure due to its susceptibility to corrosion. Uncoated construction steel used as dowels or reinforcing is vulnerable to corrosion if exposed to moisture and deicing salts, so it is critical to maintain at least 2″ between the surface of a concrete structure and all steel reinforcing. Stainless steel or coated steel should be used for dowels and tie bars. Galvanized steel WWM is used to reduce surface cracking. Steel fibers mixed throughout the concrete can offer up to a 150% increase in flexural strength, reduced potential for cracking during concrete shrinkage and increased fatigue strength (Taylor et al. 2019). The carbon footprint of steel, particularly stainless, is substantial, but can be greatly reduced with higher recycled contents. Steel reinforcing in concrete may inhibit reuse of concrete slabs and sometimes recycling of concrete for recycled concrete aggregate.
Fiber reinforced polymer dowels (FRP)	Fiber reinforced polymer dowels are made from a resin or polymer binder, reinforcing element, and inert materials. Reinforcing elements are made from fiberglass, carbon fiber, or graphite fiber. Inert fillers are calcium carbonate, clay, or hydrated alumina. FRP dowels are non-corroding and have the potential to create durable concrete structures. Embodied carbon is higher than steel. FRP may inhibit reuse of concrete slabs and sometimes recycling of concrete for recycled concrete aggregate.
Synthetic fiber reinforcing	Synthetic fiber reinforcing fibers are made from materials such as acrylic, polyethylene, polypropylene, aramid, carbon, and nylon. Some fibers are made from waste plastics. Polyethylene and polypropylene are most used in concrete structures because they are non-absorptive and not susceptible to alkali attack. Low dosages of synthetic fibers are used to reduce plastic shrinkage and settlement cracking. Higher doses of macro synthetic fibers reduce cracking, add flexural toughness, and improve surface durability and fatigue resistance. The carbon footprint of synthetic fibers is lower than steel reinforcing used to control cracking. Fiber reinforcing makes it easier to reuse concrete slabs or crush concrete for RCA (Figures 14.10–14.12).
Carbon reinforcing	Carbon materials for concrete reinforcing are relatively new products. They take the form of coated mats, mesh, and rods. Rods are made of carbon fiber reinforced plastics and can withstand higher tensile forces than steel. Mats form carbon fibers into a yarn and then are processed into a lattice structure. Carbon is four times lighter with six times more load bearing capacity than steel (Carbocon 2024). Since carbon does not rust, carbon reinforcing does not require a 2″ cover of concrete, potentially allowing for thinner concrete sections. Carbon reinforcing is often used with thin sections of ultra high-performance concrete (UHPC).

Project Delivery Processes for Low-Carbon Concrete

- Develop carbon reduction goals in pre-design with the entire design team and, if possible, concrete producers and contractors.
- Reference industry benchmarks for GWP of concrete in the region of the project or reference GSA benchmarks. Set carbon targets at a percentage below these benchmarks.
- Set a carbon budget for all the concrete used in a project's building(s), pavement, and other structures.
- Identify structures where 56-day full strength requirements can be used instead of the standard 28-day full strength requirements.
- Understand what technologies, SCMs and alternative cements are available locally.
- Use performance-based concrete specifications instead of prescriptive specifications.
- Reference the National Ready Mix Concrete Associations Guide Specification for Concrete (NRMCA 2021).

UHPC is used in precast site furnishings, wall panels, and some bridge infrastructure projects. UHPC produces products that are up to ten times stronger than conventional concrete with compressive strengths ranging from 117–200MPa (17,000–29,000psi) (Figure 14.13).

UHPC mixes contain as much as 30% portland cement – three times the cement of conventional concrete. But structures made from UHPC are usually delicate or hollow, using lower quantities of concrete overall. Some mixes incorporate silica fume as an SCM replacing a relatively

Figure 14.10 At the Livestrong Foundation Headquarters, Ten Eyck Landscape Architects stacked concrete slabs reclaimed from concrete pavement demolished on site. The wall uses a small amount of mortar to hold the slabs in place and it rests on a compacted aggregate base. Image credit: Bill Timmerman

Figure 14.11 Larger reclaimed concrete slabs at the Livestrong Campus Headquarters were used vertically for a freestanding wall at the entry. The concrete slab wall is supported with attachment to steel angles embedded in concrete footings. Image credit: Bill Timmerman

Figure 14.12 Ten Eyck Landscape Architects transformed acres of concrete and asphalt pavements at the Arizona State University Polytechnic Campus. Removed concrete slabs were repurposed for use in a garden pavement on campus. Image credit: Bill Timmerman

Figure 14.13 Sculptural UHPC picnic tables and benches. UHPC uses more cement than traditional concrete, but less UHPC is used often evening out the additional embodied carbon from the higher cement content. Image credit: Courtesy of Landscape Forms

small portion of the portland cement. Given the shift in the concrete industry toward reducing the carbon footprint of PIP and precast concrete products, some UHPC producers are exploring mixes with greater replacement of portland cement. UHPC's low porosity and water permeability imparts strong weather resistance. Its strength allows for thin section, intricate forms, perforated panels, and lattices (NPCA 2013).

Carbon Capture, Utilization, and Storage (CCUS) in Concrete

CCUS in cement, concrete, and aggregates are a relatively new technologies that can offset some of the carbon emissions from portland cement. Some of these technologies are on the market now, while others are being tested in demonstration projects. This is a rapidly changing area of the concrete industry with ongoing research on performance and carbon accounting.

Ideally carbon will be captured near where it will be stored in concrete products, fresh concrete, or aggregates as trucking carbon has a carbon footprint. Transport of the captured carbon through pipelines is in experimental phases and will require investment in new infrastructure.

Carbon Capture in Cement Manufacturing

Carbon capture in cement manufacturing is an important method of reducing carbon emissions, and the cement industry is counting on this technology to help the industry get to net-zero emissions. Post-combustion process emissions can be captured or purified from cement kiln flue gases. These technologies can be applied to existing cement plants and do not require fundamental modifications to cement kilns if space allows.

Chemical absorption is the most advanced technology and can capture up to 95% of process CO_2 emissions from a cement kiln. *Membranes*, *calcium looping*, and *direct separation* are newer technologies that are being developed (ECRA 2017). Implementation of carbon capture systems that could potentially capture virtually 100% of cement manufacturing emissions is underway in Canada.

Captured CO_2 from process emissions can be stored geologically, in enhanced oil recovery, in inert materials by reacting it with natural alkaline materials and as a feedstock to produce chemicals and fuels. It is used to cure concrete and form carbon aggregates as discussed in the next sections.

Carbon Storage in Concrete

Carbon storage in concrete is a key aspect of the concrete industry's goal of carbon neutrality by 2050. The Global Cement and Concrete Association's roadmap for net-zero concrete assigns 36% of their planned reductions to carbon sequestration in concrete (GCCA 2021).

Sequestering CO_2 in concrete traps carbon dioxide through the process of mineralization. This is accomplished in two ways: injecting CO_2 into fresh concrete during the mixing process and curing concrete in a CO_2 rich atmosphere. In both methods, carbon dioxide diffuses into the concrete and transforms into solid calcium carbonates ($CaCO_3$).

Injecting CO_2 into fresh concrete during the batching and mixing process allows for quick mineralization of the carbon, locking it into the fresh concrete and ultimately the structure. This improves the performance of concrete through enhanced compressive strength. Theoretically this strength increase could allow for use of less cement, further reducing the concrete's carbon footprint; however additional research is needed in this area.

This carbon sequestration technique is currently on the market with a retrofit technology that can be installed in any ready-mix concrete plant. Purified post-industrial CO_2 from industrial sources is used, so transportation and processing of the CO_2 must be factored into the carbon accounting (Figure 14.14).

Curing concrete in a CO_2 rich atmosphere is the other method of storing carbon in concrete. Because this process must take place in a controlled environment, it is currently

Figure 14.14 The team at CMG Landscape Architecture used a carbon sequestering technology in the concrete for this bench at the Yerba Buena Island Dog Park in San Francisco. Image credit: CMG Landscape Architecture

used only in precast concrete applications such as concrete blocks, pavers, and wall panels.

In this approach to carbon storage, CO_2 replaces steam as the method for increasing curing and strengthening. Carbonation increases strength by about 20% in the first 24 hours, so this method is good for precast concrete unit production. CO_2 curing also reduces the permeability of the concrete unit, improving durability and resistance to sulfate attack. Carbon cured block and pavers are becoming widely available. Curing concrete in a CO_2 rich atmosphere stores more carbon than the method of injecting it into a fresh concrete mix.

Carbon Storage in Aggregates

Manufactured carbon aggregate products are now on the market in a limited way. These products use a variety of technologies to produce carbon negative, lightweight aggregates for use in concrete.

One technology uses waste carbon gas to treat thermal wastes and form stable carbonates that are blended with binders and fillers to form artificial limestone. Another product uses recycled concrete and an alkaline carbonate solution to grow the carbonate to the desired size. Ideally carbon aggregate producing facilities will be co-located with a high-carbon emitting source like cement kilns or steel processing facilities. One company estimates it stores 18kg of carbon for every 45kg of aggregate produced (Blue Planet Systems 2024).

Carbon Storage in Admixtures

Biochar is a carbon rich product made by pyrolysis of organic wastes. The high temperatures in the absence of oxygen encapsulate the carbon in the biochar, storing about twice its weight in CO_2. When added to concrete, it sequesters carbon in the structure. Biochar is not a supplementary cementitious material, instead it is considered an admixture.

The benefits of adding biochar to concrete are twofold. First, the carbon is sequestered; and second, this process avoids the methane emissions associated with municipal waste treatment plants. When added to concrete, biochar can give the concrete a darker, marbled look. Biochar is not recommended to be added to concrete at more than 5% of a mixes weight, and its effect on concrete performance requires additional study (Senadheera 2023).

Concrete and Natural Carbonation

Concrete naturally sequesters CO_2 over the lifetime of a concrete structure in a process known as carbon uptake or carbonation. CO_2 penetrates the surface of hardened concrete and chemically reacts with cement hydration products to create carbonates. Carbonation is not always recognized in greenhouse gas accountings of concrete structures, but the carbon uptake can be substantial over time.

The rate of CO_2 uptake depends on several factors. Binder constituents, exposure to air, surface orientation, surface to volume ratio, surface treatment, porosity, strength, humidity, temperature, and ambient CO_2 concentration are all factors that affect the carbonation of concrete. Rates of CO_2 uptake are highest when the surface to volume ratio is high and it is exposed to air, such as with porous concrete or crushed concrete.

While the rate of carbonation decreases over time, at end of life when concrete is broken or crushed it can potentially absorb significant additional amounts of CO_2 within a few years. Leaving crushed concrete exposed to the air before it is reused is beneficial for maximum CO_2 uptake. Quantifying the rate of uptake is extremely challenging. A conservative estimate is that concrete will absorb 20% of the carbon released during manufacture of its cement (The Concrete Centre 2016).

Emerging Technologies in the Concrete Industry

In addition to CCUS, there are several emerging technologies that can improve the carbon footprint and environmental performance of cement and concrete. Several are in the early stages of development, while others are commercially available in a limited way (Table 14.10).

Table 14.10 Emerging Technologies in the Concrete Industry

Technology and Environmental Benefit	Discussion
Alternative clinker technologies and cements Potential for substantial carbon reductions	Alternative clinker technologies, also called low-carbon cements, are substances made from alternatives to portland clinker that exhibit the properties of OPC. These technologies either fully or partially avoid use of OPC. Many are produced using less energy and release fewer emissions in production. Some enhance the properties of concrete. By using alternative raw materials and reducing or eliminating the use of limestone, these clinkers can reduce the process emissions from limestone calcination that currently account for around 60% of the carbon emissions of OPC. Alternative clinker technologies are discussed in Table 14.7. Other emerging alternative clinkers are made from food waste, agricultural waste, biomass, seashells, and plastics.
Bio cements Potential reduced carbon footprint	Bio cements combine calcium with biobased materials to form or "grow" cementing materials. One technology combines calcium, carbon, a proprietary biomaterial, and aggregates (Biomason 2024). Another developed a process in which the reaction of urea with calcium ions in industrial carbide sludge forms a hard solid (Hio 2022).
High-volume SCM/PLC mixtures Potential for substantial carbon reductions	High-volume use of supplementary cementitious materials and fillers are gaining traction, particularly in Europe. High-volume fly ash concrete replaces 50–100% of portland clinker with fly ash. This produces a concrete with greater workability, later-age strength, alkai silica resistance, and reduced permeability. Challenges include variable chemical compositions of the fly ash and eventually reduced supplies. High-volume ground limestone is used in a relatively new technology called a "high-filler cement". These products used sophisticated grinding processes to optimize particle size distribution of limestone. Unlike portland clinker, they require no calcination and avoid the CO_2 emissions associated with this process. Other SCMs can be used with high-filler cement to further reduce the portland clinker. High-filler, low-clinker cements can reduce the carbon footprint of cement by 75% using as little as 20% portland clinker (Du et al. 2021).
Carbon sequestration in concrete Potential for substantial carbon reductions	Some of the most recent innovations in the concrete industry are in CCUS. Carbon is captured in cement production and used to cure concrete and create carbon aggregates. Some companies are working toward carbon negative concretes that store more carbon than is generated making them. Some of these technologies are on the market now and others are in research and testing phases. This is a rapidly changing area of the concrete industry and research on performance and carbon accounting is ongoing and necessary to ensure that the technologies are producing durable concrete with carbon benefits. CCUS is discussed previously in this chapter.
Photocatalytic cements Air pollution reduction	Photocatalytic concrete contains a catalyst, titanium dioxide, that interacts with sunlight to abate organic and inorganic air pollution through an oxidation process that converts noxious compounds to harmless ones. The titanium dioxide catalyst can be applied on a structure's surface as a coating, or it can be mixed into cement as a constituent of concrete. Owing to the cost of photocatalytic cement, it is usually used in a 9mm concrete topping layer rather than through the entire section of a concrete structure (Heidelberg Materials 2024). Photocatalytic surfaces on pavements and sound walls may someday be widespread enough to reduce some ground-level air pollution. Research, primarily in Europe and Japan, has shown concrete with a titanium dioxide (TiO2) surface layer to be a promising technology for reducing air pollutants, with reductions of NO_x ranging from 15% to 70% (Peters 2021). This wide range of results is owing to variables of wind, humidity, orientation, and urban configuration. If humidity is too high, the photocatalytic reaction may not work or may even exacerbate pollutant intensities. An ideal setting for maximum pollutant removal of photocatalytic paving or walls is one with low average humidity, low wind, and high UV intensity. Concrete is an ideal material for photocatalytic surfaces, as the reaction products can be absorbed at the surface then washed away with rain. Other applications of photocatalytic paving use a coating agent that is applied after the concrete cures. Lab tests show that these products may wear off over time, particularly with vehicular traffic, losing their effectiveness. Application on porous pavements can protect some of the coating from wear as it is slightly below the surface (Heidelberg Materials 2024).
Construction automation and efficiencies Some carbon reduction	Multiple technologies are increasing efficiencies of concrete on construction sites. On-site automation is being used for complex tasks such as excavation, mix efficiency, and concrete curing. Augmented reality (AR) and mobile interfaces are used to train builders and transmit important data to them during complex tasks. Using AR to provide visual instructions can improve the efficiency of application and embedded sensors can facilitate the use of alternative cements with real time curing data for maximizing strength and performance.

(*Continued*)

Table 14.10 (Continued)

Technology and Environmental Benefit	Discussion
3D Printed concrete Reduced waste from concrete and formwork	Three-dimensional printing of concrete structures is an emerging technology that can print complex and precise shapes in concrete structures without use of formwork. In 3D printing, concrete is extruded from a nozzle and automatically deposited in layers according to a digitally designed printing route without use of formwork (Figure 14.15–14.17). Advantages to 3D printing are high efficiency and flexibility of construction, reduced waste and dust generation, lower quantities of concrete due to hollow structures, and the ability to print complex forms (Souza et al. 2020). The typical mix for 3D printed concrete replaced larger aggregates with greater quantities of fine aggregates. Since large aggregates are essential to controlling excess shrinkage, some 3D printed concrete mixes are susceptible to cracking. Fiber reinforcements and shrinkage reducing admixtures have been tested to address this issue. Greater amounts of cement are required (up to 30%) without coarse aggregates, so testing is being done with high-volume SCMs and geopolymer cements (Xiao et al. 2021).
Energy storing concrete Potential energy storage benefits	Recent research is examining the potential for concrete to store thermal energy and electricity. A US start up mixes inexpensive carbon black into concrete in a specific way that forms a long branching network of carbon "wires" as the concrete cures. In this way, a concrete structure can act as a super capacitator – a device that stores an electric charge (Peters 2023). Developers of this technology envision roads made from this type of concrete that could charge EVs.
Luminescent concrete Potential energy savings	While concrete and cement are opaque materials, scientists have developed additives that modify the optical properties of the materials to become phosphorescent. By using additives that prevent the formation of crystals that occur in the production of cement, a material with non-crystalline structure like glass allows the passage of light inside. Phosphorescent materials absorb energy from the ultraviolet light of the sun and emit the energy later as light. This concrete technology could save energy from light fixtures by illuminating structures at night (Carreño 2016).
Self-healing concrete Increased concrete durability	Multiple emerging technologies for self-repairing or "self-healing" concrete are in the development phase. The term "self-healing" refers to a concrete material that can regain strength and heal itself after it has cracked. Some concrete can heal itself through the natural hydration of minerals present in the concrete composition, but this is not easy to control and only works for very small cracks. Technologies for self-healing concrete include built-in capsule-based, vascular, and anti-microbial healing. Biomineralization is a process where bacteria metabolically produce minerals to help repair cracks and other damage (Spinks 2015). Another technology uses carbonic anhydrase – the same enzyme that transfers CO_2 from cells in the blood stream. When added to cement, it helps turn CO_2 in the atmosphere into calcium carbonate crystals (Peters 2021). Products can be applied as a spray for existing concrete structures, a repair mortar for larger damage, and as a concrete admixture.
Reactivated cements	A very new technology has been developed that recycles cement by crushing recycled concrete into a very fine powder. This "concrete paste" is used in a steel electric arc furnace as a fluxing agent in steel production. The resulting material is a slag that chemically resembles and performs like portland cement (Dunant et al. 2024).

While the concrete sector needs innovation to meet carbon reduction targets, changes do not come easily due to several factors including standards and codes that heavily rely on OPC, reluctance of engineers and contractors to vary from well-known norms, lack of performance information, high costs of alternative products, and lack of standards for emerging technologies.

New technologies in any field can take a long time to move from the laboratory to the marketplace, but recent sustainable concrete technologies have experienced challenges in scaling up to the global concrete trade, and some companies with promising technologies have gone bankrupt before the market can embrace their products. Cost, availability, and lack of performance standards, and data are other challenges to their adoption.

Optimizing the Design of Concrete Structures

Optimized design of concrete structures is one of the most effective measures to reduce their embodied carbon and environmental impacts. Measures include optimizing

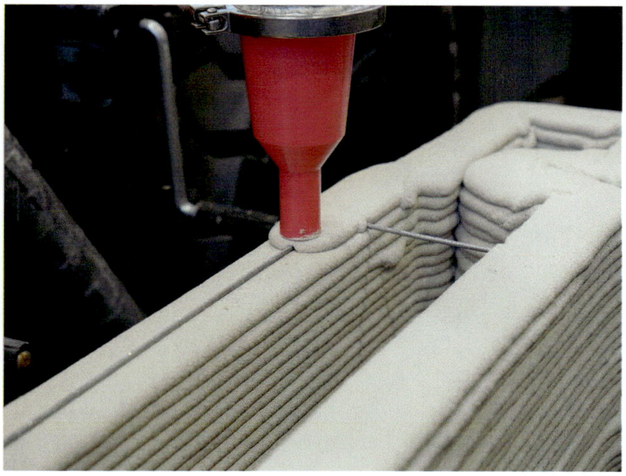

Figure 14.15, Figure 14.16 and Figure 14.17 Lindsey Heller of SKAPA Landscape Architecture + Design Visualization is using 3D printed concrete for walls, footings, benches, and even permeable pavement cells. When 3D printing concrete structures, the concrete mix is extruded from a nozzle and automatically deposited in layers achieving complex and precise shapes without formwork. SKAPA 3D prints their own nozzles for some applications. Greater percentages of cement are used; however, as many of the structures are hollow, they use lower amounts of concrete materials overall for potentially comparable embodied carbon and resource use. Image credit: SKAPA

Table 14.11 Strategies to Optimize Cement Use in Concrete Structures

Strategy	Considerations
Matching concrete strength requirements to the intended use of the structure	Higher strength concrete uses greater amounts of cement, therefore, matching the strength requirement of the concrete to the actual strength needed without over designing the structure can reduce cement quantities and reduce the carbon footprint of a concrete structure. For instance, the amount of portland cement in 3000psi/20.7MPa concrete is just 81% of the amount used in 4000psi/27.6MPa concrete. While it is simpler to just specify to the highest strength needs on a project, matching the strength of concrete to specific performance needs can reduce embodied carbon substantially.
Using sensors to avoid overdesigning structures	Digital sensors can log concrete strength and temperature on-site, eliminating the need for excess cement use. They can provide far more accurate real time information than traditional mechanical tests. Use of digital sensors is particularly important when low-carbon mixes are employed as they often have longer cure times.
Longer strength requirements (28 vs. 56 or more days)	Specifying that a concrete structure meet full strength at 56 days or more instead of 28 days allows for reduced cement content in a concrete mix. The amount of cement reduction will vary by concrete mix, but some estimates suggest that this practice could reduce the amount of cement in a mix by 8–10% (The Concrete Centre 2021). In addition to reduced cement, a longer period of time to achieve full compressive strength allows for higher percentages of SCMs, and the strength of concrete with SCMs is usually higher than that of PC concrete at all of the later intervals (56 days, 92 days, 120 days). The reduction of cement has been up to 65% in mixes with both slag and fly ash when the strength requirement was 92 days of greater (Mitchell 2024).
Optimized aggregate grading	One of the most effective ways to reduce cement quantities without reducing strength is to select aggregate sizes to optimize packing and reduce the void spaces that are filled with cement paste. This allows for a reduction in cementitious material while still achieving the required fresh and hardened concrete properties. Flow and compaction can be maintained with admixture dispersants.
Chemical admixtures	Chemical admixtures improve the performance of cement-based materials. Plasticizers and superplasticizers improve the efficiency of cement use as they reduce the amount of water needed to make concrete that is easily placed so they reduce cement use for a given strength and durability. They also allow higher use of fillers and SCMs such as calcined clays. Accelerators enable faster strength development allowing use of higher percentages of SCMs. Other chemical admixtures such as air entrainers, corrosion inhibitors, and shrinkage reducers will prolong the service life of structures.

concrete mixes for reduced use of OPC, designing durable concrete structures, and designing smaller, more efficient structures (Table 14.11).

Design decisions by designers and structural engineers impact quantities of materials used and have a direct influence on the environmental impacts and carbon footprint of concrete structures. Variables include the thickness and extent of concrete pavements, retaining walls, and freestanding seat walls. Thicker walls and pavements may require more carbon-intensive steel reinforcing as well.

Reducing pavement and wall thicknesses to the thinnest structurally sound dimension can save substantial concrete use and lower the carbon footprint of a structure. For instance, a 15cm thick concrete pavement slab has just 75% of the embodied carbon of a 20cm slab.

Durable Concrete Structures

Designing structures for durability will increase their service life, reduce replacements, and ultimately reduce their associated environmental impacts. Concrete pavements are designed to last 30 years, yet unexpected conditions such as severe weather events, heavier loads than planned, lack of maintenance, unstable bases, poor construction, and poorly designed mixes can reduce their life (Tables 14.12 and 14.13).

A major issue that can reduce the use life of concrete structures is corrosion of steel reinforcing. Corrosion results from moisture and chlorides, reaching reinforcing through cracks and fissures.

Responsible Sourcing of Concrete

The Concrete Sustainability Council (CSC) Certification System is a voluntary product certification for the concrete industry that encourages environmentally and socially responsible sourcing and practices. Ready-mix concrete plants and precast concrete plants

Table 14.12 Strategies for Durable Concrete Structures

Topic	Consideration
Reinforcing	All steel reinforcing should be held at least 50mm clear of all concrete surfaces. Fiber reinforcing can be used to control cracking of pavement slabs. Steel and FRP reinforcing may inhibit reuse and recycling of concrete after its service life.
Control joints	Control joints should be used in roughly square shapes to control cracking every 2–3 meters.
Expansion joints	Expansion joints should be employed at other structures and every 8–10 meters in pavements and walls.
Mixes	Reduction and removal of mineral salts found in the cement, aggregates, sand, and the water used in mortar and grout is important to the durability of the concrete structure. Low alkali portland cement, low water-to-concrete ratio, and clean, potable, salt free water are all helpful. Penetrating concrete sealers can also be used (Faber 2024). In frost prone regions, durability of concrete can be achieved by adding air-entraining admixtures to the mix. Reducing the maximum size of coarse aggregate will also increase durability.
Waterproofing	Waterproofing admixtures can protect concrete from water intrusion where applicable. Ensuring positive drainage across all concrete structures and avoiding standing water will minimize water damage. Good drainage behind concrete retaining walls with gravel and perforated pipe or weep holes will increase the service life of the wall.
Water to cementitious material ratio	Low water to cementitious material ratio can be used for concrete subjected to severe exposure. Too much water in a concrete mix will lower strength and increase permeability and scaling. SCMs generally increase viscosity of concrete mixes, reducing the need for water.
Construction	Moisture and humidity are critical during the curing period to keep sufficient hydration of cementitious materials and avoid shrinkage and subsequent cracking.
Subgrade	Pavement and wall footings should be placed on well-compacted subgrade or aggregate bases as required. Long-term uniformity, stiffness and drainage are critical to prevent cracking, settlement, or frost heave.

Table 14.13 Strategies for Refurbishing Concrete Pavements

Strategy	Considerations
Concrete overlays	Concrete overlays are a 5–25cm layer of concrete that is poured over an existing concrete, asphalt or composite pavement. They can be bonded (creating a monolithic structure) or unbonded to the exiting pavement underneath. Bonded overlays are 5–12cm. Joints in new overlay should be aligned with joints in existing pavement. Any cracks in the existing pavement must be repaired prior to placement of a bonded overlay. Unbonded overlays create a new structure on a stable base (the old pavement). If the overlay is unbonded, a separation layer prevents the cracks in the existing pavement from impacting the overlay. The separation layer is 25mm of hot-mix asphalt pavement. Unbonded overlays are 15–28cm thick.
Polymer-modified concrete (PMC) overlays	Polymer overlays are layers that can be added to the top of an existing concrete pavement slab or bridge. The layers are about 6.4mm thick and are a mix of a polymer binder, and natural or synthetic aggregates. Portland cement is sometimes added to the mix as well. They can seal and protect the concrete slab from deicing salts and other chemicals and can prolong the life of concrete pavements. The aggregate in the overlay can be decorative, provide texture, level uneven surfaces, and provide traction. The concrete must be structurally sound, without gaping cracks, severe delamination, or an unstable base.
Concrete resurfacers	Concrete resurfacers are a blend of no-shrink portland cement, sand, and polymer additives. They are very similar to PMC overlays, but portland cement is the main binder along with some polymer additives.
Resin bound and resin bonded surfacing	Resin bound or bonded surfaces can be installed over existing concrete slabs. They are 6mm to 25mm thick. Safety data sheets and, where available, HPDs should be consulted for human and environmental health risks and VOC contents of synthetic binders.
Epoxy stone overlays	An epoxy stone overlay is similar to PMC overlays and resin bound surfaces as it is a layer that can be placed on top of an existing concrete pavement. Epoxy is mixed with decorative gravel or river stone in a 12mm layer. The appearance is like exposed aggregate concrete. Epoxies should be UV stabilized.
Tinted sealer and concrete stains	Tinted sealers and concrete stains add color to concrete surfaces. Most tinted sealers are translucent and maintain the look of concrete. Therefore, cracks and chips in the existing concrete will still show. Tinted sealers can restore color to a slab, or they can create a completely new tint. To protect the color, a clear sealant layer is recommended.

can obtain a "CSC Certificate" and cement, slag, and aggregate suppliers can obtain a "CSC Supplier Certificate".

The Concrete Sustainability Council, founded in 2016, grew out of the Cement Sustainability Initiative (CSI), a sector project within the World Business Council for Sustainable Development. The CSC is made up of industry partners from Europe, North America, Latin America, and Asia.

Items of focus for the CSC include greenhouse gas accounting, portland cement alternatives, natural aggregate alternatives, alternative energy use, quarry restoration, air pollutant reduction, and ethical sourcing of materials. CSC certification is recognized by environmental design voluntary rating systems LEED, BREEAM, DGNB, and SITES. CSC certification can contribute to achievement of responsible sourcing credits (CSC 2024).

Decarbonizing the Concrete Industry

For over two decades, the concrete and cement sectors have been taking steps to reduce the carbon footprint of their materials and products. This has resulted in a 19.2% reduction in global average CO_2 intensity of cement production between 1990 and 2018 (GCCA 2024). This is primarily due to energy efficiency and equipment improvements in cement production, use of alternative fuels, and a reduced clinker-to-cement ratio through use of SCMs and fillers.

To meet the Paris Climate Accord's target of staying below 2°C of warming this century, the Intergovernmental Panel on Climate Change (IPCC) has mandated carbon neutrality for all sectors by 2050. The global cement and concrete sectors are engaging in several approaches to substantially reduce their carbon footprint (Table 14.14). The Mission Possible Partnership sets forth a goal for reducing concrete emissions by 50% in the mid-2030s to achieve net zero by 2050 (Mission Possible Partnership 2023).

Table 14.14 Cement and Concrete Sector Strategies for Carbon Neutrality by 2050

Decarbonization Pathway and % of Cumulative Reduction 2022–2050[1]	Discussion
Using concrete more efficiently 22%	Where possible designers and engineers can use alternative building materials to reduce use of concrete. Where concrete must be used, structures can be designed for an increased service life, refurbished and reused, and designed to use less concrete and cement. They can optimize design of structures as well as design of concrete mixes to reduce quantities of cement and concrete. Efficiencies can be found during construction with use of sensors and monitoring equipment.
Improving the energy efficiency of cement and concrete production 9%	Improving energy efficiency of cement and concrete production involves deploying the most energy efficient technologies in new cement and concrete plants and retrofitting existing facilities to improve energy performance. This involves upgrading kilns and equipment and changing plant design. Companies are increasingly using "smart" devices to track and monitor operations and machine learning to improve process control. Recovering waste heat can reduce cement factories operating costs 10–15% (GCCA 2021). Achievement of energy use reductions will require capital investment in more energy efficient plants and equipment.
Reducing process emissions 5%	The amount of portland cement clinker used can be reduced with use of clinker substitutes such as LC3 cement, fly ash, GBFS, limestone, or natural pozzolans (Table 14.2). The portland cement clinker ratio can also be reduced with use of alternative clinkers (Table 14.7). A clinker factor of 0.6 (mixes using only 60% portland cement) is the industry goal for 2050 with a 0.65 clinker factor goal for 2030 (GCCA 2021). The current global clinker factor is around 0.76. In the United States, the average clinker factor is 0.88; in the European Union, India, and China it is between 0.64 and 0.76 (Gangotra et al. 2024). The 0.6 clinker factor is estimated to mitigate almost 0.2 gigatonnes (GT) of CO_2 in 2050. If the clinker-to-cement ratio was lowered worldwide to 0.3, 1.5GT of CO_2 emissions could be saved by 2050 (Lehne and Preston 2018). Challenges include the uncertain availability of clinker substitute materials, inexperience with SCMs and alternative cements, and lack of user demand. Most of the alternative clinkers discussed in Table 14.7 are not widely used and some are not yet commercially available. Those that are commercially available are not often cost competitive with OPC, and they face raw material shortages and/or resistance from users.

(*Continued*)

Table 14.14 (Continued)

Technological innovations: –CCUS and other new technologies – Excess heat recovery for power generation – Renewable power generation and alternative fuels 45%	*Carbon capture, usage, and sequestration technologies* are being used to capture CO_2 from cement kilns post combustion. This captures process emissions from limestone calcination, which account for 60% of the carbon footprint of portland cement. Post-combustion capture technologies do not require major modifications to cement kilns and can be applied to existing facilities. Other CCUS technologies incorporate CO_2 from other industrial sources. Natural uptake of CO_2 by concrete is also included. *Excess heat recovery (EHR) technologies* for power generation capture the waste heat from cement kilns for reuse in other phases of cement manufacturing. EHR are not widely used worldwide, but they hold potential to substantially improve energy efficiency of cement production. EHR technologies are implemented most often in China with almost 90% of its domestic clinker production capacity equipped with such technologies. India, Japan, and Korea also widely use EHR technologies. *Renewable power generation* can reduce carbon emissions from the cement and concrete industry. Renewable-based options available to cement manufacturers include wind power, solar photovoltaic power, solar thermal power, and small hydropower generation.

[1] Mission Possible Partnership 2023

References

Alliance for Low Carbon Concrete and Cement. "The future depends on how we build it". https://alliancelccc.com/, accessed July 2023.

Alsadey, S. (2019). "Properties of concrete using crushed brick as coarse aggregate". *International Journal of Advanced Mechanical Civil Engineering* 6: 43–47.

American Concrete Institute (ACI). (2019). *Report on the Role of Materials in Sustainable Concrete Construction*. ACI PRC-130-19. ACI.

Antunes, M., et al. (2021). "Alternative clinker technologies for reducing carbon emissions in cement industry: A critical review". *Materials (Basel)* 15.1 (December 28): 209.

Architecture 2030 (2024). "Carbon smart materials palette: Concrete." www.materialspalette.org/concrete/, accessed July 13, 2024.

ASTM International (2023). *ASTM C595/C595M-23 Standard Specification for Blended Hydraulic Cements*. West Conshohocken, PA: ASTM International.

Bernot, Melody J., et al. (2011). "The influence of different urban pavements on water chemistry". *Road Materials and Pavement Design* 12.1: 159–176.

Biomason (2024). "Biocement grows in ambient temperatures, building with carbon to create concrete products". https://biomason.com/biocement, accessed July 13, 2024.

Blue Planet Systems (2024). "Carbon aggregates". www.blueplanetsystems.com/products, accessed July 13, 2024.

Boanda-Fuchs, A., Heierli, U., and Scrivener, K. (2024). *Low Carbon Cement: Harmonizing Environmental Goals and Housing Needs*. Swiss Agency for Development and Cooperation and Climateworks Foundation, May.

British Standards Institute (2023). *BS 8500-2:2023-TC Concrete: Complementary British Standard to BS EN 206 - Specification for Constituent Materials and Concrete*. BSI, November.

Carbocon (2024). "Carboncon: Looking to a lighter future in construction". www.carbocon.de/en/, accessed August 1, 2024.

Carreño, Berta. (2016). "Glow-hard: Luminous cement could light roads, structures". *Scientific American*.

The Concrete Centre (2021). *Specifying Sustainable Concrete: Using 56 Day Concrete Strengths*. Mineral Products Association, The Concrete Centre.

The Concrete Centre (2016). *Whole Life Carbon and Buildings*. Mineral Products Association, The Concrete Centre.

Concrete Sustainability Council (CSC) (2024). "CSC certificate of responsibly sourced concrete". https://csc.eco/certification/, accessed January 4, 2024.

CRDC Global (2024). "Introducing resin 8™". https://crdc.global/resin8/, accessed August 3, 2024.

Czigler, Thomas, et al. (2020). *Laying the Foundation for Zero-Carbon Cement*. McKinsey & Company, 9.

Debbarma, Solomon, Selvam, M., and Singh, Surender. (2020). "Can flexible pavements' waste (RAP) be utilized in cement concrete pavements? A critical review". *Construction and Building Materials* 259: 120417.

Du, Sen, Zhao, Qingxin, and Shi, XianmingXianming. (2021). "High-volume fly ash-based cementitious composites as sustainable materials: An overview of recent advances". *Advances in Civil Engineering* 2021.1: 4976169.

Dunant, C.F., et al. (2024). "Electric recycling of Portland cement at scale". *Nature* 629: 1055–1061. https://doi.org/10.1038/s41586-024-07338-8.

European Cement Research Academy (ECRA) and Cement Sustainability Initiative (CSI) (2017). *Development of State of the Art Techniques in Cement Manufacturing:*

Trying to Look Ahead. Düsseldorf and Geneva: ECRA. www.wbcsdcement.org/technology.
Faber, S. (2024). Personal communication, September 13.
Fan, Chengcheng, et al. (2021). "A comparative study on solidification/stabilization characteristics of coal fly ash-based geopolymer and Portland cement on heavy metals in MSWI fly ash". *Journal of Cleaner Production* 319.
Federal Highway Administration (FHWA) (2022). *Tech Brief: Use of Recycled Concrete Aggregate in Concrete Paving Mixtures*. FHWA-HIF-22-020. FHWA, February.
Gangotra, A., Kennedy, K., and Carlsen, W. (2024). *The US Needs to Lower Cement Emissions: Blended Cement Can Help*. World Resources Institute, May 9.
Gartner, Ellis, and Sui, TongboTongbo. (2018). "Alternative cement clinkers". *Cement and Concrete Research* 114: 27–39.
Global Cement and Concrete Industry (GCCA). (2021). *The GCCA 2050 Cement and Concrete Industry Roadmap for Net Zero Concrete*. GCCA.
Global Cement and Concrete Industry (GCCA). "Cement and concrete around the world". https://gccassociation.org/concretefuture/cement-concrete-around-the-world/, accessed May 18, 2024.
Global Cement and Concrete Industry (GCCA). "Key facts". https://gccassociation.org/key-facts/, accessed June 12, 2024.
Harrison, Edward, Berenjian, Aydin, and Seifan, MostafaMostafa. (2020). "Recycling of waste glass as aggregate in cement-based materials". *Environmental Science and Ecotechnology* 4: 100064.
Hart, J. (2022). *Cement: Raising Ambitions, Reducing Emissions*. EcoCem, November.
Heidelburg Materials. "TX active". www.heidelbergmaterials.us/products/cement/tx-active, accessed May 8, 2024.
Hio, J. (2022). "Renewable biocement made entirely out of waste material". *Industrial Equipment News*.
IIT Madras, Tara and IIT Delhi (2015). "Life cycle assessment of LC3". www.LC3.ch.
International Energy Agency. "Cement" www.iea.org/energy-system/industry/cement, accessed May 8, 2024.
International Living Futures Institute (2024). *Living Building Challenge Red List and CASRN Guide 2024*. IFLI.
Islam, Rashidul, et al. (2019). "An empirical study of construction and demolition waste generation and implication of recycling". *Waste Management* 95: 10–21.
John, Vanderley M., et al. (2018). "Fillers in cementitious materials – experience, recent advances and future potential". *Cement and Concrete Research* 114: 65–78.
Kurda, R., Silvestre, J.D., and de Brito, J. (2018). "Toxicity and environmental and economic performance of fly ash and recycled concrete aggregates use in concrete: A review". *Heliyon* 4.4 (April 23).
Lehne, Johanna, and Preston, FelixFelix. (2018). "Making concrete change". In *Innovation in Low-Carbon Cement and Concrete*. The Royal Institute of International Affairs, Chatham House.
Manjunatha, M., and Rakshith, S. (2022). "Use of waste foundry sand as fine aggregates for structural concrete – a review". *Journal of Engineering, Design and Technology* 20.6: 1468–1481.
Mission Possible Partnership (2023). *Making Net Zero Concrete and Cement Possible: An Industry Backed 1.5 °C-Aligned Transition Strategy*. Mission Possible Partnership.
Mitchell, J. (2024). Personal communication, October 8.
Moore, C. (2020). "Britain tops construction aggregates recycling figures". *New Civil Engineer*, June.
National Precast Concrete Association (NPCA) (2013). *Ultra High Performance Concrete: Guide to Manufacturing Architectural Precast UHPC Elements*. NPCA.
National Ready Mixed Concrete Association (NRMCA) (2021). *Guide to Improving Specifications for Ready Mixed Concrete*. 2PE004-21. NRMCA Publication.
National Ready Mixed Concrete Association (NRMCA) (2022). *NRMCA Member Industry-Average EPD for Ready Mixed Concrete*. NRMCA.
Natural Pozzolan Association. https://pozzolan.org/, accessed July 13, 2024.
Noller, C. (2020). *Wagners Geopolymer Concrete: Product Carbon Footprint Declaration*. The Footprint Company, February 20.
Ohemeng, Eric A., and Ekolu, Stephen O. (2020). "Comparative analysis on costs and benefits of producing natural and recycled concrete aggregates: A South African case study". *Case Studies in Construction Materials* 13: e00450.
Peters, A. (2021). "This self-healing concrete automatically fills in cracks". *Fast Company*, June.
Peters, A. (2023). "MIT researchers developed a new type of concrete that can store energy". *Fast Company*.
Pillai, Radhakrishna G., et al. (2019). "Service life and life cycle assessment of reinforced concrete systems with limestone calcined clay cement (LC3)". *Cement and Concrete Research* 118: 111–119.
Portland Cement Association (PCA) (2024). "Recycled aggregates". www.cement.org/learn/concrete-technology/concrete-design-production/recycled-aggregates#:~:text=It%20is%20generally%20accepted%20that,mechanical%20properties%20of%20the%20concrete, accessed August 2, 2024.
Provis, J. (2018). "Alkali-activated materials". *Cement and Concrete Research* 114: 40–48.
Raju, S., and Lalit, R. (2023). *Unleashing the Potential of Limestone Calcined Clay Cement LC3: Why India's New LC3*

Code Is a Game Changer for Decarbonizing Concrete. Rocky Mountain Institute, August 10.

Scrivener, Karen L., John, Vanderley M., and Gartner, Ellis M. (2018). "Eco-efficient cements: Potential economically viable solutions for a low- CO_2 cement-based materials industry". *Cement and Concrete Research* 114: 2–26.

Senadheera, Sachini Supunsala, et al. (2023). "Application of biochar in concrete – a review". *Cement and Concrete Composites* 143.

Souza, Marcelo Tramontin, et al. (2020). "3D printed concrete for large-scale buildings: An overview of rheology, printing parameters, chemical admixtures, reinforcements, and economic and environmental prospects". *Journal of Building Engineering* 32: 101833.

Spinks, R. (2015). "The concrete that can fix its own cracks". *The Guardian*, June.

Sublime Systems (2024). https://sublime-systems.com/technology/, accessed June 18, 2024.

Tam, Vivian W.Y. (2008). "Economic comparison of concrete recycling: A case study approach". *Resources, Conservation and Recycling* 52.5: 821–828.

Taylor, P.C., et al. (2019). *Integrated Materials and Construction Practices for Concrete Pavement: A State-of-the-Practice Manual*, Second Edition. Ames, IA: National Concrete Pavement Technology Center, Iowa State University, May.

UN Climate Change (2021). *Climate Action Pathway Human Settlements 2021: Vision and Summary*. UN Climate Change.

US GSA (2023). *Inflation Reduction Act Low Embodied Carbon Concrete Requirements*. US General Services Administration.

Van Dam, Thomas J., et al. (2015). *Towards Sustainable Pavement Systems: A Reference Document*. FHWA-HIF-15-002. Federal Highway Administration.

Xiao, Jianzhuang, et al. (2021). "Large-scale 3D printing concrete technology: Current status and future opportunities". *Cement and Concrete Composites* 122: 104115.

Yu, Tom, et al. (2017). "From research to reality can we implement performance-based specifications for durability and longevity of concrete? Will they work?" *Concrete International* 39: 39–46.

CHAPTER 15

Asphalt Pavement Materials

Asphalt pavement or bituminous pavement, formally called asphaltic concrete, is the most used paving material. Over 90% of new roads in the United States are surfaced with asphalt. Asphalt pavement is a mix of primarily asphaltic binder and aggregates. Other additives may be incorporated to impart specific properties to the mix. While asphalt pavements are associated primarily with vehicular applications, the continuous monolithic surface makes them suited to curvilinear paths and greenway trails. Asphalt pavement is also used as an underlying base for unit pavements and specialized surface applications.

Asphalt is inexpensive, the materials are readily available, asphalt plants are widespread, and asphalt pavements are durable and repairable. A wide variety of surface overlays and finishes can be applied to fit any design setting and extend the life of asphalt pavement (Figure 15.1).

Primary drawbacks to asphalt pavements include their contribution to urban heat islands, energy use and emissions from heating of asphalt pavement materials, and air emissions and particulates from heating at the plant. Impervious asphalt pavement, like other dense pavements, concentrates runoff and contributes to non-point source pollution.

Asphalt pavement mixes and surfaces can be specified to reduce these drawbacks with cooler mix temperatures and incorporation of high amounts of RAP. Surfaces for asphalt pavement can be employed to increase the surface reflectivity of the pavement, reducing contributions to urban heat islands (Table 15.1).

This chapter addresses the materials and construction of asphalt pavements, while Chapter 6 details asphalt pavement structures, and Chapter 7 covers porous asphalt pavement structures.

Asphalt Pavement Structures

Asphalt pavement is a flexible paving structure that tolerates minor movements from expansion and contraction of materials and soils. The mix is placed in layers called lifts and held together by tack coats. Asphalt pavement rests on a compacted aggregate base course and sometimes a subbase. In lighter applications with strong soil bearing, the asphalt pavement may rest directly on prepared subgrade.

Asphalt Pavement Course

Asphalt mixtures are heated at a plant and delivered to the site hot or warm. Mixes are placed with a mechanical paving machine or raked into place by hand, then roller compacted into 2–10cm layers. The asphalt pavement stiffens as it cools, although it will slightly soften and harden continuously while in use with temperature fluctuations. Prolonged exposure to UV radiation and/or high heat can oxidize the surface of the pavement (Marks 2024).

Asphalt pavements can be comprised of multiple layers depending on expected design loads and vehicular traffic. Generally, the heavier traffic, the greater number of layers or layer thicknesses resulting in greater overall pavement thicknesses. Asphalt pavements for pedestrian and light vehicular applications will usually be one 5–10cm layer. Where multiple layers are employed, the top layer is called the wearing course, and the lower layers are called the base courses.

Asphalt pavement can have a wide variety of surface courses and treatments that will impact the appearance and performance of the pavement as well as extend the longevity of the installation (Table 15.1 and Figure 15.1).

Prime and Tack Coats

Prime and tack coats bond the layers of asphalt pavements into one monolithic system. *Prime coats* bond layers to the base and help to waterproof it (Willis 2024). *Tack coats* bond upper layers together. Poor tack and prime coat performance can lead to compromised bonding of the layers, slippage cracking, delamination, fatigue cracking, and potholes. The tack coat is a sprayed application of an asphalt binder or emulsion on an existing asphalt, new asphalt, or an existing

DOI: 10.4324/9781003413974-17

Table 15.1 Selected Asphalt Pavement Surface Courses

Type of Surface, Thickness, and Life Expectancy	Use, Performance, and Other Considerations
Surface Courses for Improved Pavement Performance	
Open graded friction course (OGFC) 18–52mm thickness 8–12 years	An open graded friction course provides improved surface drainage to an asphalt pavement during rainfall. Water drains vertically through the surface course to the underlying dense graded asphalt pavement layer below where it drains laterally to the edge of the pavement. The OGFC layer has 15–25% air voids and is typically 18–52mm thick. Benefits include improved safety due to reduced splash and spray, reduced risk of hydroplaning, improved skid resistance, and reduced tire noise in dry conditions.
Polymer-modified asphalt (PMA) in surface courses Approx. 50mm thickness 18+ years	PMA can improve durability and longevity of surface courses. Rutting, fatigue, and fracture are also improved. Where available, recycled polymers should be used in PMA surface courses as virgin polymers will add substantially to the embodied carbon of a mix.
Rubberized asphalt (RA) in surface courses 18–52mm thickness 10–15 years	Rubberized asphalt mixes used in surface courses are durable and highly resistant to reflective surface cracking. Use of RA can allow for thinner surface courses using less material. The binder for RA surface courses contains about 20% ground rubber from recycled tires. There can be health concerns with use of recycled tires as when they are heated or chipped, they can release PAHs that could be harmful to workers. Runoff from these surfaces could contaminate adjacent environments (US EPA 2024a; Habitable 2020, 2024a, 2024b; USGBC 2019).
Stone matrix asphalt (SMA) surface courses 50–75mm thickness 20+ year	Stone matrix asphalt surface courses, also called stone mastic asphalt, are dense, gap graded mixes that have a stable stone on stone structure and a highly designed asphalt binder with filler, fibers, and/or asphalt modifiers. SMA surface courses are primarily used for high volume and heavy loading applications. This is a premium mix that has improved durability, fatigue, and rutting resistance. Wet weather friction is improved as well.
Surface Courses for Pavement Preservation	
Chip seal 6–12mm thick 3–7 years	Chip seal is a single-size crushed aggregate and binder layer that can be used over new or existing asphalt pavement, or over an aggregate base in very low traffic applications. It is not for heavy truck traffic or greater than 8% grades. Chip seal can be placed with hot, warm, or cold mix applications. The color of the aggregates in chip seal will determine the color of the surface, therefore it can be useful for improving the UHI performance of asphalt pavement. Multiple chip seal layers can be employed with coarser aggregate used in the lower layers.
Thin asphalt overlays 25–38mm thick 7–10 years	Thin asphalt overlays, sometimes called non-structural overlays, are applied to an existing pavement to preserve and rehabilitate it. They improve the pavements condition and extend its life. Thin asphalt overlays can be dense graded, open graded, or stone matrix asphalt.
Slurry seal 3–10mm thick 3–8 years	Slurry seal is a cold mix thin surface treatment of emulsified asphalt, dense graded fine aggregate, mineral filler, and water. It has a smooth texture that is good for recreational paths but offers less skid resistance for roads.
Micro surfacing 10–22mm thick 5–8 years	Micro surfacing, an enhanced slurry seal, is made of polymer-modified emulsified asphalt, crushed fine aggregate, mineral filler or additives, and water. The binder has flexibility and less aging during placement. The almost clear binder can be pigmented or used with colored aggregate offering aesthetic opportunities and better urban heat island performance. The smooth surface is good for recreational trails.
Cape seal 10–22mm thick 7–15 years	A cape seal is a slurry seal or micro surfacing on top of a chip seal. The slurry fills the voids in the chip seal. It can be used over an existing asphalt pavement or over an aggregate base in very low traffic applications.
Other sealers Thicknesses and service life vary	There are multiples type of sealers for asphalt pavements that coat the surface to protect it from wear and degradation. Products are acrylic, epoxy, coal tar, or asphalt based, and some contain fine sand and grit for added durability. Look for products with information transparency labels or certifications limiting VOC levels, harmful constituents, and worker health risks.

Sources: Heitzman 2018; Muench and Hand 2019; Van Dam 2015; Maher et al. 2005

Figure 15.1 Types of asphalt pavement and asphalt pavement surfaces. Image credit Meg Calkins

concrete slab that will receive an asphalt pavement layer. The primary material used is emulsified asphalt.

Base Course(s)

A well-compacted, unbound aggregate base course distributes the design loads over a greater subgrade area and will limit subgrade deformation. The base course will vary by thickness from 100–200mm. Heavy duty pavements or weak subgrades will also require a subbase of coarser graded aggregate that extends beyond the pavement layer and the base. Subbase thicknesses vary from 150–300mm. Porous asphalt pavement bases and base reservoirs are discussed in Chapter 7 "Porous Pavements and Green Roofs".

While asphalt pavements are most commonly supported by an aggregate base, they can be installed as a full depth pavement bearing directly on prepared and compacted subgrade. In this case, the lowest layer of the pavement will be comprised of a mix of larger aggregates and asphalt binder that distribute the load over a larger expanse of subgrade. This method is not appropriate for poorly drained soils or climates with freeze thaw cycles.

Edges

Edges or curbs are recommended to protect the edge of asphalt pavements from erosion, but it is one of the few flexible pavement types that can remain structurally stable without edging. Concrete bands or curbs at the edge of vehicular applications will protect the edge from wheel damage as well as clearly demarcate the limits of the pavement. Raised curbs will direct runoff as well. Steel or wood edges of asphalt pathways can blend with the surroundings and minimize the appearance of the edge.

Materials for Asphalt Pavements

Asphalt mixtures are comprised of aggregates, an asphalt binder, and sometimes dust or mineral filler. The aggregates and filler make up 94–96% of the mix by weight, and the remaining 4–6% is the asphalt binder.

Asphalt Binders

The asphalt binder, also called bitumen or asphaltic cement, is a semi-solid, viscous constituent of crude petroleum. Most asphalt is derived from the distillation process of selected crude oils, although it can be derived from a naturally occurring material. The asphalt binder is made flowable by heating and as it cools it hardens. The heating of binders and aggregates at the plant, the drying of aggregates, keeping the mix hot on the way to the site and during placement are significant embodied carbon impacts of asphalt pavement (Table 15.2).

There are four general types of asphalt mixtures that vary by constituents and production temperature. The range of temperatures impacts their performance, placement, energy use, and embodied carbon.

Hot-Mix Asphalt

Hot-mix asphalt (HMA) pavement is mixed at temperatures ranging from 150°C to 190°C to ensure proper aggregate coating and workability of the mixture. It is placed at temperatures between 135°C and 155°C (Corrigan 2010). Heating the mixture to high temperatures and keeping it hot as it is transported to the site and placed uses a large amount of energy. Air emissions are released throughout this process (Milad et al. 2022; Khare et al. 2020; del Carmen Rubio et al. 2013), although they have been reduced over the past two decades (NAPA 2024; Schacat et al. 2022).

Warm-Mix Asphalt

Warm-mix asphalt (WMA) pavement is produced at temperatures ranging from 85–130°C. Organic additives, chemical additives, and/or water foaming techniques are employed to reduce bitumen viscosity, ensure proper aggregate coating, enhance workability, and maintain

Table 15.2 Hot-Mix Asphalt (HMA) vs. Warm-Mix Asphalt (WMA)

	Hot-Mix Asphalt	Warm-Mix Asphalt	Comments
Production temperature	150–190 °C Placement at 135–155°C.	85–130°C, however lower temps possible with some products.	WMA has 20–75% lower energy consumption due to cooler heating temperatures (Milad et al. 2022).
Additives or foaming needed	No	Yes	WMA relies on additives or foaming processes to allow for reduced mix and placement temperatures.
Energy consumption	Higher	Lower	WMA embodied carbon is reduced by an average of 24% (Sukhija and Saboo 2021).
CO_2 emissions	Higher	25–30% lower	Emissions are directly related to the temperature of the mix as less energy is required for cooler mixes.
Other emissions	Higher	Lower	Average reductions with WMA are SO_2 (37%), NO_x (38%), CO_2 (30%), VOC (42%), dust (38+%) (Sukhija and Saboo 2021). WMA produces a healthier work environment for construction workers (Sukhija and Saboo 2021).
Cost	Lower initial cost	Higher initial cost because of additives, but lower energy and construction costs.	WMA has 25% energy cost savings with 20°C reduction in temperature (Sukhija and Saboo 2021). Fewer roller passes can save on construction time and cost.
Construction	Widely familiar	Less widely familiar and placement varies by binder additive.	WMA has enhanced asphalt mixture workability and compaction because of reduced viscosity. Extended paving season for WMA as can be placed in cooler temps.
Recycled materials	Yes	Yes	Higher quantities of RAP can be possible with WMA.
Use phase	More binder aging.	Less binder aging.	WMA results in increased density and reduced variability of asphalt. Risk of reduced binder/aggregate adhesion with WMA because of low temps and less compaction. Can be addressed with some products, recycled aggregates, and processes.

Sources: Milad et al. 2022; Sukhija and Saboo 2021; Muench and Hand 2019; Corrigan et al. 2010 desired performance. Energy use, embodied carbon, and emissions are reduced with the lower temperatures of warm mix asphalt.

Foamed asphalt avoids use of chemical additives to create warm-mix asphalt. It is a warm-mix process where cold water, along with compressed air, is injected into hot asphalt in a specially designed chamber. This results in a thin-film, high-volume asphalt foam with reduced viscosity and increased coating potential. Table 15.2 compares HMA and WMA.

Half-Warm Mix Asphalt

Half-warm mix asphalt (HWMA) pavement is produced and placed at temperatures ranging from 60–100°C. HWMA mixes use emulsions and foamed bitumen as a binder. Aggregates are heated to temperatures of 90–95°C then mixed with emulsion that has been heated to 60–65°C. Benefits of HWMA are substantially reduced energy use during manufacturing at the plant, reduced emissions at the plant and jobsite, and improved worker health and safety. HWMA consumes 50% less energy than HMA, whereas WMAs consumes 25–30% less (Milad et al. 2022).

Cold-Mix Asphalt

Cold-mix asphalt (CMA) pavement is produced and placed at temperatures ranging from 0 to 30°C. CMA is a mixture of emulsified asphalt, water, and/or a dilutant to enable mixing. CMA is limited to the rehabilitation of deteriorated pavements and roads with very low traffic loads. Asphalt emulsion can be sprayed on the aggregate and mixed in place, travel plant mixed, or pug mill mixed at the aggregate source. It can also be mixed at asphalt plants incorporating high recycled content (Marks 2024).

Some cold-mix products contain VOCs, hydrocarbons, and chemicals that pose risks to human health. Cutback asphalt is a type of cold mix that consists of asphalt binder mixed blended with solvents. Safety data sheets (SDS) and labels should be checked for risks. There are some low-VOC cold patch products available that rely on compaction, not evaporation to harden.

Aggregates

At around 95% of the mixture by weight, aggregates, including recycled asphalt pavement, are the main constituents of an asphalt pavement. Stone on stone contact within the pavement section provides structural stability while the asphalt binder holds the stones in place. The wearing surface of an asphalt pavement uses fine aggregates to provide a smooth fine textured surface, although, surface courses such as chipseal use larger stone chips to vary the surface color and reflectivity of the pavement. Lower asphalt courses, called base courses (not to be confused with an aggregate base course), are often comprised of strong coarser aggregates to add strength to the pavement.

Use of virgin aggregates in asphalt pavement poses considerable environmental impacts from their mining, crushing, and transport. But incorporation of RAP into asphalt pavement mixtures substantially reduces these environmental impacts. Chapter 12 "Aggregates and Sand Materials" discusses these impacts in greater detail.

Mineral filler consists of fine-grained mineral particles either present in or separately added to the aggregate mix. It is sometimes referred to as dust or rock dust and is defined as the portion of aggregates passing through a 0.075mm sieve (ASTM D242). Separately added fillers include baghouse fines, diatomite, hydrated lime, cement, glass powder, steel slag, and bentonite.

Environmental and Human Health Impacts of Asphalt Pavements by Life Cycle Phase

Primary environmental impacts during the product stage of asphalt pavement include GHG releases from processing and heating material, feedstock carbon from use of fossil fuel-based materials, and resource use from the comparatively high volumes of aggregates in asphalt pavements (Figure 15.2). In the use phase, standard asphalt pavement contributes to the urban heat island effect.

Health impacts include emissions of PAHs and particulate matter released from the combustion of fossil fuels or the asphalt binder during the production heating process. VOCs can be released from tack coats, asphalt surfacing products, and cold mix asphalts (Milad et al. 2022; Khare et al. 2020; Murtagh and Vallette 2017; del Carmen Rubio et al. 2013).

Cooling asphalt mixes, incorporating higher quantities of RAP, and asphalt plant management with a focus on energy efficiency are all approaches that reduce energy use and emissions. Energy use and emissions of asphalt pavement are often considered together because lowering the mix and placement temperatures of asphalt pavement will directly reduce the emissions of from producing, mixing, and placing the pavement.

Phase A1 Raw Material Extraction and Processing

Phase A1 activities include petroleum extraction and refinery operations for asphalt binder production, and

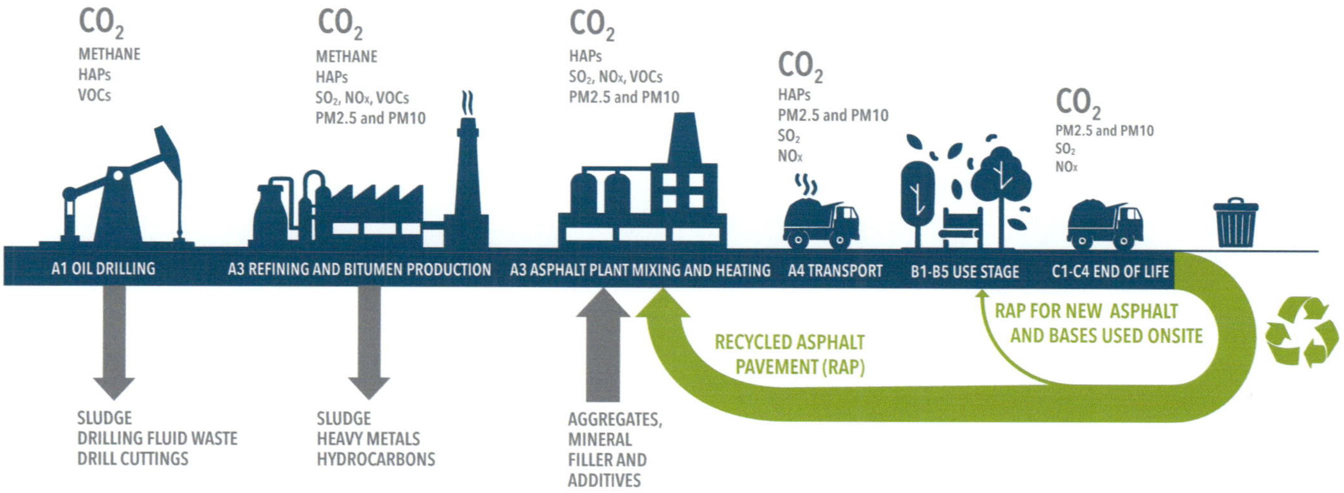

Figure 15.2 The life cycle of asphalt pavements. Asphalt pavement has a very high recycling rate with over 96% being reused in new asphalt pavement or a bass aggregate. Image credit: Meg Calkins

the quarrying and processing of aggregates and mineral fillers. Air and water emissions from petroleum refining and production of asphalt binder vary widely by facility and chemical constituents in the petroleum. Air emissions include hydrocarbons, sulfur oxides, particulates, and other HAPs in addition to fuel-related emissions such as CO_2. Water emissions include dissolved and suspended solids, phenols, oils, acids, and trace amounts of heavy metals (Milad et al. 2022; Khare et al. 2020; del Carmen Rubio et al. 2013). In asphalt pavement carbon accounting, there can be variations in the EC of petroleum extraction and refinery operations figures as asphalt cement is a co-product petroleum processing. Some LCAs may not count these activities, while others do.

Environmental and human health impacts from the rejuvenators, dilutants, and chemical additives for RAP, WMA, and CMA are not easily characterized. Safety data sheets (SDS) will be available for these products that should detail worker safety hazards and precautions. Look for products with lower VOC contents. The environmental impacts posed by these products are more challenging to identify. Some risks might be contamination of adjacent natural areas and waterways from runoff containing hydrocarbons and other chemicals.

Phases A3 Manufacturing, A4 Transportation, and A5 Construction

Phase A3 is the production of the asphalt mixtures including generation of electricity to power equipment and heat for the mix. Production involves aggregate stockpiling, heating the aggregates and asphalt binder, and mixing.

Most asphalt plants are currently powered by fuel oil and natural gas, but as the industry works toward reducing greenhouse gas emissions, some are transitioning to renewable energy sources and taking efficiency measures.

Reducing the moisture content of aggregates and RAP requires substantial energy and is a major cost as well. It takes about 25 MJ to remove 1% of moisture from one metric ton of aggregate. Measures to reduce energy use include grading and paving in stockpile areas, covering stockpiles, and using drier material from the top of the pile. For every 1% reduction in moisture content, drying cost is reduced by more than 10%. This can result in significant cost and energy savings (Muench and Hand 2019).

Other steps being employed for energy efficiency at asphalt plants are insulation of plant components, efficiencies of processes, capture of waste heat, and use of alternative fuels such as solar, biomass, biodiesel, and wind.

As HMA requires heating of the binder and aggregates to temperatures of 150–190°C and maintaining temperature of 135°C and 155°C through transportation and placement, energy use and emissions can be substantial. Emissions and fumes are released during transport and at the jobsite, although the industry has taken steps to capture and reduce them. Emissions associated with HMA production include criteria pollutants (PM2.5 and PM10), HAP metals, and HAP organic compounds (Milad et al. 2022; Schacat et al. 2022; Khare et al. 2020; del Carmen Rubio et al. 2013). The gaseous emissions associated with HMA production include the criteria pollutants sulfur dioxide (SO_2), nitrogen oxides (NOx), carbon monoxide (CO), and VOCs, as well as HAP organic compounds (Zemba 2021).

Uncaptured emissions and fumes from heating asphalt binders can pose irritation symptoms, and even more severe health impacts to workers and communities during plant mixing and placement of asphalt pavement. Prolonged exposure to PAHs and PM from asphalt emissions can contribute to lung and cardiovascular diseases. PAHs are carcinogens and can cause kidney and liver problems. Hydrocarbons have been shown to be released to the environment from pavements during the use phase as well (Khare et al. 2020; Bernot et al. 2011).

The biggest reduction of energy use and emissions at the asphalt plant and during placement comes with use of warm-mix asphalt and half-warm mix asphalt. Produced and placed at lower temperatures, use of WMA reduces energy use by an average 30%, and use of HWMA can reduce energy use by 50% or more. Hazardous emissions are substantially reduced with cooler temperatures as well.

Locally procured aggregates for an asphalt pavement mix will reduce transportation costs, energy, and emissions. Many state agencies have aggregate requirements that are the same for use in all types of asphalt pavements. This may require longer travel distances for higher quality aggregates when local aggregates could be used in lower volume roads and parking lots (Muench and Hand 2019).

B1–B5 Use, Maintenance, and Refurbishment of Asphalt Pavements

The primary impact of asphalt pavements during the use phase stems from their contributions to the urban heat island effect. The dark surface of asphalt pavement absorbs heat from the sun, then radiates it to the air leading to increased temperatures of 1°C to 6°C in urbanized areas. The surface of asphalt pavement in some climates can be up to 50° higher than a reflective white surface (Heat Island Group 2024).

Table 15.6 discusses a variety of approaches to reduce asphalt pavements contribution to the urban heat island effect. Urban heat islands and the impacts of various pavement types and surfaces are discussed in greater detail in Chapter 6 "Pavements".

While emissions to air are a lesser factor of asphalt pavements in use and maintenance, the pavement can contain measurable levels of VOCs, PAHs, and heavy metals. These compounds may be released to the soil and water in small amounts as the asphalt pavement cures, ages, and weathers (Bernot et al. 2011).

End of Life Stages C1–C4 and D

End-of-life impacts of asphalt pavements are minimal as most asphalt pavement is recycled directly back into new asphalt pavement on-site or is hauled to an asphalt plant for recycling. Where applicable, in situ recycling of the pavement will reduce the embodied carbon of hauling, but there are fewer quality control processes in the field; therefore this may not always be the best option (Willis 2024).

Embodied Carbon of Asphalt Pavements

The initial embodied carbon of standard hot-mix asphalt pavement is lower than portland cement concrete pavement, although when considering a 30-year life cycle, the repair and replacement of asphalt pavements can increase embodied carbon across the life cycle. Use of warm mix and high RAP content can improve embodied carbon performance of these materials substantially.

The two major phases of asphalt pavement production that have the highest embodied carbon occur in phase A1, where raw materials are produced and in phase A3 where the materials are mixed, and heated. For HMA in phase A3, the aggregate heating process contributes 60% or more of the total carbon emissions while the asphalt heating and mixing processes only contribute 14% and 12%, respectively (del Carmen Rubio et al. 2013).

During the use phase (B1), asphalt roads will influence fuel consumption of vehicles. On high-volume roads this may be a significant impact, while not as great on lower volume roads (Wang et al. 2012). Fuel efficiency has been shown to be better on concrete highways than on asphalt highways due to lower rolling resistance (Wang et al. 2012; Araújo et al. 2014; Mukherjee 2014; American Asphalt 2024). This will become less pronounced as vehicles move toward greater fuel efficiency.

Embodied carbon of asphalt pavements can be reduced with use of warm-mix asphalt, half-warm mix asphalt, and cold-mix asphalt. These mixes are defined by their temperatures and a comparison of embodied carbon, emissions, and other factors are shown in Table 15.2. EC can also be reduced with use of high recycled asphalt pavement contents.

Table 15.3 provides EC figures from multiple sources because the numbers vary by source, region, mix, mix temperature, and RAP content. It is recommended that asphalt pavement be specified from companies with third-party verified EPDs for their product.

Table 15.3 Embodied Carbon of One Metric Ton of Asphalt Pavement Materials (kgCO_2e/t)

Asphalt Pavement Type	Asphalt Pavement Components	Embodied Carbon (kgCO_2e/t)	Source
US GSA IRA low EC asphalt pavement	Top 20% limit	55.4	US GSA Low Embodied Carbon Asphalt Requirements 2023
US GSA IRA low EC asphalt pavement	Top 40% limit	64.8	US GSA IRA Low Embodied Carbon Asphalt Requirements 2023
US GSA IRA low EC asphalt pavement	Better than average limit	72.6	US GSA IRA Low Embodied Carbon Asphalt Requirements 2023
HMA pavement	No RAP	53.7	Schacat et al. 2024 for NAPA, assumes average transport distance to site
HMA pavement	20% RAP	47.4	Schacat et al. 2024 for NAPA, assumes average transport distance to site
HMA pavement	50% RAP	37.9	Schacat et al. 2024 for NAPA, assumes average transport distance to site
HMA with 100% RAP	100% RAP	43.91	Green Asphalt 2023
HMA pavement	2% Recycled Asphalt Shingles (RAS)	51.5	Schacat et al. 2024 for NAPA, assumes average transport distance
HMA pavement	5% RAS	48.3	Schacat et al. 2024 for NAPA, assumes average transport distance
HMA pavement	0% RAP	83.1	Average of Product EPDs listed in the Building Transparency Database 2024
HMA pavement	10% RAP	80.4	Average of Product EPDs listed in the Building Transparency Database 2024
HMA pavement	20% RAP	75.2	Average of Product EPDs listed in the Building Transparency Database 2024
HMA pavement	40% RAP	69.4	Average of Product EPDs listed in the Building Transparency Database 2024
WMA pavement	0% RAP	79.0	Average of Product EPDs listed in the Building Transparency Database 2024
WMA pavement	10% RAP	75.1	Average of Product EPDs listed in the Building Transparency Database 2024
WMA pavement	20% RAP	71.3	Average of Product EPDs listed in the Building Transparency Database 2024
WMA pavement	40% RAP	52.4	Average of Product EPDs listed in the Building Transparency Database 2024

Note: EC values represent life cycle stages A1–A3 (Raw material supply, transportation, and manufacturing) unless stated otherwise. Embodied Carbon figures will change over time as processing efficiencies are realized, mixes are cooled, and more renewable energy sources are employed..

Reducing the Impacts of Asphalt Pavement Materials

Major strategies to reduce the environmental impacts of asphalt pavements are to (1) extend the longevity of an asphalt pavement's life and (2) employ measures to reduce the embodied carbon, emissions, resource use, and other environmental impacts of the pavement (Table 15.7 and Figure 15.3).

These measures include use of recycled materials such as RAP and recycled tire rubber (Table 15.4); cooling the asphalt mix with warm, half-warm, or cold mixes; reducing the use of virgin petroleum-based binders; and changing specifications to permit use of locally available aggregates (Van Dam et al. 2015).

Recycled Materials in Asphalt Pavements

Asphalt pavements can be major consumers of recycled materials in the circular economy for both the aggregate base and the pavement layers. Use of RAP in new asphalt pavement is increasing, and it is technically possible to use 50% RAP or more, although high amounts are not always supported by standards and codes.

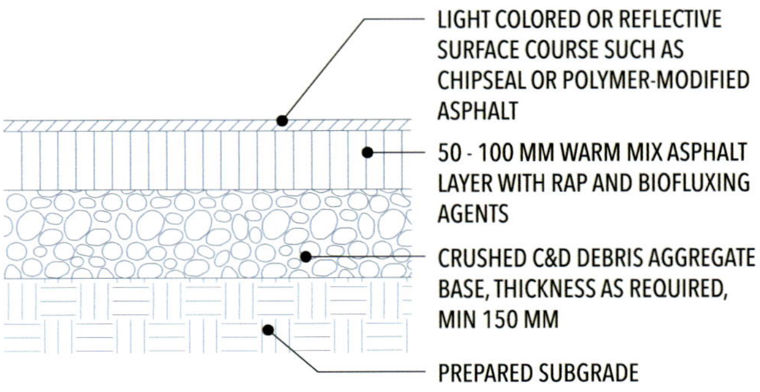

Figure 15.3 An example detail of a low carbon, resource efficient asphalt pavement. Image credit: Kathleen Elliott and Meg Calkins

Use of locally recycled materials for aggregates coupled with cooling mixture temperatures can substantially reduce the

Table 15.4 Recycled Materials for use in Asphalt Pavements

Recycled Material	Substitution Amount	Notes
Recycled asphalt pavement	15–60% (Milad et al. 2022) 100% (Green Asphalt 2023)	Improves moisture sensitivity and resilient modulus; Less rutting; Good fatigue performance. RAP should be tested and regraded to ensure quality pavement. Consider in-place recycling where applicable.
Ground tire rubber (recycled tires)	15–22% (Milad et al. 2022)	Rubber modified asphalt offers better skid resistance, extended pavement life, greater resistance to cracking and aging, noise reduction, and reduced tire wear. Lower cost and embodied carbon RMA improves moisture sensitivity and resilient modulus.
Recycled polymers	2–8% asphalt binder 0.2–1% of total mix (Willis et al. 2020)	Improved toughness, abrasion resistance, and bond strength. Not widely used, so performance is somewhat unknown, but studies show promising results. Toxicity issues with heating some recycled polymers. Not a significant way to recycle plastics.
Recycled asphalt shingles	2–10% by mass of total mixture (Muench and Hand 2019)	Mixtures must be engineered to ensure performance.
Blast furnace slag	30% (Milad et al. 2022)	Improved fatigue resistance and stiffness modulus.
Steel slag	40% (Milad et al. 2022)	Improved resilient modulus and fatigue resistance.
Glass	10% (Milad et al. 2022)	In WMA, reduced resilient modulus, creep, and moisture susceptibility.
Carbide lime and hydrated lime	Low amount	Enhanced resistance to moisture damage. Use of lime can increase the embodied carbon of an asphalt pavement.
Foundry sands	30% (Zhang et al. 2021a)	Used as a fine aggregate in asphalt pavements. Meets federal Superpave requirements. Increased moisture resistance and improved rutting resistance.
Glass fiber	Low amount (Milad et al. 2022)	With RAP improved rutting and moisture susceptibility.
Jute fiber	Low amount (Milad et al. 2022)	Enhanced fracture resistance.
Other materials	varies	Other recycled materials that have been used to a lesser extent in asphalt pavements are ceramic waste, fly ash, waste steel, and agricultural waste such as cellulose fibers.

Sources: Green Asphalt 2023; Milad et al. 2022; Zhang et al. 2021a; Willis et al. 2020; Muench and Hand 2019; Van Dam et al. 2015

embodied carbon, embodied energy, and virgin resource use of asphalt pavements. Additionally, waste materials are reused in a beneficial application.

A wide variety of waste materials can be used in asphalt pavements, and in many cases, these materials offer cost and performance advantages as well as environmental benefits. Table 15.4 details recycled materials for use as aggregates and mineral fillers and as additives in asphalt binders.

Recycled materials may impact the performance of asphalt pavements, so their use must be carefully considered and, where applicable, recycled aggregates should be tested for constituents and contaminants. Another consideration is the impact that inclusion of recycled materials will have on the future recyclability of the asphalt pavement.

Recycled Asphalt Pavement

Recycled asphalt is one of the few construction materials that is closing materials loops and approaching zero-waste status. Nearly 18% of construction and demolition waste in the United States is asphalt concrete, and over 96% of this is diverted from landfills for use in new asphalt pavement and/or base layers (US EPA 2018).

The National Asphalt Pavement Association (NAPA) survey of asphalt producers showed more than 93% of RAP being used in new asphalt pavements (estimated at 98.1 million metric metric tons), with the remaining 7% used in unbound bases and other civil engineering applications. NAPA estimated that use of RAP reduced the need for 4.9 million metric tons of asphalt binder and more than 93 million metric tons of aggregates (Williams et al. 2024).

Theoretically RAP can be used for a majority of an asphalt mix, although it is more commonly used for 15–50% of the mix with 22% being the US average (Williams et al. 2024). Most asphalt plants have the necessary hardware to incorporate RAP, and while there may be a slight cost increase for processing the RAP and the recycling agents that soften recycled asphalt binders, there are overall cost savings. One study showed cost savings estimates of 6% with use of 15% RAP, 10–20% with 25% RAP, and 29–40% with 50% RAP (Milad et al. 2022).

Reductions of more than 20% in greenhouse gas emissions and over 45% energy consumption have been found by using 50% RAP mix designs (Milad et al. 2022). A mix of 100% RAP with reduced embodied carbon of the asphalt 50% below that of standard HMA (Green Asphalt 2023).

Chemical and oil rejuvenators are added to RAP or asphalt binders to soften and rejuvenate the aged binders of the reclaimed asphalt pavement. The chemicals react with the old asphalt binder to separate and redisperse the aggregates. Biobased rejuvenator products are available.

Close inspection of RAP materials is critical to ensure the integrity of the new asphalt mixture. RAP can contain contaminants such as glass, porcelain tile, geotextiles, aggregate base materials, and PCC. In addition, RAP must be tested if it is used in high volumes to ensure qualities that will produce an appropriate asphalt mix for the new pavement. Because the performance of untested RAP in asphalt pavement may vary, it is common to use higher percentages of RAP in the lower course(s) of an asphalt pavement and lower percentages in the surface course.

RAP can contain PAHs, heavy metals, VOCs, and other chemicals from drippings on the pavement and surfaces applied during the use phase of the pavement. RAP should be tested to determine the presences of harmful constituents particularly when it is used in applications where children might have contact with it or in sensitive ecosystems (Murtagh and Vallette 2017).

In-Place Asphalt Pavement Recycling

Asphalt pavement can be recycled into new pavement or a stabilized base in place through hot, warm, or cold-mix applications. This approach is primarily performed on large expanses of pavement as bringing the equipment to the site to remove, regrade, mix, and place the asphalt pavement can be expensive. Costs of hauling the waste pavement to a plant are reduced, and embodied carbon can be substantially reduced by reduced transport and use of recycled materials (ARRA 2015).

Hot and warm in-place recycling (HIPR and WIPR) is accomplished by heating and softening the existing pavement, milling it, adding a rejuvenating agent, then placing and compacting it with traditional hot-mix equipment. The HIPR or WIPR layer(s) are then topped with a new surface layer of hot mix, warm mix, or chipseal. Disadvantages of this process are the emissions that result from the process and the lack of opportunity to significantly modify the asphalt mix.

Cold in-place recycling (CIP), the least energy-intensive method of in-place recycling, involves milling up an existing asphalt pavement, mixing in emulsified asphalt and a rejuvenating agent, then placement and compaction. A new, thin layer of HMA, WMA, or chipseal is placed over the recycled pavement. As no heating is involved in the CIP process, energy savings are estimated at 40–50% (Maher et al. 2005).

Full depth reclamation (FDR) removes and pulverizes the asphalt pavement and the underlying base layer into a new stabilized base. A small amount of cementitious material (lime, cement, kiln dust, or fly ash) or new asphalt cement is added to further stabilize the base. Foamed asphalt is used successfully in FDR. A new, thin layer of HMA, WMA, or chipseal is placed over the stabilized base. FDR is primarily used for badly cracked pavements where a surface overlay will not extend the life of the pavement for very long.

Recycled Tires

End-of-life tires (ELT) can be used as asphalt binder modifiers and/or as aggregates in asphalt pavements. Use of recycled tires in new asphalt pavement is one of the major uses for the 300 million tires in the United States and close to a billion tires globally that are discarded each year. But there can be health concerns with the use of recycled tires. When heated or chipped, they can release PAHs that could be harmful to workers. Additionally, recycled tires may add heavy metals such as lead to an asphalt mix (US EPA 2024b; Habitable 2020, 2024a, 2024b; USGBC 2019; Murtagh and Vallette 2017).

Ground tire rubber, also called recycled rubber or crumb rubber, is produced by grinding recycled tire rubber after the steel and fiber reinforcing has been stripped. It is used as an asphalt modifier and/or fine aggregate in asphalt pavement mixes. There are two sizes of ground tire rubber. "Ground rubber" is 2mm and smaller, and "coarse ground rubber" is 2mm or larger (US EPA 2016).

Rubber modified asphalt (RMA) is a blend of hot paving grade asphalt cement, reclaimed ground tire rubber (at least 15%), and additives. RMA is used in structural and surface layers of asphalt pavement and for chipseal. Performance benefits of RMA pavements include better skid resistance, extended pavement life, greater resistance to cracking and aging, noise reduction, and smoother surfaces, reducing tire wear. RMA overlay costs are reduced up to 43% below standard HMA overlays with an expected 10% longer use life. Embodied carbon of an asphalt pavement with RMA is reduced due to longer service life and low maintenance requirements (Recycled Rubber Coalition 2022).

Recycled Plastics

Recycled plastics are used in asphalt pavements in asphalt binders and as aggregates. As they are not as widely used as RAP and RMA, the impact that they have on new asphalt pavement is less well known. In a *wet process*, melted recycled plastics can be added into asphalt binders as polymer modifiers or asphalt replacements between 2% and 8% of the asphalt binder by weight. In a *dry process*, recycled polymers are added directly into the mixture as aggregate replacement, mixture modifiers, or binder modifiers. This can be 0.2–1% of an asphalt mix by weight (Willis et al. 2020).

Use of recycled polymers for asphalt modification reduces the penetration and ductility and increases the softening point, viscosity, and high temperature performance of an asphalt binder. This could lead to better rutting resistance of a pavement. Plastic coated aggregates have been shown to have better toughness, abrasion resistance, and bond strength (Willis et al. 2020). Research on potential health impacts of heating plastic aggregates for use in asphalt pavements is ongoing (Marks 2024).

Substitutes for Petroleum-Based Asphalts

As bitumen is a co-product of petroleum production, the movement away from fossil fuel use could eventually drive up the price or even limit supplies of petroleum-based asphalt binders. Polymer-based binders, biobased binders, and others are in various stages of development. Table 15.5 presents some alternatives to petroleum-based asphalt cements. Some of these binders can be mixed with traditional oil-based asphalt binders, while others are a complete substitute.

Improving the Thermal Performance of Asphalt Pavements

Standard black asphalt contributes to the UHI effect by absorbing heat from the sun, then reradiating it causing elevated air temperatures in urban and suburban areas. Standard black asphalt pavement has a surface reflectance of just 0.05–0.10 when new and 0.10–0.25 when aged. There are several asphalt surfaces discussed in Table 15.6 that can increase the surface reflectance of asphalt pavements. For greater discussion of UHI and pavements, refer to Chapter 6 "Pavements".

Durability and Rehabilitation of Asphalt Pavements

Asphalt pavements require more maintenance than concrete pavements, but regular resurfacing, repair and rehabilitation can allow for a 30-year or longer service life. Concrete pavements are designed for a 30-year use life with little maintenance, but current levels of portland cement use mean that concrete pavements are a major contributor to climate change along with concrete use in buildings.

A major environmental benefit of asphalt pavement is the ability to extend the pavements life with coatings, surfaces, and overlays while the pavement structure remains in place. Depressions, cracks, and potholes of asphalt pavements can be easily patched to match the existing surface. Another repair approach is to mill the top surface and install a new wearing course on top of the pavement.

Strategies for extending the life cycle of asphalt pavements are discussed in the box "Extending Asphalt Pavements Life Cycles: Balanced Mix Design Specifications and Perpetual Pavements". Asphalt pavement surfacing types and rehabilitation methods are discussed in Table 15.1.

Table 15.5 Substitutes for Petroleum-Based Asphalt Binders

Binder	Considerations
Bioorganic binders are made from tree, plant, and animal waste. Some exhibit properties like asphalt as they can flow when heated, coat aggregate, and be sprayed.	Some bioorganic binder products are coming on the market, but they are not in widespread use. Examples are binders derived from lignin, cellulose, corn stover (the non-food part of corn), swine waste, algae (Audo et al. 2015), and vegetable oil. Lignin is composed primarily of phenolic compounds in a hard, polymer-like structure and cellulose is a dense polymer chain of sugar molecules (Zhang 2021b).
Rubber modified asphalts with ground tire rubber can be used as an asphalt modifier for 15–22% of the asphalt binder (Van Dam et al. 2015).	Rubber modified asphalts with ground tire rubber as an asphalt modifier are discussed earlier in this chapter.
Recycled plastic binders are created by melting plastic waste, primarily polyethylenes.	Recycled plastics can be added to an asphalt mixture in a wet process as a polymer modifier at 2–8% of the binder weight. It can reduce fatigue and cracking and mitigate rutting. When plastic is added in a dry process, it is more often used as an aggregate replacement (Willis et al. 2020).

Table 15.6 Methods for Improving the Thermal Performance of Asphalt Pavements

Type	Construction and Components
Surface gritting with light-colored aggregate	Surface gritting involves spreading aggregate over newly placed HMA and pressing it in with a roller.
Chip-seals with light-colored aggregate	Chip seal applies a mix of asphalt and aggregate on an existing pavement surface. It can be a single layer as thick as the aggregate or multiple layers.
Sand seals with light-colored aggregate	Sand seals are used for preventive maintenance to fill cracks and delay pavement deterioration. A polymer-modified asphalt emulsion is broomed into the pavement surface and covered with fine sand.
Sand/shot blasting and abrading binder surface	A mechanical process to remove surface coatings using steel shot. This exposes the natural color of the aggregate.
Colorless and reflective synthetic binders used with light aggregate	Clear or colored synthetic binder is used for the surface course of the asphalt pavement. With clear binders, the color is determined by the aggregate.
Surface painting using light-colored paint	Specially formulated paint can alter the surface color of the pavement. The paint is a combination of cement fortified acrylic resins, epoxy-based polymers, and a blend of aggregates.
Micro surfacing using light-colored materials	Micro surfacing can be applied to new or existing pavements. It is a thin layer (3–7mm) of cement, sand, fillers, and a proprietary blend of emulsified polymer resin.
Grouting of open graded friction course with cementitious materials	This is a semi-rigid surfacing process that is an open graded asphalt layer with 20–25% voids filled with high strength cement grout. This wearing course is about 38mm thick.
Bonded concrete overlay on asphalt (BOCA)	A BOCA is PIP concrete over milled asphalt pavement (10–16mm). An ultra-thin bonded concrete overlay is 5–10mm.

Source: Tran and Powell 2009

Table 15.7 Approaches for Improving the Environmental and Health Performance of Asphalt Pavements

Objective	Strategy	Economic Impact	Environmental/Human Health Impact
Reduce virgin binder in asphalt pavement	Use greater quantities of RAP if performance can be realized.	Reduced cost.	Reduced energy and resource use; beneficial use of waste material; reduced need for asphalt binder.
	Use rubber modified asphalt binders.	Some increase in initial cost but could have reduced cost with thinner overlays and increased life.	Extends life of pavement; reduces use of resources; beneficial use of waste material.
	Use recycled asphalt shingles as partial replacement for asphalt binder.	Reduced cost if RAS are available.	Less use of petroleum resources; beneficial use of waste material.
	Use bio-binders (refer to Table 15.5).	Economic impacts not well known yet; limited availability.	Avoids use of petroleum-based binders; other environmental and human health impacts less known.
	Use sulfur modified asphalt.	Economic impacts not well known yet; limited availability.	Beneficial use of waste product; less use of petroleum resources.
Reduce virgin aggregate content in asphalt pavement	Use greater quantities of RAP if pavement performance can be realized.	Reduced cost; reduced impact to surrounding communities.	Reduced energy and resource use; beneficial use of waste material.
	Use other recycled materials such as: recycled rubber, recycled polymers, glass, recycled asphalt shingles, ceramic waste, and others (refer to Table 15.4).	Reduced cost if locally available; reduced impact to surrounding communities.	Reduced energy and resource use; beneficial use of waste materials.
	Use in-place recycling (full depth reclamation or partial depth recycling).	Can reduce initial costs through lower trucking and material costs and extended pavement life cycle.	Reduced use of virgin materials, carbon footprint, and emissions. Reduced trucking fuel consumption and carbon impacts.

(*Continued*)

Table 15.7 (Continued)

Objective	Strategy	Economic Impact	Environmental/Human Health Impact
	Use recycled polymers.	Some increase in cost for construction, but payback in pavement life.	Beneficial use of waste materials; potential health impacts from heating certain types of polymers.
Reduce embodied carbon, energy use, and emissions generated to produce asphalt pavement	Use WMA and HWMA to reduce mixing and placement temperatures.	Cost increases for additives can be offset by energy savings of cooler mixes.	Reduced energy and GHG to produce asphalt pavement; reduced worker exposure to fumes; additives could have some health impacts and carbon impacts to produce.
	Use renewable energy and fuel sources.	May increase cost.	Reduced embodied carbon and other emissions of asphalt concrete pavement.
	Use more efficient processes and plant designs for fuel efficiency.	Increased capital costs but greater fuel efficiency will eventually payback.	Reduced embodied carbon of asphalt concrete and other emissions of asphalt pavement.
Extend the life of asphalt pavement	Increase compaction specifications.	Some increase in cost for construction, but payback in pavement life.	Longer pavement life reduces resource use, embodied carbon, and other emissions.
	Use WMA to obtain better compaction.	Small increase in cost for construction, but payback in pavement life.	Reduced resource use, embodied carbon, and other emissions.
	Improve mixture designs.	Some cost for new equipment and training, but payback in pavement life.	Longer pavement life reduces resource use, embodied carbon, and other emissions.
	Use rubber modified asphalt for binders and aggregates.	Some increase in cost for construction, but payback in pavement life.	Longer pavement life reduces resource use, embodied carbon, and other emissions; beneficial use of waste materials; thinner pavement overlay saves resources and emissions; increased exposure of workers to fumes.
	Use lime anti-strip or other additives to decrease risk of early failure due to moisture damage.	Some increase in cost, but payback with longer life.	Longer pavement life reduces resource use, embodied carbon, and other emissions; carbon impacts of lime production.
Reduce transportation impacts of asphalt pavement	Use locally available materials.	Lower initial cost but may have higher life cycle costs if quality of local materials leads to poor performance.	Reduced resource use, carbon, and other emissions.
Increase the solar reflectance of asphalt	Use lighter colored aggregates, chip seals, surface layers, and coatings; use porous asphalt.	Potential higher initial cost with alternative surface treatments.	Greater solar reflectance will reduce asphalts contributions to urban heat islands.
Avoid harmful constituents and additives in asphalt pavements	Consult SDS prior to specifying products; look for low-VOC products with minimal diluents and solvents.	Low-VOC products may be slightly more expensive.	Lower health risks; lower environmental impacts.
Reduce the stormwater impacts of asphalt pavement	Make asphalt pavement porous.	Higher initial and life cycle cost of porous asphalt pavement may be offset by reduced gray stormwater infrastructure such as catch basins and detention ponds.	Quantity of runoff is reduced as stormwater is infiltrated and cleansed by soil. Reduced impacts on receiving waters.
	Reduce the size of paved areas with narrower roads, smaller parking lots, shared driveways, narrower paths.	Paving less will save construction and maintenance costs.	Quantity of runoff is reduced as paved areas are smaller. Reduced resource use and carbon footprint with smaller pavements.

Sources: Milad et al. 2022; Muench and Hand 2019; Van Dam et al. 2015

When selecting a surface application for asphalt pavement with the intent of minimizing environmental impacts, consideration should be given to longevity, application temperature, incorporation of recycled material, and surface color/reflectivity.

Extending the Service Life of Asphalt Pavements: Balanced Mix Design Specifications and Perpetual Pavements

With increasing incorporation of recycled aggregates, recycled binders, and lower binder temperatures, the specification of asphalt pavement mixes can be complex. The asphalt industry and the US Federal Highway Association have developed standards and processes that can result in more durable asphalt pavements using a performance-based specification system.

Balanced mix design (BMD) is defined as "asphalt mix design using performance tests on appropriately conditioned specimens that address multiple modes of distress taking into consideration mix aging, traffic, climate and location within the pavement structure" (Van Dam et al. 2015). BMD evaluates final mix properties rather than individual raw materials, allowing for optimization of recycled materials and assurance of good mix performance for producers. The BMD process incorporates mechanical tests to determine the rutting- and cracking-resistance of a mix so that the mix is "balanced" to optimize both rutting and cracking performance (Muench and Hand 2019).

Perpetual pavements are defined by the Asphalt Pavement Alliance (APA) as "an asphalt pavement designed and built to last longer than 50 years without requiring major structural rehabilitation or reconstruction and needing only periodic surface renewal in response to distresses confined to the top of the pavement" (Newcomb et al. 2020). Recent studies of perpetual pavements show a reduction in GHG impacts of around 20% (Schacat et al. 2024).

References

American Asphalt (2024). "Why use concrete for roads instead of asphalt". https://americanasphalt.com/why-use-concrete-for-roads-instead-of-asphalt/, accessed September 29, 2024.

American Foundry Society (2024). "Asphalt". www.afsinc.org/, accessed July 29, 2024.

Araújo, J.P.C., Oliveira, J.R.M., and Silva, H.M.R.D. (2014). "The importance of the use phase on the LCA of environmentally friendly solutions for asphalt road pavements". *Transportation Research Part D: Transport and Environment* 32: 97–110. https://doi.org/10.1016/j.trd.2014.07.006.

Asphalt Recycling and Reclaiming Association (ARRA) (2015). *Basic Asphalt Recycling Manual*, Second Edition. FHWA-HIF-14-1001. ARRA.

ASTM (2019). *D242 Standard Specification for Mineral Filler for Asphalt Mixtures*. West Conshohocken, PA: ASTM International.

Audo, M., et al. (2015). "Subcritical hydrothermal liquefaction of microalgae residues as a Green route to alternative road binders". *ACS Sustainable Chemistry & Engineering* 3.4: 583–590. https://doi.org/10.1021/acssuschemeng.5b00088.

Bernot, M.J., et al. (2011). "The influence of different urban pavements on water chemistry". *Road Materials and Pavement Design* 12.1: 159–176.

Corrigan, M., Newcomb, D., and Bennert, T. (2010). *From Hot to Warm*. United States Federal Highway Administration (US FHWA), July–August.

del Carmen Rubio, María, et al. (2013). "Comparative analysis of emissions from the manufacture and use of hot and half-warm mix asphalt". *Journal of Cleaner Production* 41: 1–6.

Garcia, J., and Hansen, K.R. (2001). *HMA Pavement Mix Type Selection Guide (IS-128)*. Lanham, MD: National Asphalt Pavement Association.

Green Asphalt (2023). *An Environmental Product Declaration (EPD) for Asphalt Mixtures*. LCA Performed by Ben Ciavola. Declaration No. 104.263.810 v5. Green Asphalt, March 2.

Grosenheider, Kim E., et al. (2005). *A Review of the Current Literature Regarding Polycyclic Aromatic Hydrocarbons in Asphalt Pavement*. Mn/DOT Contract 81655. University of Minnesota.

Habitable (2020). *New LEED Circularity Pilot Credit Incorporates HBN Optimize Recycling Research*. Habitable, January 30.

Habitable (2024a). *Flooring Product Guidance*. Habitable, May 8.

Habitable (2024b). *Turf Product Guidance*. Habitable, May 8.

Heat Island Group (2024). *Cool Science*. Heat Island Group, Berkeley Lab. https://heatisland.lbl.gov/coolscience/cool-pavements, accessed August 19, 2024.

Heitzman, M.A., Brown, E.R., and Hickey, J. (2018). *Thinlays for Pavement Preservation*. IS-141. Lanham, MD: National Asphalt Pavement Association.

Hughes, C.S., and Kandhal, P.S. (2002). *Designing and Constructing SMA Mixtures - State of-the-Practice (QIP-122)*. Lanham, MD: National Asphalt Pavement Association.

Kandhal, P.S. (2002). *Design, Construction, and Maintenance of Open-Graded Asphalt Friction Courses (IS-115)*. Lanham, MD: National Asphalt Pavement Association.
Khare, P., et al. (2020). "Asphalt-related emissions are a major missing nontraditional source of secondary organic aerosol precursors". *Science Advances* 6.36: eabb9785.
Legret, M., et al. (2005). "Leaching of heavy metals and polycyclic aromatic hydrocarbons from reclaimed asphalt pavement". *Water Research* 39: 3675–3685.
Maher, Michael, et al. (2005). *Context Sensitive Roadway Surfacing Selection Guide*. FHWA-CFL/TD-05-004. Federal Highway Administration, Central Federal Lands Highway Division.
Marks, H. (2024). Personal communication, October 1.
Milad, A., et al. (2022). "A comparative review of hot and warm mix asphalt technologies from environmental and economic perspectives: Towards a sustainable asphalt pavement". *International Journal of Environmental Research and Public Health* 19.22: 14863.
Muench, S.T., and Hand, A.J.T. (2019). *Sustainable Asphalt Pavements: A Practical Guide*. National Asphalt Pavement Association (NAPA).
Mukherjee, D. (2014). "Effect of pavement conditions on rolling resistance". *American Journal of Engineering Research* 3.7: 141–148.
Murtagh, C., and Vallette, J. (2017). *Optimizing Recycling: Reclaimed Asphalt Pavement (RAP in Building & Construction*. Healthy Building Network, StopWaste and the San Francisco Department of the Environment.
National Asphalt Pavement Association (NAPA) (2024). *Understanding Asphalt Mix Plant Emissions*. NAPA, February.
Newcomb, D.E. (2009). *Thin Asphalt Overlays for Pavement Preservation (IS-135)*. Lanham, MD: National Asphalt Pavement Association.
Newcomb, D.E., Timm, D.H., and Willis, J.R. (2020). *Perpetual Pavements: A Manual of Practice*. Quality Improvement Publication 130. National Asphalt Pavement Association (NAPA).
Noel, B. (2023a). "An argument for increased RAP percentages nationwide". *ForConstructionPros.com*, May 11.
Noel, B. (2023b). "Green asphalt: Be humble and kick asphalt". *ForConstructionPros.com*, December 4.
Recycled Rubber Coalition (2022). *Rubber-Modified Asphalt: Paving the Road to the Future*. Recycled Rubber Coalition.
rePave (2023). "Flexible best practices: Recommendations for the design and construction of long-life flexible pavement alternatives using existing pavements". www.pavementrenewal.org/#resources, accessed June 18, 2023.
Schacat, J., Willis, R., and Ciavola, B. (2022). *GHG Emissions Inventory for Asphalt Mix Production in the United States*. SIP-106. National Asphalt Pavement Association (NAPA).
Schacat, J., Willis, R., and Ciavola, B. (2024). *The Carbon Footprint of Asphalt Pavements: A Reference Document for Decarbonization*. SIP-109. National Asphalt Pavement Association (NAPA).
Sukhija, M., and Saboo, N. (2021). "A comprehensive review of warm mix asphalt mixtures-laboratory to field". *Construction and Building Materials* 274: 121781.
Tran, N., and Powell, B. (2009). *Strategies for Design and Construction of High-Reflectance Asphalt Pavements*. NCAT Report 09-02. National Center for Asphalt Pavement Technology (NCAT), April.
United States Environmental Protection Agency (US EPA) (2016). *Ground Rubber Applications, Asphalt Rubber*. US EPA, February 22.
United States Environmental Protection Agency (US EPA) (2020). *Advancing Sustainable Materials Management: 2018 Fact Sheet Assessing Trends in Materials Generation and Management in the United States*. US EPA, 4.
United States Environmental Protection Agency (US EPA) (2024a). *Heat Island Effect*. US EPA, June 6.
US Environmental Protection Agency (US EPA) (2024b). *Federal Research on Recycled Tire Crumb Used on Playing Fields and Playgrounds*. US EPA Safer Chemicals Research.
US General Services Administration (US GSA) (2023). *Inflation Reduction Act Low Embodied Carbon Asphalt Requirements*. US GSA.
US Green Building Council (USGBC) (2019). *LEED BD+C: Circular Products Pilot Credit, v4.1*. USGBC, November 11.
Van Dam, T.J., et al. (2015). *Towards Sustainable Pavement Systems: A Reference Document*. FHWA-HIF-15-002. Federal Highway Administration.
Von Quintus, H.L., Mallela, J., and Buncher, M.S. (2007). *Quantification of Effect of Polymer-Modified Asphalt on Flexible Pavement Performance*. Transportation Research Record: Journal of the Transportation Research Board, No. 2001. Washington, DC: Transportation Research Board of the National Academies, 141–154.
Wang, T., et al. (2012). "Life cycle energy consumption and GHG emission from pavement rehabilitation with different rolling resistance". *Journal of Cleaner Production* 33: 86–96. https://doi.org/10.1016/j.jclepro.2012.05.001.
Watson, D.E., and Heitzman, M.A. (2014). *NCHRP Synthesis 464: Thin Asphalt Concrete Overlays*. Washington, DC: Transportation Research Board of the National Academies.

Way, G.B., Kaloush, K.E., and Biligiri, K.P. (2011). *Asphalt Rubber Standard Practice Guide*, First Edition. Phoenix, AZ: Rubber Pavements Association.

West, R.C. (2015). *Best Practices for RAP and RAS Management (QIP-129)*. National Asphalt Pavement Association (NAPA).

Williams, B.A., Copeland, A., and Ross, T.C. (2018). *Asphalt Pavement Industry Survey on Recycled Materials and Warm-Mix Asphalt Usage: 2017*. No. IS 138 (8e). National Asphalt Pavement Association (NAPA).

Williams, B.A., Willis, J.R., and Shacat, J. (2024). *Asphalt Pavement Industry Survey on Recycled Materials and Warm-Mix Asphalt Usage 2022*. National Asphalt Pavement Association (NAPA) Information Series 138. National Asphalt Pavement Association (NAPA), March.

Willis, R. (2024). Personal communication, October 1.

Willis, R., Yin, F., and Morales, R. (2020). *Recycled Plastics in Asphalt Part A: State of the Knowledge*. IS-141. National Asphalt Pavement Association (NAPA).

Zemba, S. (2021). "How significant are emissions from asphalt paving materials facilities?". *Sanborn Head*, August 11.

Zhang, Y., et al. (2021a). "Investigations into stabilized waste foundry sand for applications in pavement structures". *Resources, Conservation and Recycling* 170: 105585.

Zhang, Z., et al. (2021b). "A comprehensive review of bio-oil, bio-binder and bio-asphalt materials: Their source, composition, preparation and performance". *Journal of Traffic and Transportation Engineering (English Edition)* 9.2: 151–166. https://doi.org/10.1016/j.jtte.2022.01.003.

CHAPTER 16

Earth-Based Materials

Earth-based materials have been used as a building method for millennia, and even today an estimated 8–10% of the world's population lives in earthen dwellings (Marsh and Kulshreshtha 2021). While earth-based construction is minimally used in most developed regions, it is gaining some traction worldwide as a local, low-carbon construction method. The bulk of raw materials for earth-based structures are found on or near construction sites minimizing transportation costs and emissions.

UN Habitat and other NGOs have identified earth-based construction, primarily compressed earth blocks (CEB), and rammed earth (RE), as local building methods that can address the housing crises in some rapidly developing countries in Asia and Africa. This has spurred a substantial increase in research on the structural performance of these building methods and additives to improve their performance and durability (Bredenoord and Kulshreshtha 2023; UN Habitat 2015).

Low-carbon construction goals are spurring owners, designers, and engineers to consider earth-based construction methods and materials as they can be substantially lower in embodied carbon. The embodied carbon of earth-based structures varies by the wide variety of stabilizers used. For example, a CEB stabilized with 10% portland cement is nearly the same embodied carbon as PIP concrete, while the EC of a CEB without cement is just 1/8th the EC of concrete.

As a major carbon footprint of site construction is the transport of heavy, bulky building materials, using materials from on, or near the site, will reduce the carbon footprint of site construction substantially. Soil that is available on site or locally may not be suitable without amendment for earth-based construction, but there a multiple methods to improve its potential performance.

There is a common misconception that earth-based structures are only appropriate in dry, arid climates. While earthen structures require only minimal weather protection in these regions, earthen walls can be employed in just about any setting with adequate stabilization and plaster to protect the structure from moisture. Earth-based structures will require greater attention and maintenance than typical masonry structures, so this may limit their use in public landscapes or sites without strong maintenance regimes.

Within conventional project delivery settings, the construction cost of earthen structures, such as rammed earth, CEB, or adobe will be higher than those of PIP concrete or fired brick. Earthen structures are more labor intensive and time-consuming to build, and contractors practiced in these methods are rare.

In the construction settings of developing regions where labor is inexpensive, earth construction methods can offer a cost effective alternative to expensive, imported building materials like fired clay brick, PIP concrete, and CMU construction. Portable hydraulic machines for compressing a soil mix produce compressed earth blocks that are uniform, durable, and strong. Some are interlocking and can be used by relatively unskilled builders.

Codes and Standards for Earthen Construction

Extensive research on earth-based construction methods in the past few decades has informed codes and standards for earthen construction. German building standards are currently the most well-developed to support earth blocks (CEB and adobe), earth mortar, and earth plaster. In the United States, the New Mexico building code addresses earthen construction most comprehensively while Arizona and California codes are more limited. Other states reference these codes. In 2018, the International Building Code released *Section 2109 Empirical Design of Adobe Masonry* (Table 16.1).

ASTM E2392/E2392M-10 *Standard Guide for Design of Earthen Wall Building Systems* offers technical requirements and considerations for earthen building systems including

DOI: 10.4324/9781003413974-18

Table 16.1 Selected Codes and Standards for Earth-based Construction

Source	Codes	Compressive Strength Required	Modulus of Rupture Required
International Building Code[1]	Section 2109 Empirical Design of Adobe Masonry	2.07MPa 300psi	.35MPa/ 50psi
New Mexico Building Code[2]	Title 14.7.4 Earthen Building Materials Code adobe, rammed earth, CEB, mortar, and plaster	2.07MPa 300psi	.35MPa/ 50psi
German Codes[3]	DIN 18945 Earth Blocks DIN 18946 Earth masonry mortar DIN 18947 Earth plasters DIN 18940 Standard for Load-Bearing Earth Masonry	2.5–7.5MPa–2–4MPa 1–1.5MPa	----
Source	**Standards**		
ASTM	E2392–10 Standard Guide for Design of Earthen Wall Building Systems		
ASTM	D422–63Test Particle Size Analysis Soils		
ISO	14688–1 Geotechnical investigation and testing identification and classification of soil		
Standards New Zealand[4]	NZS 4297 Engineering Design of Earth Buildings NZS 4298 Materials and construction for earth buildings		
Indian Standard	IS13827 Improving Earthquake resistance of Earthen Buildings		
Australian Standard	HB195 Australian Earth Building Handbook		

[1] IBC 2021

[2] State of New Mexico 2021

[3] Schroeder 2018

[4] Walker and Morris 2021

adobe, rammed earth, cob, cast earth, and other earth-based building types. The standard also references standards and codes from the International Building Code and countries including New Zealand, Australia, China, Germany and India.

Soils for Earth-Based Construction

Soils are made up of four main grades of materials: gravel, sand, clay, and silt. Using a sieve analysis, each type of soil material is classified according to the size of its grains. Soils for earthen construction are subsoils with a variety of particle size and mineral structures.

The properties of soil related to earthen construction are as follows:

Texture is the particle size distribution within the soil. This is expressed as percentages of gravel, sand, silt, and clay. While clay acts as a binding agent for earthen construction, too much can result in unstable structures. Soils with high percentages of clay can be amended with sand. Soils with a high percentage of coarse grains can crumble but can be amended with the addition of fine particles (Figure 16.1).

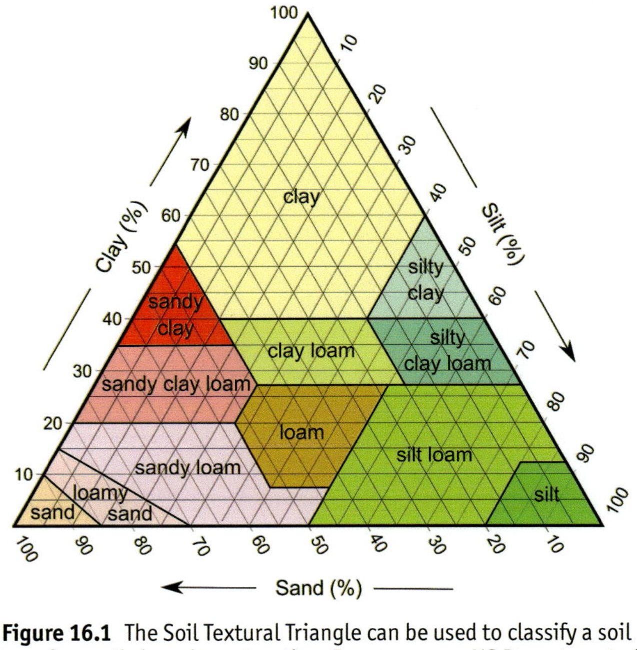

Figure 16.1 The Soil Textural Triangle can be used to classify a soil type for earth-based construction. Image source: US Department of Agriculture.

Plasticity refers to a soil's ability to deform without disintegrating or cracking. The Plasticity Index (PI) is a measure of the plasticity of the soil. Soils with high amounts of clay are highly plastic with a high PI, and soils with lower

PIs are granular. Medium plastic soils are best in earthen construction although clay can be added to soils with lower plasticity. Highly plastic soils will need to be amended with sand (Table 16.2).

Table 16.2 Soil Plasticity Index

Non-plastic soils	0
Slightly plastic soils	<7
Medium plastic soils	7–17
Highly plastic soils	>17

Source: ASTM 2020

Compactability refers to a soil's potential to be compacted thereby reducing voids and porosity. This measure is particularly important for unstabilized rammed earth and compressed earth blocks.

Cohesiveness is the ability of a soil's particles to remain together when a tensile force is applied. The mix of clay, silt, and sand as found in loamy sands, sandy loam, and sandy clay loam will usually have good cohesion. Stabilizing agents and fibers are often added to soil mixes for earth structures to improve cohesiveness of the mix (Table 16.3).

Table 16.3 Stabilizers and Binders for Earth-Based Construction

Stabilizer or Binder and Applicability	Properties and Performance
Portland cement (PC) Portland cement is used as a stabilizer in a range of percentages of 2.5–10% by weight of a CEB or RE mix. Given the high embodied carbon of PC, lower percentages of PC or cement substitutes should be used. *CEB, RE, PISE, stabilized pavement bases* *Not used in: Unfired clay brick (UCB)/adobe*	Compressive strengths of CEB/RE with PC can range from 2–20MPa with varying cement contents and compressive forces of the presses. Minimum cure time is 14 days with 28 being ideal. Higher percentages of PC can result in EC that is in the same range as concrete. PC should not be used to stabilize soils with higher than 15% plasticity index. PC offers increased weather resistance to wind and rain erosion. PC decreases CEB/RE water absorption but does not completely waterproof the structure.
Lime Quicklime, hydrated, and hydraulic limes can be used as stabilizers in soils high in clay content alone or along with other stabilizers such as portland cement or other pozzolans (Tables 16.4 and 16.5). *CEB, RE, PISE, stabilized pavement bases*	Lime works well for soils with high clay concentrations and plasticity indexes higher than 15%. Lime stabilization can result in blocks with a compressive strength of 3–18MPa. Earthen materials stabilized with lime gain more compressive strength over time. Lime stabilized blocks ca be vulnerable to moisture.
Silica fume Silica fume is a by-product of producing silicon metal or ferrosilicon alloys. It is a very reactive pozzolan and is used to produce high strength concrete. *CEB, RE*	Very high strengths can be achieved. Usually used with PC. Silica fume helps to decrease water absorption of RE and CEB. Expensive.
Fly ash Fly ash is a by-product of coal combustion. It is a reactive pozzolan and is commonly used as a portland cement substitute in concrete. *CEB, RE, PISE, stabilized pavement bases*	As it is an industrial by-product, the embodied carbon is assigned to another product, therefore embodied carbon accounting will find this to be low-carbon material in earthen structures. Usually used in combination with PC, but there is potential to use it without PC. With movement away from coal combustion, supplies could be limited. Fly ash is on the LBC Red List as it can contain heavy metals.
Geopolymer binders are made by alkali activation of aluminosilicate source materials (fly ash, GGBFS, metakaolin, calcined clays, zeolite) in a polymer formation (Chapter 14).	Competitive with PC in performance, but supplies are variable. Used in concrete in niche markets in Australia, South Africa, Brazil, India, Ukraine, and the United States. Substantially lower EC than PC. Only a few research studies in earth-based construction but slightly higher compressive strengths than PC.
Bitumen, also called asphalt emulsion, is used as a stabilizer and water proofing agent primarily in unfired clay bricks/adobe. The microscopic bitumen droplets are either mixed with solvents (cutback) or dispersed in water as an emulsion. The material differs from the type of asphalt cement or bitumen used in asphalt pavement as it is not heated, and greater quantities of bitumen are used. *Unfired clay bricks/adobe*	Semi-stabilized is 2–4% asphalt emulsion and stabilized is 4–8% asphalt emulsion. Improves the cohesive strength of granular materials and reduces their moisture sensitivity. Most effective with sandy or silty soils. Compressive strength is slightly improved but falls off after 6% bitumen content. Look for low-VOC cutback products. Bitumen content of emulsions is 55–65% so EC will be higher.

(*Continued*)

Table 16.3 (Continued)

Stabilizer or Binder and Applicability	Properties and Performance
Natural resins offer good stabilization for earthen materials. Natural resins include gum arabic, palmo-copal, rosin, vinsol, lignin, molasses, and ethyl cellulose. Natural biopolymers include starch, cellulose, chitosan, carrageenan, gelatin, and alginates. *EC, CEB and stabilized pavement*	Performance data is limited. Higher cost. Lower embodied carbon. Large quantities required for some. Most render soil impermeable to moisture, although some are water soluble. They are often used in conjunction with PC and/or lime for stabilization of EC/CEB.
Synthetic resins can be used for soil stabilization as an alternative to cement or hydrated limes. These include gypsum, calcium, and potassium salts, furfural-based and acrylic based resins, and a wide variety of polymers. *EC, CEB, and stabilized pavement*	Cause a chemical bond to form between particles resulting in a dense, water-resistant surface. Usually more expensive than cement and lime, and their performance varies widely. VOCs and health impacts can be a concern of some synthetic binders such as polyvinyl acetate (PVA) and polyvinyl chloride (PVC). Synthetic fibers for reinforcing earthen structures include polypropylene fibers, nylon fibers, and recycled tire shreds. Polypropylene fibers are often mixed with lime or cement stabilized soils to decrease their brittleness in up to 0.25% of dry weight of the mix (Hall et al. 2012).
Biological and natural stabilizers and fibers of plant and animal origin have been used to stabilize earth construction for centuries. Traditional biological stabilizers include straw, cow dung, cactus mucilage, animal blood, casein, and wool. Agricultural waste includes straw, coconut fiber, flax, hemp, wood fiber, rice husk, cassava peel and powder, sisal fiber, date palm fiber, and many other localized materials.	Performance of RE/CEB with these constituents varies widely as does their availability. Jannat et al. (2020) summarizes performance research on dozens of agricultural waste additives for unfired earth blocks. Biopolymers of guar gum and xanthum gum can increase durability and resistance to erosion of CEBs. While sometimes used alone, many are used in conjunction with lime or cement. Most are specific to a locality or region.
Non-biological waste includes pozzolanic materials such as fly ash, silica fume (both discussed previously), ground granulated blast furnace slag, and non-pozzolanic materials such as bottom ash, salvaged steel fiber, iron mine spoil, marble dust, waterworks sludge, crumb rubber, calcium carbide residue, and many other waste materials.	Some waste materials are pozzolanic and perform similarly to cement. Others are fibrous and help resist tensile forces. Performance of RE/CEB with these constituents varies widely as does their availability. Jannat et al. (2020) summarizes performance research on dozens of non-agricultural waste additives for unfired earth blocks.

Sources: Bailly et al. 2024; Idriss et al. 2022; Minke 2022; Jannat et al. 2020; Muguda 2020; Hall et al. 2012; Guillaud 1994; McHenry 1984

Soil Property Testing

Soil for use in earthen structures should be tested for texture, plasticity, compactability, and cohesiveness. Simple on-site tests can be used initially, but ultimately a lab analysis should be performed to identify specific performance potential and any required soil amendments.

Four simple tests for soil properties are as follows:

The ***ribbon test*** works a wet mass of soil in the hand to extrude a damp ribbon of soil about 15mm in diameter. The ribbon should be able to hang from the hand without breaking for at least 12cm. This test establishes the plasticity and cohesion of the earth mixture (ASTM 2020). A variation of this test starts with a moist ball of soil and presses it between the thumb and forefinger to form a hanging ribbon that is 2–3mm thick. The longer the ribbon, the more clay content in the soil. The shortest ribbon will be loamy sand (Figure 16.2).

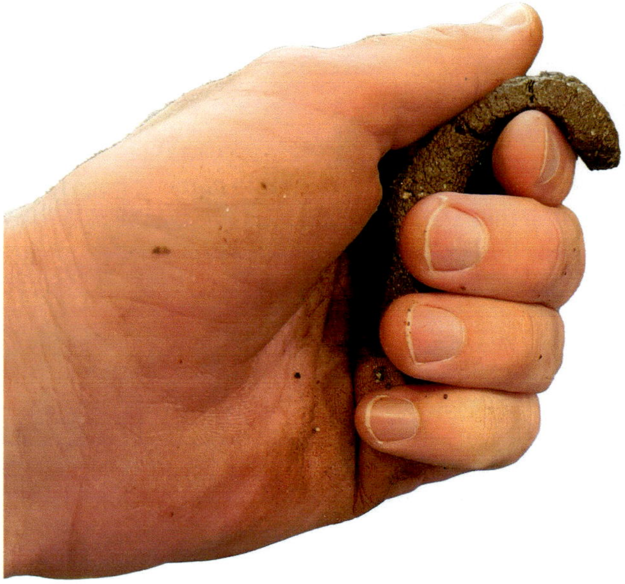

Figure 16.2 The soil ribbon test works a wet mass of sieved soil in the hand to extrude a damp ribbon of soil. The longer the ribbon the higher the clay content. This example has very high sand content and broke at only 3cm. Image credit: Meg Calkins and Jack Elvin

Figure 16.3 The shake test partially fills a jar with water then soil is added, and the jar is shaken. Gravel and sand will sink to the bottom, clay will remain suspended, and organic matter should float to the top. Image credit: Meg Calkins

Figure 16.4 The moist ball test uses a hand formed ball of moist soil. If it forms into a ball that holds together, there is a lot of clay in the soil. Image credit: Meg Calkins and Annie Elvin

The ***shake test*** involves filling a glass jar half full of a soil and then adding water to fill the jar. Shaking should be done until all particles are suspended. Particles will fall to the bottom of the jar in order of weight with gravel on the bottom, then sand, then silt. Clay will remain suspended and organic matter should float to the top. This test will reveal rough percentages of soil constituents (Houben and Guillard 1994) (Figure 16.3).

The ***ball test*** uses multiple hand formed balls of soil about 2cm in diameter. They are set aside to dry away from direct sun. After they are completely dried, none of the balls should be breakable between the thumb and fingers of one hand (ASTM 2020).

The ***moist ball test*** uses a hand formed ball of moist soil. The ball is about 25mm. With pressure from the thumb, if the ball holds completely together, there is a lot of clay in the soil. If it shatters into particles, there is a lot of sand, and if it breaks but does not shatter, there is a good range of clay and sand particles in the soil (Houben and Guillard 1994) (Figures 16.4 and 16.5).

Soil Mixes for Earth-Based Construction

Ideal soil types will vary by type of earthen structure, but generally, well-graded subsoils with minor shrinkage and medium plasticity are best. Soils with very high clay contents (>30%) are not ideal, but sand can be added to improve the mix. Soils with less than 12% clay content will need an addition of fine silt. Clay content of 12–20% is best for rammed earth and compressed earth block structures, although soils with up to 40% of clay can be used with straw in the mix (Minke 2022).

Soils with greater than 2% organic matter should not be used in earth-based construction, and even 1% organic matter is considered problematic. The addition of lime or calcium chloride at 2% can reduce the harmful effects of organic matter (Houben and Guillaud 1994).

Soils for earth-based structures can be modified by adding extra clay or sand, compacting them, and/or the addition of a stabilizing binder such as cementitious material, asphalt emulsion, or biological components including agricultural wastes. Binders are added to achieve higher compressive strengths and water resistance (Table 16.3).

Soil mixes for earthen structures containing soil, stabilizers, and other additives should be tested prior to construction for performance and quality control. Sample mixes are formed into cubes or blocks then tested for compressive

Figure 16.5 If the moist ball breaks but does not shatter, there is a good range of clay and sand particles in the soil as depicted here. Image credit: Meg Calkins and Annie Elvin

strength, modulus of rupture, resistance to water erosion, and absorption. Building codes shown in Table 16.1 require mix designs to meet standards for compressive strength, modulus of rupture, and absorption. Tests may be required if mixes vary from the code standard.

Structural Considerations for Earth-Based Structures

Earth materials for structural use are evaluated on the same criteria as other masonry structural materials such as concrete and CMU. Strength (compressive, tensile, flexural, shear), modulus of rupture, and water absorption, and tensile and shear strength are all evaluated for earthen materials and structures.

Overall, the compressive strength of earth-based construction is low, but it can be improved with stabilizers such as cement, lime, and other pozzolans. Soil's low strength is also compensated for by building thicker walls. It is generally accepted that the compressive strength of earth structures should range from 2–5MPa (Table 16.3) (Minke 2022).

Many factors influence the dry compressive strength of earthen materials. The include grain distribution, grain quality, binding strength of clay minerals, preparation (grinding, sieving, etc.), amount of water, compaction, and additives such as binders, waterproofing agents, and fibers.

Compressive strengths of earthen structures will vary by type of soil even with the same amount of a given stabilizer. For instance, the addition of 5% portland cement by weight can produce compressed earth blocks with the relatively wide compressive strength range of 4–8MPa (Houben 1994; Hall et al. 2012).

Stabilizing Additives for Earthen Materials

There are three approaches to soil stabilization that modify the physical and/or mechanical properties of a soil for earth-based construction.

1. The physical texture or plasticity of the soil can be altered by adding or subtracting sand or clay.
2. Soil can be mechanically compacted to increase its compressive strength and resistance to loading. Examples of this are rammed earth and compressed earth block.
3. Binders and stabilizers such hydraulic materials, pozzolanic materials, bituminous materials, agricultural waste, and industrial waste can be added to impart strength, cohesiveness, and weather and water resistance. (Tables 16.3–16.7).

Stabilizing additives are a very common soil mix addition to improve performance in earthen construction. They can bind soil particles together, increase compressive strength, increase modulus of rupture, reduce water penetration, and help to avoid shrinkage. They can be inorganic materials such as cement, asphalt emulsion, or lime or organic materials such as straw, fibers, or animal dung. Some additives such as portland cement can reduce the possibility of recycling earthen materials after their useful life.

The effectiveness of stabilizing binders varies by type of soil. Non-hydraulic lime should be used to stabilize cohesive soils, while cement, hydraulic limes, and sometimes asphalt emulsions should be used to stabilize granular soils (Walker and Standards Australia 2002; Houbin and Guillaud 1994).

Table 16.4 Types of Lime

Lime Binder Type	Mean Compressive Strength Range at 28 Days (MPa)	Clay Content and Set Time	Use
Hydrated lime	0.3–1.3[2]	N/A	Used for soil stabilization in pavements.
Feebly hydraulic lime, NHL 2.0	1.3–2.0[2]	Contains 5–10% clay[1]; slakes quickly; sets in 15–20 days	Used for less expensive work and in mild climates.
Moderately hydraulic lime, NHL 3.5	2.0–5.0[2]	Contains 11–20% clay[1]; slakes in an hour or two; sets in 6–8 days	Used for higher quality work and exterior walls in freezing climates.
Eminently hydraulic lime, NHL 5.0	5.0–10.0[2]	Contains 21–30% clay[1]; slakes very slowly and sets in 2–4 days	Used in harsh environments such as damp locations and near saltwater.

[1] Harrison 2005

[2] Hall et al. 2012

Table 16.5 Selected Standards for Lime in Construction

Source	Codes and Standards
ASTM	C110 Standard Test Methods for Physical Testing of Quicklime, Hydrated Lime and Limestone
ASTM	C141 Standard Specification for Hydrated Hydraulic Lime for Structural Purposes
ASTM	C207 Standard Specification for Hydrated Lime for Masonry Purposes
ASTM	C977 Standard Specification for Quicklime and Hydrated Lime for Soil Stabilization
ASTM	C1489 Standard Specification for Lime Putty for Structural Purposes
AASHTO	M216 Lime for Soil Stabilization

Table 16.6 Additives and Surface Treatments for Weather Protection of Earth-Based Construction

Integral Admixtures and Applicability	Properties and Performance
Bitumen	See discussion in Table 16.3.
Silicone water repellent admixtures minimize water movement within the capillaries of a cement stabilized earthen structure and reduce water absorption. Mixed into rammed earth, compressed earth block, adobe brick, and cob.	Bonds to matrix of soil. Very durable. Does not affect vapor permeability or surface appearance. Achieves 80% reduction in water absorption. Added at 0.05% or less by weight. Silicone admixture will slow set times when used with cement so formwork should stay on longer and/or the structure should be covered to retain moisture while curing.
Oil or fat based repellent admixtures react with calcium to form calcium stearate or oleate, insoluble salts.	0.1% by weight. Subject to biological deterioration, less durable than silicone repellents.
Surface Applied Stucco Plasters	**Properties and Performance**
Earth plaster, also called mud plaster, uses a dry soil mix with fibers and water. Ideal mix is 10% clay, 40% silt, 50% sand. 14–26mm thickness. Applied in single or multiple layers. Requires annual maintenance and repair.	Lower embodied carbon than cement and synthetic stuccos. Local materials can be used. Adheres well to earthen structures. Has a porous and permeable structure so does not protect against water penetration unless a water repellent admixture or surface applied water repellent is used.
Earth plaster with polymer and silicone incorporates a polymeric emulsion and a silicone water repellent admixture.	Dry soil mix (10 parts) should have less than 10% clay content. The polymeric emulsion (2 parts) can be acrylic, PVA, SBR latex, or other polymer that will bind ingredients and adhere to the structure. Silicone admixture is 0.4 parts. Tests show 90% of water is excluded while maintaining 80% vapor permeability.
Earth plaster with emulsified asphalt, 3–10% emulsified asphalt is added to the basic earth plaster. A variation is to paint the structure with asphalt emulsion and add sand.	Applied in two thin coats for a 10–20mm thickness. Surface of wall is darkened by asphalt.

(*Continued*)

Table 16.6 (Continued)

Integral Admixtures and Applicability	Properties and Performance
Earth plaster with cement incorporates portland cement or alternative cements in one part to 8–10 parts sandy soil.	Use on earthen structures stabilized with cement and/or lime. Has better adhesion to earthen structures than sand-based plasters. Has a porous and permeable structure so does not protect against water penetration unless a water repellent admixture or surface applied water repellent is used.
Lime plaster is made from hydrated or hydraulic lime, sand, and water. Type S is hydrated lime for plaster.	More flexible than cement stucco. Very breathable, but not waterproof. A water repellent admixture or surface applied water repellent can be used in lime plaster.
Portland cement stucco, also called cement stucco or traditional stucco, is made from portland cement, lime, sand, and water. It is a multi-layered plaster for exterior walls.	Use only on earthen structures stabilized with cement and/or lime. Has a porous and permeable structure so does not protect against water penetration unless a water repellent admixture or surface applied water repellent is used. Relatively high embodied carbon.
Polymer-modified cement stucco plaster is cement stucco with polymers added.	Adherence to surface and waterproofing is increased, and cracking is decreased. More expensive than cement stucco and less vapor permeability.
Synthetic stucco, also called acrylic stucco, contains acrylic resins mixed with sand to give the appearance of traditional stucco.	Offers a very good water resistance, but poor vapor permeability, which can trap moisture on and in the wall. Not recommended unless the earthen structure is fully cement stabilized. Synthetic stucco has greater elasticity and will expand and contract more easily with temperature extremes than traditional stucco. Relatively high embodied carbon.
Sealants and Paints	
Silicone water repellent sealers and coatings penetrate the capillaries and reduce water absorption. Can be painted on plaster or stucco.	Does not affect vapor permeability or surface appearance unless tinted. Silicone masonry paints and highly filled acrylic coatings can contain silicone water repellants.
Resin based film forming treatments are painted on and penetrate the earthen structure. Can be painted on plaster or stucco.	Can maintain permeability to water vapor. Do not affect surface appearance unless tinted. Impregnation is important to success.
Asphalt emulsion can be applied to the surface of earthen wall structures as a water proofing agent.	Inexpensive treatment. Not recommended for waterproofing the outward face of a wall as dark brown color will affect appearance and clear coatings, and plasters do not adhere well to the treated surface. Best used on the back of an earthen retaining wall, or on walls that will have a surface or veneer applied.
Plant-based sealers are brushed on the surface of the earthen structure or on mud plaster.	Performance varies. Will need regular re-application.

Sources: Minke 2022; Kebao et al. 2012; Hall et al. 2012; Houben and Guillard 1994; Norton 1997; McHenry 1984

Table 16.7 Compressive Strengths and Embodied Carbon of Earthen Construction Materials

Earthen Material	Compressive Strength (CS) Range (MPa)	Embodied Carbon (EC) Range ($kgCO_2e/kg$)	Sources and Notes
CEB/RE, unstabilized	0.76–3.62	0.011	CS: Range from multiple sources, can meet building codes for CEB EC: Average of multiple sources
CEB, 20% sand added	7.18	Not available	Sand is in addition to sand in soil CS: Saeed et al. 2020
CEB, natural fibers 1–3%	3.16	Not available	Fibers increase comp and tensile strength CS: Average of multiple sources, can meet building codes for CEB
CEB/RE, 2.5% PC	2–5.6	0.0345	CS: Houben and Guillaud 1994 EC: Cabrera et al. 2021
CEB/RE, 5% PC	4–8	0.0586	CS: Houben and Guillaud 1994 EC: Average of multiple sources
CEB/RE, 7.5% PC	6.5–11	0.0815	CS: Houben and Guillaud 1994 EC: Average of multiple sources

(Continued)

Table 16.7 (Continued)

Earthen Material	Compressive Strength (CS) Range (MPa)	Embodied Carbon (EC) Range ($kgCO_2e/kg$)	Sources and Notes
CEB/RE, 10% PC	7–15	0.104	CS: Houben and Guillaud 1994 EC: Average of multiple sources Concrete EC ranges from 0.1–0.2
CEB/RE, 2.5% hydraulic lime (HL)	1.5–2.5	0.027	CS: Houben and Guillaud 1994 EC: Average of multiple sources
CEB/RE, 5% HL	2.5–5	0.0579	CS: Houben and Guillaud 1994 EC: Average of multiple sources EC of hydraulic lime is similar to PC
CEB/RE, 7.5% HL	3–6	0.0809	CS: Houben and Guillaud 1994 EC: Average of multiple sources
CEB/RE, 10% HL	3.5–7	0.1038	CS: Houben and Guillaud 1994 EC: Average of multiple sources
Unfired clay bricks (UCB)/ Adobe, unstabilized	-	0.035	EC: Hardy V3 2024
UCB/Adobe, bitumen 2%	0.5–2.75	0.0387	CS: Houben and Guillaud 1994 2.07MPa meets IBC and NM Building codes for adobe
UCB/Adobe, bitumen 4%	0.5–3	0.0425	CS: Houben and Guillaud 1994 2.07MPa meets IBC and NM Building codes for adobe
UCB/Adobe, bitumen 8%	0.5–2	0.05	CS: Houben and Guillaud 1994 2.07MPa meets IBC and NM Building codes for adobe
Geomodular bag walls	Strength of granular fill	0.251 for one bag and one plate	EC: FlexMSE 2024 Filled bags are 300mm wide, 760mm long and 140mm high.
Other Masonry Materials for Reference	**Compressive Strength Range (MPa)**	**Embodied Carbon Range ($kgCO_2e/kg$)**	**Sources and Notes**
PIP concrete, 20% SCM	20.7	0.117	EC: NRMCA 2022
Clay brick, fired	62.0	0.237	EC: BIA 2020
Concrete masonry unit	17.5	0.105	EC: CCMPA 2020

Notes: Embodied carbon data is from the product life-cycle stage, phases A1–A3.

These EC figures do not include mortar in the CEB structure.

Interlocking CEBs can be employed without mortar. EC for mortar varies by type.

This information is for reference only as actual compressive strengths will vary by soil texture, compressive production force, and stabilizing material.

Research on performance and embodied carbon of CEB is also generally applicable to RE.

Cement and Pozzolanic Stabilization

Cementitious materials, most commonly portland cement, are used as stabilizers for compressed earth block, rammed earth, pisé, and stabilized soil bases. Cement increases compressive strength of earth structures in direct proportion to the quantity of cement used, but only up to a point. For example, compressed earth blocks with 10% portland cement have been shown to achieve compressive strengths of 7–15MPa while use of 2.5% PC has a range of 2–5.6 (Houben and Guillaud 1994).

While cement is the most common stabilizer for CEB and RE, it is a carbon-intensive material and can add substantially to the carbon footprint of an earthen structure. CEBs and RE with 2.5–5% PC can meet requirements of most building codes. Use of 10% PC has an embodied carbon that is just slightly less than that of PIP concrete, and it is not as strong (Figure 16.7).

Alternative cementitious materials can be used as stabilizers in earthen structures, increasing strength and reducing plastic shrinkage. Alternative cementitious materials such as geopolymers and alkali activated cements are discussed in detail in Chapter 14, Table 14.7.

Pozzolanic materials are "siliceous or siliceous and aluminous material that in finely divided form and in the presence of moisture, chemically react with calcium hydroxide to form compounds having cementitious properties" (ACI 2024).

Pozzolans are increasingly used as cement substitutes in concrete, and many can also be employed to stabilize earthen structures. Pozzolanic materials used as stabilizers in earth structures are fly ash, GGBFS, silica fume, calcined clay, and rice husk ash (Table 16.3). These materials are often used in conjunction with portland cement.

Lime Stabilization

Lime is a common stabilizer for soils with high clay content as the lime reacts with clay to bind the particles together with a pozzolanic reaction. It can be used as a stabilizer for compressed earth block, rammed earth, pisé, and stabilized soil bases. Lime can also reduce the shrinkage of soils and eliminate swelling (Table 16.4).

Non-hydraulic limes are produced by burning calcium carbonate (limestone) in a lime kiln above 900°C converting it into unslaked lime also called quicklime (calcium oxide). When water is added, it becomes slaked lime or hydrated lime (HL). Both slaked and unslaked lime are used as stabilizers in earth-based structures and pavements. Trass lime is a type of hydrated lime mixed with ground trass (a natural pozzolan).

Naturally hydraulic limes (NHLs) are formed by the calcination of limestone containing clay minerals or chalk parent rock. As these limes are partially hydraulic, they can be used as a higher strength, faster setting substitute for hydrated lime or quicklime.

HL and NHL set initially with water then continue to set slowly with air. They gain strength over a long period of time as they absorb carbon from the air in a process called carbonation. Types of limes used in earth-based construction are discussed in Table 16.4, and selected standards for lime in construction are listed in Table 16.5.

Bituminous Binders and Emulsions

Bitumen, also called asphalt emulsion, improves the cohesive strength of granular materials and reduces their moisture sensitivity. Unlike cement, lime, and pozzolanic materials, the use of asphalt emulsion does not cure and harden into a rigid composite, instead it disperses among the fine particles holding them together. Use of bitumen improves durability and shear strength of earth-based structures, and it introduces water repellent properties. Bituminous binders are primarily mixed into adobe bricks and stabilized soil pavements.

Surface applied asphalt emulsions are used on a wide variety of earthen walls as a waterproofing agent (Table 16.6). Most asphalt emulsions are dark colored, so this can impact the appearance of the earthen structure.

Biological Binders and Stabilizers

Biological stabilizers, including natural fibers, agricultural waste, and biopolymers, are employed to increase strength (both compressive and shear), resist erosion, and reduce shrinkage of earth-based structures. While most fibers have a lower compressive strength than soil, they hinder lateral strain during compression, leading to overall increased compressive strength. They help to resist tensile forces from the expansion and contraction of earth materials and add cohesiveness to the mix.

Natural fibers and additives are usually used in earth-based mixtures at 4–20% by volume. Biological materials can contain a lot of moisture, so some should be partially dried prior to use, and earthen structures should be allowed to dry prior to applying plaster.

Biological stabilizers can be employed alone or in conjunction with cement or lime stabilization. Research shows that biological stabilizers, sometimes along with cement, in unfired earth blocks can increase compressive strength with a wide range of results from 0.46–30MPa (Figure 16.5) (Jannat et al. 2020).

Tree resin emulsions, lignosulfonates, and enzymatic emulsions are all used as stabilizers in granular soil pavements and can be used to stabilize other earthen structures. Natural binders for soil and granular pavements are discussed in greater detail in Table 6.7.

Waterproofing and Weather Protection Additives and Surface Treatments

While earthen structures are possible in many climates, driving rain and sleet can saturate the materials leading to erosion and structural failure. Soils are permeable and hydrophilic by nature; therefore water penetration is a major vulnerability of earth-based construction.

Water proofing agents can be integral admixtures to a soil mix, or they can be applied to the surface of an earthen structure after construction. While cement and lime are common binders in earth-based construction, they provide only limited resistance to water penetration. In fact, water penetration of cement or lime stabilized structures can lead to efflorescence, calcium carbonate deposits, on the surface of the structure. Table 16.6 discusses additives and surface treatments for protection of earthen structures.

While their use is relatively new in earthen structures, silicone water repellents are very effective at protecting earth structures from water damage. They substantially reduce water absorption while maintaining vapor permeability. They can be used as admixtures or surface applied after construction, and they do not affect the color of the material (Hall et al. 2012).

Plasters will inhibit moisture penetration into earthen walls and protect them from erosion caused by wind, rain, and snow. Most earthen structures will require mud or lime plasters as cement plaster can trap moisture in the wall.

Foundations for Earth-Based Walls

Foundations for earthen walls vary by structural requirements, type of wall system, subsoil, and c limate. In areas with temperature extremes, rigid concrete footings for earthen walls using cement and/ or mortar must rest a minimum of 15cm below the frostline. In warm, dry climates, alternative footings such as stone and gravel trench foundations can be used for lower freestanding walls and unit walls without mortar.

A ***poured in place concrete*** spread footing is the most common foundation for modern earthen walls. A short concrete stem wall extending up from the footing will hold the earthen wall 15cm above finished grade to protect it from soil moisture and rain splash. The width of the stem wall will be equal to or greater than the width of the earthen wall. As concrete can conduct moisture, a waterproofing agent should be painted on the stem wall where the earth material rests.

Stabilized earth can also be used as a foundation material for freestanding earthen walls in arid climates. Stabilizers include cement, lime, pozzolanic materials, and asphalt emulsion. An integral waterproofing agent such as a silicone admixture should be added to the mix for moisture protection.

Stone foundations with cement, lime, or mud mortar can be used for low earth walls in mild climates. Stones will be staked with batter and interlock similar to a dry stack stone wall. Reclaimed concrete pavement foundations could be used in some lower earthen walls.

Well-compacted gravel trench foundations are a mixture of gravel and sand. They can be used with earthen walls that don't use cement stabilizers or cement mortars and/or in areas where frost heave is not a concern. These foundations can perform as a type of base isolation in earthquake prone areas. Earthen walls above these foundations should be waterproofed for the first few courses to avoid water wicking into the wall. They should also be placed on well-draining substrates.

Automation and Innovation in Earthen Construction

Table 16.8 Automation and Innovations in Earth-Based Construction

Robotic fabrication	Robotic fabrication can be used to produce formwork, place, and compact materials. Robotic arms have been used to compact soil mixes, feed materials, and assemble formwork (Gomaa et al. 2023).
3D printing	3D printing from CAD tools for geometry production and robotic code generation can "print" earthen structures and formwork. Freeform formwork for rammed earth construction has been fabricated using robotic 3D printing with rigid recycled plastic (Gomaa et al. 2023) (Figures 16.6 and 16.7).
Automated mechanical approaches	Automated and mechanized approaches are being developed to reduce construction costs. In Australia, a semi-automated approach uses a portable prototype system for RE that consists of a material hopper that automatically discharges the soil mix into formwork, an automatic compaction tool, and a slip formwork (Form Earth 2024). A fully automated approach uses a robotic system that provides an augmented ramming process with slip formwork controlled by a robotic arm. A robot deposits the earth material layer by layer the robotic tamping follows (Kloft et al. 2019). Another fully automated approach produces precast rammed earth in custom shapes from a CAD/BIM model using unstabilized soil (Gomaa et al. 2023).

A primary obstacle to the use of earth-based structures in many projects is construction time and the resulting labor costs. Mixing, forming, and placing earth mixtures is a time-consuming process; therefore exploring automation and other production and construction innovations can make earth-based construction competitive with other more carbon-intensive masonry materials such as concrete and concrete masonry units (Table 16.8).

Earth-Based Construction Methods

Rammed Earth Construction

Rammed earth (RE) construction, also called pisé de terre, is an earth-based building method where a soil mix is placed by 100–200mm lifts into temporary formwork and compacted either pneumatically or manually (Figures 16.8–16.10). This results in a dense, solid, and rigid monolithic wall structure that can be nearly as strong as concrete if portland cement is used as a stabilizing binder.

If stabilized and waterproofed, rammed earth structures can be used in any climate. Rammed earth has been used for centuries in China, Africa, the United Kingdom, and

Figure 16.6 and Figure 16.7 World Advanced Saving Project (WASP) and Mario Cucinella Architects designed and constructed TECLA (technology and clay) residential prototypes by 3D printing a soil mixture for a very low-carbon structure. These images show early prototype wall sections, but the finished structures are dome shaped with the walls acting as structure, roof, and external cladding. The proprietary printing system uses multiple modular 3D printers to print an entire building in 72 hours. Image credit: WASP and Mario Cucinella Architects

Figure 16.8 The soil colors and striations from placement in layers make a distinctive rammed earth wall finish. Image credit: Karoline Cullen, Dreamstime.com.

Figure 16.9 Rammed earth wall construction uses slip formwork and the soil mix is placed in layers. Pneumatic tampers are most commonly used to compact the layers, but hand tampers can be used as well. Image credit: JYB Devot

Figure 16.10 Andrea Cochran used cement stabilized rammed earth retaining walls in a private garden in California. Image credit: Andrea Cochran Landscape Architecture

the United States. While it is enjoying a renaissance for residential construction in the US southwest, Europe, Australia, and New Zealand, there is still a perception that it is a crude, inferior building method. Rammed earth freestanding site walls are detailed in Chapter 9, and construction considerations are discussed in Table 16.9.

Soils for Rammed Earth Construction

Ideal soils for rammed earth construction contain a range of particle sizes from small gravel to coarse and fine sand and clay. The following range of soil texture can be used: 5–45% clay, 15–40% silt, 20–70% sand and fine gravel. One typical ratio is 70% sand/small gravel and 30% silt and clay (Easton and Easton 2012). Other sources call for a high sand content with a maximum of 8% clay (Hall et al. 2012).

Water is added to the mix in amounts to allow the soil to bond properly when compacted but not crumble. If too much water is added, soil will be spongy and not compact properly. Proper water quantity is determined by the *soil ball drop test*. If a ball of soil is dropped from waist height and it breaks apart, the water amount is correct. If it does not break, there is too much moisture in the soil mix (Houben and Guillaud 1994).

Table 16.9 Considerations for Freestanding Rammed Earth Walls

Aspect	Considerations
Soil mix	Soil mixes for rammed earth can vary widely, but the average is 70% sand and fine gravel and 30% clay. Ranges are 5–45% clay, 15–40% silt, 20–70% sand and fine gravel. Max rock sizes are 40mm (1–½″) in diameter. Max clay lumps are 12mm (½″). Max. 2% soluble salts.[1] Use of local soil will reduce embodied carbon of transport.
Wall thicknesses	Unreinforced wall thicknesses range from 260–600mm with a minimum thickness in the New Mexico code of 260mm (10″) for a 3m (10′) wall. Thicker walls generally require lower percentages of cement for stabilization.
Stabilizers	Cement is the primary stabilizer used at percentages of 3–10% by weight of mix. Cement substitutes can be used to replace a portion or even all the cement. Alternative stabilizing materials such as fly ash, lignite ash, and hydrated or hydraulic lime can be used. White portland cement or lighter pozzolans will maintain the color of the soil in rammed earth walls.
Compressive strength	The New Mexico Earthen Building Materials Code requires rammed earth building structures to have a minimum compressive strength of 300psi (2.07MPa). Compressive strengths of unstabilized rammed earth walls range from 0.76–3.62MPa (Table 16.7).
Structural considerations	Performance research on structural performance, life cycle impacts, stabilizers, and weathering is growing for modern rammed earth construction. This data has informed codes and standards for rammed earth construction, making it easier to obtain permits for rammed earth structures in recent years.
Joints	Expansion joints should occur in walls every 4500mm. Keyways should be employed to maintain alignment of the wall segments.
Reinforcing	Low rammed earth walls may not require reinforcing. In areas of high seismic risk, earthen walls should incorporate vertical reinforcing with the soil mix rammed around the steel bars that tie the wall stem into the footing. Where there is no seismic risk, vertical reinforcing can be eliminated with generous height-to-thickness ratios providing stability of the wall stem.
Waterproofing	Rammed earth is less susceptible to water damage than adobe or brick and mortar because it is a monolithic structure, but it can be waterproofed by the addition of waterproofing agents to the mix, surface applied waterproofing agents or plasters. The NM code does not allow emulsified asphalts in the mix.
Foundations	Most foundations for rammed earth walls will be PIP concrete spread footings. For low walls, foundations could be cement stabilized rammed earth. Keyways should be employed between wall stems and spread footings.

(Continued)

Table 16.9 (Continued)

Aspect	Considerations
Lateral support	Rammed earth walls that are longer than 7 meters or 24′ should be buttressed or have some other means of lateral support per the New Mexico building code.
Plaster	Rammed earth walls can be finished with lime or earth plasters. Synthetic stucco should not be used as it is not breathable and can trap moisture. Cement plasters can be used if cement is used as a stabilizing agent. The NM code requires stucco, lime, or mud plasters for rammed earth walls that are not stabilized with at least 6% cement (Table 16.6).
Construction	Soil is dumped into the forms in layers then compacted to about half the thickness of the loose soil lift. Ramming can be accomplished either manually or by compressed air hammers with square steel heads. Layers can be thicker when ramming by machine with average thicknesses of 130–160mm once compacted. Manually compacted layers are usually around 100mm. Placement in layers gives rammed earth walls a striated appearance. If cement is used, rammed earth must be mixed in small batches, so the mix does not set while waiting for placement. Construction waste from the soil mix can be returned to the soil if portland cement is not used. If portland cement is added, extra mix can be used in soil cement pavement.
Maintenance	The plaster of rammed earth walls will require periodic maintenance. Refer to Table 16.6 for a discussion of maintenance requirement for earth plasters.

Sources: Minke 2022; Muguda 2020; Hall et al. 2012; Houben and Guillaud 1994; McHenry 1984

Stabilizers for Rammed Earth Construction

Rammed earth can be strengthened with pozzolanic or cementitious stabilizers. When these are employed, the mix is sometimes called stabilized rammed earth (SRE). Most modern rammed earth buildings incorporate 3–10% portland cement as a stabilizer lending higher compressive strength and increasing durability. Sometimes this is referred to as cement stabilized rammed earth (CSRE) (Table 16.5). Lime is another stabilizer that is used alone or in combination with cement to increase compressive strength. Given the relatively high embodied carbon of both portland cement and lime, cement substitutes, alternative stabilizers, and reduced cement contents should be considered in rammed earth structures (Figure 16.7).

If soil mixes are tested for compressive strength, modulus of rupture and water absorption prior to construction. This can reduce excessive use of portland cement.

Rammed earth site walls can use lower amounts of cement and cement substitutes and or even avoid cement stabilization entirely. Far less common, other stabilizers such as bioenzymes, agricultural and non-agricultural waste materials, industrial by-products, and volcanic sand and marble sawing waste are being tested to reduce or avoid high-carbon portland cement and lime. Recent biobased additives being explored are xanthan gum, guar gum, and linseed oil.

The New Mexico Earthen Building Materials code considers rammed earth walls to be fully stabilized with 6% or greater portland cement by weight, but research is showing that alternative stabilizers alone or combined with lower cement quantities can meet the New Mexico compressive strength requirements (Giuffrida et al. 2022; Basanna et al. 2020; Toufigh and Kianfar 2019; Damme and Houben 2018; Porter et al. 2018). As rammed earth and compressed earth block are both soil mixes compacted under pressure, they employ many of the same stabilizers (Tables 16.3 and 16.7).

Rammed Earth Wall Construction Process

One of the greatest challenges to rammed earth construction is the time it takes to construct a structure. The modern rammed earth construction process is more efficient with the use of heavy equipment such as front-end loaders for placing the soil mix in forms and pneumatic ramming tools for compacting the mix in lifts. This makes rammed earth structures more expensive to build.

The formwork for rammed earth walls is similar to that of PIP concrete formwork. Vertical posts spaced approximately every meter support horizontal wood boards or steel form panels. Rods tie the forms together at the top and bottom to resist outward forces from the ramming of the soil mix. There have been recent innovations in formwork systems for rammed earth that allow for enhanced reusability, versatility, and time efficiency (Gomaa et al. 2023).

Compressed Earth Block Construction

Like fired clay bricks and adobe, compressed earth blocks (CEB) are modular units made from soil, water, sand, and sometimes stabilizers. CEBs, also called compressed stabilized earth blocks (CSEBs), differ from fired bricks and adobe in that they are produced under high pressure lending them additional compressive strength and stability. Manual or powered hydraulic presses compress the blocks at forces ranging from 7MPa (manual presses) to 25MPa (hydraulic presses) (Figures 16.11 and 16.12). Compressive strength and bulk density of CEBs increases substantially with compressive force of the press. As opposed to other earthen materials, their use is not limited to dry, arid countries as they can withstand wider temperature and humidity fluctuations than unfired adobe bricks (Tables 16.10 and 16.11).

CEB are used throughout the world for low buildings and are growing in use as a low-cost housing material in India, Africa, Sri Lanka, Cambodia, Nepal, Mexico, El Salvador, and other rapidly developing countries (Bredenoord and Kulshreshtha 2023).

In site applications, CEBs are used in both freestanding and lower retaining wall applications. They are produced with portable equipment on or near a construction site with local soil and pozzolanic or cementitious components added for stability.

Compressed earth blocks can be produced in a wide variety of shapes and sizes that usually result in modular dimensions for ease of stacking and interlock. Holes in the block allow for vertical reinforcing and grout around the reinforcing for stability.

A recent innovation in compressed earth block is interlocking blocks called interlocking stabilized soil blocks (ISSB) (Figure 16.11). These blocks have deformations that interlock, eliminating the need for mortar and increasing the resistance to out-of-plane displacement (Figure 8.23).

Figure 16.11 Double interlocking stabilized soil blocks are formed from a soil mixture stabilized with small amounts of cement or lime. These blocks interlock vertically and horizontally forming stable walls without mortar. Image credit: Dwell Earth

Figure 16.12 A portable compressed earth block machine. Blocks can be produced with this machine on the project site if the soil type is suitable. Image credit: Dwell Earth

ISSBs could perform in retaining walls similarly to precast concrete segmental retaining wall blocks. Embodied carbon of interlocking block CEB structures can be lower without use of carbon-intensive mortar. Interlocking blocks are shown to better resist seismic forces than traditional CEB and mortar structures.

Adobe Construction

Adobe bricks, also called adobe blocks, unfired clay bricks, or sun dried bricks, are molded composite bricks made from soil, water, and organic material such as straw or other fibers. They are often stabilized with bitumen emulsion (also called emulsified asphalt) (Table 16.12). Soil suitable for unfired clay/adobe bricks is widespread, and production equipment is easily portable, making them a local, low-carbon building material.

While adobe bricks originated in dry, arid climates without trees for lumber, if fully stabilized, they can be used in just about any climate. Adobe bricks are modular like fired clay bricks, and they are produced in a wide variety of sizes (Table 16.13). In structures, adobe bricks are usually laid in a running bond pattern with clay-based or lime mortar in 12–25mm joints. The bricks can be formed with voids to receive vertical steel reinforcing bar.

Table 16.10 Considerations for Compressed Earth Block Walls

Topic	Considerations
Soil mix	A wide variety of soils can be used in CEBs making them local materials in just about any region and even on many sites. Soil should be subsoil, not topsoil. Ideal soil is 30–60% clay soil content, and 40–70% sand content with medium plasticity. Soil with a medium plasticity index of 7–17% can be used without stabilizers in applications that require low strength such as lower freestanding walls. Soils with low (<7%) or high (>17%) plasticity indexes should be stabilized.
Stabilizers	Compressed earth blocks are most commonly stabilized with portland cement or lime. But other pozzolans, resins, or industrial wastes can be used. Stabilizers increase compressive strength and reduce swelling of soil in the blocks. In some regions, stabilized compressed earth blocks are called compressed stabilized earth blocks.
Fiber reinforcing	The strength of CEB can be increased by adding fiber reinforcing to increase tensile resistance of the blocks. A network of fibers can reduce plastic shrinkage cracking and improve ductility and shear strength. Natural fibers such as date palm, banana, or quick lime fibers are preferable, but synthetic fibers such as recycled plastics can be used as well (Table 16.3).
Compressive strength	Compressive strengths of CEBs, ranging from 2–16MPa, are a factor of soil type, compressive force of press, and stabilizer type. Unstabilized MPa of CEB is dependent on soil type, sand content, and press compression, but ranges from 1–8MPa. Compressive strengths of 2–2.5MPa meets most codes for low earthen buildings and walls.
Structural considerations	Performance research about structural performance, life cycle impacts, stabilizers, and weathering is growing for CEBs. This data has informed codes and standards for rammed earth construction, making it easier to obtain permits for CEB structures in recent years.
Reinforcing	Most CEBs are formed with voids to receive vertical reinforcing if required.
Cure time	7–14 days, although they can be placed immediately in non-load bearing applications.
Moisture protection	CEBs should be held a minimum of 15cm above finished grade, and they should have an integral waterproofing agent or applied waterproofing (Table 16.6). Retaining walls should have drain rock and a perforated pipe wrapped in filter fabric placed behind the wall.
Mortar and grout	Lime or earth mortars and grouts should be used. Ideally, mortar is the same soil mix as the bricks with the addition of lime and water. Portland cement mortar should not be used. Typical mortar joint sizes are 10–20mm. ISSBs do not require mortar.
Finishes	CEBs can be finished with lime or earth plasters, or integral water proofing agents can be used. Synthetic stucco should not be used as it is not breathable and can trap moisture. Cement plasters can be used with cement stabilized CEB (Table 16.6).
Seismic considerations	In seismic areas, bond beams are required at the top of building walls and should be considered for use at the top of freestanding and retaining CEB walls as well. They can be wood, engineered lumber, or concrete. Gravel trench foundations can perform like base isolation in an earthquake.
Manufacturing	The manufacturing process for CEB involves sifting the soil, weighing, mixing, observing moisture content, casting, and compressing the mixture in a mold. After compression, the block is removed and stacked for curing for 7–14 days. Most hydraulic CEB machines can produce 1000 to 8000 blocks per day. Manual presses can produce 200–700 blocks per day (Dorado 2022).
Maintenance	The plaster of CEB walls will require periodic maintenance. Refer to Table 16.6 for a discussion of maintenance requirement for earth plasters.

Sources: Bailly et al. 2024; Bredenoord and Kulshreshtha 2023; Idriss et al. 2022; Minke 2022; Cabrera et al. 2021; Fernandes et al. 2019

Table 16.11 Comparing CEB and Adobe

Topic	Compressed Earth Blocks	Unfired Clay Brick/Adobe Bricks
General	Stronger, denser, and more uniform.	Less strong, dense, and uniform. Can have rounded edges.
Soil mix	Wide range of appropriate soils. Soils with medium plasticity are best.	More narrow range of soils. Loamy sands, sandy loam, and sandy clay loam are best.
Production equipment	Portable hydraulic press.	Wood molds.
Curing/drying	Some CEBs can be used immediately and continue to cure if stabilized; PC stabilized blocks in high load applications should cure for 7–14 days prior to use.	Blocks need to air dry for 10–14 days in a protected environment.
Stabilizers	Cement, lime, pozzolans, natural and synthetic resins, natural and synthetic fibers.	Asphalt, natural and synthetic fibers.
Compressive strengths	A wide range of compressive strengths from 2–16MPa is possible. Unstabilized MPa CEB ranges from 2–8MPa.	Compressive strengths vary from 0.5–3MPa.
Reinforcing	Natural or synthetic fibers.	Natural or synthetic fibers.

Sources: Minke 2022; Cabrera et al. 2021; Moquin 2000; Fernandes et al. 2019; Easton and Easton 2012; Houben and Guillaud 1994

While adobe bricks are local, non-toxic, low-carbon materials, their use is not without challenges. Bricks can erode or deteriorate over time if they are not properly stabilized and protected from the weather. Maintenance of adobe bricks structures is required at least annually to repair bricks and stucco plaster. Adobe bricks are labor intensive to produce on or near the site as they are molded and dried individually. This can result in relatively expensive adobe structures unless the bricks are produced by volunteers and or owners (Tables 16.12–16.15; Figures 16.13 and 16.14).

Table 16.12 International Building Code (IBC) Section 2109 Empirical Design of Adobe Masonry

Criteria	Requirement
Compressive strength	Minimum 300psi or 2.07MPa, average of multiple bricks
Modulus of rupture	Minimum 50psi 0.35MPa, average of multiple bricks
Moisture content	Not to exceed 4% by weight
Shrinkage cracks	No more than three shrinkage cracks, and no cracks are greater than 76mm (3") in length and 3.2mm (1/8") width, average of multiple bricks
Exterior building wall thickness	Minimum 254mm (10") New Mexico code is min. 14"
Absorption	Does not absorb more than 2.5% moisture by weight when submerged in water for 7 days
Foundations	Extend minimum 152mm (6") above grade
Adobe brick at grade	Fully stabilized adobe must be used for the first 102mm (4") above grade

Source: IBC 2021

Table 16.13 Common Sizes and Weights of Adobe Bricks

Size of Brick in Millimeters	Size of Brick in Inches	Approximate Weight	Notes
200x300x100mm	8x12x4in.	11kg/25lbs	US Southwest adobe brick
250x360x100mm	10x14x4in.	16kg/35lbs	New Mexico code requires min 14" walls
305x406x130mm	12x16x5in	24kg/53lbs	Arizona code requires min. 16" walls
305x381x130mm	12x15x5in	24kg/53lbs	UK unfired brick
280x280x130mm	11x11x5in	18kg/39lbs	New Zealand/ Australia adobe/ unfired brick
245x175x 115mm	9.6x7x4.5in	9kg/20lbs	European adobe/ unfired brick

Sources: Minke 2022; Easton and Easton 2012; Moquin 2000; Houben and Guillaud 1994

Table 16.14 Considerations for Adobe Brick Walls

Topic	Considerations
Soil mix	Ideal soils for adobe bricks are loamy sands, sandy loam, and sandy clay loam. Sand content should be greater than 50%. Clay content of the soil should not exceed 30%, and no more than half the clay content should be expansive clays as this can cause uneven drying through the brick and result in cracking. Too much kaolinite clay will produce weak adobe bricks.
Stabilizers	The most common stabilizer for adobe bricks is emulsified asphalt. Full stabilized bricks will contain 4–8% asphalt emulsion. Semi stabilized bricks contain 2–4% asphalt emulsion. Other stabilizers are casein and lime, water glass, linseed oil, and silicone. A wide variety of natural or synthetic fibers are used in adobe brick for binding the mix, resisting tensile forces, and controlling cracking.
Production	Adobe bricks are formed in wood or metal molds. The soil mix is rammed into molds to push the material firmly into corners. Excess material is scraped off the top of the mold. The blocks are removed from the molds after a few hours and stacked to allow for air circulation around them. Adobe bricks should be dried for 14 days away from direct sunlight to minimize shrinkage and cracking. They must be protected from rain and moisture. Thinner bricks will dry more quickly which is why they are not normally thicker than 130mm (5in). While commercial adobe brick factories can produce up to 3000 blocks per day, most on-site brick production operations can produce 300–600 blocks per day.
Mortar and grout	Typical mortar joint sizes are 10–20mm. Mortar for adobe brick walls can be a similar soil mix as the adobe with the addition of cement or asphalt emulsion for stabilization. Use of typical portland cement mortar is not recommended as it is too rigid for the relatively soft unfired clay/adobe bricks and can trap moisture against the wall. Instead, lime or mud mortar is recommended. Table 9.4 discusses types and performance of mortars and grout.
Dry time	Most unfired clay/adobe bricks can fully dry in two weeks. Thinner bricks will dry more quickly. Drying bricks in the sun can lead to greater cracking with some soil mixes.

(Continued)

Table 16.14 (continued)

Topic	Considerations
Moisture protection	Unfired clay/adobe bricks should be held a minimum of 15cm above finished grade and at least the first course of brick should be fully stabilized with emulsified asphalt. Surface and subsurface drainage should be directed away from unfired clay/adobe brick walls.
Reinforcing	Vertical reinforcing between the unfired clay/adobe bricks will tie into the stem wall and footing. Horizontal reinforcing in mortar joints can be steel horizontal web, grid, or wire reinforcing. This should be placed every three courses.
Finishes	Adobe brick walls can be finished with lime or earth plasters. Synthetic stucco should not be used as it is not breathable and can trap moisture. Cement plasters should not be used as their expansion rate is different than adobe bricks and they can trap moisture against the wall. Adobe bricks that are fully stabilized do not require protective finishes, although plaster may be desirable for a more finished appearance (Table 16.6).
Wall thickness and height	Modern unfired clay/adobe brick walls are usually 26–40cm thick with wider blocks used for earthquake prone areas and taller walls. Unfired clay/adobe brick walls should not be higher than eight times their width.
Structural considerations	Unfired clay/adobe brick walls should not be taller than eight times their width. The IBC Section 2109 Empirical Design of Adobe Masonry and other codes provide structural requirements for building walls.
Footings and foundations	In most locations, blocks are placed on a PIP concrete foundation and low stem wall per local codes and conditions. For lower walls in areas where frost heave is not a concern, unfired clay/adobe bricks can be laid directly on a well-compacted gravel and sand trench foundation. This can act as a type of base isolation in earthquake prone areas. Other foundations are constructed from stone and mud mortar to hold the adobe away from moist soil.
Seismic considerations	In seismic areas, bond beams are required at the top of building walls and should be considered for use at the top of freestanding and retaining walls as well. They can be wood, engineered lumber, or concrete. Gravel trench foundations can perform like base isolation in an earthquake.
Maintenance	The plaster of adobe brick walls will require periodic maintenance. Refer to Table 16.6 for a discussion of maintenance requirement for earth plasters.

Sources: Minke 2022; Easton and Easton 2012; Moquin 2000; Houben and Guillaud 1994

Table 16.15 Other Earthen Construction Methods

Construction Method	Performance and Construction Considerations
Cob is an earthen construction system using moist balls of sandy clay soil stacked and packed into place to form a monolithic wall.	Walls are between 30 and 60cm thick and applied in lifts up to 45cm high. Requires no formwork, ramming, or mechanical equipment but is labor intensive. Very low embodied carbon material especially if local soil is used. Soils with 5–25% clay content can be used. The soil mix should contain approximately 10% fibers for reinforcing and tensile resistance. Fibers should be 20–34cm in length. Cob material should be held 15cm min above grade and lime or earth plasters should be employed for weather protection (Table 16.6) (Smith 2000).
Pneumatically impacted sprayed earth (PISE) is a modified version of rammed earth where a moist soil mixture is sprayed onto one-sided formwork with high-pressure air delivery. Cement stabilizes the mix.	This method can the speed of construction, but it requires more expensive equipment and a large crew. The force of impact during the spraying of the mixture compacts the mix resulting in a monolithic wall. Shaving and shaping after application will produce a smooth wall. This is similar to dry-mix shotcrete and can use the same equipment. Soils should have a plasticity index of less than 10, max. clay content of 15%, and max aggregate size of 16mm. Reinforcing bar or welded wire mesh is used to resist cracking and tensile forces. Forms should be left in place for at least 12 hours after forming (Easton and Easton 2012).
Earthbag, also called superadobe, stacks plastic or cloth bags filled with soil and sometimes sand or gravel to form a wall.	Bags are laid in courses with barbed wire laid horizontally between the courses for added stability. Any kind of soil or aggregate can be used as the structural stability is derived from the weight of the particle filled bags, not the cohesiveness of the soil. Earthbags can be used for retaining walls but should be covered with plaster or vegetation as plastics are vulnerable to UV degradation and cloth is vulnerable to moisture (Kennedy and Wojciechowska 2000).

(Continued)

Table 16.15 (continued)

Construction Method	Performance and Construction Considerations
Geomodular bag wall products are filled with soil and/or sand and used for retaining walls, embankments, and other infrastructure applications. Some contain openings for vegetation.	Bags are laid in staggered courses with barbed wire or proprietary plates laid horizontally between the courses for added stability. Most systems use polypropylene bags filled with sandy soil. Some wall systems have openings for plants and call for sand with organic matter to support plant health. Some can be hydroseeded. Live staking can be incorporated into some systems. Geomodular bag walls are stacked with a batter leaning back into the hillside. Geogrid fabric tiebacks into the hillside can also be employed for taller walls (FlexMSE 2024). See Chapter 8.
Rammed earth tires offer a permanent formwork for rammed earth walls. A dry soil mix is tightly packed into a tire with a manual or pneumatic tamper.	Rammed earth tires are laid in a running bond pattern, filled with a dry soil mix, and compacted. Courses of tires are placed with a batter. Stability is dependent on the weight of the rammed earth tires, so soil type is not as important. Finished tires will weigh 135–180kg/300–400lbs. Most used for retaining walls. Aesthetics of this wall system may not be desirable, so it could be covered with cement, lime, or mud plaster (Table 16.6) (Reynolds 2000).

Figure 16.13 Adobe brick reinforced with straw. Image credit: Ccat82, Dreamstime.com

Figure 16.14 Dried adobe brick are often thinner than fired clay brick to allow them to dry more quickly. Image credit: Lluvia Morales

Figure 16.15 Trass lime rammed earth walls at Novartis Campus Park, designed by Vogt Landscape, were built by Martin Rauch in his signature style. Image credit: Christian Vogt

References

ACI (2024). www.concrete.org/tools/frequentlyaskedquestions.aspx?faqid=688.

Akipek, F.O., and Yazar, T. (2017). "Common action walls". http://potplus.org/common-action-walls/, accessed August 25, 2021.

ASTM Standard D2487-17. (2020). *Standard Practice for Classification of Soils for Engineering Purposes (Unified Soil Classification System)*. West Conshohocken, PA: ASTM International. www.astm.org.

ASTM Standard D4318-17e1. (2018). *Standard Test Methods for Liquid Limit, Plastic Limit, and Plasticity Index of Soils*. West Conshohocken, PA: ASTM International. www.astm.org.

ASTM Standard E2392-10. (2023). *Standard Guide for Design of Earthen Wall Building Systems*. West Conshohocken, PA: ASTM International. www.astm.org.

Bailly, G.C., et al. (2024). "Advancing earth-based construction: A comprehensive review of stabilization and reinforcement techniques for adobe and compressed earth blocks". *Eng* 5: 750–783. https://doi.org/10.3390/eng5020041.

Basanna, H., et al. (2020). "Role of stabilizers and gradation of soil in rammed earth construction". *Journal of Materials in Civil Engineering* 32.5: 04020072.

Brick Industry Association (BIA) (2020). *U.S.-Canada Industrywide Clay Brick: Environmental Product Declaration*. BIA.

Bredenoord, J., and Kulshreshtha, Y. (2023). "Compressed stabilized earthen blocks and their use in low-cost social housing". *Sustainability* 15.6: 5295. https://doi.org/10.3390/su15065295.

Cabrera, S., et al. (2021). "Compressed earth blocks (CEB) stabilized with lime and cement: Evaluation of both their environmental impact and compressive strength". *Revista Hábitat Sustentable* 10: 71–80.

Canadian Concrete Masonry Producers Association (CCMPA) (2020). *Environmental Product Declaration: Normal Weight and Light Weight Concrete Block Masonry Units as Manufactured by the Members of the Canadian Concrete Masonry Producers Association*. CCMPA and ASTM International.

Christoforou, E., et al. (2016). "Cradle to site life cycle assessment (LCA) of adobe bricks". *Journal of Cleaner Production* 112: 443–452. https://doi.org/10.1016/j.jclepro.2015.09.016.

Damme, H.V., and Houben, H. (2018). "Earth concrete: Stabilization revisited". *Cement and Concrete Research* 114: 90–102.

Dominguez, T. (2011). *Guide G-521 ABC's of Making Adobe Bricks*. New Mexico State University, March.

Dorado, P., Cabrera, S., and Rolón, G. (2022). "Contemporary difficulties and challenges for the implementation and development of compressed earth block building technology in Argentina". *Journal of Building Engineering* 46: 103748. https://doi.org/10.1016/j.jobe.2021.103748.

Easton, D. (2007). "Rammed earth". In *Alternative Construction: Contemporary Natural Building Methods*. Eds. L. Elizabeth and C. Adams. New York: Wiley.

Easton, D., and Easton, T. (2012). "Modern rammed earth construction techniques". In *Modern Earth Buildings: Materials, Engineering, Constructions and Applications*. Eds. Matthew R. Hall, Rick Lindsay, and Meror Krayenhoff. Elsevier.

Fernandes, J., et al. (2019). "Life cycle analysis of environmental impacts of earthen materials in the Portuguese context: Rammed earth and compressed earth blocks". *Journal of Cleaner Production* 241. https://doi.org/10.1016/j.jclepro.2019.118286.

FlexMSE (2024). "FlexMSE: Exceptionally durable, beautifully sustainable". https://flexmse.com/, accessed February 20, 2024.

Form Earth (2024). "Free form rammed earth". www.youtube.com/watch?v=4TXj5IIkUIY, accessed February 17, 2024.

Giuffrida, G., et al. (2022). "Natural and recycled stabilizers for rammed earth material optimization". In *International Conference on Sustainability in Energy and Buildings*. Singapore: Springer Nature.

Gomaa, Mohamed, et al. (2023). "Rethinking rammed earth construction for industry 4.0: Precedent work, current progress and future prospect". *Journal of Cleaner Production* 136569.

Guettala, A., Abibsi, A., and Houari, H. (2006). "Durability study of stabilized earth concrete under both laboratory and climatic conditions exposure". *Construction and Building Materials* 20.

Hall, M.R., Lindsay, R., and Krayenhoff, M. (Eds.). (2012). *Modern Earth Buildings: Materials, Engineering, Constructions and Applications*. Elsevier.

Hardy, C. (2024). *Carbon Data Set Landscape, Version 3*. Sasaki Associates, Inc.

Harrison, J. (2005). *Carbonating and Hydraulic Mortars: The Difference Is Not Only in the Binder*. Hobart, Tasmania: TecEco Pty. Ltd.

Houben, H., and Guillaud, H. (1994). *Earth Construction: A Comprehensive Guide*. London: Intermediate Technology Publications.

Idriss, Eguekeng, et al. (2022). "Engineering and structural properties of compressed earth blocks (CEB) stabilized with a calcined clay-based alkali-activated binder". *Innovative Infrastructure Solutions* 7.2: 157.

International Building Code (IBC) (2021). *Chapter 21 Masonry: Section 2109 Empirical Design of Adobe Masonry*. International Code Council, Inc.

International Standards Organization (ISO) (2017). *Geotechnical Investigation and Testing: Identification and Classification of Soil*. International Standards Organization.

Jannat, Nusrat, et al. (2020). "Application of agro and non-agro waste materials for unfired earth blocks construction: A review". *Construction and Building Materials* 254: 119346.

Kebao, R., et al. (2012). "Integral admixtures and surface treatments for modern earth buildings". In *Modern Earth Buildings: Materials, Engineering, Constructions and Applications*. Eds. Matthew R. Hall, Rick Lindsay, and Meror Krayenhoff. Elsevier.

Kennedy, J., and Wojciechowska, P. (2000). "Chapter 8: Modular contained earth, earthbag". In *Alternative Construction: Contemporary Building Methods*. Eds. L. Elizabeth and C. Adams. New York: John Wiley & Sons, Inc.

Kloft, H., et al. (2019). "Robotic fabrication of components from rammed earth" (translated from German). *Dtsch. BauZeitschrift* 54–59.

Maher, M., et al. (2005). *Context Sensitive Roadway Surfacing Selection Guide*. Lakewood, CO: Federal Highway Administration, Central Federal Lands Highway Division.

Marsh, A.T.M., and Kulshreshtha, Y. (2021). "The state of earthen housing worldwide: How development affects attitudes and adoption". *Building Research & Information* 50.5: 485–501. https://doi.org/10.1080/09613218.2021.1953369.

McHenry, P.G. (1984). *Adobe and Rammed Earth Buildings*. Tucson, AZ: University of Arizona Press.

Minke, G. (2022). *Building with Earth: Design and Technology of a Sustainable Architecture*, Fourth Edition. Birkhäuser.

Moquin, M. (2000). "Adobe". In *Alternative Construction: Contemporary Natural Building Methods*. Eds. L. Elizabeth and C. Adams. New York: Wiley.

Muguda, S., et al. (2020). "Durability and hygroscopic behaviour of biopolymer stabilised earthen construction materials". *Construction and Building Materials* 259: 119725.

Muheise-Araalia, D., and Pavia, S. (2021). "Properties of unfired, illitic-clay bricks for sustainable construction". *Construction and Building Materials* 268 (January 25): 121118.

National Ready Mixed Concrete Association (NRMCA) (2022). *NRMCA Member Industry-Average EPD for Ready Mixed Concrete*. NRMCA.

Norton, John. (1997). *Building with Earth: A Handbook*, Second Edition. London: Intermediate Technology Publications.

Nouri, Hamed, Safehian, Majid, and Mir Mohammad Hosseini, Seyed Majdeddin. (2023). "Life cycle assessment of earthen materials for low-cost housing a comparison between rammed earth and fired clay bricks". *International Journal of Building Pathology and Adaptation* 41.2: 364–377.

Porter, H., et al. (2018). "Rammed earth blocks with improved multifunctional performance". *Cement and Concrete Composites* 92: 36–46.

Reynolds, M. (2000). "Chapter 8: Modular contained earth: Earth-rammed tires". In *Alternative Construction: Contemporary Building Methods*. Eds. L. Elizabeth and C. Adams. New York: John Wiley & Sons, Inc.

Saeed, I.A., Mohammed, T.J., and Jihad, S.A. (2020). "Production of earth units compressed and stabilized by using cement and pozzolana". In *IOP Conference Series: Materials Science and Engineering*. Vol. 745. IOP Publishing.

Schroeder, H. (2018). "The new DIN standards in earth building – the current situation in Germany". *Journal of Civil Engineering and Architecture* 12: 113–120.

Smith, M. (2000). "Chapter 6: Cob". In *Alternative Construction: Contemporary Natural Building Methods*. Eds. L. Elizabeth and C. Adams. New York: John Wiley & Sons, Inc.

State of New Mexico (2021). "New Mexico earthen building materials code". In *Title 14 Housing and Construction, Chapter 7 Building Codes General*. State of New Mexico.

Toufigh, V., and Kianfar, E. (2019). "The effects of stabilizers on the thermal and the mechanical properties of rammed earth at various humidities and their environmental impacts". *Construction Building Materials* 200: 616–629.

United Nations (UN) Habitat (2015). *Urban Energy Technical Note 20: Interlocking Stabilized Soil Blocks (ISSB)*. UN Habitat.

Walker, P., and Standards Australia (2002). *HB 195: The Australian Earth Building Handbook*. Standards Australia International Ltd.

Walker, R., and Morris, H. (2021). "Development of the New Zealand earth building standards NZS 4297:2020, NZS 4298:2020 AND NZS4299:2020". SE SOC Conference, Hamilton, NZ.

CHAPTER 17

Fired Clay Bricks

Fired clay bricks are tactile, warm, and durable. Brick colors, sizes, and joint types offer a wide range of design possibilities, along with the patterns from the modularity of bricks. If locally produced, their color can reflect and blend with the local soil. Clay and shale resources are abundant around the world.

Fired clay bricks are energy intensive to produce, and as a result, relatively high in embodied carbon, yet brick is very durable, so it can be used over and over in multiple structures. The expense and embodied carbon of fired clay brick is leading to incorporation of wastes in bricks, mixes to allow lower firing temperatures, renewable energy sources and clay brick alternatives such as fly ash brick and compressed earth block (Chapter 16 "Earth-Based Materials").

Brick Statistics

Over 500 million metric tons of clay are used globally each year to produce 1.8 trillion bricks per year. The global clay brick industry, including site construction, was valued at $690 billion in 2023 (Precedence Research 2024). Growth in the brick industry is expected to increase, particularly in rapidly developing parts of the world where clay is abundant and brick structures can be built by relatively unskilled workers.

Brick production in the United States is greatest in the South Atlantic and South-Central regions of the country. And brick consumption is greatest in these areas as well, although consumption is also high in the northeast central part of the United States (BIA 2007). Given the relatively heavy weight of bricks (2120kg/m^3), they should be procured within 250km (155 miles) of the project site. A cubic meter of bricks that is transported 300km (186 miles) by truck will have a greater carbon footprint from the transportation emissions than from the production of the bricks.

While a large percentage of clay brick is produced and used in the United States, imported brick was valued at 36.2 million US dollars in 2022 with 73% imported from Mexico and 4% imported from China. The United States exported clay brick valued at 29.8 million primarily to Canada (90%) (OEC 2024). In 2023, over 90% of houses in the United Kingdom were built using brick.

Imported brick can be a concern because the Bureau of International Labor Affairs (ILAB) lists bricks as a good that may be made with child labor or forced labor in several countries (ILAB 2023; Grace Farms Foundation 2022). Forced and child labor in the global brick industry is discussed later in this chapter.

Life Cycle Impacts of Clay Brick

Brick is primarily comprised of clays or clayey shales with sand, ash, and/or oxides. It is graded, mixed with water, then formed, dried, and fired. The primary environmental impact from brick manufacture is the embodied carbon resulting from firing bricks at very high temperatures for prolonged periods (Figure 17.1).

A1 Raw Material Acquisition

Clay or shale minerals, formed from the mechanical and chemical breakdown of rocks, are abundant in many regions of the world, hence the prevalence of fired and air-dried historic brick structures. Clay minerals are primarily composed of silica and alumina with varying amounts of metallic oxides. The metallic oxides influence the color of the brick and the approach to firing.

Most clay is mined in shallow open pits using draglines, power shovels, front-end loaders, or backhoes. Some kaolin clay is extracted by hydraulic mining and dredging while fire clay is mined underground. Brick quarries are usually located near or adjacent to brick kilns to minimize the transport cost of these heavy materials.

DOI: 10.4324/9781003413974-19

The primary environmental impacts from raw material extraction for bricks are loss of topsoil, removal of land cover, air pollution, and loss of habitat. Most clay quarries do not produce toxic runoff; however dust and sediment can runoff or drift into receiving waters causing surface water pollution and negatively impacting aquatic life.

A serious human health impact from raw material acquisition is the risk of breathing in fine crystalline silica particles. Prolonged exposure can lead to multiple diseases including silicosis, an incurable lung disease, lung cancer, kidney disease, and chronic obstructive pulmonary disease (COPD) (US OSHA 2024).

Wastes as Raw Materials for Bricks

Bricks offer an opportunity for beneficial use of pre- and post-consumer waste. A variety of industrial and agricultural wastes can be substituted for some clay or shale in bricks, reducing environmental impacts from clay and shale quarrying (Table 17.1). Research has shown positive impacts to the brick strength and durability from certain waste materials, while other waste materials allow bricks to be fired at lower temperatures.

Brick Quarry Restoration Efforts

Restoration practices for spent brick quarries vary widely by country, state, and municipal regulations. The relatively shallow depth of brick quarries makes them simpler to restore than deeper stone and metal ore quarries. In the United Kingdom, the Biodiversity Net Gain law has influenced quarry restoration plans and activities with an estimated 84% of quarries having a restoration plan. These plans often promote bio and geodiversity after the quarry is closed.

Table 17.1 Wastes Used in Brick Manufacture

Type of Waste	
Sludges	Water treatment plant sludge and sludge ash
	Sludge from industrial wastewater
	Sludge from brewing
	Coffee processing sludges
	Textile sludges
	Red mud from aluminum manufacture
	High-biopolymer sludge
Quarry wastes and slurries	Stone quarry and processing sludges
	Granite-basalt fines
	Boron mine tailing
	Kaolin fine quarry residues
	Granite and marble slurry
Ashes	Waste incineration ash
	Bottom ash
	Fly ash
Steel processing wastes	Steel slag
	Ground granulated blast furnace slag
	Foundry sands
	Mill scale
Agricultural wastes	Sugarcane bagasse
	Rice husk ash
	Olive husk
Other waste	Waste glass
	Brick dust
	PFAS contaminated soils

Sources: Murmu and Patel 2018; Baghel et al. 2020; RMIT University 2024; Aouf 2023

A3 Brick Manufacturing

Manufacture of clay brick is a multi-step energy-intensive process, resulting in substantial energy use and relatively high embodied carbon.

Clay Preparation: The first step in the manufacturing process is to transform the clay and shale minerals into a plastic state that can be molded and shaped. Minerals are crushed to the appropriate size and mixed to impart different properties such as strength or color. Next the materials are ground, water is added, and the mixture is rolled and squeezed through a sieve to ensure the proper gradations and plasticity of material.

Forming: Bricks are primarily extruded then wire cut, but they can also be pressed into molds or hand formed. Wire cutting produces a smooth finish, but bricks may be mechanically textured or patterned. Pressed bricks are mechanically pressed into molds. They will have greater variability in the edges. These forming methods can produce 40,000 bricks in an hour at some facilities (BDA 2023). Hand molded bricks are still produced and valued for their variable character.

Drying: Wet bricks are stacked and dried to less than 5% moisture content primarily with tunnels, chambers, or kilns. Some manufacturers use waste heat from the firing kilns for this process.

Firing: During firing, reactions occur within the clay transforming weak "green" bricks into strong, durable units. Firing times and temperatures vary widely by type of kiln, raw materials and finished strength. Firing temperatures range from 900–1250°C (1650–2280°F), and bricks are in the kiln for 10–40 hours. Kilns are primarily fueled with

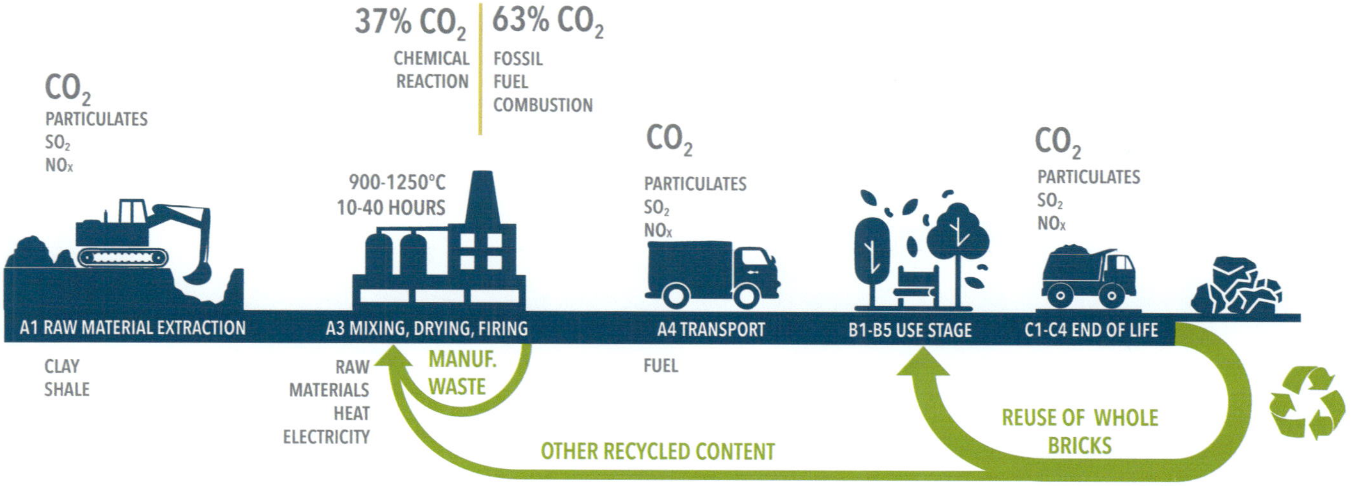

Figure 17.1 The life cycle of fired clay bricks. Image credit: Meg Calkins

natural gas, although some use liquified petroleum gas (LPG), coal, or oil. The energy-intensive process of firing generates greenhouse gas emissions as well as other air pollutants, particularly if coal is firing the kilns.

Reducing the Environmental Impacts of Brick Manufacture

While the global brick industry has not mapped out a path to net zero carbon like other industries such as cement, concrete, and steel, some national brick associations are establishing measures to reduce environmental impacts of brick manufacture. Primary actions being taken in brick manufacturing include energy, water, and resource efficiency measures.

Embodied carbon reductions: Efforts to reduce fossil fuel energy use and improve the carbon footprint of brick focus on use of alternative fuels, production efficiencies, and incorporation of waste materials (Table 7.1). Alternative fuels used include renewables such as wind and solar energy, low-carbon fuels such as hydrogen and electric firing, and methane captured from adjacent landfills. Many modern brick manufacturing facilities reuse as much energy and heat as possible for both financial and environmental reasons.

Water efficiency: Water is used in the brick making process to help shape bricks before they are dried and fired with 300–700kg of water used to produce one metric ton of bricks (BDA 2023). Some manufacturers are working to reduce the quantity of water they consume while also using captured rainwater or process water.

Waste reductions: Waste that is sent to the landfill from brick production is quite minimal as processing wastes are primarily returned to the production stream and reused in new brick. In the United Kingdom, waste per metric ton sent to the landfill has declined by 50% from 1.2kg per metric ton in 2011 to 0.6kg/t in 2022 (BDA 2023).

Circular economy: Brick can contribute to a circular economy because it is a durable product that can be recovered and reused multiple times. Brick that is 200 years old is still in use or reuse in pavements, walls, and buildings around the world. The mortar in brick structures can be repointed to extend the life of a structure, and brick structures can be deconstructed and the bricks reclaimed (Figures 17.2 and 17.3). Brick pavements on sand setting beds are easily repaired and/or deconstructed. The lack of mortar allows for easy dismantling of the pavement and reuse of the bricks elsewhere. When bricks are mortared in a pavement or wall, the labor cost of removing the mortar can be high. Instead, these bricks are often sent to a landfill or crushed for pavement base.

The Embodied Carbon of Brick Materials

The CO_2 emissions from brick manufacture stem from three sources:

- ***Direct emissions*** from fuel consumption primarily result from the combustion of natural gas to dry green bricks, then heat kilns to over 1000°C.
- ***Indirect emissions*** come from the use of electricity to power equipment and lights in the manufacturing facility,

Figure 17.2 and Figure 17.3 The Bai'etan Exhibition Center landscape by Sasaki transforms a former industrial port into a public waterfront on the Pearl River in Guangzhou, China. This SITES Platinum project used bricks salvaged on-site and repurposed as special pavement in the plaza. Image credit: Savor Photography

although some companies are transitioning to electricity from renewable sources. In the United Kingdom, 95% of electricity used comes from renewable sources or is generated on site (BDA 2023).

- ***Process emissions*** are released from clay or additives as they are heated during firing. These are the most challenging emissions to abate, although research is underway on several approaches to capture and store them.

The embodied carbon from industry-wide EPDs in the United States, Canada, and the United Kingdom is relatively high compared with alternative materials such as concrete, concrete blocks, unfired clay bricks, fly ash bricks, compressed earth block, and local stone (Tables 17.2 and 17.3). However, when the entire life cycle of a clay brick is considered, the exceptionally long service life of bricks makes their embodied carbon relatively low per year of service. The service life of bricks is estimated to be over 150 years even if the brick structure will not last that long.

Human Health Impacts and Equity Issues in Brick Manufacture

Human health impacts of brick manufacture include worker health impacts from air emissions and airborne particulates during raw material extraction, manufacture, and construction. Lack of safe and fair working conditions can also be an issue in some countries.

Human Health Impacts of Brick Manufacture

Fluorine is present in the raw materials for bricks and, when fired, hydrogen fluoride is released along with other fluorine compounds. Chlorine is also present in raw materials and

Table 17.2 Embodied Carbon of Fired Clay Bricks and Their Alternatives

Brick or Brick Product	Embodied Carbon (EC) Range ($kgCO_2e/kg$)	Sources and Notes
Fired clay brick	0.237	Industry wide EPD from the Brick Industry Association (BIA 2020)
Unfired clay brick/adobe	0.035	Hardy 2024
Compressed earth block, 5% PC	0.0586	Average of multiple sources
Fly ash brick	0.16	Comprised of fly ash, stone dust/sand, lime, gypsum, and cement (Levi and Raut 2021)
Precast concrete paver	0.27	Drycast (Singh et al. 2020)

Notes: Embodied carbon data is from the product life-cycle stage, phases A1–A3. Embodied Carbon figures will change over time as processing efficiencies are realized, and more renewable energy sources are employed.

These EC figures do not include mortar in the structure.

Table 17.3 Alternatives to Fired Clay Brick

Alternative Brick or Block	Embodied Carbon	Notes
Geopolymer bricks Aluminosilicate materials are combined with alkaline actives to form geopolymer binders. The geopolymer binder, mixed with water and clay, cures to form a strong, durable brick without the need for firing.	20–34% of the embodied carbon of fired clay bricks.[1]	Geopolymer bricks can incorporate waste materials such as fly ash, brick dust, and other construction, industrial and agricultural wastes.[1]
Fly ash bricks are comprised of fly ash (50–70%), stone dust/sand (15–20%), lime, gypsum (15–20%), and sometimes a bonding agent (5–8%). They are compressed at a very high pressure and cured for 24 hours in a steam bath at around 66°C.	40–52% of the embodied carbon of fired clay bricks.[2]	ASTM Standard C1790 offers a standard for fly ash bricks without portland cement. Min. compressive strength is 3500psi. Fly ash is a by-product of coal combustion, therefore heavy metals such as lead, cadmium, or mercury are likely present. Proponents argue that they are encased in the brick matrix and are inert. The fly ash supply will eventually become limited with a move away from coal. These smooth, gray bricks have a low moisture absorption rate of 6–12% in contrast to a clay face brick with a max. allowable absorption rate of 20%[2].
CEBs Compressed earth blocks are modular units made from soil, water, sand, and sometimes stabilizers such as cement and/or lime. CEBs are produced with a hydraulic press at forces ranging from 2–40MPa.	5–44% of the embodied carbon of fired clay bricks.[3]	Compressive strength is primarily related to the quantity of cement and/or lime used. Force of press also influences strength. Increased use of cement and/or lime increase the embodied carbon of the CEB product.[3] Interlocking CEB reduce the need for mortar. Refer to Chapter 16 "Earth-Based Materials" for more information.
Unfired clay brick/adobe brick Unfired clay bricks, also called adobe, sun dried bricks, or earth blocks, are molded composite blocks made from soil, water, and organic material such as straw or other fibers. They are often stabilized with bitumen emulsion.	5–21% of the embodied carbon of fired clay bricks.[4]	Compressive strengths of unfired clay brick are substantially lower than fired clay bricks. Semi stabilized adobe bricks contain 3–5% asphalt emulsion and fully stabilized bricks contain 5–10% asphalt emulsion. Fully stabilized bricks can be left exposed. Other waterproofing stabilizers are casein, lime, water glass, linseed oil, and silicone. Refer to Chapter 16 "Earth-Based Materials" for more information.
Waste bricks A variety of waste materials are crushed and formed into bricks with different binders. Some binders are cementitious or polymer-based, while others are ceramic and require firing to solidify.	Reductions in embodied carbon vary widely.	Some product information is vague about the type and amount of recycled content and the binder materials. Questions to ask of the manufacturers of these products are: • What percentage of post-consumer waste do you use? What are the materials? • What is the binder? • What is the embodied carbon of the product?

[1] Hodhod et al. 2020; Maaze and Shrivastava 2023; Nagaraju and Bahrami 2024

[2] Paul Levi and Raut 2021

[3] De Villiers and Mbewe 2016; Cabrera et al. 2021

[4] Christoforou 2016

becomes hydrogen chloride when fired. Health impacts of these emissions include eye, nose, and respiratory irritation, pulmonary edema, bronchial spasms, and eye and skin burns (US CDC 2024).

Particulate matter (PM) is released during the brick manufacturing process from grinding, drying, and screening operations. Water is used to hold down dust as well as pollution control equipment, but some is released and can pose a hazard for workers and communities around the brick plant. Silicosis is a risk during raw material extraction, manufacturing, and cutting brick during construction.

Worker Safety and Equity in the Global Brick Industry

Design for Freedom characterizes bricks as "one of the most at-risk construction materials made by forced and bonded labor" (Grace Farms Foundation 2022). This is a global issue with more than 20 countries on the US Department of Labor's List of Goods Produced by Child Labor or Forced Labor (ILAB 2022) (Table 17.4; Figures 17.4 and 17.5).

While most bricks used in European countries, the United States and Australia are produced within the country, a percentage are imported with some from countries on the ILAB 2022 list. In 2022, 4% of brick used in the United

Table 17.4 Human Rights Issues and Bricks

Country	Child Labor	Forced Labor
Afghanistan	x	x
Argentina	x	
Bangladesh	x	
Bolivia	x	
Brazil	x	
Burma	x	x
Cambodia	x	x
China	x	x
Colombia	x	
Ecuador	x	
Egypt	x	
India	x	x
Iran	x	
North Korea		x
Nepal	x	x
Pakistan	x	x
Paraguay	x	
Peru	x	
Russia		x
Uganda	x	
Vietnam	x	

Source: ILAB 2022

Figure 17.4 A child laborer in the brick industry in India. Image credit: ZMS

States and nearly 3% of brick used in Europe and the United Kingdom was imported from China. Thirty-four percent of clay brick produced in India is exported to the United Kingdom with 13% exported to Australia (OEC 2024). Both China and India are on the ILAB 2022 list.

Figure 17.5 Teenage laborers in the brick industry in Bangladesh. Image credit: Kartik

Properties and Types of Clay Brick

There are several types of brick that are designed for specific uses (Table 17.5 and Figure 17.6). For instance, paving brick have very high compressive strengths to support vehicular traffic. Both thin brick and hollow brick use quite a bit less material than a solid standard sized brick at 215mmx102.5mmx65mm (7–5/8″ x 3–5/8″ x 2–1/4″).

Brick Grades

The specification of brick with the appropriate properties is key to ensuring the durability and performance of a brick structure. Building bricks are classified by their resistance to freezing and thawing. Grades of bricks, classified by compressive strength, water absorption, and other criteria are shown in Table 17.6. Paving bricks are classified

Table 17.5 Types of Brick

Facing brick *ASTM C216*	Used in both structural and non-structural applications where at least one face of the brick requires a finished appearance. Facing bricks may have voids that are less than 25% of the cross sectional are of the brick. Brick voids allow for steel reinforcing and may also be filled with mortar. Facing brick are graded by weather exposure and appearance.
Hollow brick *ASTM 652*	Hollow bricks have voids in the units that are greater than 25% of the gross cross-sectional area but less than 60%. Brick void shapes vary widely and can be filled with steel reinforcing and/or mortar. Hollow brick are graded by weather exposure and appearance.
Building brick *ASTM C62*	Building brick are used in both structural and non-structural applications where a finished appearance is not a requirement. Building brick may have voids that are less than 25% of the cross sectional area of the brick. Building brick are graded by weather exposure.
Thin brick *ASTM C1088*	Thin brick are used as a veneer on wall structures. They have normal face dimensions but a reduced thickness. Thin brick are classified by interior or exterior use.
Paving brick *ASTM C902 and C1272*	Paving brick, used as a wearing surface, usually have minimal absorption and higher compressive strength as they are subjected to traffic loads. Paving brick are graded by intensity of traffic, weather, abrasion resistance, and appearance.
Glazed brick *ASTM C1405*	Glazed brick have a ceramic glaze fused to the face of the brick. The glaze can be applied prior to firing or after firing. Glazed brick can be used in structural or non-structural applications. Glazed brick are graded by appearance.

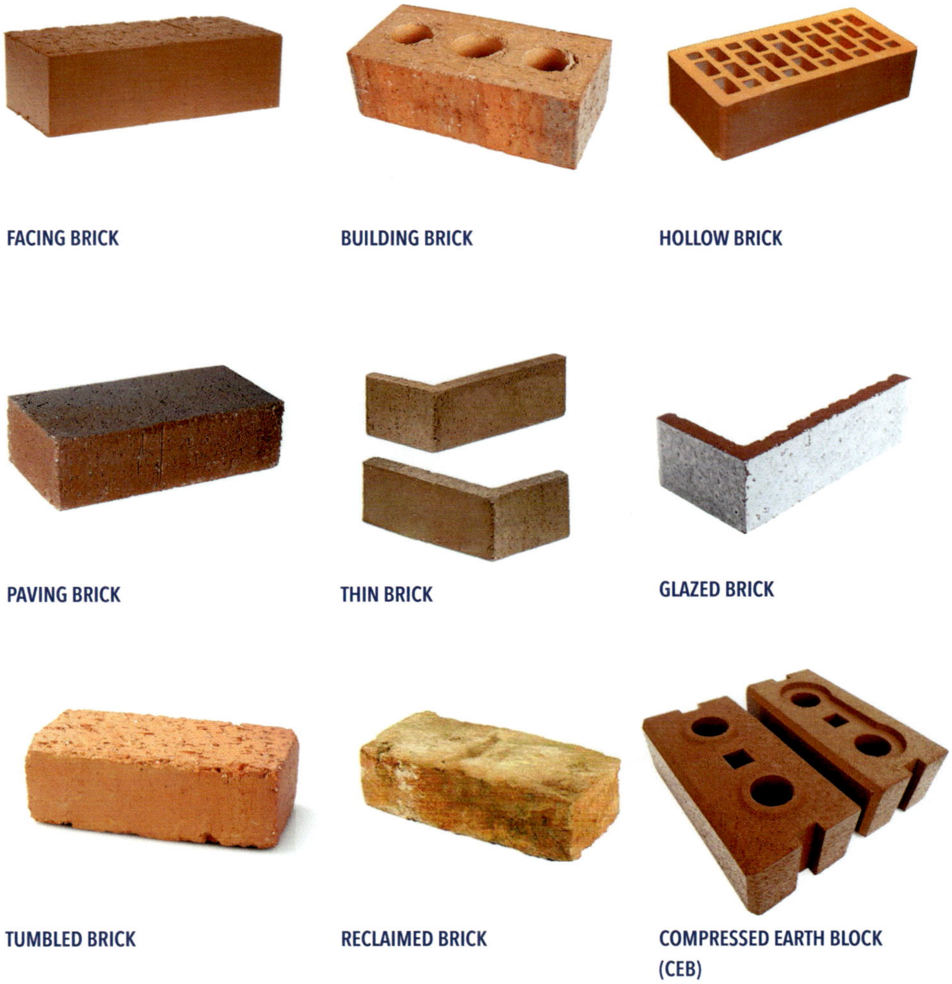

Figure 17.6 Types of brick used in site construction. Image credit: Meg Calkins

according to weather and expected traffic load and are shown in in Table 17.7.

Compressive strength varies by the intended use of the brick and the brick grade. It is measured by a force placed perpendicularly to the brick plane.

Water absorption affects the weathering of the brick and its bond with mortar. It is expressed as a maximum percentage with severe weather bricks allowing lower rates of absorption.

Durability is a combination of compressive strength and water absorption. It is also a factor of resistance to freezing and thawing in the presence of moisture, abrasion, and staining.

Table 17.6 Brick Grades

Brick Grade	Min Compressive Strength	Water Absorption (%)
Grade SW (severe weathering) is used in applications where high resistance to damage caused by freeze thaw cycles is required.	20.7MPa, 3000psi	17%
Grade MW (moderate weathering) is used is used in applications where some resistance to damage caused by freeze thaw cycles is required.	17.2MPa, 2500psi	22%
Grade NW (negligible weathering) is used in applications where minimal resistance to damage caused by freeze thaw cycles is acceptable.	10.3MPa, 1500psi	No limit

Source: ASTM C62–23 Standard Specification for Building Brick

Table 17.7 Paving Brick Grades

Brick Grade	Min Compressive Strength	Water Absorption (%)
Pedestrian and Light Traffic Paving Brick		
Grade SX (severe weathering) is used in applications where bricks may be frozen while saturated with water.	52.2MPa, 8000psi	8%
Grade MX (moderate weathering) is used is used in applications where resistance to freezing is not an issue.	20.7MPa, 3000psi	14%
Grade NX (negligible weathering) is used in interior applications.	20.7MPa, 3000psi	No limit
Type I brick is used in applications with extensive abrasion.		
Type II brick is used in applications with moderate abrasion.		
Type III brick is used in applications with low abrasion.		
Heavy Vehicular Paving Brick		
Type R is brick set in a mortar or bituminous setting bed supported by a concrete or asphalt pavement base.	52.2MPa, 8000psi	6%
Type F is brick set on an aggregate setting bed with aggregate joints set on an aggregate base.	69MPa, 10,000psi	6%
PS brick pavers are for general use.		
PX brick pavers are for applications where a higher degree of precision and lower variation is required.		
PA brick pavers are intended to produce architectural effects resulting from color, texture, and size.		

Sources: ASTM C1272–22 Standard Specification for Heavy Vehicular Paving Brick

ASTM C902–22 Standard Specification for Pedestrian and Light Vehicular Paving Brick

Figure 17.7 MVVA used brick pavement at the Max Family Garden at St. Ann's Warehouse in Brooklyn. Existing brick walls of the warehouse were retained and frame views of the Brooklyn bridge and other landmarks. Image credit: Conor Coghlan

Figure 17.8 The herringbone brick paving pattern accommodates the flowing paths and the transitions to the irregularly spaced bricks. Image credit: Michael Van Valkenburgh Associates

Clay Brick Structures

Clay bricks are commonly used for retaining walls, free standing walls, stairs, and vehicular and pedestrian pavements (Tables 17.9 and 17.10; Figures 17.7–17.9). Brick wall structures are primarily rigid structures that

Table 17.8 Brick Shaping Techniques

Brick Shaping Technique	Qualities of Fired Brick
Extruded bricks	Even, smooth cut surfaces with sharp edges. Some contain cutting marks or grooves. Shallow, irregular marks on the surface resulting from materials that are burned out during firing.
Pressed bricks, semidry	Mostly smooth surfaces and structured brick.
Pressed bricks, dry	Dense brick with smooth surfaces on all sides. Sharp edges and accurate dimensions.
Water struck bricks	Machine or handmade. Handmade are smooth dense brick with water or clay streaks. Machine made have piston print on only one side.
Hand struck bricks	Machine or handmade. Machine made have irregular sand covered surfaces and softer edges than pressed or extruded bricks. Handmade are sharper edged than machine made with smooth surfaces and pressing creases.

employ mortar and grout to form a composite assembly on a concrete footing (Figure 17.9). There are a few interlocking brick products that are designed for brick walls without mortar. Brick walls are discussed in Chapter 8 "Retaining Walls and Bioengineered Embankments" and Chapter 9 "Freestanding Walls".

Clay bricks are also used in pavements. Sand set bricks on a rigid or flexible base are primarily recommended for more durable, repairable pavement structures. Rigid brick pavement using mortared bricks on a concrete base is not recommended as they are vulnerable to freeze thaw cycles that can damage mortar joints and are difficult to repair. Brick pavement structures are discussed in Chapter 6 "Pavements".

Mortar and Grout for Clay Brick Structures

Bond strength, workability, and water retentivity are the most important properties to consider in mortar selection (Figure 17.10). Refer to Table 9.4 for a discussion of mortar and grout types. Grout is used in brick masonry structures to fill cells of hollow units, voids around reinforcing, and spaces between wythes. Partially grouted walls can reduce use of carbon-intensive grout while still achieving structural performance.

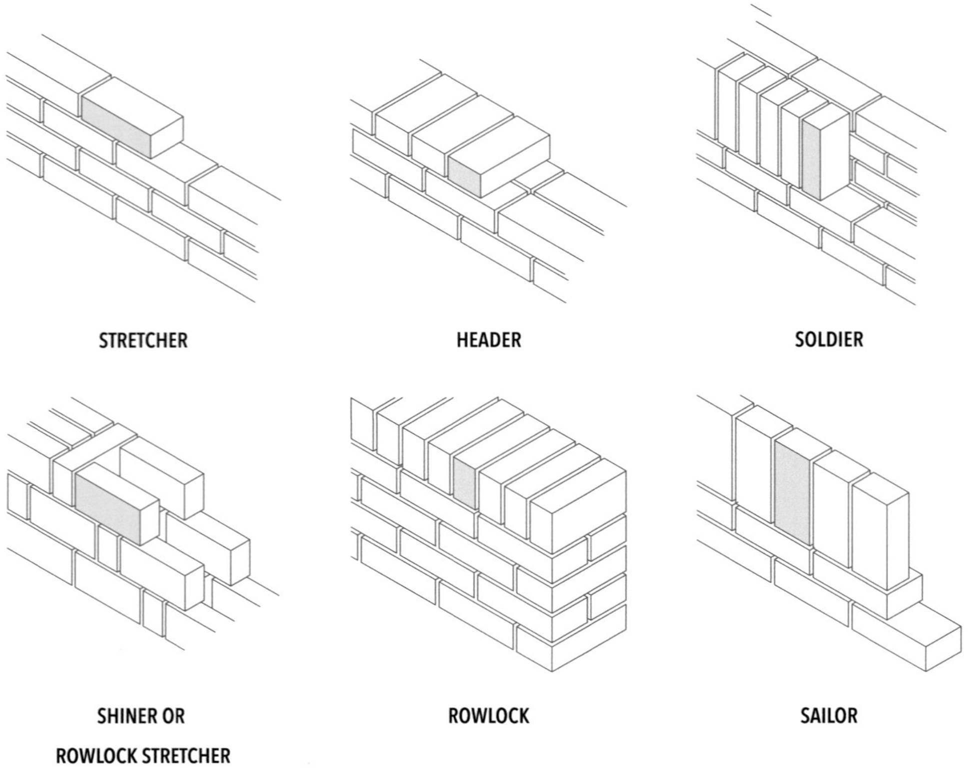

Figure 17.9 Nomenclature for the orientation of bricks in a wall. Image credit: Kathleen Elliott and Meg Calkins

Table 17.9 Detailing Durable Brick Structures

Topic	Considerations
Brick grades	Specify the appropriate grade of brick for the climate and structure in which it will be used, but do not over specify brick, e.g., don't use SW brick in a tropical climate.
Mortar joints	Specify compression and tooling of mortar into joints to reduce permeability. Avoid types of joints where water can be trapped on a shelf, where mortar protrudes from the joint, or make its way between the mortar and the brick (Figure 17.10). Do not apply mortar during rain or temperatures under 5°C. Mortar joints should be checked periodically for cracks between the brick and mortar. Cracks can allow water to seep into the wall or pavement, and in freezing climates can expand, damaging the brick or mortar.
Wall caps	Use capstones and flashing on rigid cavity walls, veneer walls, and multiple wythe brick walls.
Moisture protection	Examine walls periodically for moisture penetration. If it is occurring, address the source of moisture immediately through wall repair measures. In high moisture of hurricane regions, favor masonry headers over metal wall ties to bond solid brick wall wythes or brick cavity walls together as the metal ties can corrode. Hold all steel reinforcing in joints back from the mortar face by at least 20mm.
Using reclaimed bricks	Reclaimed bricks formerly placed with portland cement mortar may have traces of mortar that can compromise the bond with mortar in the new structure. Reclaimed historic bricks may have highly variable durability, strength, and absorption characteristics. Design of the new structure should accommodate these potential irregularities. Labor costs may be higher when using reclaimed brick, as cleaning of bricks prior to placement may be required, and placement may not be as fast as with new bricks. Some building codes require a 50% reduction in working stresses from those of new masonry when using salvaged bricks. Salvaged brick may have a more worn or rugged appearance.
Additional items	Refer to Chapters 6, 8, and 9 for more considerations for durable brick structures.

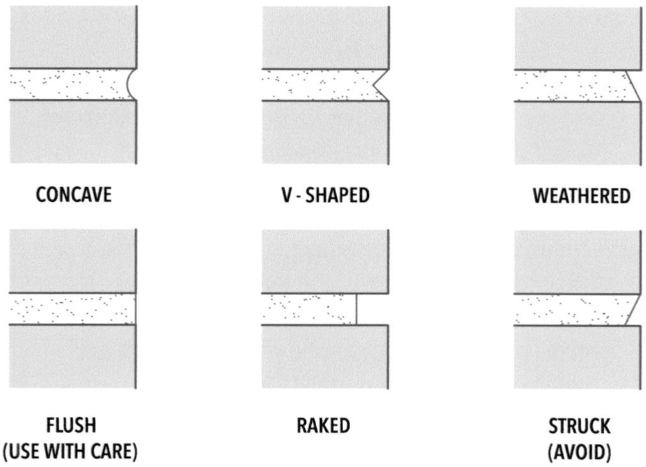

Figure 17.10 Mortar joints should be designed to shed water. Struck and raked joints could allow water to penetrate the bottom of the joint. Image credit: Kathleen Elliott and Meg Calkins

Table 17.10 Maintenance of Masonry Structures[1]

Cleaning Approach	Considerations for Use	Notes
General cleaning notes	Test the cleaner on a small, inconspicuous area of the masonry before broad application.	Avoid high VOC content cleaners or those with toxic constituents that may impact the environment around the structure.
Pressure washing Removes dirt, smoke, pollutants, oil, and efflorescence	Avoid driving water into masonry joints. Maximum intensity will vary by brick type. Low pressure washing should be safe and gentle to the masonry surface. Medium and high pressure can damage softer masonry and penetrate joints.	Not effective for heavy stains. Can use a lot of water. One of the least expensive methods of cleaning masonry. Avoid use in cold weather.
Steam cleaning Removes dirt, smoke, and pollutants	Steam cleaning is safe for all but the most porous masonry. Performs well on rough surfaces.	Safest method of cleaning masonry. Relatively inexpensive. Avoid use in cold weather.
Steam cleaning with chemicals Removes dirt, smoke, pollutants, oil, efflorescence, paint, and organic stains	Steam cleaning with chemicals enhances general. steam cleaning by removing tougher stains.	Consult product literature for applications, VOC levels, and hazardous chemicals.
Acids Removes dirt, smoke, pollutants, oil, efflorescence, brown stains, and organic stains	Avoid use of acids on high lime content masonry or mortars. Avoid use on light colored, soft, sandy brick. Clean entire surface to keep color uniform. Rinse structure thoroughly with water after application.	Protect surrounding vegetation and receiving waters by containing runoff. Consult product literature for applications, VOC levels, and hazardous chemicals.
Cleaners and detergents Removes dirt, smoke, pollutants, oil, efflorescence, brown stains, green stains, and organic stains	Household cleaners such as dish soap and other products can be used. Clean entire surface to keep color uniform. Rinse structure thoroughly with water after application.	Protect surrounding vegetation and receiving waters by containing runoff. Consult product literature for applications, VOC levels, and hazardous chemicals.
Proprietary cleaners Removes dirt, smoke, pollutants, oil, efflorescence, brown stains, green stains, white bleed, paint, rust, and organic stains	Clean entire surface to keep color uniform. Rinse structure thoroughly with water after application.	Protect surrounding vegetation and receiving waters by containing runoff. Consult product literature for applications, VOC levels, and hazardous chemicals.
Sandblasting Removes dirt, smoke, pollutants, paint	Avoid use on polished and very smooth surfaces. Pressures and abrasives vary. Sand blast entire surface to keep color uniform. Protect mortar joints.	Good for heavier staining. Workers should be protected from silica inhalation.
Dry brush Oil, tar, efflorescence, and paint	Labor intensive, but easy to control pressure. Brush entire surface to keep color uniform.	Good for lighter staining. Workers should be protected from silica inhalation.
Sanding discs, wet aggregate Removes dirt, smoke, pollutants, and paint.	Removes outer surface of masonry. Avoid use on polished or carved surfaces. Less abrasive than sandblasting. Easier to control.	Good for heavier staining. Workers should be protected from silica inhalation.

[1] This table applies to brick, stone, precast concrete, and PIP concrete surfaces. It does not address other earth-based construction.

References

Aouf, S.A. (2023). "Emy Bensdorp takes on PFAS pollution by turning 'toxic' soil into bricks". *Dezeen*, October.

ASTM Standard C62 (2023). *Standard Specification for Building Brick*. West Conshohocken, PA: ASTM International.

ASTM Standard C216 (2023). *Standard Specification for Facing Brick*. West Conshohocken, PA: ASTM International.

ASTM Standard C652 (2022). *Standard Specification for Hollow Brick*. West Conshohocken, PA: ASTM International.

ASTM Standard C902 (2022). *Standard Specification for Pedestrian and Light Vehicular Paving Brick*. West Conshohocken, PA: ASTM International.

ASTM Standard C1088 (2023). *Standard Specification for Thin Brick Veneer Units Made from Clay or Shale*. West Conshohocken, PA: ASTM International.

ASTM Standard C1272 (2022). *Standard Specification for Heavy Vehicular Paving Brick*. West Conshohocken, PA: ASTM International.

ASTM Standard C1405 (2023). *Standard Specification for Glazed Brick (Single Fired, Brick Units)*. West Conshohocken, PA: ASTM International.

ASTM Standard C1790 (2021). *Standard Specification for Fly Ash Facing Brick*. West Conshohocken, PA: ASTM International.

Baghel, R., Pandel, U., and Vashistha, A. (2020). "Manufacturing of sustainable bricks: Utilization of mill scale and marble slurry". *Materials Today: Proceedings* 26, Part 2: 2136–2139.

Brick Development Association (BDA) (2023). *The UK Clay Brickmaking Process*. London; BDA.

Brick Industry Association (BIA) (2007). *Brick Overview*. Reston, VA: BIA.

Brick Industry Association (BIA) (2020). *U.S.-Canada Industrywide Clay Brick: Environmental Product Declaration*. BIA.

Cabrera, S., et al. (2021). "Compressed earth blocks (CEB) stabilized with lime and cement: Evaluation of both their environmental impact and compressive strength". *Revista Hábitat Sustentable* 10: 71–80.

Christoforou, E., et al. (2016). "Cradle to site life cycle assessment (LCA) of adobe bricks". *Journal of Cleaner Production* 112: 443e452.

De Villiers, Wibke, and Mbewe, Peter. (2016). *Environmental Sustainability Evaluation of Low-Cost Buildings: The Case of Buildings Under the JENGA Project*.

Grace Farms Foundation (2022). *Design for Freedom Toolkit*. New Canaan, CT: Grace Farms Foundation.

Hodhod, O.A., Alharthy, S.E., and Bakr, S.M. (2020). "Physical and mechanical properties for metakaolin geopolymer bricks". *Construction and Building Materials* 265.

Maaze, M.R., and Shrivastava, S. (2023). "Selection of eco-friendly alternative brick for sustainable development; a study on technical, economic, environmental and social feasibility". *Construction and Building Materials* 408.

Murmu, A.L., and Patel, A. (2018). "Towards sustainable bricks production: An overview". *Construction and Building Materials* 165: 112–125.

Nagaraju, T.V., and Bahrami, A. (2024). "Sustainable geopolymer bricks manufacturing using rice husk ash: An alternative to fired clay bricks". In *Sustainable Structures and Buildings*. Ed. Alireza Bahrami. Springer International Publishing AG.

Observatory of Economic Complexity (OEC). "Building bricks 690410 exporters and importers". https://oec.world/en/profile/hs/building-bricks, accessed May 17, 2024.

Paul Levi, K., and Raut, Ashwin. (2021). "Embodied energy analysis to understand environmental impact of brick industry in West Godavari region". *Materials Today: Proceedings* 47: 5338–5344.

Precedence Research (2024). "Bricks market size, share, and trends 2024 to 2034". www.precedenceresearch.com/bricks-market, accessed November 8, 2024.

RMIT University (2024). "Energy smart bricks keep waste out of landfill". *RMIT University News*, April.

Singh, A., Vaddy, P., and Biligiri, K. (2020). "Quantification of embodied energy and carbon footprint of pervious concrete pavements through a methodological lifecycle assessment framework". *Resources, Conservation & Recycling* 161: 104953.

Think Brick Australia (2024). *Clay Brick and Roof Tile: Sustainability Strategy and Road Map*. Think Brick Australia.

US Bureau of International Labor Affairs (ILAB) (2022). *List of Goods Produced by Child Labor or Forced Labor*. US Department of Labor, ILAB.

US Centers for Disease Control. 2024. "Emergency preparedness and response: Facts about hydrogen fluoride (hydrofluoric acid)". https://emergency.cdc.gov/agent/hydrofluoricacid/basics/facts, accessed May 8, 2024.

US Occupational Safety and Health Administration (US OSHA) (2024). "Silica, crystalline: Health effects". www.osha.gov/silica-crystalline/health-effects, accessed May 8, 2024.

CHAPTER 18

Wood Materials and Products

Wood can be a renewable, relatively low-carbon construction material if it comes from a sustainably managed forest and is preserved, detailed, and maintained to increase its longevity. Wood is tactile, biophilic, and generally easy to construct with; however, logging of some wood species can have devastating effects on sensitive ecosystems.

The species of wood and, if required, the preservation or modification treatment is of utmost importance to sustainable wood construction. Maximizing the service life of lumber is key to renewability. Lumber should be in use for the amount of time it takes to regrow a tree to yield a comparable size wood member. For instance, it takes 90 to 120 years to grow a *Handroanthus heptaphyllus* tree to yield Ipe lumber for deck boards, but the deck boards will likely not be in use for 90 years even if they are reclaimed, refinished, and reused. Conversely, southern yellow pine trees are grown in 20–30 years to yield deck boards that can be thermally modified and will last until new trees can be grown.

This chapter will discuss the embodied carbon considerations of wood products, environmental and human rights issues of current forestry practices, forestry certification programs, types of wood and their characteristics in lumber, designing with and specifying reclaimed wood, wood preservative treatments, and wood finishes. Chapter 10 "Decks and Boardwalks" and Chapter 11 "Rails, Fences, and Screens" discuss design and detailing of wood structures.

Forest Carbon, Ecosystems, and Indigenous Populations

Degradation of forests for logging threatens biodiversity and contributes to climate change. Even when forests are replanted, logging can irreversibly erode soil carbon, contribute to species loss through the removal of food sources and shelter from predators, and reduce plant diversity. Forests are home to more than 80% of terrestrial biodiversity, and they provide ecosystem services of clean air, food and medication, habitat and water regulation.

Forests are major regulators of the global climate and water cycles. Forests and their soils are the largest terrestrial carbon sink as they absorb about 16 billion metric tons (gigatons GT) of carbon annually. Currently global forests hold nearly 900 GT of carbon in their leaves, branches, roots, and soils. Maintaining forest land is a vital strategy in combating climate change.

Logging impacts the climate in three ways. First, vegetation is removed that has been sequestering carbon. Second, slash and roots are left to decay and release carbon after logging, and third, soil is disturbed when trees are cut and removed, and again when trees are replanted. The release of vast soil carbon stores that have been captured over millennia can be substantial but are not well quantified.

Forestry was the largest driver of global tree cover loss with 155 million hectares lost between 2001 and 2023 (WRI 2024). An estimated 15 billion trees are cut down annually resulting in significant forest loss (Ritchie 2021). The terms *forest loss* and *tree cover loss* broadly capture the impacts of permanent deforestation and forest degradation.

Permanent deforestation is the conversion of forest to non-forest land and is the outcome of several pressures including agriculture, cattle ranching, land grabbing, infrastructure development, and extractive industries. Certain human activities such as logging, poaching, and fires, as well as climate change, contribute to forest degradation and to deforestation by fueling land use change and exploitation.

Forest degradation represents a series of alterations to the health and integrity of an ecosystem that result in a loss of diversity, coverage, connectivity, and resilience. It also jeopardizes the livelihoods, and therefore the survival, of Indigenous peoples. A degraded forest is

DOI: 10.4324/9781003413974-20

one that has been logged and has recovered naturally or artificially but has lost, through human activities, the structure, function, species composition, or productivity normally associated with a natural forest type expected on that site. Degradation increases greenhouse gas emissions.

Destruction refers to those processes that totally transform ecosystems, that is, when they are entirely converted to other uses such as agriculture and livestock – generally at odds with the suitability of the soils – or are obliterated altogether through catastrophic events such as fires, deforestation, and land use changes.

Disconnection relates to the fact that in a healthy ecosystem both terrestrial and freshwater elements must be interconnected. This is particularly true for those species that move around a lot in search of food and mates. Habitat alteration results in the isolation of these elements, leaving many species with no chance of survival (Living Amazon Report 2022).

Tropical Forests

Tropical rainforests are some of the planet's richest ecosystems providing habitat for a diverse array of plants and animals. They store an estimated 375 gigatons of carbon in their trees and soil and regulate climate patterns globally. When a tree dies or falls, it is consumed very quickly by microorganisms and carbon is released by decay. Because of the heat and humidity, vegetation grows so rapidly that decaying organic matter is almost immediately incorporated into new growth. This results in a larger amount of carbon stored in vegetation than in any other type of ecosystem (Table 18.1). Tropical forests store about 120 metric tons of CO_2 per hectare in their vegetation and 123 metric tons of CO_2 per hectare in the first meter of soil (Noon et al. 2022; Ometto et al. 2022).

In tropical forest regions there is a tension between the forests and agricultural expansion. While several countries have moved to protect their tropical forests, 96% of human-caused permanent deforestation occurs in the tropics with a loss of 4.1 million hectares globally in 2022. This forest loss produced 2.7 gigatons (Gt) of CO_2 emissions – an amount equivalent to all of India's annual fossil fuel emissions (Searchinger et al. 2023).

The Amazon, the largest rainforest in the world, regulates the climate and affects weather patterns for a large part of North, Central, and South America. It also contributes to global climate stabilization by storing 150–200 billion metric tons of carbon in its soils and vegetation. It is in a critical situation with threats to its health, size, biodiversity, and over 500 distinct Indigenous communities

Table 18.1 Carbon Storage in Earth's Ecosystems, Metric Tons of Carbon Stored per Hectare

Type of Ecosystem	Soil tCO_2/ha	Plants tCO_2/ha
Boreal forests	344	64
Tropical forests	123	120
Temperate forests	96	57
Temperate grasslands	236	7
Deserts and semi deserts	42	2
Tundra	127	6
Wetlands	643	43
Tropical savannas	117	29
Croplands	80	2

Source: Canadell et al. 2023

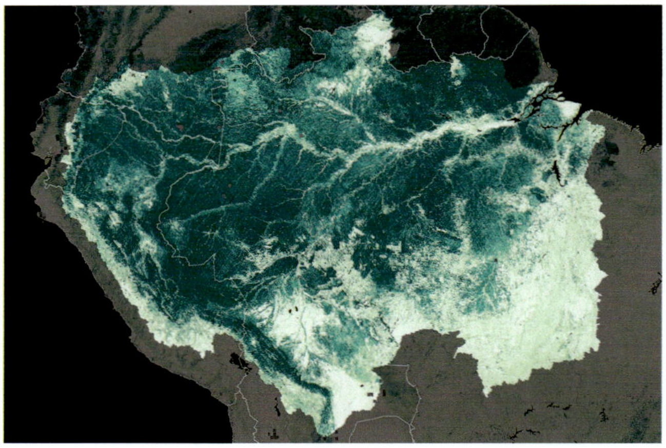

Figures 18.1 This image from NASA's Earth Observatory shows the above ground biomass density in the Amazon in 2022. The darker green indicates greater biomass density. Image credit: NASA Earth Observatory

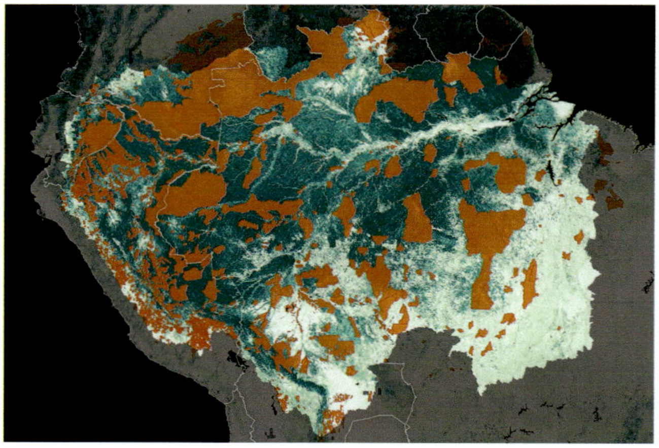

Figure 18.2 The brown overlays depict the Indigenous territories for 1.5 million people. While their communities have been threatened and Indigenous people even murdered by illegal logging activities, overall, the study shows that Indigenous communities are helping to protect some of the most intact parts of the Amazon rainforest. Image credit: NASA Earth Observatory

Figure 18.3 An Eastern European boreal forest that has been clear cut. Image credit: Max5128

(Figures 18.1 and 18.2). Climate change and the magnitude of deforestation have increased the dry season length and drought frequency. The ability of the forest to recover from dry events has slowed (Boulton et al. 2022). Total annual Amazon forest loss has declined by 55% from 2020 to 2023. In 2023, forest loss was 911,740 hectares (Finer et al. 2023).

Boreal Forests

Boreal forests ring the Arctic Circle with primarily coniferous trees storing an estimated 703 gigatons of carbon in the trees and soil (Mackey et al. 2024) (Figure 18.3). They stretch across Canada, Alaska, Scandinavia, and Russia and comprise ⅓ of the world's forests. As the world's largest remaining intact forest, Canada's boreal forest is a vital resource for Indigenous and local communities, threatened species, and international efforts to stabilize the global climate. However, boreal forests are being steadily harvested for lumber and other wood-based products with an area the size of Ohio logged between 1996 and 2015 (Lewis and Jordan 2021).

The global boreal forest is very effective at storing atmospheric carbon for long periods (even thousands of years) because of its short summers and high soil acidity, which slows rates of decomposition. In contrast to tropical forests, boreal forests store as much as 95% of their carbon in soils, wetlands, and peatlands. Boreal forests store about 64 metric tons of CO_2 per hectare in their vegetation and 344 metric tons of CO_2 per hectare in the first meter of soil (Noon et al. 2022; IPCC 2022).

More than 600 Indigenous communities depend on Canada's boreal forest for their livelihoods and cultural identities. The boreal is important for iconic species such as moose, boreal caribou, and Canada lynx, as well as hundreds of migratory bird species. In addition to extensive carbon stores and climate mitigation, the boreal provides ecosystem services such as fresh water, habitat, and clean air.

More than 90% of the logging in Canada's boreal forest is done by clearcutting with removal of nearly all trees in an area. Replanting of clear-cut areas creates tree stands that are less biologically and structurally diverse and do not have the same habitat value or ecological function. They are also less resilient to future disturbances such as extreme weather, fires, and climate change than the original forest was.

Logging in Canada's Ontario and Quebec provinces has removed 35.4 million acres since 1976 with only 56 million acres of well-established trees remaining in the region. This has left a patchwork of isolated tree stands that are unable to support the wildlife (Mackey et al. 2024).

Temperate Forests

Forests in the temperate zone occur in North America, Europe, Australia, New Zealand, Asia, and South America, and they contain some of the largest and oldest trees in the world. Types of temperate forests include deciduous, coniferous, mixed forest, and rainforest.

Carbon sinks in temperate forests are attributed to the annual increase of "live biomass" due to the yearly growth of living trees including the new leaves in spring and their loss in fall (Table 18.1). While older, larger trees are a small fraction of all trees in temperate forests, they store 41–84% of the total vegetative carbon stock. Removing a mature forest creates a "carbon debt" that can take decades or even centuries to recover (Birdsey et al. 2023). Tropical forests store about 57 metric tons of CO_2 per hectare in their vegetation and 96 metric tons of CO_2 per hectare in the first meter of soil (Noon et al. 2022).

Permanent deforestation of temperate forests has occurred from urbanization and transition to agricultural land. In the United States, 90% of temperate forests in the lower 48 states were cleared away in the 17th to 19th centuries, although some reforestation has taken place (Barnes et al. 2024). Forestry practices in temperate forests vary widely from large-scale plantations to clearcutting to selective harvesting of natural forests.

Protections for Forests and Biodiversity

The Global Biodiversity Framework

One million plant and animal species are now threatened with extinction – some will disappear within decades (IPBES 2019). At the United Nations Biodiversity Conference in December 2022, 188 nations signed the Kunming-Montreal

Global Biodiversity Framework (GBF). The GBF aims to address biodiversity loss, restore ecosystems, and protect the rights of Indigenous people. The plan consists of several measures to stop the loss of nature with the most important being to conserve and manage 30% of the world's lands, inland waters, coastal areas, and oceans and to restore 30% of degraded ecosystems by 2030 (UNEP 2022).

Glasgow Leaders Declaration on Forests and Land Use

Leaders from over 130 countries with 90% of the world's forests committed to work together to stop and reverse forest loss and land degradation by 2030 in the *Glasgow Leaders Declaration on Forests and Land Use*. In addition to agreeing to protect and restore forests, the leaders recognized the critical importance and value of knowledge and forest guardianship provided by Indigenous peoples and local communities. The agreement emphasizes their inclusion in decision-making about forest management of their lands. Perhaps most importantly, 12 countries pledged 12 billion for forest related climate finance before 2025.

There are two international entities that list restricted and endangered woods.

The Convention on International Trade of Endangered Species (CITES) is the result of an agreement between many governments around the world. CITES has three different levels of protection for wood species:

Appendix I lists species threatened with extinction. Trade in specimens of these species is permitted only in exceptional circumstances.
Appendix II lists species not necessarily threatened with extinction, but in which trade must be controlled to avoid utilization incompatible with their survival.
Appendix III lists species that are protected in at least one country, which has asked other CITES parties for assistance in controlling the trade.

If wood is listed in Appendix I or Appendix II of CITES, its trade is restricted and sometimes illegal depending on the species and the country of import.

The International Union for the Conservation of Nature (IUCN), founded in 1948, publishes the IUCN Red List of Threatened Species. Species on the Red List fall into one of three categories:

Critically endangered: Reportedly facing an extremely high risk of extinction in the wild in the immediate future.
Endangered: Not critically endangered, but reportedly still facing a very high risk of extinction in the wild in the near future.
Vulnerable: Not endangered, but reportedly still facing a high risk of extinction in the wild in the medium-term future.

The IUCN offers four additional assessments for species:

Near threatened: Technically doesn't meet the Red List criteria of a vulnerable or endangered species but is close to qualifying and/or may qualify in the near future.
Conservation dependent: Currently the focus of a species or habitat-specific conservation program. Cessation of such conservation programs would shortly result in the species becoming listed on the Red List.
Data deficient: Either not enough data to make an accurate assessment or a species' listing has been disputed or challenged.
Least concern: Species that aren't near threatened and are not dependent on conservation efforts.

The IUCN Red List only considers risk of extinction; it does not consider habitat destruction or deforestation issues surrounding tree species.

Wood Products and Their Carbon Impacts

Through photosynthesis, CO_2 is removed from the air and the carbon is distributed in stems, roots, and leaves of a tree. As trees and other plant matter decay, the carbon is stored in the soil of forests (See the box "The Carbon Cycle" in Chapter 2 "Decarbonization of Materials and Structures"). Carbon that is sequestered during biomass growth is called *biogenic carbon*. When a tree is cut for wood products, the lumber continues to store some of this carbon (*stored carbon*) for decades until it decomposes or burns (Table 18.2).

Wood and wood products are often relatively low embodied carbon construction materials as compared with concrete and steel. This, along with their ability to be renewed, and their biophilic nature makes wood a strong choice of construction material.

But quantifying the true embodied carbon of wood materials can be challenging and the methods of calculation are sometimes disputed. Variables that impact the embodied carbon calculations of wood products include forest management practices, transportation distances, scope of the LCA, and how biogenic carbon storage is treated.

Table 18.2 Global Warming Potential and Embodied Carbon of Wood Products

Material Name **Density** (kg/m³)	**Description**	**Global Warming Potential (GWP)**[1], **Fossil A1–A3**[2] (kg CO_2e/kg)	**Global Warming Potential GWP**[1], **Biogenic**[3] (kgCO_2e/kg)	**Comments and Sources**
ACQ pressure-treated pine 600kg/m³	Surfaced, pine	0.297	-1.83	Hardy, Version 3, September 2024
Copper azole treated softwood 486kg/m³	Surfaced, Radiata pine	0.204	-1.64	Wood Processors and Manufacturers Association 2021
LOSP treated softwood 486kg/m³	Surfaced, Radiata pine	0.311	-1.64	Wood Processors and Manufacturers Association 2021
Milled hardwood 1050kg/m³	Tropical hardwoods such as Ipe, Cumaru, Jatoba	0.490	-1.83	Hardy, Version 3, September 2024
Milled softwood 460kg/m³	Untreated softwood decking – add stain/paint separately	0.137	-1.55	Carbon Leadership Forum Baseline 2023; carbon stored from ICE 2019
Thermally modified hardwood 562kg/m³	Decking and cladding; ash, poplar, oak	0.648	-1.28	Average of multiple product EPDs
Thermally modified softwood 420–472kg/m³	Decking and cladding; radiata pine, pine and spruce	0.673	-1.81	Average of multiple product EPDs
Acetylated lumber, softwood 515kg/m³	Decking and cladding; radiata pine	0.633	−1.56	Agrodome 2022
Furfurylated lumber, softwood 625kg/m³	Decking and cladding; radiata pine and Scots pine	0.588	−1.21	Kebony 2022
Glue laminated timber, untreated 621kg/m³	Softwood glulam, untreated	0.244	-1.41	Carbon Leadership Forum Baseline 2023; carbon stored from ICE 2019
Glue laminated timber, pressure treated 621kg/m³	Softwood glulam, copper-based treatment	0.397	-1.41	Wood Solutions 2017
Fused bamboo composite decking 680kg/m³	Fused bamboo with binder	1.060	-1.65	Dasso 2020
Composite decking 930kg/m³	48% recycled content LDPE, 50% fiber, 2% new LDPE binder	1.49	-0.83	Average of multiple product EPDs

[1] Embodied carbon (EC) is characterized in LCAs and EPDs as global warming potential (GWP). Units are given in CO_2 equivalents (CO_2e).

[2]All figures are carbon equivalents for phases A1 (Raw Material), A2 (Transport), and A3 (Manufacturing) for one kilogram of wood product.

[3] Figures are for biogenic carbon stored until the wood product burns or decomposes. When burned or decomposed, the carbon is released resulting in a zero balance for biogenic carbon.

Note: Embodied Carbon figures will change over time as processing efficiencies are realized, and more renewable energy sources are employed.

Forest Management Practices and Embodied Carbon

Forest management practices impact the true carbon footprint of timber and wood products. Many life cycle assessments don't account for differing forest practices even though a clear-cut forest left to reseed on its own will ultimately result in a substantially higher carbon footprint for the lumber than if it was taken from a forest that was selectively harvested and/or revegetated. In a sustainably managed forest, new trees are planted to replace those that are cut resulting in an offset or even net gain of carbon sequestration. The US Forest Service has shown that carbon stores can increase as more and more trees are cut for long lasting wood products, and the forests they were harvested from are revegetated (Prestemon et al. 2022).

The main factors affecting the quality of forestry practices are rotation time between harvesting, the percentage of trees retained in the forest, and the protection of riparian buffers.

Short rotation forestry vs. long rotation forestry: The rotation frequency of the forest impacts both carbon storage and habitat value. Forests on a longer rotation cycle (e.g., 60 years vs. 40 years) will store more carbon than a young forest and the ecosystem will have more time to re-establish and function.

Percentage of trees harvested: Higher percentages of tree retention in forests will maintain higher forest carbon. Some jurisdictions have regulations about the maximum percentage of trees that can be harvested from a forest in any one rotation, while other jurisdictions allow clear cutting. The downside of higher retention is that loggers must build more roads, transport longer distances, and disturb more forests to yield a comparable amount of lumber for their business.

Soil carbon: LCAs don't traditionally consider the carbon released from soil disturbance during logging procedures. As more carbon is stored in soils than in trees, this carbon release can be substantial but is not well researched. Soil disturbance resulting from clear cutting is likely to release more carbon from the soil than selective harvesting (Lacroix et al. 2016).

Biogenic Carbon in LCAs

Wood products are heavily marketed as carbon neutral or even carbon positive due to the biogenic carbon stored while they are in use. ISO 21930 offers a standard methodology for determining biogenic carbon storage credits for wood and biobased products. Stored biogenic carbon is expressed as a negative value representing CO_2 pulled from and not yet released back to the atmosphere. Yet there are important and unresolved concerns with the way that biogenic carbon is quantified in LCAs and reported in EPDs.

Some important factors such as carbon released from soils during logging and from roots, slash and mill waste are not currently quantified in LCAs. Additionally, the amount of carbon sequestration lost from the cut tree and the lag time for newly planted trees to absorb significant atmospheric CO_2 are also not considered. Research is ongoing in these areas, and incorporation of these impacts in LCA standards is being considered.

Scope of the LCA

Wood products can appear carbon positive if an LCA only covers the product stage (A1–A3) of a product. The GWP for biogenic carbon stored may be higher than the GWP for fossil fuel combustion. LCAs that consider a full cradle-to-grave or cradle-to-cradle system boundary will also reflect the carbon emissions released from burning or decomposition of the wood. This results in a net-zero figure for biogenic carbon, plus fossil fuel emissions to manufacture, process and transport the lumber (for more information see the box "Biogenic Carbon and Carbon Stored in Construction Materials" in Chapter 2).

A product stage LCA will not quantify transportation of the wood product from the factory gate to the construction site. In the global wood products trade, this embodied carbon can be substantial. Wood can be harvested in Africa, shipped to China for processing, and then sold to a lumber company in Australia. Some wood modification treatments are only done in Europe, so timber harvested in New Zealand or Canada is shipped to Europe for treatment then shipped elsewhere for use.

Climate Smart Wood

The Climate Smart Wood Group was formed in 2019 in response to concerns about current wood LCA practices oversimplifying forestry practices when accounting for the carbon footprint of lumber and wood products. The group, including representatives from the Forest Stewardship Council and the Carbon Leadership Forum, emphasizes climate resilience, climate justice, and climate mitigation.

The Climate Smart Wood Group offers the following definition of Climate Smart Forestry "Climate Smart Forestry (CSF) is forestry that increases forest resilience in the face of climate change and sequesters and stores more carbon over time compared to conventional practices" (CSWG 2024).

> *An overarching concept of CSF is to support forests that are growing more than they harvest. Forests whose carbon stocks are growing are a sink for atmospheric carbon and it generally means that forest growth exceeds timber harvest and natural mortality.*

Obtaining lumber grown and harvested with climate smart practices can involve extra work as wood must be traced through several owners from the forest of origin and through all processing and distribution steps along the way. The path of most lumber is not a simple forest to sawmill to distributer to consumer path; instead, a single order of wood might come from multiple forests, through multiple sawmills, treatment facilities, and distributors. Using certified lumber with rigorous practices that require a chain of custody for all steps will help with this aim.

Practices associated with climate smart forestry include (CSWG 2024) the following:

- Reduced size of harvest openings
- Longer harvest rotations
- Managing forests for a diversity of tree sizes and ages
- Thinning fire prone stands
- Protecting habitat, water quality, and aquatic habitat
- Restoring habitat
- Reduced use of fertilizers, pesticides, and herbicides
- Protecting high value forests
- Protecting the rights of Indigenous peoples and communities to steward their own land.

The Social and Cultural Impacts of Wood Products

Over 25% of global land area is traditionally owned, managed, used, or occupied by Indigenous peoples (UNEP 2023). These ecosystems are under pressure but are generally declining less rapidly than other lands. While some of the nature managed by Indigenous peoples and local communities is formally protected, other areas are increasingly under pressure for resource extraction activities including mining and logging. This is threatening local livelihoods and, in some cases, the safety of Indigenous peoples.

Despite their intimate understanding of the ecosystems on their land, Indigenous people often fight for a voice in the decisions made about their lands. The *Glasgow Leaders Declaration on Forests and Land Use* and other policies support the rights of Indigenous peoples to participate in decisions about the land that is designated as their territory, but this does not happen with regularity in practice. Several Indigenous territories are lost each year to logging and mining, despite some being protected lands.

Illegal Logging and Human Rights Violations

Tropical hardwoods are often the foundational trees for their ecosystem and when they are removed, the system rapidly declines. Many countries issue permits to control the number of trees that are harvested each year; but illegal logging removes far more trees than is healthy for the ecosystem, stressing the forest and causing decline. Often within a few years the land is transitioned to agriculture or ranching (Boulton et al. 2022).

An estimated 10–30% of timber in the global trade is harvested illegally with high percentages of tropical hardwood originating from illegal activities. Illegal logging is made easier by a lack of transparent and regulated systems of enforcements; the relative ease of falsifying documents for permits, quotas, and locations; and theft from protected areas and Indigenous lands. In the Brazilian Amazon, illegal logging supplies far more timber to the market than from legal sources with an estimated 90% of deforestation resulting from illegal activities such as logging, mining, or clearing land for agricultural use (Vergara 2022).

Human right violations often occur during illegal logging (Table 18.3). As much as half of illegal logging depends on forced laborers working in unsafe and unregulated conditions (Grace Farms Foundation 2022). A major source of both forced and child laborers are human trafficking victims.

It is not just tropical hardwood that is illegally logged with forced labor. The People's Republic of North Korea (DPRNK) has forced thousands of North Korean workers into logging, sawmills, and other work in the boreal and temperate forests of Russia. It is estimated that over half of the $12 billion in lumber exported from Russia was illegally harvested by these forced laborers (Grace Farms Foundation 2022).

Indigenous communities are intimidated and displaced for illegal harvest of trees. From 1985 to 2016, The Pastoral Land Commission (Spanish name, CPT) recorded 18,012 land conflicts in the Brazilian Amazon that resulted in 1,722 killings. A number of these were from disputes with Indigenous peoples and guardians of the forests (Greenpeace Brazil 2017).

It is very difficult to track wood products through multiple processing phases and distributers. China is the largest importer of illegal timber from Africa and Asia where it is processed and then distributed to other countries (GFF 2022). This complicates the supply chain to the point that buyers may not be aware they are purchasing illegal timber.

To stop the trade of illegal timber to Europe, the European Union has enacted several pieces of legislation that prohibit placing of illegally harvested timber on the EU market. Beginning in 2025, the European Union will require rigorous traceability of imported wood products (EC 2024).

The United States Lacey Act was broadened in 2008 to include illegally trafficked plants in addition to wildlife. The Act bans trade in timber that was acquired in violation of US, foreign, or international laws. It has strict liability provisions holding buyers responsible if any timber purchased proves to be illegal. Forgery in

Table 18.3 Human Rights Issues and Timber

Country	Child Labor, ILAB	Forced Labor, ILAB	Forced Labor, DFF	Human Trafficking, DFF
Brazil		x	x	x
Cameroon				x
Cambodia	x			
China				x
Czech Republic				x
Guyana				x
Kyrgyz Republic				x
Laos				x
Madagascar			x	x
North Korea		x	x	x
Papua New Guinea				x
Peru		x	x	x
Russia		x	x	
Sweden				x
Solomon Islands				x
Tanzania				x
Uganda				x
Vietnam	x			x

Sources: ILAB 2022; DFF 2022

some chains of custody for tropical hardwoods could put purchasers at risk (Vergara et al. 2022; Earthsight 2021; Conniff 2018).

Forest Certification Programs

There are multiple certification programs that aim to protect forest health and workers along the forest products supply chain. Standards, level of detail, prescriptiveness, and required documentation vary among programs.

Forest Stewardship Council

The Forest Stewardship Council is a rigorous, voluntary forest certification used around the world. Currently, it is the only certification formally recognized by the Living Building Challenge (LBC).

FSC certifies the entire chain of custody of lumber from forest management, through processing and finishing, to the final purchase of the lumber. FSC aims to provide assurance that lumber comes from well-managed forests, is not illegally harvested, is worked by fairly paid and treated workers, and is not from protected lands or lands of Indigenous peoples.

FSC has a policy of zero deforestation meaning forests where lumber is harvested will not be converted to non-forest uses such as real estate development of agriculture. Trees taken from FSC certified forests are replaced with an equal or greater number of trees; and clear-cut sizes are limited to protect ecosystems. Safeguarding of ancient and endangered forest and biodiversity preservation are other important goals of the Forest Stewardship Council.

The FSC requires fair wages and a fair work environment, and they protect the rights of communities, including Indigenous peoples, in and around the forests they certify. This is particularly important as Indigenous groups are not often included in decisions by logging companies.

In 2023, over 400 million acres of forest were FSC certified accounting for about 10% of working forests worldwide. Around 170 million acres of working forests are certified in North America including 25% of commercial logging operations in the Canadian boreal forest (FSC 2024).

The ten principles listed next are the foundation of the FSC standards and have been developed to be relevant to different forest ecosystem and a wide diversity of cultural, political, and legal settings. They require any managers of certified forests to (FSC 2023)

1. comply with all applicable laws
2. maintain or improve the social and economic well-being of workers
3. uphold the rights of Indigenous peoples
4. maintain or improve the social and economic well-being of local communities
5. manage their products and services in a way that maintains or improves their long-term economic viability, social benefits, and environmental benefits
6. maintain, conserve, and/or restore the ecosystem services and environmental values of managed forests and also avoid, repair, or mitigate negative environmental impacts
7. establish a management plan that outlines their economic, environmental, and social policies and objectives
8. demonstrate progress toward meeting these objectives
9. maintain or improve high conservation values
10. ensure that all management activities comply with FSC principles and criteria.

Programme for the Endorsement of Forest Certification

The PEFC is an umbrella program that sets standards for and certifies local and regional forest certification systems. PEFC members include private companies, non-governmental organizations, and trade associations. In the United States, PEFC recognizes the Sustainable Forestry Initiative.

Sustainable Forestry Initiative

The Sustainable Forestry Initiative (SFI) is a voluntary certification system that offers standards and certifies forest management, sourcing and chains of custody, and harvesting practices in North America. SFI standards are developed by a wide variety of forest industry stakeholders, but logging companies originated the standard. While 159 million hectares of working forests are certified by the SFI, most environmental groups feel as if the policies and practices are not rigorous enough. Some do not even go beyond the forestry regulations of the local governing bodies. SFI is criticized for not requiring the free, prior, and informed consent of Indigenous peoples; not protecting primary forests and threatened tree species; and for their misleading labeling and marketing (Gutierrez Garzon et al. 2023; Jordan 2022).

American Tree Farm System

The American Tree Farm System (ATFS) was developed in 1941 by private landowners to improve forest management practices on private land in the United States. Around 74,000 family forest owners are part of the program encompassing nearly 8 million hectares of land. ATFS is endorsed by PEFC and has 8 standards along with 14 performance measures and 22 indicators. Like SFI, the ATFS is less prescriptive and detailed than Forest Stewardship Council Certification (Gutierrez Garzon et al. 2023).

Reclaimed and Reused Wood

Use of reclaimed wood offers many benefits including an extended service life, a reduction in the embodied carbon of a structure, reduced waste, and reduced pressure on forests for new lumber. It can provide the opportunity to use high quality, unique wood in creative ways. But use of reclaimed wood can pose challenges and is not always cost effective. Reclaimed wood comes from a variety of sources.

Typical ***architectural salvage*** sources are buildings and bridges. The most common buildings to be deconstructed are warehouses, military facilities, and industrial buildings because larger structural members are used and are not usually covered by finishes such as lath and plaster. Most of the costs for architecturally salvaged wood come from the labor required to deconstruct the building and to clean the wood of nails and other fasteners.

Horticultural salvage is lumber from fallen or timbered trees. Many large street, yard, and orchard trees are taken down each year due to lightning strikes, development nuisance, or life-cycle reasons. A good number of these are very large hardwoods or fruit trees that can yield large, unique pieces of lumber. This practice has grown to the point that the cost of taking down a tree may be offset by the sales value of the wood (Figures 18.4–18.10).

Sinker logs are retrieved from river bottoms. Many old-growth forest cypress and yellow pine logs harvested around the turn of the 20th century in the US Southeast sank to the bottom of waterways and have been preserved due to a lack of oxygen underwater. While this is a good source of high-quality reclaimed lumber, the practice has ecological drawbacks. Sensitive aquatic ecosystems may be disturbed as logs are salvaged from the river floor.

Jobsite salvage or construction and demolition waste is a source of reclaimed lumber from deconstructed structures.

Figure 18.4 and Figure 18.5 CMG Landscape Architecture used eucalyptus timbers from trees removed on-site for a Bay Area tech campus. The repurposed eucalyptus replaced concrete seats and retaining walls with new amphitheater style wooden seating. All of the timber was milled and dried locally without treatments or protectants, further reducing environmental impacts. Image credit: CMG Landscape Architecture

Figure 18.6 Reclaimed white oak from Dorthea Dix Park will be used in custom furnishings for the redesigned park in Raleigh, North Carolina. MVVA worked with local fabricator and supplier Raleigh Reclaimed on the wood reuse. Image credit: MVVA, Inc.

Figure 18.7 At Presidio Tunnel Tops Park, designed by Field Operations, the Fallen Tree is a centerpiece of the playground. The white oak, estimated to be 250 years old, was felled by a storm in south-western Ontario and reclaimed and refurbished by Earthscape Play. A child leaps across the gap between two sections of the Fallen Tree. The spacing and uneven tree surface create an added challenge, one that mimics what children might find in the woods. Image credit: Earthscape Play

Figure 18.8 The Fallen Tree's root flare is a cozy spot for children to hide or hang out. Wood's thermodynamic properties mean it is pleasant to the touch in all weather. Image credit: Earthscape Play

Figure 18.9 It took more than 500 hours to carefully sand and hand-carve the tree to maintain its true shape but remove all bark, sapwood, and rot. This careful finishing is critical to the longevity of reclaimed trees in exterior settings, as unfinished fallen trees are more attractive to insects and decay organisms. Image credit: Earthscape Play

Figure 18.10 An Earthscape craftsperson looks through the largest section of the Fallen Tree. Originally this section was expected to remain a solid piece like the other two sections, but root rot necessitated that it be hollowed out. This created an even more unique play experience. Image credit: Earthscape Play

Sources include wood members from deconstructed buildings, decks, timber retaining walls, or wood construction waste. Often reclaimed wood is still good, but the design needs have changed, so the structure is being removed. This is particularly true of structural members from a deck. The deck boards weather to the point that the owner may decide to completely replace the deck, when the structural members are still viable.

Contractors can be wary of jobsite salvage for several reasons. First, if the reclaimed wood will be used in a structural situation, it may need to be regraded. This can add expense and delay construction schedules. Second, there may not be enough salvaged lumber for the new structure, making it difficult to match with new lumber. The wood removed on-site must be stored in a dry place, sometimes for months before it is needed for the new structure.

Major benefits of building with reclaimed wood lie in the wood's quality and uniqueness compared with the limited variety and quality of newly harvested wood that is on the market today. Some wood salvaged from deconstructed buildings was cut from old-growth forests and is far denser, closer grained, and freer of defects than the wood that is currently grown quickly and harvested young. Some old buildings used large solid structural members that are not even available in the lumber trade today. Last, some unique woods like heart cypress, heart redwood, or Port Orford cedar can be found on the salvage market (Table 18.4).

Table 18.4 Considerations for Use of Reclaimed Wood

Topic	Consideration
Design	Let the materials inspire the design Discuss using reclaimed wood with the owner, contractor and other consultants early in the project design process. Make owners aware of aware of nail holes and other age marks from previous use. Keep big wood members whole where possible, rather than ripping them down, then they can be salvaged again.
Sourcing	Source and reserve the wood before design work is complete to avoid design revisions and/or procurement delays. Specify wood from local sources where possible to avoid the resource consumption of ex-tensive trucking. Request a chain of custody for any questionably sourced wood. Make sure there is some extra wood in the order. It may be difficult to find more of the same later.
Specifications	Include appearance standards in specifications. Use less pristine wood for rougher applications like concrete formwork or under deck structure. Specify drilled and bolted connections rather than nailed, as salvaged wood is drier than "green" wood and may spilt. Preserve reclaimed wood that is not already preserved or naturally decay resistant.
Finishing	When using fallen trees, remove all bark, sapwood, and rot. This careful finishing is critical to the longevity of reclaimed trees in exterior settings. Unfinished fallen trees will be attractive to insects and decay organisms.
Regrading	Have wood in structural situations regraded if required. Plan for this in construction schedules.
Design for deconstruction	Design new wood structures for ease of deconstruction. Use bolted or screwed connections. Do not use glue or nails.
Rating systems	Obtain reclaimed wood from sources that are recognized by LEED, SITES, and LBC standards.

Table 18.5 Type and Considerations of Engineered Wood for Exterior Use

Type	Considerations
Glue laminated wood "Glulams" are made by bonding together layers of smaller lumber into a larger member with moisture resistant adhesives.	Wood should be treated with preservatives or modifications. Adhesives should be rated for exterior use. Capable of supporting substantial loads with smaller wood pieces (Figures 11.11 and 18.12). Should be coated with UV resistant finishes that are periodically reapplied. Ensure drainage of water away from the glulam. Common glue is phenol resorcinol formaldehyde (PRF) adhesive. Formaldehyde is an IARC and US NIH NTP[1] known carcinogen. Some glulams are treated with IPBC/PER treatments that contain permethrin, a substance classified by the US EPA as "likely to be carcinogenic to humans" (US EPA 2006).
Composite lumber Wood composite materials are a mix of wood fibers such as sawdust and a plastic binder.	They are used as decking and lumber. Composite lumber encapsulates the wood fibers in a polymer, so they don't biodegrade like other wood products. Some composite lumber is 100% plastic or partially bamboo, rice hull, or other biobased materials. Glues, adhesives, and plastics and their health impacts vary widely.
Exterior plywood Thin layers of wood veneer that are treated with chemicals and preservatives to resist moisture, then bound with an adhesive.	Used in sheds, boardwalks, concrete formwork, and outdoor furniture. Usually water resistant, but not waterproof. Should be finished with exterior paint or water protective stain or sealant. Common glue is phenol resorcinol formaldehyde adhesive. Formaldehyde is an IARC and US NIH NTP[1] known carcinogen.

[1]International Agency for Research on Cancer (IARC)

US National Institutes of Health, National Toxicology Program (US NIH NTP)

The process for designing with and specifying reclaimed lumber is different than that of new lumber. When using new lumber, structures are designed, detailed and specifications are written, and then new lumber is procured by the contractor just before construction. As reclaimed lumber sizes, types, and quantities are highly variable, it is recommended that the design of a structure be conceptualized, then the necessary lumber be procured, then the structure design adapted to the available timber where the design of the new structure grows out of the uniqueness of the wood itself.

Challenges to this different process are the need for long-term storage of the material and bid structures that

Figure 18.11 An oriole's nest inspired the design of the large play nest. Its pear shape frames the structure that is fully climbable inside and out. The frame is constructed of Alaskan yellow cedar glue laminated timbers and the horizontal climbing boards are acetylated lumber. Image credit: Earthscape Play

Figure 18.12 The nest is surrounded by black locust posts. The same black locust lumber was used to create the Woodland Wall that represents the tightly planted historic forest of the Presidio. Black locust has an extremely long in-ground life span when it has been fully debarked, and the sapwood has been removed. Image credit: Earthscape Play

may not allow funds for purchase of reclaimed wood early in the design process. An alternative is to approach the procurement of the wood like tagging plant material. If the reclaimed wood can be "pretagged" with the source and price written into the specs, then the wood could be "reserved" for a project.

The Forest Stewardship Council certifies reclaimed wood under its FCS Recycled label. FSC offers a standard for reclaimed wood called "Sourcing reclaimed material for use in FSC Product Groups or FSC Certified Projects". It recognizes post-consumer wood material coming from commercial transport packaging, C&D debris, damaged stock and rejected products, and other used wood from a variety of post-consumer residential, commercial, or industrial sources. The Forest Stewardship Council does not recognize sinker logs as reclaimed lumber as they consider them to be part of the waste from logging operations. The FSC awards the Recycled label after an independent third-party audits the chain of custody of the wood all the way back to where it was originally used.

Biodeterioration of Wood Products

Biological organisms that degrade wood are fungi, insects, bacteria, and marine borers. Wood products are vulnerable when exposed to conditions such as mild temperatures, moisture, and oxygen that allow growth of wood-decaying organisms. Insects such as termites can damage wood products in some regions.

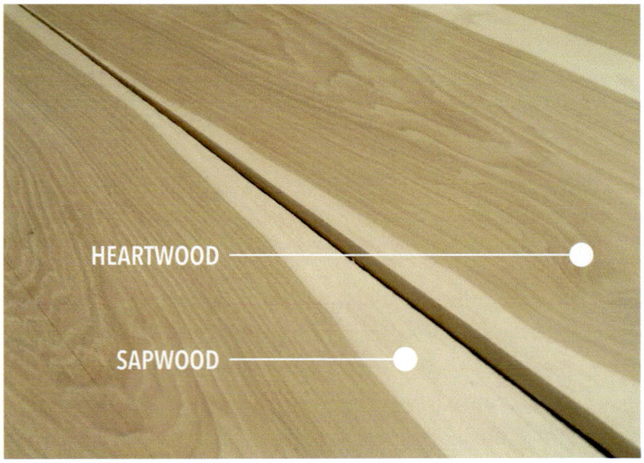

Figure 18.13 Heartwood is more decay resistant than sapwood. Image credit: Meg Calkins

Serious decay of wood can occur when its moisture content is above the fiber saturation point, which is 30% on average. This can happen with water standing on the wood, water penetrating unfinished end grain, or wood that is in contact with the soil. Wood in water-soaked conditions deprived of oxygen will not support the development of decay fungi.

Wood in a tree is divided into two zones: the heartwood and the sapwood (Figure 18.13). The actively conducting part of the wood where the cells are still alive is the *sapwood*. This is the band of lighter colored wood adjacent to the bark. The *heartwood* is the darker colored wood found in the inner part of the tree. Decay producing fungi will attack either the

Table 18.6 Average Heartwood Decay Resistance of Common Domestic and Imported Lumber Species[1]

Very Resistant	Resistant	Moderately Resistant	Slightly or Nonresistant
US Domestic			
Black locust	Cedar, Atlantic white	Douglas-fir	Ashes
Cedar, Alaskan yellow	Cedar, Eastern red	Larch, Western	Firs, true
	Cedar, Northern white	Pine, eastern white, old growth	Hemlocks
	Cedar, Western red including sap wood	Redwood, young growth	Pines
	Redwood, old growth		Spruces
	Walnut, black		
Imported			
Ipe (lapacho)	Balau	Mahogany, African	Meranti, light red
Jarrah	Mahogany, American	Meranti, dark red	
Teak, old growth	Spanish-cedar	Teak, young growth	
Cumaru	Garapa		
Massaranduba			
Tigerwood			
Manilkara			
Jatoba			

[1] 90% of wide side of lumber pieces should be heartwood (ICC 2021).

Source: Ross 2021

heartwood or sapwood of a tree, but the heartwood of some species contains various biochemicals called extractives that lend it decay resistance and strength.

Table 18.6 classifies the heartwood decay resistance of commonly used lumber species. It is important to note that the heartwood of old-growth trees will generally offer more decay resistance than second growth trees of the same species. However, there are multiple environmental reasons to avoid lumber from old-growth trees discussed earlier in this chapter. Species listed in the "moderately resistant" and "slightly or non-resistant" categories will require some type of preservative treatment.

Wood Durability Ratings

Wood durability ratings are used to estimate how long a wood product will last in contact with the ground. The ratings do not consider site specific factors such as moisture, temperature, weather, installation competency, and physical stresses (Tables 18.7–18.9).

Wood Preservation Methods

Wood that is not naturally decay resistant, also called naturally durable, must be treated or modified to protect it from decay

Table 18.7 Durability Classes of Wood-Based Materials to Attack by Fungi[1]

Durability Class	Designation	Expected Wood Life in Contact With the Ground
1	Very durable	25+ years
2	Durable	15–25 years
3	Moderately durable	10–15 years
4	Slightly durable	5–10 years
5	Not durable	0–5 years

[1] Refers to heartwood (not sapwood), modified or treated wood

Source: BS EN 350 2016

Table 18.8 Durability Classes of Wood-Based Materials to Attack by Insects[1]

Durability Class	Designation
DC D	Durable
DC M	Moderately durable
DC S	Not durable

[1] Refers to heartwood (not sapwood), modified or treated wood

Source: BS EN 350 2016

Table 18.9 Use Classes of Wood Materials

EN 335 Use Class	AWPA Use Category	Description
Use class 1	UC1	Interior, dry
Use class 2	UC2	Interior, damp
Use class 3.1	UC3A	Exterior, no ground contact, protected
Use class 3.2	UC3B	Exterior, no ground contact, exposed
Use class 4.1	UC4A	Exterior, ground contact, general use
Use class 4.2	UC4B	Exterior, ground contact, heavy duty
Use class 5	UC5A, B, C	Marine Use, (A) Northern waters, (B) Central waters, (C) Southern waters

Sources: BS EN 335 2013; AWPA 2024

fungi, insects, and marine borers. Wood modification or preservative treatments will increase the service life of wood allowing for more efficient use of forest resources.

Wood Modification Treatments

With concern for the health of tropical forests, alternatives to tropical hardwoods in site construction are increasingly used. Wood modification treatments use physical, biological, or chemical processes to improve the properties of wood. Thermal modification, acetylation, and furfurylation improve durability of the wood by providing resistance to decay and termites and improve properties of the lumber such as dimensional stability and hardness (Table 18.10).

Wood modification can be applied to both softwoods and temperate hardwoods resulting in a wood product that performs competitively with tropical hardwoods. Designing

Table 18.10 Modified Wood Comparison Chart

	Janka Hardness (lbf[1]) Density[2] kg/m^3 Ideal Moisture Content Fire Rating	Durability Class[3] Use Class Warranty	Comments[4]
Thermally modified lumber Ash, red oak, poplar, pine, spruce	Janka Hardwoods (HW)1320–1400lbf Softwoods (SW) 870–1129lbf Density HW 581–628kg/m^3 SW 400–420kg/m^3 4–8% moisture content Fire rating HW Class A SW Class B	Durability HW 1 SW 1 Use class HW 3.2 SW 3.2	Moisture content is about half that of the untreated species. Wood grain is pronounced, and some is dark brown. Surface can be treated with natural oil products to delay graying and weathering. FSC certified is available. Fully recyclable and biodegradable.
Acetylated lumber Radiata pine	Janka 922lbf Density 515kg/m^3 3–5% moisture content Fire rating Class C	Durability 1 Use classes 1–4 Warranty: 25 years in ground contact; 50 years above ground	Increased density and dimensional stability. Swelling and shrinkage are reduced by 75% or more. Fully recyclable and biodegradable. FSC certified available.
Furfurylated lumber Radiata pine Scots pine	Janka 1619lbf Density 570–670kg/m^3 4–12% moisture content Fire rating Class B	Durability 1 Use class 3.2–4 Warranty 30 years	FSC certified available. Scots pine end grain cuts should be treated with a wood preservative in the field. Close to 50% more weight is added with treatment. The concentration of furfuryl alcohol monomer in the product is below the limit for classification as hazardous waste.

[1] Janka hardness is a measure of the side grain harness of wood. The pound-force (lbf) measure is commonly used to express Janka hardness, even in some countries using the metric system.

[2] Assumes 12% moisture content of wood.

[3] Based on EN 350 Durability of wood and wood-based products.

[4] Refer to Table 10.6 for design and construction considerations and Table 10.9 for maintenance considerations

Sources: Thermory 2024; Accoya 2024; Kebony 2024

with, detailing, and maintaining modified wood structures is discussed in detail in Chapter 10 "Decks and Boardwalks" and Chapter 11 "Rails, Fences and Screens".

Thermally Modified Wood

Thermally modification (TM) heats wood to temperatures above 200°C in a process called torrefaction, permanently modifying the carbohydrates (hemicelluloses) and rendering the wood inedible to decay fungi and insects. The process limits oxygen and uses different methods such as steam or hot water to prevent the wood from burning. During the modification process, the wood is dried to lower its moisture content and to reduce expansion and contraction owing to moisture changes over the life of the lumber.

The resulting decay and insect resistant lumber is dimensionally stable and darker in color. There is some loss of strength, but testing shows that it is not significant.

No chemicals are introduced in the non-toxic thermal modification process, so the lumber is easily recyclable. Energy to heat the wood and transport it from forest to processing to the jobsite are the main environmental impacts; some companies take steps to capture waste heat and use renewable fuel sources.

Both hardwoods and softwoods can be thermally modified. White Ash, yellow poplar, oak, Scots pine, and spruce are the most common thermally modified woods. While the largest amount of thermally modified wood is treated in Europe, some companies have modification facilities in the United States. Some of the ash, oak, and pine are sourced from the United States as well. FSC certified TM lumber is widely available.

TM lumber can be priced at a premium, like tropical hardwoods, but cost savings come in the ease of constructability. These products are easier to work with and require no special tools, which cuts installation time and waste. In a deck application, the understructure is typically built with conventional pressure-treated lumber, and the decking is thermally modified lumber. Because thermally modified woods are not as dense as many tropical hardwoods, it does not hold heat, making sitting on the bench or walking barefoot on the deck more comfortable in the heat of summer (Figure 18.14).

Acetylated Wood

Acetylated wood is modified by reacting the lumber with acetic anhydride (produced from agricultural crop waste) and heat to modify the composition. This process greatly reduces the ability of the wood to absorb water and renders it more dimensionally stable and durable. Acetylated

Figure 18.14 Michael Van Valkenburgh Associates used thermally modified lumber for a deck at Wellesley College. They found radial cuts for this deck far easier with TM lumber than with hard, dense tropical hardwoods. Image credit: MVVA Inc.

Figure 18.15 Move Art used acetylated wood for this play structure at Gascoigne Park in London. Image credit: Move Art

wood's swelling and shrinking are reduced by 75%, and its durability is substantially increased (Accoya 2024). As with TM wood, microorganisms and fungi are no longer able to recognize the wood as a food source. Stake tests show that acetylated wood performs better in-ground than CCA treated lumber (Accoya 2020).

Like thermal modification, the acetylation process alters the full cross section of the wood, so cuts and machine holes do not need to be treated with a preservative. Because there is minimal shrinking and swelling of the wood, paints and finishes are expected to last longer (Figure 18.15).

Furfurylated Wood

Furfurylated wood is impregnated with furfural alcohol (a monomer and agricultural by-product) then heated to penetrate and surround the cell walls of the wood. Furfurylated lumber shares its durability (European Durability Class 1 – "very resistant" to decay), dimensional stability, and color traits with thermally modified and acetylated lumber. But because of the polymerization, it is heavier and harder. Given the hardness and increased brittleness, the manufacturer recommends predrilling screw holes and using self-sealing stainless steel screws.

Figure 18.16 and Figure 18.17 The Olin Studio used furfurylated lumber for decking and structures throughout Pier 26 in NYC. Kebony slats articulate a viewing shed structure in the park. The Janka hardness of furfurylated lumber is similar to some tropical hardwoods, but not as hard as Ipe. Image credit: Meg Calkins

The furfurylation process adds an additional 50% to the dry weight of the wood (Figures 18.16 and 18.17).

Naturally Durable Woods

Naturally durable woods, also called decay-resistant woods, produce extractives that discourage insects and decay and mold fungi. Extractives and their effectiveness vary by tree species and are most prevalent in the heartwood of trees (Table 18.11). The International Residential Building Code for Deck Construction (2024) requires that at least 90% of the wide side of natural durable wood boards be heartwood.

Common trees native to North America that yield durable heartwood lumber include Black locust (*Robinia pseudoacacia*), white oak (*Quercus alba*), black walnut (*Juglans nigra*), Alaskan yellow cedar (*Chamaecyparis nootkatensis*), Western red cedar (*Thuja plicata*), Easter red cedar (*Juniperus virginiana*), Port Orford cedar (*Chamaecyparis lawsoniana*) redwood (*Sequoia sempervirens*), and Western juniper (*Juniperus occidentalis*) (Meier 2024).

Trees native to Europe that yield very durable heartwood lumber are English, Holm and Sessile oaks (*Quercus robur, ilex, petraea*), cedar (*Cedrus brevifolia*), sweet chestnut (*Castanea sativa*), and the European yew (*Taxus baccata*) (Meier 2024).

Several tropical hardwoods are very durable, yet the harvest of many types of commercially available species poses substantial environmental and human rights impacts. Refer to the "Forest Carbon, Ecosystems, and Indigenous Populations" section of this chapter for greater detail.

Black Locust Lumber

Lumber from the heartwood of the black locust tree is hard, dense, and decay resistant. Native to the central-eastern United States, they are fast-growing pioneer species. They are considered invasive in some parts of the United States. Black locust trees grow large enough to yield high quality, dense, decay-resistant heartwood lumber in 25 to 40 years – just 20% to 40% of the time needed to grow comparably decay-resistant tropical hardwoods. Black locust's Janka hardness rating, a measure of how well the wood can withstand dents and dings, is 1700 pounds-force (lbf). The Janka hardness number can also predict challenges in nailing, screwing, sanding, or sawing, and although black locust is hard compared to North American softwoods, it is only half that of Ipe and Cumaru – notoriously challenging woods to build with.

Visually, black locust is known for its knots, mineral streaks, surface checks, and bark insertions. Some proponents of black locust appreciate the natural character of the wood; however, clients should be prepared for the appearance – and changing appearance – of the wood. The lumber grain can open and look like it's splitting, then close again within one season. Premium grade has fewer imperfections, and No. 1 grade is less expensive but has more imperfections.

The greatest construction challenge when using black locust lumber stems from its irregular, winding tree form that can result in lumber that slightly bends and warps wanting to return to its original shape. Black locust lumber manufacturers suggest attaching deck boards at 40cm and avoiding extremely long lengths in structures.

Specifications for black locust lumber should require use of heartwood and address graining, imperfections, strength, and moisture content. Contractors should be made aware of potentially long lead-times to source and dry black locust lumber (Figures 18.18–18.20).

Oak

Oak is increasingly used in exterior settings as it is durable, hard, and abundant in North America and Europe. White oak and red oak are the two main types of lumber used in North America with European oak used in the United Kingdom and Europe. Kiln-dried European oak is the most resistant to UV degradation and decay causing fungi, with white oak also relatively durable when untreated. Red oak should be thermally modified if exposed to the elements.

All three types of oak have densities ranging from 675kg/m^3 to 759kg/m^3 with Janka harness measures of 1120–1360lbf (pounds-force). When building with oak, nail and screw holes should be predrilled. Like any other wood, it can develop checks with moisture fluctuations (Figure 18.21).

Preservative Treatments for Wood

Pressure-treated softwood lumber is the most common material used for structural members of wood decks,

Table 18.11 Naturally Durable (Decay Resistant) Wood Comparison Chart

Wood	**Janka Hardness (lbf)**[1] **Density**[2] **kg/m**^{3} **Ideal Moisture Content** **Fire Rating**	**Durability Class**[3] **Use Class** **Warranty/Life Expectancy**	**Comments** **IUCN Status**
Black locust *Robinia pseudoacacia* East Central United States	Janka 1700lbf Density 770kg/m^3 11% moisture content Fire Rating Class A	Durability 1–2 Use class 3–4 Life expectancy 30–50 years	Only heartwood is decay resistant. Heavy, hard, moderately low shrinkage. Lumber grain can open and look like it's splitting, then close. Tendency to warp. Extra fasteners should be used.[4] IUCN Red List: *Least concern*
White oak and European oak *Quercus alba, Quercus robar* Middle and Eastern United States, Europe	Janka 1120–1360lbf Density 675–759kg/m^3 12% moisture content Fire Rating Class C	Durability 2 Use class 3 Life expectancy 15–40 years	Good stability, hard so should be predrilled, European oak has a curving grain. IUCN Red List: *Least concern*
Alaskan yellow cedar *Chamaecyparis nootkatensis* Northwestern North America	Janka 580lbf Density 495kg/m^3 12% moisture content Fire Rating Class B	Durability 1–2 Use class 3 Life expectancy 20–30 years	Durable, but easy to work. Stands up to wear without splitting or splintering. Smooth straight grain and close rings. Very slow growing tree. FSC certified available. IUCN Red List: Chamaecyparis nootkatensis are *Near Threatened*
Redwood *Sequoia sempervirens* Western United S	Janka 450lbf Density 415kg/m^3 10–14% moisture content Fire Rating Class B	Durability 3 Use class 3 Life expectancy 5–30 years	Very soft. Easy to work. Not durable in high traffic settings. Only heartwood is decay resistant. Decay resistance of second growth trees has low to moderate decay resistance. IUCN Red List: Sequoia sempervirens are *Endangered*
Western red cedar lumber *Thuja plicata* Western North America	Janka 350lbf Density 370kg/m^3 10–14% moisture content Fire Rating Class B	Durability 2 Use class 3 Life expectancy 10–20 years	Very soft. Not durable in high traffic settings. IUCN Red List: *Least concern*
Easter red cedar *Juniperus virginiana* Eastern North America	Janka 900lbf Density 530kg/m^3 10–14% moisture content Fire Rating Class B	Durability 1–2 Use class 3 Life expectancy 15–20 years	Durable, but easy to work. Stands up to wear without splitting or splintering. Smooth straight grain with knots. Heartwood is reddish. FSC certified available. IUCN Red List: *Least concern*
Ipe *Handroanthus, Roseodendron, Tabebuia spp.* Northern South America, Amazon	Janka 3600lbf Density 1100kg/m^3 14–18% moisture content Fire rating Class A	Durability 1 Use class 5 Life expectancy 30–40 years	Challenging to build with as it is dense and hard, oily wood does not take finishes easily. IUCN Red List: *Handroanthus spp.* are characterized as: *Endangered, Vulnerable* and *Near Threatened* CITES: Appendix II
Cumaru *Dipteryx spp.* Northern South America, Amazon	Janka 3300lbf Density 1085kg/m^3 14–18% moisture content Fire rating Class A	Durability 1 Use class 4 Life expectancy 30–40 years	Challenging to build with as it is dense and hard, oily wood does not take finishes easily. IUCN Red List: *Dipteryx spp.* are *Near Threatened* CITES: Appendix II
Jarrah *Eucalyptus marginata* Southwestern Australia	Janka 1860lbf Density 835kg/m^3 14–18% moisture content Fire rating Class B	Durability 2 Use class 4 Life expectancy 15–40 years	Challenging to work with because of high density and irregular grain. Western AU ban IUCN Red List: Eucalyptus marginata are *Near Threatened*

(*Continued*)

Table 18.11 (Continued)

Wood	Janka Hardness (lbf)[1] Density[2] kg/m³ Ideal Moisture Content Fire Rating	Durability Class[3] Use Class Warranty/Life Expectancy	Comments IUCN Status
Jatoba *Hymenaea courbaril* Central America, northern South America	Janka 2690lbf Density 910kg/m³ 14–18% moisture content Fire rating Class A	Durability 2–3 Use class 3 Life expectancy 25+ years	Challenging to work with because it is dense and hard. Susceptible to attack from marine borers. IUCN Red List: *Hymenaea spp* are *Stable*
Merbau *Intsia species* Malaysia, Phillipines, New Guinea, Pacific Islands	Janka 1840lbf Density 815kg/m³ Fire rating Class A 14–18% moisture content	Durability 1–2 Use class 4 Life expectancy 30–40 years	Good durability and resistance to insects. Stains black in the presence of moisture. Difficult to saw. IUCN Red List, Intsia spp. are *Near Threatened*
Garapa or **Grápia** *Apuleia leiocarpa* South America	Janka 1650lbf Density 820kg/m³ 14–18% moisture content Fire rating Class A	Durability 3 Use class 2 Life expectancy 20–30 years	Also known as Brazilian oak. Durable although vulnerable to termites. IUCN Red List: Apuleia leiocarpa listed as *Declining*
Massaranduba *Manilkara spp.* Caribbean, Central and South America	Janka 3130lbf Density 1080kg/m³ 14–18% moisture content Fire rating Class B	Durability Use class Life expectancy 50+ years	Also known as Brazilian redwood, South Pacific redwood, torem, bulletwood. Very durable with good insect resistance, susceptible to marine borers IUCN Red List: *Manilkara spp.* are listed as Endangered, *Vulnerable* and *Declining*
Tigerwood *Astronium spp.* Mexico to Brazil	Janka 2170lbf Density 905kg/m³ 14–18% moisture content Fire Rating Class A	Durability Use class Life expectancy 30+ years	Also called Goncalo Alves, Brazilian koa, zebrawood, mura, kingwood IUCN Red List: *Astronium spp.* are listed as *Endangered* and *Least Concern*
Teak *Tectona grandis* Southern Asia	Janka 1070lbf Density 655kg/m³ 14–18% moisture content Fire Rating Class C	Durability 1 Use class 4 Life expectancy 50+ years	Plantation grown in tropical regions of Asia, Africa, and South America. Slow growing and dense. IUCN Red List: *Tectona grandis* are listed as *Endangered* and *Decreasing*
Meranti *Shorea spp.* Southeast Asia	Janka 800–1600lbf Density 590–850kg/m³ 14–18% moisture content Fire Rating Class A	Durability 2–4 Use class 2 Life expectancy 15–25 years	Also called white meranti, dark red meranti, yellow meranti, balau, Philippine mahogany Susceptible to insect attack IUCN Red List: *Shorea spp.* are listed as *Endangered, Vulnerable, Near Threatened, Least Concern* and *Decreasing*

[1] Janka hardness is a measure of the side grain harness of wood. The pound-force (lbf) measure is commonly used to express Janka hardness, even in some countries using the metric system.

[2] Assumes 12% moisture content of wood.

[3] Based on EN 350 Durability of wood and wood-based products.

[4] Refer to Table 10.7 for design and construction considerations and Table 10.10 for maintenance considerations

Sources: Meier 2024; Gerard et al. 2017

Figure 18.18 and Figure 18.19 The Nest is an interactive play structure designed and constructed by Tri-Lox at a park in Long Island City. The modular structure has a stainless steel frame with black locust lumber slats. The black locust lumber was sourced from regional working forests. Image credit: Tri-Lox.

Figure 18.20 The regionally sourced black locust wood for the benches at Little Island was sourced and milled by Tri-Lox and fabricated by SITU. The grain of black locust lumber can open and look like it's splitting, then close again within one season. This is a unique characteristic of this durable natural material, and it does not mean there is a problem with the wood. Image credit: Meg Calkins

pergolas, guardrails, and fences, even when deck boards may be modified woods, composite lumber, or durable wood. Pressure-treated lumber is relatively fast growing, comparatively inexpensive and often easier to procure from regional sources (Table 18.12).

The most commonly pressure-treated softwoods are pines (southern yellow pine, Eastern white pine, red pine, ponderosa pine), firs (hem fir, Douglas fir), and spruce (black spruce, white spruce). FSC certified pressure-treated lumber is widely available.

Pressure treatments are chemicals that are biocides or insecticides. Most are regulated by federal or state environmental protection agencies. Pressure treatments fall into two categories: oil borne and waterborne. The type of preservative used depends on the type of exposure and the end use. Waterborne preservatives have minimal odor

Figure 18.21 MVVA used reclaimed white oak for a nature play pole climber at the Port Lands in Toronto. The white oak is treated with a low-VOC clear sealant. This photo was taken during the construction process. Image credit: MVVA, Inc.

Table 18.12 Pressure Treatments for Softwood Products

Pressure Treatment Use Class	Environmental and Human Health Risks[1]	Notes[2]
Water-Based Preservatives		
Alkaline copper quat (ACQ-A, ACQ-B, ACQ-C, ACQ_D) Use class 3.1–4.1	Copper is acutely toxic to fish and other marine life (US EPA 2007) Copper oxides: US CWA Toxic Pollutants CERCLA Hazardous Substance	50–67% copper oxide and 33–50% quaternary ammonium compound (quat). Used for decking and dimensional lumber. Widely Available. Slight greenish tone when new.
Micronized copper azole and copper azole (CA-B, CA-C, MCA, MCA-C) Use class 3.1–4.1	Copper is acutely toxic to fish and other marine life (US EPA 2007) Tebuconazole: Aquatic toxicity Copper oxides: US CWA Toxic Pollutants CERCLA Hazardous Substance	96% copper oxide and 4% triazole compounds. Used for decking and dimensional lumber. Widely available. Less corrosive to copper fasteners. Slight greenish tone when new.
Ammoniacal copper zinc arsenate (ACZA) Use class 3.1–4.1	Copper is acutely toxic to fish and other marine life (US EPA 2007) Arsenic Pentoxide: IARC Carcinogen REACH Restricted OSHA Reg. Substance CERCLA Hazardous Substance Copper oxides: US CWA Toxic Pollutants CERCLA Hazardous Substance	50% copper oxide, 25% zinc oxide, and 25% arsenic pentoxide. Utility poles, railroad ties, foundation pilings, structural timbers, and permanent wood foundations. Slight brown to bluish green tone when new. For wood placed in seawater.
Chromated copper arsenate (CCA) Use class 4.1–5	Copper is acutely toxic to fish and other marine life (US EPA 2007) Arsenic Pentoxide: IARC Carcinogen REACH Restricted OSHA Reg. Substance CERCLA Hazardous Substance	47.5% chromium oxide, 18.5% copper oxide, and 34% arsenic pentoxide. Voluntarily removed from commercial and residential construction by the preserved wood industry in 2003. Still used in industrial and class 5 applications.
EL2 Use class 3	Non-metallic wood preservative has rapid degradation in seawater and is toxic to aquatic life (ECHA 2024) 4,5-Dichloro-2-N-octyl-4-isothiazolin-3-one: REACH Annex III.	98% Fungicide 4, 5-dichloro-2-N-octyl-4- isothiazolin-3-one (DCOI) 2% insecticide imidacloprid. Carbon-based wood preservative. Contains an integrated stabilizer to reduce checking and splitting. Used for decking and dimensional lumber.
Propiconazole-tebuconazole-imidacloprid (PTI) Use class 3	Propiconazole: US EPA C Poss. Carcinogen Tebuconazole: US EPA C Poss. Carcinogen	Two fungicides (propiconazole and tebuconazole) and the insecticide imidacloprid. Non-metallic, carbon-based preservative.
Oil-Based Preservatives		
Creosote (CR, CR-S, CR-PS) Use class 3–4	Creosote: IARC Carcinogen OSHA Carcinogen REACH Restricted OSHA Reg. Substance CERCLA Hazardous Substance	Used in telephone poles, railroad ties and marine pilings. Rail ties treated with creosote should not be used to construct raised vegetable beds.
Pentachlorophenol (PCP-A, PCP-C) Use class 4–5	Pentachlorophenol: POP severely restricted IARC Carcinogen CERCLA Hazardous Substance OSHA Reg. Substance	Penta is preservative primarily used in industrial uses. Rail ties treated with penta should not be used to construct raised vegetable beds.
Copper naphthenate (CuN, CuN-W) Use class 3–4	Classified by the EPA as acceptable for consumer use	Used to treat field cuts and holes in pressure-treated wood.

(Continued)

Table 18.12 (Continued)

Pressure Treatment Use Class	Environmental and Human Health Risks[1]	Notes[2]
DCOI Use class 3–4	Non-metallic oil-based wood preservative is toxic to aquatic life (ECHA 2024) 4,5-Dichloro-2-N-octyl-4-isothiazolin-3-one: REACH Annex III	Uses the same active ingredient (4, 5-dichloro-2-N-octyl-4-isothiazolin-3-one) as the waterborne preservative EL2.

[1]US Clean Water Act Toxic Pollutants

US Clean Water Act Priority Pollutants

US Clean Air Act (CAA) Hazardous Air Pollutants (HAPs) List

EU REACH 2024 Restricted Chemical List (Annex XVII)

US Environmental Protection Agency Group C Possible Carcinogen

US Occupational Safety and Health Administration Specifically Regulated Substances

Persistent Organic Pollutant (POP), Stockholm Convention banned or severely restricted substances

[2] Ross 2021

and leave the wood with a dry, stainable surface. Oil borne treatments impart water repellency to the wood and are commonly used in industrial and marine settings.

Pressure treatments are applied in treatment plants where bundles of wood products are placed into large pressure cylinders. A combination of vacuum, pressure, and sometimes heat force the preservative deeply into the wood, although not usually through the entire wood member.

Many waterborne preservatives contain copper as an active ingredient as it is effective against most types of decay fungi and major insect pests. It has low toxicity for humans and other mammals; however, it can negatively impact aquatic organisms as discussed in Table 18.12. Most formulations of copper-based preservatives also contain a co-biocide as some copper tolerant fungi can sporadically cause damage in wood treated with copper.

Selection of the type of pressure treatment depends on how and where the lumber will be used. Based on moisture exposures, the American Wood Protection Association (AWPA) and European Standard Use Classes are shown in Table 18.9.

Treated Wood Waste

Some states prohibit the disposal of treated wood waste (TWW) in landfills due to potential toxicity issues and leaching of dangerous chemicals. While CCA treatment is used only in special uses today, it was the standard treatment for wood preservation in the United States until 2004. As a result, there may be a good chance that an exterior wood structure demolished today was constructed with CCA treated lumber.

CCA treated wood can be hard to identify. If the structure was built before 2004 and it is not a naturally decay-resistant wood like redwood, cedar, or a tropical hardwood, it is probably CCA treated lumber. Other signs of pressure treatment are a green tinge, rot on the inside but not the outside of the lumber, or incising marks where the pressure treatment was pushed into the lumber.

Other treated woods of concern are those with copper naphthenate, creosote, and pentachlorophenol (PCP) treatments. The State of California has categorized all pressure-treated lumber – even the more benign new copper treatments – to be hazardous waste requiring its disposal in hazardous waste facilities (DTSC 2021).

References

Accoya (2020). *Accoya Performance Testing Summary V4.0*. Accoya, July.

Accoya (2024). *Accoya Wood Information Guide V3.9.1*. Accoya.

Agrodome B.V. (2022). *Accoya Wood Environmental Product Declaration*. Agrodome Advice & Development, February 14.

American Wood Protection Association (AWPA) (2024). *2024 AWPA Book of Standards*. AWPA.

Barnes, M.L., et al. (2024). "A century of reforestation reduced anthropogenic warming in the Eastern United States". *Earth's Future* 12.2.

Birdsey, R.A., et al. (2023). "Assessing carbon stocks and accumulation potential of mature forests and larger trees in U.S. federal lands". *Frontiers in Forests and Global Change* 5.

Boulton, C.A., Lenton, T.M., and Boers, N. (2022). "Pronounced loss of Amazon rainforest resilience since the early 2000s". *Nature Climate Change* 12.3: 271–278.

BS EN 350: 2016 (2016). *Durability of Wood and Wood-Based Products – Testing and Classification of the Durability to Biological Agents or Wood and Wood-Based Materials*. European Standard.

BS EN 335: 2013 (2013). *Durability of Wood and Wood-Based Products – Use Classes: Definitions, Application to Solid Wood and Wood-Based Products*. European Standard.

California Department of Toxic Substances Control (DTSC) (2021). *AB-332 Hazardous Waste: Treated Wood Waste: Management Standards*. CA DTSC, August.

Canadell, Josep G., et al. (2023). "Intergovernmental panel on climate change (IPCC): Global carbon and other biogeochemical cycles and feedbacks". In *Climate Change 2021: The Physical Science Basis. Contribution of Working Group I to the Sixth Assessment Report of the Intergovernmental Panel on Climate Change*. Cambridge University Press, 673–816.

Climate Smart Wood Group (CSWG) (2024). "Climate smart forestry: A working definition of climate-smart forestry". www.climatesmartwood.net/forests/, accessed August 24, 2024.

Conniff, R. (2018). *Greenwashed Timber: How Sustainable Forest Certification Has Failed*. Yale Environment 360, February 20.

Earthsight (2021). *FSC Hall of Shame: The Ethical Wood Label's Long Line of Scandals*. Earthsight, October 25.

EN 15804 + A2 (2021). *EN 15804:2012+A2:2019/AC:2021 – Sustainability of Construction Works – Environmental Product Declarations – Core Rules for the Product Category of Construction Products*. EN, June 2.

European Chemicals Agency (ECHA). "Substance infocards". https://echa.europa.eu/substance-information/-/substanceinfo/100.058.930, accessed July 13, 2024.

European Commission (EC) (2024). "Regulation on deforestation-free products". https://environment.ec.europa.eu/topics/forests/deforestation/regulation-deforestation-free-products_en, accessed August 24, 2024.

Finer, M., Ariñez, A., and Mamani, N. (2023). *Amazon Deforestation & Carbon Update for 2023*. MAAP, 201.

Forest Stewardship Council (FSC) (2023). *FSC Principles and Criteria for Forest Stewardship*. FSC.

Forest Stewardship Council (FSC) (2024). "Facts & figures". https://connect.fsc.org/impact/facts-figures, accessed August 23, 2024.

Gatti, L.V., et al. (2021). "Amazonia as a carbon source linked to deforestation and climate change". *Nature* 595: 388–393.

Gerard, J., et al. (2017). *Tropical Timber Atlas*. Editions Quae.

Grace Farms Foundation (2022). *Design for Freedom Toolkit*. New Canaan, CT: Grace Farms Foundation.

Greenpeace Brazil (2017). *Bloodstained Timber: Rural Violence and the Theft of Amazon Timber*. Greenpeace Brazil.

Gutierrez Garzon, A.R., et al. (2023). "A comparative analysis of five forest certification programs". *Forests* 11.

Hardy, C. (2024). *Carbon Dataset Landscape, Version 3*. Sasaki Associates, Inc.

IPBES (2019). *Global Assessment Report on Biodiversity and Ecosystem Services of the Intergovernmental Science-Policy Platform on Biodiversity and Ecosystem Services*. Eds. E. S. Brondizio, J. Settele, S. Díaz, and H. T. Ngo. Bonn, Germany: IPBES secretariat.

International Code Council (ICC) (2021). *2021 Deck Construction Based on the International Residential Code*. ICC, April.

Jordan, A. (2022). *SFI Under Scrutiny for Misleading Sustainability Claims*. Natural Resources Defense Council, December.

Kebony (2024). "Product guidelines". https://us.kebony.com/wp-content/uploads/2024/06/Kebony_Product-Guidelines_V9.pdf, accessed May 8, 2024.

Lacroix, E.M., Petrenko, C.L., and Friedland, A.J. (2016). "Evidence for losses from strongly bound SOM pools after clear cutting in a Northern Hardwood forest". *Soil Science*, 181.5: 202–207.

Lawrence, D., and Vandecar, K. (2015). "Effects of tropical deforestation on climate and agriculture". *Nature Climate Change* 5.1: 27–36.

Lewis, C., and Jordan, A. (2021). *By a Thousand Cuts: How Powerful Companies' Wood Sourcing Is Degrading Canada's Boreal Forest*. R:21-02-A. Natural Resources Defense Council (NRDC).

Lovejoy, T.E., and Nobre, C. (2018). "Amazon tipping point". *Science Advances* 4.2.

Mackey, B., et al. (2024). "Assessing the cumulative impacts of forest management on forest age structure development and woodland caribou habitat in boreal landscapes: A case study from two Canadian Provinces". *Land* 13.6.

Meier, E. (2024). "The wood database 2024". www.wood-database.com/, accessed January 1, 2024.

Noon, M.L., et al. (2022). "Mapping the irrecoverable carbon in earth's ecosystems". *Nature Sustainability* 5: 37–46.

Ometto, J.P., et al. (2022). "Cross chapter paper 7: Tropical forests". In *Climate Change 2022: Impacts, Adaptation and Vulnerability. Contribution of Working Group II to the*

Sixth Assessment Report of the Intergovernmental Panel on Climate Change. IPCC.
Prestemon, J.P., Nepal, P., and Sahoo, K. (2022). "Housing starts and the associated wood products carbon storage by county by shared socioeconomic pathway in the United States". *PLoS One* 17.8.
Ritchie, H. (2021). *Not All Forest Loss Is Equal: What Is the Difference Between Deforestation and Forest Degradation*. Our World in Data, February.
Ross, R. (2021). *Wood Handbook: Wood as an Engineering Material*. FPL-GTR-282. US Department of Agriculture, US Forest Service, Forest Products Lab.
Searchinger, T., et al. (2023). *The Global Land Squeeze: Managing the Growing Competition for Land*. World Resources Institute (WRI).
Thermory (2024). "Thermory: Decking boards and modular systems". https://thermory.com/technical-area/?filter=gallery_category_taxonomy%3A111, accessed August 24, 2024.
UNEP (2022). "Kunming-Montreal global biodiversity framework". *Convention on Biological Diversity*. UN Environment Programme.
UNEP (2023). "Indigenous peoples and the nature they Protect". *Nature Action*. UN Environment Programme, September 19, 2023.
US Bureau of International Labor Affairs (ILAB) (2022). *List of Goods Produced by Child Labor or Forced Labor*. US Department of Labor, ILAB.
US Environmental Protection Agency (US EPA) (2006). *Permethrin Facts*. EPA 738-F-06. US EPA.
US Environmental Protection Agency (US EPA) (2007). *Aquatic Life Ambient Freshwater Quality Criteria - Copper*. EPA 822-R-07. US EPA.
Vergara, A., et al. (2022). *Living Amazon Report 2022*. World Wildlife Fund Amazon Coordination Unit, December.
World Resources Institute (WRI) (2019). *How Much Forest Was Lost in 2019*. WRI, April.
World Resources Institute (WRI) (2024). *Indicators of Forest Extent: Forest Loss*. WRI, April.

CHAPTER 19

Biobased Materials and Products

Prior to the mid-20th century, biological materials were the primary resource for construction materials and consumer products. In 1900, biological materials such as agricultural and forest products were used for about 41% by weight of all material resources. But by 1995 only 8% of the material stream was biobased as industries shifted to highly processed technical materials such as plastics, concrete, and metals (Matos and Wagner 1998).

In 2022, biobased materials accounted for 12% of the materials used in the building industry, and this number is expected to grow. This reflects the beginning of a shift toward use of renewable, biobased materials and products in construction. Mass timber and lumber are replacing some steel and concrete building structures; sheathing and siding are using engineered wood and agricultural waste board; and interior finishes are increasingly biobased (Lecompte and Picandet 2022) (Figures 19.1 and 19.2).

In site construction, biobased materials are used in a surprising variety of products. Some of the more high-profile uses of biobased products are in erosion and sedimentation control, and composite lumber. But biobased content is also used in pavement stabilization, concrete, asphalt, reinforcing bar, deicers, powder coating, paints, stains, sealants, adhesives, cleaners, and more products.

Biobased Defined

The US Department of Agriculture defines biobased products as "derived from raw materials such as plants and other renewable agricultural, marine, and forestry materials" (USDA 2016). Some products will be fully biobased, while others combine biological content with technical content such as adhesives, waterproofing agents, petroleum-based plastics, or synthetic binders (Figure 3.6). While the terms *biobased materials* and *biobased content materials* are distinguished here, the term "biobased product" is commonly used in the construction industry to refer to both.

Biobased materials refer to products that primarily consist of a substance (or substances) derived from renewable biomass. Most biobased materials go through processing to convert the biological materials into products or fuels. They are made from renewable resources and that can be replenished within a relatively short period of time.

Biobased content materials and products are composite products that combine biological material and fossil fuel-derived polymers and/or minerals. Sometimes they are called "biocomposites" but often the biobased content is marketed and the technical content not revealed. Expressed as a percentage, the "biomass content" of the product is divided by the total organic carbon content by weight (ASTM 2022).

Biobased feedstocks come from three main sources:

By-products of agricultural and forestry operations such as straw, slash, rice hulls, and coconut coir are the largest source of feedstocks for biobased materials.

Recycled and reprocessed biobased waste such as cellulose fiber from paper and textile recycling.

Materials that are specifically cultivated specifically for biobased products and plastics include algae, sugarcane, jute, and bamboo. Wood for lumber falls into this category, but it is covered in Chapter 18 "Wood Materials and Products".

Benefits and Impacts of Biobased Materials

Biobased and biobased content products are commonly perceived to be environmentally preferable, and they can be, but not always. Even though they are made at least partially from renewable materials, their biological feedstocks can pose environmental impacts. And biocomposite materials

DOI: 10.4324/9781003413974-21

Figure 19.1 and Figure 19.2 The BioMat Pavilion LightPRO Shell in Stuttgart is a load bearing shell structure constructed from "bamboo-like" 25mm pultruded profiles made of natural flax and hemp fibers. The profiles are bent and lashed together with cables to form the sculptural structure for the fabric covering. The pavilion was designed and constructed by a team at the Institute of Building Structures and Structural Design at the University of Stuttgart. Image credit: Institute of Building Structures and Structural Design

incorporating synthetic chemicals and constituents can pose both environmental and human health impacts. Sustainability claims of biobased materials and products abound; therefore it is critically important to fully understand the constituents of these products and their impacts.

Environmental Benefits of Biobased Materials and Products

Biobased products can offer several environmental benefits with the most beneficial being renewability of at least a portion of their feedstock. Most are rapidly renewable within five years. Some are non-toxic and do not create hazardous waste by-products. Some conserve fossil fuel resources because of their biological feedstock, but they are not always lower embodied carbon.

Biobased products store some of the carbon that was sequestered in plant cell walls until the biobased product decomposes or burns. For a greater discussion of this, refer to the Chapter 2 box "Biogenic Carbon and Carbon Stored in Construction Materials".

The 140 gigatons of by-products from agricultural and forestry operations generated worldwide each year offer ample feedstock for biobased construction materials and products (Tripathi et al. 2019). And the beneficial reuse of this waste preserves carbon stores longer than if the waste decomposed in the field or forest. However, the supply chain for some biomass by-products is limited, supplies are geographically and seasonally variable and processing them into products can be more expensive than use of standard non-renewable feedstocks.

Coordination of the supply chain must be improved to avoid increased carbon emissions from collection of the biomass, treatment, and processing. Additionally, agricultural fields and forests must be managed responsibly to avoid use of fossil fuel-based fertilizers and herbicides and maintain the health of surrounding ecosystems.

Currently the supply of agriculture and forestry residues outweigh both current and projected demand for biobased products and materials, so their end-of-life pathways contribute greenhouse gases to the atmosphere within a short time of harvest. Most of the biomass from agricultural by-products is burned, releasing CO_2 directly into the atmosphere, or it is abandoned on land where it releases greenhouse gases from decomposition. Excess forest biomass residues will also release CO_2 as they decompose, and some can exacerbate wildfires with a buildup of brush (Sahoo et al. 2021).

Environmental Impacts of Biobased Materials and Products

Environmental impacts of biobased materials stem from production of their feedstocks as well as their processing.

Some biobased products, such as bioplastics, are used as substitutes for synthetic compounds and are very highly processed. They may incorporate chemical constituents that are harmful to human health and the environment, and their processing is fossil fuel intensive.

There are multiple environmental impacts associated with agricultural feedstocks. Many are grown using fertilizers, herbicides, and pesticides. Air and water pollution results from fertilizer use, and pesticide application can harm plants and animals around fields. Use of these fossil fuel-based products increases the embodied carbon of a biobased product but are not always included in LCAs.

The area required to produce agricultural biomass can be land consumptive with natural ecosystems removed for cropland. Land is redirected from other functions such as wildlife habitat or waterway protection, and biodiversity is reduced with growth of a single crop. Soil erosion is a consequence of some cropping practices leading to decreased fertility, air pollution, water turbidity, and chemical pollution.

While some biobased products, such as erosion control blankets and fiber rolls, are designed to be biodegradable, most are protected from decomposition by chemical constituents that inhibit decay and insects. These preservatives and binders often prohibit recycling of the product after its service life.

Embodied Carbon Impacts of Biobased Materials

The embodied carbon of biobased materials varies widely and, like most other materials, is partially related to their degree of processing. The embodied carbon of some bioplastics and biocomposites can be higher than some fossil fuel-based plastics. In contrast the embodied carbon of a coir fiber log is far lower than a polypropylene silt fence.

It is challenging to quantify the entire carbon life cycle of biomaterials as it can be nuanced and complex. Product stage (A1–A3) EPDs do not tell the whole story of stored biogenic carbon in these materials. Some EPDs show a negative quantity for stored biogenic carbon below a positive number for GWP emissions from for growing the plant materials, powering processing equipment, transport, and mineral-based components. When "totaled" this can substantially reduce the total GWP or even show a "carbon positive" figure, but it does not account for the biogenic carbon that will be released back to the atmosphere at the end of the product's life if it decomposes or is burned. The standard EN15804 + A2 2019 requires accounting of biogenic carbon storage in the product stages, packaging stage, and the end-of-life stages resulting in a zero-carbon balance for biogenic carbon and true GWP emissions for the actual processing impacts (EN 15804 + A2 2019).

If the product is disposed of in a landfill, the carbon originally removed from the atmosphere and stored in the biomass may be anaerobically degraded and released as methane. Instead of "zero-ing out" with a simple release of carbon stored, this would increase the carbon impacts of the product as methane is 25 times more potent than CO_2.

Last, both direct and indirect land use changes are only accounted for in some EPDs. Direct land use change for biobased materials is the clearing of land to grow the crops. Indirect land use changes occur where a crop displaces another land use causing it to relocate and disturb another area of land. Both of these will impact the carbon footprint of a product.

Challenges to Widespread Use of Biobased Products

With plenty of biomass by-product, the construction material industry is poised to substantially scale up use of biobased construction materials, but there are some challenges that will only be overcome with supportive policies, codes, standards, and financial incentives. Major challenges to the incorporation of waste biomass in construction materials stem from the lack of continuous, year-round supplies, geographic variations of crops, and lack of standard quality. Crops are harvested once or twice per year; therefore, biomass by-products are only produced at those times. Storage of these heavy, bulky biobased materials can be challenging as they will need conditioned storage to not biodegrade.

The availability of biomass by-products varies by geographic location. Coconut coir will be readily available in parts of the world that grow and process coconuts, while corn husks will be available in a completely different geographic location. Resulting biobased products will be somewhat limited by location as well. And while some products can be shipped, some are heavy, and shipping long distances will increase their carbon footprint.

Biomass by-product feedstocks can be of variable quality and the non-standardized nature of biological materials can impact production techniques and quality. The production techniques for many biobased products are expensive and involve complex processing.

Biobased Materials and Products in Site Construction

Current uses of biobased materials and products in site construction vary widely, and they are far more ubiquitous than is obvious. Biobased materials are easily seen in applications such as erosion and sedimentation control products and engineered lumber and decking; but biobased content is also found in aggregate pavement stabilizers, plastics, concrete, cement, asphalt, bricks and blocks, and coatings and adhesives. Biobased compounds in materials like concrete, asphalt, or coatings often only make up a small percentage of the overall product.

Applicability of biobased materials and products to site construction varies by the properties of the product. Some of the most common biobased products used in building construction such as panels made from agricultural residues cannot be used outside as they will be compromised by moisture and sunlight. Other biobased lumber products can be used outdoors but may need to be treated with preservatives, stains, and sealers.

Standards and Policies Supporting Biobased Construction Materials

Policies that financially incentivize biobased building materials have recently been adopted in the United States and Europe with the intent to scale up the market for biobased products and reduce disposal of agricultural and forestry residues. But more policy support is needed along with new standards for specific types of biobased construction products. Other policy incentives and regulations are needed to coordinate residue management, manufacturing, and quality control (Table 19.1).

The USDA administers the single attribute BioPreferred Program to incentivize increased use of biobased products. The program offers Biobased Product Purchasing Requirements that set minimum biobased contents for federal purchasing for hundreds of products from office supplies to building materials (Table 19.2). This program maintains a database of certified biobased products in over one hundred categories.

As biobased materials and products are often biocomposites, all constituents must be identified by percentage. Key issues to consider when specifying these products are as follows:

- Hazard profile of the material or product (Chapter 4)
- Percentage of biobased content
- Origin and production of biobased content
- Environmental impacts of all constituents
- Embodied carbon of the product across the full life cycle (Chapter 5).

Certifications and Ecolabels for Biobased Materials and Products

There are multiple certifications and ecolabels for biobased materials and products. Some certify the feedstocks, while others certify biobased content. Still others certify the end-of-life options for biobased materials (Table 19.3).

For biobased materials to be truly environmentally preferable, they must be sourced from agricultural or forest product wastes or be sustainably grown and produced. A biological material can't be called renewable if it is overharvested or if it depletes soil or plant resources.

Bioplastics and Biocomposites

The feedstocks for bioplastics and some biocomposites are biopolymers. Biopolymers are renewable, organic polymers that are produced by the cells of living organisms.

There are three classifications of biopolymers:

1. Polymers that are *directly extracted or removed from biomass* including polysaccharides and proteins.

Table 19.1 Standards for Biobased Materials

	Standards for Biobased Materials
	Standards for Biodegradable Materials
EN 16785	Bio-based products – Bio-based content – Part 1: Determination of the bio-based content using the radiocarbon analysis and elemental analysis
ASTM D6866	Standard Test Methods for Determining the Biobased Content of Solid, Liquid, and Gaseous Samples Using Radiocarbon Analysis
ISO 22404	Plastics – Determination of the aerobic biodegradation of non-floating materials exposed to marine sediment – Method by analysis of evolved carbon dioxide
ISO/DIS 22403	Plastics – Assessment of the inherent aerobic biodegradability and environmental safety of non-floating materials exposed to marine inocula under laboratory and mesophilic conditions – Test methods and requirements
EN 17033	Biodegradable mulch films for use in agriculture and horticulture – Requirements and test methods
EN 13432	Standard for biodegradable and compostable packaging

Table 19.2 USDA Minimum Biobased Contents for Materials Used in Site Construction

Use in Site Construction	Product Categories	Minimum Biobased Content[1]
Adhesives, sealants, paints, stains, and coatings are made wholly or partially from vegetable oils and natural resins. The hazard profile of the material or product should be considered as they can contain hazardous chemicals. VOC levels should not exceed the State of California, Air Resources Board, VOC Limits (California ARB 2021).	Adhesives	24%
	Blast media	94%
	Epoxy systems	23%
	Exterior paints and coatings	83%
	Graffiti and grease removers	34%
	Metal cleaners	56%
	Powder coatings	34%
	Paint removers	41%
	Traffic and zone marking paints	30%
	Wood and concrete sealers – membrane concrete sealers	11%
	Wood and concrete sealers – penetrating liquids	79%
	Wood and concrete stains	39%
Wall and pavement materials A wide variety of biological materials are used as stabilizers or reinforcing in concrete and earth-based construction materials. Usually, these materials are selected and incorporated based on their local availability. They are often by-products of other processing. Other biobased wall and pavement materials are "grown" by microorganisms, such as algae of fungi. Still other materials are made from bioplastics.	Fiber reinforcing in concrete	Not specified
	Blocks and bricks wholly or partially from biobased materials such as mycelium, algae, cork, and food waste	Not specified
	Stabilizers and reinforcing for earth-based structures	Not specified
	Biobased asphaltic binders	Not specified
	Lightweight aggregate in concrete	Not specified
Pavement products Biobased pavement products are produced from vegetable oils, lignin, and other biological constituents. The hazard profile of the material or product should be considered. There are several very effective biobased aggregate pavement stabilizers available (Table 6.7).	Asphalt restorers	68%
	Asphalt and tar removers	80%
	Concrete curing agents	59%
	Concrete and asphalt cleaners	70%
	Concrete and asphalt release fluids	87%
	De-icers	93%
	Dust suppressants (and stabilizers)	85%
	Masonry and paving systems	25%
	Playground and athletic surface materials	22%
Erosion control products are a large area of minimally processed biobased materials for site construction. Many products incorporating jute, coir, straw, and cellulose will biodegrade over time, leaving nutrients in the soil.	Erosion control materials	77%
Intermediates are materials that are used in the production of another product that is sold to consumers. Most biobased intermediates for site construction materials are bioplastics and natural rubbers.	Intermediates – binders	47%
	Intermediates – plastic resins	22%
	Intermediates – rubber materials	96%
Engineered lumber and decking products combine biobased fibers and a binder that may be biobased or synthetic. Binder materials should be identified and hazard profiles checked. Many of these products are used in interior building applications, so if the product will be used outside it should be rated for exterior use.	Composite panels – plastic lumber	23%
	Lumber, engineered wood products	25%
Miscellaneous	Mulch and compost materials	95%
	Trash cans and waste receptacles	87%
	Synthetic turf with plant-based infill	Not specified

[1] USDA 2024

Table 19.3 Selected Certifications and Ecolabels for Biobased Products

	Country	Certification or Ecolabel
Sustainable feedstocks	International	Forest Stewardship Council, more information in Chapter 18
	International, in US it is SFI	Program for the Endorsement of Forest Certification (PEFC), more information in Chapter 18
	International	International Sustainability and Carbon Certification (ISCC); Biomaterials and Bio-circular Materials
	International	Roundtable on Sustainable Biomaterials (RSB)
	EU	REDCert
	EU	Better Biomass
	Scandinavia	Nordic Swan Ecolabel
	North America	USDA Organic label
	International	International Federation of Organic Agriculture Movements
	EU	European Union Organic logo
Bio-based content	North America	USDA Certified Biobased Product Label, BioPrefered Program
	North America	UL 9798 Bio-based Product Content Verification
	North America	US EPA Safer Choice Label
	North America	Green Circle Certification for Biobased Content
	North America	Sustainable Biomaterials Collaborative
	EU	EU Biobased Content Certification Scheme
	EU	EU Ecolabel
	EU, based in Belgium	OK Biobased by Vincotte
	EU, based in Germany	Din-Geprüft Biobased

2. Polymers that start with *biobased monomers* such as polylactide (PLA) and are produced by chemical synthesis.
3. Polymers *produced by microorganisms* or genetically modified bacteria including bacterial cellulose and polyhydroxyalkanoates.

Biodegradable polymers are capable of clean biodegradation by microorganisms into simple molecules such as carbon dioxide, water, or methane. In industrial composting processes, they will break down by 90% in six months. Under European Standards (EN 13432) they are marked with a "compostable" symbol.

Bioplastics

Bioplastics are made from biopolymers produced from renewable biomass sources such as sawdust, woodchips, straw, corn starch, food waste, and vegetable fats and oils. Some are obtained from processing biopolymers such as starch, cellulose, chitosan, soy protein, gluten, and alginate. Others are chemically synthesized from lipids from plant or animals and sugar derivatives.

Upstream impacts may be lower than petroleum-based plastics as they are renewable, and they avoid fossil fuel feedstocks. Some will have a lower carbon footprint than petroleum-based plastics, but not always.

Currently about 52% of bioplastics such as PLA, polybutylene succinate and polyhydroxyalkanoates are technically biodegradable, while about 48% of others with a different molecular structure are durable. Durable bioplastics include Bio-PET and biopolyethylene (Bio-PE). Bio-PE can be recycled in most regions as it is almost chemically identical to fossil fuel based polyethylenes (Figure 19.3).

Environmental and Human Health Impacts of Bioplastics

Bioplastics have been marketed as a "green" alternative to fossil fuel-based plastics, but research shows serious downstream environmental and human health impacts. Biodegradable bioplastics that end up in landfills will not break down if they don't have access to oxygen. They can last for centuries and slowly release methane (Robbins 2020).

Currently, bioplastic alternatives exist for nearly all conventional plastics (PP, PET, PE, PVC). And while they use renewable feedstocks, they often share similar

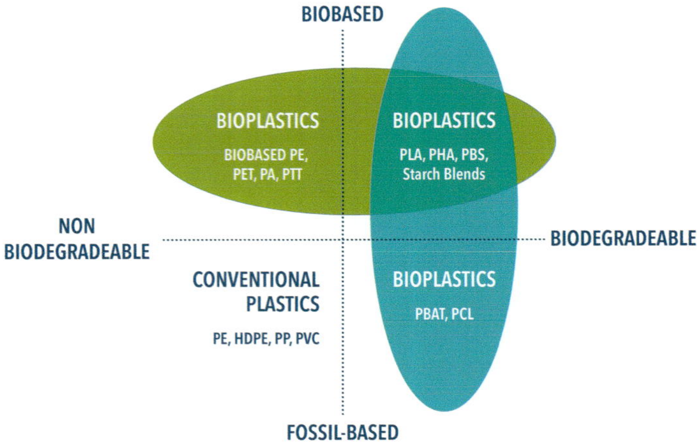

Figure 19.3 Bioplastics are classified based on their biodegradability and their feedstock (biobased vs. fossil-based). Adapted from European Bioplastics 2016. Image credit: Meg Calkins

environmental and human health impacts with their conventional plastic counterparts. The end products of most bioplastics contain some of the same synthetic chemicals as petroleum-based plastics.

Additional studies of the health and environmental impacts of bioplastics are urgently needed. A 2020 study of bioplastics found that most bioplastics contain toxic chemicals. Two-thirds of 43 bioplastics studied were found to induce baseline toxicity, and 42% induced oxidative stress, an imbalance of free radicals and antioxidants in the body (Zimmermann et al. 2020).

A 2010 study found that bioplastics production resulted in greater amounts of pollutants released due to both the chemical inputs to grow the crops and the chemical processing required to convert the organic material to plastic. B-PET was found to pose the greatest toxic effects to ecosystems and release the most carcinogens (Tabone et al. 2010).

Use and Growth in the Bioplastics Industry

The global bioplastics industry accounted for only 0.5% of the 400 million metric tons of plastics produced in 2023, but the bioplastics market is projected to grow to 7.43 million metric tons in 2028 (European Bioplastics 2023). Polyethylene furanoate (PEF), the biobased counterpart to fossil fuel-based PET, is expected to grow 163% by 2030 (Ghi et al. 2022).

Most bioplastics are used in packaging and fibers with only around 14,000 metric tons used in construction materials. In the construction industry, bioplastics are used in

- concrete mixtures and dry premixed mortars
- cement substitutes
- biobased manufactured lightweight aggregates
- paints, coatings, and stucco
- stabilizers in earthen construction
- materials and reinforcing for 3D printing
- geotextiles
- lighting and furnishings.

"Growing" bioplastics directly from organic waste is a promising new manufacturing technology. Organisms such as algae, fungi, and microorganisms can process waste biomass by transforming it into bioplastic. Fungi mycelia grow by consuming the lignocellulosic content of agricultural waste. The cellulose, both food and structure, solidifies within a week, uniting the smaller portions of the tissue and hardening into a form that is light and strong. Bio bricks are a product that results from this process. They exhibit properties such as good fire resistance, compressive and flexural strength, and insulation (Anjaly et al. 2023).

Biocomposites

Biocomposites are made up of two or more materials where at least one is naturally derived (Figures 19.4 and 19.5). They are currently more common in the construction industry than bioplastics, although this could change with the increasing emphasis on phasing out use of fossil fuels. In the site construction industry, biocomposites are used in

- lumber
- decking
- fencing
- concrete reinforcing bar
- fiber reinforced concrete
- asphalt pavement
- cisterns
- fiber reinforced plastic products.

Figure 19.4 and Figure 19.5 The Arbo Skin Pavilion in Stuttgart demonstrates the capabilities of a new bioplastic called Arboblend containing over 90% renewable materials, primarily lignin and other reinforcing fibers, developed for the construction industry. The Pavilion was designed and constructed by a team from the Institute of Building Structures and Structural Design (ITKE). The pyramidal modules were produced through the extrusion of bioplastic granules into sheets that were then thermoformed into shapes. CNC milling removed some sections creating openings in the façade. Image credit: Institute of Building Structures and Structural Design

A principle for the circular economy involves avoidance of mixing biological nutrients with technical nutrients within a material or product to allow for recycling (Chapter 3). It is preferable for biocomposites to mix one biobased material with another rather than to mix a biobased material with a synthetic material.

Completely biobased composites are biopolymers such as starch plastics, polylactides (PLA), soy plastics, cellulosic plastics, or microbial produced polymers reinforced with natural fibers such as flax, hemp, or kenaf. These are usually biodegradable.

Other biocomposites are a mix of biobased material with fossil fuel-based plastics or other non-biological material. These mixed biological and technical materials are not biodegradable and are often not even recyclable.

In 2022, 12.7 million metric tons of biocomposites were produced worldwide with 50% of this production in Asia, and around 22% in both the United States and Europe. The construction industry was the second largest consumer of biocomposites behind the auto industry. Exponential growth is expected in the biocomposites industry with the market size predicted to triple by 2030 (Polaris Market Research 2022).

The costs of bioplastic and biocomposite products can be higher, but some prices are coming down with increased demand. Factoring in the intangible impacts of climate change and the associated health and weather impacts, fossil fuel-based plastics are costing our planet. The United Nations Environment Program (UNEP) estimates that petroleum-based plastic production costs the world $75 billion in natural capital each year. These impacts stem from greenhouse gas emissions, pollution of the marine environment, and air pollution from incineration (Valuing Plastic 2014).

Natural Fiber Products in Site Construction

Biobased products made from natural fibers can replace some synthetic plastic products used in site construction. Some are cultivated specifically for products such as jute, while others are by-products of agricultural processing such as coir and straw (Tables 19.4 and 19.5; Figures 19.6–19.8).

Some natural fiber products are minimally processed and can be left to decompose in place. For instance, jute, coir, and straw are used in erosion control matting, blankets, and fiber rolls that are intended to eventually decompose by the time plants are established enough to control erosion. As these products biodegrade, they provide nutrients and organic matter to the soil.

Minimally processed natural fiber products generally have a lower embodied carbon than their synthetic counterparts. For instance, polypropylene erosion control fabric has an embodied carbon over ten times that of jute (National Jute Board 2024). The embodied carbon from the jute and coir life cycle comes from their transport as they are primarily produced in Southeast Asia and South America.

Table 19.4 Natural Fibers Used in Site Construction Materials

Natural Fiber and Environmental Impacts	Use in Construction Products
Hemp is a type of cannabis plant with very low levels of psychoactive THC. It grows very quickly and can capture atmospheric carbon twice as effectively as forests at 8–15 metric tons of CO_2 per hectare. Forests typically capture 2–6 metric tons per hectare depending on the type of forest. Hemp farms produce more usable fibers per acre than forestry (Müssig et al. 2020).	The woody inner part of the stem is used to make erosion control blankets. Hemp fibers are needle punched into a biodegradable mat made from rayon or other natural fibers. Look for blankets that are 100% biodegradable. The stiff fibers on the outside of the stem are used to produce bioplastic products such as cladding panels. Hempcrete is another use for hemp fibers. When the hemp is mixed with lime, water, and sand, it creates a material that can be precast or formed for non-structural applications. Hempcrete can absorb and release moisture, but it is vulnerable to microorganisms, so it should be plastered if used in exterior settings.
Jute is a long, rough fiber collected from the phloem of dicot plants. It is called bast fiber and is composed primarily of cellulose and lignin. Jute must be grown in very moist environments with warm humid climates and annual monsoon seasons. Jute stalks are soaked in water for several days to separate the fibers from the stalk. This can impact water quality by increasing organic loads in rivers, lakes, and estuaries. Chemical fertilizers and pesticides are not used as often to produce jute as they are for other agricultural crops. One hectare of jute will absorb up to 15 metric tons of CO_2 (National Jute Board 2024).	Jute can be spun into coarse, strong threads, and it is the second most common fiber after cotton. Most of the jute trade is in South Asia, so jute and jute products will likely have a carbon footprint from their transport to North America or Europe. Jute is commonly used in temporary erosion control matting and blankets. Sometimes it is called jute netting. It is a woven mesh made from spun mesh yarns. It reduces soil moisture loss due to its ability to absorb up to 450% of its weight. It is usually a very loose weave (70–75% open) that allows for plant growth. Most jute matting can be used on slopes up to 3:1. It will biodegrade over time. Jute fibers are added to concrete to increase its strength and resist cracking, and they are used as a stabilizer in asphalt pavement. Jute fibers are used in earth-based masonry applications such as rammed earth, compressed earth block, and sun dried brick. Jute is also used in composite panels and building products.
Coir, also called coconut fiber, is extracted from the outer husk of coconuts. The fibers are found between the hard internal shell and the outer coat. Coir fiber is extremely water resistant, but this means it must be soaked in freshwater for several months prior to processing. This can impact water quality from the sodium and nutrient laden wastewater. Other techniques of processing employed to speed the process use higher amounts of energy.	Most coir and coir products are produced in Southeast Asia and South America. In site construction, its primary use is in geotextiles such as erosion control matting and blankets. Like jute, it retains moisture and facilitates seed germination. It will biodegrade in a few years, but it will take more time than jute or straw products. Coir is used as a stabilizer in bricks, concrete, earth masonry, and asphalt. It is also used in wall board and flooring.
Straw is an agricultural by-product from the growth and processing of grain crops such as barley, oat, rye, and wheat. Straw sequesters a lot of carbon, but it decomposes relatively quickly. Straw is readily available in North America and Europe.	Straw is used in several construction materials. Straw is used in whole form as fiber reinforcing in concrete (rarely), earthen walls and earth and lime plaster. Straw bales are used in sediment control and straw fibers are used in erosion control matting and rolls. Compared to coir, straw decomposes in a much shorter time. Straw is used in many applications in building construction such as straw bale construction and agriboards.
Rice hulls are the hard outer shells of rice. They contain around 20% opaline silica combined with lignin and cutin. As most rice is produced in Asia, rice hulls or the products made from them will need to be transported for use in North America.	Rice hulls are resistant to moisture penetration and decomposition, making them appropriate in products for exterior use. They are combined with a binder material and used in engineered lumber and decking products. At least one product line uses PVC as the binder material. Use considerations, care, and maintenance of these products is discussed in Tables 10.8 and 10.11 in Chapter 10. Rice hulls are also used to produce lightweight concrete as they have high silica content.
Wood fibers and chips are a by-product of the lumber and demolition industries.	In site construction, wood waste is incorporated whole into mulches for planting beds, playgrounds, and pathway surfaces. It is also pressed into a wide variety of engineered wood products. Exterior grade MDF, plywood, fiberboard, and other engineered lumber products are available.
Bamboo fibers are produced by stripping and shredding bamboo culms. Bamboo characteristics and performance are discussed later in this chapter.	Bamboo fibers are used in fused bamboo decking and lumber products. Use considerations, care and maintenance of these products is discussed in Tables 10.8 and 10.11 in Chapter 10.

Table 19.5 Other Biobased Material Feedstocks for Construction Materials

Biobased Material and Environmental Impacts	Use in Construction Products
Algae are a large, diverse group of organisms that live in water. Algae, including kelp and seaweed, are estimated to contribute about 50% of the CO_2 removal and storage on earth (Fujita et al. 2022). Algae grow 10–50 times more quickly than plants.	Algae are used in the production of bioplastics. Algae are used to produce a biogenic limestone that is substituted for quarried limestone traditionally used to make cement. This bio-cement is produced by biomineralizing cyanobacteria grown using seawater, sunlight, and CO_2. The process produces biogenic limestone (calcium carbonate) that is burned to produce cement. Producers estimate that it is carbon neutral because it releases carbon sequestered by the algae, and if some of the biogenic limestone is added directly to the cement without burning it, it could be carbon positive (Living Materials Laboratory 2024). Bio-cement and aggregate are combined in a biocomposite to form precast concrete products such as wall blocks, interlocking retaining wall blocks, and pavers.
Mycelium is a network of fungal threads called hyphae. The rootlike hyphae feed on low-grade agricultural waste and grow very quickly. Mycelium can be grown into a water, mold, and fire-resistant building material in many forms. It can be grown in a mold to form blocks or sheet material taking around five days to fully form. It is also toxic to termites. Production of mycelium uses energy and has associated carbon impacts. To obtain the biomass, it must have high levels of refrigeration and drying (Figures 19.9 and 19.10).	Mycelium is used in packaging applications as an alternative to polystyrene. It is used in composite boards and blocks. Its compressive strength is only 0.207MPa, so it is not ready to be a structural material, but it is stronger than concrete relative to its weight (Bonnefin 2022). Because mycelium products are biodegradable, they are usually only appropriate for temporary applications in site construction.

Figure 19.6 and Figure 19.7 The livMatS Pavilion in the Botanic Garden at the University of Freiburg is a load bearing structure made from robotically wound flax fiber. The structure was designed and fabricated by a team from the Institute of Building Structures and Structural Design at the University of Stuttgart in collaboration with biologists from the Cluster of Excellence "Living, Adaptive and Energy-autonomous Material Systems (livMatS)" at the University of Freiburg. Advanced robotic fabrication techniques were employed to weave the latticework of the structure. Image credit: Institute of Building Structures and Structural Design

Natural fibers are used in earth-based construction as stabilizers and reinforcing. These fibers are discussed in Table 16.5.

Natural fibers are also incorporated in lumber, board, and decking products. These are technically biocomposite materials as binder materials such as glues are employed. These binders can be synthetic or biobased. Some of these products are marketed as green products because of their biobased content, but it is important to identify the binder material as well. At least one natural fiber composite lumber product uses PVC to bind the rice hulls.

Bamboo

Bamboo is used in site construction in whole form for fencing and other structures; and bamboo fibers are incorporated into engineered lumber and decking products. As the fastest growing plant in the world, it is a grass that can be harvested every 1–5 years depending on the species.

Figure 19.8 and Figure 19.9 The BIOPOD Team from Rhode Island School of Design (RISD) created tessellated floating plant beds made from a mycelium-based material intended to cleanse water bodies though bioremediation. The team fabricated the Reishi mycelium BIOPODs by tightly packing mycelium into a hexagonal plexiglass mold. Placed in the Providence River to create "miniaturized wetlands", the mats will slowly degrade over time to release plants into the surrounding waterway. The plant roots, microbes on the roots, and the mushrooms will break down some pollutants in the river. Image credit: Avantika Velho

Bamboo forests comprise about 3% of all global forested land, mostly in China, India, Japan, Indonesia, Thailand, and the Himalayas. They sequester about 1.46 times the carbon of fir forests and 1.33 times that of tropical forests (UNEP 2023).

Bamboo is farmed on forest plantations in Asia, Africa, and South America, with a small amount of production in North America. Most bamboo and bamboo products used in North America are imported from China and East Africa.

Around 70% of bamboo is harvested from natural forests and 30% is grown on forest plantations. The sustainable harvest of bamboo, even from natural forests, is quite different from the harvest of timber. Only 20–25% of the stalks and culms are harvested each year, and bamboo forest health does not decline as easily as that of the boreal or rainforest ecosystems that are harvested for timber (Beena and Seethalakshmi 2011).

Pole Bamboo in Construction

Bamboo poles, called culms, have been used for millennia to produce light flexible structures, and the material is enjoying a resurgence with the greater emphasis on biobased materials in the building fields. Bamboo poles can be used in walls, roofing, flooring, and fencing.

Bamboo has good tensile strength, compressive strength, and elasticity making it useful in structural applications such as columns, beams, and framing. Pole bamboo construction offers particularly good performance in seismic zones as it has flexibility and resilience. Basic bamboo connections and considerations for pole bamboo in site structures are discussed in Chapter 11 and shown in Figures 11.26 and 11.27.

Treatment and Protection of Bamboo Poles

Bamboo weathers similarly to wood, so when it is used in exterior settings it must be treated. It is also vulnerable to termites, insects, and fungal attacks. Mature culms (3–5 years old) must be used because the sugar levels are stabilized and the culm is strong.

Bamboo poles are typically treated by immersing them in a solution of borax, boric acid, and water. Borax is sodium borate or tetraborate. The borate solution used is about 3% borax, 2% boric acid, and 95% freshwater (Farrugia and Goutham 2022).

Borax and boric acid can pose health impacts if ingested or inhaled in large amounts. The US EPA classifies boric acid as moderately toxic and sodium tetraborate as acutely toxic. It may disrupt hormones and harm the male reproductive system (US EPA 1993). In the environment, borax can damage broadleaf plants. Grasses are more tolerant of exposure to borax. The impact of borax on aquatic species is not well studied.

The three methods of borax treatment of bamboo culms are as follows:

Full immersion submerges the entire pole in the borax treatment solution. It soaks into the bamboo tissues completely. Prior to submersion, all of the nodes of the bamboo are pierced to allow the solution to enter the inside of the culm.

Cold soak immersion uses an open tank system to soak the poles for 7–14 days in a warm environment or even longer in a cold environment.

Hot soak immersion heats the soak to encourage the borates to enter the woody mass more quickly. The bamboo soaks for 24 hours or even less time with higher temperatures.

Other treatments of bamboo include the following:

Heating and smoking bamboo poles reduces the moisture content and modifies the sugars. This approach is similar to thermal modification for lumber and could be scaled for larger production of bamboo poles for construction.

Water leaching submerges the bamboo in clean water for 4–8 months slowly washing away the sugars and starches.

Fermentation buries the bamboo in mud and leaves for 3–4 months. Bacteria and microorganisms convert the starches and sugars into acids. This method is inconsistent and may result in only partially treated bamboo.

Saltwater immersion of bamboo for one to three months allows the sugars to absorb the salt solution and release sugary sap.

Engineered Bamboo Construction Products

Engineered bamboo products for site construction include lumber and decking. Mechanical performance of these products can be comparable to tropical hardwoods. Use considerations, care and maintenance of these products is discussed in Tables 10.8 and 10.11.

Engineered bamboo products have embodied carbon that is similar to steel products, although the carbon stored can offset the carbon impacts of its production at least until it decomposes. Bamboo decking product binders add to the product's carbon footprint. Many of these products are imported from China, further increasing the embodied carbon for their transportation.

Binders for engineered bamboo products are typically formaldehyde-based resins, which can pose health impacts during manufacture and use. One formaldehyde-free synthetic resin that is sometimes used is EPI (emulsion polymer isocyanate) (Adhesives + Coatings 2024).

Treatment of engineered bamboo products varies widely. In many cases, the binder protects the bamboo fibers from insect and fungal attack. In other cases, hydrophobic coatings are used that may contain harmful chemicals or VOCs. Other products treat the bamboo fibers with preservatives prior to mixing them with the binder.

These practices can pose health and environmental hazards and increase the carbon footprint of engineered bamboo products. While the embodied carbon of the products will be provided in product EPDs, the health hazards will not be enumerated. It is very difficult to obtain information about the binders used as many are proprietary.

References

Adhesives + Coatings (2024). "EPI glue for wood". www.adhesivesandcoatings.com/, accessed March 31, 2024.

Anjaly, P.T., et al. (2023). "Synthesis and commercialization of bioplastics: Organic waste as a sustainable feedstock". *Science of the Total Environment* 904: 167243. ISSN 0048-9697. https://doi.org/10.1016/j.scitotenv.2023.167243.

ASTM Standard D6866 (2022). *Standard Test Methods for Determining the Biobased Content of Solid, Liquid, and Gaseous Samples Using Radiocarbon Analysis*. West Conshohocken, PA: ASTM International. www.astm.org.

Beena, V.B., and Seethalakshmi, K.K. (2011). *Reproductive Biology and Biochemical Changes Associated with Flowering of Dendrocalamus Stocksii and Ochlandra Travancorica*. Diss. Kerala Forest Research Institute.

Bonnefin, I. (2022). "Emerging materials: Mycelium brick". *Certified Energy*. www.certifiedenergy.com.au/emerging-materials/.

California Air Resources Board (ARB). (2021). *Summary of Architectural Coating Rules in California: VOC Limits*. State of California, ARB, September 2.

EcoChain (2024). "Does 'biobased' always mean 'more sustainable'?". https://ecochain.com/blog/, accessed July 4, 2024.

EN 15804 + A2 (2021). *Sustainability of Construction Works - Environmental Product Declarations - Core Rules for the Product Category of Construction Products*. EN 15804:2012+A2:2019/AC:2021. EN, June 2.

European Bioplastics (2016). *Fact Sheet: What Are Bioplastics?* European Bioplastics, January.

European Bioplastics (2023). *Bioplastics Market Development Update 2023*. European Bioplastics. www.european-bioplastics.org/news/publications/.

Fairs, M. (2021). "Hemp 'more effective than trees' at sequestering carbon says Cambridge researcher. *Dezeen*.

Farrugia, M., and Goutham, S. (2022). *The 3 Stages for Turning Bamboo into a Reliable Construction Material*. Bamboo University, March 19.

Fujita, R.M., et al. (2022). *Carbon Sequestration by Seaweed: Background Paper for the Bezos Earth Fund*. EDF Workshop on Seaweed Carbon Sequestration. New York: Environmental Defense Fund (EDF).

Ghi, T., et al. (2022). *Why the Bio-Based Materials Market Is Finally Poised for Growth*. Arthur D. Little, September.

Keena, N., Duwyn, J., and Dyson, A. (2022). *Biomaterials Supporting the Transition to a Circular Built Environment in the Global South*. New Haven: Yale Center for Ecosystems + Architecture and United Nations Environment Programme. https://globalabc.org/resources/publications/.

Keena, N., et al. (2022). "A life-cycle approach to investigate the potential of novel biobased construction materials toward a circular built environment". *Energies* 15.19. https://doi.org/10.3390/en15197239.

Lecompte, T., and Picandet, V. (2022). *Bio-Based Materials Improve the Comfort and Carbon Footprint of Buildings*. Polytechnique Insights. Institut Polytechnique de Paris. www.polytechnique-insights.com/en, accessed October 4, 2022.

Living Materials Laboratory (2024). *Living Materials Laboratory: Projects*. Boulder: University of Colorado. https://livingmaterialslab.org/, accessed July 4, 2024.

Matos, G.R., and Wagner, L.A. (1998). "Consumption of materials in the United States, 1900–1995". *Annual Review of Energy and the Environment* 23: 107–122.

Müssig, J., et al. (2020). "Transdisciplinary top-down review of hemp fibre composites: From an advanced product design to crop variety selection". *Composites. Part C: Open Access* 2: 100010, ISSN 2666-6820. https://doi.org/10.1016/j.jcomc.2020.100010.

National Jute Board (2024). *Jute Geotextiles and Green Jute*. Government of India, Ministry of Textiles, National Jute Board. www.jute.com/web/guest/green-jute, accessed March 31, 2024.

Polaris Market Research (2022). "Biocomposites market share, size, trends, industry analysis report, by fiber type; by polymer type (synthetic, natural); by product (hybrid, green); by end-use; by region; segment forecast, 2022–2030". *Polaris Market Research PM2412*.

Robbins, J. (2020). *Why Bioplastics Will Not Solve the World's Plastic Problem*. Yale Environment 360, Yale School of the Environment, August 31.

Sahoo, K., et al. (2021). "Life-cycle assessment and techno-economic analysis of biochar produced from forest residues using portable systems". *The International Journal of Life Cycle Assessment* 26: 189–213. https://doi.org/10.1007/s11367-020-01830-9.

Tabone, M.D., et al. (2010). "Sustainability metrics: Life cycle assessment and green design in polymers". *Environmental Science & Technology* 44.21: 8264–8269.

Tripathi, N., et al. (2019). "Biomass waste utilisation in low-carbon products: Harnessing a major potential resource". *NPJ Climate and Atmospheric Science* 2.1: 1–10. https://doi.org/10.1038/s41612-019-0093-5.

United Nations Environment Programme (UNEP) (2014). *Valuing Plastic: The Business Case for Measuring, Managing, and Disclosing Plastic Use in the Consumer Goods Industry*. UNEP.

United Nations Environment Programme (UNEP) (2023). *Building Materials and the Climate: Constructing a New Future*. Nairobi: UNEP.

US Department of Agriculture (USDA) (2016). *Fact Sheet: Overview of USDA's BioPreferred Program*. USDA Press Release No. 0047.16. USDA.

US Department of Agriculture (USDA) (2024). "BioPreferred product categories". www.biopreferred.gov/, accessed June 16, 2024.

US Environmental Protection Agency (US EPA) (1993). *Prevention, Pesticides and Toxic Substances: Boric Acid*. US EPA-738-F-93-006. USDA.

Zimmermann, L., et al. (2020). "Are bioplastics and plant-based materials safer than conventional plastics? In vitro toxicity and chemical composition". *Environment International* 145 (December): 106066. https://doi.org/10.1016/j.envint.2020.106066.

CHAPTER 20

Metals and Metal Products

Metals play an important role in site construction from reinforcing concrete, to guardrails, and long span overhead structures. Metals are distinguished by their strength-to-size ratio, durability, and versatility. Yet they can be among the most carbon and resource intensive materials responsible for 8-9% of global greenhouse gas emissions each year (Climate Watch 2023).

Two billion metric tons of metals are produced each year using 10% of global energy consumption and 3.2 billion metric tons of minerals (Raabe 2023). Additionally, several billion metric tons of soil and rock result as rock-to-metal ratios (RMR) vary widely from 9 metric tons of material to produce one metric ton of steel, to 125 metric tons of material removed to produce one metric ton of copper (Nassar et al. 2022).

There are multiple actions that can be taken to reduce the environmental and human health impacts of metals in site construction. These include detailing durable metal structures with connections that allow for deconstruction, optimizing the size and shape of metal members to use less material, and specifying high recycled content metals that are produced regionally with lower carbon manufacturing processes.

Metal types, products, and finishes are discussed along with detailing considerations for durable metal structures in this chapter (Figure 20.1). Tradeoffs to reduce embodied carbon and health burdens are addressed as well. Metal structures such as rails, screens, and fences are discussed in Chapter 11.

Types of Metals in Site Construction

Steels and aluminums are the two most common types of metals used in site construction. These metals are rarely used in their pure form, instead additional metals, called alloys, are added to optimize properties of the base metal. The most common alloys are nickel and chromium that are added to mild steel to create corrosion resistant stainless steel.

Metals fall into two general categories:

Ferrous metals are iron alloys including construction steel, weathering steel, stainless steel, cast iron, and wrought iron. Ferrous metals are characterized by their strength and low rate of thermal expansion.

Non-ferrous metals used in site construction include aluminum, copper, bronze, titanium, and zinc. They have a higher rate of expansion and can change shape slightly after construction.

Steel

Steel is an iron alloy with 2% or less carbon (Table 20.1). It is the most used metal in site construction as it is strong, durable, and relatively inexpensive compared with other metals. It is characterized by its ability to form high strength and relatively thin structures. Nearly all steel contains alloying elements to impart various properties. The product range of steels is broad from structural I-beams to perforated sheets and wire rope. Custom steel shapes and casts are also possible. Types of steel used in site construction include stainless steel, weathering steel, and mild steel with a protective finish.

Mild Steel

Mild steel, also called low-carbon steel, has a carbon content of 0.05–0.25%. This leads to an increase in strength, ductility, and ease of forming and welding. Construction steel, also called structural steel, is a type of mild steel with small amounts of manganese, phosphorous max, sulfur max, and silicon used in applications that require high tensile strength. Construction steel is available in several grades, the two most common of which are detailed in ASTM Standards A36 and A572. Construction steel can't be used

DOI: 10.4324/9781003413974-22

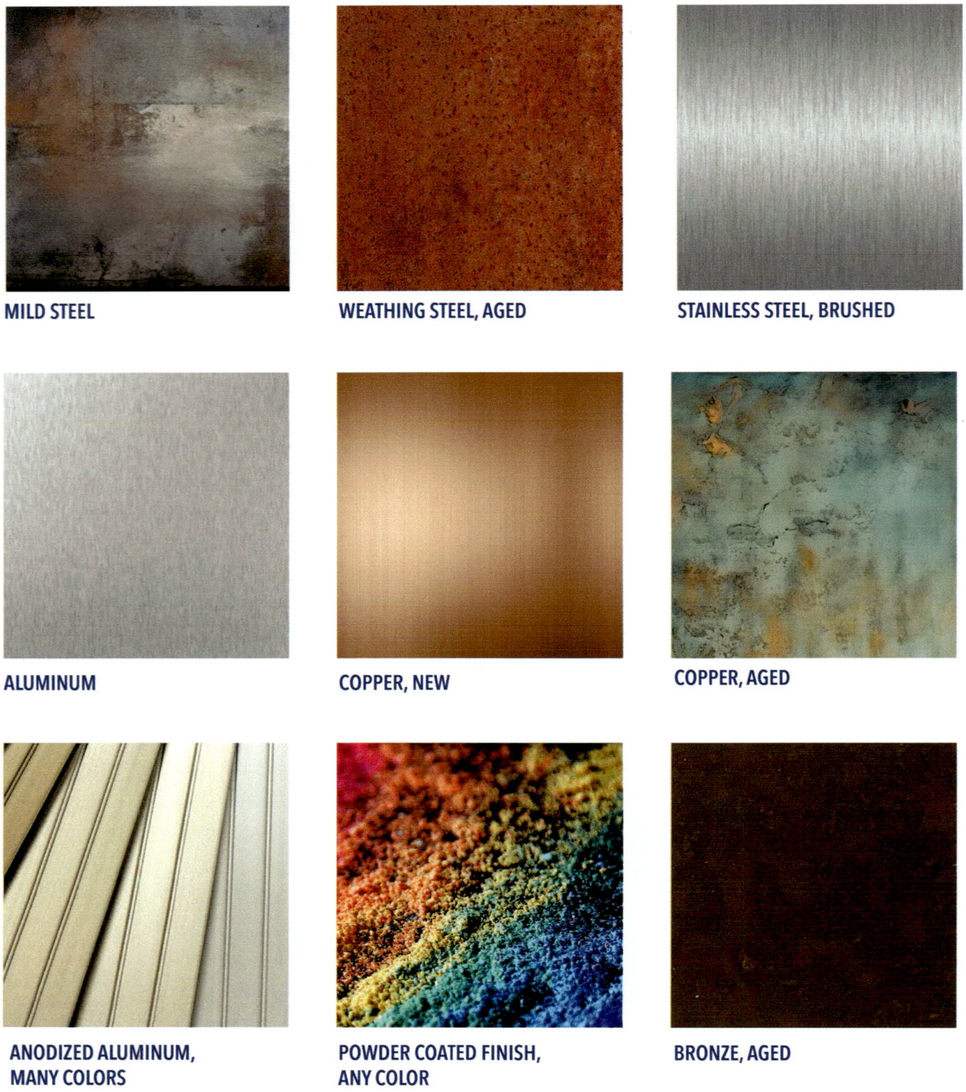

Figure 20.1 Some typical metals and metal finishes used in site construction. Image credit: Meg Calkins

Table 20.1 Hot Rolled Steel vs. Cold Rolled Steel

	Hot Rolled (HR) Steel	Cold Rolled (CR) Steel
Production	Steel is rolled at high temperatures allowing it to be easily shaped and formed. As it cools, it shrinks slightly with less control over the uniformity as compared to cold rolled steel.	Cold rolled steel is hot rolled steel that has been further processed by being allowed to cool then annealed and temper rolled. Closer dimensional tolerances are possible.
Strength	Very strong	20% stronger than HR
Finish	Scaly finish that can be removed by pickling, grinding, or sandblasting.	More finished surface
Malleability	More malleable than CR	Harder than HR
Cost	Less expensive	More expensive
Applications	Structural components	Structural components, decking, and framing
Types of steel	Construction steel, SS, weathering steel	Construction steel, SS, weathering steel
Embodied carbon	Lower	Higher because it is cooled then reheated

Sources: Chesterfield Steel 2024; GROB Inc. 2020

in exterior applications without a protective coating as it will corrode. Protective coatings for construction steel are detailed in later in this chapter.

Weathering Steel

Weathering steel, also called Corten steel, is a type of steel alloy that forms a protective patina of a rust-like material when exposed to moisture. Its alloying elements are copper, chromium, nickel, silicon, and phosphorus added in very small amounts. Weathering steel is available in sheets, shapes, tubes, plate, and bar stock (Figures 20.2–20.4).

Weathering steel's patina has no effect on the strength or other properties of the steel. The warm, textured, dark brown layer is a barrier to further corrosion, making the material resistant to moisture and chemical damage. This patina will self-heal if scratched.

The color of weathering steel darkens as it ages with the final color dependent on the site and climate conditions. Sulfur dense atmospheres speed oxidation and darkening. In very dry climates, the patina may not naturally form, or it may be patchy. In those cases, a chemical application can be employed to form an even patina.

While weathering steel's patina is a protective layer, water runoff can carry small amounts of the patina onto adjacent surfaces leading to stains (Figure 20.2). Sealing the weathering steel after the patina has fully formed is recommended.

Figure 20.2 If not sealed, water runoff from weathering steel can carry small amounts of the patina onto adjacent surfaces leading to stains. Image credit: Meg Calkins

Figure 20.3 The China State Construction Engineering Corporation used weathering steel for pavilions and a seat wall at Yellowhorn Farm Park in Inner Mongolia. Image credit: Wenzhong Gao

Figure 20.4 Weathering steel plate with a sculptural connection was used by Heatherwick Studio and MNLA for retaining walls at Little Island in NYC. Image credit: Meg Calkins

Construction with Weathering Steel ASTM A588 and DIN EN 10025-5 offer standards for weathering steels. In addition to typical considerations for detailing and construction of steel structures discussed later in the chapter and in Chapter 11, specific considerations for construction with weathering steels include the following:

- Avoid using weathering steel in ground contact. If used in ground contact, it should be well sealed with a waterproofing agent.
- Vegetation should be trimmed back from the structure.
- Specify thicker material sections in humid climates.
- Seal the weathering steel if in contact with surrounding hardscape such as walls and pavement to avoid staining.
- Avoid overlapping joints where water can be trapped.

Stainless Steel

Stainless steel (SS) is an alloy steel with chromium and nickel added to impart corrosion resistance. SS forms a chromium-rich oxide layer that protects the material from corrosion and self-heals with any damage.

Austenitic stainless steels from the 300 series are the types of SS used in site construction. *Type 304* is the most common, containing 18% chromium and 8% or 10% nickel. *Type 316* offers greater corrosion resistance in environments with heavy air pollution or localized chemicals that could coat it. The composition of Type 316 is 16–18% chromium, 10–14% nickel, 2–3% molybdenum, and 0.08% carbon.

Stainless steel is available in sheets, plates, bars, wires, and tubing. ASTM and DIN EN standards for specification of stainless steels are offered by form of metal (Table 20.2). Multiple surface finishes can be applied to SS from polished to textured and patterned. Generally, rougher finishes are not as easily cleaned with rainwater, but they will not show dirt and dust as much as polished finishes.

Environmental and Human Health Burdens of Steel Production

Making steel is an energy and carbon-intensive process (Figure 20.5). With nearly 2 billion metric tons of primary steel produced annually, the global iron and steel industry is responsible for about 7% of greenhouse gas emissions from energy production and industrial processes globally (Mission Possible Partnership 2022). As the market for steel is expanding with growth in the building sector worldwide, substantive changes must be made to bring the iron and steel industry to net-zero emissions by 2050.

There are two sources of raw materials for steel: iron ore and recycled steel scrap. Most steel is a combination of these two inputs with 90–99% recycled content for steel produced in EAF and around 30% in a basic oxygen furnace (BOF) (Steel Manufacturers Association 2024).

Iron Ore Mining

Iron ore is an abundant material as it makes up 5% of the earth's crust. Environmental impacts from iron ore mining include resource use, water pollution, air pollution, and land degradation.

Resource use and waste: The rock to rock-to-metal ratio (RMR) of iron is among the lowest of any metal but it still averages about 9:1 (Nassar et al. 2022). This means that around 18 billion metric tons of rock are excavated for the nearly two billion metric tons of iron produced annually.

Table 20.2 Standards and Certifications for Responsible, Low-Carbon Metals

Source	Standard or Certification
Initiative for Responsible Mining Assurance	The Initiative for Responsible Mining Assurance offers a standard for best practices for industrial scale mining. Two standards are available: *Mining and Mineral Processing* and the *Chain of Custody Standard for Responsibly Mined Minerals*.
ResponsibleSteel	The multi-stakeholder ResponsibleSteel organization offers an *International Production Standard* for responsible production, sourcing, and stewardship of steel. The standard covers responsible sourcing, labor rights, GHG emissions, effluents and waste, water stewardship, and biodiversity. Steel producers can become certified by third-party auditors.
Global Energy Monitor	Global Steel Plant Tracker (GSPT) provides information on individual crude iron and steel production plants worldwide.
Alliance for Sustainable Building Products	Delivering Innovative Steel ReUse Toolkit (DISRUPT) provides business considerations for steel reuse, case studies, scenario mapping, supply chain models, guidance, and policy notes.
Aluminum Stewardship Initiative	The Aluminum Stewardship Initiative offers standards for responsible production, sourcing and stewardship of aluminum. Aluminum product producers can become certified through a self-assessment and a third-party audit.

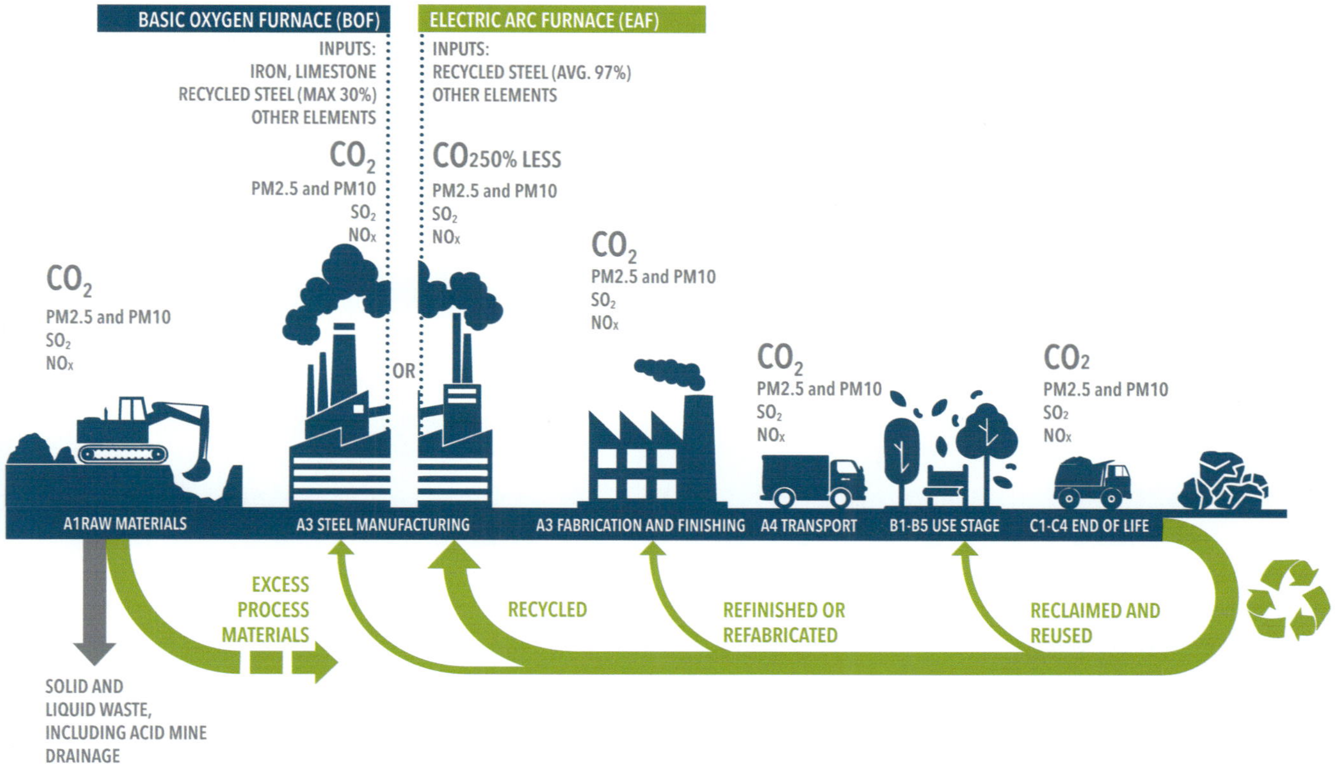

Figure 20.5 The life cycle of steel. Steel produced in a BOF is twice as carbon intensive as that produced in an EAF. A majority of the steel made in the United States is produced in an EAF. Image credit: Meg Calkins

Water pollution: Iron ore tailings, solid wastes resulting from the mining process, can contaminate receiving waters with heavy metals and harmful chemicals. Acid mine drainage (AMD) results from a combination of low pH and high levels of heavy metals that can negatively impact both water and soil quality.

Air pollution: Mining equipment including diesel generators and trucks release air pollutants such as CO_2, carbon monoxide, nitrous dioxide, and sulfur dioxide. Sulfur dioxide can react with water vapor in the air to form acid rain harming vegetation and, ironically, corroding metal structures.

Primary Steel Production

Primary steel is produced by processing iron ore in either a blast furnace-basic oxygen furnace system (BF-BOF) or in a DRI plant. The two processes reduce the ore to iron with different techniques (Table 20.3).

Blast Furnace-Basic Oxygen Furnace Steel Production: In 2021, 1.95 billion metric tons of steel were produced in BF-BOFs around the world. This carbon-intensive process releases over 2 metric tons of CO_2 for every metric ton of steel produced and accounted for 70% of global steel production (Kinch 2022).

The first step in the BF-BOF steel making process is to produce coke through a process called coking, which burns coal at temperatures of 900–1100°C for 15–36 hours. Carbon is emitted along with other chemicals and heavy metals.

In the BF-BOF furnace, the coke is burned at temperatures as high as 1300°C to chemically reduce iron ore to create pig iron. The next steps in the process combine iron ore, limestone, and steel scrap to produce steel. The maximum amount of recycled steel scrap used in blast furnaces is 30%. Substantial carbon emissions are released both from energy to heat the blast furnace, and process emissions from the chemical reaction of reducing iron ore and burning limestone.

China, by far the largest producer of steel, makes most of it in BF-BOFs with many new BF plants in the pipeline. In the United States less than 30% of steel is produced in BF-BOFs. Reliance on the carbon-intensive BF-BOF steel making process could impede the steel industry's efforts to achieve net-zero by 2050 (Swalec and Grigsby-Schulte 2023).

Typical steel products produced in BF-BOFs include hollow structural sections (HSS) and metal decks as they need an additional step to roll them into form using primary steel. Plates may also be produced in BF-BOFs.

Table 20.3 Embodied Carbon of Primary Metals

Primary Metal	Embodied Carbon, $kgCO_2e/kg$	Source and Notes
Global average of steel production	1.91	Global average; World Steel Association (WSA) 2023
Steel produced in BF-BOF, average RC is 23%[1]	2.33	Global average; WSA 2023
Steel produced in EAF, average RC is 90–93%[1]	0.68	Global average; WSA 2023
Steel produced in DRI-EAF	0.85	Mission Possible Partnership 2022
Steel produced in hydrogen-based DRI	0.08	Mission Possible Partnership 2022 Assumes 100% renewable energy sources
Stainless steel, primary, 30% scrap	6.82	World Stainless 2023; not third party verified
Stainless steel, primary, 50% scrap	3.28	World Stainless 2023; not third party verified
Stainless steel, primary, 75% scrap	2.42	World Stainless 2023; not third party verified
Stainless steel, primary, 85% scrap	2.08	World Stainless 2023; not third party verified
Copper, primary	4.1	International Copper Association 2023
Aluminum global average	10.4	Yang 2021
Aluminum, North America	8.46	Produced from primarily hydroelectric power; The Aluminum Association, Wang 2022
Aluminum produced in China	12.8	Yang 2021
Aluminum, 100% recycled content, North America	0.527	Produced from hydroelectric power; The Aluminum Association, Wang 2022

[1] AISI and SMA 2021

Note: Embodied Carbon figures will change over time as greater amounts of recycled content is used, processing efficiencies are improved, and more renewable energy sources are employed.

Direct Reduced Iron – Electric Arc Furnace (DRI-EAF): Direct reduced iron is a process of the chemical removal and reduction of oxygen from iron ore in its solid form producing primary steel in an EAF. The current practice, accounting for about 5% of primary steelmaking, is to use fossil resources such as natural gas or coal to produce the iron without melting it. Carbon combines with the oxygen in the iron ore producing metallic iron called sponge iron (IIMA 2018). This process is accomplished at much lower temperatures than the production of pig iron in a BOF, therefore it is lower carbon. Steel produced with this traditional DRI route releases about 1.5 metric tons of CO_2 per metric ton of steel produced (as opposed to $2.4tCO_2$ in BOF-BF).

A very promising new DRI technology reduces iron ore with hydrogen instead of carbon, therefore the waste gas is water, not CO_2. This process is called *hydrogen-based direct reduced iron*. There are two main low-carbon sources of hydrogen for this new DRI process. "Green" hydrogen is produced by combining renewable energy with electrolysis, while "blue" hydrogen is produced from fossil fuels and the carbon released is captured and stored (CCS) (WSA 2022).

Secondary Steel Production

Secondary steel is produced in an *electric arc furnace (EAF)*, which uses an electrical current to melt steel scrap and recycle it into new steel. EAFs use much less energy than BOFs and since they are powered by electricity, they could theoretically be 100% powered by renewable sources. While more than 70% of steel produced in the United States is made in EAFs, there is not currently enough scrap steel to meet global demand, so some primary steel will always be produced. Worldwide, only 25% of global production is produced with EAFs and 43% of planned steel plants are EAFs (MPP 2022).

The carbon footprint of secondary steel produced in an EAF is about 25% of that produced in a BOF. Increased steel recycling and use of EAFs will help lower steel industry emissions, but to fully decarbonize, substantive changes will need to be made in the primary steel making industry.

Typical steel products produced in EAFs include hot rolled shapes such as angles, channels tubes, wide flange members, rebar, and some plates. Specifications should require EAF production for these products.

Getting to Net Zero in Steel Production

Like other manufacturing industries, the transition away from coal-based steel production is underway with a multi-pronged approach to net-zero in the industry by 2050. Strategies include reducing demand for primary steel, material efficiencies, renewable energy sources, new technologies such as hydrogen-based direct reduced iron, and carbon capture, utilization, and storage (Swalec and Grigsby-Schulte 2023; MPP 2022) (Tables 20.2 and 20.3).

Reduced Demand for Primary Steel and Increased Scrap Recycling

Demand for primary steel must be reduced despite projected growth in the construction sector worldwide and increasing global scrap recycling is a key approach to reduce the demand. The US steel recycling rate is around 70%, and the global recycling rate just 60% (AISI and SMA 2021). The lower global average is largely due to immature or non-existent recycling infrastructures in some countries.

Designers and engineers play an important role in reducing demand for primary steel. They can employ material optimization strategies, design structures for deconstruction, and detail durable structures as discussed later in the chapter.

Hydrogen-Based Direct Reduced Iron or Iron Ore Electrolysis

The most substantial shift toward net zero in the steel industry will occur with increased use of the relatively new hydrogen-based DRI or iron ore electrolysis primary steel making processes. Instead of coking coal, the DRI process uses hydrogen that reacts with iron ore to make sponge iron at lower temperatures without melting the ore. The waste product released is water, not CO_2. The hydrogen-based DRI process can take place in an EAF plant, therefore these plants can produce enough primary steel to supplement recycled steel. The International Energy Agency estimates that at least 42% of primary steelmaking must use these processes by 2050 (IEA 2023). If "green" hydrogen is used and the EAF uses 100% renewable energy sources, this new steel will have a substantially lower carbon footprint at just 0.08tCO_2e/t steel produced. For context, primary steel produced in a BF-BOF releases about 2.4tCO_2e/t steel (MPP 2022).

Carbon Capture, Utilization, and Storage

CCUS technologies can be installed at coal powered BF-BOFs and natural gas DRI plants. While the industry must transition to hydrogen-based DRI, these carbon capture technologies can address residual process emissions during the transition. Ambitious CCUS technologies on some BF-BOFs are expected to result in zero-carbon primary steel by 2028 (IEA 2023; MPP 2022).

Zero-Carbon Electricity

Renewable energy sources such as wind, solar, geothermal, and hydropower are key to decarbonizing the steel industry. Generating zero-carbon energy for steel production will require a rapid scale up of renewable sources and transmission capacity.

Responsible Steel Standard

The multi-stakeholder ResponsibleSteel organization offers an *International Production Standard* for responsible production, sourcing, and stewardship of steel. The standard covers responsible sourcing, labor rights, GHG emissions, effluents and waste, water stewardship, and biodiversity (ResponsibleSteel 2024). Steel producers can become certified by third-party auditors.

Iron

Iron, the base metal for steel alloys, is not widely used in construction because of the superior performance of mild and stainless steels. Due to higher amounts of carbon, iron has better corrosion resistance than steel so cast iron products are still used primarily in infrastructure applications for drainage and piping products.

Cast Iron

Cast iron is a group of iron alloys with a carbon content greater than 2% and a silicon content of 1–3%. They have a low melting temperature and can be brittle. They are very durable when cast and are commonly used for drain grates, sewer caps, and pipes. Because this form of iron is cast into molds, custom designs can be achieved relatively inexpensively, particularly if molds are produced through CNC technology. Cast iron is corrosion resistant, but coatings are used to improve its appearance.

Gray cast iron contains 2–4% carbon and 1–3% silicon. It is brittle and susceptible to impact cracking and fatigue if overloaded, but it has compressive strength comparable to low-carbon steel. It is comprised of flake graphite.

White cast iron contains less silicon and is harder than gray cast iron making it more difficult to machine. It is brittle, but more abrasion resistant than gray cast iron and it can be welded.

Ductile cast iron, also called spheroidal cast iron, has nodular graphite and greater tensile strength than gray and white cast iron. It also has excellent abrasion resistance.

Wrought Iron

Wrought iron is not produced today except in bespoke artisan applications. Products that are sold as "wrought iron" today are usually steel that is galvanized or powder coated black. True wrought iron was historically used for gates, fences, guardrails, planters, and gazebos. It has a fibrous structure that makes it easily identifiable.

Aluminum

Aluminum is a non-ferrous metal that is characterized by its versatility, corrosion resistance, and relatively high strength to weight ratio. Aluminum products can be molded, extruded, or rolled into a wide variety of shapes. While not as commonly used as steel, aluminum has a place in site construction in overhead structures, enclosure systems, handrails, benches, tree grates, trellises, lighting structures, and more. About a quarter of all aluminum produced is used by the construction industry.

Aluminum's ability to withstand many elements and resist corrosion makes it useful in exterior applications. It naturally builds a thin layer of whitish gray powdery oxidation that protects the metal from corrosion, but in the most urban environments, this layer should be formed through the electrolytic oxidation process of anodization. Anodizing results in a thicker, more consistent protective layer that has greater corrosion resistance.

Other protective surfaces for aluminum include powder coating, painting and interference color discussed later in the chapter. Uncoated and anodized aluminum surfaces can be ground, brushed, or polished to produce a wide variety of visual effects, although fine patterns in exterior settings could be compromised over time due to the metal's relative softness.

Alloys are generally employed for aluminum construction products to impart hardness and greater durability. Aluminum products are either wrought or cast.

Wrought Aluminum

Wrought aluminum has very good mechanical strength and can be formed into many shapes. It is produced by smelting aluminum ingots with an alloying metal, then extruding, rolling, or forging it into a form. Extruded profiles such as bars, rods, and shapes are the strongest aluminum products. Wrought aluminum has good structural integrity and a smooth, even surface. It is very formable and can be shaped, machined, or extruded into just about any shape.

The primary types of wrought aluminum alloys used in construction are:

5000 series alloys, composed of aluminum with magnesium (0.5–5.5%), are known for their strength and high corrosion resistance. They are suitable for harsh environmental conditions such as marine environments.

6000 series alloys, composed of aluminum with magnesium (0.35–1.5%) and silicon (0.2–1.8%), are the most common aluminum alloys used in construction. Aluminum 6061 is resistant to corrosion and stress cracking and is weldable and easily formable. It is used for aluminum structural shapes such as angles, beams, channels, I-beams, and T-shapes.

Cast Aluminum

Cast aluminum is made from aluminum billets that are melted, mixed with alloying elements, and poured into molds to produce products or shaped for additional processing. Cast aluminum is not widely used in site construction although it can be employed for decorative applications.

Environmental and Human Health Burdens of Aluminum Production

In 2023, 70 million tons of primary aluminum was produced globally with nearly 60% in China. Only 5.6% was produced in North America (International Aluminum 2024). Manufacturing location is important because production technologies and energy sources vary among countries (Figure 20.6).

The production of primary aluminum is very carbon intensive with the industry responsible for about 0.35 GT of greenhouse gas emissions worldwide in 2021. Aluminum demand is expected to grow 40% by 2030 (IEA2022). The carbon emissions from production of one ton of aluminum range widely from 3–12 tons of CO_2e.

Raw materials Extraction

Primary aluminum production is a resource intensive process with 4–9 tons of bauxite, the main source of aluminum, required to produce just one ton of aluminum (Nassar et al. 2022). Eighty-four percent of bauxite is extracted from open-pit mines in Australia, China, Guinea, and Brazil (International Aluminum 2024).

Bauxite mining involves removal of land cover, habitat destruction and loss of biodiversity. Washing of minerals into receiving waters can pollute and degrade them, and air pollution from fossil fuel powered extraction machinery and trucks impacts air quality.

Primary Aluminum Production

After mining, bauxite is sent around the globe to refineries where the bauxite is heated, separated, and filtered to extract alumina in a multi-step, energy intensive process. The waste from this process, called red mud for its pinkish red color from iron, is substantial with 1.2 tons of red mud for every ton of alumina produced.

There are over three billion tons of red mud stored in ponds, and dried mounds around the world with few opportunities to use as it contains toxic heavy metals such as arsenic, lead,

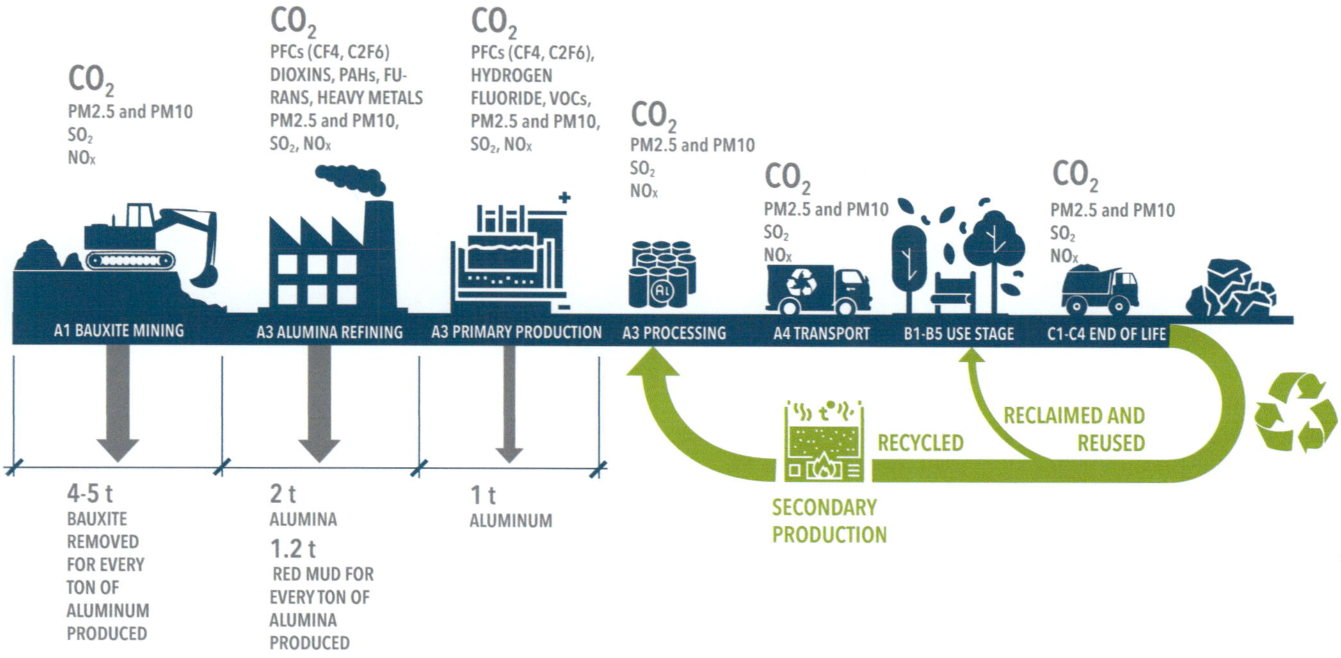

Figure 20.6 The typical life cycle of aluminum. Image credit: Meg Calkins

cadmium, mercury, and antimony. Red mud also contains radioactive constitutents and is extremely corrosive. New research is exploring the use of red mud in steel production (Max-Planck-Gesellschaft 2024).

Aluminum smelting is the process of extracting aluminum from alumina, its oxide. Electrolytic reduction is used to convert alumina into aluminum with hundreds of cells filled with molten cryolite. The result is pure liquid aluminum, which is then cast into various shapes and forms. The electrolytic reduction process is the most energy-intensive step in the process, accounting for 65% of the carbon footprint of primary aluminum. The alumina refining process accounts for 33% (Wang 2022).

Small amounts of perfluorocarbons (PFCs) are released during the smelting process as the electrolytic bath falls below certain levels. The PFCs, tetrafluoromethane (CF_4) and hexafluoroethane (C_2F_6), are some of the most potent greenhouse gases at 6500 times stronger than CO_2. CF_4 is estimated to last in the atmosphere for 5000 years. In the United States, the aluminum industry's self-reported releases of PFCs have decreased 88% since 1990 (US EPA 2024a). But despite the reductions, the largest US producer of aluminum emitted 24 tons of PFCs in 2021 (McKenna 2022).

Aluminum and Embodied Carbon

Despite relatively robust recycling rates, the embodied carbon of aluminum products is still substantial, although it is declining. The carbon footprint of primary aluminum in North America has declined by 49% since 1991. This is primarily due to technological advancements such as improved process controls, the phase out of old smelting technologies, and increased use of renewable hydroelectric energy sources (Wang 2022) (Table 20.3).

Around 85% of the emissions from aluminum production stem from energy sources. In 2021, over 95% of the primary aluminum smelting in North America was accomplished with renewable hydroelectric power. In Europe, over 92% was produced with hydroelectric power, but in China, where around 60% of all aluminum is produced, coal is currently the primary energy source for smelting. As a result, the embodied carbon of primary aluminum produced in China can be twice as high as that produced in North America (Aluminum Association 2021).

Projects with low-carbon goals should source aluminum products produced in North America, Europe, or other countries that use high percentages of renewable energy to produce their aluminum. Given the global nature of the aluminum industry, it may be challenging to determine the location of the primary aluminum production.

The Aluminum Stewardship Initiative offers standards for responsible production, sourcing, and stewardship of aluminum. Aluminum product producers can become certified through a self-assessment and a third-party audit (ASI).

Aluminum Recycling

While the process for producing primary aluminum is carbon and resource intensive, producing a new aluminum product with 100% recycled content has just 6% (520kg CO_2e/ton) of the carbon footprint of new aluminum and uses just under 7% of the energy inputs (Wang 2022). The aluminum recycling rate varies around the world with the aluminum can collection rate being about 50% in the United States and over 98% in Germany (MPP 2023). The recycling of aluminum from construction and auto parts in the United States is over 90% (AEC 2024).

Because aluminum recycling has been occurring since the material became widespread, 75% of all aluminum ever produced is estimated to be still in use today. Globally, the recycled input average in new aluminum is 32%, but it is 57% in North America (AA 2024). Unfortunately, the global demand for aluminum outweighs the recycling rate, and even if all aluminum were recycled, the scrap would only replace half of the current demand (UNEP 2023).

The Aluminum Industry's Path to Net-Zero

Like the steel sector, the aluminum industry is working toward net-zero carbon by 2050. The North American and European aluminum industries have already made some progress with technological improvements and use of renewable energy sources. The responsibility for reaching net-zero lies with a broad array of stakeholders including mining operations, refineries, smelting facilities, designers, engineers, and aluminum product specifiers (Table 20.2).

The key steps toward decarbonization of aluminum products are as follows:

Transitioning to low-carbon power sources will have the single biggest impact on decarbonization of the industry. The challenge will be to transition to renewables in some countries as China, the largest aluminum producing country, still relies on coal and natural gas sources with only 16% of smelting energy from hydroelectric sources in 2021 (AEC 2021).

Increasing aluminum recycling rates will substantially reduce the carbon intensity of the production of aluminum. As part of the industry's move to net-zero carbon by 2050, average scrap collection rates for aluminum products will need to increase to 90% or 81 million metric tons worldwide. (MSS 2023).

Technological efficiencies in production such as heat recovery, low-carbon anodes in smelters, and decarbonized thermal energy in refineries will contribute to lower carbon aluminum (Figures 20.7 and 20.8).

Figure 20.7 and Figure 20.8 Nebbia Works used low-carbon aluminum to construct a temporary pavilion in the water feature at the Victoria & Albert Museum in London. The aluminum producer, EN+ used a revised production process that resulted in a product with embodied carbon of 0.01tC02e/t. This is just a fraction of the typical US aluminum EC of 8.46tC02e/t. The design of the structure turned 2D sheets into a 3D space in an effort to minimize material use and waste. When the structure is removed, it will be fully recycled into new low carbon aluminum. Image credit Nebbia Works

Optimization of aluminum products include resource efficiency measures such as keeping aluminum products in use for a long time and using less aluminum by building smaller structures or specifying smaller members. This points to the importance of specifying appropriate aluminum finishes for the environment in which the aluminum is used.

Copper and Copper Alloys

Copper is used in site construction in flashing, cladding, roofing, lighting, and wiring, and it is alloyed to create bronze and brass. Copper is an important component of wood preservation treatments as well. Copper resists corrosion by forming a protective patina in shades ranging from golden brown to dark brown and eventually blue green. The patina can be hastened with chemical applications.

The two most common copper alloys used in site construction are *bronze* and *brass*. Bronze is a copper-tin alloy, and brass is an alloy of copper and zinc. They are both harder than copper and resist corrosion well. Bronze is most often cast as it is very hard. It can resist seawater and salt air well.

Environmental and Human Health Impacts of Copper and Copper Alloys

Copper is a very resource intensive metal with anywhere from 40–1000 tons of material excavated to produce one ton of copper (Nassar et al. 2022). Copper is mined in open pits with 2 tons of overburden removed for every ton of copper ore mined. Then the amount of copper derived from the copper ore is just 0.7% on average creating substantial amounts of waste. Copper deposits occur primarily in Chile, Australia, Peru, Mexico, and the western United States.

Water runoff from copper tailings can carry heavy metals into receiving waters. Copper is toxic to aquatic organisms and soil microorganisms (Nor 1987).

The resource intensive mining and copper concentration processes contribute 65–80% of copper's embodied carbon. These emissions come from electricity use at mine sites and diesel

Table 20.4 Embodied Carbon of Metal Products

Product	Embodied Carbon Phases A1–A3 $kgCO_2e/kg$	Source and Notes
Steel reinforcing bar for concrete, fabricated	0.854	98% average recycled content (RC); CLF 2023 Baseline from the Concrete Reinforcing Steel Institute EPD 2022
Steel wire and mesh, fabricated	1.7	CLF 2021 Baseline
Steel hot rolled steel shapes, fabricated – wide flange and other beams, channels, angles, tees	1.22	93% average RC; CLF 2023 Baseline from American Institute of Steel Construction EPD 2021
Plate steel, fabricated, thicker than 6mm (¼″)	1.73	80% average RC; CLF 2023 Baseline from American Institute of Steel Construction EPD 2021
Steel merchant bar, fabricated round, square, rectangular	1.48	Average RC; CLF 2023 Baseline from American Institute of Steel Construction EPD 2021
Steel hollow structural sections, fabricated, square, rectangle, circle tube stock	1.99	Average RC; CLF 2023 Baseline from American Institute of Steel Construction EPD 2021
Steel cold-formed steel framing, galvanized, studs and track	2.44	Average RC; CLF 2023 Baseline from Steel Framing Industry Association EPD 2021
Steel open-web steel joists, prefabricated steel joists and girders with open middle web and top and bottom chords	1.43	Average RC; CLF 2023 Baseline from Steel Joist Institute EPD 2021
Steel decking, galvanized or painted	2.32	Average RC; CLF 2023 Baseline from Steel Deck Institute EPD 2022
Galvanized steel hot rolled structural sections, plates and hollow structural sections	1.71	RC not specified; American Galvanizers Association 2020
Stainless steel 304, hot rolled austenitic	1.91	84% recycled content; Aperam 2022
Stainless steel 304, cold rolled austenitic	2.02	84% recycled content; Aperam 2022202
Stainless steel, rebar	4.99	Acerinox 2023
Aluminum extrusions, mill finish	10.25	CLF 2023 Baseline from Aluminum Extruders Council EPD 2022
Aluminum extrusions, painted	11.67	CLF 2023 Baseline from Aluminum Extruders Council EPD 2022
Aluminum extrusions, anodized	10.76	CLF 2023 Baseline from Aluminum Extruders Council EPD 2022
Aluminum sheet	7.46	The Aluminum Association, North America, Wang 2022
Aluminum, cast	4.75	The Aluminum Association, North America, Wang 2022
Aluminum, extruded	6.21	The Aluminum Association, North America, Wang 2022
Copper sheet	2.72	Assumes 90% recycled content; Copper Development Association 2019
Copper tube	3.37	Assumes 90% recycled content; Copper Development Association 2019

truck hauling. Twenty to thirty-five percent of the total emissions occur in smelting and refining of copper cathodes (RMI 2024).

Globally, the average recycled content of copper is slightly more than 30%. The International Copper Association estimates that two-thirds of the 690 million metric tons of copper produced over the last hundred years is still in use (ICA 2023).

Human Health and Equity Issues in Metals Industries

The mining, processing, and finishing of metals can pose substantial human health impacts through release and exposure of heavy metals and toxic chemicals.

Table 20.5 Metals and Metal Compounds Substance Priority Lists[1]

Metal and Metal Compounds Presence in Building Materials Listings	Health Burdens and Risks	Releases to the Environment in the United States2·3
Hexavalent chromium (CR VI) and CR VI compounds *stainless steel production, electroplating, pigments and wood preservation* #17 on ATSDR Substance Priority List On LBC Red List	Hexavalent chromium compounds, also called chromium VI, are forms of the metallic element chromium. Workers in industries processing CR VI can be exposed to the compound, and it has been found in the water and soil of fence-line communities. Hexavalent compounds have been shown to cause lung cancer when inhaled. Other health impacts include respiratory irritation, kidney damage, liver damage, pulmonary congestion, and edema (ATSDR 2012a).	The US Toxic Release Inventory does not distinguish between CR III and CR VI. A rough estimate is that one-third of CR releases are CR VI. In 2022, 1,738 metric tons of CR VI were released to land from metal mining. Another 2,051 metric tons were released to land and air from primary metals production.
Nickel and nickel compounds *stainless steel, aluminum and copper alloying element, electroplating* #57 on ATSDR Substance Priority List Some nickel compounds on the LBC Red List	Long-term exposure to nickel can cause lung and nasal cancers. Some compounds can cause kidney and liver damage and negative reproductive effects (ATSDR 2022).	In 2022, 5,022 metric tons of nickel and nickel compounds were released from industries (TRI 2022a, 2022b).
Zinc and zinc compounds *galvanized coatings for steel, electroplating* #74 on ATSDR Substance Priority List Some compounds on the LBC Red List	Zinc is an essential nutrient for humans, animals, and some plants in very small amounts. Long-term exposure to high amounts of zinc can lead to liver damage and neurological issues (ATSDR 2022c).	In 2022, 67,665 metric tons of zinc dust and zinc compounds were released from industries (TRI 2022a, 2022b).
Copper and copper compounds *pipes, flashing, specialty screens, wood preservatives, pesticides* #120 on ATSDR Substance Priority List Some compounds on the LBC Red List	Copper is an essential nutrient for humans, animals, and some plants in very small amounts. Copper is toxic to aquatic organisms and species. Ingesting high amounts of copper over a long period of time can lead to liver and kidney damage (ATSDR 2022b).	In 2022, 22,938 metric tons of copper and copper compounds were released from industries (TRI 2022a, 2022b).
Lead and lead compounds *Flashing, soft solder, roofing, linings, electrical conduits, fly ash*. Used in PVC products, particularly those produced outside the US and Europe, but no longer used in paint.* #2 on ATSDR Substance Priority List On LBC Red List	Lead is a persistent bioaccumulative toxin that accumulates in the fatty tissues of people and wildlife. Exposure increases up the food chain. Children exposed to lead have decreased cognitive performance, slower growth, and other developmental problems. Adults exposed to lead experience neurological, renal, hematological, and reproductive issues (ATSDR 2020).	Lead disposal by metal mines is responsible for 88% (around 204,343 metric tons) of all lead releases in 2022. Primary metal production is responsible for an additional 10,943 metric tons of lead. The primary metals sector was responsible for the largest releases of lead to the air at 32% (around 54,000kg) in 2022.
Arsenic and arsenic compounds *Metal alloy for some bronze, wood preservation, pesticides, fly ash** #1 on ATSDR Substance Priority List On LBC Red List	Arsenic is a naturally occurring element found as an impurity in metal ores such as copper. Arsenic is a known carcinogen through respiratory or gastrointestinal exposure. It can cause cardiovascular disease and neurological and renal problems (WHO 2022). CCA treated lumber is largely phased out as a wood treatment except in utility poles, pilings, docks, and retaining structures.	In 2022, 104,140 metric tons were released to land, air, and water from metals mining with an additional 1438 metric tons released from the primary metals sector.

(Continued)

Table 20.5 (Continued)

Metal and Metal Compounds Presence in Building Materials Listings	Health Burdens and Risks	Releases to the Environment in the United States2,3
Mercury and mercury compounds *Fly ash*, fluorescent and HID lighting, batteries, septic tanks,* #3 on ATSDR Substance Priority List On LBC Red List	Mercury is a persistent bioaccumulative toxin neurotoxin that is present in the environment from coal combustion emissions. Mercury exposure, even in small amounts, can damage the nervous system, kidneys, liver, and immune system. Mercury vapors pass easily from the lungs to the bloodstream (ATSDR 2022d).	In 2022, the metals mining sector released 1,482 metric tons to land. The primary metals sector was responsible for the largest releases of mercury to the air at 36% (around 6,000kg) (EPA Mercury). Releases to water were 984kg.
Cadmium and cadmium compounds *PVC products, electroplating materials, anti-rust paints, fly ash** #7 on ATSDR Substance Priority List On LBC Red List	Cadmium is a persistent bioaccumulative toxin. Acute exposure to cadmium can damage the lungs, while chronic exposure can result in kidney, bone, and lung diseases (ATSDR 2012b).	In 2022, 584 metric tons of cadmium compounds were released to land from metals mining and 487 metric tons were released from the primary metals sector.

*Fly ash, a by-product of coal combustion, may contain heavy metals such as mercury, cadmium, and lead depending on the chemical composition of the coal that was burned.

[1] IFLI 2024; ATSDR 2022a

[2] TRI 2022a

[3] TRI 2022b

Workers during phases A1 and A3 are impacted as well as fence-line communities around mines, manufacturing, and finishing facilities and waste disposal points (Table 20.5).

Additionally, human right abuses occur primarily in metals mining through use of forced and child labor in some countries. They also occur in the metals precursor industries of limestone mining and coal mining (Table 20.6).

Table 20.6 Potential Human Rights Abuses of Metals, and Their Precursors

Metal	Country	Child Labor	Forced Labor
Charcoal	Brazil	x	x
Charcoal	Uganda	x	
Coal	Afghanistan	x	
Coal	China		x
Coal	Columbia	x	
Coal	Mongolia	x	
Coal	North Korea		x
Coal	Pakistan	x	x
Cobalt ore	Democratic Republic of the Congo	x	
Copper	Democratic Republic of the Congo	x	
Iron	North Korea		x
Pig iron	Brazil	x	x
Limestone	Egypt	x	
Limestone	Paraguay	x	
Tin	Bolivia	x	
Tin	Democratic Republic of the Congo	x	x
Tin	Indonesia	x	
Zinc	Bolivia	x	
Brass	India	x	

Sources: ILAB 2022; Grace Farms Foundation 2022

Using Metals in Site Construction

The remainder of this chapter addresses use of metals in site structures (Figures 20.9–20.13; Tables 20.7, 20.8, and 20.10). The focus is on optimizing metal members in structures, specifying durable metals and finishes, and general considerations for metal connections and detailing (Tables 20.16 and 20.17). Embodied carbon is provided for selected steel and aluminum product sizes and forms (Table 20.9).

Several other chapters in this book discuss metals. Chapter 14 "Concrete Materials and Products" covers steel reinforcing for concrete structures, while Chapters 6, 8, and 9 discuss steel reinforcing in pavements and walls. Chapter 11 "Rails, Fences, and Screens" details these metal structures and offers additional specifics on approaches to metal durability.

Figure 20.9 Typical metal shapes used in site construction. Image credit: Meg Calkins

Weathering and Durability of Metals

Given the environmental impacts and embodied carbon of metals, they must be specified and detailed for durability. Selecting the appropriate type of metal and finish is a key consideration for a metal structure if it will last for several decades.

Metal Corrosion Stresses

Corrosion of metals is intensified by the environment in which they are placed, the design or the structure, and maintenance practices (Table 20.11). The following conditions contribute to weathering and corrosion of metals.

Atmospheric Corrosion

Atmospheric corrosion occurs with a film of moisture on the metal surface. It is increased by humidity, the occurrence of condensation, and the amount of pollution in the atmosphere. High humidity, fog, and other precipitation can combine with corrosive compounds to form an oxide layer on metals that may eventually lead to corrosion. Significant corrosion can take place if the relative humidity is about 80% and the temperature above 0°C. Pollutants and salts cause corrosion at much lower humidity levels. Heat also contributes to corrosion. Driving rain can rinse structures and reduce corrosion.

Environmental Pollutants

Environmental pollutants in air or rainwater can cause corrosion and oxidation of metals. Common sources of

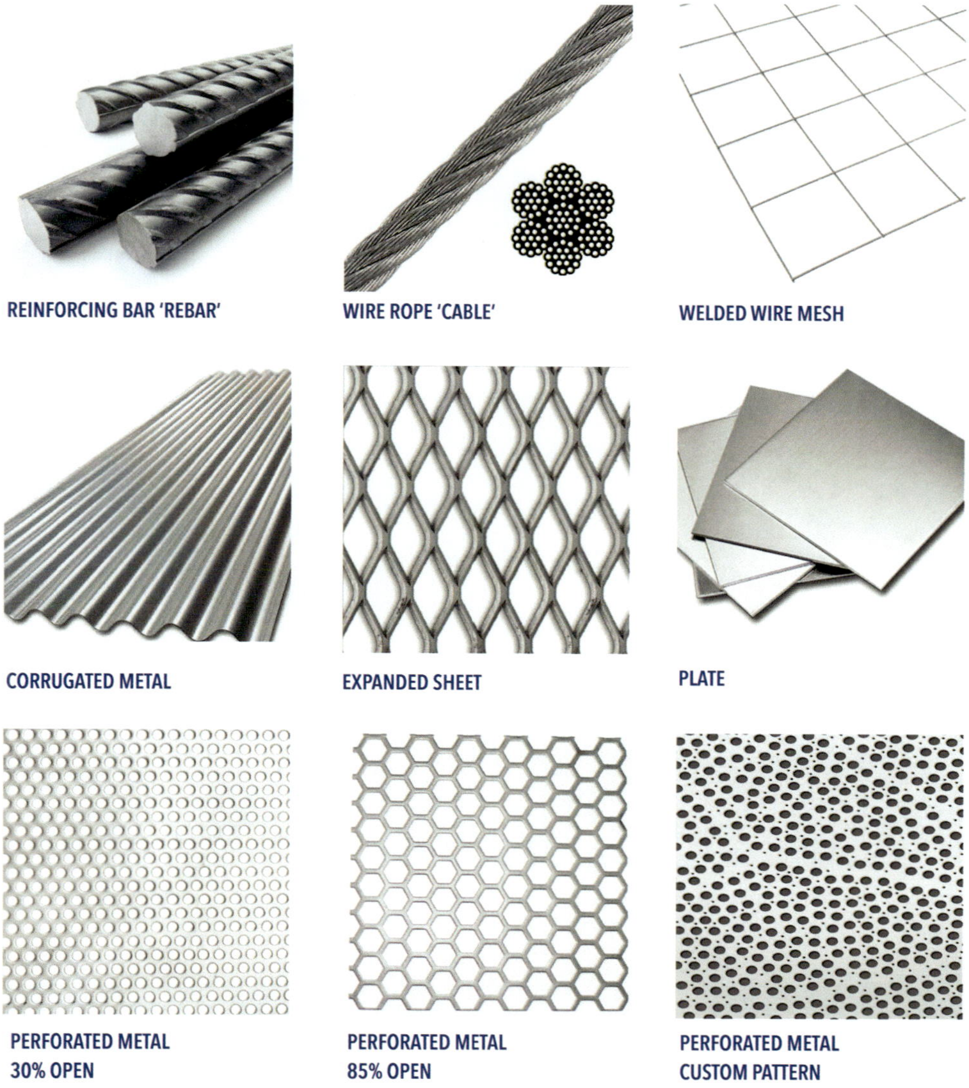

Figure 20.10 Typical metal products used in site construction. Image credit: Meg Calkins

environmental pollutants are exhaust from vehicles, industrial pollution, fossil fuel combustion, and acid rain. Sulfur dioxide and nitrous oxides are particularly corrosive to metals.

Coastal and Deicing Salts

Metals can be vulnerable to salt exposure either from coastal environments and/or deicing salts. Humidity and heat can concentrate salts on the surfaces of metals, so it is best to orient metals to allow rainwater to wash off the salts.

Coastal environments are generally considered to be within 8–16km of a coast, however strong winds can carry salts in the air even greater distances. Stainless steel grade 316 is recommended for coastal environments. Aluminum is vulnerable to pitting and should be avoided in coastal environments.

Deicing salts containing sodium chloride, calcium chloride, and magnesium chloride can corrode construction steels and stainless steels adjacent to streets and other pavements. High speed traffic on pavements can cause deicing salts to spread well beyond highways. Aluminum is not as vulnerable to deicing salts.

Galvanic Corrosion

Galvanic corrosion, also called bimetallic corrosion, can result from contact between two dissimilar metals on the galvanic series in the presence of moisture (Figure 20.14). Dissimilar metals should be separated by neoprene washers, coatings, or other non-metallic elements. It is important to pay attention to both metals and metal coatings when specifying fasteners that will be in contact with both.

Figure 20.11 Clayton Korte Architects exploited the strength to size ratio of metals with relatively small member sizes for the frame and steel cables and lights forming the overhead plane in the outdoor dining area at Cosmic Saltillo in Austin. Image credit: Erika Rich

Figure 20.12 McGregor Coxall converted the site's largest and most visually dominant oil storage tank into an overhead structure in a central gathering place in Ballast Point Park. The structure is a sculptural reinterpretation of the tank. Image credit: Christian Borchert

Figure 20.13 The overhead structure incorporates eight 1kW hour wind turbines, standing in contrast to the structure's previous use as a container for fossil fuels. Image credit: Christian Borchert

Design of the Structure and Its Connections

Crevices and joints in metal structures that allow water to collect can be problematic, as can deep surface textures on metals (Figures 11.15–11.17). Surface textures and grain of finishes should be oriented vertically and not sheltered so rainwater can wash off corrosive compounds.

Metal Treatments, Finishes, and Coatings for Durability

Metal finishes, coatings, and treatments are applied to the surface of metals to achieve a wide variety of effects, but one of the primary reasons for the application of metals coatings or treatments for metals in exterior settings is to protect a metal from corrosion due moisture or chemical

Table 20.7 Metals and Their Available Forms

+ = Common / = Less common X = not produced c = coating	Construction Steel	Weathering Steel	Stainless Steel	Aluminum	Copper or Copper Alloys	Galvanized Steel	Cast Iron
Profiles (angles, channels, T, etc)	+	+	+	+	/	+	x
Tubes (round, square, rectangular)	+	+	+	+	+	+	x
Pipes	+	/	+	/	+	+	+
I-beams and W beams (wide flange beams)	+	+	+	+	x	+	x
Plates	+	+	+	+	+	+	x
Bar (merchant bar)	+	+	+	+	/	+	x
Sheets	+	+	+	+	+	+	x
Wire and mesh	+	/	+	/	+	+	x
Wire rope	+	x	+	/	/	+	x
Cast forms	+	x	x	+	+	x	+
Hardware and fasteners	+	x	+	+	+	+	x

Table 20.8 Connections for Metal Structures

+ = Common / = Less Common	Bolts/ Screws	Welds	Adhesives	Rivets	Fittings	Solder/ Brazing	Performance Considerations
Construction steel	+	+	/	/	/	/	Welded connections between major structural members, such as beams and columns, make it very difficult to remove the members for reuse. When designing for deconstruction and reuse, bolts and screws are the easiest connections to disassemble. Welds should be ground smooth, so moisture does not collect in the imperfections and corrode the weld surface or the adjacent metal.
Stainless steel	+	+	/	/	/	/	
Weathering steel	+	+	/	/	/	x	
Aluminum	+	+	+	/	+	/	Mechanical fasteners such as bolts, screws, and rivets are the most common methods of joining aluminum. These fasteners require an overlap of material, and they are often hidden. To avoid galvanic corrosion, steel fasteners should be separated from the aluminum by a plastic or neoprene washer. Coated steel fasteners are recommended. Adhesives can join multiple aluminum members, but it is difficult to deconstruct glued connections. Aluminum welds use gas tungsten arc welding (GTAW), also known as tungsten inert gas (TIG) welding or metal inert gas (MIG) welding.
Copper	/	/	/	/	+	+	Soldering, brazing, press fittings, and flared fittings are most common, but glued and welded connections can be used.
Titanium	+	+	/	/	/	+	The most common way to join titanium is through GTAW or gas metal arc welding (GMAW). Adhesives, soldering, and bolting are also used. Brazing is used to join titanium and stainless steel.
Cast iron	/	+	+	/	/	/	Braze welding is the most common way to join cast iron components, but fittings, couplings and epoxies are also used.

Table 20.9 Embodied Carbon of Steel Shapes, Unfinished Construction of Steel and Stainless Steel (SS)

Shape and EC $kgCO_2e/kg$ (Table 20.4)	Typical Dimensions (mm) Height/Depth x Base x Thickness	Weight kg/m (m^2 for Sheet Metal)	Embodied Carbon[1] $kgCO_2e/m$
Construction Steel			
Angles 1.22kgCO_2e/kg	20x20x3 40x40x4 60x60x5 100x100x10	0.88 2.42 4.57 15.02	1.08 2.95 5.58 18.30
T section 1.22kgCO_2e/kg	20x20x3 40x40x4 60x60x5	0.90 2.49 4.68	1.09 3.74 5.71
Square tube 1.99kgCO_2e/kg	20x20x3 40x40x4 60x60x5	1.48 4.31 8.31	2.95 8.58 16.52
Rectangular tube 1.99kgCO_2e/kg	30x20x1.5 40x30x2.6 50x30x2.6 60x40x2.6 60x40x4	1.08 2.55 3.00 3.81 5.64	2.15 5.07 5.97 7.58 11.22
Round tube 1.99kgCO_2e/kg	21.3 dia., 2.3 wall 33.7 dia., 2.6 wall 33.7 dia., 4 wall 42.4 dia., 3.2 wall 59.4 dia., 5 wall	1.08 1.99 2.93 3.09 6.66	2.15 3.96 5.83 6.15 13.25
Pipe 1.99kgCO_2e/kg	22 dia., 4 wall 34 dia., 4.5 wall 38 dia., 5 wall 42 dia., 5.5 wall	1.74 3.22 4.00 4.86	3.46 6.41 7.96 9.67
I-beam 1.22kgCO_2e/kg	152.4x88.7x4.5 (web) 7.7 (flange) 203.2x101.8x5.4 (web) 9.3 (flange) 260.4x102.2x6 (web) 10 (flange) 304.4x123.4x7.1 (web) 10.7 (flange)	16.00 23.12 28.30 37.00	19.52 28.18 34.53 45.14
Channel 1.22kgCO_2e/kg	40x20x5 60x60x5 100x50x6 220x80x9	2.87 6.62 10.60 29.40	3.50 8.08 12.93 35.87
Plate 1.73kgCO_2e/kg	1000x1000x6 1000x1000x8 1000x1000x10	47.12 62.80 78.552	81.48 108.64 135.81
Merchant bar, flat 1.48kgCO_2e/kg	20x5 30x6 40x6 40x8	0.79 1.41 1.88 2.51	1.17 2.09 2.78 3.71
Merchant bar, square 1.48kgCO_2e/kg	15x15 20x20 25x25 30x30	1.77 3.14 4.91 7.07	2.62 4.65 7.27 10.46
Merchant bar, round 1.48kgCO_2e/kg	15 22 25 30	1.39 2.98 3.85 5.55	2.06 4.41 5.70 8.21
Sheet 1.22kgCO_2e/kg	1 2 3 4	7.85 15.72 23.55 31.40	9.58 19.15 28.73 38.31

(Continued)

Table 20.9 (Continued)

Shape and EC $kgCO_2e/kg$ (Table 20.4)	Typical Dimensions (mm) Height/Depth x Base x Thickness	Weight kg/m (m^2 for Sheet Metal)	Embodied Carbon[1] $kgCO_2e/m$
Welded wire mesh 1.7$kgCO_2e/kg$	102x102 Metric Wire (MW)18.7 102x102 MW25.8 152x152 MW13.1 152x152 MW25.8	2.96 4.06 1.47 2.73	5.03 6.90 2.50 4.64
Stainless Steel			
SS wire rope, 1.7$kgCO_2e/kg$	2.5 3.2 5 8	0.03 0.05 0.12 0.33	0.08 0.13 0.30 0.79
SS T section, $kgCO_2e/kg$	20x20x3 40x40x4 60x60x5	0.90 2.43 4.31	2.18 5.88 10.43
SS sheet, $kgCO_2e/kg$	1 2 3 4	7.92 15.83 23.75 31.67	19.17 38.31 57.48 76.64
SS sheet, perforated $kgCO_2e/kg$	1mm, 30% open area 1mm, 50% open area 2mm, 30% open area 2mm, 50% open area 3mm, 30% open area 3mm, 50% open area	5.80 4.32 11.61 8.70 17.30 13.00	14.04 10.41 28.07 21.05 41.87 31.46

Note: Embodied Carbon figures will change over time as greater amounts of recycled content is used, processing efficiencies are improved, and more renewable energy sources are employed.

Table 20.10 Standards for Metals in Site Construction

Standard Source and Number	Title
Steel standards	
ASTM A36	Standard Specification for Carbon Structural Steel
ASTM A276	Standard Specification for Stainless Steel Bars and Shapes
ASTM A492	Standard Specification for Stainless Steel Wire Rope
ASTM A500	Standard Specification for Cold-Formed Welded and Seamless Carbon Steel Structural Tubing in Rounds and Shapes
ASTM 514	Standard Specification for High-Yield Strength, Quenched and Tempered Alloy Steel Plate, Suitable for Welding
ASTM 572	Standard Specification for High Strength Low-Alloy Columbium-Vanadium Structural Steel
ASTM A580	Standard Specification for Stainless Steel Wire
ASTM A588	Standard Specification for High-Strength Low-Alloy Structural Steel, up to 50ksi [345MPa] Minimum Yield Point, with Atmospheric Corrosion Resistance. ASTM International.
DIN EN 10025–5	Hot rolled products of structural steels – Part 5: Technical delivery conditions for structural steels with improved atmospheric corrosion resistance.
ASTM A709	Standard Specification for Structural Steel for Bridges
ASTM A992	Standard Specification for Structural Steel Shapes
Aluminum standards	
ASTM B209	Standard Specification for Aluminum and Aluminum-Alloy Sheet and Plate
ASTM B221	Standard Specification for Aluminum and Aluminum-Alloy Extruded Bars, Rods, Wire, Profiles, and Tubes
ASTM B308	Standard Specification for Aluminum-Alloy 6061-T6 Standard Structural Profiles

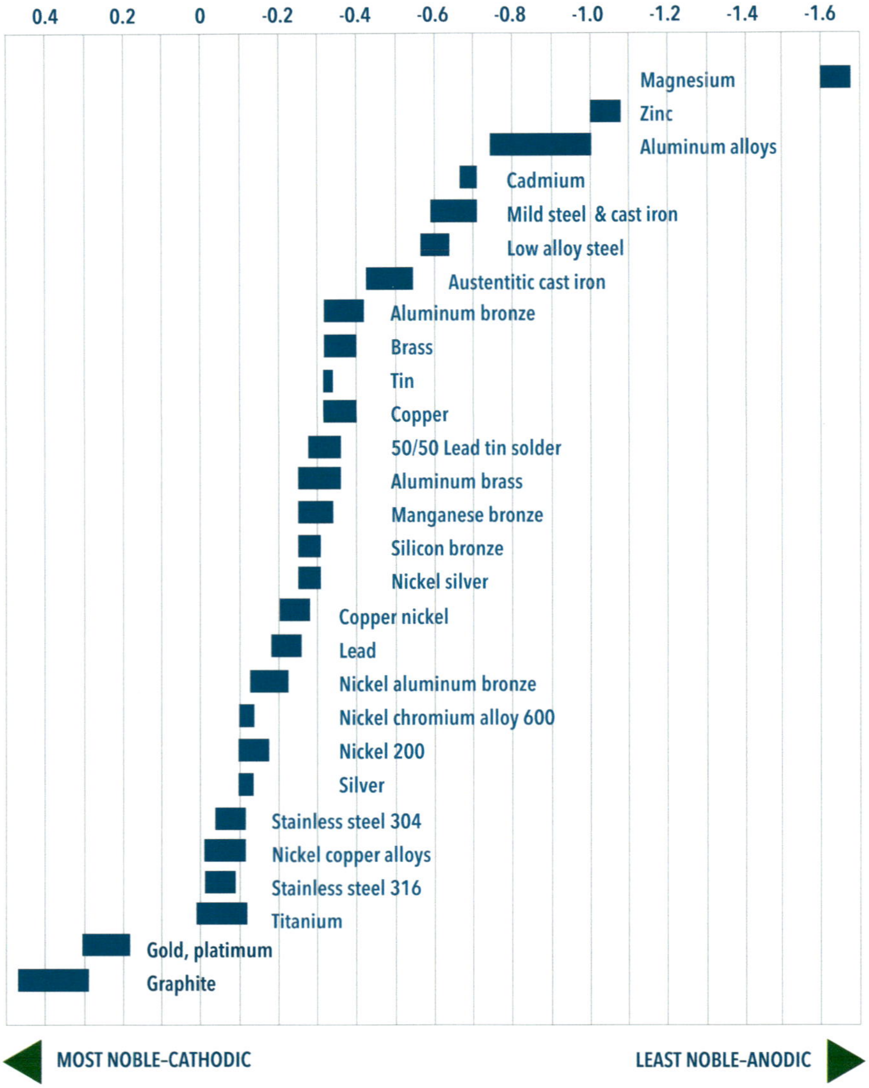

Figure 20.14 Galvanic corrosion, also called bimetallic corrosion, can result from contact between two dissimilar metals on the galvanic series in the presence of moisture. Image credit: Meg Calkins

Table 20.11 Metal Corrosion Categories

Corrosion Classification[1]	Thickness Loss μm/year[2] GA = Galvanized Steel ST = Unalloyed Steel	Appropriate Metals for Environment
C1 Very Low Not exposed to corrosion in normal atmospheric air. Heated buildings with clean atmosphere.	GA ≤ 0.1 ST ≤ 1.3	All metals.
C2 Low Rural areas with little or no pollution and low sulfur dioxide levels. Indirect moisture exposure with good ventilation. Small amounts of auto emissions and varying levels of ammonia, nitrogen, fertilizers, diesel exhaust, animal fluids (waste).	GA 0.1–0.7 ST 1.3–25	Weathering steel, powder coated steel, galvanized steel, 304 SS, unfinished or finished aluminum.
C3 Medium Urban areas with moderate pollution and some atmospheric pollutants such as SO_2. Coastal areas with some salinity. Deicing salt splash, auto emissions, building heating fuel combustion emissions that elevate sulfur dioxide and nitrogen oxide.	GA 0.7–2.1 ST 25–50	Weathering steel, heavy powder coated or galvanized steel, 304 and 316 SS, thick anodized or heavy powder coated aluminums.

(Continued)

Table 20.11 (Continued)

Corrosion Classification[1]	Thickness Loss µm/year[2] GA = Galvanized Steel ST = Unalloyed Steel	Appropriate Metals for Environment
C4 High Industrial areas and coastal areas with moderate salinity. Salt air, acid rain, atmospheric pollutants such as SO_2, NO_2, acid rain, or low pH dew from corrosive pollutants released into the atmosphere.	GA 2.1–4.2 ST 50–80	Weathering steel, heavy powder coated or galvanized steel, 316 SS, thick anodized or heavy powder coated aluminums used with care and well maintained. Clean structures regularly (Table 20.16).
C5-I Very high (Industrial) Industrial areas with high humidity and aggressive atmosphere. Atmospheric pollutants such as SO_2, NO_2, acid rain, or low pH dew from corrosive pollutants released to the atmosphere.	GA 4.2–8.4 ST 80–200	Heavy powder coated steel, heavy galvanized steel, 316 SS, only anodized or powder coated aluminum with minimum coating thickness of 18µm² clean structures regularly (Table 20.16).
C5-M Very high (Marine) Coastal and offshore areas with high salinity. Salt air, acid rain, sulfur water, animal fluids (waste).	GA 4.2–8.4 ST 80–200	Heavy powder coated steel, heavy galvanized steel, 316 SS, only powder coated aluminum with minimum coating thickness of 18µm² machined holes or cut end of aluminum are vulnerable in these saline environments. Clean structures regularly (Table 20.16).

[1] ISO 2012

[2] ISO 2017

Sources: PBZ 2024; Archinect 2020; Schroeder 2023

Table 20.12 Finishes and Coatings Deposited on Metals

Metal Finish or Coating	Environmental and Human Health Burdens	Considerations
Polyester powder coating (PPC) is a sprayed-on polymer or resin powder that adheres to the metal electrostatically. The powder liquifies, flows, and coats the metal. It takes place in a controlled environment where overspray is captured and reused. Typical polymers used for powder coating are polyester, polyurethane, polyester-epoxy, fusion bonded epoxy, and acrylics. *Steel, aluminum* Thickness 20–100µm	Durable, relatively non-toxic coating with long use life. PPC does not require a solvent base so poses less toxicity and is lower embodied carbon. Avoid PPC that contain PVC. Recycling of PPC coated metal requires extra processes to clean the metal adding EC. When heated, emissions from the polymer coating may pose a worker health risk.	PPC produces a more durable and smoother finish than paint because of the electrical adherence bond with the metal. Wide variety of colors like paint. PPC cannot be applied in the field so nicks and scratches will need to be sanded to the base metal then repaired with paint. As they are thermoplastics, they will soften slightly when hot then harden again. See AAMA 2604 or 2605 standards.
Fluoropolymer powder coatings (FPC) are a blend of resins and fluoropolymer lubricants. They are softened and applied to a metal in a heated environment in a similar process to PPC. Coating takes place in a controlled environment where overspray is captured and reused. *Aluminum* Thickness 2–50µm	Chemical and abrasion resistant. Very durable. Long use life. PPC does not require a solvent base, but FPC does. Involves chemicals on the LBC Red List such as polyvinylidene fluoride (PVDF) and fluoroethylene vinyl ether (FEVE). Recycling of FPC coated metal requires extra processes to clean the metal adding EC. When heated, emissions from the polymer coating may pose a worker health risk.	Wide range of colors and textures available. Durable coatings with excellent wear and abrasion resistance. As they are thermoplastics, they will soften slightly when hot then harden again. See AAMA 2604 or 2605 standards. FPC produces a smoother finish and is more durable than paint because of the electrical adherence bond with the metal.
Thermal cure epoxy powder coatings (TCEPC) are high molecular weight resin or polymer coatings that are applied in a heated environment. The coating forms a bond to the metal. Sometimes this is a primer coat with powder coating applied as a surface layer. *Steel, aluminum* Thickness 25–152µm	Resistant to solvents, acids, and most chemicals, but not as resistant to UV. Some may contain chemicals on the LBC Red List or the California Banned Solvents list. Avoid **TCEPC** that contains PVC. Recycling of **TCEPC** coated metal requires extra processes to clean the metal adding EC. When heated, emissions from the polymer coating may pose a worker health risk.	Wide range of colors available. Durable coatings with excellent wear and abrasion resistance. Some are thermoplastics that will soften slightly when hot then harden again. See AAMA 2604 or 2605 standards. **TCEPC** produces a smoother finish and is more durable than paint because of the electrical adherence bond with the metal.

(Continued)

Table 20.12 (Continued)

Metal Finish or Coating	Environmental and Human Health Burdens	Considerations
Hot dip galvanizing is a process where steel is dipped into a bath of molten zinc at 450°C. When exposed to air, the zinc reacts with oxygen and carbon dioxide to form zinc oxide and zinc carbonate. This protective coating forms a metallurgical bond with the steel as it cools. When exposed to weather, the zinc layer sacrifices itself to protect the steel by cathodic protection. *Steel, cast iron* Thickness 30–100μm	Zinc is a toxic material to humans if inhaled/ingested in high amounts (Table 20.5). Zinc is toxic to aquatic organisms. Zinc coated steel is not as easily recycled as uncoated steel. Technologies exist to separate and recycle the zinc but are not available everywhere. Life expectancy of galvanized coatings is 10–20 years in rural exposures, 8 years in coastal exposures, and 4 years in industrial exposures.	Individual components can be hot-dipped or multi member constructions can be dipped once assembled. Many baths are 7–17.5m in length and 1.3–2m in width. Depths are 1.8–3.2m. Longevity of the coating is related to its thickness. Damage can be field repaired with cold dip galvanizing paint. Resistant to solvents.
Cold dip galvanizing is a zinc-rich paint that is applied to steel with brushes, rollers or sprays cold. It is not a chemically bonding process like hot dip galvanizing, instead it is a mechanical bond. *Steel, cast iron* Thickness 5–30μm	Some cold dip galvanizing paints contain solvents that are on the California Banned Solvents list. This coating is not as durable as powder coating or hot dip galvanizing.	Cold dip galvanizing is good for field repairs of hot dip galvanized structures. Resistant to solvents.
Aluminized steel coats carbon steel with an aluminum-silicon alloy through a hot-dipping process that is like galvanizing. The aluminum diffuses into the steel.	As aluminized steel is aluminum diffused with steel, this treatment renders the metal non-recyclable.	Very corrosion resistant and resistant to pitting. Welded connections are more complicated.
Vapor deposition (VD) is a coating process where materials in a vapor state are condensed with a chemical conversion to form a solid, very thin layer on the surface of a metal. Metals that are deposited on the base metal include aluminum, chromium, cadmium, cobalt, copper, nickel, titanium, and zinc. *Steel, stainless steel* Thickness 2–10μm	VD avoids the electrolytic solutions and plating baths that are used for electroplating, anodizing, and hot dip galvanizing; therefore it may pose fewer toxicity risks and substantially reduce the quantity of hazardous wastes that these other processes generate. It does use large amounts of energy.	VD finishes are very thin, so they may not resist nicks and scratches. Match fasteners to metal coating type to minimize galvanic reactions and resulting corrosion. Not a common coating for site construction products.
Thermal spray (TS), also called metalizing, applies a thin, molten metal coating on a metal surface. The metal coating is heated using a flame or electric arc then sprayed at high velocity. Most common metals in the coating include aluminum, zinc, and their alloys. *Steels* Thickness 25–635μm	TS avoids the electrolytic solutions and plating baths that are used for electroplating, anodizing, and hot dip galvanizing; therefore it may pose fewer toxicity risks and substantially reduce the quantity of hazardous wastes that these other processes generate. It does use large amounts of energy.	Match fasteners to metal coating type to minimize galvanic reactions and resulting corrosion. Not as commonly used in site construction products as powder coating.
Electroplating, also called electrodeposition, is the process of plating a thin layer of metal onto another metal through the process of hydrolysis. *Steels, copper, zinc, tin* Thickness 2.5–508μm	Metals used in electroplating include nickel, cadmium, chromium, and other heavy metals that post health burdens and are persistent bio accumulative toxins (Table 20.5).	Metallic appearances can be achieved with a variety of textures.
Silicone modified polyester coating is a two-coat system for metal panels and roofing. *Steels* Thickness 16–23μm	Some may contain chemicals on the LBC Red List or the California Banned Solvents list. Some may contain chemicals on the IFLI Red List or the California Banned Solvents list.	Wide range of colors and textures available.
Epoxy air-dry coatings are a mix of epoxy and a hardener that is self-priming. They are applied in a factory environment. They are not heated but allowed to air dry. Thickness 102–152μm per coat	Resistant to solvents, acids, and most chemicals. Some may contain chemicals on the LBC Red List or the California Banned Solvents list.	Multiple coats can be applied.

(*Continued*)

Table 20.12 (Continued)

Metal Finish or Coating	Environmental and Human Health Burdens	Considerations
Metal paints, liquid coatings with an organic base, can be applied in a factory setting through sprays or they can be applied in the field. Paints contain pigments and resins such as epoxies, acrylics, alkyds, polyurethanes, or others. Paints use either solvents or water that transform the paint into a liquid.	Some paints contain solvents that release VOCs and impact air quality. Other paints contain heavy metals such as cadmium and chromium. Some may contain chemicals on the LBC Red List or the California Banned Solvents list.	Wide range of colors and finishes available. Multiple coats can be applied. Can be touched up in the field.
Siloxanes are protective metal coatings made up of silicon, oxygen, and carbon containing compounds. The metal is submerged in a solution of organosilicon compounds where hydrolysis and condensation reactions occur bonding the metal and the siloxane. *Steels, aluminum* Thickness 30–60µm	Good chemical and UV resistance. Some may contain chemicals on the LBC Red List or the California Banned Solvents list.	Clear. The chemical bond makes siloxanes long lasting with good wearing ability. Can be applied in the field, but there is a long curing period of up to 72 hours.
Clear coats are a wide range of field or factory applied finishes for metal protection. They are also applied over other corrosion protective finishes. *Steels, aluminum, bronze, copper, brass.* Thickness 38–50µm	Some may contain chemicals on the IFLI Red List or the California Banned Solvents list.	Should not be the only protective finish applied on construction steel as they are not protective enough. Can be field applied over weathering steel, copper, or bronze after the desired patina is reached. Must be regularly reapplied every 2–10 years.

Sources: ILFI 2024; Zahner 2021, 2024; PBZ 2024; Metal Coatings 2024; Milošev and Frankel 2018; Zimmermann 2015

Table 20.13 Chemical Finishes and Treatments on Metals

Metal Finish or Coating or Treatment	Environmental and Human Health Burdens	Considerations
Anodizing is an electrochemical process for aluminum that increases the corrosion resistance, wear resistance, and aesthetics of the metal's surface. The metal is submerged in an acid electrolyte bath and an electrical current is passed through it forming a tightly adhering oxide layer. *Aluminum, aluminum alloys, magnesium, titanium* Thickness 5–25µm	An anodized coating will last between 10 and 20 years. Most anodized metals can be recycled. Anodizing uses a sulfuric acid bath. Air emissions of sulfuric acid can impact workers and fence-line communities. Unsafe disposal of the bath could impact surrounding environments. Recent process modifications have extended the life of anodizing baths. Aluminum hydroxide sludge is a toxic by-product.	Anodizing can achieve shades of silver-gray to dark gray and black. Other colors are achieved by immersing the anodized metal in a liquid solution containing dye or pigment where the anodic coating absorbs the dye. Anodizing may not offer enough resistance to salt air and other chemical damages. The anodized layer has a porous, ordered structure that makes it easy to add color or other finishes. For this reason, anodizing can be a first step in other treatments.
Conversion coatings (CC) use a chemical or electrochemical conversion treatment that converts the native oxide to a new oxide with desired properties. *Aluminum, steel* Thickness 0.2–1µm	Some CC coated metals can be recycled. Coatings include black oxide, phosphate, chromate. Chromate can pose health burdens.	CC is used as a surface coating or a base for other finishes.
Interference color (IC) manipulates the oxide layer on a metal surface with a deposition of a thin metallic layer that creates an interplay of light. It is produced by the destructive or constructive interference of light reflected from the two surfaces. IC adds color to anodized aluminum. *Aluminum, aluminum alloys, SS, titanium* Thickness 5–50µm	IC uses an acid bath. Air emissions of sulfuric acid can impact workers and fence-line communities. Unsafe disposal of the bath could impact surrounding environments.	IC offers a wide range of colors to the anodizing process. The colors do not fade with time because they are impervious to the effects of UV radiation.

Sources: Zahner 2021, 2024; PBZ 2024; Metal Coatings 2024; Milošev and Frankel 2018

Table 20.14 Mechanical Finishes for Metals

Metal Finish or Coating or Treatment	Environmental and Human Health Burdens	Considerations
Mill finish is the surface texture of the metal after going through the extrusion, drawing, or rolling process.	Mill finishes are the lowest in embodied carbon as they are the least processed. The recycling potential of the metal is not impacted.	Mill finish is rough and lacks luster. It may have spots of oxidation or mill oil.
Polishing and buffing smooth the surfaces of metals with abrasives, belts, and buffers that vary by type of metal. They remove oxidation and prevent contamination and oxidation. The surface of the metal can become stronger.	Environmental and human health risks are limited beyond the energy use to achieve them. The recycling potential of the metal is not impacted.	Polished and buffed surfaces range from reflective to satin to matte finishes. Multiple steps and abrasives are used. Polished surfaces will show water spots, dirt, and pollution that lands on the surface.
Other mechanical finishes create textures and patterns on the surface of the metal. Finishes include stamping, pressing, engraving, rolling, dimpling, shot peening, media blasting, and many others. They roughen the surface of the metal creating a desired appearance or preparing the metal for another type of coating.	Environmental and human health risks are limited beyond the energy use to achieve them. The recycling potential of the metal is not impacted.	Wide variety of visual effects. When the surface of a metal is roughened, surface area is increased and more metal is exposed to oxidation and possibly corrosion. Rougher finishes on SS should be avoided in marine and industrial environments.

Sources: Zahner 2020, 2021, 2024; PBZ 2024; Miloševand Frankel 2018

Figure 20.15 and Figure 20.16 The steel exoskeleton of the Mill 19 building is exposed and celebrated in a loggia and full height corridors for exterior circulation among outside spaces at the Mill. Steel members were clear-coated in locations where visitors might touch them, but the large overhead members were left uncoated to weather, adding a patina to the pavement below with their rust flakes, and drips. New metal structures in the space are yellow powder coated steel to contrast with the existing structures. Image credit: Gaffer Photography

compounds. Metal finishes are also used to impart color, texture, or other visual qualities to a metal product or structure (Table 20.15; Figures 20.15 and 20.16).

Deposition finishes are metals, metal alloys, polymers, or resins that coat the surface of a metal to protect the metal from corrosive elements, harden the surface, and impart color or textural properties. Many deposition finishes are heated to form a durable electrostatic or metallurgical bond with the base metal product (Table 20.12).

Chemical finishes for metals are treatments applied to the surface using chemical solutions to alter its properties. These are typically used to impart corrosion resistance, achieve a desired aesthetic, or enhance adhesion of further coatings (Table 20.13).

Mechanical finishes for metals use physical methods to alter the surface of a metal resulting in a variety of textures, patterns, and reflectivity levels. Typical finishes are buffing, brushing, tumbling, and bead blasting (Table 20.14).

Table 20.15 Weathering and Finishes by Metal Type

Metal, Life Span, and Color When New	Ten-Year Old Character	Appropriate Finishes
Mild steel, gray/dark blue, 10–20 years if uncoated	Orange red rust and eventual failure	Powder coating, vapor deposition, galvanizing, aluminizing
Weathering steel, gray/dark blue, 20–100 years	Dark red brown	Clear coat
Galvanized steel, 5 years in C5, 11+ years in C2 and C3, light gray, bright	Gray white (white rust), less reflective	Powder coating, thermal spray, siloxanes, paint
Aluminized steel, 8–10 years in C5, gray white bright	Light gray, less reflective	Mostly unfinished
Stainless steel, 304, 20–50 years, chrome silver	Some staining in urban, industrial, or coastal environments	Smoother mechanical finishes will resist staining better in harsh environments
Stainless steel, 316, 50+ years, chrome silver	No change	Smoother mechanical finishes will resist staining better in harsh environments
Titanium, soft gray with slight gold cast	Little change over time	Can be mechanically finished
Aluminum, mill finish, silver white	Dull gray, some mottling	Anodizing, powder coating, painting
Aluminum, anodized, clear or colored, 20–70 years depending on thickness and environmental conditions	Minimal color change over time	Powder coating, painting
Zinc, gray blue tint	Dark bluish-gray color	No finish required
Copper, reddish pink	Gray-green patina	Clear coat or clear finishes
Bronze, reddish gold	Gray-green patina	Clear coat or clear finishes

Sources: Zahner 2021, 2020

Table 20.16 Design and Construction of Durable, Lower Carbon Metal Structures

Issue	Considerations
Metal production	Specify US or European produced *steel* products that have been made in EAFs or with the new hydrogen-based DRI process. Where possible, avoid steel products made in BF-BOFs. If steel is imported from China or India, it is likely to have been produced in BF-BOFs. Specify *aluminum* produced in North America, Europe, or other countries using high amounts of renewable energy for smelting. Metal producers may need to be contacted directly to ensure type of furnace for steel and energy sources for aluminum.
Designing durable metal structures	The most low-carbon, resource efficient approach is to keep metal structures in use for as long as possible. In exterior settings, durability is dependent on the appropriate type of metal, grade of stainless steel, steel, or aluminum finish for the environment surrounding the structure. Refer to Tables 20.12 and 20.13 for metal finishes and surface treatments. Isolate dissimilar metals from each other through use of inert washers, protective coatings (e.g., powder coatings and paint), and other physical barriers.
Design for moisture shedding and rinsing	Member orientation and connections to avoid prolonged exposure to moisture will avoid trapped moisture (Figures 11.15–11.17). Orienting metal panels and members to allow for rainwater rinsing will remove potentially corrosive air pollutants and dust. Consider specifying a higher grade of SS or thicker protective coatings on metal elements that will not be cleaner regularly or exposed to rain washing.
Metal finishes for recycling	Specify weathering steel, stainless steel, anodized aluminum, or metal finishes such as power coating that will not inhibit recycling of the metal after its useful life. Coatings that contain hazardous materials such as PVC, may off gas toxic air pollutants as they are heated during recycling.
Salvaged and reclaimed metal products	Reusing metal members or assemblies in new installations will extend the use life of the metal and substantially reduce the embodied carbon. Reclaimed metals can be sourced directly from local or regional demolition projects by contacting demolition or deconstruction companies. Metal scrap yards may also be good sources of reclaimed steel or aluminum. Procure reclaimed metal in the schematic design phase as new structures can be designed around the available scrap.
Designing metal structures for deconstruction	Metal structures should be designed for disassembly and reuse of members, panels, or modules. Bolted connections are the easiest to disassemble as welds require cutting the members. Welds could be used for panels or modules if they can be reused in whole form.
Structural considerations	Steel joists and trusses for decks and overhead structures are lighter weight than steel shapes such as I and W beams. They will have lower embodied carbon as less metal is used.

(*Continued*)

Table 20.16 (continued)

Issue	Considerations
Optimized use of metal shapes	Metal angles and Ts have the same structural capability as steel tubes but weigh less, use less steel and have a lower embodied carbon (Tables 20.4 and 20.9). Work with fabricators and/or structural engineers to achieve other efficiencies in a steel structure such as using a higher grade of steel and smaller members, increased amounts of recycled/reclaimed steel, or stronger aluminum alloys.
Fasteners	Use metal fasteners with an equal or greater corrosion resistance than the metal used for the structure (Figure 20.14). Isolate fasteners from the metal structure with plastic or neoprene if they are too far apart on the galvanic series to avoid bimetallic corrosion.
Welds	Steel welds that are 6mm or heavier in corrosive environments should use low-carbon stainless steels. Weld imperfections should be removed to avoid pockets for contaminants or moisture to gather.
Concrete reinforcing	Use steel reinforcing bar in concrete structures only when required. Work with structural engineers to specify the smallest diameter bar with the largest spacing, although make sure that the thickness of concrete structures is not increased to compensate for the lack of rebar. Consider using recycled polymer or other fiber reinforcing in concrete pavements where appropriate.

Table 20.17 Maintenance Considerations for Metal Structures

Activity	Considerations
Maintenance plans and instructions	Detail maintenance techniques that will protect and maximize the service life of the metal structure. In maintenance plans, identify cleaners to use and other types to avoid. Identify a regular cleaning schedule. List paint colors that will closely match powder coating or factory applied paint, so nicks and scratches can be repaired.
Cleaning	Metal structures should be regularly cleaned to wash away pollutants, salts, and substances that could lead to corrosion. Mild detergents should be used and structures rinsed well. Steam cleaning and light power washing can be used. Avoid acids and acidic substances such as vinegar. Avoid abrasive cleaning tools.
Rust removers	Diluted phosphoric acid, citric acid, and oxalic acid will remove rust and should not harm painted, galvanized, or powder coated surfaces.
Graffiti removal	Graffiti removal techniques will vary by type and finish of metal. Soft brushes, sponges, or light abrasive pads can be used, but test to make sure they don't scratch the finish. *SS:* Pressure washing, rubbing alcohol, non-acetone nail polish remover, or paint thinners such as mineral spirits or acetone. *Powder coated metal:* Use graffiti removal products designated for "sensitive surfaces" as solvent based products could damage the polymer. *Anodized aluminum:* Anodized surfaces can be sensitive to pH levels and can darken or stain with harsh acidic cleaners. Alkaline cleansers can break down the coating. pH neutral cleansers should be used. *Mill finish aluminum:* Penetrating oil, pressure washing, or paint thinners such as mineral spirits or acetone.
Powder coating repairs	If scratched, nicked, or compromised in some way, powder coating can be repaired in the field. Grind the repair area with a power hand grinder or wire brush down to the bare metal. Remove any dust from the grinding. Use tape to mark off the area to apply powder coat touch up paint. Prime with zinc-rich primer in thin layers. Apply touch up powder coat paint.
Field painting aluminum	Aluminum with mill finish, paint finish, or anodized finish can be painted if these original finishes are compromised or if a color change is desired. Oil-based metal paints should be used. First the structure should be thoroughly cleaned with a mild household laundry detergent. Coat with metal primer then with paint made for aluminum.
Field painting galvanized steel	Galvanized steel can be painted in the field. Remove "white rust", which is the natural oxide of zinc. It may require scraping but take care not to damage the galvanized finish. For tough layers, use an acidic product designed to remove it. Rinse off very well prior to priming and painting. The primer coat should be a zinc-rich paint. The finish coat should be compatible with the primer coat.

Sources: Zahner 2021, 2020, 2024; PBZ 2024; ISO 2017

References

The Aluminum Association (AA) (2024). "Infinitely recyclable". www.aluminum.org/Recycling, accessed March 31, 2024.

Aluminum Extruders Council (AEC) (2024). "Sustainability: Reducing the impact". https://aec.org/sustainability, accessed March 31, 2024.

Aluminum Stewardship Initiative (ASI) (2024). "ASI standards overview". https://aluminium-stewardship.org/get-certified/overview, accessed March 31, 2024.

American Iron and Steel Institute (AISI) and the Steel Manufacturer's Association (SMA) (2021). *Technical Report: Determination of Steel Recycling Rates in the United States*. AISI and SMA, July 27.

Archinect (2020). *Finish Options for Exterior Steel*. Archinect Discussion Forum, September 14.

ASTM (2019). *ASTM 588 Standard Specification for High-Strength Low-Alloy Structural Steel, up to 50 ksi [345 MPa] Minimum Yield Point, with Atmospheric Corrosion Resistance*. ASTM International.

California Air Resources Board (ARB) (2021). *Summary of Architectural Coating Rules in California: VOC Limits*. State of California, ARB, September 2.

Chesterfield Steel (2024). "Differences between hot rolled steel and cold rolled steel". www.chesterfieldsteel.com/differences-between-hot-and-cold-rolled-steel/, accessed March 31, 2024.

Climate Watch (2023). *World Greenhouse Gas Emissions in 2020 by Sector, End Use and Gases (Static)*. Key Visualizations, Climate Watch. www.climatewatchdata.org/.

European Standard (2019). *DIN EN 10025-5 Hot Rolled Products of Structural Steels - Part 5: Technical Delivery Conditions for Structural Steels with Improved Atmospheric Corrosion Resistance*. European Standard.

Grace Farms Foundation (2022). *Design for Freedom Toolkit*. New Canaan, CT: Grace Farms Foundation.

GROB Inc. (2020). *Hot Rolled Steel vs. Cold Rolled Steel*. GROB Inc., July 30.

International Aluminum (2024). *Primary Aluminum Production*. International Aluminum.

International Copper Association (2023). *Copper Environmental Profile*. ICA and Copper Alliance.

International Energy Agency (IEA) (2022). *Tracking Report - Aluminum*. IEA. www.iea.org/reports/aluminium.

International Energy Agency (IEA) (2023). *Net Zero Roadmap A Global Pathway to Keep the 1.5 °C Goal in Reach, 2023 Update*. IEA.

International Iron Metallics Association (IIMA) (2018). *Use of Direct Reduced Iron (DRI) in the Electric Arc Furnace (EAF) for Steelmaking*. Iron Metallics Association, Ltd.

International Standards Organization (ISO) (2012). *ISO 9223 Corrosion of Metals and Alloys - Corrosivity of Atmospheres - Classification, Determination and Estimation*. ISO.

International Standards Organization (ISO) (2017). *ISO 12944-2 Paints and Varnishes - Corrosion Protection of Steel Structures by Protective Paint Systems - Part 2: Classification of Environments*. ISO.

Jones, C., and Hammond, G. (2019). *Inventory of Carbon and Energy (ICE) Version 3.0*. Circular Ecology and University of Bath, November.

Kinch, D. (2022). "Direct-reduced iron becomes steel decarbonization winner". In *Energy Transition: Metals*. S&P Global, June 22.

Living Building Challenge (LBC) (2024). *LBC Red List, Priority List and Watch List*. International Living Futures Institute (IFLI).

Max-Planck-Gesellschaft (2024). *Green Steel from Toxic Red Mud: An Economical Process with Green Hydrogen Can Be Used to Extract CO_2-Free Iron from the Red Mud Generated in Aluminum Production*. Max-Planck-Gesellschaft Newsroom, January 30.

McKenna, P. (2022). "Why American aluminum plants emit far more climate pollution than some of their counterparts abroad: Short answer: They're old". *Inside Climate News*, December 6.

Metal Coatings (2024). "Industrial coating services". www.metcoat.com/industrial-coatings.htm, accessed March 31, 2024.

Milošev, I., and Frankel, G.S. (2018). "Review – conversion coatings based on zirconium and/or titanium". *Journal of the Electrochemical Society* 165.3: C127–C144. https://doi.org/10.1149/2.0371803jes.

Mission Possible Partnership (MPP) (2022). *Making Net-Zero Steel Possible: An Industry-Backed, 1.5 °C Aligned Transition Strategy*. Mission Possible Partnership, RMI, Energy Transitions Commission, and World Economic Forum.

Mission Possible Partnership (MPP) (2023). *Making Net-Zero Aluminum Possible: An Industry-Backed, 1.5 °C Aligned Transition Strategy*. Mission Possible Partnership, RMI, Energy Transitions Commission, We Mean Business Coalition, World Economic Forum and International Aluminum.

Nassar, N.T., et al. (2022). "Rock-to-metal ratio: A foundational metric for understanding mine wastes". *Environmental Science & Technology* 56.10: 6710–6721. https://doi.org/10.1021/acs.est.1c07875.

Nor, Y.M. (1987). "Ecotoxicity of copper to aquatic biota: A review". *Environmental Research* 43.1: 274–282. https://doi.org/10.1016/S0013-9351(87)80078-6.

PBZ (2024). "7 types of metal finishes and when to use them". https://pbzmfg.com/blog/types-of-metal-finishes-when-to-use-them/.

Raabe, D. (2023). "The materials science behind sustainable metals and alloys". *Chemical Reviews* 123: 2436–2608.

ResponsibleSteel (2024). "ResponsibleSteel international production standard, version 2.1". www.responsiblesteel.org/certification/.

Rocky Mountain Institute (RMI) (2024). *Decarbonizing the Copper Sector: Discussion Topics and Considerations for a 1.5 °C-aligned Trajectory and Target-setting Methodology*. Rocky Mountain Institute (RMI) and the Copper Mark, April 18.

Schroeder, T. (2023). "Selecting aluminum finishes for coastal areas". *The Construction Specifier*, March 1.

Steel Manufacturers Association (2024). "Steel is the new green". https://steelnet.org/sustainability/, accessed March 31, 2024.

Swalec, C., and Grigsby-Schulte, A. (2023). *Pedal to Metal: It's Time to Shift Steel Decarbonization into High Gear*. Global Energy Monitor and Global Steel Plant Tracker, July.

United Nations Environment Programme (UNEP) (2023). *Building Materials and the Climate: Constructing a New Future*. Nairobi: UNEP.

US Agency for Toxic Substances and Disease Registry (ATSDR) (2012a). *Chromium – Tox FAQs*. Comprehensive Environmental Response, Compensation, and Liability Act (CERCLA), ATSDR, October.

US Agency for Toxic Substances and Disease Registry (ATSDR) (2012b). *Cadmium – Tox FAQs*. Comprehensive Environmental Response, Compensation, and Liability Act (CERCLA), ATSDR, October.

US Agency for Toxic Substances and Disease Registry (ATSDR) (2020). *Lead – Tox FAQs*. Comprehensive Environmental Response, Compensation, and Liability Act (CERCLA), ATSDR, August.

US Agency for Toxic Substances and Disease Registry (ATSDR) (2022a). *2022 Substance Priority List*. Comprehensive Environmental Response, Compensation, and Liability Act (CERCLA), ATSDR.

US Agency for Toxic Substances and Disease Registry (ATSDR) (2022b). *Copper – Tox FAQs*. Comprehensive Environmental Response, Compensation, and Liability Act (CERCLA), ATSDR, April.

US Agency for Toxic Substances and Disease Registry (ATSDR) (2022c). *Zinc – Tox FAQs*. Comprehensive Environmental Response, Compensation, and Liability Act (CERCLA), ATSDR, April.

US Agency for Toxic Substances and Disease Registry (ATSDR) (2022d). *Mercury – Tox FAQs*. Comprehensive Environmental Response, Compensation, and Liability Act (CERCLA), ATSDR, April.

US Bureau of International Labor Affairs (ILAB) (2022). *List of Goods Produced by Child Labor or Forced Labor*. US Department of Labor, ILAB.

US Environmental Protection Agency (US EPA) (2024). *Aluminum*. Electric Power Systems Partnership. www.epa.gov/eps-partnership/aluminum-industry, accessed March 31, 2024.

US Occupational Safety and Health Association (OSHA) (2024). *Hexavalent Chromium Health Effects*. US Department of Labor, OSHA. www.osha.gov/hexavalent-chromium/health-effects, accessed June 9, 2024.

US Toxics Release Inventory (TRI) (2022a). *TRI On-site and Off-site Reported Disposed of or Otherwise Released, for Facilities in NAICS 2122 – Metal Mining, All Chemicals*. US TRI.

US Toxics Release Inventory (TRI) (2022b). *TRI On-site and Off-site Reported Disposed of or Otherwise Released, for Facilities in NAICS 331 – Primary Metals, All Chemicals*. US TRI.

Wang, M.J. (2022). *The Environmental Footprint of Semi-Fabricated Aluminum Products in North America*. The Aluminum Association, January.

World Health Organization (WHO) (2022). *Arsenic: Key Facts*. WHO, December 7.

World Stainless (2023). *CO_2 Emissions Report: Industry Emissions and Related Data*. World Stainless Association.

World Steel Association (WSA) (2022). *Fact Sheet: Hydrogen (H2)-Based Ironmaking*. WSA, June.

World Steel Association (WSA) (2023). *Sustainability Indicators 2023 Report: Sustainability Performance of the Steel Industry 2004–2022*. WSA, November.

Yang, M. (2021). *China's Aluminum Production Grew by 5% in 2020, Producing More Emissions Than Indonesia*. EMBER, February 7.

Zahner, L.W. (2020). *Aluminum Surfaces: A Guide to Alloys, Finishes, Fabrication and Maintenance in Architecture and Art*. Hoboken, NJ: Wiley.

Zahner, L.W. (2021). *Steel Surfaces: A Guide to Alloys, Finishes, Fabrication and Maintenance in Architecture and Art*. Hoboken, NJ: Wiley.

Zahner, L.W. (2024). "Surfaces: Architectural metal finishes". www.azahner.com/products/surfaces#specifications, accessed March 31, 2024.

Zimmermann, A. (Ed.). (2015). *Constructing Landscape: Materials, Techniques, Structural Components*, Third Edition. Birkhäuser.

CHAPTER 21

Plastics and Plastic Products

Plastics are lightweight, moldable, water resistant, and inexpensive to manufacture. For these reasons they have come to serve an important role in several applications of site construction. As most plastic products are water resistant, they can be used in ground contact. They have a relatively high strength-to-weight ratio and good flexibility and offer many design possibilities. Polymer-based coatings protect and waterproof structures made from other materials increasing their durability. Some plastics can incorporate high percentages of post-consumer recycled content.

Yet widespread use of plastics, particularly single use from packaging, has led to a global plastic waste crisis (Figure 21.1). Over 400 million metric tons of plastics were produced worldwide in 2022. Nearly 40 million metric tons of plastic waste was generated in North America but only 5–6% of this was recovered for recycling (Garside 2024). The amount of plastic in oceans is estimated to be about 150 million metric tons with 11 million tons of plastic waste flowing into it each year (Palardy 2022).

The primary feedstock for most plastics is oil or natural gas, so the embodied carbon of plastic products includes the feedstock burden as well as the energy and chemicals used to produce the plastic. When the full life cycle of plastic resins is considered, they account for about 3.4% of all GHG emissions worldwide, and global demand is expected to increase by as much as 30% in the next decade (IEA 2018).

While the carbon footprint of plastics is substantial, the human health burden of producing, consuming, and disposing of plastics must also be considered. Chemical constituents, additives, and emissions from some plastics pose health risks to humans and animals. Eliminating use of these harmful constituents is important, however identifying them in plastic products can be very challenging given their complicated chemistry.

On the supply side, primary approaches to reducing environmental and health burdens of plastics include decarbonizing production processes in the plastics industry, eliminating toxic constituents and by-products, scaling up safer biobased plastics from sustainably produced feedstocks, reusing packaging and plastics through takeback programs, improving plastics recycling, and incorporating higher recycled content into new plastics.

On the demand site, the primary approaches to reducing the burdens of plastics include reducing the quantity of plastics used, specifying safer plastics, and keeping plastic products in use for as long as possible. Longer service lives will reduce replacements and the related supply side burdens.

This chapter discusses types of plastics used in site construction by performance, environmental and human health impacts, and prospects for recycling. Characteristics of natural and synthetic rubbers are addressed along with the human rights issues of rubber production in some countries. Bioplastics and biobased plastics are covered in Chapter 19 “Biobased Materials and Products”.

Types of Plastics Used in Site Construction

Plastics are polymers that are long chains of molecules comprised of repeating links called monomers. The most common feedstocks for plastics are natural gas and petroleum, while common biobased plastic feedstocks include corn and sugarcane. Intermediate petrochemicals used to produce plastics include ammonia, methanol, and high value chemicals (HVCs) such as ethylene, butadiene, propylene, benzene, toluene, and xylene (COMET 2023).

Human and environmental health risks vary widely by plastic product, and they can occur during all phases of a plastic product’s life cycle. Emissions into air, land, and water occur during petrochemical processing and plastic manufacturing.

DOI: 10.4324/9781003413974-23

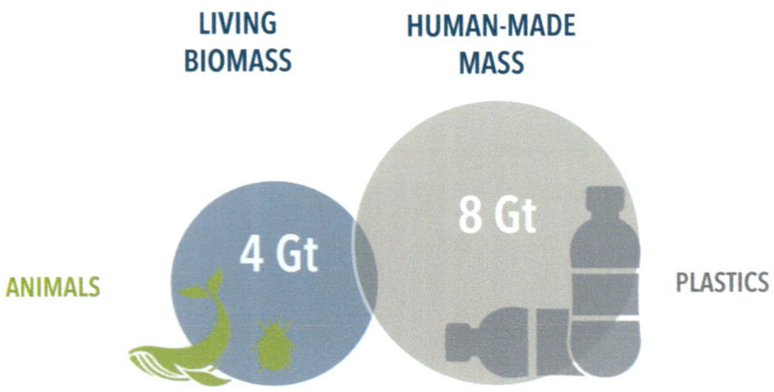

Figure 21.1 The weight of all plastic and plastic waste on earth exceeds the weight of all living animals. Adapted from source: Elhacham et al. 2020. Image credit: Drake Bruner and Meg Calkins

These emissions impact workers and fence-line communities around plants. During use, plastics can off gas VOCs and other harmful chemicals.

In the disposal phase, plastic waste that is incinerated can impact air quality for long distances. Hazardous air emissions and solid waste burdens on fence-line communities around disposal sites are increased as chemical compounds are gasified (COMET 2023).

Thermoplastics

Thermoplastics, about 80% of plastics in use today, soften in heat and become more rigid when they cool. This allows them to be recycled and reformed into new products, yet it also makes them vulnerable when exposed to extreme heat in use. Additives are employed to resist heat and UV degradation where necessary. Other additives are used to maintain flexibility, add color, foam the plastic, and impart strength.

The most common additive compounds are plasticizers such as phthalates and bisphenols, flame retardants, cadmium and lead compounds, curing agents like formaldehyde, biocides such as arsenic compounds, organic tin compounds and triclosan, alkyl phenols, and colorants such as azocolorants and cadmium compounds. Many of these compounds are endocrine disruptors or pose other health risks in manufacture use and disposal (Flaws et al. 2020).

These additives are added to plastics in varying amounts from small to very large percentages of the finished product. For example, up to 80% of a PVC product could be added plasticizers such as phthalates (Flaws et al. 2020).

Polyethylene

Polyethylene (PE), a polyolefin, is a relatively benign thermoplastic created by the polymerization of ethylene. The precursor to PE is a fossil fuel like most other plastics, and the carbon and other emissions from its production are not insignificant. Recycling infrastructure for PE is widespread and it is the most recycled family of plastics.

High-density polyethylene (HDPE), resin code #2, is relatively easy to form, low-cost, hard, and durable. It can withstand most solvents and is resistant to temperatures up to 110°C (230°F). Non-contaminated HDPE can be recycled up to ten times (Goldsberry 2018).

Low-density polyethylene (LDPE), resin code #4, is a softer PE that is used for flexible products. It will soften at lower temperatures, is less dense, and has lower tensile strength than HDPE. It is recyclable and can be recycled into new products with HDPE, although some curbside programs do not collect it.

Polyethylene terephthalate (PET), also called polyester, resin code #1, is one of the most widely produced and recycled plastics. Clear PET has one of the highest recycling market values. It is made from ethylene glycol whose precursor is the carcinogen and mutagen, ethylene oxide (US EPA 2024). In 2022, over 160,000lbs of ethylene oxide emissions were released in the United States, but emissions have been reduced by 80% reduction since then (ACC 2024).

Polypropylene

Polypropylene (PP), resin code #5, is a lightweight but tough plastic with very good heat resistance. It is another type of polyolefin and is a by-product of petroleum refining. It is a safe plastic like most polyethylenes, and it can be recycled multiple times.

Polyvinyl Chloride

PVC, also called vinyl, is lightweight, inexpensive, durable, and resistant to UV degradation, many chemicals, and corrosion. It can be made flexible or rigid. Numerous construction products are made from PVC, but it poses serious environmental and human health burdens throughout its life cycle (NCI 2024; Material Research L3C 2024; WHO 2023; Flaws et al. 2020; HCWH Europe 2021; ATSDR 2014) (Table 21.1).

PVC is produced by polymerization of the vinyl chloride monomer (VCM). Over 50% of VCM is chlorine, which is derived from rock salt (sodium chloride). Chlorine is mixed with ethylene from natural gas refining to produce ethylene dichloride (EDC), which is then cracked to become the vinyl chloride monomer. Chlorinated PVC (CPVC) is PVC that has gone through an additional chlorination process to improve its fire resistance and corrosion resistance.

Table 21.1 PVC Constituents, Additives, and By-Products

Constituent	Health and Environmental Burdens and Risks
Vinyl chloride monomers (VCMs)	VCM, the chemical intermediate for PVC, has been classified as a Group 1 Human Carcinogen by the International Agency for Research on Cancer since 1979. Additional exposure risks include impairment of the central nervous system; irritation of eyes, mucous membranes, and respiratory tract; and acute lower respiratory issues. Chronic exposure can cause kidney, lung, and brain cancers as well as lymphoma and leukemia (ATSDR 2023; NCI 2024).
Ethylene dichloride (EDC)	EDC exposure risks include nausea, vomiting, dizziness, internal bleeding, bluish-purple discoloration of the mucous membranes and skin (cyanosis), rapid but weak pulse, unconsciousness, and even death (NIOSH 2014).
Chlorine gas	Chlorine gas is listed on the US OSHA *List of Highly Hazardous Chemicals, Toxics and Reactives* "which present a potential for a catastrophic event at or above threshold quantity" (OSHA 2019). Chlorine gas exposure risks include skin and eye burns; respiratory irritation, pulmonary edema, and permanent lung damage; and headache and dizziness (NIOSH 2018).
Dioxins	Dioxins are a by-product of PVC production or incineration that are released into air, water, and land. Dioxins are a group of chemical compounds that are persistent environmental pollutants (POPs) accumulating in the food chain in the fatty tissue of animals and fish. They are highly toxic and can cause reproductive and developmental issues, immune system damage, hormone disruption, and cancer (WHO 2023).
Heavy metals	Metals such as zinc, organotin, and, to a lesser extent, cadmium and lead are added to PVC to extend the life of some products. These metals are highly toxic and can cause cancer. They target the body's cardiovascular, renal, gastrointestinal, neurological, reproductive, and respiratory systems (OSHA 2024). Lead is no longer used in PVC pipes in North America or Europe.
Phthalates	Phthalates are added to PVC products in amounts up to 80% by weight to make them more flexible (Flaws et al. 2020). TPE plastics also contain phthalates. Diethylhexyl phthalate (DEHP), the most common PVC plasticizer, is a semi-volatile organic compound (SVOC). DEHP can be emitted to air or by attaching to dust for years while the product is in use. In disposal, it can leach into groundwater. DEHP is an endocrine disruptor and has been linked to reproductive issues, birth defects, and delayed mental development as well as liver and kidney damage (ATSDR 2022a).
Organotins	Organotins are used as heat stabilizers in PVC and CPVC pipes. Organotins are endocrine disrupting chemicals that also pose reproductive and neuro toxicity impacts. Mono/dibutyl tin and mono/dimethyl tin are most reported in the United States (Material Research, L3C 2024).
Alkylphenols	Alkylphenols are added to PVC for heat stabilization in pipes and other products. They are also used in cleaners, degreasers, and adhesives. Alkylphenols are endocrine disruptors and reproductive toxins. They accumulate in fatty tissue and are found in breast milk (Acir and Guenther 2018). They are toxic to aquatic organisms.
Tetrachloride	Tetrachloride is released from chlor-alkali facilities to the air and environment. It is a global warming and ozone depleting gas. Long-term exposure to tetrachloride can cause liver, kidney, and central nervous system issues (US EPA 2023).

During production of PVC, vinyl chloride is released to the environment through discharges of industrial plant waste to land, air, and water. In 2021, total self-reported releases from vinyl chloride and PVC plants in the United States were 194,221kg (HCWH Europe 2021).

Vinyl chloride is a known human carcinogen and liver toxicant. It is ranked only behind lead, mercury, and arsenic on the US Agency for Toxic Substances and Disease Registry (ADTSR) Substance Priority List (ATSDR 2023).

Vinyl chloride poses health risks to multiple populations throughout the life cycle of PVC. Workers in PVC factories have been found to have unusually high rates of liver cancer (Material Research L3C 2024; Flaws et al. 2020; HCWH Europe 2021; ATSDR 2014). Fence-line communities around PVC plants, primarily comprised of low-income people of color, are exposed to toxins with multiple communities having been forced to relocate due to toxins in the environment.

Nearly 1000 unplanned releases of vinyl chloride have occurred in the United States since 2010. Train derailments have released 2268 metric tons of vinyl chloride to communities along rail lines (Material Research L3C 2024; Flaws et al. 2020). In landfills, the vinyl chloride monomer can leach from PVC products in landfills and microbes in the environment degrade other chlorinated substances into vinyl chloride.

PVC often contains additives to impart flexibility, resistance to UV degradation, or other qualities. Phthalates are added to PVC products in amounts up to 80% by weight to make them more flexible (Flaws et al. 2020). Metals such as zinc, organotin, and, to a lesser extent, cadmium and lead are added to PVC to extend the life of some products.

When PVC products are incinerated or accidentally burned, they release dioxins, phthalates, and heavy metals. If they are mixed into municipal waste, incineration can release these toxins. During building fires, first responders are exposed as well (Material Research L3C 2024). Refer to Table 21.1 for more complete accounting of the exposure risks related to PVC, its by-products, and added constituents.

Polyvinyl chloride is resin code #3, although it is minimally recycled in the United States and Europe due to the release of dioxins, heavy metals, and phthalates when heated. PVC in the recycling stream can contaminate other plastic recycling as well.

Numerous countries are considering PVC phase-outs. PVC and many of its additives are on the IFLI Living Building Challenge Red List. Safer substitutes exist for all PVC products; therefore, these alternative plastics or other materials should be specified (Table 21.2).

Table 21.2 Common Plastic Site Construction Products and Their Safer Alternatives

Product Type	Common Plastic and/or Harmful Constituents	Lower Carbon and/or Healthier Alternative Plastic and Minimum Recycled Content[1] PC=Post-Consumer Recycled Content RC=Total Recovered Content	Non-Plastic Alternatives
Pressure pipe	PVC, CPVC, PVDF, PEX	Heat fused joints: PP Mechanical joints: PE, HDPE, LDPE, HDPP	Cast iron, carbon steel, galvanized steel, stainless steel
Non-pressure pipe	PVC, ABS, PEX	Heat fused joints: PP Mechanical joints: PE, HDPE (100% RC), LDPE, HDPP	Vitrified clay, cast iron, galvanized steel
Garden hose	PVC	PU, thermoplastic elastomers (TPE) (phthalate-free), thermoplastic vulcanites (TPVs), rubber 60–65% PC	Stainless steel
Irrigation drip lines	PVC	HDPE, recycled rubber	
Soaker hose	PVC	PU, recycled rubber 60–70% PC	
Fencing and guardrails	PVC	HDPE, plastic lumber (PVC free), composite lumber (PVC free) 60–100% PC, 90–100% RC	Modified lumbers, pressure-treated lumbers, naturally decay-resistant lumbers (avoid tropical hardwoods), bamboo, wattle fencing, living fences, 100% recycled powder coated steel
Composite lumber	Wood/PVC Rice hull/PVC	HDPE, plastic lumber (PVC free), composite lumber (PVC free) Rice hull/non-PVC plastic 60–100% PC, 90–100% RC	Modified lumbers, pressure-treated lumbers, naturally decay-resistant lumbers (avoid tropical hardwoods)
Plastic lumber	PVC	HDPE 25–100% PC, 75–100% RC, Mixed resins 50–100% PC 95–100% RC	Modified lumbers, pressure-treated lumbers, naturally decay-resistant lumbers (avoid tropical hardwoods)
Outdoor furniture	PVC	HDPE	Modified lumbers, pressure-treated lumbers, naturally decay-resistant lumbers (avoid tropical hardwoods)
Edging	PVC	HDPE, composite lumber (non-PVC), plastic lumber (non-PVC) 30–100% PC	Wood, 100% RC steel, wattle edging, concrete, brick
Pavers and patio blocks	PVC	Rubber/rubber blends Plastic/plastic blends (PE, mixed resins) 90–100% RC	Domestic stone, precast concrete, brick
Pavement reinforcing grids	PVC	HDPE, PE, PPE, non-PVC mixed resins 90–100% PC	

(Continued)

Table 21.2 (Continued)

Product Type	**Common Plastic and/or Harmful Constituents**	**Lower Carbon and/or Healthier Alternative Plastic and Minimum Recycled Content**[1] PC=Post-Consumer Recycled Content RC=Total Recovered Content	**Non-Plastic Alternatives**
Tree grates	PVC	HDPE 50–100% PC	100% recycled content steel
Parking stops		Plastic or rubber 100% PC	Concrete
Benches	PVC, tropical hardwoods	Plastics (HDPE, PPE) 90–100% PC, 100% RC Plastic composites 50–100% PC, 100% RC	Wood, stone, 100% recycled content steel, concrete
Playground equipment	PVC and PVC coated steel	Plastics (HDPE, PPE) 90–100% PC, 95–100% RC Plastic composites (non-PVC) 50–75% PC, 95–100% RC	Wood, stone, 100% recycled content steel
Playground surfaces	Synthetic turf with PVC, PFAS; Crumb rubber playground surfaces (PAHs and lead)[2]	90–100% PC EPDM rubber, PVC free synthetic turf	Mulch, engineered wood fiber, sand, pea gravel, grass
Artificial turf	PVC, PFAS, Crumb rubber (PAHs and lead)[2], antimicrobials	Nylon, PP, PE	Natural grass, plant-based or mineral-based infills over synthetic infills
Non-woven geotextile		PP, TPO, EPDM, LDPE	
Woven geotextile		PP, TPO, EPDM, LDPE	
Geomembrane liners	PVC	RPE, TPO, HDPE, EPDM	Bentonite clay (pond liners)
Fabrics – furnishings, umbrellas and fabric structures	PVC	LDPE, PP, Nylon	Cotton canvas
Lightweight structural fill	EPS (expanded polystyrene, Styrofoam), XPS (Extruded polystyrene)		Expanded shale, clay, slates, slag; foamed glass aggregate
Conduit	PVC, CPVC	HDPE, PP	Steel
Expansion joint filler	PVC,	PP, PUR, silicone	Wood, cork
Expansion joint sealant		PUR, silicone	
Paints	High VOC paints, oil-based paints with solvents	Post-consumer latex paint White 20% PC Gray/Dark colors 50–99% PC Consolidated latex paint 100% PC	Lime wash
Powder coating	Fluoropolymer powder coatings	Polyester, polyurethane, polyester-epoxy, fusion bonded epoxy and acrylics	
Adhesives	Glues and epoxies containing formaldehyde, vinyl compounds, solvents and high levels of VOCs	Low VOC, solvent-free epoxy resins, polyester resins, polyurethane resins	
Sealers/ waterproofing agents	Sealants containing solvents, vinyl compounds, and high levels of VOCs	Low VOC, solvent-free sealers; RPE, HDPE, EPDM membranes	Bentonite clay pond liners

[1]EPA Comprehensive Procurement Guidelines

[2] Habitable 2024a, 2024b

Thermoplastic Elastomers

TPE, sometimes referred to as thermoplastic rubbers, are mixed plastic and rubber that have both thermoplastic and elastomeric properties. They can stretch and elongate then return to their original shape. TPE plastics are used for liners, roofing membranes, and pipes and in weatherstripping. Some TPE products contain phthalates to make them more flexible, posing health risks (Table 21.1).

Acronitrile Butadiene Styrene

Acronitrile butadiene styrene (ABS) is a thermoplastic used in pipes and other architectural products. It is used as a substitute for PVC in piping applications as it has fewer life-cycle impacts, but it can be damaged by sunlight and when burned, it gives off hydrogen cyanide. Like PVC, ABS is technically recyclable, but the market is very limited, so it is not usually recycled.

Polystyrene

Polystyrene (PS), resin code #6, is produced by the polymerization of styrene and benzene. It is transparent, rigid, brittle, and moderately strong. Styrene has been linked to liver cancers and neuro effects (OSHA 2024).

When foamed (XPS) or expanded (EPS), PS produces a foam that is used as insulation and lightweight fill. Both XPS and EPS contain the brominated flame retardant HBCD, a known persistent bioaccumulative toxin, endocrine disruptor, and neurotoxin (NCI 2024). EPS is being phased out as a single use plastic in takeaway containers. It may be phased out as an insulation as well.

Thermosets

With heating, pressurization, or reactions with a hardening agent, thermoset plastics undergo an irreversible chemical reaction forming cross-linked bonds or polymers. These bonds are durable and heat resistant and render the plastic non-recyclable. Thermoset polymers are found in paints, coatings, and adhesives as well as other plastic construction products. Types of thermosets include epoxies, phenolic polyurethanes, urea formaldehyde, and cross-linked polyethylene.

Polyurethane

Polyurethane (PU or PUR) is a thermoset resin that is primarily made with toluene diisocyanate (TDI) or methylene diphenyl diisocyanate (MDI). During production, TDI and MDI can be released into the air, water, and land and result in moderate to severe respiratory impacts to workers (ATSDR 2022b). Polyurethane is considered stable when cured and exposure risk is limited. In site construction, polyurethane is primarily used in adhesives, coatings, sealants, and elastomers.

Cross-Linked Polyethylene

Cross-linked polyethylene (PEX) is low-density polyethylene that is cross linked with an agent such as dicumyl peroxide. It is a semi-thermoset polymer. PEX has good mechanical properties at relatively high temperatures and is used as a pipe substitute for PVC hot water pipes. It can't be melted down or recycled.

Studies of PEX piping have identified chemicals such a toluene, a neurotoxin, and MTBE, a carcinogen (Kelley et al. 2014). Other research found that PEX pipes leached more chemicals and released more odors than PE pipes (Stoiber 2017).

Bioplastics and Biocomposites

Biobased plastics are derived from renewable, natural raw materials. Some are obtained from processing biopolymers such as starch, cellulose, chitosan, soy protein, gluten, and alginate. Others are chemically synthesized from lipids from plants, animals, and sugar derivatives.

Biocomposites are made up of two or more materials where at least one is naturally derived. Completely biobased composites are biopolymers reinforced with natural fibers, while other biocomposites are a mix of biobased material with fossil fuel-based plastics or other non-biological material. These mixed biological and technical materials are not biodegradable and are often not even recyclable.

Improved agricultural practices such as reduced use of fertilizers and pesticides can reduce the environmental burdens of biobased plastics. Other improvements include use of wastes, including food waste, as feedstock; optimization of energy use and renewables; use of otherwise fallow lands; and mechanical recycling of bioplastics.

Bioplastics can have a lower embodied carbon than other plastics as the feedstocks are not fossil fuel based, however, there are still serious environmental burdens such as deforestation, excessive fertilizer and water use, and excessive chemical use. Not all bioplastics and biocomposites are biodegradable, so some can offer durable products for site construction. Bioplastics and biocomposites are discussed in greater detail in Chapter 19 "Biobased Materials and Products".

Plastics and Greenhouse Gas Emissions

The production of plastics accounted for about 3.4% of global carbon emissions in 2022 with additional emissions resulting from their disposal (US EPA 2024c). The primary feedstock

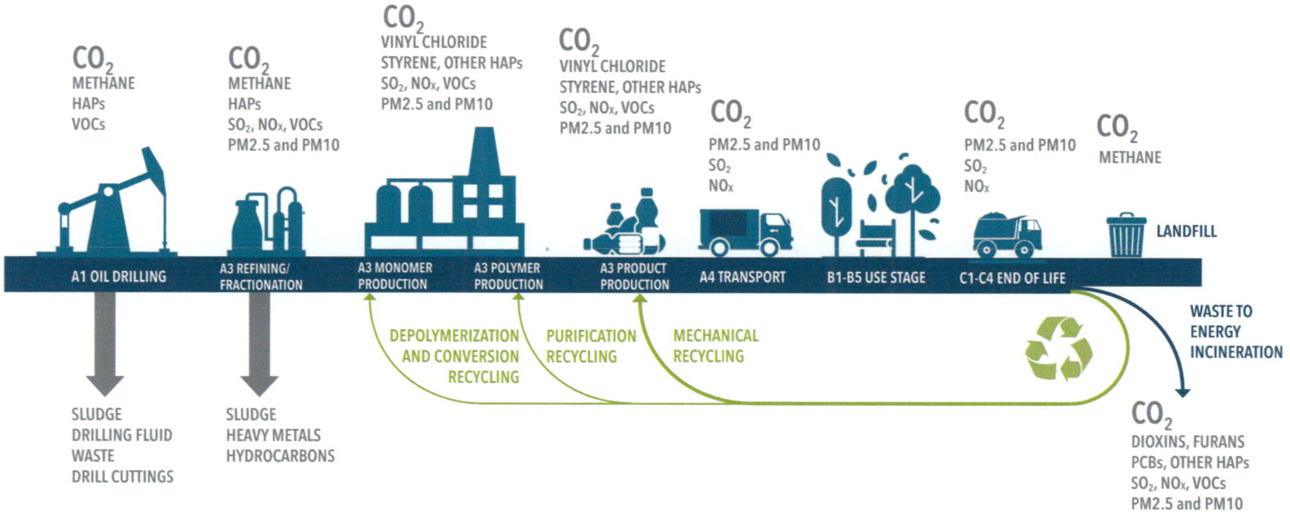

Figure 21.2 The typical plastics production life cycle. Unlike metals, asphalt, and concrete, plastic recycling industries are not robust as costs are high, infrastructure is not in place everywhere, and some plastic products are not safe to recycle. Image credit: Meg Calkins

Table 21.3 Emissions Factors for Plastics, Well to Resin, kgCO_2e/kg

Resin Code	Plastic	Range kgCO_2e/kg	50th Percentile kgCO_2e/kg
1	PET	1.70–3.63	2.47
2	HDPE	1.04–1.91	1.80
3	PVC	1.31–2.69	2.20
4	LDPE	1.22–2.20	1.96
5	PP	1.18–2.01	1.76
6	PS	2.34–3.51	3.20
	PC	3.40–3.40	3.40

Source: COMET 2022

Note: Embodied Carbon figures will change over time as processing efficiencies are improved, and more renewable energy sources are employed.

for many plastics is oil or natural gas, so their embodied carbon includes the feedstock burden as well as energy and chemicals used to produce the plastic (Figure 21.2). Generally, simpler plastics like polypropylene and high-density polyethylene offer safer, lower carbon alternatives in many applications (Tables 21.3 and 21.4). PVC is more processed than PP or HDPE; however its feedstock burden is lower as it is over 50% chlorine by weight.

Carbon emissions of plastics vary widely based on ingredients and production methods. Even within the same resin, EC can vary due to differing accounting methods and variable data. Production of plastic resins is responsible for around 60% of the carbon emissions of plastics on average. At the extrusion and fabrication level, an average of 30% of the carbon footprint, procurement distances, energy source, and processing equipment result in even greater variations.

Plastics have different densities, so it is important to look at the weight of a product when comparing its EC to the same product made from a different plastic. For example, PEX has a similar embodied carbon as PVC, yet it is one-seventh the weight for a similar size product, so the actual EC of a PEX product is much less than the comparable PVC product.

Emissions from the end-of-life phase of plastics account for an average of 9% of their carbon footprint. These are discussed in the "End-of-Life Considerations for Plastics" section of this chapter.

Future Embodied Carbon Reductions for Plastics

There are multiple places within the plastics production and recycling industries to reduce their embodied carbon. Perhaps the largest reductions can be made with a shift to *renewable energy sources* in the production of plastic resins. One study found that with a shift to 100% renewable energy sources, the EC of a PET product can be reduced by 56% and a PVC product by 75% (Zheng et al. 2019) A shift from fossil fuel feedstocks to biobased waste feedstocks for plastics could also cut emissions.

Biobased plastics often are lower embodied carbon in part because the feedstock is not fossil fuel based. One study estimated a 93% reduction from the baseline with 100% renewable energy use and the use of bioplastics (Carbon Tracker 2020). Use of bioplastics could shift the environmental burden from the climate to water and land resources if biobased feedstocks are not grown sustainably.

Table 21.4 Embodied Carbon of Common Plastic Products Used in Site Construction

Plastic Product	Density kg/m^3	Embodied Carbon kgCO_2e/kg[1]	Notes and Source
PVC pipe, pressure	1450	2.81	Average of multiple product EPDs
PVC pipe, non-pressure	1530	2.67	Average of multiple product EPDs
PE pipe, pressure	960	2.88	Average of multiple product EPDs
PE pipe, non-pressure	960	2.61	Average of multiple product EPDs
PP Pipe	920	3.14	Average of multiple product EPDs
Geotextile, PP	900	3.36	Average of multiple product EPDs
PVC membrane	1300	3.47	Average of multiple product EPDs
EPDM membrane	938	3.77	Average of multiple product EPDs
Polyurethane resin	1313	4.99	Hardy Carbon Dataset V3 2024; Range is 4.1–5.87kgCO_2e/kg
Epoxy resin	1100	4.75	Hardy Carbon Dataset V3 2024; Range is 3.8–5.7kgCO_2e/kg
Butyl rubber	1350	5.77	Hardy Carbon Dataset V3 2024
Styrene butadiene rubber	1600	5.5	Hardy Carbon Dataset V3 2024
Isoprene rubber	681	5.31	Hardy Carbon Dataset V3 2024
Fiberglass	1051	1.93	Hardy Carbon Dataset V3 2024; Range is 1.35–2.5kgCO_2e/kg
Fiberglass and plastic 50/50	996	2.44	Hardy Carbon Dataset V3 2024; Range is 1.91–2.97kgCO_2e/kg
Playground surfacing, EPDM rubber	938	3.71	Hardy Carbon Dataset V3 2024; Range is 3.41–4.01kgCO_2e/kg 70% RC is 0.752kgCO_2e/kg
Synthetic turf, nylon and PP	107	5.50	Hardy Carbon Dataset V3 2024; Range is 3.0–10.0kgCO_2e/kg
Lightweight structural fill: EPS geofoam	45.7	7.54	Hardy Carbon Dataset V3 2024; Range is 2.5–12.5kgCO_2e/kg
Lightweight structural fill: UL-FGA foamed glass aggregates	180	0.62	Hardy Carbon Dataset V3 2024; Range is 0.043–1.2kgCO_2e/kg
Rice hull composite lumber (50% rice hull, 50% recycled HDPE)	1200	1.65	Hardy Carbon Dataset V3 2024; Range is 0.9–2.4kgCO_2e/kg
Wood/plastic composite lumber (48% recycled LDPE and 50% wood fiber)	930	1.08	Hardy Carbon Dataset V3 2024; Range is 1.01–1.15kgCO_2e/kg
Wood/plastic composite lumber (50% PVC and 50% wood fiber)	1300	3.22	Hardy Carbon Dataset V3 2024; Range is 2.48–3.97kgCO_2e/kg

Note: Embodied Carbon figures will change over time as greater amounts of recycled content is used, processing efficiencies are improved, and more renewable energy sources are employed.

End-of-Life Considerations for Plastics

About 400 million metric tons of plastic waste are produced globally each year (UNEP 2024). In 2019, an estimated 73 million metric tons of plastics waste were generated in the United States at about 220kg per person. This amount is five times the global average per capita and accounts for about 12% of municipal solid waste (MSW) (Alves 2023).

Over 80% of the US generated plastic waste in 2019 was sent to landfills while only 5–6% was recovered for recycling as compared with an average of 9% worldwide. Of the 5–6% recovered in the United States, 1–2% is sent out of the country for recycling primarily to Mexico and Canada (Alves 2023).

A large amount of global waste ends up in our oceans killing thousands of mammals, sea birds, and turtles every year or breaking into microplastics and harming smaller forms of aquatic life. Without action, 29 million metric tons of plastic waste are expected to flow into our oceans annually by 2040 (Palardy 2022).

Managing plastic waste is a global crisis that is exacerbated by limited, inconsistent, and economically challenged recycling systems. Some plastics such as PVC can release toxins in recycling, while others are comingled and not economically viable to separate. Last, it is less expensive to produce new plastics, creating a disincentive to procure and use recycled materials.

Recycling Plastic Waste

Plastic recycling can be more expensive than creating virgin plastics from fossil fuels, recycling infrastructure is not robust everywhere, and technical challenges inhibit high levels of recycling. In the US, primary plastics recycled were PET (38%) and HDPE bottles (17%), non-bottle rigids (22%), film (22%), PP and other bottles (0.5%), and other plastics (0.2%). These figures are down 11% from a high in 2016 (Circularity in Action 2024). Currently, polyethylenes (#1, #2, #4) and polypropylenes (#5) are the most economically viable plastics recycled in the United States with relatively robust infrastructures in place, markets for reuse of the recycled materials, and safer recycling processes (Figure 21.3).

Site construction products can be a good destination for plastic waste that can't be recycled back into its original products. For instance, plastic milk jugs made from HDPE are recycled into plastic and composite lumber. The US EPA has set both post-consumer and total recovered content thresholds for several site construction products in their *Comprehensive Procurement Guidelines* (Table 21.2).

Molecular recycling, also called advanced recycling, uses heat, enzymes, solvents, or sound waves to break down and purify plastic waste. Resulting products include polymers, monomers, oligomers, or hydrocarbon products. The three main technologies of molecular recycling are purification, depolymerization, and conversion (Table 21.5).

Figure 21.3 Plastics recycling symbols have created some confusion because studies show that people recycle products if they have one of these symbols when, in reality, most recycling programs only take 1 and 2 plastics. Some will also take plastics numbered 4 and 5. When other types of plastics are mixed into the recycling stream, batches can be contaminated and sometimes compromised. Image credit: Drake Bruner and Meg Calkins

Table 21.5 End-of-Life Emissions for Plastics

End-of-Life Process	Average Emissions $kgCO_2e/kg$ Plastic
Mechanical recycling, the most common method of plastic waste recycling, breaks down plastic waste through shredding or chipping without significantly altering the material's chemical structure. The plastic flakes are converted into pellets that can be used to make new products. HDPE and PET are the most common plastics that are mechanically recycled, but theoretically any thermoplastic can be. Mechanical recycling permits multiple cycles of reuse or recycling. It is expensive and labor intensive but relatively low in embodied carbon.	0.30
Purification, also called solvent recycling, uses solvents to break down the plastic waste and remove additives such as dyes or stabilizers. It produces a virgin grade polymer. There are emissions with this process, and possible human health impacts depending on the type of plastic waste or solvents used.	1.60
Depolymerization processes use heat or enzymes to reduce polymers to their monomers. Repolymerization converts the monomer into new plastics. There are substantial emissions with this process, and potential human health impacts depending on the type of plastic waste.	2.50
Conversion processes use gasification or pyrolysis (heat) and pressure to convert polymers to hydrocarbon molecules. These are then reprocessed into new monomers, then polymers. There are substantial emissions with this process, and potential human health impacts depending on the type of plastic waste.	2.80
Incineration of plastic waste, also called waste-to-energy, can be an alternative energy source. It is generally agreed that plastics-to-fuel and plastics-to-energy should not be considered recycling. There are substantial emissions with this approach and documented human health impacts depending on the type of plastic waste (Tables 4.3 and 21.1).	3.15
Landfill	0.02

Source: Luu et al. 2023

Note: Emissions figures will change over time as processing is improved and more renewable energy sources are employed.

Molecular recycling technologies are much newer than mechanical recycling and less widespread. Benefits include a greater purity in the polymer or monomer than mechanical recycling can offer, the ability to produce new plastic products that are just as high quality as the original product, and the opportunity to recycle a much greater variety of plastics (molecular recycling).

Drawbacks of molecular recycling include higher costs, higher EC, limited recycling infrastructure, and unproven technologies at scale. The embodied carbon of molecular recycling is very high, sometimes higher than the production of products from virgin resources.

Some plastic products cannot be recycled. Multi-material composites such as composite lumber or products produced with mixed resins are not recyclable because it is impossible to separate the materials, and the chemical makeup is uncertain. Sometimes, mixed resin products can be recycled into other mixed resin products, but this is not common as the mixed resins may contain PVC or other plastics that could release toxins when heated.

While PVC waste is technically recyclable and heavily marketed as such by PVC product manufacturers, it is not commonly recycled in the United States as recyclers consider it difficult and expensive to recycle. This is due to the many hazardous constituents in PVC and the high chlorine content. PVC plastic waste must be separated from other plastic waste adding cost to the process.

End-of-Life Emissions from Plastics

There are multiple end-of-life scenarios for plastic waste with varying environmental and human health burdens. Both carbon and toxic emissions from plastics continue in the end-of-life phase of plastics. CO_2 emissions for plastics in the end-of-life stage account for an average of 9% and as much as 70% of the total life-cycle carbon burden of plastics (Figure 21.4 and Table 21.5) (Luu 2023).

Emissions from plastics recycling: The carbon emissions burden from plastics recycling varies by recycling technology. Mechanically recycled rigid plastics average about 0.3kgCO_2e/kg while chemical/molecular recycling methods can be substantially higher. The GHG intensity of virgin PP is about 2.0kgCO_2e/kg resin while pyrolysis recovered PP is 2.3kgCO_2/kg resin. Mechanically recovered PP is 1.3kgCO_2e/kg resin (Blysma et al. 2023).

Landfilling plastics: From a carbon standpoint, landfilling plastics has been found to be the least intensive option at just 0.017kgCO_2e per kg of plastic waste, although this approach uses more non-renewable fossil fuel resources to produce new plastic products. Additionally, accidental landfill fires or leaking effluent may release toxins from plastics.

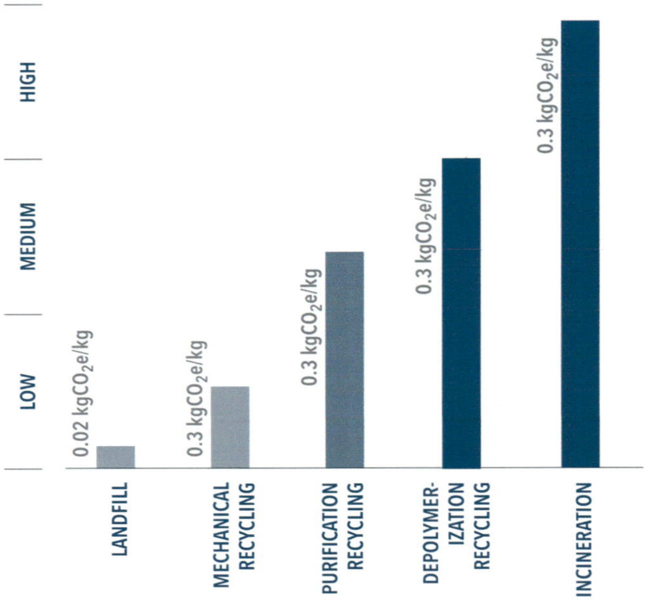

Figure 21.4 The embodied carbon of various end of life scenarios for plastics. Image credit: Meg Calkins

Incinerating plastics: One quarter of all plastic waste worldwide is incinerated, producing substantial carbon emissions even when pollution control equipment is employed. Incineration of plastic waste can decrease air quality for long distances and increase hazardous solid waste burdens on the facility and fence-line communities. Of particular concern is the incineration of PVC, which releases dioxins and phthalates. Incineration of PVC also results from landfill fires, building fires, and backyard burning.

Natural and Synthetic Rubbers

Rubber is a durable, waterproof, and elastic material that comes from natural or synthetic latex. About 32 million metric tons of rubber are produced annually worldwide with about 2/3 of this being synthetically produced (ECRA 2024) Most natural rubber is produced in Southeast Asia with China with leading production in synthetic lumber.

Primary products made from rubber are tires, medical supplies such as latex gloves, flooring, footwear, toys, and tubes. In site construction, rubber is used for tubing, washers, expansion joint filler, adhesives, joint sealants, caulks, and other products.

Natural rubber is derived from the liquid tree sap of tropical plants such as *Hevia brasiliensis* or *Landolphia owariensis*, while synthetic rubber is made from petroleum and natural gas. Synthetic rubbers are polyisoprene, neoprene, nitrile rubber, and styrene butadiene rubbers (SBR).

Polyisoprene, a polymer of isoprene is a relatively benign synthetic rubber.

Neoprene is made from chloroprene molecules. It is more resistant to oil and chemicals. Chronic exposure to chloroprene can lead to cardiovascular disorders and liver function abnormalities (Chemical Safety Facts 2024).

Styrene butadiene rubbers are derived from the copolymerization of styrene and 1,3-butadiene. 1,3-butadiene is a potential carcinogen, teratogen and reproductive hazard (US EPA 2005). Styrene has been classified by the IARC as a probable carcinogen.

Rubber Waste

Globally, an estimate 1 billion tires are discarded each year causing an escalating rubber waste crisis (Dabic-Miletic et al. 2021). End-of-life tires (ELT) contribute to air, water, and soil pollution, and the sheer volume of ELT means that this is a serious environmental burden. Many waste tires are burned for energy or fuel, but incineration of ELT releases toxic air pollutants and heavy metals even with pollution control equipment. The waste from tire incineration results in potentially hazardous solid waste as well. Unintentional fires in waste tire piles are not uncommon and are an uncontrolled source of air and water pollution.

Rubber Recycling

Rubber is a recyclable material but there are challenges that inhibit its widespread recycling. Rubber's durability makes it energy intensive to process into tire chips and crumb rubber. Recycling technologies have not been fast enough and infrastructure not broad enough to reduce the large volume of waste tires. Newer technologies such as pyrolysis and devulcanization will speed recycling, but can release toxic chemicals and carbon emissions.

There are health concerns with recycled rubber fill used in artificial turf applications, recycled rubber playground surfacing, recycled tire mulch in playgrounds, and crumb rubber use in hot-mix asphalt pavement. Concentrations of PAHs have been found to be as high as 1% of recycled crumb rubber by weight. This is a concern for workers manufacturing and installing these products as well as individuals playing on these surfaces, particularly in higher temperatures (US EPA 2024; Habitable 2020, 2024a, 2024b; Gomes et al. 2020).

Recycled rubber could be safely used in paving blocks, parking stops and soaker hoses. Refer to Table 21.2 for recommended recycled rubber contents for these and more products. Half-warm mix asphalt and cold-mix asphalt are other applications for recycled crumb rubber.

Composite Plastics

Composite plastics are polymers strengthened with fibers, filler, powders, particulates, and other constituents that impart properties such as strength or stiffness. Fibers in composite plastics may be fabric, metal, glass, fiberglass, wood, or agricultural waste. Powders are made of carbon, graphite, silicates, ceramics, or other organic or inorganic materials.

In site construction, plastic composites are used in piping, conduit, lumber, composite rebar, fencing, and edging. Some plastics composites comingle polymers with agricultural or wood processing waste for high recycled content products. Composite materials can be reused, but they are not recyclable due to their comingled constituents.

Composite Lumber

Composite lumber is one of the most common plastic composites in site construction used for decking, edging, and fencing and even some structural applications. Types of resins in plastic lumber are commonly HPDE or PVC mixed with agricultural waste such as rice hulls, wood processing waste, or fiberglass. HDPE-based composite lumber often uses post-consumer HDPE, while PVC-based composite lumber does not usually incorporate a recycled polymer. These products claim some recycled content, but it is usually from the agricultural or wood waste.

Major considerations for specifying composite lumber are discussed in Table 21.7. Detailed considerations for use and maintenance of composite lumber are discussed in Tables 10.8 and 10.11 in Chapter 10 "Decks and Boardwalks".

Ethical Sourcing of Plastics and Rubber

Human rights violations can occur during all life cycle phases of plastics and rubber. A primary human rights violation from plastics are the health impacts discussed throughout this chapter and in Tables 4.3 and 21.1. Chemical and plastics industry workers and fence-line communities are primarily impacted by plastics health burdens with these populations being primarily lower income people of color.

Lesser-known human rights violations in the production of plastics and rubber occur using forced labor and child labor

Table 21.6 Strategies to Reduce the Environmental and Human Health Burdens of Plastics

Issue	Consideration
Reducing plastic waste	Maximize the use life of plastic products in site construction as this will reduce demand for new plastics. Detail structures that incorporate plastic products to maximize their use life.
Facilitating recycling	Specify products that are single polymer plastic products that are HDPE, LDPE, PET, or PP as they can be more readily recycled.
Plastic alternatives	Where appropriate, consider non-plastic materials such as biobased or mineral-based materials with lower environmental and human health burdens. Some will have a greater embodied carbon but may be more durable (Table 21.2).
Air quality and human health impacts	Specify low-VOC paints, sealants, adhesives, and coatings. Refer to VOC limits by from the State of California, Air Resources Board, VOC Limits.
Health impacts	Specify safer plastics. Avoid plastic products with IFLI Red List constituents such as vinyl chloride, phthalates, formaldehyde, PFAS, and California Banned Solvents. Refer to Tables 4.3 and 21.1.
Information transparency	Request environmental product declarations, health product declarations, and safety data sheets from manufacturers to check all constituents. Prioritize these products and those with Declare labels.
Recycled content	Specify plastic products with at post-consumer recycled content and at least 50% total recovered content. Refer to the EPA's Comprehensive Procurement Guidelines for minimum recycled content for a specific product type.
Reducing packaging waste	Specify products that avoid or minimize single use plastic packaging. Look for product manufacturers that take back and reuse/recycle packaging.

Table 21.7 Considerations for Specifying Composite Lumber

Aspect	Consideration
Recycled content	Specify composite products with a minimum of 50% post-consumer recycled content and 95–100% total recovered content. Avoid composites with less than 50% recovered content.
Polymer type	Avoid composite lumbers containing PVC or polystyrene. It can be very difficult to identify the polymer in composite products. If the manufacturer advertises that their product is made from "salt" this means that it is likely PVC based.
Local sources	Specify locally or regionally sourced and produced composite lumber.
Lighter weight composites	Specify composite lumber shapes that maximize strength and performance while minimizing weight. This can lower the embodied carbon of the product.
Design and installation	Refer to Table 10.8 in Chapter 10 for composite lumber decking design and installations considerations.
Maintenance	Refer to Table 10.11 in Chapter 10 for composite lumber decking maintenance considerations.

outside North America and Europe. China's Xinjiang Uyghur Autonomous Region (XUAR) was recently listed by the US Customs and Border Protection for their common practice of forced labor in the PVC industry there. Ten percent of the world's supply of PVC comes from this region of China. PVC and PVC products made in this region are banned from US import (US CBP 2024).

Forced and child labor are a concern in rubber industries in several countries in Southeast Asia and West Africa. The US Bureau of International Labor Affairs (ILAB) highlights rubber produced by child labor in Cambodia, Indonesia, Liberia, Philippines, and Vietnam. Rubber produced in Malaysia and Burma could use forced laborers (ILAB 2022). Design for Freedom has also highlighted these countries, and they add one more for child labor concerns, Cote d'Ivoire (DFF 2022).

There are three organizations that verify the sources and production of natural rubbers. The Forest Stewardship Council ensures that rubber is sustainably sourced and produced. The PEFC offers Sustainable Forest Management Certification for rubber growers.

The Fair Rubber Association administers the Fair Rubber Standard that addresses labor practices, worker health and safety issues, and environmental standards to produce rubber. Products that adhere to the standard are certified.

The Future of Plastics

Global plastics production is expected to increase to 579 million metric tons by 2028 and as much as 30% in the next decade (IEA 2018). This rate of growth is largely due to the inexpensive cost of plastics as compared with alternative materials.

Efforts are being made on the world stage to address the widespread environmental and human health burdens of plastics. With the sheer volume of single use plastics, millions of metric tons of plastics finding their way into oceans each year, low recycling rates, and toxins in some polymers and additives, policies must be put in place to reduce the impacts.

An international plastics treaty could scale back or eliminate single use plastics, PVC, and other harmful plastics and additives. It could curb plastics production and address the plastics waste crisis, particularly the plastics waste in our oceans.

The International Living Future Red List of toxic substances to avoid is growing more robust and transforming product markets with thousands of chemicals listed, and even more on their "Watch List". Several major corporations have eliminated PVC and other harmful plastics from their products and/or packaging, and several countries are exploring bans and phase-outs.

The UN Environment Programme and the Regional Activity Centre for Sustainable Consumption under the Stockholm Convention call for four main approaches to address issues in the plastics industry (UN Environment 2020).

Safer plastics: Innovations must be developed for safer plastics that avoid hazardous constituents and by-products.
Collaboration among industries: Collaboration among industries related to plastics must safely close materials loops to support the circular economy.
Increased recycling: Innovations must be made to recycling systems to increase recycling rates safely.
Transparent information: Plastics producers must provide the public with transparent information on the risks and exposures of hazardous chemicals and by-products.

Innovations in Plastics

There are hundreds of new types of polymers in development with increasingly complex chemistries. Some innovations address human health and environmental concerns, while others are exploring performance improvements. Innovations are numerous and constantly evolving.

Multiple "safer" plastic constituents have been developed as substitutes for those restricted by the EU REACH program, the US Toxic Substances Control Act, or California's list of banned substances. However, some of these substitutes may pose serious health impacts but are too recently developed for the impacts to have been fully tested and verified.

While bioplastics have been produced for a few decades, climate change and waste reduction imperatives are driving innovations in bioplastic materials made from food and agricultural wastes. Bioplastics are addressed in Chapter 19 "Biobased Materials and Products".

Sequestering carbon in plastics is another recent innovation. One product mixes biochar with a binder made from sugarcane to produce a "carbon negative" plastic. Biochar, a nearly pure carbon charcoal-like material, is produced from burning forestry and agricultural waste in an oxygen-free furnace.

A Circular Economy Vision for the Plastics Industry

Several governments and NGOs have put forth ambitious visions for a circular economy in the plastics industry. A circular economy is a systems solution framework where materials are designed to be used and reused in a closed loop. Virgin resource use is eliminated from the system and waste is not created that is not reused for a new process or product at its highest value.

The UNEP estimates that a shift to a circular economy in the plastics industry could reduce the volume of plastics entering oceans by 80% by 2040 and reduce virgin plastic production by 55%. Greenhouse gas emission could be reduced by 25% (UNEP 2022).

True circularity is a tall order, especially for the plastics industry where many products have such short use lives. Business models must shift, policies must be established and enforced, and innovations must be made.

The UNEP vision for a circular plastics industry designs out waste, keeps materials in use, and protects human health and the environment. The vision states the following aims:

- ***Unnecessary use of plastics should be eliminated.*** Reducing use of plastics is the single most important goal of the plan, particularly plastic packaging.
- ***Takeback of plastic products and packaging by manufacturers should be required.*** Reuse of plastics is preferred over recycling as this can reduced embodied carbon and transportation. A takeback requirement could spur innovations in design and use of safer materials.
- ***All plastic should be reusable, recyclable, or compostable.*** Plastics that are harmful or contaminate recycling processes would be eliminated from use. Plastic products would be designed to fit within recycling systems.
- ***Virgin plastic feedstocks should be avoided***. New plastic products would only be from recycled feedstocks.
- ***All plastics are free of hazardous materials***. This could strengthen recycling markets making it easier and more economically viable to recycle.

References

Acir, I.H., and Guenther, K. (2018). "Endocrine-disrupting metabolites of alkylphenol ethoxylates – a critical review of analytical methods, environmental occurrences, toxicity, and regulation". *Science of the Total Environment* 635 (September 1): 1530–1546. https://doi.org/10.1016/j.scitotenv.2018.04.079.

Agency for Toxic Substances and Disease Registry (ATSDR) (2014). *Medical Management Guidelines for Vinyl Chloride*. US Centers for Disease Control. https://wwwn.cdc.gov/TSP/MMG/MMGDetails, accessed May 20, 2024.

Agency for Toxic Substances and Disease Registry (ATSDR) (2022a). *ToxFAQs for Di(2-Ethylhexyl)phthalate (DEHP)*. ATSDR.

Agency for Toxic Substances and Disease Registry (ATSDR) (2022b). *ToxFAQs for Toluene Diisocyanate (TDI)*. ATSDR.

Agency for Toxic Substances and Disease Registry (ATSDR) (2023). *ATSDR's Substance Priority List*. US Centers for Disease Control. www.atsdr.cdc.gov/spl/index.html.

Alves, B. (2023). "Per capita plastic waste generation worldwide 2019, by region". *Statista*, December 20.

American Chemistry Council (2024). "Ethylene oxide is a versatile chemical building block". www.americanchemistry.com/chemistry-in-america/chemistries/ethylene-oxide, accessed May 21, 2024.

BCC Publishing (2023). *Market Trends in the Plastics Industry: An Analysis of Developments by Key Plastics Manufacturers*. Report Code: PLS115B. BCC Publishing, September.

Blysma, S., Conway, T.J., and Fallurin, J. (2023). *We Can Cut Petrochemicals Use Today: Framing the Challenge and Opportunity*. Rocky Mountain Institute.

California Air Resources Board (ARB). (2021). *Summary of Architectural Coating Rules in California: VOC Limits*. State of California, ARB, September 2.

Carbon Tracker (2020). *The Future's Not in Plastics*. Carbon Tracker.

Chemical Safety Facts (2024). "Neoprene". www.chemicalsafetyfacts.org/chemicals/neoprene/, accessed May 8, 2024.

Circularity in Action (2024). *Data Summary*. US Post Consumer Plastic Recycling Dashboard 2022, March.

Coalition on Material Emissions Transparency (COMET) (2023). *Making Plastic Emissions Transparent*. COMET. www.cometframework.org.

Dabic-Miletic, S., Simic, V., and Karagoz, S. (2021). "End-of-life tire management: A critical review". *Environmental Science and Pollution Research* 28: 68053–68070. https://doi.org.prox.lib.ncsu.edu/10.1007/s11356-021-16263-6.

Design for Freedom. (2022). *Design for Freedom Toolkit*. New Canaan, CT: Grace Farms Foundation.

Elhacham, E., Ben-Uri, L., Grozovski, J. et al. (2020). "Global human-made mass exceeds all living biomass". *Nature* 588, 442–444. https://doi.org/10.1038/s41586-020-3010-5.

European Rubber Chemicals Association (ERCA) (2024). "How is rubber made?" https://erca.cefic.org/how-is-rubber-made/, accessed May 10, 2024.

Flaws, J., et al. (2020). *Plastics EDCs and Health: A Guide for Public Interest Organizations and Policy Makers on Endocrine Disrupting Chemicals & Plastics*. Endocrine Society and ICPN, March.

Garside, M. (2024). "Annual production of plastics worldwide from 1950 to 2022". *Statista*. www.statista.com/statistics/282732/global-production-of-plastics-since-1950/, accessed May 23, 2024.

Goldsberry, C. (2018). "Scientific tests prove HDPE can be recycled at least 10 times". *Plastics Today*. www.plasticstoday.com/packaging/scientific-tests-prove-hdpe-can-be-recycled-at-least-10-times.

Gomes, Filipa, et al. (2020). "A review of potentially harmful chemicals in crumb rubber used in synthetic football pitches". *Journal of Hazardous Materials* 409: 124998. https://doi.org/10.1016/j.jhazmat.2020.124998.

Habitable (2020). *New LEED Circularity Pilot Credit Incorporates HBN Optimize Recycling Research*. Habitable, January 30.

Habitable (2024a). *Flooring Product Guidance*. Habitable, May 8.

Habitable (2024b). *Turf Product Guidance*. Habitable, May 8.

Health Care Without Harm (HCWH) Europe (HCWH) (2021). *The Polyvinyl Chloride Debate: Why PVC Remains a Problematic Material*. Health Care Without Harm Europe, June.

International Energy Agency (IEA) (2018). *The Future of Petrochemicals: Towards More Sustainable Plastics and Fertilisers*. IEA and Organization for Economic Cooperation and Development (OCED), October.

Kelley, K.M., et al. (2014). "Release of drinking water contaminants and odor impacts caused by green building cross-linked polyethylene (PEX) plumbing systems". *Water Research* 67: 19–32.

Living Building Challenge (LBC) (2024). *LBC Red List, Priority List and Watch List*. International Living Future Institute (IFLI).

Luu, P., et al. (2023). *Transitioning to a Circular System for Plastics: Assessing Molecular Recycling Technologies in the United States and Canada*. Closed Loop Partners.

Material Research, L3C (2024). *Toxic Substances Control Act Technical Report: Exposure to Vinyl Chloride*. Material Research, L3C, Earthjustice and Beyond Plastics, March.

National Cancer Institute (NCI) (2024). *Vinyl Chloride*. National Institutes of Health. www.cancer.gov/about-cancer/causes-prevention/risk/substances/vinyl-chloride, accessed May 19, 2024.

The National Institute for Occupational Safety and Health (NIOSH) (2014). *Current Intelligence Bulletin 25: Ethylene Dichloride (1,2-Dichloroethane)*. US Centers for Disease Control. www.cdc.gov/niosh/, accessed May 19, 2024.

The National Institute for Occupational Safety and Health (NIOSH) (2018). *Chlorine*. US Centers for Disease Control. www.cdc.gov/niosh/topics/chlorine/default.html, accessed May 19, 2024.

Occupational Safety and Health Administration (OSHA) (2019). *List of Highly Hazardous Chemicals, Toxics and Reactives (Mandatory)*. U.S. Department of Labor.

Occupational Safety and Health Administration (OSHA) (2024). "Cadmium". www.osha.gov/cadmium, accessed May 19, 2024.

Palardy, J. (2022). *Breaking the Plastic Wave: Solving the Plastic Pollution Problem*. The PEW Charitable Trust, March 10.

Stoiber, T. (2017). *Amid Pipe Wars, Researchers Wary of Plastic Pipes Leaching Chemicals*. Environmental Working Group (EWG), November 14.

UN Environment Program (UNEP) (2022). *Historic Day in the Campaign to Beat Plastic Pollution: Nations Commit to Develop a Legally Binding Agreement*. UNEP, March 2.

UN Environment Program (UNEP) (2024). "Our planet is choking on plastic". www.unep.org/interactives/beat-plastic-pollution, accessed August 4, 2024.

UN Environment Regional Activity Center (RAC) for Sustainable Consumption and Production (2020). *Plastic's toxic additives and the circular economy*. UN Environment RAC, September.

US Bureau of International Labor Affairs (ILAB) (2022). *List of Goods Produced by Child Labor or Forced Labor*. US Department of Labor, ILAB.

US Customs and Border Protection (US CBP) (2024). *Uyghur Forced Labor Prevention Act*. US CBP, October 16.

US Environmental Protection Agency (US EPA) (2005). *An Updated Study of Mortality Among North American Synthetic Rubber Workers*. Record ID 258324. US EPA, March.

US Environmental Protection Agency (US EPA) (2023). *Carbon Tetrachloride*. Report 56-23-5. US EPA.

US Environmental Protection Agency (US EPA) (2024a). *TRI National Analysis: Ethylene Oxide*. US EPA. www.epa.gov/trinationalanalysis/ethylene-oxide, accessed May 21, 2024.

US Environmental Protection Agency (US EPA). (2024b). *Federal Research on Recycled Tire Crumb Used on Playing Fields and Playgrounds*. US EPA Safer Chemicals Research.

US Environmental Protection Agency (US EPA) (2024c). *Impacts of Plastic Pollution*. www.epa.gov/plastics/impacts-plastic-pollution, accessed June 14, 2024.

World Health Organization (WHO) (2023). *Dioxins*. WHO.

Zheng, J., and Suh, S. (2019). "Strategies to reduce the global carbon footprint of plastics". *Nature and Climate Change* 9: 374–378.

Index

For Product Safety Concerns and Information please contact our EU representative GPSR@taylorandfrancis.com Taylor & Francis Verlag GmbH, Kaufingerstraße 24, 80331 München, Germany

Batch number: 10801187

Printed by Printforce, the Netherlands